Current Progress in Engineering Sciences

Uday Shanker Dixit · Nipjyoti Bharadwaj ·
Sumit Kumar · Debabrata Sikdar
Editors

Current Progress in Engineering Sciences

Select Papers of RIC 2024, Volume 1

Springer

Editors
Uday Shanker Dixit
Department of Mechanical Engineering
Indian Institute of Technology Guwahati
Guwahati, Assam, India

Nipjyoti Bharadwaj
Department of Civil Engineering
Indian Institute of Technology Guwahati
Guwahati, Assam, India

Sumit Kumar
Department of Chemical Engineering
Indian Institute of Technology Guwahati
Guwahati, Assam, India

Debabrata Sikdar
Department of Electronic and Electrical Engineering
Indian Institute of Technology Guwahati
Guwahati, Assam, India

ISBN 978-981-96-8752-7 ISBN 978-981-96-8753-4 (eBook)
https://doi.org/10.1007/978-981-96-8753-4

This Springer imprint is published by the registered company Springer Nature Singapore Pte Ltd.
The registered company address is: 152 Beach Road, #21-01/04 Gateway East, Singapore 189721, Singapore

Foreword

The Research and Industrial Conclave (RIC) is an annual event organized by the Students' Academic Board of the Indian Institute of Technology Guwahati, aimed at fostering collaboration between industry and academia. The ninth edition of the Conclave was held from August 9 to 11, 2024, at IIT Guwahati, drawing over 800 participants who actively engaged in various activities. A key highlight of the Conclave was the presentation of research papers by emerging researchers, reflecting the enthusiasm and innovation of young minds.

For the first time, the organizers have decided to publish selected papers from the Conclave as Proceedings, in collaboration with Springer Nature. The papers are compiled into three volumes:

Volume 1: Current Progress in Engineering Sciences
Volume 2: Current Progress in Physical Sciences
Volume 3: Current Progress in Interdisciplinary Research.

We are currently witnessing the era of Industry 4.0, where the integration of digital technologies with the physical world is at the forefront of innovation. With advancements in the Internet of Things (IoT) and Artificial Intelligence (AI), creating digital twins of industrial systems has become increasingly feasible. However, achieving the vision of Industry 4.0 requires interdisciplinary research and collaboration, as isolated efforts are insufficient to address the complexities involved.

Even as Industry 4.0 continues to evolve, the concept of Industry 5.0 has emerged, emphasizing a more human-centric approach that prioritizes sustainability, resilience, and collaboration between humans and advanced technologies. Unlike its predecessor, Industry 5.0 recognizes the importance of contributions from not only engineering and scientific disciplines but also from the arts and humanities. This recognition aligns well with the objectives of the National Education Policy (NEP) 2020, which rightly emphasizes the significance of interdisciplinary education and research.

The articles published in these three volumes of the Proceedings of RIC 2024 reflect the growing trend toward interdisciplinary research. Furthermore, many of the works are aligned with the United Nations Sustainable Development Goals (SDGs), indicating a broader vision to address global challenges. Given that most of the lead authors belong to Generation Z, it is not surprising that computational tools have been enthusiastically and effectively utilized throughout their research.

The selection of articles involved a rigorous peer-review process, with each paper reviewed by at least two experts to ensure high-quality standards. Each volume has been meticulously edited by four domain experts, supported by a dedicated team of student volunteers.

We hope that these volumes will serve as valuable archival resources and inspire future research that continues to bridge the gap between industry and academia.

April 2025

Prof. Devendra Jalihal
Director
Indian Institute of Technology Guwahati
Guwahati, Assam, India
director@iitg.ac.in
dj@iitg.ac.in

Preface

Research and Industrial Conclave 2024 (RIC 2024) with the thematic name as "Integartion'24" was held during August 9–11, 2024, marking the ninth edition in the series. Volume 1 of Proceedings of RIC 2024 contains the select papers highlighting the *Current Progress in Engineering Sciences*. The volume contains 43 papers covering Civil Engineering, Mechanical Engineering, Electrical Engineering, Electronic Engineering, Computer Engineering and Chemical Engineering.

Historically, Civil Engineering is the oldest branch of engineering. John Smeaton (1724–1792), who designed several bridges, canals, harbors, and lighthouses, is considered the father of Civil Engineering. Mechanical Engineering emerged with the development of steam engines; many regard James Watt (1736–1819) as the father of Mechanical Engineering. This discipline significantly contributed to the First Industrial Revolution (circa 1750–1840). Similarly, Electrical Engineering played a major role in the Second Industrial Revolution (1870–1914), with notable figures like Michael Faraday (1791–1867), Thomas Alva Edison (1847–1931), Nikola Tesla (1856–1943), and James Clerk Maxwell (1831–1879) making groundbreaking contributions.

Electronics Engineering evolved from Electrical Engineering with the invention of the thermionic valve, essentially a diode, by Sir John Ambrose Fleming (1849–1945) in 1904. By properly controlling the flow of electrons, electronics has given birth to numerous revolutionary products, including the first electronic computer, ENIAC, in 1946, which employed about 18,000 thermionic valves and consumed approximately 150 kW of electrical power. The year 1947 was significant for India as it achieved independence from British colonial rule, but it was also noteworthy for the invention of the transistor by John Bardeen (1908–1991), Walter Brattain (1902–1987), and William B. Shockley (1910–1989) at Bell Laboratories, USA. Transistors have transformed the field of Electronics Engineering, influencing nearly all engineering disciplines and even giving rise to the interdisciplinary field of Mechatronics in the early 1970s.

During the Second Industrial Revolution, Chemical Engineering emerged from Industrial Chemistry. The first Chemical Engineering course was offered at the University of Manchester in 1887 by George Edward Davis (1850–1907), often regarded as the father of Chemical Engineering. Today, apart from Chemistry, Chemical Engineers require knowledge of Mechanical, Electrical, Electronics, and Computer Engineering. Simultaneously, Chemical Engineers have significantly influenced other engineering disciplines. For example, the American multinational chemical company DuPont developed the technique of Critical Path Method (CPM) in the late 1950s for managing chemical plants.

In the era of the Fourth Industrial Revolution, there is considerable overlap among various engineering disciplines. Engineering techniques are now grounded in sound scientific principles. Artificial Intelligence (AI), modeling, simulation, and optimization techniques are increasingly applied across all engineering branches. The interdependence among engineering disciplines is evident, and the articles in this volume reflect this trend. The lead authors of these articles are young researchers who have attempted to address traditional problems using modern tools while focusing on sustainability. We hope this volume will contribute to enhancing understanding across disciplines, promoting interdisciplinary research.

We are deeply grateful to our authors and reviewers for their valuable contributions. While we cannot list the names of the reviewers to preserve anonymity, their efforts are sincerely appreciated. We welcome feedback from readers on this book.

Guwahati, India

Uday Shanker Dixit
Nipjyoti Bharadwaj
Sumit Kumar
Debabrata Sikdar

Acknowledgements

We extend our deepest gratitude to **Prof. Devendra Jalihal**, Director of the Indian Institute of Technology Guwahati, for his gracious patronage of the Research and Industrial Conclave 2024. We are immensely thankful to **Prof. K. V. Krishna**, Dean of Academic Affairs, for his invaluable guidance and unwavering support. Our appreciation also goes to **Prof. Vimal Katiyar**, the then Dean of Industrial Interactions and Special Initiatives, and **Prof. Rupam Barman**, Chairman of the Students' Academic Board, for their significant contributions to the success of this event.

The Conclave's triumph is a testament to the outstanding leadership of **Prof. Pankaj Kalita**, Convener, and **Prof. R. B. Sharmila**, Co-convener. We deeply appreciate the relentless dedication of our students—**Mr. Shubham Singh** (Student Convener), **Mr. Nikhil Kumar Mishra** (Student Co-convener, PG), and **Mr. Rishi Patel** (Student Co-convener, UG). Their tireless efforts and exceptional coordination were instrumental in making the event a resounding success. We also acknowledge the remarkable commitment of **Mr. Hitesh Ramrakhiyani**, General Secretary of the Students' Academic Board, whose contributions were invaluable.

We are particularly grateful to **Dr. Faladrum Sharma**, Project Officer at the Pandit Hemchandra Mission for the Popularization of Science and Assamese Culture, for his commendable support. Our sincere thanks also go to **Mr. Mangal Singh**, Research Scholar, Department of Chemical Engineering, and **Ms. Suchetna Kushwaha**, Research Scholar, School of Energy Science and Engineering, for their valuable assistance throughout the event.

We owe a special debt of gratitude to all the authors, reviewers, and invited speakers for enriching the Conclave with their knowledge and expertise. We also warmly thank **Springer Nature** for their generous sponsorship and collaboration in publishing the Proceedings of Selected Papers. We are especially grateful to **Ms. Swati Meherishi**, Editorial Director (Engineering, Materials Science, Energy, Water, Climate) at Springer Nature, New Delhi, and her dedicated team for their steadfast support and cooperation.

The success of the Research and Industrial Conclave 2024 would not have been possible without the dedication, collaboration, and guidance of everyone involved. We are truly thankful to all.

Acknowledgement

Research and Industrial Conclave—Integration 2024

We are excited to share the highlights from the 9th edition of the Research and Industrial Conclave—Integration 2024, which took place on August 9–11, 2024. This year's event was a landmark occasion, celebrating the vibrant spirit of research and innovation within our community.

Distinguished Guests

We were honored to welcome Shri Rajiv Ratan Chetwani, Director of DISM at ISRO, as our Chief Guest. His presence underscored the significance of our efforts in advancing scientific and industrial research. We also had the pleasure of hosting

Prof. Pedro from Alto University, Finland, who delivered an enlightening keynote address. Additionally, Dr. Purnima Devi Barman, a renowned Wildlife Biologist and Green Oscar Winner, enriched our program with her invaluable insights.

Record Attendance

This year, we welcomed over 800 participants—the highest attendance ever for the Scientifique (Poster/Oral Presentation) event. For the first time, selected papers were published in Springer across three comprehensive book volumes, marking a significant milestone in our publication efforts.

Workshops and Activities

The Conclave featured a diverse range of workshops, including:

- **ML/AI Workshop** by Prof. Debanga Neog and Prof. Chiranjib Sur
- **MATLAB Workshop** by Prof. Prabeer Baruah
- **IPR Workshop for Tech Transfer** by Lalit Ambastha (IP Strategist, IP Baazaar) and Dr. Sanjeev Kumar Majumdar (Sr. Manager NRDC)
- **Scientific Writing Workshop** by Swati Mehrishi (Editorial Director, Springer)

We also hosted Emprendimiento, a startup pitching event featuring over 50 startups, creating a unique platform for innovation and investment. Our Institute Open Day welcomed over 1600 school students, offering engaging workshops on Mathematics and Physics, interactive sessions from the design department, and an exciting drone demonstration by the mechanical department.

Hackathon and Tech Talks

The Conclave included a Hackathon for UG/PG students, with more than 400 participants, providing excellent opportunities for practical problem-solving and collaboration. Our Tech Talks by DST Scientists and Professors explored practical solutions through patents, showcasing cutting-edge research and its applications.

Thank You!

As we reflect on this remarkable journey, we extend our heartfelt gratitude to all contributors, participants, and supporters who made this event possible. We look forward to more inspiring and impactful gatherings in the future!

Dr. Pankaj Kalita
Faculty Convener

Prof. Uday S. Dixit
Chief Editor, RIC'24

Mr. Shubham Singh
Student Convener

Mr. Nikhil Kumar Mishra

Mr. Hitesh Ramrakhiyani

Contents

About the Editors

Dr. Uday Shanker Dixit is presently Professor in the Department of Mechanical Engineering and Head of Center for Indian Knowledge System at Indian Institute of Technology (IIT) Guwahati. He received B.E. degree in Mechanical Engineering from erstwhile University of Roorkee (now IIT Roorkee) in 1987, M.Tech. degree in Mechanical Engineering from IIT Kanpur in 1993, and Ph.D. in Mechanical Engineering from IIT Kanpur in 1998. He has worked in two industries—HMT, Pinjore and INDOMAG Steel Technology, New Delhi, where his main responsibility was designing various machines. Dr. Dixit joined the Department of Mechanical Engineering, IIT Guwahati, in 1998, where he is currently a Professor. He was also the Officiating Director of Central Institute of Technology, Kokrajhar from February 2014 to May 2015. Dr. Dixit is actively engaged in the research in various areas of design and manufacturing since last three decades. Most of his research work is focused on Modelling of Manufacturing Processes and involves finite element analysis and soft computing. He has also contributed in design optimization. He has extensively used finite element and soft computing tools for the design and manufacture of mechatronic systems. He has authored several papers and books. Presently he is an Associate Editor of the Journal of Institution of Engineers (India), Series C, Regional Editor Asia of the International Journal of Mechatronics and Manufacturing Systems and Associate Editor of the Journal of Micro manufacturing. He has guided several doctoral and masters' students. Dr. Dixit has investigated a number of sponsored projects and developed several courses. He is a Fellow of Indian Welding Society, National Advisory Committee member of International and All India Manufacturing, Technology and Design Conference. He was a Board Member of IIT Kanpur from 2018 to 2021. He is a reviewer of several journals and has organized several conferences. He has executed several socially relevant projects, such as technology development to upgrade the skills of brass metal artisans of Hajo, Guwahati and development of machine for making plates out of areca nut.

Dr. Nipjyoti Bharadwaj is currently an Assistant Professor in the area of Transportation Engineering in the Department of Civil Engineering at the Indian Institute of Technology (IIT) Guwahati. Before joining IIT Guwahati, he worked as an NRC research associate at Turner-Fairbank Highway Research Centre, USA. He earned his Ph.D. in civil engineering from the University of Missouri-Columbia, USA. He received his B.E. and M.Tech. degrees in civil engineering from Assam Engineering College, Guwahati, India, and the S. V. National Institute of Technology, Surat, India. His research interests include traffic operation, modeling, safety, naturalistic driving, driving simulator, ITS, and simulation. Dr. Bharadwaj has published over 20 research papers in international and national journals as well as conference proceedings.

Dr. Sumit Kumar is currently working as an Assistant Professor in the Department of Chemical Engineering at the Indian Institute of Technology (IIT) Guwahati. He obtained his Ph.D. in Petroleum Engineering from IIT (ISM) Dhanbad in 2020, B.Tech. in Petroleum Engineering with M.Tech. in Petroleum Management (Dual Degree) from IIT (ISM) Dhanbad in 2015. His key research areas include Flow through porous media, Modelling and Simulation, Adsorption and Pyrolysis, CO_2 sequestration, Shale gas research, Enhanced oil recovery, Reservoir flow simulation and Rock mechanics. He has taught several courses including Reservoir Engineering, Reservoir Simulation, Production Engineering, Well Testing, Enhanced Oil recovery, Engineering Economics and Energy Management.

Dr. Debabrata Sikdar is an Associate Professor at the Department of Electronics and Electrical Engineering, Indian Institute of Technology (IIT) Guwahati. He received dual degrees in B.E. (Hons.) Electronics and Instrumentation Engineering and M.Sc. (Hons.) Physics in 2008, master degree M.E. Communication Engineering in 2012 from BITS, Pilani, India, and then Ph.D. in 2015 from Monash University, Australia. He worked with General Electric from 2008 to 2010. After Ph.D., he joined Imperial College London, as a postdoctoral Research Associate in 2015. His research interests include nanophotonic and plasmonic devices, metamaterials, modern display technologies. He has pioneered many novel techniques in theory, modeling and simulation of light interaction with guided and scattering media at the nanoscale, proposed new tuneable metadevices, and published several research articles and patents. He is the recipient of several prestigious scholarships and awards such as Victoria India Doctoral Scholarship, Douglas Lampard Electrical Engineering Research Prize and Medal (by Monash University), Marie Curie Individual Fellowship (by European Commission). Recently, the esteemed Journal of Physics: Condensed Matter—IOP Science recognized him as an EMERGING LEADER 2023 for his remarkable contribution in the areas of Nanophotonics and Metamaterials.

Performance Evaluation of Water Treatment Sludge-Based Geopolymer for Road Subgrade Application

Alok Bijalwan, Bitupan Sonowal, and Anil Kumar Mishra

Abstract This article explores the utilisation potential of water treatment sludge (WTS) along with fly ash (FA) and ground granulated blast furnace slag (GGBS) as binding material for road subgrade application. The strength and durability of WTS-FA-GGBS geopolymers are examined through a succession of unconfined compressive strength (UCS), California bearing ratio (CBR) and alternate wetting–drying tests, respectively. The mechanical properties of WTS-FA-GGBS geopolymers were evaluated using a variety of factors, including binder proportions, curing period, curing conditions, (ambient curing and heat curing) and sodium hydroxide (NaOH) concentration. The results revealed that increasing the binder proportions, NaOH concentration and curing time improved the strength of the WTS-FA-GGBS geopolymer. Furthermore, heat curing was also conducted to examine the effect of heat on the mechanical properties of the geopolymer samples. The heat-cured geopolymer samples showed better unconfined compressive strength (UCS) values as compared to that of the ambient cured samples. The investigation indicated that all geopolymer samples satisfied the minimum strength standards for cement-stabilised mix, with a maximum mass loss of 8.17%. This is within the stipulated limitations (<14%) as per IRC:37-2018. The California bearing ratio (CBR) values of WTS-FA-GGBS geopolymers exceed the minimum requirements for cement-stabilised subgrade or subbase material given by IRC: SP:89-2010. Additionally, FESEM and XRD studies validated the alterations that occurred during geopolymerisation. Finally, a leaching study was conducted to investigate the geo-environmental impacts of different geopolymer specimens, and it showed that the heavy metal concentrations in the geopolymer leachates are well within the limits set by USEPA.

A. Bijalwan (✉) · B. Sonowal · A. K. Mishra
Department of Civil Engineering, Indian Institute of Technology, Guwahati, Assam, India
e-mail: b.alok@iitg.ac.in

B. Sonowal
e-mail: bitupansonowal@alumni.iitg.ac.in

A. K. Mishra
e-mail: anilmishra@iitg.ac.in

U. S. Dixit et al. (eds.), *Current Progress in Engineering Sciences*,
https://doi.org/10.1007/978-981-96-8753-4_1

1 Introduction

The demand for sustainable and eco-friendly construction materials is becoming crucial due to the growing global requirement for infrastructure development. Portland cement is the major binding agent utilised in the concrete industry. The projected global annual production is anticipated to increase from around 2.54 billion tonnes in 2006 to 4.38 billion tonnes in 2050, presuming a yearly growth rate of 5% [1]. However, the projected 2015 cement production estimate was met by 2011. Cement production alone accounts for 4–8% of global anthropogenic CO_2 emissions [2, 3]. The construction sector worldwide is currently at a critical juncture, as it must deal with the challenge of fulfilling growing infrastructure needs while also complying with strict environmental requirements. Conventional building materials, especially Portland cement, are linked to substantial carbon emissions and depletion of resources, which contribute to climate change and environmental damage. This has prompted an intensified quest for more sustainable and environmentally friendly solutions. Geopolymers have attracted significant interest as a potential solution due to their reduced impact on the environment, enhanced mechanical characteristics, and efficient utilisation of industrial waste products.

An enormous amount of industrial waste material is produced each year due to the world's rapid industrialisation and urbanisation. Recent studies have mostly focused on identifying methods to extract valuable materials from different forms of industrial waste that are commonly discarded in landfills, with specific emphasis on their desirable thermal, mechanical, and physicochemical properties. [4, 5]. Industrial waste management costs rise annually due to the substantial land requirements for its disposal. Recycling industrial waste materials in certain field applications is a mutually beneficial solution to this issue. In addition to reducing the need for traditional, resource-intensive materials like Portland cement, geopolymers contribute to addressing waste disposal problems by turning industrial by-products into useful building materials.

The worldwide production of WTS has risen significantly in response to the growing demand for potable water. Every day, approximately 100,000 tonnes of sediment are produced by water treatment plants worldwide [6]. India, with the highest population in the world, is expected to raise its water consumption from 710 billion cubic metres (BCM) in 2010 to 1180 BCM by 2050, with industrial and domestic demand estimated to be 2.5 times higher (CPCB 2011). Although there are ways for treatment and disposal, they are often costly and have a significant negative impact on the environment. The WTS can be used in some field applications to bridge the gap between efficient water treatment and sustainable sludge management. Traditionally, the most common ways of getting rid of WTS have been through landfilling and soil application [7]. Landfill disposal is characterised by inefficiency due to the need for more stringent rules and increasing costs. While it offers agronomic benefits, it also poses concerns surrounding the accumulation of heavy metals, which can result in toxicity problems for human beings, aquatic life, and terrestrial conditions.

Water treatment sludge is typically considered waste and poses disposal challenges. Its incorporation as a key ingredient in geopolymers promotes a circular economy by diverting it from landfills and addressing environmental concerns. Fly ash (FA) and ground granulated blast furnace slag (GGBS), both industrial by-products, are used as binding materials, further reducing reliance on virgin resources and contributing to waste valorisation. This study demonstrates the potential of WTS-FA-GGBS geopolymers as a sustainable alternative to conventional cement in road subgrade applications. Cement production is a significant source of CO_2 emissions, whereas these geopolymers utilise industrial by-products, reducing waste and environmental impact. The materials exhibit excellent strength and durability, meeting IRC standards for subgrade and subbase materials. The geopolymerisation process offers a low-carbon solution compared to energy-intensive cement production. Additionally, the study confirms the environmental safety of these materials through leaching tests. This research promotes eco-friendly construction practices while addressing waste management challenges.

2 Materials and Methodology

2.1 Material Characterisation

The WTS utilised in this research came from the drinking water treatment facility at IIT Guwahati in Assam, India. This water treatment plant draws its raw water supply from the Brahmaputra River. Fly ash was procured from Jalandhar, Punjab, India. The GGBS utilised in this investigation was procured from Jindal Saw Ltd. in Gujarat, India. After being removed from the bottom of the clariflocculator tank, the water treatment sediment was dried in an open area. The desiccated sludge was subsequently crushed into fine particles and screened through a 425 μm sieve. The raw WTS's UCS has been calculated in compliance with ASTM D698 at the maximum dry density MDD and optimum moisture content OMC. The particle size distribution of the water treatment sludge was determined using the combined method of sieve analysis and hydrometer analysis, following the guidelines provided by ASTM, 2007b [8] and ASTM, 2021[9], respectively. It was noted that the water treatment sediment was primarily composed of sand 10.6% (4.75–0.075 mm), silt 74.5% (0.075–0.002 mm) and clay 14.9% (<0.002 mm) and was categorised as MI (silt of intermediate plasticity) according to the Unified Soil Classification System (USCS). According to ASTM, 2007a [10], WTS had an OMC of 27.0% and an MDD of 1.441 g/cc. The WTS was found to have a specific gravity of 2.40 after being evaluated in accordance with ASTM D854 [11]. The differential free swell index of WTS is found out to be 7.7% which lies in the low category as classified by Holtz and Gibbs [12]. The physical parameters of WTS are listed in Table 1.

Scanning electron microscopy (SEM) pictures, as illustrated in Fig. 1, revealed that the appearance of FA is spherical with a few irregular features, while the shape

Table 1 Physical properties of WTS

Serial no.	Properties	Confirming to code	Result
1	Liquid Limit	ASTM D4318–10	41.9%
2	Plastic Limit	ASTM D4318–10	30.2%
3	Plasticity Index	ASTM D4318–10	11.7
4	% Clay	ASTM D422–63	14.9%
5	% silt	ASTM D422–63	74.5%
6	Classification	ASTM D2487 (USCS)	MI
7	Differential free swell index	IS 2720: Part XL: 1977	7.7%
8	Optimum moisture content (OMC)	ASTM D698	27%
9	Maximum Dry density (MDD)	ASTM D698	1.441 g/cc
10	Specific gravity	ASTM D854	2.40
11	Unconfined Compressive strength (UCS)	ASTMD2166	285 kPa
12	Soaked CBR values	IS 2720: Part 16: 1987	4.45%

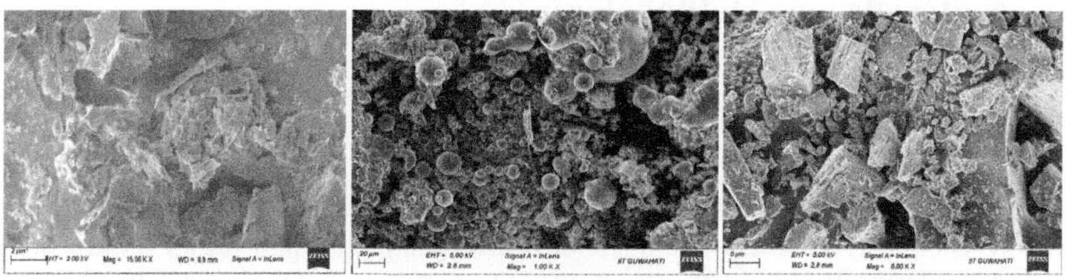

(a) WTS (b) FA (c) GGBS

Fig. 1 SEM image of **a** WTS **b** FA **c** GGBS

of WTS particles was found to be irregular and flaky. GGBS, on the other hand, had an angular shape. An X-ray fluorescence spectrometer (XRF) study was employed to determine the chemical composition of the WTS, FA and GGBS, as illustrated in Table 2. The primary chemical oxides in WTS are silicon dioxide (SiO_2) (55.51%), aluminium oxide (Al_2O_3) (16.03%) and ferric oxide (Fe_2O_3). Since geopolymer is mostly composed of silica and alumina. It is not feasible to use WTS as the only raw material for geopolymerisation due to its low amorphous character. To solve this problem, WTS geopolymer is synthesised using FA and GGBS as binder materials. The fly ash utilised in the study can be classified as Class F fly ash conforming to ASTM C 618 [13] criteria. Alkali activators are commonly employed in the production of geopolymers, and they are amorphous and semi-crystalline polymers made of aluminosilicate [14, 15]. It's more likely for aluminosilicate materials to break down when they are in an alkaline environment. This is caused by alkaline activators like sodium or potassium hydroxide or their silicates. This process releases silicon and aluminium ions within the solution when Si-O and Al-O molecules dissolve in the raw materials. Dissolved ions can recombine with the alkali activator and produce the geopolymer framework, a three-dimensional aluminosilicate network. This polymerisation process incorporates the creation of newly formed Si-O-Al and Si-O-Si bonds. Na_2SiO_3 not only enhances the dissolution of aluminium (Al) and silicon (Si) but also aids in the bonding mechanism during the geopolymerisation process [16]. Hence, the study employs a mixture of NaOH and Na_2SiO_3 as two alkaline liquids. The sodium hydroxide initially arrived in pellet form by Merck Life Science Pvt. Ltd., India, while sodium silicate was manufactured by Loba Chemie Pvt. Ltd., India, and was in a liquid state. Liquid sodium silicate is 8.0% Na_2O and 27% SiO_2. For these experimental investigations, NaOH concentrations of 4, 6, 8 and 10 M were implemented. A NaOH solution was produced for the test sample simply 1 day prior to its casting. The ratio of Na_2SiO_3 to NaOH (i.e. SS/SH) was set to be 1. In order to facilitate the formation of gels as well as the precipitation of silicates during geopolymerisation, the aqueous solution of Na_2SiO_3 was combined [17].

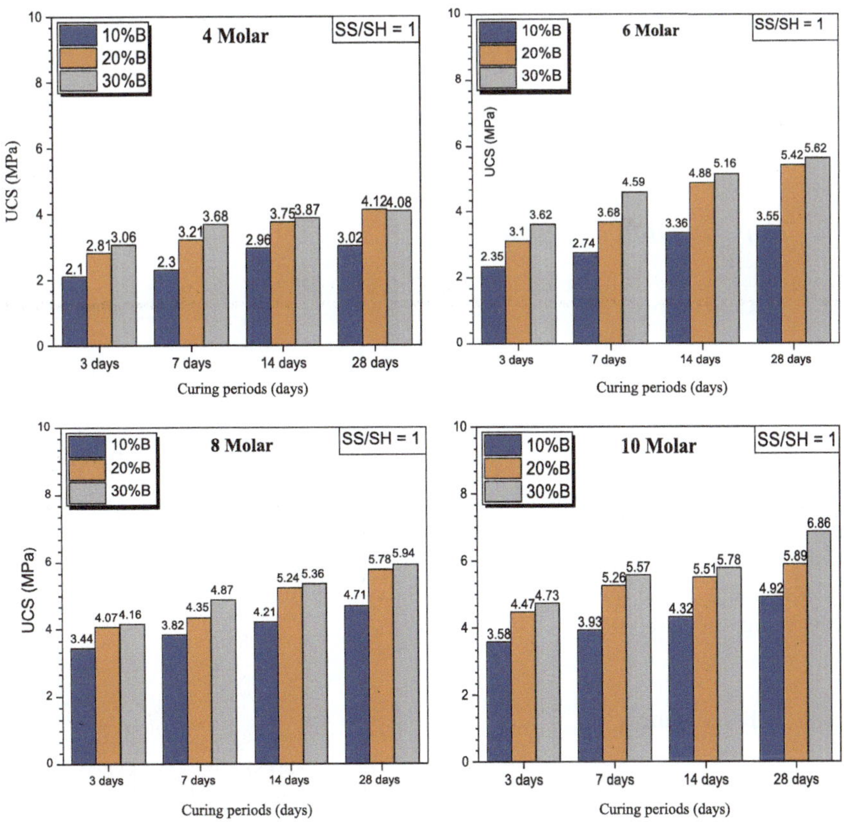

Fig. 2 UCS of WTS-FA-GGBS geopolymers cured at ambient condition for 3, 7, 14 and 28 days, with 4, 6, 8 and 10 molar concentration of alkaline activator, for SS/SH ratio of 1

Table 2 XRF displays the chemical composition of WTS, FA and GGBS

Major oxides	SiO_2	Al_2O_3	Fe_2O_3	MnO	MgO	CaO	Na_2O	K_2O	TiO_2	P_2O_5	Si-O	LOI
WTS (%)	55.51	16.03	7.27	0.086	2.62	2.07	1.71	2.97	1.54	0.17	0.21	9.82
FA (%)	55.49	29.27	4.62	0.040	0.62	0.87	0.00	1.48	2.17	0.49	0.05	4.90
GGBS (%)	33.15	17.67	1.43	0.319	5.52	35.56	0.34	0.62	0.66	0.01	1.25	3.47

2.1.1 Sample Preparation

The research involved the synthesis of WTS-based geopolymers with FA and GGBS as binding ingredients, with an FA to GGBS ratio of 1:1. To explore how binder content affects WTS-based geopolymer strength, samples were formed with 10, 20 and 30% binder. The alkaline activator used is a mixture of NaOH and Na_2SiO_3. Four NaOH concentrations (4, 6, 8 and 10 M) were utilised to examine the effect of activator solution concentration on the strength of the WTS-based geopolymer composite.

2.2 *Experimental Programme*

2.2.1 UCS Testing

The unconfined compressive strength, or UCS values of the samples, were evaluated following the guidelines of ASTMD2166 [18]. At least, three specimens with a height of 76 mm and a diameter of 38 mm were made for each combination of mixes. These specimens were compacted at the maximum dry density (MDD) and optimum moisture content (OMC). The mean value of the three specimens will be recorded as the (UCS) of the specimen. The samples undergo two types of curing: ambient curing and heat curing at a temperature of 100°C. The duration for the curing process under ambient conditions is 3, 7, 14 and 28 days, whereas for samples subjected to heat curing, the duration is 1, 3 and 7 days.

2.2.2 CBR Testing

The California Bearing Ratio (CBR) experiments were conducted for various samples in accordance with the guidelines provided by IS 2720-16 [19]. According to the guidelines laid down by IRC-37 [20], the samples were compacted statically to 97% of the MDD in a 150-mm-diameter, 175-mm-high mould. Samples were soaked for 4 days. After the soaking period load penetration curve is plotted.

2.2.3 Durability Tests

Using a series of alternating wetting–drying cycles, the durability of different samples was evaluated in compliance with IS 4332-4 (1968) criteria [21]. The production of samples for durability testing is comparable to that of UCS tests. The wetting–drying periods were initiated by immersing standard samples of a variety of mix formulations in water (wetting) for 5 hours after 7 days of ambient curing. After being dried in an oven at 70 °C for 42 h, the specimens were allowed to cool at ambient temperature for 1 hour. Subsequently, the dried samples were scratched with a metal brush, applying 18–20 vertical strokes, prior to being re-weighed and re-moistened. Consequently, each wetting–drying cycle lasted for a duration of 48 h, and the samples underwent a total of 12 wetting–drying cycles. The mass loss percentage and unconfined compressive strength of different combinations were assessed after 12 wetting–drying cycles.

2.2.4 Microstructure and Mineral Composition Analysis

FESEM and EDS analyses were conducted on different mix combinations of the samples to examine the microstructural and elemental level changes that occur during the geopolymerisation reaction. Images captured by a field emission scanning electron microscope (FESEM) (Zeiss, model: Gemini) were used for the microstructural investigation and a combination of FESEM and OXFORD EDS (energy-dispersive X-ray spectroscopy was used to assess the elemental composition. Specimens that had already undergone UCS testing were used to obtain samples for FESEM and EDS analyses. Following the unconfined compressive strength tests, the samples were ground into small particles and oven-dried overnight at 60° to eliminate any moisture before gold-coating for additional analysis. Then, the necessary levels of magnification were used to acquire the images. Using energy-dispersive X-ray spectroscopy (EDS-Oxford Instruments), the composition of elements is determined. In addition, a constant angular speed ranging from 5° to 80° (2θ) was used to generate the XRD patterns of the geopolymer samples.

2.2.5 Leaching Test

Using a USEPA 1311, the WTS-FA-GGBS geopolymer matrix with 30% binder content underwent toxicity characteristics leaching process (TCLP) testing to evaluate its potential toxicity at 4 and 10 M concentrations of NaOH with an SS/SH ratio of 1[22]. In order to get the sample ready for TCLP tests, 5 gm of the sample (with a particle size less than 2 mm) was combined with 100 mg of extraction fluid (a mixture of acetic acid and deionised water) at a pH of 4.93 ± 0.05 (pH was adjusted with 1N NaOH). The sample-to-solution ratio was 1:20, and it was shaken at room temperature for 18 h at 30 ± 2 rpm through a bath shaker. The solution was then passed through a glass fibre filter. Finally, Inductively Coupled Plasma Mass Spectrometry (ICP-MS) was used to measure the concentration of the leachate.

3 Results and Discussions

3.1 Unconfined Compressive Strength (UCS)

The UCS values of geopolymerised materials are impacted by a variety of parameters, including binder content, molar concentration, of alkali activators, curing period and curing conditions. The effects of these variables are addressed in the following sections.

3.1.1 Effect of Binder Content on Unconfined Compressive Strength

The graph in Fig. 2 clearly shows that the strength of the WTS-FA-GGBS-based geopolymers improves with increasing binder content, regardless of alkaline activator concentration and curing period. The degree of amorphousness in WTS is insufficient to be employed as the sole material for geopolymerisation. Hence, additional binder material, which is a combination of FA and GGBS, is used in the study because both binders are amorphous; these react immediately in the alkaline environment and produce the geopolymer matrix. A robust geopolymer matrix is formed, which increases strength when the binder content goes from 10 to 30%. That is because reactive components, such as alumina and silica, are more readily available in FA and GGBS, which promotes the geopolymerisation reaction.

3.1.2 Effect of Curing Period on Strength

The curing period influences the strength growth of geopolymer materials. Figure 2 shows the UCS values of samples with varying molarities and binder contents that were ambient cured or air-dried for 3, 7, 14 and 28 days. It has been noted that the UCS values increase as the curing period increases from 3 to 28 days, regardless of the binder content and concentration of the alkaline activator. In the initial 3 days of alkali activation, the precursor materials' alumina and silica are leached off, leading to the formation of geopolymeric gels. The geopolymerisation reaction begins in the first 3 days of curing and continues to progress. Figure 3 shows a sharp increase in strength as the curing period extends from 3 to 7 days. The geopolymeric gels developed by the alkali activation undergo additional polymerisation and cross-linking by the completion of the 7-day curing period. This results in a denser and more linked geopolymer network, which increases strength. In comparison to the early phases, the geopolymer's strength increases as the condensation process

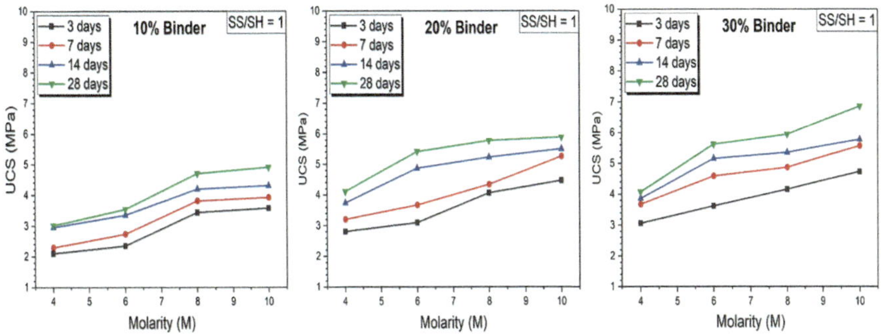

Fig. 3 Variation of UCS values of WTS-FA-GGBS geopolymer with different curing period and molarity for 10, 20 and 30% Binder and SS/SH ratio of 1

nears completion, and a more robust geopolymer matrix is generated at the end of the 28 days. A common benchmark for evaluating the strength performance of cementitious materials, including geopolymer, is the strength of the 28-day-cured samples.

3.1.3 Effect of NaOH Concentration

Alkaline solution concentration can significantly impact geopolymerised sample strength. Figure 3 show WTS-FA-GGBS geopolymer strength with varied binder percentages, curing periods, and alkaline solution NaOH concentrations. When the concentration of sodium hydroxide was raised from 4 to 10 M, the strength of 28-day ambient-cured samples with 10% binder rose from 3.02 to 4.92 MPa. A higher concentration of NaOH makes aluminosilicate source materials more soluble and speeds up the dissolution of free $[SiO4]^-$ and $[AlO4]^-$ tetrahedra, which then polymerise to create a Si-O-Al-O link between them. This Si-O-Al-O bonding is negatively charged; therefore, a higher proportion of SH can ensure that there are enough Na^+ cations to compensate for this charge imbalance. According to Singh et al. [23], increasing the concentration from 5 to 10 M improves the strength of geopolymer samples more significantly. Wang et al. [24] found that the unconfined and flexural strength of metakaolin-based geopolymers improved when the NaOH concentration increased from 4 to 12 M.

3.1.4 Effect of Heat Curing

Temperature and curing conditions have a significant impact on how a geopolymer material behaves in terms of strength development. This study looked at how heat curing affected the strength of WTS-FA-GGBS geopolymers with varying binder contents (10, 20 and 30%), concentrations of NaOH (4, 6, 8 and 10 M), and curing periods. The series of prepared WTS-FA-GGBS geopolymer samples had been initially permitted to be set for 24 h at ambient temperature. They were then cured for 1, 3 and 7 days at 100 °C in an oven. Figure 4 shows the UCS values of WTS-FA-GGBS geopolymers with varying binder contents, and NaOH concentrations after curing at 100 °C for 1, 3 and 7 days. For all NaOH concentrations and binder contents, the UCS values are found to be notably impacted by the heat-curing condition. For instance, when cured for 1–3 days at 100 °C, strength was improved from 3.92 to 4.84 MPa at a 4 M concentration of NaOH and 10% binder content. This suggests that heat curing can promote early strength growth in geopolymer samples at ambient curing conditions, aluminosilicate dissolution and polymerisation rates are relatively slow, which reduces strength. High curing temperatures easily dissolve silica and alumina from source materials, forming substantial aluminosilicate gels with high strength at an early curing age. Temperature curing enhances the mobility of Na^+ and OH^- ions, resulting in increased dissolution of SiO_2 and Al_2O_3 [25, 26]. However, after further curing for 3–7 days, the samples lost some of their strength. Prolonged heat at 100 °C caused significant moisture loss and microcracks in geopolymer samples, which reduced their strength.

3.1.5 California Bearing Ratio (CBR)

Raw water treatment sludge (WTS) has a soaked CBR value of 4.45%, which is below the minimal standard for materials used for constructing roads. Materials used in subgrade construction must have a required minimum CBR value of 8% in order to comply with the criteria set out by IRC:37-2012 [27]. Samples were set to cure at room temperature for 3 days before being immersed in water for 4 days while carrying a 5 kg surcharge load. Geopolymers with a 10% binder exhibited an increase in CBR values from 64.2 to 161.8% as the molarity was raised from 4 to 10 M. At greater NaOH concentrations, the dissolution of silica and alumina from the source materials speeds up the synthesis of geopolymer gels. Raw material particles are bonded by this gel, therefore improving bearing resistance, strength, as well as CBR characteristics. Moreover, higher binder content resulted in a higher CBR value. With a NaOH concentration of 4 M, the CBR went up from 64.2 to 137.1% as the binder content increased from 10 to 30%. A significant quantity of binder guarantees sufficient silica and alumina for geopolymerisation, producing a material with enhanced bearing resistance.

The highest CBR value was 260.32% with 30% binder, 10 M NaOH, and 1 SS/SH ratio. The IRC: SP:89 [28] specifies that a cement-stabilised mix employed as subgrade material must have a minimum CBR value of 15%. Figure 5 demonstrates that all WTS-FA-GGBS geopolymer samples soaked CBR values exceed subgrade or subbase material criteria, regardless of binder quantity and NaOH concentration.

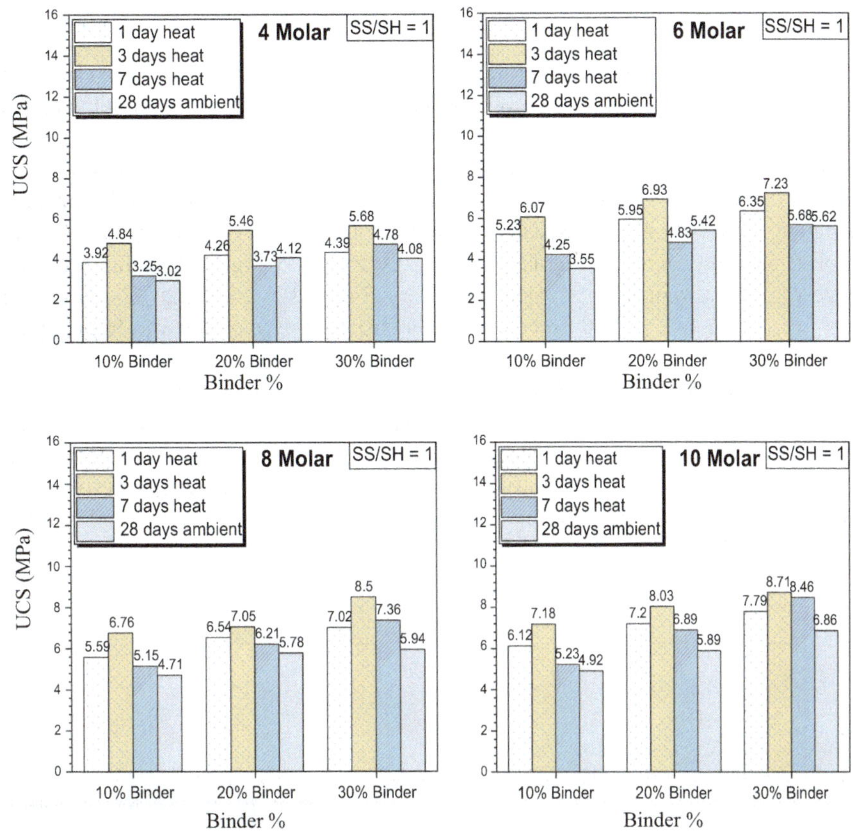

Fig. 4 UCS of heat cured WTS-FA-GGBS geopolymers for a curing period of 1, 3 and 7 days at a temperature of 100 °C, with 4, 6, 8 and 10 M alkaline solution, for SS/SH ratio of 1

3.1.6 Durability Performance of Geopolymerised Sample

The wetting–drying experiments were conducted in accordance with the parameters specified by IS 4332-4 (1968) [21]. Figure 6a shows that when the concentration of NaOH and the percentage of binder increases, the mass loss percentage tends to decrease. When the molarity is raised from 4 to 10 M, mass loss declines from 5.88 to 2.41% at 30% binder content. Furthermore, with a rise in binder content from 10 to 30%, the percentage of mass loss dropped from 8.17 to 5.88% at a NaOH concentration of 4 M. A greater concentration of NaOH causes the precursor materials' alumina and silica to dissolve more quickly, forming extensive geopolymer gels and a denser, more linked geopolymer matrix. Thus, it leads to decreased mass loss during wetting and drying cycles. The plot in Fig. 6b illustrates the residual unconfined compressive strength (UCS) of geopolymer samples after undergoing 12 cycles of wetting and drying. We compare the UCS values with the UCS of samples

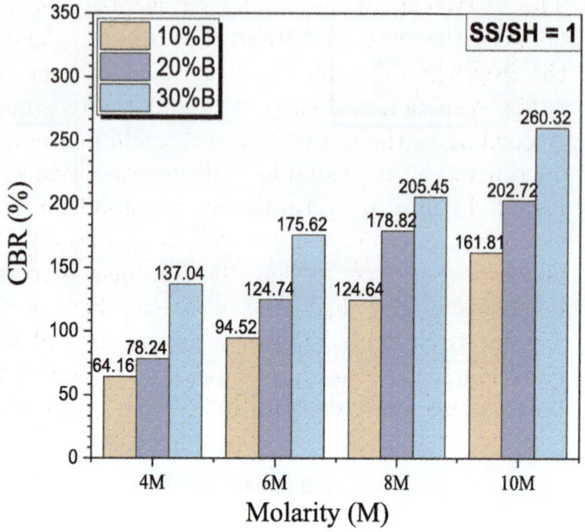

Fig. 5 Soaked CBR values of WTS-FA-GGBS geopolymer samples at different concentrations of SH, Binder contents and SS/SH ratio of 1

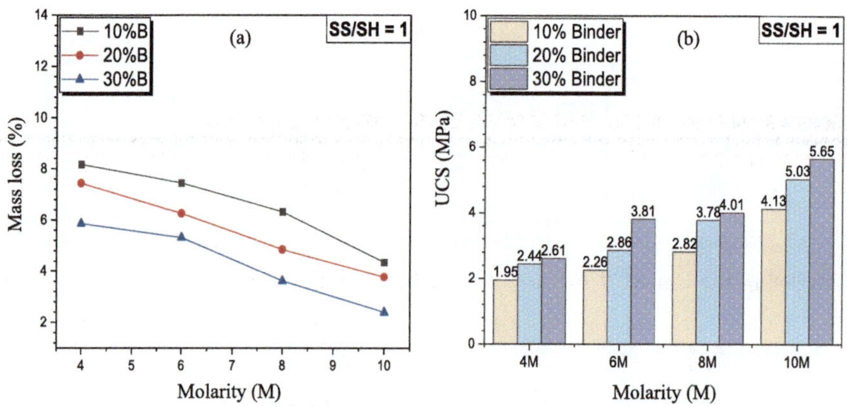

Fig. 6 **a** Mass loss percentage and **b** UCS of WTS-FA-GGBS geopolymers after 12 wetting–drying cycles, evaluated at different molarities of SH, binder contents and SS/SH ratio of 1

cured under ambient conditions for 28 days, considering varying molarities of SH and binder content. The term "residual" refers to the remaining strength of the samples after they have undergone 12 cycles of wetting and drying.

3.1.7 Microstructure and Mineral Composition Analysis

Figure 7 shows the XRD pattern of WTS-FA-GGBS geopolymer specimens that were cured at ambient temperatures for a duration of 28 days with 30% binder and for NaOH concentrations of 4 and 10 M. XRD patterns showed that all geopolymer samples contained geopolymer reaction products like calcium aluminosilicate hydrate (CASH), calcium silicate hydrate (CSH) and sodium silicate hydrate (NSH), which contribute towards their strength. According to several studies [29–32], the CASH gel is the main reaction product that develops when the calcium in GGBS is activated by an alkali. Alkali dissolves calcium in GGBS, forming free Ca^{2+} cations. At the same time, the activator breaks down the aluminosilicates to release [SiO4]- and [AlO4]- tetrahedral units. These units then join together to form silica and alumina monomers. In this way, the negatively charged alumina and silica monomers are balanced out by the free Ca^{2+} cations. This creates CASH gel, which helps make the geopolymer composite stronger, denser and more linked.

Figure 8 illustrates the microstructure of WTS-FA-GGBS geopolymer samples that were cured at room temperature. These samples had a binder content of 30% and an SS/SH ratio of 1. The SH concentrations tested were 4 and 10 M. In Fig. 8, FA and GGBS particles have smoother surfaces, indicating that low concentrations of NaOH (4 M) were insufficient to leach silica and alumina from precursor materials, limiting reaction product production. As the molarity increased from 4 to 10 M, the FA and GGBS particles were almost completely dissolved, resulting in a considerable fraction of reaction products (CASH, NASH, CSH). This resulted in a denser and more compact structure, which in turn increased the compressive strength.

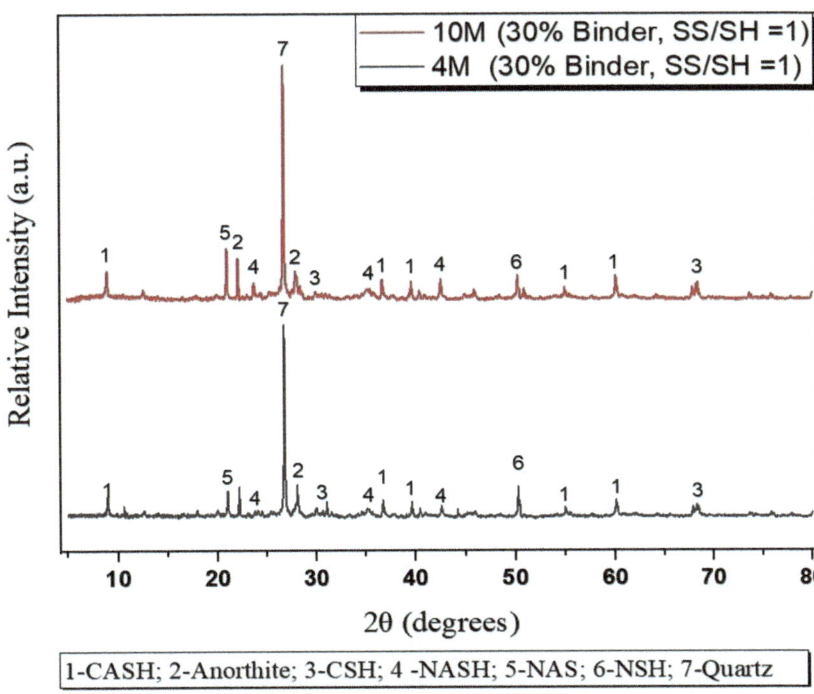

Fig. 7 XRD pattern of 28 days ambient cured WTS-FA-GGBS geopolymers with 30% Binder and SS/SH ratio of 1 for two different concentrations of SH (i.e. 4 and 10 M)

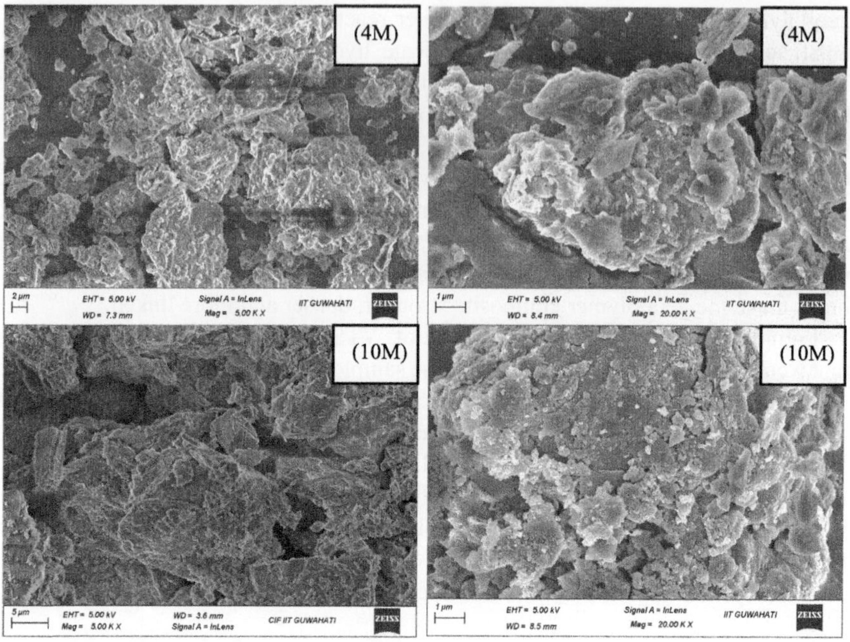

Fig. 8 SEM image of WTS-FA-GGBS geopolymers with 4 and 10 M SH concentration at 5KX and 20KX magnification

In order to evaluate the aluminosilicate components' solubility more accurately, it is essential to examine the geopolymer matrix's elemental composition using EDS analysis. Figure 9 illustrates the WTS-FA-GGBS geopolymer's EDS patterns. These samples were cured at the ambient temperature by utilising 30% binder, an SS/SH ratio of 1 and at 4 and 10 M NaOH concentrations. A higher compressive strength was observed in the geopolymerised specimens with increasing Si/Al and Na/Al ratios [33]. Greater Si/Al and Na/Al ratios suggest enhanced dissolution of aluminosilicates, resulting in increased compressive strength in geopolymer samples due to the formation of more prominent geopolymer gels.

3.1.8 Leaching Potential of the Geopolymer Samples

Table 3 summarises the heavy metal concentrations in the leachate as indicated by the TCLP analysis. According to Table 3, the levels of heavy metals in the leachate are significantly lower than the regulation limits set by the USEPA. Geopolymerisation in very alkaline conditions forms small and large aluminate and silicate compounds from aluminosilicate minerals. By means of a physical binding mechanism, these structures efficiently capture and encapsulate various heavy metals, therefore enabling the ongoing development of polycondensation reactions [34].

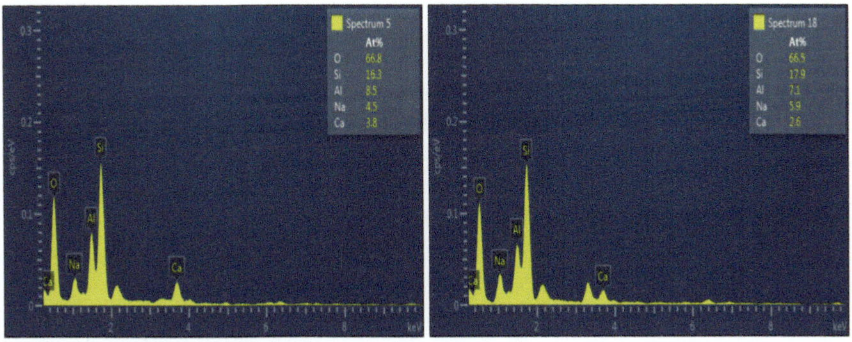

Fig. 9 EDS result of the WTS-FA-GGBS geopolymers with 30% binder and SS/SH ratio of 1 for two different concentrations of SH **a** 4 M and **b** 10 M

Heavy metal immobilisation in geopolymers depends on factors such as the metal type, raw material characteristics, activator type and pore structure. Therefore, it is clear that the use of WTS-FA-GGBS-based geopolymers in subgrade construction does not pose any potential environmental hazards, involving the discharge of heavy metals that could lead to groundwater pollution.

Table 3 Toxicity characteristics of the geopolymerised sample with 30% Binder (FA + GGBS) and SH concentration 4 and 10 M keeping SS/SH ratio 1

Analytes	70%WTS-30%B	4 M–30%B 1 (g/ml)	10 M–30%B 1 (g/ml)	USEPA Limits for hazardous waste (g/ml)
Cd	0.000	0.000	0.000	**1.0**
Zn	0.159	0.197	0.059	**25.0**
Pb	0.036	0.030	0.010	**5.0**
Cu	0.128	0.120	0.000	**25.0**
Cr	0.141	0.123	0.009	**5.0**
Ni	0.056	0.042	0.032	**25.0**
Mn	1.064	0.789	0.431	**10.0**
Ba	0.235	0.303	0.085	**100**
Se	0.002	0.001	0.001	**1.0**

where 4 M–30%B 1 denotes 4 M is NaOH concentration, 30%B is the binder percentage of (FA + GGBS) and SS/SH ratio = 1

4 Conclusion

This study examined the strength development properties of geopolymer produced from (WTS) using (FA) and (GGBS) as binding agents. An alkaline activator has been made by combining SH and SS solutions with an SS/SH ratio of 1. The mechanical characteristics of WTS-FA-GGBS geopolymers were evaluated, considering several factors including different binder content, curing period, conditions of curing and SH concentration.

The main conclusions and observations of the present research based on the analysis and experimental test results are as follows:

- The increase in molarity of alkali activator and binder content improved the UCS and CBR of the treated WTS.
- An increase in the curing temperature and curing period had a positive impact on the compressive strength of the geopolymer samples. Heat-cured samples' UCS values increased from 1 to 3 days and then dropped up to 7 days due to moisture loss and microcracks. All geopolymer samples satisfied IRC (2014) cement-stabilised subgrade strength requirements (>1.2 MPa).
- For all geopolymer samples, mass loss levels were significantly lower than the limitations (<14%) established by IRC:37-2018 for the cement stabilised mix.
- The improved dissolution of the raw materials caused by higher concentrations of SH resulted in widespread aluminosilicate gel formation, which in turn contributed to higher strength, as confirmed by scanning electron microscopy (SEM).
- The TCLP results show that heavy metal levels in the geopolymer leachate are within USEPA guidelines.

References

1. M. Schneider, M. Romer, M. Tschudin, H. Bolio, Sustainable cement production—present and future, Cement and concrete research 41(7) (2011) 642–650.
2. E. Gartner, Industrially interesting approaches to "low-CO2" cements, Cement and Concrete research 34(9) (2004) 1489–1498.
3. C. Chen, G. Habert, Y. Bouzidi, A. Jullien, Environmental impact of cement production: detail of the different processes and cement plant variability evaluation, Journal of cleaner production 18(5) (2010) 478–485.
4. F. Colangelo, R. Cioffi, F. Montagnaro, L. Santoro, Soluble salt removal from MSWI fly ash and its stabilization for safer disposal and recovery as road basement material, Waste Management 32(6) (2012) 1179–1185.
5. S.A. Bernal, E.D. Rodríguez, A.P. Kirchheim, J.L. Provis, Management and valorisation of wastes through use in producing alkali-activated cement materials, Journal of Chemical Technology & Biotechnology 91(9) (2016) 2365–2388.
6. A.O. Babatunde, Y. Zhao, Constructive approaches toward water treatment works sludge management: an international review of beneficial reuses, Critical Reviews in Environmental Science and Technology 37(2) (2007) 129–164.

7. M. Sanchez-Monedero, C. Mondini, M. de Nobili, L. Leita, A. Roig, Land application of biosolids. Soil response to different stabilization degree of the treated organic matter, Waste Management 24(4) (2004) 325–332.
8. D. ASTM, Standard test method for particle-size analysis of soils (2007).
9. ASTM, D7928-21e1; Standard Test Method for Particle-Size Distribution (Gradation) of Fine-Grained Soils Using the 527 Sedimentation (Hydrometer) Analysis, ASTM International: West Conshohocken, PA, USA (2021).
10. D. ASTM, 698; Standard Test Methods for Laboratory Compaction Characteristics of Soil Using Standard Effort, ASTM International: West Conshohocken, PA, USA (19428-2959) (2007).
11. ASTM, Standard test methods for specific gravity of soil solids by water pycnometer, Designation (2000).
12. W.G. Holtz, H.J. Gibbs, Triaxial shear tests on pervious gravelly soils, Journal of the Soil Mechanics and Foundations Division 82(1) (1956) 867-1-867-22.
13. C. ASTM, 494. Standard Specification for Chemical Admixtures for Concrete, West Conshohocken, PA: ASTM International, 1999.
14. J. He, G. Zhang, Geopolymerization of red mud and rice husk ash and potentials of the resulting geopolymeric products for civil infrastructure applications, Developments in Strategic Materials and Computational Design II: Ceramic Engineering and Science Proceedings 32 (2011) 45–52.
15. J. Provis, J. Van Deventer, Geopolymerisation kinetics. 2. Reaction kinetic modelling, Chemical engineering science 62(9) (2007) 2318–2329.
16. J. Kiventerä, P. Perumal, J. Yliniemi, M. Illikainen, Mine tailings as a raw material in alkali activation: A review, International Journal of Minerals, Metallurgy and Materials 27 (2020) 1009–1020.
17. D. Khale, R. Chaudhary, Mechanism of geopolymerization and factors influencing its development: a review, Journal of materials science 42 (2007) 729–746.
18. ASTM, ASTM D2166/D2166M-16: Standard Test Method for Unconfined Compressive Strength of Cohesive Soil, ASTM, 2016.
19. IS, Laboratory determination of CBR, Bureau of Indian Standards New Delhi, India, 1987.
20. IRC-37, Guidelines for the design of flexible pavements, Indian code of practice, IRC 37 (2001).
21. IRC, IS 4332-4: Methods of test for stabilized soils, Part 4: Wetting and drying, and freezing and thawing tests for compacted soil-cement mixtures, Bureau of Indian Standards New Delhi, India, 1968.
22. USEPA, Waste leachability: The need for review of current agency procedures (1999).
23. S. Singh, A. Kumar, T. Sitharam, Experimental study on strength, durability, hydraulic and toxicity characteristics of soil treated with mine tailings based geopolymers for sustainable road subgrade application, Construction and Building Materials 414 (2024) 134894.
24. H. Wang, H. Li, F. Yan, Synthesis and mechanical properties of metakaolinite-based geopolymer, Colloids and Surfaces A: Physicochemical and Engineering Aspects 268(1) (2005) 1–6.
25. P. Sajan, T. Jiang, C. Lau, G. Tan, K. Ng, Combined effect of curing temperature, curing period and alkaline concentration on the mechanical properties of fly ash-based geopolymer, Cleaner Materials 1 (2021) 100002.
26. S. Singh, A. Kumar, T. Sitharam, Mechanical, Microstructural, and Durability Properties of Soil Stabilized with Alkali-Activated Jarofix for Road Applications, Journal of Hazardous, Toxic, and Radioactive Waste 27(4) (2023) 04023020.
27. IRC-37, Tentative guidelines for the design of flexible pavements, Indian Road Congress, Kama Koti Marg, Sector-6, R.K. Puram, New Delhi-110 022, 2012.
28. IRC, Guidelines for soil and granular material stabilization using cement, lime and fly ash, IRC: SP 89 (2010).
29. S.R. Abdila, M.M.A.B. Abdullah, R. Ahmad, S.Z.A. Rahim, M. Rychta, I. Wnuk, M. Nabiałek, K. Muskalski, M.F.M. Tahir, Syafwandi, Evaluation on the mechanical properties of ground granulated blast slag (GGBS) and fly ash stabilized soil via geopolymer process, Materials 14(11) (2021) 2833.

30. M. Hoy, S. Horpibulsuk, A. Arulrajah, A. Mohajerani, Strength and microstructural study of recycled asphalt pavement: Slag geopolymer as a pavement base material, Journal of Materials in Civil Engineering 30(8) (2018).
31. S. Lekshmi, J. Sudhakumar, An assessment on the durability performance of fly ash-clay based geopolymer mortar containing clay enhanced with lime and GGBS, Cleaner Materials 5 (2022) 100129.
32. H. Miraki, N. Shariatmadari, P. Ghadir, S. Jahandari, Z. Tao, R. Siddique, Clayey soil stabilization using alkali-activated volcanic ash and slag, Journal of Rock Mechanics and Geotechnical Engineering 14(2) (2022) 576–591.
33. M. Yaghoubi, A. Arulrajah, M.M. Disfani, S. Horpibulsuk, M.W. Bo, S. Darmawan, Effects of industrial by-product based geopolymers on the strength development of a soft soil, Soils and foundations 58(3) (2018) 716–728.
34. K. Taki, S. Mukherjee, A.K. Patel, M. Kumar, Reappraisal review on geopolymer: A new era of aluminosilicate binder for metal immobilization, Environmental Nanotechnology, Monitoring & Management 14 (2020) 100345.

Effectiveness and Challenges of Local Retrofitting Methods Used in Low- and Medium-Rise Non-ductile RC-OGS Frames

Bajrabahu D. N. Deo, Romanbabu M. Oinam, and Sanjeev Kumar

Abstract Retrofitting old buildings to make them seismically resilient and durable is a better option rather than demolishing them. It is also a good method for safeguarding the structure from dynamic loads. There are numerous local retrofitting options, most of which include jacketing. According to literature surveys, carbon fibre-reinforced polymers (CFRP) for jacketing are growing in popularity. This study focuses on CFRP jacketing as a local retrofitting solution, examining its benefits and drawbacks for low- to medium-rise reinforced concrete open-ground storey (RC-OGS) buildings. Several parameters are employed to define structural behaviour, including base shear, drift ratios, inter-storey drift (IDR) and so on. Numerical validation of the experimental work, which includes a half-scale single-bay two-storey experimental CFRP retrofitted frame specimen, was performed to carry out these works numerically, and the results were in good agreement with the experiment results. The dynamic study demonstrates that the local retrofitting technique performs better when base shear, drift ratios, IDR and other factors are taken into account, but it is ineffective when it comes to the total structure's energy dissipation capacity at the global level. The nonlinear static analysis demonstrates that in medium-rise structures, the retrofitted non-ductile structure just moves damage from the ground level to the first floor rather than fully following a pre-defined failure mechanism. Furthermore, removing these issues in conjunction with other retrofitting solutions is a future scope of our work.

Keywords Non-ductile RC-OGS frame · CFRP jacketing · Retrofitting · Numerical modelling · OpenSees

---

B. D. N. Deo · R. M. Oinam (✉) · S. Kumar
Department of Civil Engineering, IIT Guwahati, Guwahati, India
e-mail: romanbabu@iitg.ac.in

B. D. N. Deo
e-mail: d.bajrabahu@iitg.ac.in

S. Kumar
e-mail: sanjeev.ku@iitg.ac.in

U. S. Dixit et al. (eds.), *Current Progress in Engineering Sciences*,
https://doi.org/10.1007/978-981-96-8753-4_2

1 Introduction

Many existing reinforced concrete (RC) structures in India were built prior to the establishment of seismic design guidelines. These buildings are very prone to earthquakes. Existing open-ground storey (OGS) structures are more vulnerable than other RC buildings because of the high shear and rotational demand in the ground columns [1]. Some of these buildings sometimes referred to as non-ductile reinforced concrete structures, lacked sufficient strength to withstand lateral loads and had a brittle failure mode [2]. This is evident from the incidence of the Bhuj earthquake, where one hundred fifty OGS buildings collapsed [3]. Also, the majority of uncollapsed OGS buildings have been damaged heavily. Such issues become an inherent property of OGS structures. The poor structural safety of these structures sometimes demands demolition and rebuilding. However, this approach is not economically viable and can significantly affect the nation's economy.

The local retrofitting processes primarily refer to jacketing procedures such as concrete jacketing, steel caging or jacketing, CFRP jacketing and so on. Jacketing procedures are successful and can improve mechanical qualities such as stiffness and strength, minimise inter-storey drift, remove sources of weakness (stress contraction) locally in low- to moderately damaged structures and are cost-efficient and impactful in component-level retrofitting [4]. Retrofitting with a CFRP jacket boosts the strength, especially for square specimens, mainly columns that are extensively damaged [5]. However, the presence of pre-damage in members may have a detrimental impact on confinement efficiency, resulting in a minimal increase in strength when compared with unconfined strength [6]. The CFRP jacketing is capable of decreasing residual displacements and oblique cracks between the shear and flexural zones in beams under both impact and dynamic loads [7]. Benzaid et al. experimented with the effect of CFRP on the structural parameters such as strength and ductility of plane concrete columns and reinforced concrete columns while considering various test parameters such as column slenderness, number of wrap layers and concrete strength, and they concluded new peak stress equations and effective strain coefficients that can easily predict the axial strength and corresponding strain of FRP confined column [8]. This model considers the effects of effective circumferential failure strain and effective lateral confinement failure. They also discovered that all CFRP-wrapped specimens failed, producing a characteristic sound and that the fibre rupture begins in the centre and spreads to the end. The smaller the area in which the CFRP bursts, the greater the slenderness ratio. The deformation capacity decreases as the quantity of CFRP increases. On the other hand, the members in compression reveal that the dilation ratio grows after the axial strain reaches an identical value, at which point the CFRP jackets begin to alter their behaviour noticeably. The strength gain was improved because the CFRP was activated after a significant axial strain. It is also likely that some strength deterioration occurs before the jacket contributes effectively to strength [9]. Samy et al. used CFRP jacketing with an elliptical column profile and found that it increased its strength by about 60% compared to the typical CFRP wrapping process and by 80% over the original column specimen [10]. Even in eccentric load conditions, the fully wrapped designs of CFRP sheets produce better outcomes for columns [11].

2 Research Significance

It is clear that component-level testing less precisely forecasts the behaviour of the actual building under dynamic loads, and a minimum of work is performed while considering the overall structure of the non-ductile type. Considering this literature-based study, CFRP jacketing is chosen for future investigation in this study. However, retrofitting ground floor columns in OGS frames is a traditional method used in this numerical study. To simulate low- and medium-rise buildings, two different frames of three-storey as low-rise buildings and six-storey as medium-rise buildings are considered. Both are designed for gravity loading. These structures are assessed using the performance-based plastic design (PBPD) approach and retrofitted with the ACI 440 [12]. The primary goal of this work is to identify the benefits and drawbacks of the local retrofitting method using RC-OGS frames.

3 Structural Evaluation and Retrofitting

3.1 PBPD Method

The energy balance concept guides the performance-based plastic design (PBPD) technique, which also considers the inelastic behaviour and structural component properties during the design process to reach a desired goal, and it was successfully applied to the RC moment frames [13, 14]. It can also be used to retrofit RC frames [15, 16]. According to the PBPD method, the design yield base shear of a frame structure can be expressed as follows [17]:

$$\frac{V_y}{W} = \frac{-\alpha + \sqrt{\alpha^2 + 4S_A^2\left(\frac{\gamma}{\eta}\right)}}{2}; \text{ Here, } \alpha = \left(\frac{h_1 \theta_p 8\pi^2}{T^2 g}\right) \tag{1}$$

where V_y is the design base shear, W represents the total seismic weight of the study frame, γ is the energy modification factor and η represents the factor used for pinched hysteretic behaviour. The dimensionless parameter, α and period-based spectral acceleration, S_A both depend on the time period, T value. h_I represents the height of the ground floor and θ_p is the plastic rotation. Here, it was assumed that the θ_p of the frame is the same as inelastic drift that depends on the yield drift, θ_y and ultimate drift, θ_u through the relation ($\theta_p = \theta_u - \theta_y$). The values of T (for RC-MRF building) and S_A for a target DBE spectrum can be calculated using code provisions

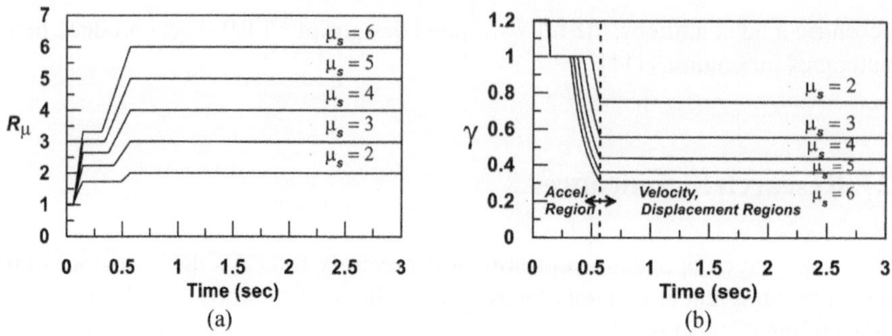

Fig. 1 Variation of **a** ductility reduction factor and **b** structural ductility factor with natural time periods

[18] for the evaluation of structure as given below:

$$S_A = \frac{Z}{2}\left(\frac{S_a}{g}\right); \text{ and, } T = 0.075h^{0.75} \tag{2}$$

where Z represents the zone factor, s_a/g is the design acceleration coefficient for different soil types for the time period, T and h is the height of the structure. The seismic input energy, E_i and the dissipated energy, E_d can be calculated by the equations given below:

$$E_i = \frac{\gamma W S_A^2 T^2}{8\eta} \tag{3}$$

$$E_d = (2\theta_p + \theta_y)M_{pc}n_c \tag{4}$$

Here, M_{pc} represents the moment capacity of the column and n_c= number of columns. Equation (4) is derived based on the assumption that all flexural plastic hinges formed in all the ground floor column ends, indicating the soft storey mechanism of the OGS structure. In this study, the relationship suggested by Newmark and Hall [19] is used for calculating the ductility reduction factor. Figure 1 shows the relationship between μ_s, R_μ and T is the structural period, and Fig. 3 shows the resulting values of γ.

3.2 *Case Study Frames and Evaluation*

This study uses two interior frames for three-storey and six-storey buildings, which are considered to have varied dimensions[20]. Figure 2 shows the used frame elevation for both frames, in which C1, C2, C3 and B1, B2, B3 represent the columns and beams used on the ground floor, 1st–2nd floors, 3rd–4th–5th floors, respectively, for the six-storey frame and C4, B4 represents column, beam for the three-storey frame.

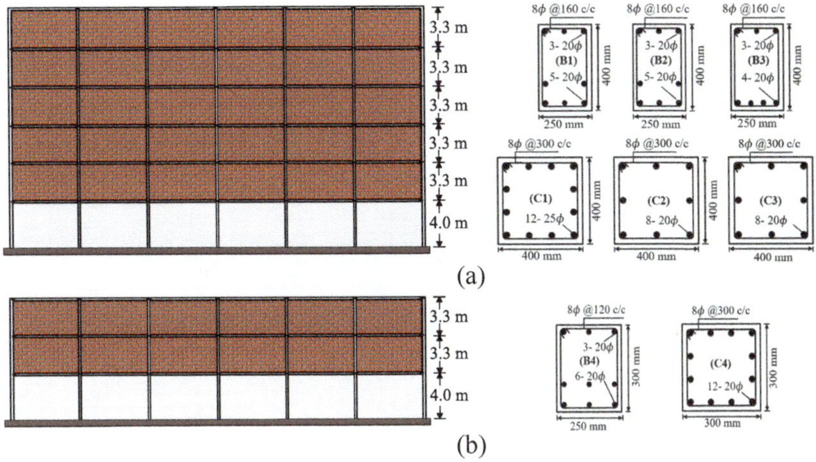

Fig. 2 Elevation and section details of study frames: **a** six-storey and **b** three-storey

Both the structures are used as moment-resisting RC-OGS frames and were designed under gravity loads utilising the Indian standard codes [21, 22]. The building has a 150 mm slab thickness and 115 mm wall thickness. The unit weight of the masonry infill wall is 18 kN/m^3. The concrete strength of 25 MPa and Fe415 steel reinforcement bars are used. The values of yield stress, ultimate stress and elongation percentage for Fe415 grade steel bars are 415 MPa, 485 MPa and 14.5%, respectively. The column and beam cross-sections are 400 mm × 400 mm, 250 mm × 400 mm for the six-storey frame and 300 mm × 300 mm, 250 mm × 300 mm for the three-storey frame, respectively. Moderate exposure conditions and earthquake zone V are also considered. All roofs except the top were first subjected to a live load of 3 kN/m^2. A load factor of 0.25 has been used to calculate the seismic weight for this live load. The columns were designed for axial load with biaxial moment considering dead and live loads. The clear covers for columns and beams are 40 and 25 mm, respectively.

For the six-storey frame, for a target DBE value, the evaluated θ_y is obtained as 0.78%. However, the used CFRP jacket is designed for a target ultimate drift (θ_t) of 2%. The energy modification factor, γ can be determined from the structural ductility factor, μ_s, and the ductility reduction factor, R_μ using the relationship mentioned in Fig. 1. The yield base shear (V_y/W) the unstrengthened frame is obtained as 0.293 after including the P-delta effect in the frame. The evaluation results show that the structure leads to the deficiency of a bending moment of 282 kN-m and a shear force of 110 kN for each column. This deficiency will be countered with the help of retrofitting at the ground floor level. A similar evaluation approach has been used for the three-storey frame, as it shows a deficiency of a bending moment of 66 kN-m and a shear force of 30 kN for each column. Table 1 describes the evaluation results for both frames using the PBPD method. Both of the frames were evaluated for a target drift of 2% under the DBE value. For these deficiency values of bending moment and shear force, the CFRP jackets are designed as per the provisions of ACI 440 [12].

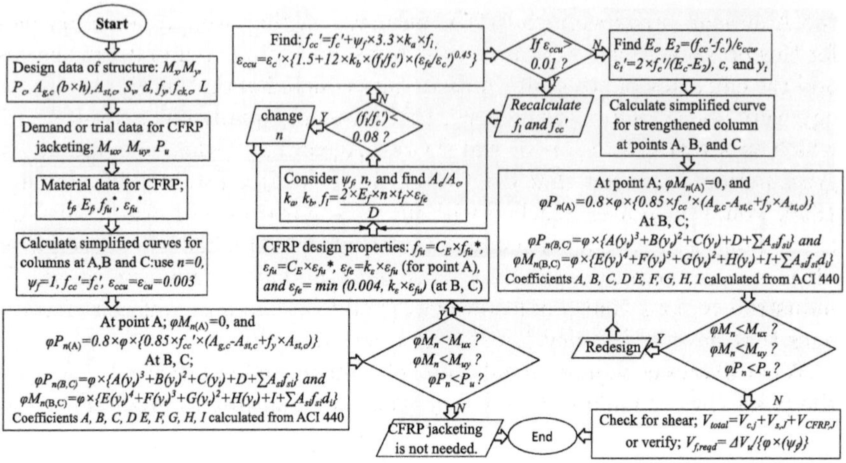

Fig. 3 Design procedure for CFRP jacketing as per ACI 440 [12]

Table 1 Structural evaluation using the PBPD method

Sl. No.	Parameters	Values	
		Six-storey	Three-storey
1	Time period, T (s)	0.7225	0.44059
2	Importance factor, I (Residencial building)	1.2	1.2
3	Response reduction factor, R (Categorised as OMRF)	3	3
4	Seismic coefficient, S_a/g (soft soil)	2.31121	2.5
5	Spectral response, S_A (g)	0.41	0.45
6	Yield drift, θ_y (%)	0.78	1.15
7	Length of plastic hinge, L_p (% of column length)	9.7	9.7
8	Target drift for DBE, θ_t (%)	2	2
9	Required structural ductility, μ_s (θ_t/θ_y)	2.564	1.7391
10	Ductility reduction factor, R_μ (From Fig. 1a)	2.564	1.3913
11	Energy modification factor, γ (From Fig. 1b)	0.65	1
12	The factor for the pinched hysteretic parameter, η	0.4	0.4
13	Seismic weight of the structure, W (kN)	7792	3572
14	Tributary gravity load at the first floor (With P-Δ effect) (kN)	155.85	71.44
15	Required base shear, V_y (kN)	2292.46	1131.954
16	Ratio, V_y/W	0.29	0.3169

3.3 Design of CFRP Jacket

In the case of CFRP jacketing, initially, it is necessary to calculate the simplified curve for the unstrengthen column and then evaluate the amount of CFRP required, as described by ACI 440 [12]. As per the code provisions, three points are the column under pure compression, the neutral axis present at a distance of effective depth and the neutral axis between compression and tension zone considered for evaluation. Figure 3 demonstrates the step-by-step procedure for CFRP jacket design. Considering the six-storey OGS frame for CFRP jacketing, the total thickness of 4.5 mm CFRP used for retrofitting purposes, which has an ultimate tensile strength of 3793 N/mm^2, modulus of elasticity of 227527 N/mm^2 and rupture strain of 0.0167 mm/mm is used for design with the procedure of ACI 440 and guidelines. The obtained bending and shear capacity improved to 374 kN-m and 576 kN, respectively. The CFRP wrapping scheme follows six CFRP strips with a width of 100 mm in the middle positions and two strips with a width of 150 mm wrapped at the plastic hinge location, i.e. at the top and bottom of the column section. Similarly, for the three-storey frame, CFRP's total thickness was 1.65 mm. Below, Fig. 4 summarises the P-M and M-ϕ response of normal and retrofitted frames in both three-storey and six-storey frames, where T-USF represents three-storey unstrengthened frame, T-SF represents three-storey frames strengthened with CFRP, S-USF represents six-storey unstrengthened frame and S-SF represents six-storey frames strengthened with CFRP.

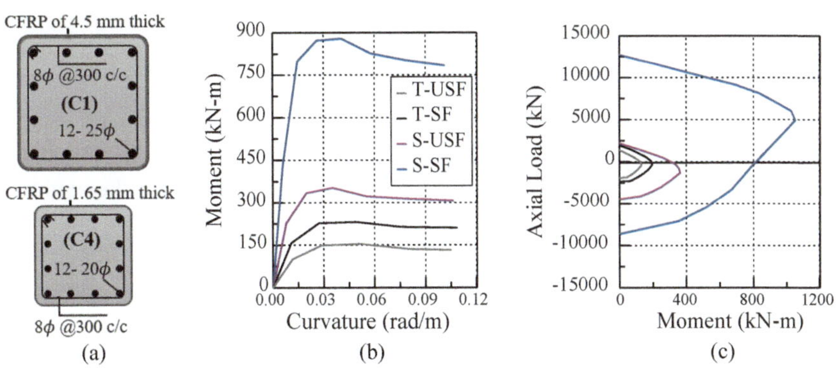

Fig. 4 Retrofitted column cross-sections and M-ϕ and P-M responses

4 Numerical Modelling

The computer software OpenSees is frequently used to simulate the response of structural structures because of its simplicity and realistic performance. Since all of the frames were numerically modelled, the frame members are represented as P-Delta-fibre-type force-based nonlinear members. The weight from the roof and the live load have been provided as a UDL on beams. Concrete02, a Uniaxial material command in OpenSees, was used to account for the compression and tensile behaviour of concrete. Concrete strength in beam and column members is divided into two categories: confined and unconfined. The properties of confined concrete can be calculated by using the confinement model. In the beam and column elements, a material model, the modified Ibarra–Medina–Krawinkler degradation model with the pinched hysteretic response (ModIMKPinching), was used with section aggregator. The mentioned material model was calibrated for shear behaviour using the column and beam members' shear capacity and shear deformation, allowing the structure to display appropriate shear and flexure behaviour under lateral loads. The Pinching4 uniaxial material command was used to model the pinching behaviour of concrete and longitudinal reinforcing steel. For the purposes of this study, the infill wall is modelled with a single strut model element in both diagonal directions and the concrete01 material model. The equivalent struts used in the model for unreinforced masonry infill walls were calculated using the following equations [18]. The width of the equivalent strut can be computed using the following equation:

$$w_{ds} = 0.175\alpha_h^{-0.4} L_{ds};\ \alpha_h = h\left(\sqrt[4]{\frac{E_m t \sin 2\theta}{4E_f I_c h}}\right) \tag{5}$$

Here, t is the thickness of the wall, θ is the angle of the diagonal strut with the horizontal, I_c is the moment of inertia of the adjoining column and h is the clear height of the URM (unreinforced masonry) infill wall. w_{ds} is the width of the equivalent strut and L_{ds} is the length of the equivalent strut model. E_f and E_m are the modulus of elasticity of RC MRF building and URM infill, respectively, which can be calculated as follows:

$$E_f = 5000\sqrt{F_{ck}} \tag{6}$$

$$E_m = 550 f_m;\ \text{and}\ f_m = 0.433 f_b^{0.64} f_{mo}^{0.36} \tag{7}$$

The F_{ck} represents the characteristic compressive strength, f_m is the compressive strength of a masonry prism. f_b is the compressive strength of brick in MPa and f_{mo} is the compressive strength of mortar in MPa. For this study, the compressive strength of mortar and brick is considered to be 4 and 15 MPa, respectively. Figure 5 below represents the fiber section with the section aggregator and the equivalent strut model used in the frames.

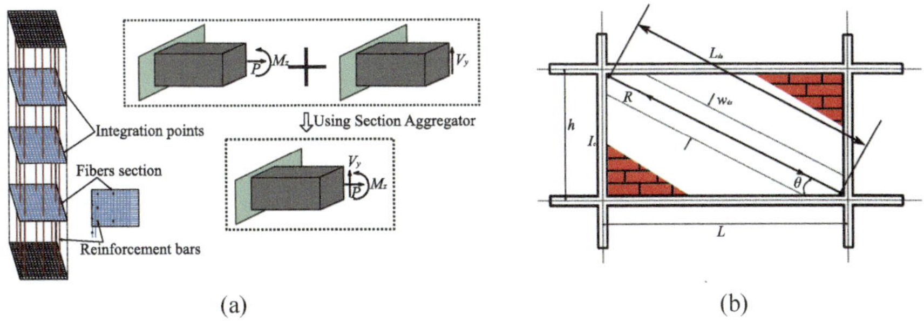

Fig. 5 Schematic representations of **a** fibre section with section aggregator, and **b** URM strut model

There are two techniques to provide CFRP in numerical modelling in opensees: providing an equivalent confinement model to the column and simply providing an extra patch on the outer side of the column with CFRP material properties. In the case of an equivalent confinement model, the confinement value can be computed using ACI 440, which adheres to the model proposed by Lam and Teng [23, 24]. However, in order to demonstrate an extra patch outside of the column, a proper material model and damage to the CFRP patch must be provided, along with adequate calibration.

5 Numerical Verification of an Experimental Study

A two-storey and single-bay RC frame with CFRP wrapping experiment conducted by Karaiya et al. [25] was considered for numerical validation. The overall dimension of the test specimen was 4000 mm in width and 3800 mm in depth. The column and beam cross sections are 200 mm × 200 mm and 200 mm × 280 mm, respectively. The material properties are used as mentioned in the experimental work for numerical modelling purposes. The CFRP used in the experimental work has a modulus of elasticity of 240 kN/mm^2, tensile strength of 3800 N/mm^2 and ultimate strain of 1.55%, respectively. The thickness of CFRP sheets is 0.117 mm. The equivalent confinement value was calculated using design procedures provided by [12] for numerical implementation. The material models used in the numerical model are concret02 for concrete and pinching4 material for steel. The loading protocol was used, as mentioned in the experimental work. Figure 6 summarises the comparison of numerical and experimental results, as these numerical results show a good agreement with the experimental results.

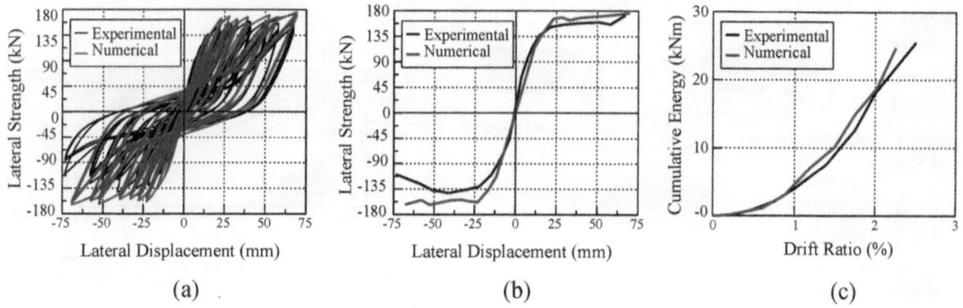

Fig. 6 Comparison of experimental and numerical results; **a** hysteretic response, **b** backbone curve and **c** energy curves

6 Results and Discussions

The case study structures have been modelled using the design values as described above. Using proper material constitutive models in OpenSees, the numerical models and CFRP jacketing processes are developed for the case study. The six-storey frame has a seismic weight of 7792.5 kN and the three-storey frame has a seismic weight of 3572 kN. In OpenSees, the gravity analysis of the case study structure yields a similar result.

For numerical investigation, nonlinear static, as well as dynamic analysis, are performed for all four frames, such as the three-storey unstrengthened frame (T-USF), three-storey frame strengthened with CFRP (T-SF), six-storey unstrengthened frame (S-USF) and six-storey frame strengthened with CFRP (S-SF). Nonlinear dynamic analysis was used to investigate the seismic response of the study frames. For this purpose, seven different real-time ground motion records, such as the ground motions at Chamoli (EQ1), El Centro (EQ2), Northridge (EQ3), Whittier Narrows (EQ4), Superstition Hills (EQ5), Kobe (EQ6) and Jiashi (or Northwest, China) (EQ7) are selected. For the design spectrum, soft soil type, earthquake zone V, was considered, and IS code provisions [18] were used. The design spectrum is used to scale all the ground motion data, and that scaled data of ground motion is used for further time-history analysis. Table 2 shows the details of selected ground motion data. Figure 7 summarises the scaled ground motion data and design spectrum with all seven ground motions. Furthermore, for the three-storey frames, the scaled acceleration data used 0.75 g (acceleration due to gravity), whereas for the six-storey frame, 1 g is used for numerical analysis.

6.1 *Effectiveness and Challenges*

After designing and numerically modelling all of the frames, the nonlinear static analysis was performed. The investigation shows that the base shear values for the various frames are 699 kN for T-USF, 1306.42 kN for T-SF, 1319.7 kN for S-USF and

Table 2 Details of selected ground motions

Sl no	Earthquake	Country	Year	Station	Component	Magnitude (Mw)	Distance (km)	PGA (g)	Time interval	Total time	Source
1	Chamoli (EQ1)	India	1999	Gopeswar (IITR)	20	6.6	17.3	0.36	0.02	24.34	SMC
2	El Centro (EQ2)	USA	1940	El Centro	180	6.9	12.2	0.35	0.02	53.76	PEER
3	Northridge, CA (EQ3)	USA	1994	Santa Monica (City Hall)	360	6.7	27.4	0.37	0.02	60	CSMIP
4	Whittier Narrows, CA (EQ4)	USA	1987	Los Angeles (Obregon Park)	360	6.1	14.2	0.43	0.02	40	CSMIP
5	Superstition Hills, CA (EQ5)	USA	1987	El Centro (Parachute Test)	315	6.6	1	0.38	0.01	22.31	PEER
6	Kobe (EQ6)	Japan	1995	Kobe University, Japan (CEOR station)	90	6.9	0.9	0.306	0.01	32	PEER
7	Northwest, China (EQ7)	China	1997	Jiashi, China (CSB station 19,001)	270	6.1	27.7	0.273	0.02	60	PEER

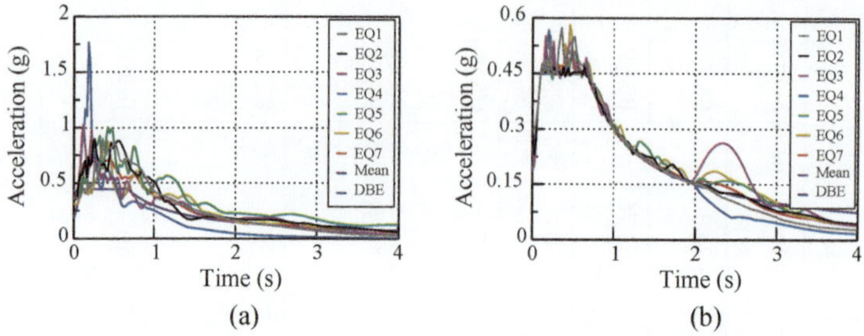

Fig. 7 Used ground motions, **a** unscaled spectra, and **b** scaled spectra (5% damping)

2054.54 for S-SF, respectively. Figure 8 illustrates the hinge mechanisms and capacity curves for each frame. The hinge mechanism of T-SF has a correct distribution of hinges and efficiently overcomes the typical open-ground storey characteristic. T-USF columns have a shear capacity of 100 kN as the frame experiences fewer shear hinge forms. Also, in the case of T-USF, the frame evaluation (Table 1) shows that the frame has a yield drift of 1.15%. In comparison to the S-USF frame, the T-USF frame is more ductile. The S-USF frame has a shear capacity of 115 kN and requires higher structural ductility than T-USF, which describes a typical non-ductile RC-OGS structure with shear-deficient columns. Figure 8 indicates that the S-USF frame's capacity curve is shear-dominated. The CFRP retrofitted frame (T-SF) does, however, confront some shear hinge development on the top floors, as the ground floor columns have been modified with CFRP. Table 3 shows the demand versus design capacity (Design value for moment and reduced shear force value used in numerical analysis) ratio for various columns and beams in all frames, as well as the critical force values that the members experienced during nonlinear static analysis.

Dynamic analysis was used to determine structural behaviour for both three- and six-storey frames. Some key factors for the frames include drift ratio, inter-storey

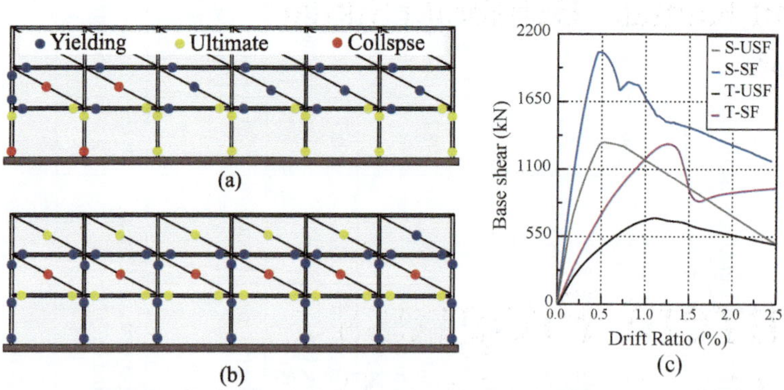

Fig. 8 Obtained hinge mechanisms for **a** T-USF, **b** T-SF and **c** capacity curves for all frames

Table 3 Demand versus capacity ratio in frames for different members

Sl. No.	Member type	Frame type	Capacity		Demand				D/C	
			SF (kN)	BM (kNm)	SF (kN)	Col. No.	BM (kNm)	Col. No.	SF	BM
1	Ground floor columns	T-USF	80	81	97.5	C1	152	C1	1.21	2.53
		T-SF	262	186	180	C1	236	C1	0.687	1.26
		S-USF	115	140	217	C5	305	C5	1.887	2.17
		S-SF	576	374	347	C7	381	C6	0.602	1.018
2	First-floor columns	T-USF	80	81	82.5	C1	74	C1	1.03	1.23
		T-SF	80	81	110.1	C1	145	C1	1.37	2.41
		S-USF	118	116	78	C1	252	C1	0.661	2.172
		S-SF	118	116	197	C3	331	C6	1.67	2.853
3	Ground floor beams	T-USF	133	136	106.5	B6	172	B1	0.8	1.264
		T-SF	133	136	143	B6	142	B4	1.075	1.044
		S-USF	142	162	146	B1	250	B1	1.028	1.54
		S-SF	142	162	157.1	B4	205	B4	1.106	1.265

drift (IDR), residual drift (RDR), base shear, dissipated energy and maximum plastic rotation due to earthquakes.

Figure 9 compares floor level versus drift and IDR responses for all frames. In the case of T-USF, the maximum drift obtained for EQ5 was 1.011% on the ground floor. The average drift of all earthquakes is 0.727% at ground floor level. After retrofitting, i.e. for the T-SF, the maximum drift for the Kobe earthquake (EQ6) is 0.568%, whereas the average drift for all earthquakes is 0.473% at ground floor level. Similarly, in terms of inter-storey drift (IDR), the average IDR achieved for T-USF and T-SF was 1.927 and 1.2%, respectively, representing a 60.5% reduction in IDR at the ground floor level. In the case of T-USF, EQ2 and EQ5 have a maximum IDR of 2.586 and 2.67%, respectively, after retrofitting with CFRP, which is decreased by 50.34 and 44.16%. The overall comparison of results demonstrates the effectiveness of a local retrofitting scheme in low-rise buildings.

In the case of a three-storey frame, the maximum plastic rotation is very large in EQ2 and EQ5, as the frame is most damaged at the ground-level columns. The three-storey frame is flexure-dominated, as shown in static pushover curves. However, in the case of a six-storey frame with shear-deficient columns, the S-USF fails under EQ5. CFRP retrofitting significantly reduces plastic rotation values while increasing base shear capacity. The average base shear values for T-USF, T-SF, S-USF and S-SF frames are 588.06, 1118.4, 1314.1 and 2256.1 kN, with base shear to seismic weight ratios of 0.165, 0.313, 0.168 and 0.289, respectively. Table 4 summarises the maximal plastic rotation and base shear of all the frames under different earthquakes.

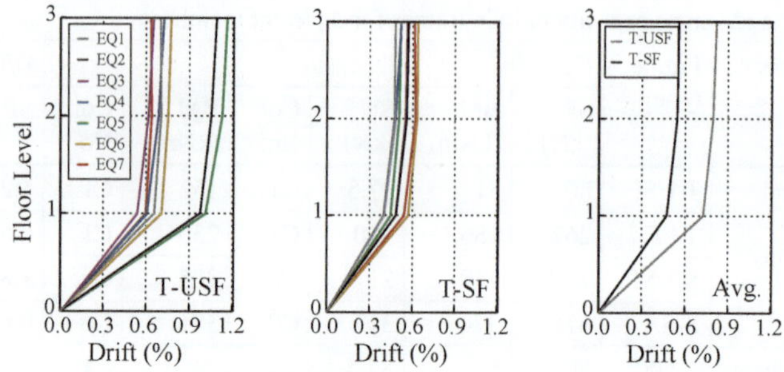

(a) Floor level v/s drift responses (three-story frames)

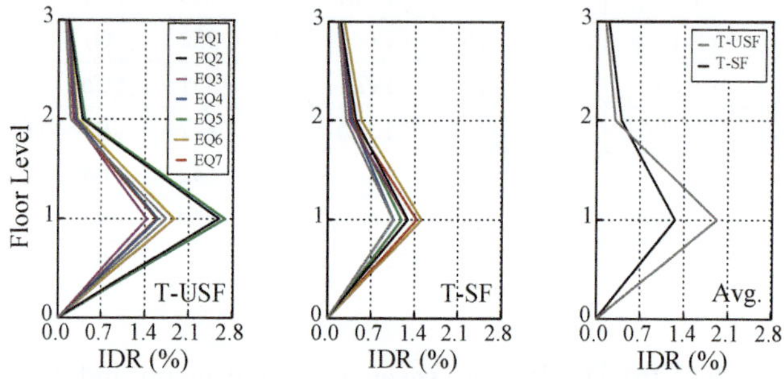

(b) Floor level v/s IDR responses (three-story frames)

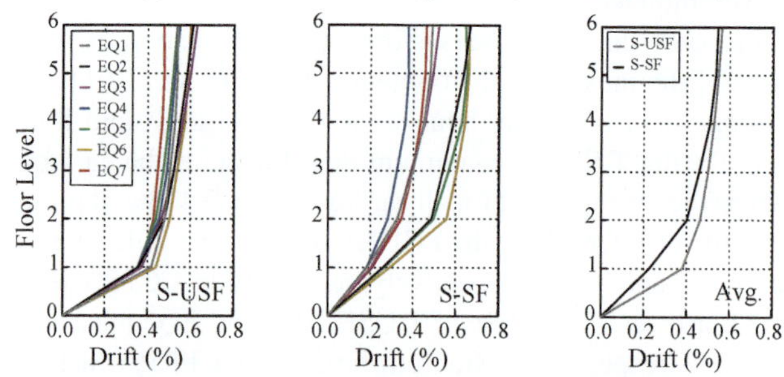

(c) Floor level v/s drift responses (six-story frames)

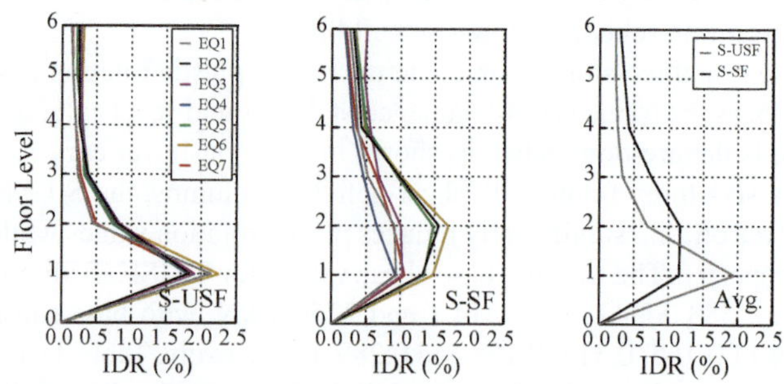

(d) Floor level v/s IDR reponses (six-story frames)

Fig. 9 Retrofitted and non-retrofitted frame responses under dynamic analysis

Table 4 Maximum plastic rotation and base shear of all frames under different earthquakes

	Max. plastic rotations (rad) for unstrengthened frames				Base shear (kN)			
	T-USF	Time (s)	S-USF	Time (s)	T-USF	T-SF	S-USF	S-SF
EQ1	0.00506	5.02	0.00557	5.58	542.12	970.75	1315.2	2042.6
EQ2	0.02075	12.98	0.00504	13.06	694.26	1138.2	1306.6	2368.4
EQ3	0.00485	14.74	0.0068	15.9	532.15	1112.8	1311.6	2287.5
EQ4	0.00573	8.66	0.00518	5.1	565.18	1087.3	1316.4	2035.3
EQ5	0.02636	19.55	0.00887	19.88	691.82	1070.3	1310.8	2349.6
EQ6	0.00612	12.77	0.00704	18.4	575.66	1257.7	1319.1	2442.0
EQ7	0.00442	6.14	0.00518	7.3	515.22	1192.2	1318.6	2267.5
Avg	–	–	–	–	588.06	1118.4	1314.1	2256.1

In general, the non-uniform MDOF system has a significantly larger maximum residual storey drift than the uniform MDOF system, and most current trend studies do not account for residual drift, which is also necessary for structural inspections. In the case of the six-storey frame, due to local retrofitting of ground floor columns, the CFRP jacket effectively strengthens the ground floor. Still, the damage concentrated on the first floor rather than following the predicted pattern for hinge formation. As shown in Fig. 10a, the ground-level columns are severely damaged, which is characteristic of OGS frame behaviour. Figure 10b illustrates that the first-level columns are more damaged than the ground floor, indicating that the damage has shifted. Figure 10c further indicates that, rather than decreasing the structure's residual drift at the weak storey, the local retrofitting system shifted it to the upper level. This problem may not occur in low-rise buildings but is common in medium- and high-rise buildings. It is also worth mentioning that local retrofitting schemes are inadequate regarding global energy dissipation capacity and proper reduction of residual drifts, particularly in medium to high-rise buildings.

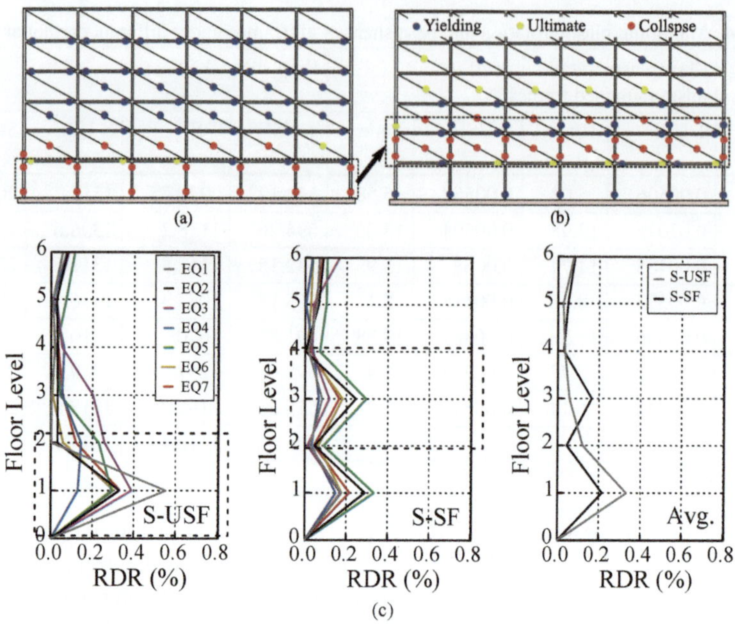

Fig. 10 Formation of hinges during nonlinear static analysis in **a** S-USF, **b** S-SF, and **c** residual drift response for six-storey frames

7 Summary and Conclusions

This study considers two frames of three- and six-storey buildings and is designed under gravity loads. To examine the inadequacy of these frames, the performance-based plastic design (PBPD) method was applied, which considers the inelastic condition by applying a pre-selected yield mechanism for target drift. As the OGS frames showed the maximum damage at the ground floor columns, these columns are retrofitted using a CFRP jacket based on the design procedure of ACI 440. The commercial software OpenSees uses different material models for the numerical modelling of concrete, steel and masonry struts. Also, a numerical validation of an existing study has been carried out, which justifies the modelling technique of the frame and CFRP jacket. The scaled earthquakes to a target DBE have been used for the nonlinear dynamic analysis. The conclusion obtained from this study shows that:

- The CFRP jacket intended for three- and six-storey buildings improves lateral strength, reduces drift ratio and improves inter-storey drift ratio (IDR) following retrofit. This demonstrates the efficacy of the local retrofitting system.
- The six-storey OGS frame's hinge mechanism shows damage sifting from the ground to the first floor only, which does not fulfil the desired mechanism considered for retrofitting. This could disadvantage local retrofitting procedures using jacketing in medium-rise RC-OGS frames of non-ductile type.

- Due to inadequate energy dissipation capacity, local retrofitting solutions are ineffective for reducing residual deformation in six-storey non-ductile RC-OGS frames.

Following this information, it is necessary to develop a hybrid seismic upgradation scheme of retrofitting that combines local retrofitting with energy dissipation devices or find a new retrofitting scheme to improve energy dissipation while reducing or controlling other parameters, which is one of the future scopes of this work.

References

1. Murty, C., Greene, M., Jain, S. K., Prasad N. P., Mehta, V. V.: Earthquake Rebuilding in Gujarat, India An EERI Recovery Reconnaissance Report. Earthquake Engineering Research Institute (2005).
2. Oinam, R. M., Sahoo, D. R., Sindhu, R.: Cyclic Response of Non-ductile RC Frame with Steel Fibers at Beam-Column Joints and Plastic Hinge Regions. Journal of Earthquake Engineering 18(6), 908–928 (2014).
3. Jain S. K., Jaiswal, A.: Post-earthquake handling of buildings and reconstruction issues emerging from the 2001 Bhuj earthquake. In: Proceedings of the Seventh US National Conference on Earthquake Engineering, Boston, USA (2002).
4. Sengupta, A. K., Reddy, C. S., VT, V. N., A. A.: Seismic Analysis and Retrofit of Existing Multi-storeyed Buildings in India- An Overview With Case Study. In: 13th World Conference on Earthquake Engineering, Vancouver, B.C., Canada (2004).
5. Al-Nimry, H., Neqresh, M.: Confinement Effects of Unidirectional CFRP Sheets on Axial and Bending Capacities of Square RC Columns. Engineering Structures 196 (2019).
6. Dalgic, K. D., Ispir, M., Ilki, A.: Cyclic and Monotonic Compression Behavior of CFRP-Jacketed Damaged Noncircular Concrete Prisms. Journal of Composites for Construction (2015).
7. Zhang, J., Wu, J., Du, W., Tong, C., Zhu Z., Jing, Y.: Residual load-carrying performance of CFRP strengthened RC beam after drop hammer impact. International Journal of Impact Engineering 175, p. 104547 (2023).
8. Benzaid R., Mesbah, H. A.: Strength Model for Square Concrete Columns Confined by External CFRP Sheets. Structural Engineering and Mechanics 46(1), 111–135 (2013).
9. Demir, U., Ispir, M., Sahinkaya, Y., Arslan, G.,Ilki, A., Axial Behavior of Noncircular High-Performance Fiber-Reinforced Cementitious Composite Members Externally Jacketed by CFRP Sheets. Journal of Composites for Construction 23(4) (2019).
10. Samy, K., Fouda, M. A., Fawyz, A., Elsayed, T.:Enhancing the Effectiveness of Strengthening RC columns with CFRP sheets. Case Studies in Construction Materials 17, p. e01588 (2022).
11. Wang, J., Lu, S., Yang, J.: Behavior of eccentrically loaded rectangular RC columns wrapped with CFRP jackets under different preloading levels. Journal of Building Engineering 34, p. 101943 (2021).
12. ACI 440.: Guide for the Design and Construction of Externally Bonded FRP Systems for Strengthening Concrete Structures. Farmington Hills (2017).
13. Leelataviwat, S., Goel, S. C., Stojadinovic, B.: Energy-based Seismic Design of Structures using Yield Mechanism and Target Drift. Journal of Structural Engineering 128 (8), pp. 1046–1054 (2002).
14. Liao, W. C.: Performance-based plastic design of earthquake resistant reinforced concrete moment frames (2010).
15. Karki, P., Oinam, R. M., Sahoo, D. R.: Evaluation of seismic strengthening techniques for non-ductile soft-story RC frame. Advances in Concrete Construction 9(4), 423–435 (2020).

16. Sahoo D. R., Rai, D. C.: Design and evaluation of seismic strengthening techniques for reinforced concrete frames with soft ground story. Engineering Structures 56, 1933–1944 (2013).
17. Goel, S. C., Chao, S. H.: Performance-based plastic design: earthquake-resistant steel structures (2008).
18. IS 1893: Criteria for Earthquake Resistant Design of structures. Part I: General Provisions and Buildings. Sixth Revision ed., New Delhi, Bureau of Indian Standards (2016).
19. Newmark N. M., Hall, W.: Earthquake Spectra and Design. Engineering monographs on earthquake criteria (1982).
20. Haselton, C. B., Goulet, C. A., Mitrani-Reiser, J., Beck, J. L., Deierlein, G. G., Porter, K. A., Stewart, J. P., Taciroglu, E.: An Assessment to Benchmark the Seismic Performance of a Code-Conforming Reinforced Concrete Moment-Frame Building. Pacific Earthquake Engineering Research Center (2008).
21. IS 456: Indian Standard Code of Practice For Plain and Reinforced Concrete. New Delhi, Bureau of Indian Standards (1978).
22. IS 1893: Criteria for Earthquake Resistant Design of Structures. New Delhi, Bureau of Indian Standards (1984).
23. Lam, L., Teng, J.: Design-oriented stress–strain model for FRP-confined concrete. Construction and Building Materials 17, 471–489 (2003a).
24. Lam, L., Teng, J.: Design-oriented Stress–Strain Model for FRP-confined Concrete in Rectangular Columns. Journal of REINFORCED PLASTICS AND COMPOSITES 22(13), 1149–1186 (2003b).
25. Karaiya, B. K., Oinam, R. M., Sahoo, D. R., Gupta, A.: Lateral cyclic performance of a CFRP retroftted two-story RC frame with open ground story. *Bulletin of* Earthquake Engineering 18, 5919–5939 (2020).

Sustainable Mobility: Evaluating the Utility of E-Rickshaw in Medium-Sized City Using a Binary Logit Model

Danial Doley, N. Mohamed Hasain, and Mokaddes Ali Ahmed

Abstract Various paratransit modes are prevalent in urban areas, influenced by the diverse needs and characteristics of commuters. In developing countries, the absence of adequate public transportation is commonly observed in medium-sized cities, leading many urban residents to rely on paratransit for their daily travel, especially within central business district (CBD) areas. Paratransit modes are often favored due to their affordability and convenience, effectively filling the gap left by public transport systems. Among these, e-rickshaws have recently gained prominence in urban areas of developing nations. Their appeal lies in their affordability and eco-friendly nature, making them a potential replacement for traditional fossil fuel-based paratransit vehicles. This study focuses on assessing the utility of E-Rickshaws and understanding commuter preferences in medium-sized cities where public transportation is inadequate. The research was conducted along the CBD corridor of Silchar, a medium-sized city in India. A commuter survey was carried out using questionnaires targeting users of various transport modes along the selected corridor. The utility of E-Rickshaws was analyzed using a binary logit model, and the statistical validity of the models was confirmed through various goodness-of-fit tests. Sensitivity analysis was also performed to identify the factors that most significantly influence the choice of E-Rickshaws. The study presents key findings and policy recommendations, offering valuable insights into the role of E-Rickshaws in achieving sustainable transportation in medium-sized cities. These findings highlight the potential of E-Rickshaws to serve as a viable, eco-friendly mode of transport in urban areas where public transportation options are limited.

D. Doley (✉)
Department of Civil Engineering, Indian Institute of Technology Guwahati, Guwahati, Assam, India
e-mail: d.danial@iitg.ac.in

N. M. Hasain
Department of Civil Engineering, Indian Institute of Technology Bombay, Mumbai, Maharashtra, India
e-mail: mohamedhasaincivil@gmail.com

M. A. Ahmed
Department of Civil Engineering, National Institute of Technology Silchar, Silchar, Assam, India
e-mail: maa@civil.nits.ac.in

U. S. Dixit et al. (eds.), *Current Progress in Engineering Sciences*,
https://doi.org/10.1007/978-981-96-8753-4_3

Keyword E-rickshaw · Utility · Sensitivity analysis · Binary logit model

1 Introduction

The remarkable growth in demand for mobility services, driven by rapid urbanization in non-industrialized nations, has significantly strained urban transport infrastructure and services. In many cities with existing public transit, the formal services are frequently insufficient and unreliable, leading to a reliance on personal mobility and informal public transit alternatives to fulfill the mobility requirements of the population [29]. The operation of the transportation infrastructure heavily influences the economy and social development of a region. It moves people, goods, and services to their final destination. The transportation infrastructure generally connects several places, which need improvement in most developing countries. Further, the traffic in the area and transportation structure should be appropriately addressed for long-term progress for sustainable transportation infrastructures. India, being one of the fastest-growing economies, is urbanizing, with the urban population growing at a median annual rate of roughly 3%. With the pattern of urbanization, the necessity for flexible transportation increases, and paratransit mode provides flexible and dynamic transportation. The public transportation systems in the country are among the most heavily used in the world, serving the mobility demand of the second-largest population [25]. The automobile business in India is quickly developing, with a yearly production of over 4.6 million vehicles, and the vehicular volume is supposed to rise further in the future [24]. Indian streets are becoming congested every year, and more private vehicles are being used, especially two- and four-wheelers. However, there is also a significant share of paratransit mode prevailing in medium-sized cities due to the lack of affordability by middle-class people. Instead of using public transportation, commuters frequently opt for paratransit mode because of the availability, accessibility, and flexibility of the destination. Further, medium-sized cities lack an efficient public transport system unlike major and large cities. In India, more than 2.5 million motors are used as paratransit domestically, and this quantity is growing, with more than 2,50,000 newer auto-rickshaws (motorized three-wheelers) bought each year [7]. This increase is mainly because of the higher demand and mode choice behavior of the commuters.

With rising global pollution levels, India has seven of the world's ten most polluted cities. India suffers from severe air pollution, long-term health ailments and related deaths, particularly in urban and semi-urban areas. Vehicle emissions mainly cause this poor air quality [19]. Auto-rickshaw is the most common mode of public transportation in the country. Because of the pollution problems, many cities are initiating the prohibition of using petrol-powered rickshaws within the city limits. As a result, various automobile manufacturers have initiated the redesign of the traditional petrol-powered rickshaw into a more energy-efficient and environmentally friendly design. Many leading auto-rickshaw producers, like Bajaj Auto, are establishing expansions in deals toward a sustainable and non-polluting energy source such as electricity or renewable energy. Over the past few years, several existing automobile manufacturers and newly dedicated companies have made remarkable efforts to transform conventional vehicles into electric vehicles (EVs), providing reliable and environmentally friendly solutions [15]. The Smart Cities Mission of the Government of India has stressed promoting sustainable modes of transportation. The Smart City concept is proposed to improve the quality of life of citizens by enabling comprehensive city (re)development and upgrading programs [33]. The practice of an attractive and environment-friendly paratransit mode for last-mile connectivity is one of the aims of the urban mobility plan of the "Smart Cities Mission" [8].

Compared to the current alternative fuel-powered rickshaws, a renewable source would make it a better option. One approach to accomplish this is to employ an energy system, such as an electric system, that can utilize many renewable energy sources [28]. The battery-operated electric rickshaw is an alternative to the auto rickshaw since it is environmentally friendly. Also, the whole system cost, which includes reduced maintenance and fuel savings, is predicted to be significantly lower than the traditional auto rickshaw [31]. Electric Rickshaw (also locally known as tuk-tuk or toto) is a three-wheeled battery-operated paratransit vehicle shown in Fig. 1. This E-Rickshaw caught the attention of policymakers and automobile industries in recent years to provide e-mobility to commuters and keeping control of air pollution in medium-sized cities where proper public transportation does not exist. Most E-Rickshaw drivers in the country are below the poverty line, unemployed and some are illiterate individuals. Additionally, it is observed that a growing number of people from rural areas are migrating to semi-urban areas to operate an E-Rickshaw in an underdeveloped city due to its lower costs. Nowadays, every household has electricity, and E-Rickshaw operators can easily recharge E-Rickshaw batteries at home without any trouble. This 4-seater vehicle works on a DC motor ranging from 650 to 1400 W and is equipped with a 4 Li-ion/lead acid battery of 12 V connected in series [17]. It is important to note that the specific features and regulations related to E-Rickshaws can vary by region and country due to their speed limitation of 25 km/h.

Fig. 1 Pictorial view of E-rickshaw

One of the important aspects of any economy is its efficient transportation within the cities. Further, developing economies mainly depend on public transportation. In India, the major cities usually have well-functioning public transportation, including the metro rail system. However, medium-sized cities do not often have functional public transportation in developing economies. The paratransit mode within the CBD area often serves the role of public transportation. Therefore, it is important to focus on these paratransit modes in medium-sized cities and assess their role as a substitute for public transportation. Further, the rising trend of EVs is also visible in medium-sized cities. This study focuses on assessing the utility of E-Rickshaws and the mode choice behavior of commuters in a medium-sized city in India. Hence, the objectives of the study are (a) to measure the revealed modal share of E-Rickshaw in the medium-sized city corridor, (b) to assess the effect of different characteristics of commuters on the utility of Rickshaw using Binary Logit model.

2 Literature Review

As urbanization has accelerated in the past few years, the economy has grown and the family incomes and living standards in urban regions have increased. This led to a notable surge in the number of privately-owned vehicles. Due to concerns about pollution, there is a need for an eco-friendly mobility solution nationwide.

A study conducted in Patna, India, analyzes the travel patterns and perceptions of safety for users of E-Rickshaw paratransit. The study employed statistical models to analyze how socio-demographic factors and travel circumstances influence trip frequency, mode substitution, and the prospective adoption of E-Rickshaws. The results indicate that E-Rickshaws largely substitute traditional auto-rickshaw trips [21]. A study conducted by Majumdar and Jash [17], through surveys, indicated that E-Rickshaws have the potential to mitigate carbon emissions and enhance environmental conditions due to their minimal energy consumption compared to other motorized public transport modes. The authors also discussed effective implementations and suggested improvements in vehicle design and appropriate regulation. In a separate study, Shandilya et al. [28] examined the historical, current and future aspects of E-Rickshaws. Their analysis suggested that E-Rickshaws are a promising mode of present and future transport, making strides in the Indian transport system. They have also pointed out issues such as slow speed, the disposal of lead-acid batteries and limited seating capacity. Another study investigated the attitudes of drivers of three popular paratransit modes, namely E-Rickshaws, auto rickshaws, and cycle rickshaws. The study also provides a deeper knowledge of the socio-economic backgrounds of these paratransit drivers. The study uses data from 620 paratransit drivers collected during a face-to-face survey in Delhi, India. Grey relation analysis was used to analyze how the three paratransit driver groups perceive E-Rickshaws. The results show that paratransit drivers are predominantly from low-income and educated backgrounds. A key E-Rickshaw service hurdle is a lack of amenities, such as designated parking, fixed charging stations, and fare regulation [22].

Kumar and Chakrabarty [13] stated that EVs involve a high capital cost but lower operating costs. This paper introduces a model designed to assess the total cost of ownership (TCO) between electric vehicles (EVs) and various fuel types (petrol, diesel, and compressed natural gas (CNG)) used in internal combustion engines (ICE). The analysis revealed that, on average, the TCO per kilometer for electric two-wheelers and E-Rickshaws is lower than that of their ICE counterparts under typical daily usage conditions in Indian urban settings. Secondly, it is inferred that the TCO per km of E-Rickshaws is less than that of their ICE counterparts when the vehicle usage exceeds 110 km per day. In Indian cities, three-wheelers are mainly used for first- and last-mile connectivity, where daily travel distances range between 100 and 150 km. In this scenario, E-Rickshaw drivers can easily attain the break-even point with their current daily travel distance. Singh et al. [30] present an overview of the introduction of E-Rickshaws in Delhi (NCR) and examine their impact on sustainability and the challenges associated with e-mobility. The study employed three types of surveys: survey research, non-participant observations, and questionnaire surveys. Results indicated that E-Rickshaws are socially accepted, economically viable, and environmentally sustainable in the region.

The rapid expansion of the three-wheeler E-Rickshaw in India has created jobs for both semi-skilled and unskilled individuals. These manufacturing facilities, ancillary units, and distribution centers employed many semi-skilled workers. Some agencies contributed to expansion by purchasing many E-Rickshaws and renting them to the underprivileged. However, there are still prospects for additional growth if Central/State Governments and manufacturers take some of the actions outlined in the study [27]. With a different approach, a study conducted in Hyderabad, India, developed a driving cycle for E-Rickshaws in rural and urban traffic conditions. The developed driving cycle is further used to determine the driving characteristics of an E-Rickshaw and is compared with other typical driving cycles. Random selection and k-means clustering techniques are used to develop driving cycles for E-Rickshaws in rural and urban traffic conditions. The analytical results show that k-means clustering outperforms random selection and accounts for traffic conditions systematically. The findings can help simulate the energy consumption and driving characteristics of E-Rickshaws compared to comparable fossil-fuel vehicles [4]. Ansari and Sinha [2] conducted a comparative analysis of service quality perception among users of Intermediate Public Transport (IPT) modes, specifically auto-rickshaws and E-Rickshaws in Patna. Users express high satisfaction with cleanliness and travel time for both IPT modes but share concerns about security, convenience, and safety for female users, particularly during night hours. The researchers assert that these findings hold significance for policymakers and the development of transportation infrastructure in mid-sized cities.

The study on the importance of EVs seeks to give solutions for increasing the cost advantage of EVs by lowering their payback period by investigating the potential of a vehicle-to-home (V2H) system, hence increasing the willingness of customers to pay. Implementing V2H can address two major technological issues: the cost disadvantage of EVs and the peak leveling of the Indian power grid. It can reduce the payback period of EVs, making them a reality rather than a distant future [12].

A study conducted by Pramanik and Ashrafuzzaman Rahman [20] focused on the operational characteristics of paratransit, mainly E-Rickshaws, in a medium-sized city in Bangladesh. The paper provides substantial data emphasizing the importance of E-Rickshaws, offering valuable insights for policymakers in recognizing the significance of E-Rickshaws and promoting e-mobility. A case study conducted in West Bengal, India, delves into the practical performance of electric two-wheeler and E-Rickshaws across varied traffic conditions. While E-Rickshaws are primarily utilized in developing regions, electric two-wheelers enjoy a broad user base. Nevertheless, the majority of these vehicles are designed with a maximum speed limitation of 25 km/h. The investigation found that electric two-wheelers and E-Rickshaws exhibited specific energy consumption rates of 28.67 and 43.25 W/km, respectively. By considering the operational scenarios, the study suggests that rerouting and optimizing the usage of low-speed electric vehicles could be explored as prospective solutions for enhancing last-mile connectivity options in the future [16].

Various models can be combined to streamline, analyze, and comprehend intricate phenomena, aiding decision-making processes by illustrating their fundamental mechanisms, interactions, and behaviors. Models support hypothesis testing by replicating scenarios and evaluating their alignment with observed data. They are employed to assess the potential effects of policy interventions, regulations, or actions, assisting policymakers in evaluating the repercussions of different policy alternatives and making well-informed decisions to attain desired results. Rahman and Baker evaluate induced mode switch behavior using flyovers as a case study in the building of transportation infrastructure using multinomial regression approach. This study looks at whether building flyover infrastructure causes people to change their travel preferences. Compared to flyover users who did not change their travel mode, the study indicated that flyover users were three times more likely to switch to private transport to save travel time [23]. Saxena and Shrivastava [26] performed study to quantify the degree of satisfaction of autorickshaw operator with the chosen parameters and to identify and investigate the elements that contribute to the adaptation of vehicles in India. Using structural equation modeling, the study discovered that characteristics connected to vehicles and the economy have the greatest impact on the adaptability of E-Rickshaw, as seen from the perspective of 478 autorickshaw operator. Using a different approach linking mode choice with travel behavior by using a logit model based on utility function was done. Al-Salih and Esztergár-Kiss [1] examined how an individual's choice of transport mode and travel behavior interact, and more precisely, to determine which characteristic has the biggest impact on mode choice. To determine the highest utility of the mode option, the utility function considers the activity features, trip characteristics (such as travel duration, cost, and distance between activity places), and individual characteristics.

A similar study type is performed by Weng et al. [32], on binary logit modeling Beijing public transport commuters' mode choice behaviors. This study employed an examination of the factors influencing the mode choice behaviors of public transport commuters, focusing on subway and bus usage. A questionnaire was devised using a combination of revealed preference and stated preference techniques, with both online and paper surveys conducted to collect data on commuters' travel preferences between subways and buses. A binary logit model was suggested to analyze public transport commuters' travel decisions, with regression coefficients estimated through maximum likelihood estimation. Furthermore, utilizing data from Beijing public transport smart cards, a support vector machine classification model was constructed to discern commuters' mode choices. Lashari et al. [14] performed a binary logistic regression and regression tree from 1500 valid survey responses on users' attitude and perception affecting their intention to buy EVs. The findings indicated that within the realm of attitudinal characteristics, the most influential predictors for the purchase of an EV were perceptions related to environmental impact and economic considerations associated with EV usage.

The summary of the literature on E-Rickshaws and related topics leads to the conclusion that these paratransit vehicles are widespread in numerous developing cities. E-Rickshaws effectively offer last-mile connectivity, bridging the gap between major transportation hubs and final destinations. Their ability to navigate through narrow and congested urban streets allows them to reach areas that might pose challenges for larger vehicles. While several studies have highlighted the significance, performance, safety, and operational characteristics of E-Rickshaws, limited research has delved into their role in urban mobility, particularly in medium-sized cities where comprehensive studies on their versatility still need to be conducted. Further, the discussed studies have been conducted in the cities where the proper functional public transportation exists. Notably, more research needs to be done specifically to focus on the comparative utility of E-Rickshaws in medium-sized cities in developing countries. Researchers have identified a growing interest in this field due to its increasing demands and the associated economic and social benefits. Since E-Rickshaws are proving to be a viable mode of transport, it is important to assess their modal share, utility, and the mode choice behavior of commuters in medium-sized cities where there is no proper functioning public transportation for commuters.

3 Case Study Area

Silchar is the headquarters of the Cachar district of Assam, India and is ranked as the second-largest urban center in population and municipal area [9]. It is also one of the major cities in the Northeastern region of the country, which has an airport, an intra-state connecting railway station, a national-level university (NIT Silchar), a state-level medical college (SMC), and a central university (Assam University). It is also the starting point of a major highway corridor called the East–west corridor connecting the western region of the country. Geographically, it is situated in south Assam, covering a space of 257.5 km^2. Silchar fills in as the doorway to the North-eastern states of Manipur, Mizoram, Tripura, and different locales of Barak Valley in terms of economy and business relations. As per the 2011 census of India, Silchar town has a population of 172,830 [18]. The gender ratio stands at 989 females for every 1,000 males, significantly exceeding the national average of 940 females for

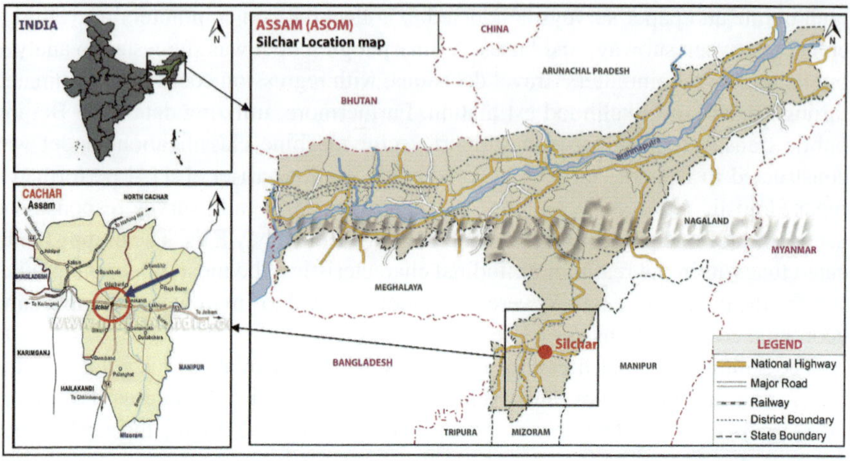

Fig. 2 Assam state map with the study location Silchar city

every 1,000 males. Also, Silchar city is one of the most densely populated regions of Assam and is a business place encompassing numerous locales. However, the city needs more stable public transportation within the city, which is being substituted by the privately owned paratransit mode. The map view of Silchar is shown in Fig. 2.

4 Survey and Data Collection

4.1 Study Corridor

The main Corridor connecting the Central Business District (CBD), Assam University, National Institute of Technology Silchar, Silchar Medical College, Apollo Hospital and a major local market has been selected for the study. The Corridor serves as a spine for intercity travel, and the E-Rickshaws play a major role as paratransit in this Corridor. Six locations with higher passenger boarding and onboarding from E-Rickshaws were selected for the data collection. The Corridor is roughly 3.6 km from the Ranghirkhari Point to the Capital Point.

4.2 Questionnaire Survey

The primary objective of this survey is to collect information on the travel data of the commuters and general characteristics influencing the choice of mode. The factors that generally affect the choice of a mode and the perceived utility of a mode can be broadly classified into the following four categories: (a) Socio-economic information like age group, gender, educational qualification, employment status and average monthly family income, (b) Household information like average household size, Vehicular ownership (e.g., Bicycle, 2-wheeler and 4-wheeler), (c) Trip-based information like trip purpose of respondents, last trip mode used, average trip length of respondents, travel time and travel cost, and (d) Service quality attributes like comfort, reliability, safety, convenience and accessibility. All the respondents are asked to provide the travel data based on their last performed trip. This also provides the stated modal split. In the service quality parameters, respondents were asked to rate the mode used, like comfort, reliability, safety, convenience, and accessibility, with the help of a 5-point Likert scale.

4.3 Sample Size

The stratified sampling technique was selected to determine the sample size of the questionnaire survey for trip makers because of its feasibility in selecting data randomly from different subgroups or strata proportionately. The empirical formula given by William G. Cochran is used to determine the sample size, n_0 [6]. A confidence level (Z is 1.96 for 95% confidence interval) of estimated proportion of an attribute (p) that is present in the population is adopted as 0.5 in this study and desired margin of precision (e) is adopted as 5%. As per the sample size, the sample size is found to be 384.16. Now the adjusted sample size for the population size of Silchar was estimated with the help of following empirical.

$$\text{adjustedsamplesize}\ (n) = \frac{\text{samplesize}}{1 + \frac{\text{samplesize}-1}{\text{populationsize}}} \tag{1}$$

For the trip makers survey, the adjusted sample size for the Silchar with the population size of 172,830 from Eq. (1) is 384. Through the questionnaire survey, 416 responses were collected along the selected study corridor.

4.4 Description of the Collected Data

The data collection was divided into four categories: socio-economic information, household information, trip-based information, and service quality attributes. Table 1 presents the summarized data and coding pattern used for each response in this study. N represents the number of observations under each category. The marginal percentage represents the proportion of observations under the category. For the last trip mode used in Private vehicle + Other paratransit mode includes bicycle, two-wheeler, car, cycle rickshaw, autorickshaw, and taxi. E-Rickshaw is considered as the dependent variable for the Binary logit model.

The data collected from the survey were categorized and analyzed in Table 4 as follows:

Socio-economic Characteristics. Majority were between 18 and 45 years (75.2%), followed by under 18 (13%) and over 45 (11.8%). About the gender 70% of respondents were male, and 30% were female. Most had education up to 10 + 2 (53.8%), followed by those with a degree (35.8%) and illiterate (10.4%). Regarding employment status, the majority were students (64.4%), followed by employed (20.4%) and unemployed (15.2%). From the point of family income 42.6% earned less than ₹25,000 monthly, 33.9% between ₹25,000–₹50,000, and 23.5% over ₹50,000.

Trip-based Information. Regarding the trip purpose work trips (37.2%), shopping trips (31%), and school/college trips (31.8%). Based on the last trip mode used about 70.2% used private vehicles and other paratransit modes, while 29.8% used E-Rickshaws. The majority trips were under 2 km (62.8%). Most trips took 10 min or less (32.7%). About the travel cost 39.9% is the highest of trips cost ₹20 or less.

Service Quality Ratings. Comfort, reliability, safety, convenience, and accessibility were rated, with a significant portion rating these aspects as good.

Regarding age group, it was observed that younger people prefer using E-Rickshaws more than older people. On the other hand, private vehicles + other paratransit mode in the study corridor have been preferred by older people as they are more convenient. In terms of employment status, employed people use private vehicles for transport in the corridor, and unemployed people are found to use E-Rickshaws comparatively more than others. Further, commuters earning more prefer private vehicles with lesser family income utilizing the E-Rickshaws. Males are found to be using private vehicles, and females are found to prefer E-Rickshaws and autorickshaw more, while the genders of the commuters are examined. This data highlights the socio-economic diversity and travel behaviors of the commuters, providing essential insights into the factors affecting their mode choice, especially regarding E-Rickshaws.

Table 1 Summary of collected response under each category along with adopted coding pattern

Category	Parameters	Categories	Coding pattern in SPSS	N	Marginal percentage (%)
Socio-economic information	Age group	<18	1	54	13.0
		18–45	2	313	75.2
		>45	3	49	11.8
	Gender	Male	1	291	70.0
		Female	2	125	30.0
	Educational qualification	Illiterate	1	43	10.4
		Up to 10 + 2	2	224	53.8
		Degree	3	149	35.8
	Employment status	Employed	1	85	20.4
		Student	2	268	64.4
		Unemployed	3	63	15.2
	Average monthly family income	<₹25,000	1	177	42.6
		₹25,000–₹50,000	2	141	33.9
		>₹50,000	3	98	23.5
Household Information	Average household size	≤2	1	273	65.6
		3–5	2	120	28.8
		≥6	3	23	05.5
	Vehicular ownership (bicycle)	–	–	416	100.0
	Vehicular ownership (2-wheeler)	–	–	416	100.0
	Vehicular ownership (4-wheeler)	–	–	416	100.0
Trip based information	Trip purpose	Work trip	1	155	37.2
		Shopping trip	2	129	31.0
		School/College trip	3	132	31.8
	Last trip mode used	Private vehicle + other paratransit mode	1	292	70.2
		E-rickshaw	2	124	29.8
	Average trip length of respondents	<2 km	1	261	62.8
		2–4 km	2	23	05.5
		>4 km	3	132	31.7
	Travel time (min.)	≤10	1	136	32.7

(continued)

Table 1 (continued)

Category	Parameters	Categories	Coding pattern in SPSS	N	Marginal percentage (%)
		11–20	2	92	22.1
		21–30	3	105	25.2
		31–40	4	13	03.2
		>40	5	70	16.8
	Travel cost (INR)	≤20	1	166	39.9
		21–40	2	57	13.7
		41–60	3	70	16.9
		61–80	4	18	04.3
		≥81	5	105	25.2
Service quality attributes	Comfort	Extremely poor	1	35	08.4
		Poor	2	53	12.7
		Neutral	3	75	18.0
		Good	4	232	55.8
		Very good	5	21	05.1
	Reliability	Extremely poor	1	30	07.2
		Poor	2	44	10.6
		Neutral	3	61	14.6
		Good	4	241	57.9
		Very good	5	40	09.6
	Safety	Extremely poor	1	53	12.8
		Poor	2	83	20.0
		Neutral	3	145	34.8
		Good	4	123	29.6
		Very good	5	12	02.8
	Convenience	Extremely poor	1	42	10.0
		Poor	2	41	09.9
		Neutral	3	61	14.7
		Good	4	259	62.3
		Very good	5	13	03.1
	Accessibility	Extremely poor	1	29	06.9
		Poor	2	49	11.8
		Neutral	3	57	13.7
		Good	4	246	59.2
		Very good	5	35	08.4

5 Statistical Analysis

Data was analyzed using IBM SPSS statistics software version 26 to develop models for the mode choice assessment of E-Rickshaw. The logit model is used to determine the utility function of the E-Rickshaw. The level of significance (α) for this study is set as 0.05 with a confidence interval of 95%. A total of twenty independent variables and one dependent variable are considered in the model from trip makers, which were obtained from the questionnaire survey. The dependent variable is taken as the choice of mode.

5.1 Estimation of Utility Function of E-Rickshaw

A utility function is a representation to define individual preferences for goods or services beyond the explicit monetary value of those goods or services. The utility maximization rule states that the user will select the alternative from his/her set of available alternatives that maximizes the user's utility [10]. Mathematically, in the field of transportation engineering, it is a function of characteristics of all the available modes and characteristics of an individual's valuation for each mode. Demand forecasting of any mode requires knowledge of the preferences of trip makers. A utility function can describe the preferences of trip makers. Therefore, using questionnaires or Google Forms, such a utility function can be mathematically constructed for the modes based on stated preferences. According to Kroes and Sheldon, the term "stated preference method" refers to a family of techniques that use individual respondents' statements about their preferences in a set of transport options to estimate utility function [11]. An individual trip maker is assumed to choose the mode of travel which offers maximum utility from all the available modes. The modal utility is mathematically expressed, as shown in Eq. 2.

$$U_{ij} = V_{ij} + \varepsilon_{ij} \tag{2}$$

where V_{ij} is function of measured variables/attributes for trip maker i and mode j; ε_{ij} is the non-measurable systematic error or disturbance; and V_{ij} can be expressed through Eq. 3.

$$V_{ij} = X_{1j}\beta_{1j} + X_{2j}\beta_{2j} \cdots X_{nj}\beta_{nj} \tag{3}$$

where X_1, X_2, ………X_n are the attributes/variables considered for mode j and β_1, β_2… β_n are their estimated coefficients. Choice behavior is a complex decision-making process; many variables impact the demand for people's travel choices [5]. The mode preference varies from person to person and from place to place, depending on the characteristics of the transport facilities at that place.

Binary Logit model. The binary logit model is the simplest mode choice model, as the choice is to be made between two modes of travel only. For at least thirty years, researchers have been working on logit models to analyze mode choice decisions [3]. In the present study, this dependent variable is considered a mode choice between E-Rickshaw and other modes. The logit model was used to determine the utility function of E-Rickshaw in this city (Peng et al. 2002). The logit model takes the form as shown by Eq. 4.

$$\mathrm{Logit}(Y) = \ln\left(\frac{P_i}{1 - P_i}\right) = \beta_1 X_1 + \beta_2 X_2 + \cdots \tag{4}$$

where P_i = Utility of the mode i; Y = Dichotomous dependent variable representing the choice of mode; β_1 and β_2 are the regression coefficients; and X_1, X_2, which can be categorical or continuous, are predictor/independent variables.

Parameter Estimates of Logit Model. The questionnaire data obtained from the trip makers survey was fed into the SPSS software. Each of the variables or parameters of the model, qualitative or quantitative, dependent, or independent, is coded as given in Table 1. Not all the independent variables significantly impact the dependent variables, so statistical tests are carried out to determine the statistically significant variables. The logit model is estimated using the maximum likelihood (ML) method. The estimated coefficients, standard errors, significance, and odds ratio of the variables are given in Table 2.

From the estimates given in Table 2, only eleven variables are found to be statistically significant as their p-value is less than the significance level of 0.05. Among the significant variables, age, educational qualification, employment status (job and student), average monthly family income, household size, vehicle ownership (bicycle, 2-wheeler and 4-wheeler), trip length, travel time and travel cost have negative coefficient value, and the rest has a positive sign in their coefficients.

The parameter estimates from the Logit model provide valuable insights into the factors influencing the choice between E-Rickshaws and private vehicles (including other paratransit modes). Among socio-demographic variables, age (B = −0.976) has a significant negative impact, indicating that younger individuals are more likely to prefer E-Rickshaws, potentially due to affordability and a greater willingness to adopt unconventional transport modes. Gender (B = 2.874) has a strong positive effect, suggesting that males are more inclined to opt for private vehicles, likely due to social norms or the perceived status and control associated with private vehicle ownership. Educational qualifications (B = −1.085) negatively influence E-Rickshaw preference, possibly reflecting concerns about comfort or the status associated with private vehicles. Employment status also plays a role, with both jobholders (B = −1.583) and students (B = −1.199) being less likely to choose E-Rickshaws, likely due to a preference for faster or more comfortable transport options.

Economic factors further highlight the divide, with higher average monthly family income (B = −1.573) significantly reducing the likelihood of choosing E-Rickshaws, as wealthier households may favor private vehicles for convenience and

Table 2 Parameter estimates of logit model using maximum likelihood method

Parameters	B	Std. error	Exp(B)
Age (X_1)	−0.976**	0.474	0.377
Gender (male)(X_2)	2.874	0.621	17.716
Educational qualification (X_3)	−1.085**	0.463	0.338
Employment status (job) (X_4)	−1.583**	0.718	0.205
Employment status (student)(X_5)	−1.199**	0.627	0.301
Average monthly family income (X_6)	−1.573***	0.367	0.207
Household size (X_7)	−0.683*	0.385	0.505
Bicycle ownership (X_8)	−0.979**	0.436	0.376
2-Wheeler ownership (X_9)	−2.352***	0.649	0.095
4-wheeler ownership (X_{10})	−1.281***	0.470	0.278
Trip purpose (work trip)(X_{11})	2.220	0.678	9.209
Trip purpose (shopping/social trip)(X_{12})	2.237	0.722	9.368
Trip length (X_13)	−0.448	0.283	0.639
Travel time (X_{14})	−0.114	0.275	0.892
Travel cost (X_{15})	−0.564***	0.240	0.569
Comfort (X_{16})	0.606*	0.357	1.832
Reliability (X_{17})	0.885**	0.382	2.422
Safety (X_{18})	1.233***	0.340	3.430
Convenience (X_{19})	1.599	0.369	4.947
Accessibility (X_{20})	0.349	0.335	1.418

*$p < 0.1$; **$p < 0.05$; ***$p < 0.01$

status. Similarly, larger household sizes (B = −0.683) are less likely to choose E-Rickshaws, likely due to logistical challenges or the need for more versatile transport options. Vehicle ownership whether bicycles (B = −0.979), two-wheelers (B = −2.352), or four-wheelers (B = −1.281) also strongly discourages E-Rickshaw use, as individuals tend to rely on their existing vehicles.

Trip characteristics show a clear preference for E-Rickshaws in non-essential trips like shopping or social visits (B = 2.237) and work trips (B = 2.220), possibly due to their affordability and convenience for short distances. However, longer trip lengths (B = −0.448) discourage their use, reflecting practical constraints such as range limitations. Travel cost (B = −0.564) and time (B = −0.114) negative effects align with rational travel behavior, as higher costs and longer times discourage selection. On the other hand, positive perceptions of comfort (B = 0.606), reliability (B = 0.885), safety (B = 1.233), and convenience (B = 1.599) significantly increase the likelihood of choosing E-Rickshaws, with safety and reliability emerging as particularly important factors.

Overall, these results underscore the interplay of demographic, economic, and perceptual factors in transportation mode choice. While E-Rickshaws are seen as affordable and suitable for short trips, barriers such as range limitations, comfort, and perceptions of status must be addressed to make them more competitive against private vehicles and other paratransit modes. Among the parameters, the gender and the purpose of the trips, along with the accessibility, convenience, and comfort of the E-Rickshaw, do not affect the utility of the mode.

Goodness of Fit Test. The −2 Log-likelihood and R^2 values, which are statistical measures of the model, are given in Table 3. It is found that this model can explain 52.9–70.6% of the variance of the dependent variable. Since the present study attempts to predict the human mode choice behavior, lower R^2 values are likely. A good model may have a lower R^2 value, whereas a biased model may have a higher R^2 value. This depends on the field of study. Hence, R^2 cannot be the only parameter based on which the adequacy of the model can be decided. An additional performance measure of the Hosmer and Lemeshow Test was performed to determine the goodness of fit of the model. This test infers the prediction quality of the developed logistic regression model. It uses the Chi-square test and indicates a poor fit if the significance of the Chi-square value is less than 0.05 at the 95% significance level. The test statistics are given in Table 4, and it was observed that the significance of the chi-square value is greater than 0.05.

Sensitivity Analysis. Sensitivity Analysis was performed to assess the influence of each parameter on the utility of E-Rickshaw in the selected corridor. It was performed through the One-Factor-At-a-Time (OFAT) method, in which a single variable is gradually increased while keeping the other parameters the same and assessing the effect on the utility function (Pannell 1997). All the significance variables are increased gradually, and the corresponding utility of E-Rickshaw was calculated from the Eq. (4). The result of sensitivity analysis for socio-economic, household, travel related, and service quality characteristics are shown in Fig. 3.

The sensitivity in the E-Rickshaw's utility is assessed due to the changes in the independent variables, and the average utility of the E-Rickshaw was found to be 34.72%. In terms of socio-economic characteristics, all the significant variables, except the unemployed employment status, hurt the utility of the E-Rickshaw on the selected corridor. This is obvious as the educational qualification increases and the overall socio-economic status of the users improves, which reduces the utility of E-Rickshaws as they are a paratransit mode. On the other hand, unemployment has a positive effect on E-Rickshaw's utility. In terms of household characteristics, the utility of E-Rickshaw has been affected as vehicle ownership increases. The

Table 3 Model summary

−2 Log likelihood	Cox & Snell R square	Nagelkerke R square
178.326	0.529	0.706

Table 4 Hosmer and Lemeshow test statistics

	Chi-square	df	Sig.
Final model	11.392	8	0.180

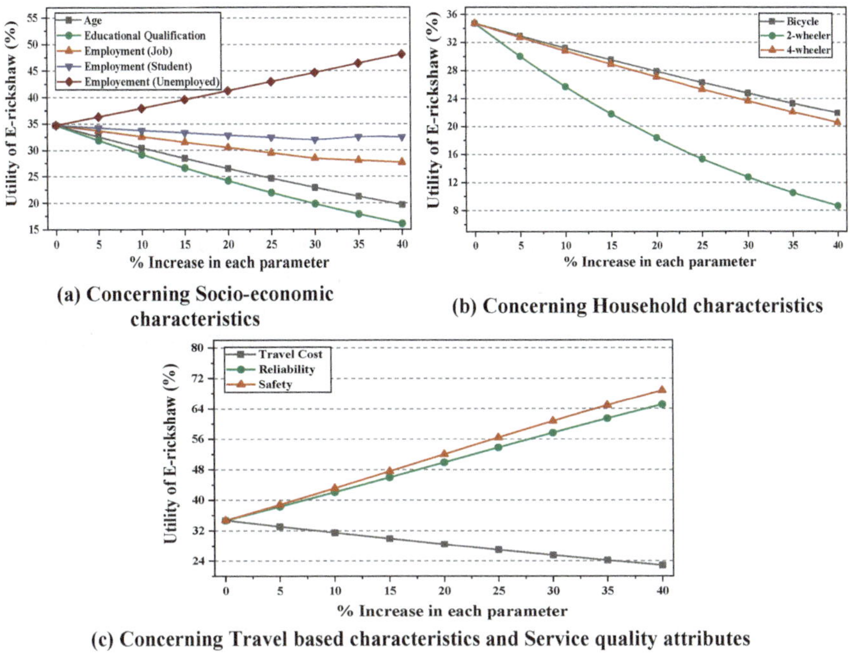

Fig. 3 Sensitivity analysis through OFAT for each characteristic

two-wheeler ownership has the largest impact on vehicle ownership, as it is more appropriate for congested traffic conditions. Further, increasing the reliability and safety of the E-Rickshaw increases its utility. Finally, increasing the travel cost of an E-Rickshaw significantly impacts its utility.

6 Conclusion

The E-Rickshaw has good quality attributes to become a sustainable paratransit mode, which benefits the environment, operators, and users. In medium-sized cities without proper public transportation, E-Rickshaws could be viable as a paratransit mode within the CBD area. From the stated modal revealed that the share of E-Rickshaw, the utility of E-Rickshaw and the mode choice behavior of commuters in medium-sized cities where proper functioning transportation is lacking. A CBD corridor has been selected in Silchar, and the stated modal share of E-Rickshaw was found to be 29.80% with respect to all the available modes, which is still lack compare to all other modes. This indicates that paratransit operators in Silchar impact on a preference for E-Rickshaws in the CBD area. The utility of E-Rickshaw has been assessed using the binary logit model. From the sensitivity analysis, the effect of each parameter on the utility of E-Rickshaw has been assessed and it found to 34.72%.

The binary logit model was applied to analyze mode choice decisions between E-Rickshaw and other transport modes. The parameter estimates indicate that several socio-economic and travel-related factors significantly impact the utility of choosing E-Rickshaw. Age, educational qualification, employment status (job and student), family income, household size, vehicle ownership (bicycle, 2-wheeler, 4-wheeler), trip length, travel time, and travel cost negatively affect the likelihood of choosing E-Rickshaw. Gender, trip purpose, accessibility, convenience, comfort, reliability, and safety have varying impacts on the utility of E-Rickshaw, with some parameters showing a positive influence.

The goodness of fit for the model was assessed using −2 Log-likelihood, Cox & Snell R Square, and Nagelkerke R Square. The model explained between 52.9 and 70.6% of the variance in the dependent variable, indicating a moderate fit. The Hosmer and Lemeshow test, which evaluates the prediction quality of the logistic regression model, showed a significance value greater than 0.05, suggesting an acceptable model fit.

Sensitivity analysis, performed using the One-Factor-At-a-Time (OFAT) method, revealed the impact of individual parameters on the utility of E-Rickshaw. Higher education and better employment status negatively impact E-Rickshaw utility, while unemployment has a positive effect. Increased vehicle ownership, especially two-wheelers, reduces E-Rickshaw utility. Higher travel cost negatively affects E-Rickshaw utility, whereas improving reliability and safety enhances it.

Overall, the utility of E-Rickshaw is influenced by multiple factors, with socio-economic improvements and increased vehicle ownership generally reducing its utility, while reliability and safety improvements enhance it. Ensuring the safety and reliability of E-Rickshaw could improve the mode choice behavior of commuters toward E-Rickshaw as they express more weightage to these service quality attributes among all. The shift of commuters from private vehicles to E-rickshaws and other paratransit modes to E-Rickshaws could be observed in cities which lack proper public transportation. This shift is necessary since E-Rickshaw is an eco-friendly mode of transport within the CBD area. E-rickshaws will continue to play a critical role in providing last-mile connectivity, complementing the E-bus system rather than competing with it. With strategic route planning, E-rickshaws can serve areas that are not directly covered by E-bus routes, ensuring inclusivity in urban mobility. From a regulatory perspective, policies promoting the E-rickshaws, E-buses and IPT modes, such as route rationalization and fare integration, will help create a seamless and efficient urban transport system. For further study, the mode choice behavior of commuters can be modeled using the Multinomial logit model. Such that multiple mode choice can comprehend more.

Acknowledgements The authors wish to thank Mr. Saurabh Kumar and Mr. Suman Kumar, M. Tech Scholars of Department of Civil Engineering, National Institute of Technology Silchar, for their further support.

Funding The author(s) received no financial support for the research, authorship, and publication of this article.

References

1. Al-Salih, W.Q., Esztergár-Kiss, D.: Linking mode choice with travel behavior by using logit model based on utility function. Sustain. 13 (2021). https://doi.org/10.3390/su13084332
2. Ansari, M.W., Sinha, S.: Comparative assessment of service quality of ipt modes in Urban India. Civ. Eng. Archit. 8, 1436–1450 (2020). https://doi.org/10.13189/cea.2020.080626
3. Ashalatha, R., Manju, V.S., Zacharia, A.B.: Mode choice behavior of commuters in Thiruvananthapuram city. J. Transp. Eng. 139, 494–502 (2013). https://doi.org/10.1061/(ASCE)TE.1943-5436.0000533
4. Chandrashekar, C., Agrawal, P., Chatterjee, P., Pawar, D.S.: Development of E-rickshaw driving cycle (ERDC) based on micro-trip segments using random selection and K-means clustering techniques. IATSS Res. 45, 551–560 (2021). https://doi.org/10.1016/j.iatssr.2021.07.001
5. Chen, H., Gan, Z.X., He, Y.T.: Choice model and influencing factor analysis of travel mode for migrant workers: Case study in Xi'an, China. Discret. Dyn. Nat. Soc. 2015 (2015). https://doi.org/10.1155/2015/236216
6. Cochran, W.G.: Sampling Techniques. John Wiley & Sons, New York (1977).
7. Cochrane, C., Muneer, T., Fraser, B.: Design of an electrically powered rickshaw, for use in India. Energies. 12 (2019). https://doi.org/10.3390/en12173346
8. Delhi, P.: IMPLEMENTATION OF SMART CITIES MISSION (2023) https://pib.gov.in/PressReleaseIframePage.aspx?PRID=1984980#:~:text=100 Smart Cities have been, through application of smart solutions.
9. District Disaster Management Authority: City Disaster Management Plan Silchar, District Disaster Management Authority, Cachar, Assam (2012) https://asdma.gov.in/pdf/Silchar%20city%20DM%20plan.pdf
10. Koppelman, F.S., Bhat, C.: A Self Instructing Course in Mode Choice Modeling: Multinomial and Nested Logit Models (2006)
11. Kroes, E.P., Sheldon, R.J.: Stated Preference Methods: An Introduction. J. Transp. Econ. Policy. 22, 11–25 (1988)
12. Kumar, A.G., Anmol M., Akhil V.S.: A Strategy to Enhance Electric Vehicle Penetration Level in India. Procedia Technol. 21, 552–559 (2015). https://doi.org/10.1016/j.protcy.2015.10.052
13. Kumar, P., Chakrabarty, S.: Total Cost of Ownership Analysis of the Impact of Vehicle Usage on the Economic Viability of Electric Vehicles in India. Transp. Res. Rec. 2674, 563–572 (2020). https://doi.org/10.1177/0361198120947089
14. Lashari, Z.A., Ko, J., Jang, J.: Consumers' Intention to Purchase Electric Vehicles: Influences of User Attitude and Perception Sustainability 13(12), (2021). https://doi.org/10.3390/su13126778
15. Mahmoudi, C., Flah, A., Sbita, L.: An overview of electric Vehicle concept and power management strategies. 2014 Int. Conf. Electr. Sci. Technol. Maghreb, Cist. 2014. 1–8 (2014). https://doi.org/10.1109/CISTEM.2014.7077026
16. Majumdar, D., Dutta, A., & Jash, T.: Study on real-world performance of electric two-wheelers and three-wheelers under heterogeneous traffic conditions: a case study in West Bengal State, India. *Clean Technologies and Environmental Policy*, *24*(8), 2419–2439 (2022).
17. Majumdar, D., Jash, T.: Merits and Challenges of E-Rickshaw as An Alternative form of Public Road Transport System: A Case Study in the State of West Bengal in India. Elsevier B.V. (2015)

18. Ministry of Home Affairs, G. of I.: Census of India 2011 - Assam - Series 19 - Part XII B - District Census Handbook, Cachar, https://censusindia.gov.in/nada/index.php/catalog/201
19. Nambiar, A. V., John, J., Philip, N. V., Thomas, K. P., & Samuel, E. R. (2019, July). Economical Electric Rickshaw from Conventional Engine Rickshaw. In *2019 2nd International Conference on Intelligent Computing, Instrumentation and Control Technologies (ICICICT)* (Vol. 1, pp. 560–565). IEEE.
20. Pramanik, Ashrafuzzaman Rahman, M.S.: Operational Characteristics of Paratransit in Medium-sized City: A Case Study on E-rickshaws in Rangpur City, Bangladesh. J. Bangladesh Inst. Planners. 12, 7 (2019).
21. Priye, S., Manoj, M.: Exploring usage patterns and safety perceptions of the users of electric three-wheeled paratransit in Patna, India. Case Stud. Transp. Policy. 8, 39–48 (2020). https://doi.org/10.1016/j.cstp.2020
22. Priye, S., Manoj, M., Ranjan, R.: Understanding the socioeconomic characteristics of paratransit drivers and their perceptions toward electric three-wheeled rickshaws in Delhi, India. IATSS Res. 45, 357–370 (2021). https://doi.org/10.1016/j.iatssr.2021.03.002
23. Rahman, M.L., Baker, D.: Modelling induced mode switch behaviour in Bangladesh: A multinomial logistic regression approach. Transp. Policy. 71, 81–91 (2018). https://doi.org/10.1016/j.tranpol.2018.09.006
24. Ramachandra, T. V., Shwetmala: Emissions from India's transport sector: Statewise synthesis. Atmos. Environ. 43, 5510–5517 (2009). https://doi.org/10.1016/j.atmosenv.2009.07.015
25. Sahu, S., Shanker, S., Kamat, A., & Barve, A.: India's public transportation system: the repercussions of COVID-19. Public transport, 15(2), 435–478 (2023). https://doi.org/10.1007/s12469-023-00320-z
26. Saxena, A., Shrivastava, B.: Examining Factors Affecting the Willingness of Rickshaw Operators to Adopt Battery Operated Rickshaws: The Case of Bhopal, India. Transp. Res. Rec. 2677, 325–340 (2023). https://doi.org/10.1177/03611981221130339
27. Saxena, S.N.: Revolution in growth of three-wheeler electric vehicles in India. J. Glob. Tour. Res. 4, 117–126 (2019). https://doi.org/10.37020/jgtr.4.2_117
28. Shandilya, N., Saini, V., Ghorpade, A.R.: Handbook: E-rickshaws Deployment in Indian Cities. ICLEI South Asia (2019).
29. Shimazaki, T., Rahman, M.: Physical characteristics of Paratransit in Developing countries of Asia. J. Adv. Transp. 30, 5–24 (1993). https://doi.org/10.1002/atr.5670300203
30. Singh, D.R., Mishra, S., Tripathi, K.: Analysing acceptability of E-rickshaw as a public transport innovation in Delhi: A responsible innovation perspective. Technol. Forecast. Soc. Change. 170, 120908 2021(a). https://doi.org/10.1016/j.techfore.2021.120908
31. Singh, S., Priyadharshni, B., Prathyusha, C., Moses Santhakumar, S.: Assessment of the Electric Rickshaws' Operation for a Sustainable Ecosystem. Springer Singapore 2021(b).
32. Weng, J., Tu, Q., Yuan, R., Lin, P., Chen, Z.: Modeling Mode Choice Behaviors for Public Transport Commuters in Beijing. J. Urban Plan. Dev. 144, 1–9 (2018). https://doi.org/10.1061/(asce)up.1943-5444.0000459
33. Zubizarreta, I., Seravalli, A., Arrizabalaga, S.: Smart City Concept: What It Is and What It Should Be. J. Urban Plan. Dev. 142, 1–8 (2016). https://doi.org/10.1061/(asce)up.1943-5444.0000282

Traffic Safety Evaluation of School Zones in Guwahati City Using Microsimulation Modeling

Danial Doley, Vhanamane Ganesh Sukahdeo, Akhilesh Kumar Maurya, and Rahul Raoniar

Abstract "School are the doors to entirety of ocean of the education" this quote aptly describes the importance of schools in a society. With ever increasing population increase in number of schools is mandatory which in turn increase the number of school zones. Traffic safety evaluation of school zones is crucial for ensuring the safety of students, teachers, and other individuals who frequent these areas. The high volume of vehicular and pedestrian traffic during school hours, school zones are prone to accidents and other traffic-related incidents. A thorough evaluation of the traffic safety measures in place, including signage, traffic signals, crosswalks, and speed limits, can identify areas of improvement and help prevent accidents. Moreover, a comprehensive evaluation of the traffic patterns and behavior in and around school zones can help local authorities develop effective strategies to mitigate traffic risks and promote safer driving habits. Investing in the evaluation and enhancement of traffic safety in school zones is crucial to ensuring a secure and supportive learning environment for students and teachers. This study utilizes microsimulation, a computational method that intricately models traffic flow and behavior, to assess the safety of school zones. It examines current traffic conditions and compares them against various scenarios, including reduced speed limits, traffic calming devices, and pedestrian-friendly infrastructure. By analyzing these scenarios, the study identifies the most effective strategies for improving traffic safety in school zones. Microsimulation offers a precise depiction of traffic dynamics, enabling a more accurate evaluation of different safety measures and their potential effects. The findings from this project can guide local authorities and school administrators in making well-informed decisions regarding the implementation of traffic safety improvements in school zones.

D. Doley (✉) · V. G. Sukahdeo · A. K. Maurya
Department of Civil Engineering, Indian Institute of Technology Guwahati, Guwahati, Assam, India
e-mail: d.danial@iitg.ac.in

A. K. Maurya
e-mail: maurya@iitg.ac.in

R. Raoniar
The Center for Transportation Research, Western Kentucky University, Bowling Green, Kentucky, Kentucky, USA

U. S. Dixit et al. (eds.), *Current Progress in Engineering Sciences*,
https://doi.org/10.1007/978-981-96-8753-4_4

Keyword School zones · Traffic safety · Vissim · Microsimulation

1 Introduction

Commuting to and from school is an integral part of the education system. However, ensuring the safety of students during their commute poses significant challenges, especially in areas with high traffic volumes, such as school zones. While the responsibility of ensuring student safety rests with school administration, teaching staff, and parents still the local road agencies and traffic management play a vital role in mitigating the risks associated with commuting. Urbanization has contributed to increased traffic congestion, exacerbating the challenges associated with keeping children safe as they travel to and from school. These challenges are particularly significant for young children walking or biking to school. The Traffic Injury Research Foundation (TIRF) reports that children aged 14 and under are the most at-risk group when it comes to pedestrian collisions [1]. As per recent MoRTH statistics there were 28,142 pedestrian fatalities in India, and many more injuries, across all age groups of about 3,938 total fatalities for children aged 0–14 years, including all types of road accidents, including pedestrian collisions [2]. To address these concerns, proper management and scientific implementation of safety measures are crucial. Modeling and simulation techniques can be employed to test the efficacy of various safety measures and their combinations, potentially reducing conflicts and preventing casualties in school zones.

Children walking in school zones are particularly vulnerable to road accidents, which significantly contribute to road casualties. To tackle this problem, it is crucial to deeply understand the various risk factors present in school zones. This study considers various factors identified from existing research papers and studies to calculate the risk involved in school zones using modeling and simulation techniques. Safety measures can then be implemented based on the results obtained from surrogate safety assessment models. This approach allows for the implementation of appropriate safety measures to make school zones safer without causing significant disruptions to traffic flow or compromising the level of service for traffic in these areas. Ultimately, this study can help reduce the risk of accidents and promote safer commuting for students, teachers, and other individuals in and around school zones.

The remainder of this paper is formed as follows: Sect. 2 is the literature review which reviews the safety in school zones, microsimulation and its calibration, and use of surrogate safety assessment measures. Strategies to fulfilled the objectives the methodology and data collection is explained in Sect. 3. The modeling and simulation is shown in Sect. 4. Section 5 shown the analysis results. Finally, a conclusion with future research direction is given in Sect. 6.

2 Literature Review

The review covers various studies that provide insights into the behavioral traits of pedestrians and the key factors that need to be considered when assessing safety in school zones. Additionally, the literature review includes references to studies on microsimulation, calibration techniques, and safety analysis.

2.1 Safety in School Zones

Numerous studies have investigated how effective school zones are in lowering accident risks and enhancing safety. A study by Sun et al. [3] found that school zones in Hong Kong were effective in reducing the number of accidents involving pedestrians and cyclists. The study analyzed data from 48 school zones and found that the implementation of school zones reduced the number of accidents by 34%. Similarly, a study by Rahman et al. [4] found that school zones were effective in reducing the number of crashes and fatalities in and around schools. The study analyzed data from 14 cities in the United States and found that school zones reduced the risk of pedestrian fatalities by 37% and the risk of crashes involving young drivers by 29%. In addition, a study by Strawderman et al. [5] examined the impact of school zone speed limits on driver behavior and found that drivers were more likely to slow down in school zones than in non-school zones. The study also found that drivers who were aware of the school zone were more likely to comply with the speed limit than those who were not aware. Another study by Park [6] analyzed the effect of the installation of LED speed feedback signs on drivers' behavior in school zones. The study found that the LED speed feedback signs were effective in reducing the average speed of drivers in school zones, resulting in a lower risk of accidents. A Canadian study found that lowering speed limits from 50 to 30 km/h in school zones resulted in a significant decrease in fatal and injury collisions by 45.3%, and a 55.3% reduction in injuries to vulnerable road users [3]. Study conducted by Rahman [4] aimed to evaluate the safety effects of various roadway countermeasures in school zones. Using crash rates data, the most crash-prone school zone was identified. A microsimulation network was then utilized to assess the effectiveness of three countermeasures: two-step speed reduction, decreasing driveway access, and replacing the two-way left-turn lane (TWLTL) with a raised median. The results indicated that implementing the two-step speed reduction and decreasing driveway access significantly reduced crash risks, while replacing the TWLTL with a raised median showed higher crash risks compared to the base condition. These findings provide valuable insights for transportation planners and decision-makers in improving safety measures for school zones. The study conducted by Katan et al. [7] found out that reduced speed school zones are crucial for enhancing the safety of young children during their commute to and from school. This study aimed to identify effective methods for improving speed compliance in these school zones. Through a literature review, a public opinion survey, and a field study using speed monitoring displays (SMDs) in Utah, it was concluded that a combination of appropriate traffic controls, public education, and law enforcement is necessary for achieving better speed limit compliance. The SMDs proved to be effective in increasing speed compliance and improving school zone safety by reducing speeds and decreasing the percentage of vehicles exceeding the designated speed limit. Traffic volume plays an important role in vehicle and pedestrian interaction. Study conducted by Kloeden et al. [8] shows that with increase in traffic the number of conflicts of vehicle and pedestrian tend to increase slightly but the severity of collision if happens is drastically reduced due to overall reduction in traffic stream speed. But pedestrian behavior tends to become riskier as the time required to walk or cross roads increases with increase in traffic volume.

2.2 Microsimulation and Its Calibration

VISSIM calibration has been an active research area in India due to its significance in the simulation of mixed traffic conditions prevalent in the country. A study by Raju et al. [9] focused on the calibration of VISSIM models for heterogeneous traffic flow by adjusting various model parameters such as driver behavior and road network features. They used statistical tools such as the one-way ANOVA and multi-parameter sensitivity analysis to identify the significant variables and calibrate the model to minimize the error between the simulated and field data. Another study conducted in India by Arkatkar et al. [10] aimed to improve the toll plaza operation using VISSIM simulation. The study involved the calibration of the VISSIM model for the toll plaza, using field data collected from the site. The authors calibrated the model using gap acceptance parameters, lane distribution, vehicle types, and speeds, and validated the model using field data. The results of the study showed that VISSIM could be effectively used to improve the toll plaza operations. Another study the VISSIM model was calibrated by adjusting driver behavior and signal timings, then validated using field data in the context of a high-density mixed traffic corridor in Kolkata [11]. The study found that the calibrated model provided reliable results and could be used for the assessment of traffic safety measures. Similarly, a study by Zhandong et al. [12] focused on the calibration of VISSIM for an urban arterial road in China. The study used a combination of field data and simulation to calibrate the VISSIM model and validated the model by comparing it with the field data.

In conclusion, VISSIM calibration has been extensively studied by researchers in India due to its relevance in simulating mixed traffic conditions prevalent in the country. The studies mentioned above have successfully employed statistical tools and model calibration techniques to minimize the error between simulated and field data. The calibration of VISSIM models has also proved useful in improving traffic operations, which can have significant implications for the efficiency and safety of the transportation system.

2.3 Use of Surrogate Safety Assessment Measures

The use of Surrogate Safety Assessment Measures (SSAMs) in Indian conditions has gained significant attention in recent years. Several studies have explored the application of SSAMs to assess the safety performance of various transportation systems and environments in the country. In a study conducted using SSAMs were utilized to evaluate the safety performance of Bus Rapid Transit (BRT) systems in India [13]. The authors employed measures such as time-to-collision and deceleration rate to assess the safety of BRT systems and compared them with non-BRT corridors. The study highlighted the effectiveness of SSAMs in evaluating the safety outcomes of BRT systems and provided insights for improving their design and operation. A another study focused on the use of SSAMs to assess the safety performance of urban arterials in India [14]. The researchers collected traffic data and utilized measures such as time-to-collision and post-encroachment time to evaluate the safety performance of intersections. The study demonstrated the applicability of SSAMs in assessing the safety of urban road networks and identified potential areas for improvement. Furthermore, a study find by Avinash et al. [15] explored the use of SSAMs in the context of pedestrian safety in Indian cities. The researchers utilized measures such as time-to-collision and post-encroachment time to assess the risk of pedestrian-vehicle conflicts at mid-block cross walk. The study highlighted the importance of SSAMs in identifying potential safety concerns and guiding the development of pedestrian-friendly infrastructure. The study conducted by Nikolaou et al. [16] found that SSMs have the potential to serve as sustainable indicators of road safety, complementing or even substituting traditional crash analyses. The review examined 34 studies, discussing the types of SSMs collected in real road environments, their collection methods, their relationship with specific crash types, and the statistical models developed. The study Conducted by Yang et al. [17] proposed a new safety performance measure called Risk Status (RS) by integrating crash data and SSMs. By combining these two sources of information, RS provides a more comprehensive assessment of traffic safety. The study utilized real-world connected vehicle data to model the interrelationship between RS, crash frequency, SSMs, and contributing factors, highlighting the importance of considering spatial autocorrelation and unobserved heterogeneity in safety analysis. According to a research report by Gelder et al. [18], SSMs for real-time risk assessment in traffic conflicts can be derived using a data-driven methodology. By utilizing Monte Carlo simulations and regression models, the PRISMA method calculates the probability of specific events, such as crashes, without relying on assumptions about future trajectory evolution. The study demonstrated the effectiveness of the derived SSMs, particularly in longitudinal traffic interactions, and highlights the potential for future applications in other types of traffic conflicts and interactions with vulnerable road users.

Multiple studies have shown that school zones are effective in reducing accidents and improving safety. They have been found to reduce the number of accidents involving pedestrians and cyclists, decrease the risk of crashes and fatalities, and improve driver behavior and speed compliance. Researchers have focused on the calibration of VISSIM models, a traffic simulation software, to accurately represent mixed traffic conditions and improve toll plaza operations. Calibration techniques involving field data and statistical tools have been successful in minimizing errors between simulated and field data, providing reliable results for traffic safety assessments. The use of SSAMs in India has gained attention. Studies have utilized SSAMs to evaluate the safety performance of Bus Rapid Transit (BRT) systems, urban arterials, and pedestrian safety in Indian cities. SSAMs such as time-to-collision and deceleration rate have proven effective in assessing safety outcomes and identifying areas for improvement. This study mainly focuses on evaluating safety at school zones using simulation techniques by considering different factors contributing to the conflicts and traffic behavioral characteristics in school zone. Additionally, study aims at proposing safety measures to enhance the safety of school zones and measure their effectiveness with modeling and micro-simulation.

To achieve this study some objectives have been addressed: (1) Identification of suitable school zone and collection of required traffic parameters of vehicular traffic and pedestrians. (2) Modeling existing conditions of school zones in microscopic simulation software. (3) Evaluating safety of existing school zones from field data and simulated data. (4) Evaluation of different safety measures to improve the overall safety of school zones.

3 Methodology and Data Collection

There are three schools that constitute the school zone around Kamrup Higher Secondary School, St. Mary's Higher Secondary School, and T.C. Government Girls' Higher Secondary School. As shown in Fig. 1, the yellow circles from top to bottom represent these school zones respectively. To study the safety of school zone, the "Kamrup Higher Secondary School" and "TC Girls School" in Guwahati were selected as school zone for traffic study. Being in heart of Guwahati city this school makes a good school zone for traffic study located in urban area. Located near Pan Bazar area of Guwahati city which is one of the busiest of the city areas. This makes "Kamrup High School" and "TC Girls School" a crucial school zone. Most of Streets surrounding school zone are four lane and are divided. During peak hours, i.e., from 2:30 pm to 5:30 pm there is significant congestion in school zone near the school. Most of the roads have provision of footpaths of 1.5 m wide on both sides of road but footpaths are seldom available in their full capacity due to encroachments. The inner roads are unsignalized and of proper signages like yield signs and speed limits signs. The intersection of M.C. Road and G.N.B. road is unsignalized intersection which is main entry point of traffic in the school zone. These conditions warrant and have scope for lot of improvement of school zone which could be translated into safety effects in school zone.

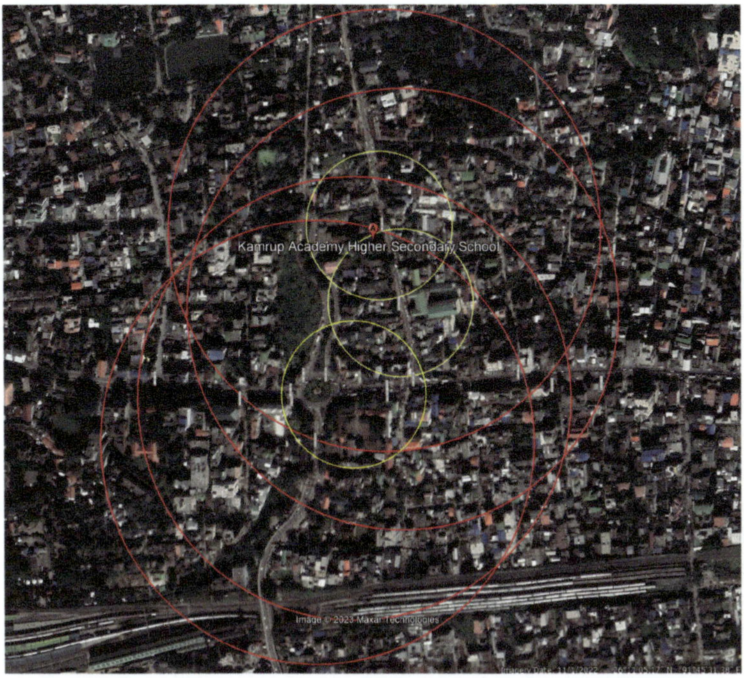

Fig. 1 Map showing extents of school zone (yellow circle covered the main school)

3.1 Data Collection

Data from the site was collected using a video graphic survey. This survey was conducted in three phases. In total five video cameras were used at a time to record video at same time and different location in school zones showed in Fig. 3a–d locations wise respectively. The proportion of various vehicle types was used as input in VISSIM. The average proportion traffic composition observed from data was shown in Fig. 4. The collected video graphic data was used to find the reasonable and real ranges of inputs of values to be given while simulating the school zone in simulation software. The parameters required were using "Datafromsky" software with accuracy of 10 frames per second. Different characteristics were obtained by analyzing footage frame by frame in Datafromsky software.

Inputs collected for the simulation are (i) Traffic volume and Traffic speed on various links (ii) Pedestrian volume and classification (iii) Classification of traffic (iv) Desired speed distribution of vehicles (v) Peak volumes of traffic and corresponding time of day.

4 Modeling and Simulation

The model of selected site using data obtained from data collection and extraction was created in VISSIM. The traffic compositions and pedestrian compositions as obtained from the collected data were fed to the VISSIM. To create the road alignment around school zone, traffic map provided by PTV VISSIM was used showed in Fig. 2. The interval of 1 s was used for simulation while simulating the modeled site.

The output collected from VISSIM as follow: (i) Vehicle records (Trajectory) (in form of.trj file) (ii) Traffic delays (iii) Pedestrian records (iv) Pedestrian delays (v) Vehicle and pedestrian travel times.

4.1 Output Analysis and Calculations

Outputs obtained from VISSIM were analyzed in SSAM to find all possible conflicts from the trajectories of vehicles and pedestrians. Potential conflicts when mapped spatially give idea about the places of high risk which can be rectified with safety measures. The following Data was collected to give inputs for simulation.

Traffic Volumes. To accurately simulate traffic conditions, it is crucial to provide the software with accurate and traffic flow data. One way to achieve this is by computing traffic flow values at regular intervals of time. In this particular case, traffic flow values were calculated every 5 min (Fig. 5). This means that the number of vehicles passing through a given point on the road network was recorded every

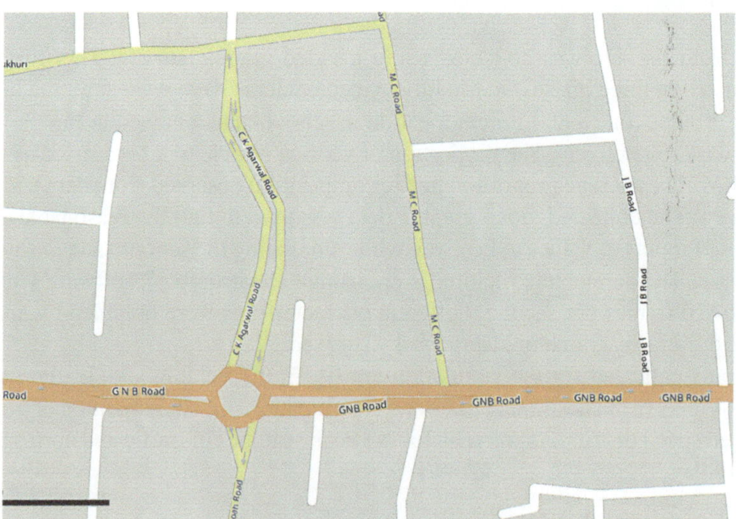

Fig. 2 PTV VISSIM map showing road alignment in school zone

a. Location 1

b. Location 2

c. Location 3

d. Location 4

Fig. 3 Different location on site of data collection

5 min. This data was then used as input to simulate traffic conditions in the modeled environment. Using a time interval of 5 min to calculate traffic flow values can help to improve the accuracy of the simulation results. This is because traffic conditions can change rapidly, and using a longer interval could result in inaccurate data being used as input. By calculating traffic flow values every 5 min, the simulation software can more accurately model real- world traffic conditions.

Desired Speed Distribution. To find desired speed distribution the video of traffic was observed manually and those vehicles which seem unaffected by other vehicles

Fig. 4 Traffic composition

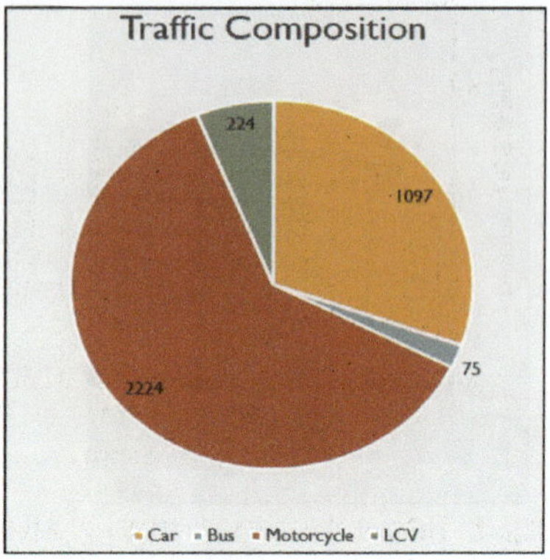

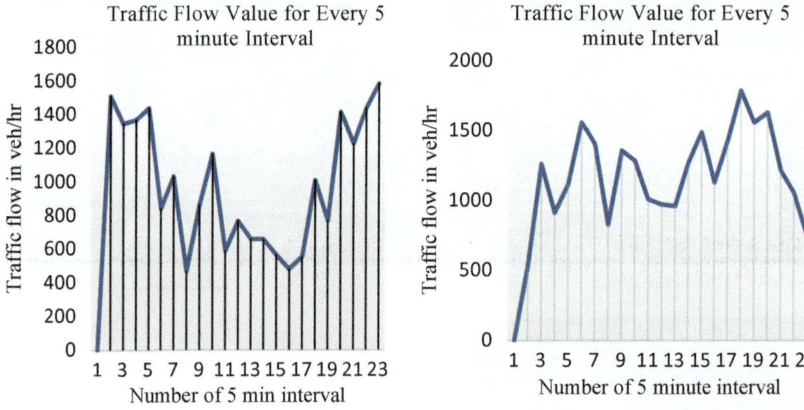

Fig. 5 Traffic flow values for each 5 min interval

were selected as vehicles in free flow. Distribution of speeds were calculated for each vehicle type separately namely motorcycles, cars, buses, light commercial vehicles (LCVs) as shown in Fig. 6.

Pedestrian Flow. Similar to traffic flow, pedestrian flow data is also important for simulating and analyzing pedestrian behavior in a virtual environment. To provide accurate input data for pedestrian simulation software, it is necessary to calculate pedestrian flow values at regular intervals as shown in Fig. 7. For this purpose, pedestrian flow values were calculated every 5 min. By collecting this data, it is possible to simulate pedestrian behavior in a virtual environment with a high degree of accuracy.

Vehicle Route Selection. One of the crucial factors in developing efficient transportation systems is the ability to accurately predict the flow of traffic at intersections or road segments. Video data collection has been increasingly used as a reliable and effective method to gather this information. By analyzing the footage, the proportion of vehicles going straight, turning left, and turning right was determined. This data

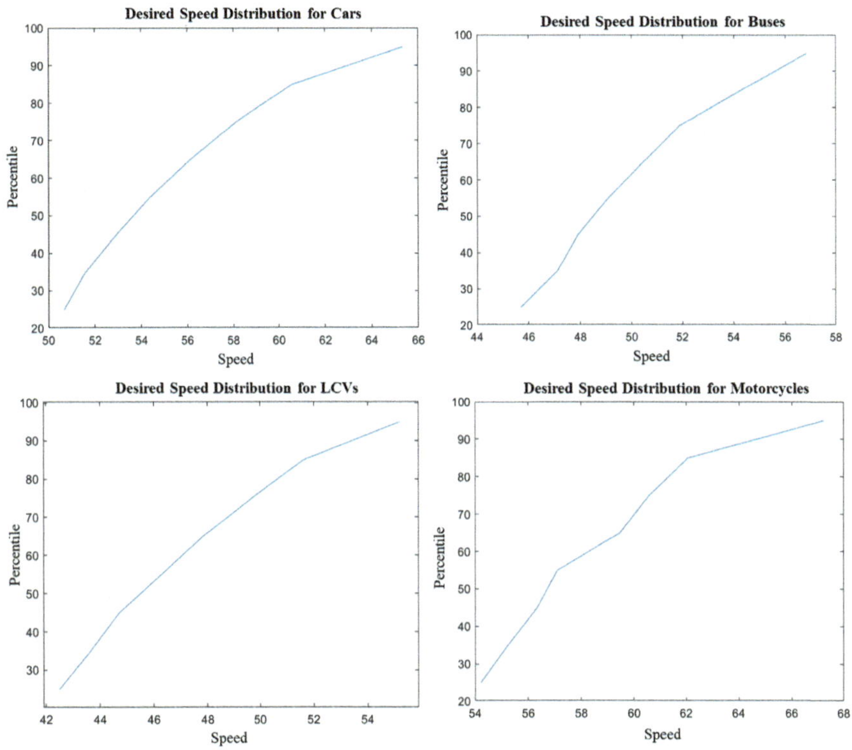

Fig. 6 Desired speed distribution for different vehicle types

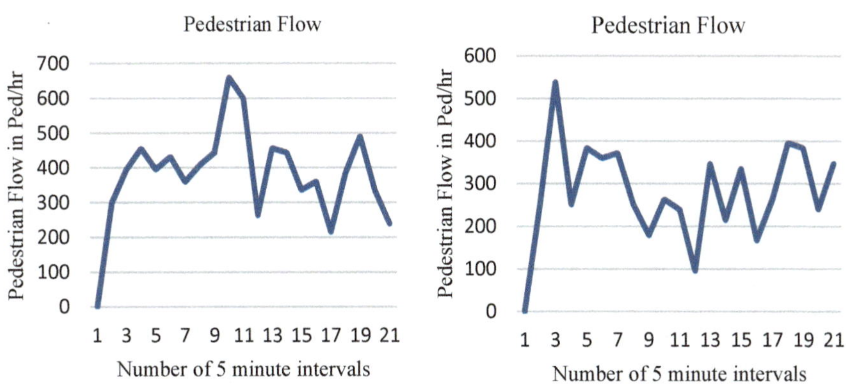

Fig. 7 Pedestrian flow for each 5 min interval

was then used to optimize vehicle routes using simulated models. By incorporating the proportion of vehicles going straight, turning left, and turning right into the simulated model, the accuracy of the model is improved, resulting in more reliable and efficient vehicle routing.

4.2 Development of Model

To develop model of school zone in VISSIM, the road alignment and geometry was created in VISSIM using collected information and with use of traffic maps provided in PTV VISSIM. The roadway width was measured on-site at various sections and incorporated into the simulation model accordingly. The sections of the road were parked vehicles were present were considered, and only the available width for vehicle passage was considered in the model.

Calibration of Model. For simulated model to depict real-life conditions with adequate accuracy calibration of model is required. The parameters obtained from data collected were used as input for simulated model as shown in Fig. 8.

The parameters given as input were (i) 5 min Traffic flow values (ii) Traffic composition (iii) Vehicle Routes (4) Desired Speed Distribution and (5) 5 min Pedestrian Flow.

To calibrate the simulation model, it was necessary to determine the appropriate values for driver behavior and pedestrian's accepted gap while crossing the road. As the road under consideration was an urban road, the "Wiedemann 74" Driver Behavior model was utilized for calibration purposes. To identify the significant variables that affect the calibration of the model, a one-way ANOVA (Analysis of Variance) test was conducted on each variable while keeping all other values constant. This statistical method allowed for the identification of the variables that have a significant impact on the calibration process. Variables that had a p-value less than 0.3 in the ANOVA test

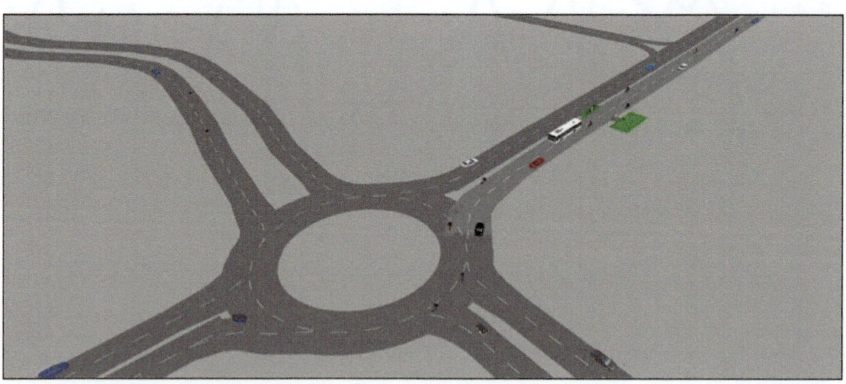

Fig. 8 Images from simulated scenarios

were considered statistically significant and were included in the calibration process. Because simulation is complicated process lot of factors are at play, so we decided to go with $p < 0.3$ and 50% significance at another place.

The use of statistical tests such as ANOVA enables the identification of important variables that affect the simulation model and its calibration. By identifying these variables, the calibration process can be optimized to ensure that the simulation accurately reflects real-world conditions. This approach enhances the reliability and accuracy of the simulation. From Table 1 ANOVA test, it was determined that the following variables had a significant impact on the calibration of the model: (i) Average Standstill Distance (ii) Additive Part of Safety Distance and (iii) Multiplicative Part of Safety Distance.

In order to account for pedestrian-vehicle interactions, "The Gap Accepted by Pedestrian" was also identified as a significant parameter. Once these significant variables were identified, their ranges were determined based on past studies and literature showed in Table 2. This allowed for a more accurate calibration of the model and a better understanding of the impact of these variables on the simulation outcomes. By selecting appropriate ranges for the significant variables, the model was able to reflect real-world conditions more accurately, resulting in more reliable simulation results for the chosen scenario.

To evaluate the accuracy of the simulation, headway distribution of the traffic stream was calculated based on the inputs and parameters used. This calculated distribution was then compared with the observed headway distribution from real-world conditions. To ensure that the model was accurately calibrated, the error between the two distributions was minimized using a Genetic Algorithm. This involved minimizing the chi-square error between the observed and simulated headway distributions. The Genetic Algorithm was run for 50 generations with a population size of 40 showed in Fig. 9. Through this calibration process, the simulation was refined to accurately reflect the real-world traffic conditions.

Table 1 P values for ANOVA test to determine significant variable

	1	2	3	4	5
Minimum look ahead distance	0.994	0.996	0.999	0.900	0.994
Minimum look back distance	0.988	0.800	0.992	0.988	0.987
Average standstill distance	**0.243**	0.978	**0.302**	**0.032**	**0.033**
Additive part of safety distance	**0.001**	0.837	**0.007**	0.815	**0.056**
Multiplicative part of safety distance	0.999	**0.0317**	**0.010**	0.990	0.850

Table 2 Range of values of variables for gap accepted

Parameter	Default	**Range**
Average standstill distance	2	1 - 3
Additive part of safety distance	2	0.1 - 3
Multiplicative part of safety distance	3	0 - 3

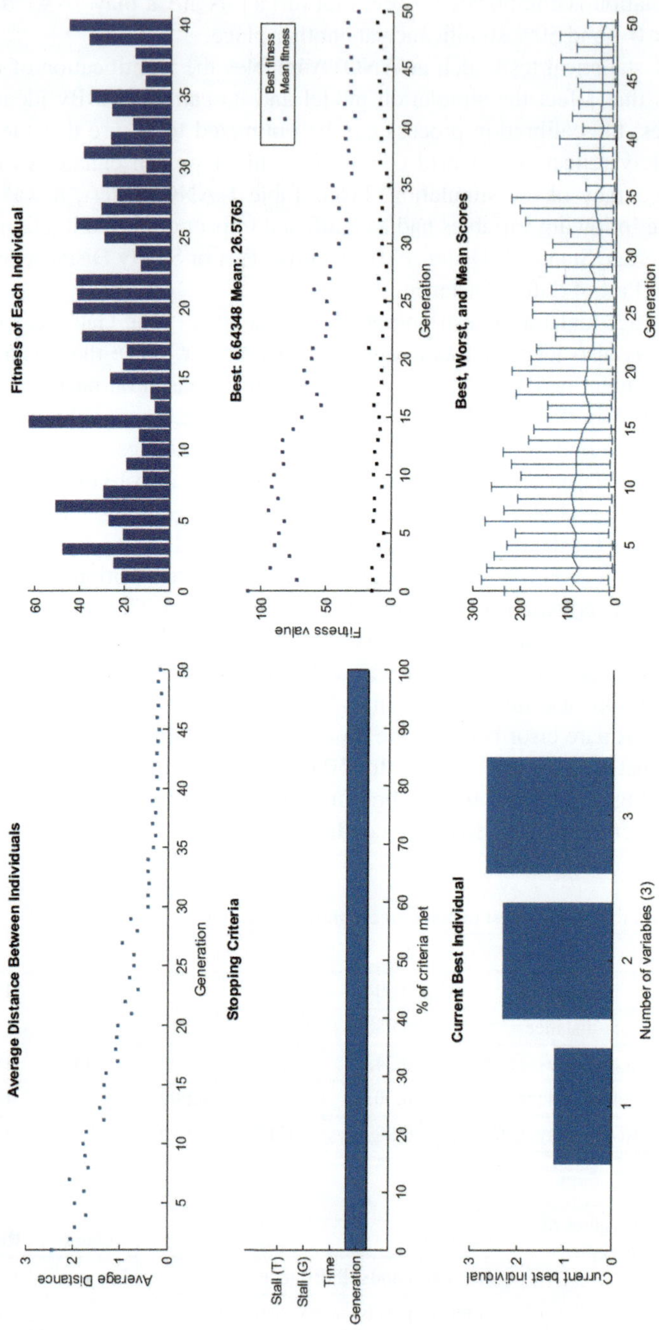

Fig. 9 Result of genetic algorithm optimization

4.3 *Validation of Model*

The parameters obtained through calibration were evaluated by testing the error in headway distributions for an additional 40 min of data that was not used previously (Fig. 10).

The average value for chi-square error between simulated and actual headway distribution was 11.859 which is less than critical value of 13.339 at 50% confidence level. Based on the results of the chi-square test conducted at a confidence level of 50%, we failed to reject the null hypothesis that the two distributions were the same, indicating that the obtained values satisfactorily fit the observed data. Therefore, the model is assumed to be calibrated and validated for the given traffic conditions.

5 Analysis and Results

5.1 *Analysis*

This section analyzes different scenarios created in simulated and calibrated model of school zone. The past studies were taken as reference for the analysis of present school zones and their existing conditions. The existing conditions were modeled and simulated in PTV VISSIM software. The study compares two aspects, namely traffic safety in school zones and delay. Traffic safety in school zones is evaluated using Surrogate Safety Assessment Models (SSAMs) such as Time-to-Collision (TTC) and Post-Encroachment Time (PET). The criteria used for potential conflicts in the study are TTC less than 1.5 s or PET less than 5 s. These metrics are commonly used in traffic safety analysis to assess the risk of potential collisions and evaluate the safety of traffic systems. A TTC less than 1.5 s indicates a high risk of collision, while a PET less than 5 s indicates a low margin of safety after a potential conflict. By comparing the safety and delay aspects of traffic in school zones, the study aims to identify potential trade-offs between safety and efficiency.

5.2 *Evaluation for Different Safety Measures*

Based on the analysis conducted, it was observed that implementing a subway system had the most positive impact on travel time and delay compared to the existing condition and the two-step speed reduction approach (Table 3). The travel time for the subway system was significantly lower than both the existing condition and the two-step speed reduction approach. Additionally, the delay for the subway system was the lowest compared to the other two approaches. In terms of potential conflicts, the subway system had the least number of potential conflicts compared to the other

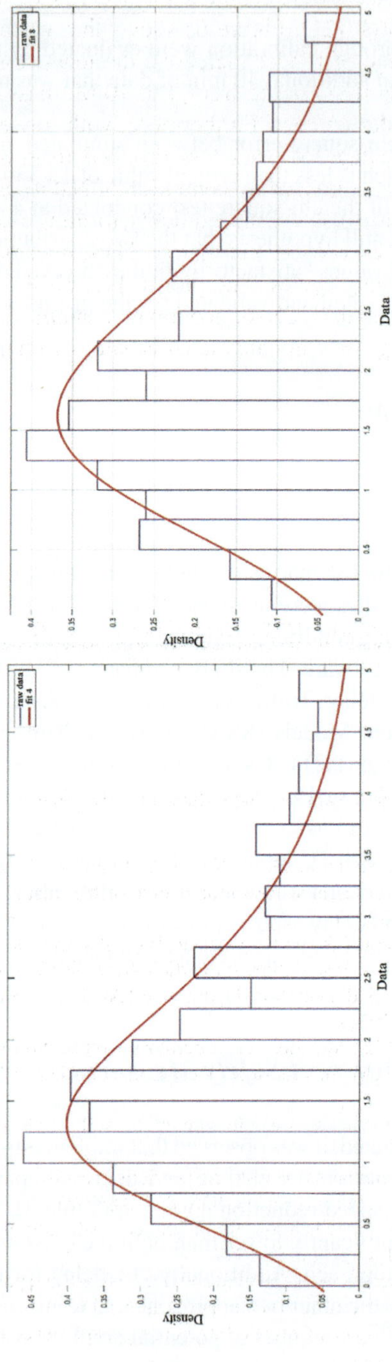

Fig. 10 Simulated and actual headway distribution

Table 3 Comparison of delay and potential conflicts for different safety measures

	Existing condition	Two step speed reduction	Subway
Travel time (s) (% change)	259.7 (0)	288.6 (+11.13%)	205.3 (–20.95%)
Delay (s) (% change)	140.7 (0)	169.6 (+20.06%)	86 (–38.87%)
Potential conflicts (nos.) (% change)	1205 (0)	948 (–21.33%)	473(–60.75%)

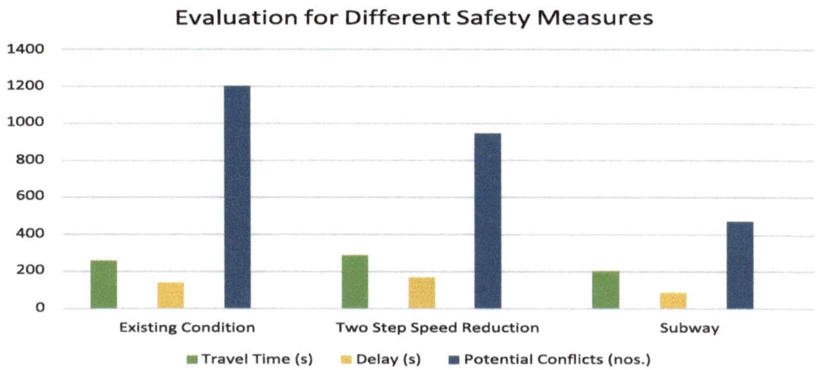

Fig. 11 Evaluation of different safety measures

two approaches. It reduced delay by 38.87% while reducing potential conflicts by 60.75%. This indicates that the subway system had a positive impact on traffic safety.

Overall, the analysis suggests that implementing a subway system can be an effective approach to reduce travel time, delay, and potential conflicts. It is important to note that the specific conditions and characteristics of the roadway and traffic should be considered when determining the most appropriate approach for a given situation (Fig. 11).

5.3 Evaluation for Different Compliance Level

Table 4 shows the results of simulations conducted for different levels of compliance rates in a given scenario. The compliance rates for this simulation were 70%, 80%, 90%, and 95%. Additionally, a compliance rate of 95% with lane discipline was also considered. The simulation results indicate that as the compliance rate increases, the travel time (in second) increases (see Fig. 12). This means that if the compliance rate is high, the traffic moves slowly, resulting in higher travel times. For instance, the travel time for a compliance rate of 70% is 259.7, whereas for a compliance rate of 95%, it is 296.3. Similarly, the delay for a compliance rate of 70% is 140.7, whereas for a compliance rate of 95%, it is 177.3. Moreover, as the compliance rate increases, the potential conflicts decrease. This means that higher compliance rates result in safer driving conditions with fewer potential conflicts. For instance, the potential conflicts for a compliance rate of 70% are 1205, whereas for a compliance rate of 95%, it is 495. Finally, the results for a compliance rate of 95% with lane discipline show the best performance in terms of travel time, delay, and potential conflicts. This indicates that having a high compliance rate with lane discipline can significantly improve the traffic flow and safety conditions.

Table 4 Comparison of delay and potential conflicts for different compliance levels

	Compliance of 70%	Compliance of 80%	Compliance of 90%	Compliance of 95%	Compliance rate of 95% with lane discipline
Travel time (s) (% change)	259.7 (0)	277.9 (+7%)	294.5 (+13.41%)	296.3 (+14.09%)	239.9 (–7.62%)
Delay (s) (% change)	140.7 (0)	158.9 (+12.94%)	175.5 (+24.74%)	177.3 (+26.01%)	120.9 (–14.07%)
Potential conflicts (nos.) (% change)	1205 (0)	1072 (–11.04%)	868 (–27.97%)	827 (–31.36%)	495 (–58.92%)

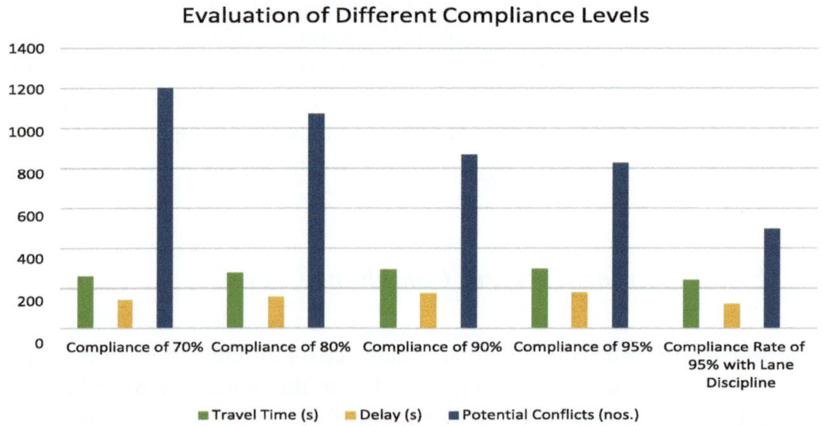

Fig. 12 Evaluation of different compliance levels

6 Conclusions and Future Research Directions

6.1 Conclusions

This study effectively investigated the traffic flow and pedestrian activity at a school zone in Guwahati city using video data collection and simulated modeling. The existing conditions were modeled and simulated in PTV VISSIM software. The study compares two aspects, namely traffic safety in school zones and delay. Traffic safety in school zones is evaluated using SSAMs such as TTC and PET. The criteria used for potential conflicts in the study are TTC less than 1.5 s or PET less than 5 s. These metrics are commonly used in traffic safety analysis to assess the risk of potential collisions and evaluate the safety of traffic systems. A TTC less than 1.5 s indicates a high risk of collision, while a PET less than 5 s indicates a low margin of safety after a potential conflict. By comparing the safety and delay aspects of traffic in school zones, the study aims to identify potential trade-offs between safety and efficiency.

It was observed that implementing a subway system had the most positive impact on travel time and delay compared to the existing condition and the two-step speed reduction approach. The travel time for the subway system was significantly lower than both the existing condition and the two-step speed reduction approach. Additionally, the delay for the subway system was the lowest compared to the other two approaches. In terms of potential conflicts, the subway system had the least number of potential conflicts compared to the other two approaches. It reduced delay by 38.87% while reducing Potential Conflicts by 60.75%. This indicates that the subway system had a positive impact on traffic safety. Overall, the analysis suggests that implementing a subway system can be an effective approach to reduce travel time, delay, and potential conflicts. However, concern raising on frequent rainfall and flooding, making subway systems near school zones may not be feasible. The possible alternative can be implemented, (i) Elevated Pedestrian Bridges: Ensure safe movement above flood-prone areas. (ii) Traffic Calming Measures: Use speed bumps, raised crosswalks, and narrowed lanes to enhance safety. (iii) Traffic Management: Introduce dedicated school transport lanes and optimized signal timings. (iv) Enhanced Drainage Systems: Pair new infrastructure with improved drainage to prevent waterlogging. (v) Pedestrian Infrastructure: Develop dedicated footpaths and cycle tracks.

The compliance rates for this simulation were 70, 80, 90, and 95%. Additionally, a compliance rate of 95% with lane discipline was also considered. The simulation results indicate that as the compliance rate increases, the travel time increases. This means that if the compliance rate is high, the traffic moves slowly, resulting in higher travel times. For instance, the travel time for a compliance rate of 70% is 259.7, whereas for a compliance rate of 95%, it is 296.3. Similarly, the delay for a compliance rate of 70% is 140.7, whereas for a compliance rate of 95%, it is 177.3. Moreover, as the compliance rate increases, the potential conflicts decrease. This means that higher compliance rates result in safer driving conditions with fewer potential conflicts. For instance, the potential conflicts for a compliance rate of 70% are 1205, whereas for a compliance rate of 95%, it is 495. Finally, the results for a compliance rate of 95% with lane discipline show the best performance in terms of travel time, delay, and potential conflicts. This indicates that having a high compliance rate with lane discipline can significantly improve the traffic flow and safety conditions.

Overall, the results of this study provide valuable insights into the effectiveness of different strategies in optimizing traffic flow and improving pedestrian safety. This study effectively integrated different methods and approaches to analyze traffic and pedestrian flow, which could lead to more effective and efficient transportation systems. The combination of video data collection and simulated modeling proved to be a powerful approach in analyzing traffic and pedestrian flow, providing a comprehensive understanding of the site conditions. The findings of this study can serve as a basis for future transportation planning and management, with potential applications in developing efficient and safe school zones. As found in this study increasing traffic rule compliance rate has huge impact on school zone safety. This can be achieved by education and awareness, clear signage and road markings, law enforcement, community engagement, driver education and training, traffic calming measures, public transport and alternative modes, and technology solutions. It is important to adopt a multi-faceted approach that combines education, enforcement, infrastructure improvements, and community involvement to foster a culture of compliance with traffic rules. By implementing these strategies, road safety can be improved, traffic congestion can be reduced, and overall compliance with traffic rules can be increased.

Expanding the research to various locations with different traffic patterns and pedestrian behaviors would enable a broader understanding of traffic dynamics and the effectiveness of various strategies in diverse urban settings for future research direction.

7 Limitations

Video data was collected from a specific site, which might not reflect the general traffic patterns and pedestrian behavior in other areas. The study's findings and conclusions may not be widely applicable and need to be validated at different locations. The accuracy of the extracted parameters largely depends on the quality of the video data. Issues such as low resolution, occlusions, and lighting variations can introduce errors or biases in the data collected. The outcomes of the simulations and the effects of different scenarios and compliance levels may be sensitive to the input parameters used in the model. Small variations or uncertainties in these parameters could lead to significant differences in the results, potentially affecting the validity of the findings. While the simulated model was tested with various scenarios, compliance levels, and their effects, it is important to note that real-world validation was not explicitly mentioned in the provided information. The accuracy and effectiveness of the model's predictions need to be validated through field studies or comparative analysis with actual observed data.

Funding The author(s) did not receive any financial support for the research, writing, or publication of this article.

References

1. https://tirf.ca/wpcontent/uploads/2017/01/TIRF_Toolkit_Factsheets_Pedestrian-Issue.pdf
2. https://morth.nic.in/sites/default/files/RA_2022_30_Oct.pdf
3. Sun, D., El-Basyouny, K., Ibrahim, S., Kim, A. M.: Are school zones effective in reducing speeds and improving safety?. *Canadian Journal of Civil Engineering, 45*(12), 1084–1092 (2018).
4. Rahman, M. H., Abdel-Aty, M., Lee, J., Rahman, M. S.: Enhancing traffic safety at school zones by operation and engineering countermeasures: A microscopic simulation approach. *Simulation Modeling Practice and Theory, 94*, 334–348 (2019).
5. Strawderman, L., Rahman, M. M., Huang, Y., Nandi, A.: Driver behavior and accident frequency in school zones: Assessing the impact of sign saturation. *Accident Analysis & Prevention, 82*, 118–125 (2015).
6. Park, J., Abdel-Aty, M., Lee, J.: School zone safety modeling in countermeasure evaluation and decision. *Transportmetrica A: transport science, 15*(2), 586–601 (2019).
7. 8. Kattan, L., Tay, R., Acharjee, S.: Managing speed at school and playground zones. *Accident Analysis & Prevention, 43*(5), 1887–1891 (2011).
8. Kloeden, C., Woolley, J., McLean, A. J.: A follow-up evaluation of the 50km/h default urban speed limit in South Australia. In *Proceedings of* (2007).
9. 10. Raju, N., Kumar, P., Arkatkar, S., Joshi, G.: Determining risk-based safety thresholds through naturalistic driving patterns using trajectory data on expressways. *Safety Science, 119*, 117–125 (2019).
10. 11. Bains, M. S., Arkatkar, S. S., Anbumani, K. S., Subramaniam, S.: Optimizing and modeling tollway operations using microsimulation: case study Sanand Toll Plaza, Ahmedabad, Gujarat, India. *Transportation Research Record, 2615*(1), 43–54 (2017).
11. Maheshwary, P., Bhattacharyya, K., Maitra, B., Boltze, M.: A methodology for calibration of vehicle class-wise driving behavior in heterogeneous traffic environment. In *World Conf Transp Res 00* (pp. 1–15) (2017).
12. 13. Zhandong, Z., Shaohui, C., Yanquan, Y., Aixiu, H., Xinyi, Z.: VISSIM simulation based expressway exit control modes research. *Procedia engineering, 137*, 738–746 (2016).
13. 14. Kathuria, A., Parida, M., Ravi Sekhar, C., Sharma, A.: A review of bus rapid transit implementation in India. *Cogent Engineering, 3*(1), 1241168 (2016).
14. Golakiya, H., Chauhan, R., Bari, C. S., Dhamaniya, A.: Pedestrian safety analysis at urban midblock section under mixed traffic conditions using time to collision as surrogate safety measure. *Current Science (00113891), 123*(9) (2022).
15. 16. Avinash, C., Jiten, S., Arkatkar, S., Gaurang, J., Manoranjan, P.: Evaluation of pedestrian safety margin at mid-block crosswalks in India. *Safety Science, 119*, 188–198 (2019).
16. 17. Nikolaou, D., Ziakopoulos, A., Yannis, G.: A review of surrogate safety measures Uses in historical crash investigations. *Sustainability, 15*(9), 7580 (2023).
17. 18. Yang, D., Xie, K., Ozbay, K., & Yang, H.: Fusing crash data and surrogate safety measures for safety assessment: Development of a structural equation model with conditional autoregressive spatial effect and random parameters. *Accident Analysis & Prevention, 152*, 105971 (2021).
18. 19. de Gelder, E., Adjenughwure, K., Manders, J., Snijders, R., Paardekooper, J. P., den Camp, O. O., Tejada, A., De Schutter, B.: PRISMA: A novel approach for deriving probabilistic surrogate safety measures for risk evaluation. *Accident Analysis & Prevention, 192*, 107273 (2023).

Extraction of Flow-Power from VIV of Two Cylinders in Diagonal Arrangement at Low Reynolds Number

Abhijit Pal and Atul K. Soti

Abstract Flow around the bluff body has piqued researchers' interest for over a century. Vortex-induced vibration (VIV) is one of many nonlinear phenomena where body motion is governed by the flow around the bluff body. Present numerical work illuminates the VIV phenomenon of rigidly coupled two cylinders. The cylinders are circular and of equal diameter. The Reynolds number, Re, is 150, and the flow is in the laminar region. The two cylinders are arranged diagonally, and the horizontal and vertical spacing are fixed at 2.0 and 3.0, respectively. The effects of input parameters such as reduced velocity, U^*, and damping ratio, ζ, are observed on vibration amplitude, phase difference, frequency, and average power generation. The damping ratio is varied between 0.04 and 0.24. Increasing damping reduces the oscillation amplitude. Maximum amplitude is noticed at the lowest damping in the range $3.0 \leq U^* \leq 8.5$. The variation of oscillation amplitude with U^* is observed to be non-monotonic. A peak power is observed at a particular U^* at every damping. Average power depends on the damping and velocity of the cylinders. The opposing nature of variation of velocity and damping as the damping increases governs the nature of harnessed power. We have obtained an optimal damping for peak power generation, which is $\zeta = 0.10$. For the present case, the peak of maximum average extracted power is close to 0.283 at $\zeta = 0.10$. The maximum harnessed is more than two times the power generated by a single cylinder and four times that of the rigidly coupled cylinder in tandem.

Keywords Power harvesting · Vortex-induced vibration · Rigidly coupled cylinder · Diagonal arrangement

A. Pal (✉) · A. K. Soti
Indian Institute of Technology Guwahati, Guwahati, Assam, India
e-mail: abhijitpal@iitg.ac.in

U. S. Dixit et al. (eds.), *Current Progress in Engineering Sciences*,
https://doi.org/10.1007/978-981-96-8753-4_5

1 Introduction and Literature Review

Depletion of fossil fuel reserves forces humankind to find alternative ways to produce energy. Apart from that, the energy source should leave a low carbon footprint. The kinetic energy of a moving body is used to produce energy like gas, steam, and wind turbines. VIV can replace the need for a rotating turbine to produce electricity by harnessing the fluid power. VIV can induce flow motion as there is a resultant fluctuating force from the interaction of body and fluid. Unlike the rotary motion of the turbine, the translational motion of the body due to VIV can produce energy.

VIV is crucial for the stability of long, thin structures like marine risers and structures tethered to the ocean floor. Also, VIV analysis is needed to design bridges and long heat exchanger tubes. However, in 2008, Bernitsas et al. [3] proposed a device called VIVACE (Vortex-induced vibration aquatic clean energy) that can utilize flow power to generate electric power. This device can harness energy from low flow velocity and run small off-shore platforms. Mass ratio and damping also have an effect on a cylinder, forced to vibrate at a particular amplitude and frequency [2]. They have found the parameter mass-damping, which governs efficiency. Efficiency attains a maximum at particular mass-damping and then falls as mass-damping increases. Sun et al. [13] found that power changes as the input parameters, i.e., spring stiffness, mass ratio, and damping at high Re. As they increase those parameters, power also increases. Soti et al. [10] have also found an optimal damping ratio ζ where maximum power will be maximum like [2]. Their numerical study follows up with the observation of strong dependence Re on power in the low Re region.

An increase in the number of bluff bodies enhances output characteristics of VIV, such as amplitude. Han et al. [7] observed that three cylinders in a triangular arrangement, moving as a unit, have higher amplitude than a single cylinder. Ding et al. [5] also obtained augmentation on amplitude using independently moving three and four cylinders. Zhao et al. [16] works involve the study of the effect of horizontal and vertical spacing of two cylinders on VIV characteristics at low Re. As horizontal spacing increases, the amplitude of vibration increases. However, vertical spacing does not significantly impact amplitude as the amplitude values are very close to the single cylinder values. The power output also significantly increases as we increase the number of bluff bodies. As we observe for amplitude in Zhao et al. [16], the distance between bluff bodies has relevance on power output. Kim and Bernitsas [8] found normalized power generation from multiple cylinders, called "synergy," increases as the number of cylinders increases. They got as much as two-fold more synergy than a single cylinder. There is also optimal spacing between cylinders where synergy peaks. Sun et al. [14] showed with the help of surface roughness patches that power output in the VIV regime can be increased for two cylinders.

From the previous discussion, it is implied that the study on the effect of the rigidly-coupled cylinder at a staggered arrangement is not explored. In a practical sense, rigidly coupled cylinders have the advantage of fewer moving parts over independently moving cylinders and reduced maintenance. Our study discusses the effect of damping on two cylinders, moving as a unit and placed at a fixed gap diagonally.

Here, the diagonal distance or gap, G/D, is kept at 3.605 as the vertical spacing T/D is 3.0 and horizontal spacing L/D is 2.0. The timing of vortices impinging on another body is fixed at a particular Re. By carefully orchestrating the distance between bodies on either side, it can be desirable to generate high lift [1]. In our study, two cylinders are synchronized. Vortices are similar signs shed periodically and in an almost sinusoidal path. So, the second cylinder, in our case, is placed in such a sinusoidal path. In our previous study, using two cylinders in tandem at a horizontal distance of 2.0 proved not feasible for effective power generation [9]. But Griffith et al. [6] show that at a vertical distance of 3.0, the increased lift can be achieved for the stationary cylinders. Also, at this vertical distance with no horizontal distance, the displacement of the front and rear cylinders is higher than at other distances. However, the rigidly-coupled cylinders show different kinds of response [16]. In practical terms, a rigidly coupled cylinder means two cylinders are attached to a single moving platform such that their relative position does not change, i.e., they move as a unit. It is observed at only tandem distance, $L/D = 2.0$, the maximum displacement is the same as that of a single cylinder, and for only a side-by-side distance of $T/D = 4.0$, the oscillation amplitude closely matches that of a single cylinder. The spacing is fixed at $L/D = 2.0$ and $T/D = 3.0$ for our present case to observe if the power extraction value from this arrangement is higher or lower than the single cylinder value. Suppose the power for the present arrangement is observed to be higher than that of a single cylinder, where the oscillation amplitude is almost the same as that of a single cylinder. In that case, we can conclude that for other arrangements where amplitude will be higher than this staggered arrangement can produce more power. The mass ratio, m^*, of the total arrangement is fixed at 2. The simulations are carried out for low Re, Re = 150. The range of reduced velocity, U^*, and the damping ratio are 3.0–8.5 and 0.04–0.24, respectively. This study on lower Re can be extended to higher Re in experiments. The relation between maximum extracted power with Re for the single cylinder is established in a previous study [12]. From the relation, it is assumed that at high Re, we can also get enhanced power from the staggered arrangement, as power is an increasing function of Re. But there is one question answered here: Is extracted power from two cylinders placed diagonally more than the single-cylinder?

2 Methodology

For the analysis of fluid–structure coupling, the equation for resolving flow and motion of structure is numerically solved in a segregated coupling manner in OpenFOAM (Open Field Operation And Manipulation). Navier–Stokes equation, along with the continuity equation, has to be solved to resolve the flow, and both are non-dimensionalized as follows:

$$\nabla^*.\mathrm{u}^* = 0 \tag{1}$$

$$\frac{\partial \mathrm{u}^*}{\partial t} + (\nabla^*.\mathrm{u}^*)\mathrm{u}^* = -\nabla^* p^* + \frac{1}{\mathrm{Re}}(\nabla^{*2}\mathrm{u}^*) \tag{2}$$

where $\mathbf{u}^* = \begin{pmatrix} u/U \\ v/U \end{pmatrix}$ and $\nabla^* = \begin{pmatrix} \partial/\partial x^* \\ \partial/\partial y^* \end{pmatrix}$. x^* and y^* can be written as $x^* = x/D$ and $y^* = y/D$. Also $t^* = tU/D$, $p^* = p/\rho U^2$ and Re is denoted as $\mathrm{Re} = UD/\nu$, where U is the uniform flow velocity, D is the diameter of both the cylinders and ρ and ν are the density and kinematic viscosity of the fluid. Velocity and pressure are coupled through a mixture of PISO (Pressure-Implicit with Splitting of Operators) and SIMPLE (Semi-Implicit Method for Pressure Linked Equations) scheme, called PIMPLE, available in OpenFOAM. PIMPLE is a robust solver that can handle large displacements. For the displacement, the classic spring-mass-damper equation is solved.

$$m\ddot{y} + c\dot{y} + ky = F(t) \tag{3}$$

where m, c, and k are the mass, damping constant, and spring constant of the whole two-cylinder arrangement. $F(t)$ denotes fluctuating fluid force on the body with time. The structural equation is non-dimensionalized as follows:

$$\ddot{y}^* + \frac{4\pi\zeta}{U^*}\dot{y}^* + \frac{4\pi^2}{U^{*2}}y^* = \frac{2}{\pi}\frac{C_L}{m^*} \tag{4}$$

where lift coefficient, $C_L = \frac{F(t)}{\frac{1}{2}\rho U^2 D}$. U^* is reduced velocity expressed as $U/f_n D$, where f_n is the natural frequency of the system. m^* is mass ratio represented as $m/(2{*}\pi/4\rho D^2 L)$. There is also damping ratio ζ defined as the ratio between damping coefficient, c and critical damping coefficient, c_c of the structure, where c_c is represented as $c_c = 4\pi f_n m$. Fluctuating fluid force induced by the fluid on the body is due to the change in pressure distribution with time in the stationary cylinder [4]. After a critical Re, the boundary layers around the cylinder cannot stay attached due to the curvature of the body. The boundary layers separate symmetrically from two points behind the cylinder and form shear layers that roll into one another into the wake behind the cylinder. Due to the rolling up of the shear layers, it creates a rotating structure on the wake, which is called vortex [15]. The change in pressure distribution leads to fluctuating force on the body. When the body is elastically mounted to move in the transverse direction only, the component of fluctuating force transverse to the fluid motion imparts motion to the body. This transverse fluid force is called fluctuating lift force, as its value and direction change with time. Due to the motion of the body, the wake behind the cylinder is altered, and it impacts the shed vortices as a part of the fluid mass moves with the cylinder. This, in turn, changes the pressure distribution pattern around the body and affects the fluid force. Therefore, there is a nonlinear interaction between the body motion and the fluid force, and both are coupled. From the point of view of the numerical study, the viscous and pressure force around the body is calculated from the momentum equation and is fed into the spring-mass-damper Eq. 3, and the body motion is calculated according to the

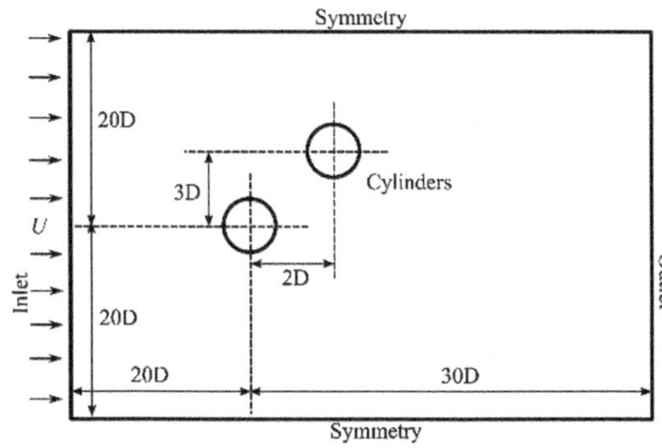

Fig. 1 Schematic of the computational domain of staggered arrangement

Table 1 Boundary conditions

Inlet	Outlet	Top and bottom	Cylinder
$u = U$	$\partial u/\partial x = 0$	$\partial u/\partial y = 0$	$u_{rel} = 0$
$v = 0$	$\partial v/\partial x = 0$	$v = 0$	$v_{rel} = 0$
$\partial p/\partial x = 0$	$p = 0$	$p = 0$	

fluctuating force. The momentum and continuity equations are solved for the new cylinder position. So, the body motion affects the fluid force, and fluid force also affects the body motion. Therefore, Eqs. 1, 2, and 3 are coupled. Here m is the mass of the total arrangement, and L is the length of one cylinder.

As all the simulations are in 2D, L is taken as 1. The computational domain is rectangular in the size of 50D × 40D. The U^* is varied in the range 3–8.5 and ζ is varied from 0.04 to 0.24. The schematic of the computational domain is shown in Fig. 1. The boundary conditions are described in Table 1.

2.1 Validation

Figure 2 shows the validation of amplitude with Zhao et al. [16]. The previous study discusses the validation details [9]. Mesh independence is showed in Table 2: From Fig. 3, we see that after mesh M4, we see there is no significant change in A/D. So, we choose M4 mesh for our numerical study.

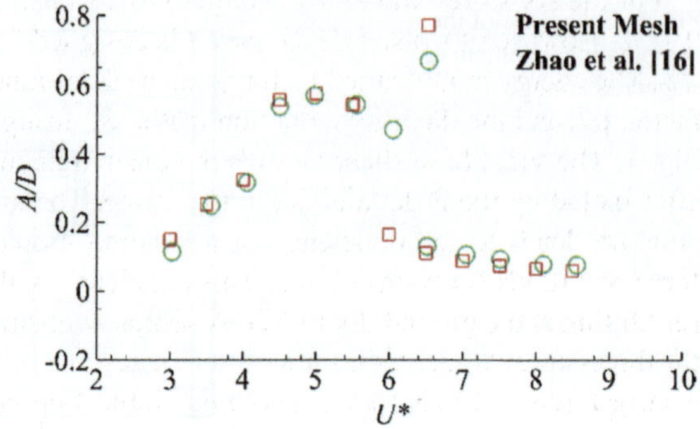

Fig. 2 Numerical validation of amplitude data with Zhao et al. [16] at L/D = 2.0, ζ

Table 2 Mesh independence

Mesh	Δx_{min}	No of elements	A/D
M1	0.04	11,139	0.762563
M2	0.02	22,197	0.823475
M3	0.015	37,440	0.831046
M4	0.012	54,590	0.823823
M5	0.010	77,058	0.840451

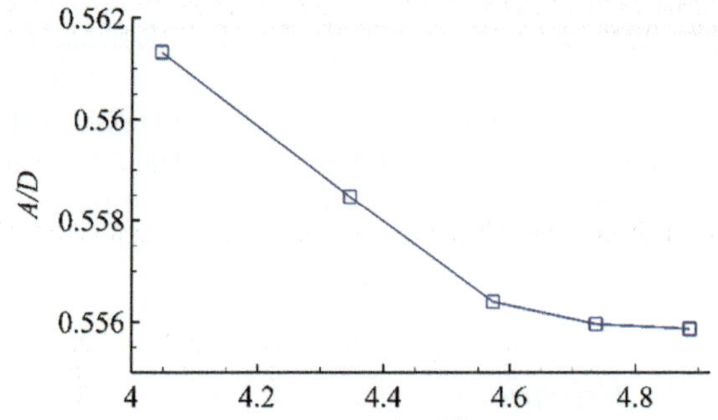

Fig. 3 Mesh independence test

3 Results and Discussion

3.1 Amplitude Response

Non-dimensional temporal displacement with U^* for $\zeta = 0.04$ is shown in Fig. 4. At $U^* = 3.0$, the displacement has a beating type of phenomenon, and there are several frequencies. Maximum displacement is almost 0.1 in this case. We observe an increase in amplitude and periodic nature of transient displacement as we increase U^* to 4.5. But at lower $U^* = 3.0$, the time period of oscillation, T, is higher than $U^* = 4.5$. At $U^* = 6.0$, though the maximum displacement decreases, the periodic oscillation nature is maintained. The time period of oscillation remains almost close to 1.0 for $U^* = 4.5$ and 6.0. As we increase U^* to 7.0, we not only see a decrease in amplitude and period but also several frequencies in response data. However, the frequencies are distinct, and five dominant frequencies are observed. We see again a beating type of phenomenon at $U^* = 8.5$ and a lot chaotic from the nature of several frequencies embedded in the signal.

The maximum displacement or non-dimensional amplitude with U^* and ζ is consolidated in Fig. 5. The amplitude response at $\zeta = 0.04$ is already discussed, and we see a rise and fall of amplitude with U^*. This rise and fall are the same for all the other ζ cases: a jump around $U^* = 3.5$ attain the highest amplitude around $U^* = 4.0$, then a gradual fall in amplitude with U^*. It is noticeable that amplitude also

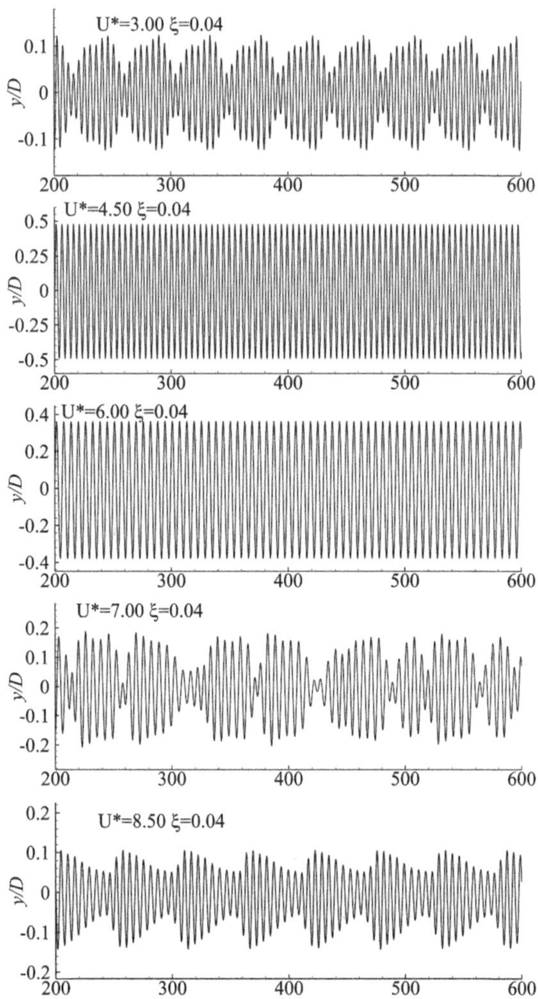

Fig. 4 Transient amplitude response at different U^* at $\zeta = 0.04$

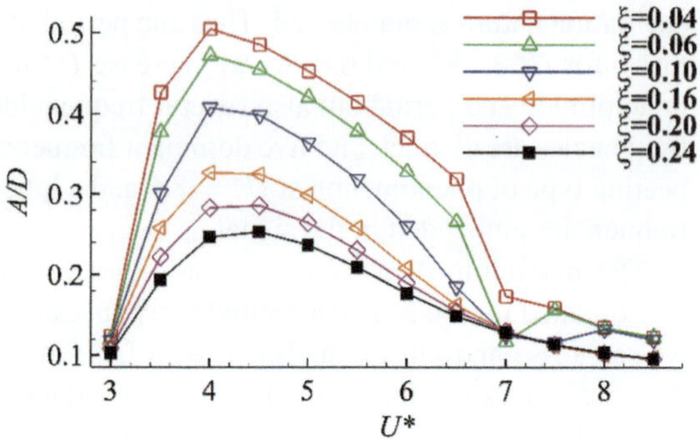

Fig. 5 Variation of amplitude response with U^* at particular damping ratios

decreases with damping, and the amplitude fall is smeared out. After $U^* = 7.0$, the amplitude almost goes to a negligible value, nearly the vibration amplitude value at $U^* = 3.0$. So, the vibration amplitude is bounded between a range of U^*. Here, we are changing U^* by changing the natural frequency of the system or precisely changing the stiffness of the springs. However, in reality, U^* changes with changing the flow velocity. So, the limiting nature of the amplitude proves advantageous at higher flow velocity as oscillation ceases if a flood-like condition ever happens.

3.2 *Frequency Response*

The dominant frequency of oscillation, f at every U^*, is shown in Fig. 6. The frequency is non-dimensionalized with respective natural frequency f_n at each U^*. In the region where f gets very close to f_n or $f/f_n = 1$, we see a significant increase in amplitude. This region of large amplitude is called the "lock-in' region. We observe a gradual increase in frequency response as U^* increases. The dotted green line denotes the frequency of shed vortices, also called Strouhal frequency, from two stationary cylinders at the same diagonal position. In theory, for a single cylinder at higher U^*, when the frequency matches the Strouhal frequency, the amplitude dips to a negligible value. This phenomenon is when the vortex-shedding frequency of oscillating mass desynchronizes from the system's natural frequency, and this phenomenon is called desynchronization. But in our case, though the amplitude almost goes to a low value at higher U^*, f/f_n does not match the stationary cylinders vortex-shedding frequency. So, we can say it is like a desynchronization region.

3.3 *Phase Difference*

The phase difference (ϕ) is the phase shift between lift force and displacement and the variation of φ with U* at different ζ is shown in Fig. 7. At zero damping, ϕ

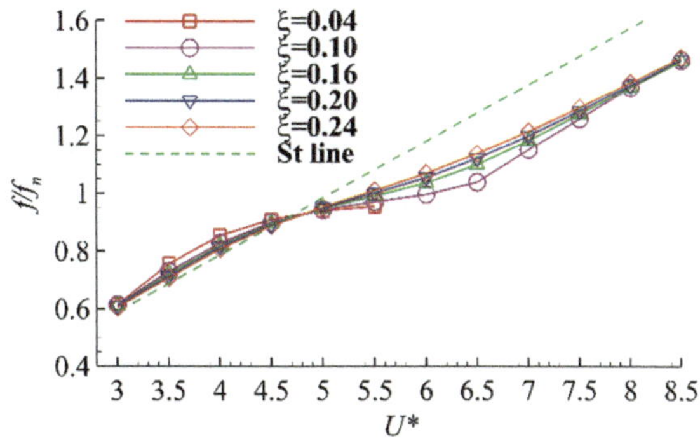

Fig. 6 Frequency response at different ζ with the variation of U^*

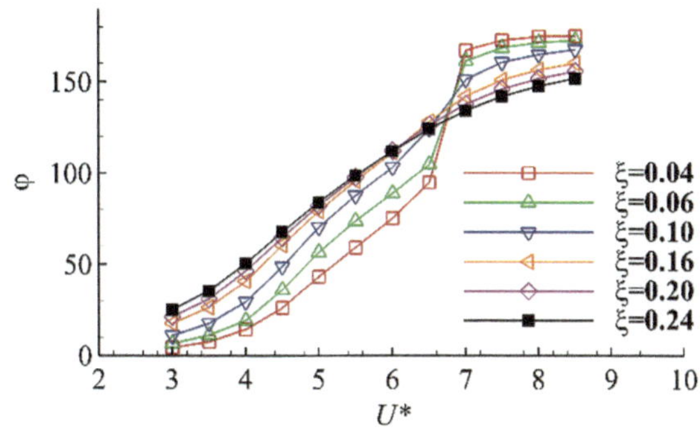

Fig. 7 Variation of phase difference at different ζ

jumps from 0° to 180° when the force is out of phase with the displacement. From the discussion in Soti et al. [10], without the loss of generality, the cylinder motion and lift force are assumed to be periodic, and lift is phase-shifted from displacement. They have shown that with the introduction of damping, ϕ will be 0° when $f/f_n < 1$ and 180° when $f/f_n > 1$. In this case, ϕ reaches from 0° to 180° gradually. At $\phi = 90°$, the amplitude is not maximum. But ϕ at which maximum amplitude is observed increases as ζ increases. We see a jump in phase around $U^* = 6.5$. But as we increase ζ, the sudden increase in ϕ gradually diminishes. This jump in ϕ indicates not only a change in amplitude response but also frequency response. We observe a sudden decrease in amplitude and the departure of oscillation frequency from natural frequency (Figs. 5 and 6).

3.4 Flow Patterns

The vorticity in z-direction ω_z defines as $(\partial v/\partial x - \partial u/\partial y)$. The contours of vorticity are presented in Fig. 8 at $U^* = 4.5$ at $\zeta = 0.14$. We picked two instances when the cylinder is at the top and bottom positions at $t = T/4$ and $T/2$, respectively, where T is the time-period of oscillation. We chose $U^* = 4.5$, as, in this instance, the

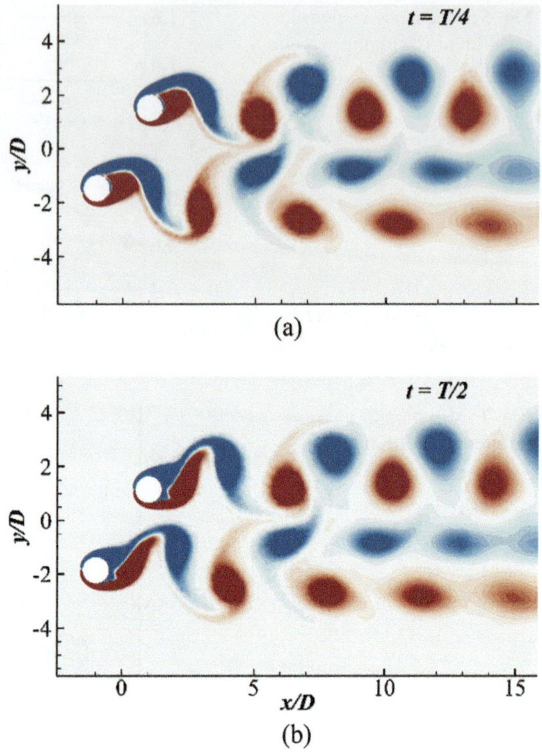

Fig. 8 Vorticity contour at timestep $t =$ **a** $T/4$ and **b** $T/2$ at $U^* = 4.5$ and $\zeta = 0.14$

response is periodic. In one cycle, two single vortices are shed from both cylinders. Two cylinders make separate vortex streets named 2S type of vortices. When the cylinders in the topmost position vortex-streets interfere at a distance $x/D = 3.0$. However, when the cylinders are at the bottom-most position, vortices interact at $x/D = 6.0$. At $\zeta = 0.14$, the interaction is not very vigorous. Only the bottom street of vortices is stretched in the axial direction as if it is convected away by the upper vortices.

3.5 Power Extraction

Extracted power can be expressed as the power dissipated by the damper, which is $c\dot{y}^2$. The total power available in the flow is $1/2\rho U^3 D$. So, one can express the instantaneous power as

$$P(t) = \frac{c\dot{y}^2}{\frac{1}{2}\rho U^3 D} \quad (5)$$

This power is non-dimensionalized and can be thought of as efficiency. However, the above ratio is not a ratio of input and output; it is the fraction of total power that can be extracted from the flow. The non-dimensional average power can be written as

$$\overline{P} = \frac{2\pi^2\alpha}{U^*}U_{rms}^2 \tag{6}$$

where $\alpha = m^*\zeta$ and U_{rms} is defined as $U_{rms} = \sqrt{\frac{1}{T}\int_T^0 \dot{y}^* dt}$. The kinetic energy of the body can be converted to electrical energy by attaching a rake-gear-pulley system to a generator. The power can be derived from the total work done on the cylinder and divided by the time period of oscillation T. The work done is nothing but the force on the cylinders, $F(t)$ times velocity integrated over a cycle of oscillation. This can be equated by multiplying the left-hand side of Eq. 3 and averaging over T. If we assume the cylinder motion and $F(t)$ are periodic, only the damping portion contributes to the power. The other terms become zero if we average over one cycle of oscillation. This is why power is modeled as the multiplication of damping force $c\dot{y}$ and $\dot{y}$. From Eq. 6 it is clear that power depends on m^*, ζ, and U^*. Variation of average power with U^* is shown in Fig. 9 for different ζ. Power also depends on the square of the maximum amplitude, $\overline{P} \propto (A/D)^2$ as evident from Bernitsas et al. [3].

The noticeable feature from Fig. 9 is the similarity between amplitude and power with the change in U^* values at a fixed ζ. Power also increases to a maximum value at $U^* = 4.0$ and then gradually comes to zero value at high damping. If we closely look at power extraction at $U^* = 4.5$ for various ζ, we see that at lowest ζ, the power is lowest. As power depends upon ζ, the contribution of low ζ to the power can not overcome the effect of high amplitude at low ζ. Again, at large ζ, the amplitude is very low, and high ζ can not overcome the effect of low amplitude. Thus, there will be an optimal ζ in between where both can maximize the power output. The optimal ζ at $U^* = 4.0$ is 0.10. So, if we can show maximum power at each ζ value, then we can deduce a relation between maximum average power, $\overline{P}_{max}$ and ζ.

Figure 10 shows the variation of $\overline{P}_{max}$ vs ζ. We see the optimum ζ is 0.10, where we obtain a peak of $\overline{P}_{max}$ which is 0.283. This peak power is recorded at $U^* =$

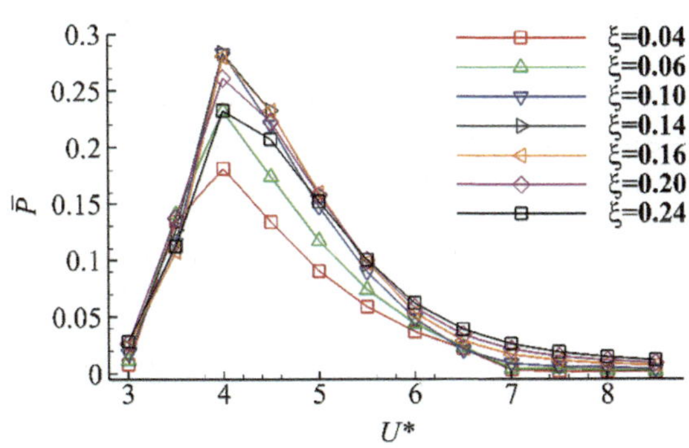

Fig. 9 Power extraction at different ζ. The maximum power obtained is 0.283 at $U^* = 0.40$

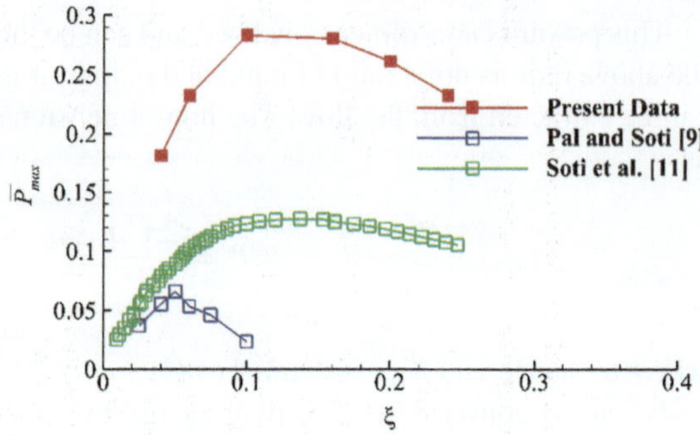

Fig. 10 Variation of power extraction with ζ. Comparison of power extraction of cylinders in diagonal arrangement with tandem arrangement [9] and single cylinder [11]

4.0. However, if we notice every U^* value, the maximum power is not obtained at $\zeta = 0.10$ for all U^*. Figure 10 also shows the comparison of extracted power of the single-cylinder from Soti et al. [11] and tandem arrangement at $L/D = 2.0$ from Pal and Soti [9] with the current results. The maximum power at a diagonal arrangement is more than four times what we can achieve in the tandem arrangement when, in both cases, $L/D = 2.0$. Also, the optimal ζ shifted to a higher value for the diagonal arrangement. Also, $\overline{P}_{max}$ for single cylinder is 0.131 [11], whereas diagonal arrangement is 0.283, more than two-fold. So, we can conclude that this diagonal arrangement with $L/D = 2.0$ and $T/D = 3.0$ two cylinders moving synchronously can be an excellent choice over tandem and single-cylinder arrangements for power extraction.

4 Conclusions

This work presented VIV of two cylinders diagonally arranged at $L/D = 2.0$ and $T/D = 3.0$. Two cylinders move like a single body. The amplitude, frequency, and phase difference with U^* for each ζ values are discussed. Amplitude peaks at a particular U^* value and then reduces due to the desynchronization of vortex shedding frequency from natural frequency. The maximum amplitude is 0.505D, which is 9% lower than the single cylinder maximum amplitude. The distinction between amplitude regimes or branches seems to be blurred in this case, as opposed to three branches observed in single-cylinder at low $m^*\zeta$. The frequency of oscillation increases as U^* increases. The synchronization region spans from $U^* = 4.5$–6.5. After the oscillation, the frequency departs from the natural frequency. The increase in damping decreases amplitude and also smoothes out the phase shift from 0° to 150°. The jump in ϕ at higher U^* not only decreases but also the value of ϕ decreases with ζ. 2S-type vortices are observed in both the vortex streets. The upper cylinder vortices help lower cylinder vortices to convect away.

The power also peaks like amplitude response and decreases at higher U^* region. The peak amplitude and peak power are observed at the same U^* value, $U^* = 4.0$;

the maximum non-dimensional power 0.283 is observed at $\zeta = 0.10$. For a U*, we observe the opposing effect of amplitude and damping on power. The peak power for every U^* value is not happening at the same ζ. Also, for a single U^*, the power peaks at a certain ζ and then falls with higher ζ. Variation of maximum power with ζ concludes an optimal ζ where we get peak power. The peak power 0.283 is observed at optimal $\zeta = 0.10$. Also, power from this diagonal arrangement is far superior to the single cylinder and cylinders in tandem at the same $L/D = 2.0$. We get more than four folds and two folds more power than single and tandem arrangements, respectively. Also, the lowest peak power is 28% more than the peak power from a single cylinder. The reason may be that the careful cylinder placement enhances the fluid force on the body. The forces may be related to how the vortices from the bottom cylinder convected away efficiently due to the presence of the upper vortex street.

Acknowledgements The authors would like to acknowledge the financial support from Grant No. SRG/2020/001082, provided by Science & Engineering Research Board (SERB), India. We are also grateful to the Param-Ishan HPC facility at IIT Guwahati for providing the required CPU hours to carry out the necessary computations.

References

1. Alam, M.M., Meyer, J.P.: Two interacting cylinders in cross flow. Physical Review E 84(5), 056304 (2011)
2. Barrero-Gil, A., Pindado, S., Avila, S.: Extracting energy from vortex-induced vibrations: a parametric study. Applied mathematical modelling 36(7), 3153–3160
3. Bernitsas, M.M., Raghavan, K., Ben-Simon, Y., Garcia, E.: Vivace (vortex induced vibration aquatic clean energy): A new concept in generation of clean and renewable energy from fluid flow (2008)
4. Blevins, R.D.: Flow-induced vibration. New York (1977)
5. Ding, L., Zhang, L., Kim, E.S., Bernitsas, M.M.: Urans vs. experiments of flow induced motions of multiple circular cylinders with passive turbulence control. Journal of Fluids and Structures 54, 612–628 (2015)
6. Griffith, M.D., Jacono, D.L., Sheridan, J., Leontini, J.S.: Flow-induced vibration of two cylinders in tandem and staggered arrangements. Journal of Fluid Mechanics 833, 98–130 (2017)
7. Han, P., Pan, G., Tian, W.: Numerical simulation of flow-induced motion of three rigidly coupled cylinders in equilateral-triangle arrangement. Physics of Fluids 30(12), 125107 (2018)
8. Kim, E.S., Bernitsas, M.M.: Performance prediction of horizontal hydrokinetic energy converter using multiple-cylinder synergy in flow induced motion. Applied energy 170, 92–100 (2016)
9. Pal, A., Soti, A.K.: Power harvesting from viv of rigidly-coupled cylinders in tandem arrangement. In: Conference on Fluid Mechanics and Fluid Power. pp. 313–318. Springer (2021)
10. Soti, A.K., De, A.: Vortex-induced vibrations of a confined circular cylinder for efficient flow power extraction. Physics of Fluids 32(3), 033603 (2020)
11. Soti, A.K., Thompson, M.C., Sheridan, J., Bhardwaj, R.: Harnessing electrical power from vortex-induced vibration of a circular cylinder. Journal of Fluids and Structures 70, 360–373 (2017)

12. Soti, A.K., Zhao, J., Thompson, M.C., Sheridan, J., Bhardwaj, R.: Damping effects on vortex-induced vibration of a circular cylinder and implications for power extraction. Journal of Fluids and Structures 81, 289–308 (2018)
13. Sun, H., Kim, E.S., Nowakowski, G., Mauer, E., Bernitsas, M.M.: Effect of mass ratio, damping, and stiffness on optimal hydrokinetic energy conversion of a single, rough cylinder in flow induced motions. Renewable Energy 99, 936–959 (2016)
14. Sun, H., Ma, C., Kim, E.S., Nowakowski, G., Mauer, E., Bernitsas, M.M.: Hydrokinetic energy conversion by two rough tandem-cylinders in flow induced motions: Effect of spacing and stiffness. Renewable Energy 107, 61–80 (2017)
15. Williamson, C.H., Roshko, A.: Vortex formation in the wake of an oscillating cylinder. Journal of fluids and structures 2(4), 355–381 (1988)
16. Zhao, M.: Flow induced vibration of two rigidly coupled circular cylinders in tandem and side-by-side arrangements at a low reynolds number of 150. Physics of Fluids 25(12), 123601 (2013)

Effect of Phase Change Material on Effective Battery Thermal Management Under Various Ambient Temperatures and High Discharge Rate

Deevanapally Siri, Pendyala Sudeepthi, Mude Hrushitha, Padala Poojitha, Dumpeta Srivalli, Varshitha Balusu, Sk. Mohammad Shareef, and G. Amba Prasad Rao

Abstract Lithium-ion batteries (LIBs) have emerged among many types of rechargeable batteries as a potential candidate due to their high energy density, longer cycle life and range. The performance of Lithium-ion batteries is very sensitive due to high temperatures arising out of high charging/discharging and high ambient temperatures. Despite numerous efforts, effective thermal management solutions that ensure uniform temperature distribution and prevent thermal runaway remain a significant

---

D. Siri · P. Sudeepthi · M. Hrushitha · P. Poojitha · D. Srivalli · V. Balusu · Sk. M. Shareef · G. Amba Prasad Rao (✉)
Department of Mechanical Engineering, National Institute of Technology Warangal, Hanamkonda, Telangana, India
e-mail: gap@nitw.ac.in

D. Siri
e-mail: ds22meb0b48@student.nitw.ac.in

P. Sudeepthi
e-mail: ps22meb0b43@student.nitw.ac.in

M. Hrushitha
e-mail: mh22meb0b68@student.nitw.ac.in

P. Poojitha
e-mail: pp22meb0b47@student.nitw.ac.in

D. Srivalli
e-mail: ds22meb0b53@student.nitw.ac.in

V. Balusu
e-mail: bv22meb0b26@student.nitw.ac.in

Sk. M. Shareef
e-mail: sm23mer2p08@student.nitw.ac.in; shareefshaik4@cvr.ac.in

Sk. M. Shareef
Department of Mechanical Engineering, CVR College of Engineering, Ibrahimpatnam, Hyderabad 501510, Telangana, India

U. S. Dixit et al. (eds.), *Current Progress in Engineering Sciences*,
https://doi.org/10.1007/978-981-96-8753-4_6

challenge, especially for prismatic batteries. In the present work, thermal management through passive mode employing phase change materials (PCM) with optimized cell pack geometries has been undertaken. Numerical studies through ANSYS have been done on a single cell and extended to a battery module with six cells. Computational fluid dynamics (CFD) based simulations were conducted exploring various PCMs, both organic and inorganic types, with varying thicknesses and under different ambient temperatures. Results indicate that for a pack of 6 cells under 8C discharge rate, optimal PCM thickness of 4 mm significantly reduces peak temperatures and enhances thermal uniformity across the battery pack. The optimized thickness with an inorganic PCM at an ambient temperature of 300 K demonstrated up to a 10% reduction in maximum temperature compared to conventional setups. Among the organic and the inorganic PCMs, the inorganic was found to be more effective. Furthermore, the impact of ambient temperature was found to be more severe.

Keywords Lithium-ion batteries · Battery thermal management system · Phase change material · Organic · Inorganic · Computational fluid dynamics

Nomenclature

ρ (kg/m^3)	Density of the cell material
T (K)	Temperature of the cell
t (s)	Time
k (W/m K)	Thermal conductivity of the cell material
I (A)	Current flowing through the battery
V (V)	Terminal voltage of the battery under load
L	Latent heat of PCM
P	Pressure of liquid PCM
g	Acceleration due to gravity
S	Momentum source
v	Velocity vector of liquid PCM
C_p (J/kg.K)	Specific heat capacity of the cell material
Q_{gen} (W/m^3)	Heat generation rate within the cell
R_{int} (Ω)	Internal resistance of the battery
V_{oc} (V)	Open-circuit voltage of the battery
ρ_{PCM}	Density of the PCM
k_{PCM}	PCM's thermal conductivity
hs	Sensible heat
h_r	Reference enthalpy
T_r	Reference temperature
C_p	Specific heat capacity
T_s	Solidus temperature which represents the start of the PCM melting
T_l	Liquidus temperature which represents the completion of PCM melting

k_l	Thermal conductivity of liquid PCM
ρ_l	Liquid PCM density

Subscripts

p	Constant pressure
gen	Generation rate
Int	Internal
oc	Open-circuit
s	Sensible
r	Reference
s	Solidus
l	Liquidus

Greek symbols

μ	Thermal dynamic viscosity of liquid PCM
β	Expansion coefficient of liquid PCM
η	Liquid fraction of PCM
ΔH	Enthalpy of phase change process

Acronyms

BTMS	Battery Thermal Management System
CFD	Computational Fluid Dynamics
C-rate	Battery charge/discharge rate expressed as a multiple of its nominal capacity
EVs	Electric Vehicles
ICEVs	Internal Combustion Engine Vehicles
LFP	Lithium Iron Phosphate
LIBs	Lithium-ion Batteries
MSMD	Multi-Scale Multi-Domain
NTGK	Non-equilibrium Thermal Galvanostatic Kinetics
PCM	Phase Change Material
RT-45	Organic PCM with a melting point of 45 °C-(Rubitherm 45)

1 Introduction

In today's world, electric vehicles (EVs) are increasingly seen as a viable and attractive choice because of their significant advantages [1]. Internal combustion engine vehicles (ICEVs) have been the backbone of transportation for over a century, providing reliable performance and a broad range of vehicle options. EVs have replaced internal combustion engine vehicles in many aspects, such as reducing emissions, lowering operating costs, and decreasing reliance on fossil fuels. Although alternative energy sources like biofuel exist for IC engines, EVs have proven to be the best option due to their superior efficiency. While biofuels are considered renewable, their production and combustion still emit greenhouse gases. But still, there are some challenges for EVs, such as limited range and long charging times. So, these challenges were addressed with Li-ion batteries as the power source.

The reason for choosing Li-ion batteries for EVs is because of their higher energy density, longer cycle life, faster charging times, and improved efficiency. But LIB generates heat during charging and discharging, which can potentially lead to thermal runaway [2]. Therefore, effective cooling methods are implemented to maintain the battery within the optimal temperature range as suggested is 15–40 °C [3]. Additionally, to ensure effective balancing and consistent discharge/charge across the entire pack, it is crucial to maintain a temperature difference of no more than 5° among the cells [4]. Therefore, maintaining the ideal temperature of the battery is facilitated with battery thermal management system (BTMS). At their core, active and passive cooling are two fundamentally different methods for BTMS. Each approach offers its own benefits and constraints, often necessitating a blend of both for optimal thermal management. Active cooling mechanisms utilize external energy sources to dissipate heat away from the battery as in air cooling and liquid cooling. Advantages of active cooling systems include enhanced cooling capacity, precise temperature control, and adaptability to real-time conditions. However, they are also more complex, consume more energy, and require more maintenance due to additional components [5]. Passive cooling systems can function independently, without the need for an external mechanism. Passive cooling systems are preferred due to their energy efficiency, low maintenance, high reliability, environmental benefits, noise reduction, long-term cost savings, scalability, and independence from the power grid [6]. Hybrid cooling combines either an active and a passive BTMS or two passive BTMS systems.

Due to its minimal power consumption and eco-friendly characteristics, employing Phase change material (PCM) is considered as a favorable approach for BTMS, aligning it with passive cooling methods. PCMs are materials that can store and release significant amounts of heat when they undergo phase transitions [7]. PCM exhibits characteristics such as higher energy storage density, good stability, and minimal supercooling [8]. To mitigate the increase in temperature and ensure the battery remains within safe operating limits, PCM can be employed. As the C-rate of the battery increases, the peak temperature rises, often exceeding the optimal temperature range [9]. The results demonstrate that the use of PCM significantly reduces the average temperature and improves temperature uniformity within the lithium-ion battery pack, offering a superior solution compared to other methods [10]. Generally, PCM is composed of three types: organic, inorganic, and eutectic. Organic PCMs, such as paraffins and fatty acids, are known for their noncorrosive properties, high heat of fusion, and good chemical and thermal stability. However, they have low thermal conductivity and lack high phase change enthalpy. Inorganic PCMs, such as salt hydrates, are recognized for their high thermal conductivity and significant phase change enthalpy but they have drawbacks such as undercooling and corrosion. Eutectic PCMs, whether inorganic-organic, organic-organic, or inorganic-inorganic combinations, offer benefits such as high heat capacity but are susceptible to corrosion at elevated temperatures [11].

Despite significant advancements in BTMS, several gaps remain. Most reviews on BTMS have explored either PCM or liquid cooling. However, there is a lack of comprehensive studies that examine both organic and inorganic PCMs. Additionally, many studies assume a constant ambient temperature, which does not reflect real-world conditions where temperatures vary throughout the year. This limitation reduces the practical applicability of the findings.

To address these gaps, the research employed a systematic approach. The study began with a numerical model of a single cell, which was validated by comparing it with experimental data from an existing model. Following successful validation, the scope was expanded to include a pack of six cells, allowing for the investigation of thermal management performance across larger cell configurations. To ensure the findings were reliable and relevant under diverse conditions, a series of simulations and experiments were conducted under varying ambient temperatures, thoroughly assessing the performance of different PCMs.

LFP batteries were chosen for this study due to their long cycle life, offering up to 2500 cycles, and deep cycle capability, making them highly reliable for long-term energy storage and electric vehicle applications. They offer excellent safety and thermal stability, with low internal resistance and a wide operating temperature range (−20 to 50 °C), reducing the risk of thermal runaway [12].

The study also incorporated the Multi-Scale Multi-Domain Non-Equilibrium Thermal Galvanostatic Kinetics (MSMD-NTGK) model. The MSMD approach offers a scalable and integrated framework for modeling thermal, electrical, and electrochemical phenomena at both cell-level and pack-level, making it suitable for large-scale battery simulations [13].

Furthermore, the NTGK model was employed. It is chosen for its accurate thermo-electrochemical coupling, enabling precise temperature prediction under constant discharge rates (C-rates) while maintaining computational efficiency. It performs well across various cooling techniques, such as natural convection, forced convection, and liquid cooling [13].

2 Numerical Modelling

To achieve the optimal thermal performance, Lithium Iron Phosphate (LFP) is chosen for its high thermal stability reducing the risk of thermal runaway. Prismatic batteries are selected due to their flat design, space efficiency, and simple integration in various applications. The key cell parameters are shown in Table 1 (Fig. 1).

2.1 *Material Properties*

The thermal behavior of the LFP prismatic battery is influenced by several material properties which include the specific heat capacity that determines the ability to store the heat and thermal conductivity which affects heat dissipation rate. Additionally, density and electrical conductivity plays a key role in overall thermal behaviors. The equivalent thermo-physical properties used are shown in Table 2.

Table 1 $LiFePO_4$ cell parameters

Parameter	Value
Type	Lithium Iron Phosphate (LFP)
Anode material	Graphite
Cathode material	$LiFePO_4$
Electrolyte	Carbonate polymer
Dimensions of cell	7.5 mm (t) × 160 mm (w) × 227 mm (h)
Nominal capacity (Ah)	20
Maximum voltage (V)	4
Minimum voltage (V)	2.1

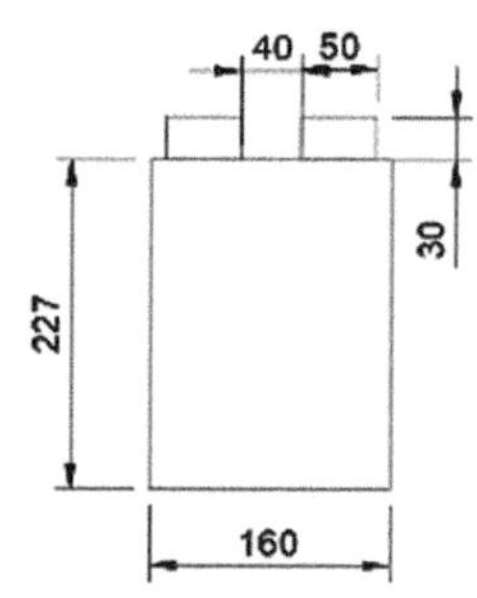

Fig. 1 Dimensions of a single cell LIB

Table 2 Properties used for the numerical study

Properties	Active material	N-tab material	P-tab material
Density (kg/m^3)	2092	8960	2700
Specific heat capacity (J/kg. K)	678	385	900
Thermal conductivity (W/m. K)	18.2	398	238

2.2 Governing Equations

Effective thermal management of Lithium Iron Phosphate (LFP) batteries is crucial for improving performance and safety. Using ANSYS Fluent and the MSMD-NTGK model allows accurate simulation of electrochemical and thermal interactions in batteries. Assumptions made in the analysis include [14]:

1. Constant specific heat capacity and thermal conductivity.
2. Uniform initial temperature for battery and PCM.
3. Negligible PCM volume change during phase change.
4. Uniform current density within the LIB.
5. Laminar PCM flow.
6. Homogeneous PCM properties.
7. Negligible heat generation in PCM.

Thermal behavior of lithium-ion batteries, especially prismatic cells, is highly influenced by the convective heat transfer coefficient and ambient temperature [15]. The thermal analysis considered natural convection with a low heat transfer coefficient of h=5 W/m^2.K applied to exposed surfaces. Ambient temperature was varied to evaluate the system's performances under different conditions.

Energy Balance in the Battery Cell. Heat conduction and internal heat generation due to electrochemical reactions, Joule heating, and overpotentials:

$$\rho C_p \frac{\partial T}{\partial t} = k\nabla^2 T + Q_{gen} \tag{1}$$

Heat Generation Rate. Joule heating from internal resistance and overpotential heat are key heat sources during charge–discharge cycles, crucial for thermal management.

$$Q_{gen} = I^2 R_{int} + I(V_{oc} - V) \tag{2}$$

PCM Energy Equation. PCM absorbs heat from the cell, melting between the solidus and liquidus temperatures, and dissipates it via conduction and convection for uniform thermal management.

$$\rho_{PCM} \frac{\partial H}{\partial t} = k_{PCM} \nabla^2 T \tag{3}$$

$$H = h_s + \Delta H \tag{4}$$

$$h_s = h_r + \int_{T_r}^{T} C_{p(PCM)} dT \tag{5}$$

$$\Delta H = \eta * L \tag{6}$$

$$\eta = \left\{ 0T < T_s \frac{(T - T_s)}{(T_l - T_s)} T_s \le T \le T_l 1T > T_l \right\} \tag{7}$$

2.3 Governing Equations for PCM

Energy Equation. The energy equation governs heat transfer within the PCM and between the PCM and battery, accounting for sensible and latent heat. This ensures safe battery temperatures, prevents overheating, extends lifespan, and improves performance.

$$\frac{\partial(\rho_l H)}{\partial t} + \nabla.(\rho_l v H) = \nabla.(k_l \nabla T) \tag{8}$$

Momentum Equation. The momentum equation describes liquid PCM fluid dynamics, essential for predicting flow patterns and ensuring efficient heat transfer.

$$\rho_l \frac{\partial v}{\partial t} + \rho_l (v.\nabla) v = -\nabla.P + \mu \nabla^2.v + \rho_l \beta g (T - T_r) + S \tag{9}$$

Continuity Equation. The continuity equation ensures mass conservation in the PCM, crucial for accurately modeling its behavior under varying thermal loads.

$$\frac{\partial \rho_l}{\partial t} + \rho_l.\nabla.v = 0 \tag{10}$$

For incompressible flows, this simplifies to:

$$\nabla.u = 0 \tag{11}$$

$$\nabla.v = 0 \tag{12}$$

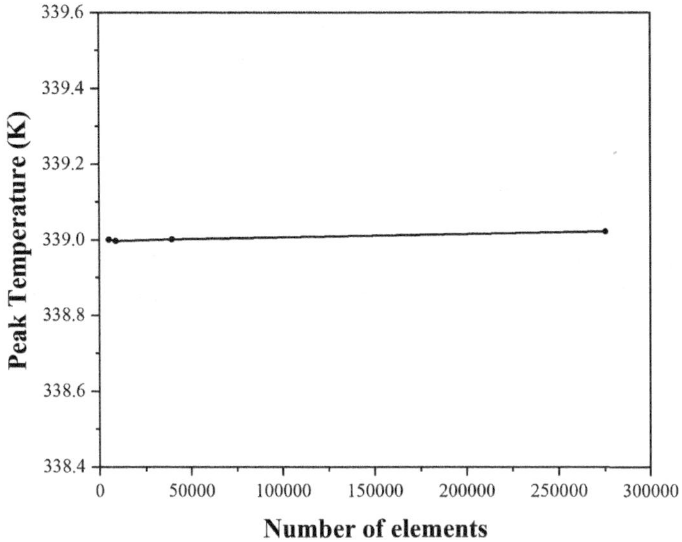

Fig. 2 Grid independence of mesh structure of a single cell

2.4 *Grid Independence*

A hex-dominant mesh was employed for the simulation, ensuring efficient division of the domain. The mesh size was set to an optimum value to ensure a balance between computational efficiency and accuracy. The MSMD was utilized to implement the NTGK model using ANSYS fluent CFD. Mesh is applied to the LFP battery pack as shown in figure. An ambient temperature of 27 °C is used during the grid independence test. Natural convection at 8C discharge rate is utilized for the analysis. The grid independence test is conducted to ensure that the simulation results remain consistent regardless of the grid employed. As the grid number gets progressed, the allowable tolerance level is less than 0.1%. From the figure, we can say that as the grid number is increased from 4976 to 8736, the temperature is decreased from 339.0014 to 338.9976 K. The change in temperature was observed to be 0.0038 K (less than 0.1%). Based on the grid independence test, the size corresponding to 39,480 elements were chosen for subsequent numerical solutions to expedite simulations (Figs. 2 and 3).

2.5 *Validation*

The simulation results of the numerical model developed were compared with previous studies for validation, as shown in Fig. 4. The LIB temperature variation over time was compared with the experimental results obtained by Ugur [16] for

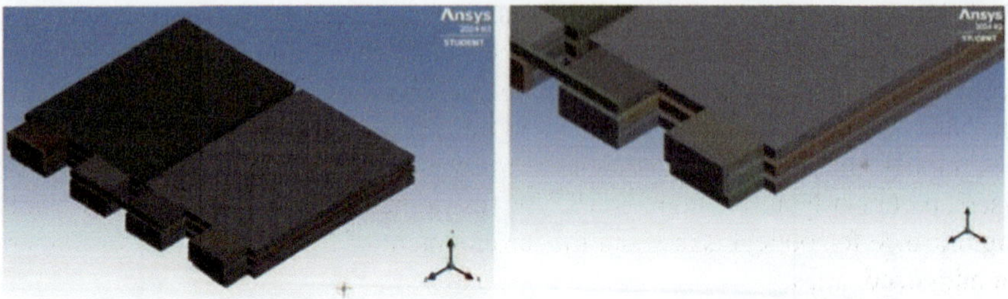

Fig. 3 Mesh structure of LIB pack

validation. The validation was conducted under a constant 5C discharge rate at an ambient temperature of 300 K. The simulations were performed using ANSYS Fluent and ran up to 800 s. The temperature obtained from the present study's numerical model closely matches the results of Ugur [16], though a slight deviation in the curve is recorded in certain regions with time.

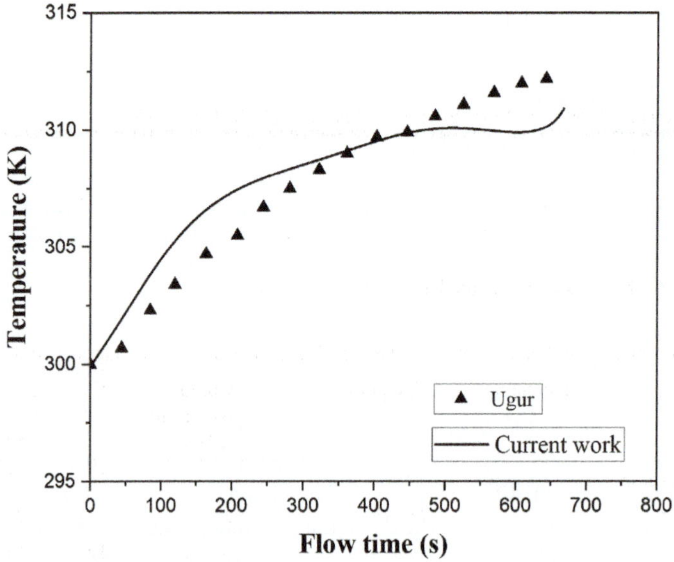

Fig. 4 Validation of present model with results of previous studies

3 Results and Discussion

Based on the numerical experiments carried out on both single and battery module of six cells for observing the effect of C-rating with flow time, the temperature profiles have been captured. The effect of different types of PCMs and ambient temperatures are presented.

3.1 Studies on a Single Cell and a Module of Six Cells at Ambient Temperature of 300 K

The contour shows uneven heat distribution in Fig. 5, indicating insufficient heat dissipation without PCM, which could lead to thermal stress and reduced cell performance.

In Fig. 6, the application of PCM shows a lower and more uniform temperature distribution across the cell. The maximum temperature is reduced, indicating better thermal regulation during high discharge rates.

Figure 7 illustrates the cell temperature distribution within the battery module. The cell temperature obtained at 8C discharge for this module of 6 cells is 340.11523 K. The simulation highlights how heat is distributed across the cells, which is crucial for assessing the battery's thermal performance and safety.

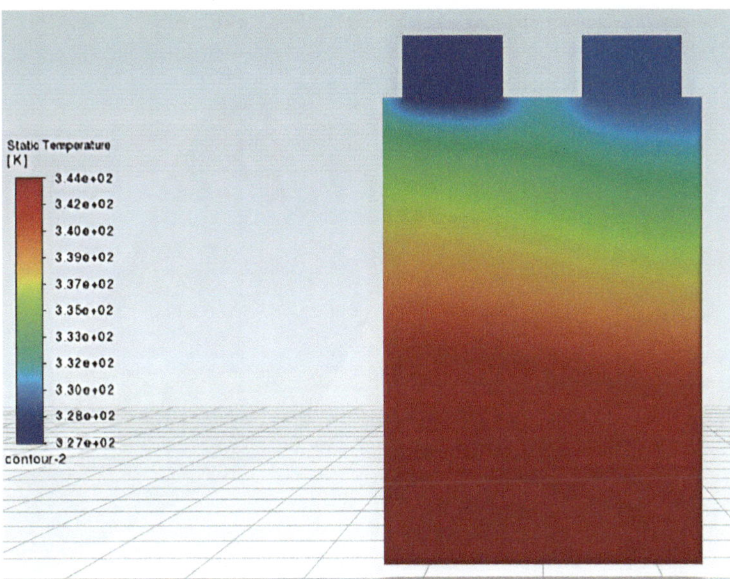

Fig. 5 Temperature effect of 8C rating on a single cell without PCM

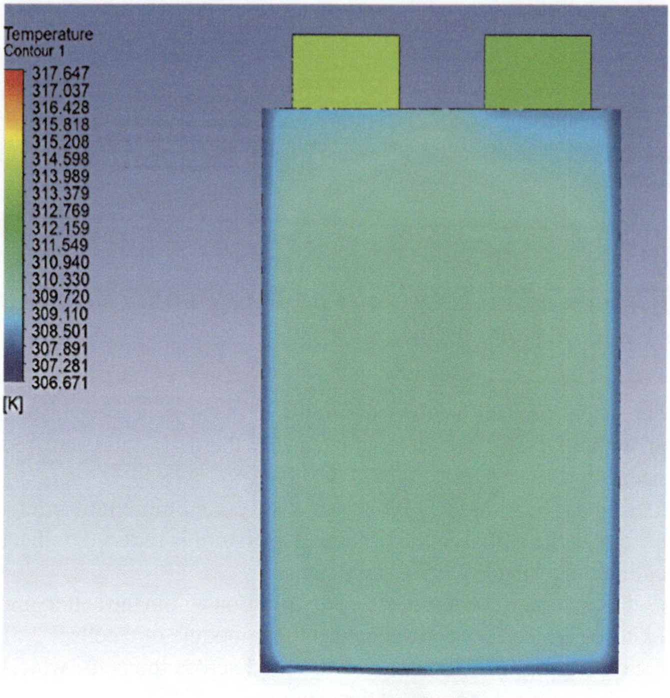

Fig. 6 Temperature contour of a single cell with RT45 PCM at 8C rating

Fig. 7 Temperature contour of a module of 6 cells at 8C rating

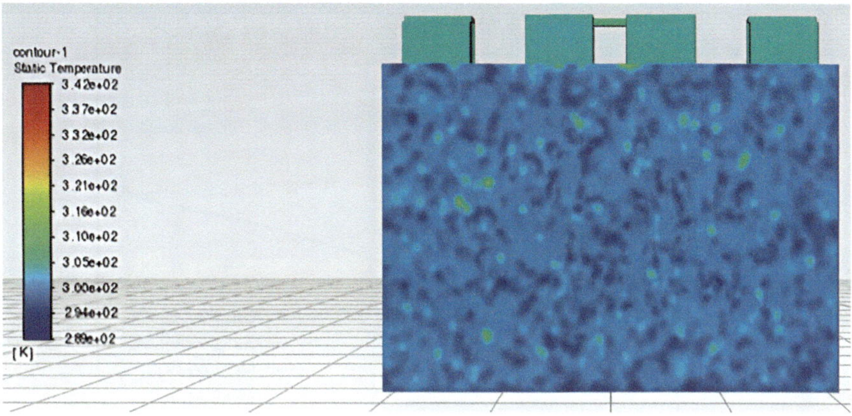

Fig. 8 Temperature contour of a module of 6 cells with $CaCl_2.6H_2O$ as PCM at 8C rating

Figure 8 contour shows the static temperature distribution in the battery module with PCM, with the cell temperature recorded as 304.86 K. Additionally, the phase change material (PCM) temperature is recorded as 300.24 K, reflecting the thermal regulation provided by the PCM.

Comparison between a single cell and module.

Figure 9 represents the variation of cell temperature under two different ratings. It can be noted that as the rating is increased the time for discharge/charge gets reduced and also the temperature throughout the flow time increases. For 5C and 8C peak battery temperatures are 325 and 338.5 K respectively, an increase 13.5 K is observed from 5 to 8C.

From the observed data, there is minimal temperature difference between single cell and module at 8C discharge. This leads to an inference that the design of battery pack is optimal. These temperatures are outside the safe range of working of LIB cell. So, it can be inferred from this that a cooling technique is required to bring down the peak temperature within the safe range. Therefore, it necessitates the effective thermal management.

3.2 Studies on a Single Cell and a Module of Six Cells at Different Ambient Temperatures

Effect of ambient temperature on a module under discharge ratings of 3C, 5C, 8C respectively (without PCM). To observe the variation of ambient temperature on a module, 3 different ambient temperatures have been considered.

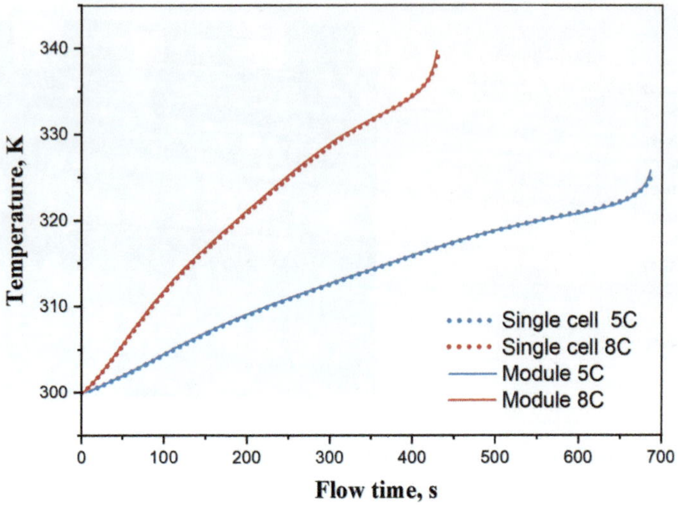

Fig. 9 Variation of cell temperature for a single cell and module under 5C, 8C

From Fig. 10, it is obvious that, Increase in discharge rate affects the cell temperatures, with higher rates leading to increased cell temperatures at all ambient conditions. For instance, at 300 K ambient temperature, the cell temperatures are 311.21, 325.76, and 339.66 K for 3C, 5C, and 8C discharge rates, respectively. This trend is consistent across other ambient temperatures, indicating that higher discharge rates generate more heat within the cells.

As the ambient temperature increases, the cell temperatures also rise at each discharge rate. At an 8C discharge rate, the cell temperature goes from 339.66 K at 300 K ambient temperature to 351.45 K at 320 K ambient temperature. This indicates that the ambient temperature significantly influences the thermal behavior of the cells. When compared the effect of ambient temperature and discharge, it is observed that the effect of ambient temperature is more severe than an increase in discharge.

The combination of high ambient temperatures and high discharge rates leads to the highest cell temperatures. At 320 K ambient temperature and an 8C discharge rate, the cell temperature reaches a maximum of 351.45 K. This combined effect suggests that without any BTMS, the cells can reach potentially unsafe temperatures, especially under high load conditions and in hot environments.

Elevated temperatures can accelerate battery degradation, reduce lifespan, and increase the risk of thermal runaway. The observed temperatures at higher discharge rates and ambient conditions are concerning from a safety and performance perspective. The results highlight the necessity of implementing an effective BTMS to manage the thermal conditions within the battery module, especially under high discharge rates and in warmer climates.

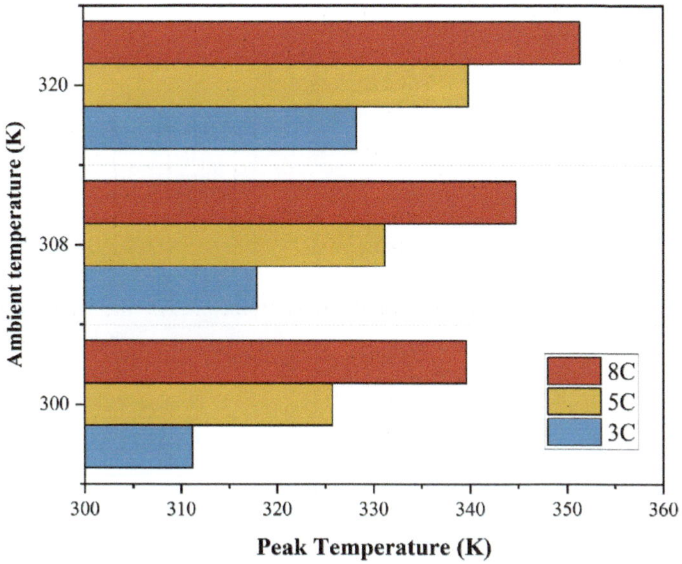

Fig. 10 Variation of temperature without BTMS at various discharge rates and ambient temperatures

3.3 Studies on a Module Without and with PCM

As described in the previous sections, phase change materials as a means of thermal management have been employed. To begin with, an organic PCM, RT-45 has been employed. Simulations have been carried out with two different types of organic PCMs and two different inorganic PCMs. The choice of PCMs is done in such a way that there is a clear variation in the thermal conductivities as well as the transition temperatures with respect to phase change.

Module without and with PCM RT45 4 mm thickness.

With RT-45 as PCM, the module cell temperature is brought down to 309.7 K from 339.6 K. In an attempt to further decrease the temperature another organic PCM-Capric acid with thermal conductivity 0.36 is employed. The thermal conductivity of Capric acid is about 50% higher than that of RT-45 (Fig. 11).

Comparison of temperature profiles with four different types of 4 mm thick PCMs.

Figure 12 illustrates the relationship between the thermal conductivity of the PCM and the peak temperature of the battery cell during operation. The graph shows a clear trend where increasing the thermal conductivity of the PCM results in a significant reduction in the peak temperature.

At a comparatively low thermal conductivity value of 0.2415 W/m K for RT-45, the peak temperature reaches approximately 310 K. As the thermal conductivity

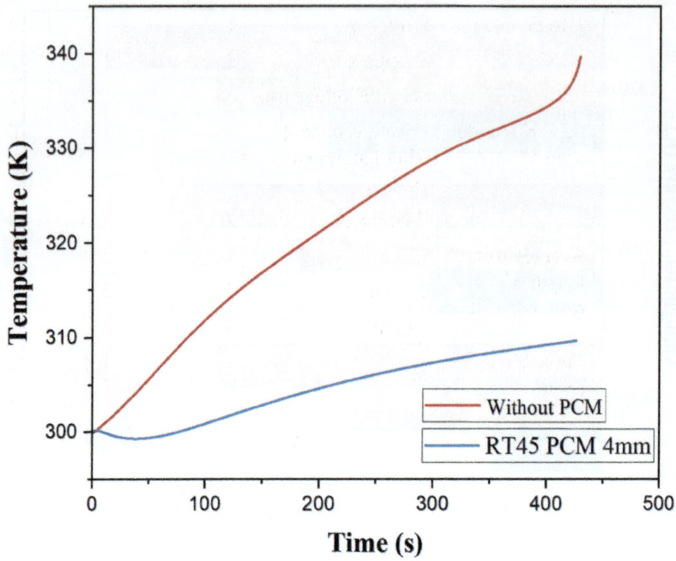

Fig. 11 Variation of temperature without BTMS and 4 mm thick RT-45 at 300 K under 8C discharge

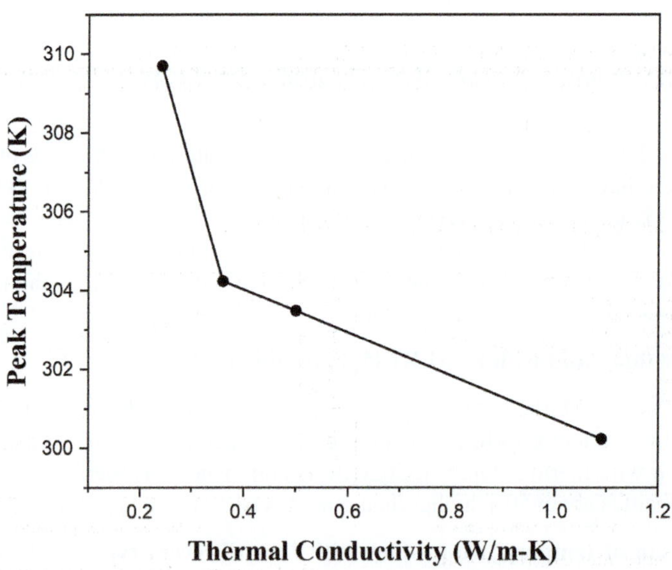

Fig. 12 Variation of Peak Temperature with Thermal Conductivity at 300 K under 8C Discharge

increases to 0.36 W/m K for Capric acid, there is a notable decrease in the peak temperature to 304 K. This trend continues, with further increments in thermal conductivity resulting in a gradual reduction in peak temperature. At a comparatively higher thermal conductivity value of 1.0 W/m K for $CaCl_2.6H_2O$, the peak temperature is observed to be 300.2 K.

This data underscores the importance of selecting PCMs with higher thermal conductivity for effective thermal management in battery systems.

Temperature variations of best organic and inorganic PCMs at 300 K under 8C discharge.

It can be inferred from Fig. 13 that in case of the organic PCMs, adoption of Capric acid demonstrated superior heat dissipation whereas inorganic PCM, $CaCl_2 \cdot 6H_2O$ exhibited more effective thermal management, establishing the fact that thermal conductivity plays a crucial role in enhancing the cooling performance of PCMs.

Temperature variations of different PCMs at different ambient temperatures.

It is certain from Fig. 14 that, at an ambient temperature of 300 K, $CaCl_2 \cdot 6H_2O$, with the higher thermal conductivity of 1.09 W/m K, achieves the lowest battery temperature. This highlights the role of thermal conductivity in thermal management. However, at higher ambient temperatures (308 and 313 K), RT45 unexpectedly exhibits superior temperature regulation compared to the other PCMs, despite having the lower thermal conductivity (0.2415 W/m K). This indicates that factors beyond thermal conductivity influence thermal management. The melting range, defined by the solidus and liquidus temperatures, significantly impacts PCM performance. RT45, with a melting range from 308 to 322 K, overlaps closely with the

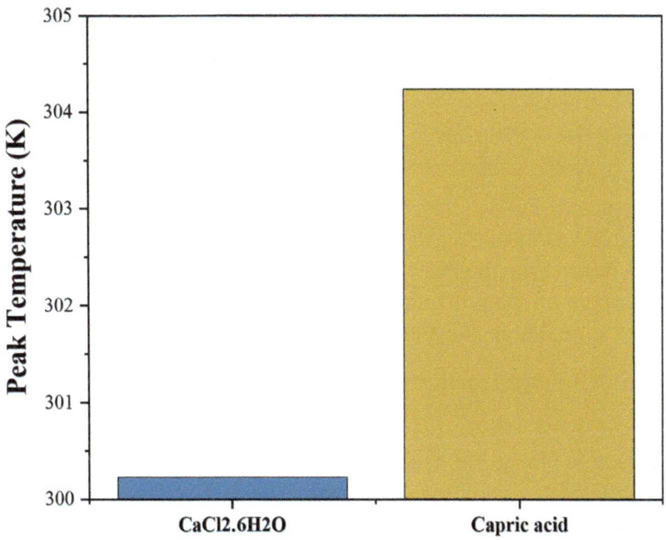

Fig. 13 Temperature variations of best organic and inorganic PCMs at 300 K under 8C discharge

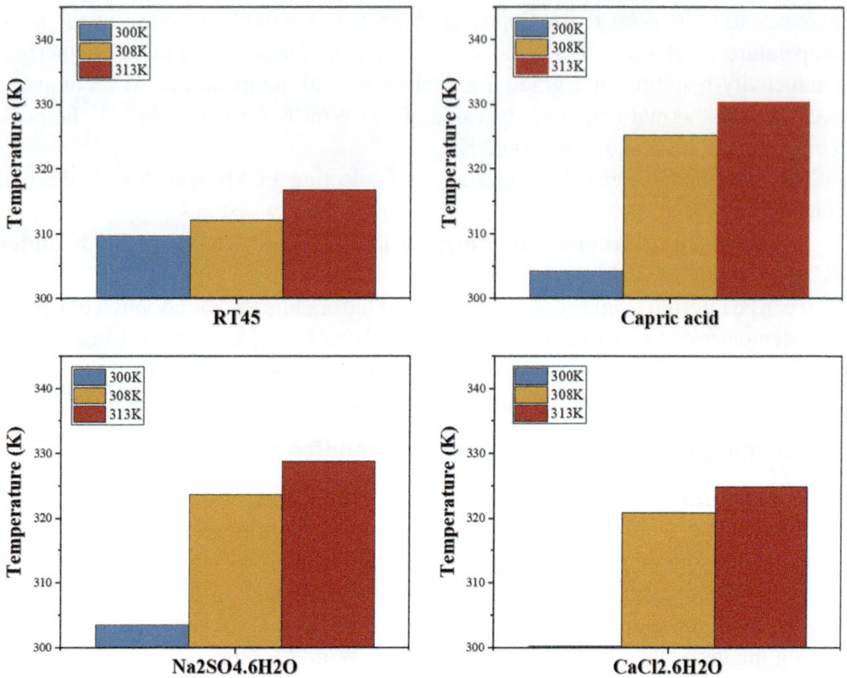

Fig. 14 Peak temperature variation of PCMs under three different ambient temperatures

ambient temperatures, enabling effective latent heat absorption and stabilizing the temperature during phase transitions.

3.4 Studies on a module without and with PCM.

The performance of PCM-based BTMS is influenced by the liquid fraction of the PCM, which reflects the extent to which the material has melted. The effectiveness of the BTMS depends on the PCM's ability to absorb heat through latent heat during melting. An optimal range of liquid fraction ensures effective heat absorption and solidification before the next cycle.

Comparison of Liquid fractions: RT45 and $CaCl_2.6H_2O$ at 313 K under 8C discharge.

The liquid fraction plot at 313 K shows RT45 starting with a low liquid fraction and gradually increasing to about 0.4, becoming progressively more liquid. In contrast, $CaCl_2.6H_2O$ begins with a high liquid fraction close to 0.6 and maintains a relatively constant value, remains predominantly in the liquid state. This suggests that RT45 absorbs heat over time, causing a notable increase in its liquid fraction, whereas

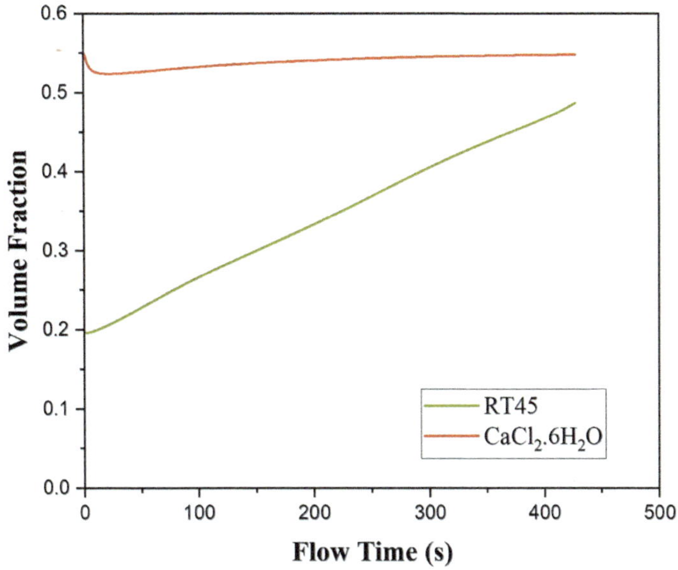

Fig. 15 Variation of liquid fraction of RT45 and $CaCl_2.6H_2O$ at 313 K under 8C discharge with flow time

$CaCl_2.6H_2O$ is already mostly in the liquid state and exhibits little change. The increased liquid fraction and higher final temperature (324 K), indicating less effective cooling as the PCM melts excessively, reducing its ability to solidify before the next cycle (Fig. 15).

Liquid Fraction of $CaCl_2.6H_2O$ at Various Temperatures.

$CaCl_2.6H_2O$'s liquid fraction increases with temperature, peaking at 313 K and lowest at 300 K. Higher temperatures result in faster and higher liquid fractions, while 300 K shows a slower and lower increase. At the higher temperature, the PCM remains mostly in a liquid state, leading to ineffective solidification (Fig. 16).

4 Conclusions

Among the rechargeable batteries, for use in EVs, LIB serves as the purpose of both energy source and storage and have occupied prominence due to its inherent properties. It is observed through the present numerical investigations that a battery module experienced high peak temperatures beyond the safe zone of operation, necessitating the use of effective thermal management under both high discharge and high ambient temperatures. At a discharge rate of 5C, the peak temperature of a single cell reaches 325 K, which rises to 338.5 K at an 8C discharge rate, indicating an increase of 13.5 K.

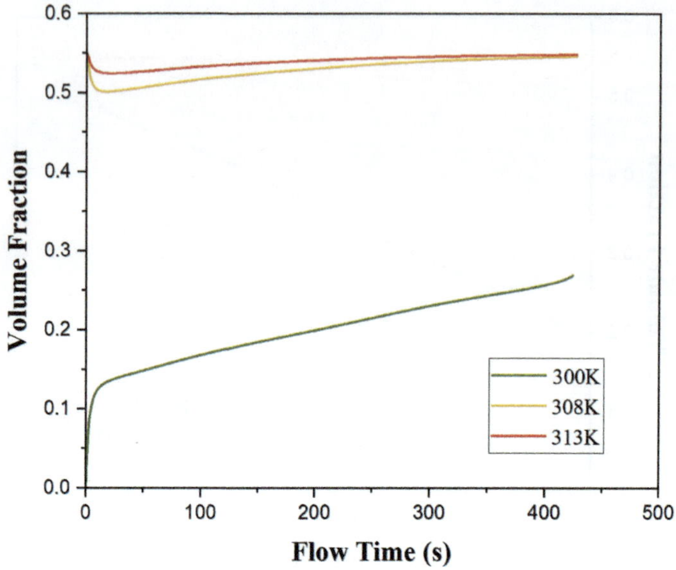

Fig. 16 Variation of liquid fraction of $CaCl_2.6H_2O$ at various ambient temperatures under 8C discharge

To mitigate the ill effects of thermal runway, associated with the exposure to peak temperatures, a passive mode of BTMS is employed using both the organic and inorganic PCMs. It is observed that thermal conductivity of a PCM plays a crucial role in effective thermal management by providing high rate of heat dissipation irrespective of organic or inorganic PCM. With the highest thermal conductivity of 1.09 W/m·K, $CaCl_2 \cdot 6H_2O$ achieved the best results for a module of six cells, reducing the temperature by 11.6%.

Further, the studies are conducted to observe the effect of high ambient temperatures, also establishing the vital role of PCMs high thermal conductivity. It is noted that effect of batteries exposed to high ambient temperatures is more severe than high discharge rates. As seen, the temperature rise for capric acid, from an ambient temperature of 300–313 K, is approximately 26.22 K for a module of six cells.

The high thermal conductivity of PCM undoubtedly reduced the peak temperature when switched over from organic PCM to inorganic PCM. However, the liquid fractions are noted to be sensitive to temperature increase. If liquid fraction increases, leading to increased flowability of PCM and flammability affects when it leaks eventually. It is observed that not only thermal conductivity alone plays a crucial role, but in addition, specific heat, and transition temperature of PCMs have a significant role to play. For instance, despite its lower thermal conductivity, RT-45 performed better at elevated temperatures for a module due to its phase change temperature aligning well with the ambient conditions with reduced temperature of 316 K.

On the other hand, inorganic PCM has drastically reduced the peak temperatures for a module of 6 cells from 344.8 to 320.9 K which is about 7%, at an ambient temperature of 308 K, but the corrosion effects associated with its use will allow judicious selection. It is concluded that during the summer season, in certain places, when the temperature is high due to either discharge or ambient condition, an operation of EVs at lower discharges is recommended.

Acknowledgements The authors gratefully acknowledge the Department of Mechanical Engineering, NIT Warangal, for providing the CFD lab facilities essential for this research. We also wish to express our appreciation to the IITG RIC team for the opportunity to present our work.

References

1. Chan CC.: The state of the art of electric, hybrid, and fuel cell vehicles. In: Proceedings of the IEEE (2007).
2. Yalçın S., Panchal S., Herdem MS.: A CNN-ABC model for estimation and optimization of heat generation rate and voltage distributions of lithium-ion batteries for electric vehicles. Int. J Heat Mass Transfer (2022).
3. Wencan zhang., Zhicheng Liang., Weixiong Wu., Guozhi Ling., Ruixin Ma.: Design and optimization of a hybrid battery thermal management system for electric vehicle based on surrogate model. International Journal of Heat and Mass Transfer (2021).
4. Bandhauer TM., Garimella S., Fuller TF.: A critical review of thermal issues in lithium-ion batteries. J Electrochem Soc. (2011).
5. Tehmina Ambreen., Arslan Saleem., Carlos E. Ugalde-Loo., Tauseef-ur-Rehman., Muhammad Tanveer, Cheol Woo Park.: 3 - Active cooling techniques for battery thermal management. Thermal Management for Batteries, 71–92 (2024)
6. Chuan-Rui Yu., Han-Sen Guo., Qian-Cheng Wang., ORCID., Rui-Dong Chang.: Revealing the Impacts of Passive Cooling Techniques on Building Energy Performance: A Residential Case in Hong Kong.
7. Sivakrishnamraju Rallabandi., Rajay Vedaraj Issac Selvaraj.: Advancements in Battery Cooling Techniques for Enhanced Performance and Safety in Electric Vehicles.: A Comprehensive Review
8. G. Amba Prasad Rao., SR Shravan Kumar.: A review of integrated battery thermal management systems for lithium-ion batteries of electric vehicles. e-Prime - Advances in Electrical Engineering, Electronics and Energy (2024).
9. S. Panchal., I. Dincer M. Agelin-Chaab R. Fraser., M. Fowler.: Thermal modeling and validation of temperature distributions in a prismatic lithium-ion battery at different discharge rates and varying boundary conditions. Applied Thermal Engineering (2016).
10. Wenjian Wang., Xuan Zhang., Chengyun Xin., Zhonghao: An experimental study on thermal management of lithium-ion battery packs using an improved passive method. Applied Thermal Engineering (2018).
11. Ling Xie., Liu Tian., Lulu Yang., Yifei Lv.: Review on application of phase change material in water tanks. Advances in Mechanical Engineering (2017).
12. Anindita Roy., Suraj Meshram., Rajkumar Bhimgonda Patil., Sreelekha Arun., Abhijeet Kore.: Life cycle testing and reliability analysis of prismatic lithium-ion-phosphate cells. International Journal of Sustainable Energy (2024).
13. Mazhar Hussain., Mohd. Kaleem Khan., Manabendra Pathak.: Thermal analysis of phase change material encapsulated li-ion battery pack using multi-scale multi-dimensional framework. Journal of Energy Storage (2023).

14. Seyed Saeed Madani., Erik Schaltz., Søren Knudsen Kær.: Thermal simulation of phase change material for cooling of a Lithium-ion battery pack. Electrochem (2020).
15. Ugur morali.: Influence of convective heat transfer coefficients on thermal behaviour of a lithium-ion cell: a numerical study. Turk J Chem (2023).
16. Ugur.: A numerical and statistical implementation of a thermal model for a lithium-ion battery. Energy (2021).

Wake Shedding Characteristics of an Airfoil in Low Reynolds Number Regime

Anand Verma, Bastav Borah, Amrita Pathak, and Vinayak Kulkarni

Abstract This study investigates the wake characteristics of the low-speed E214 airfoil within a Reynolds number (*Re*) range relevant to micro air vehicles and small-scale wind turbine applications. The experiments are conducted in an open channel wind tunnel and tested for *Re* range of 3×10^4–7×10^4. The effects of probe location, *Re*, and angle of attack (*AoA*) on wake behavior are examined. Results reveal that a single dominant frequency is observed at streamwise probe locations close up to $x/c = 2.0$, indicating a stable wake structure near the trailing edge region. Beyond this point, multiple frequencies emerge, signifying increased vortex complexity and interactions. The amplitude of the dominant frequency decreases as the probe moves downstream, reflecting energy dissipation. The shedding frequency increases with *Re* at a fixed *AoA*, whereas it decreases with an increase of *AoA* at a constant *Re*. The Strouhal number analysis shows it remains nearly constant with increasing *Re* for a given *AoA* due to the proportional relationship between shedding frequency and freestream velocity. However, the Strouhal number decreases with increasing *AoA* at a fixed *Re*, attributed to enhanced flow separation and larger, slower-moving vortices.

Keywords Wind tunnel · E214 airfoil · Hot-wire anemometer · Vortex shedding frequency · Strouhal number

A. Verma (✉) · B. Borah · A. Pathak · V. Kulkarni
Department of Mechanical Engineering, Indian Institute of Technology Guwahati (IITG), Guwahati, India
e-mail: anand2016@iitg.ac.in

B. Borah
e-mail: bastav.borah@iitg.ac.in

A. Pathak
e-mail: amrita.pathak@iitg.ac.in

V. Kulkarni
e-mail: vinayak@iitg.ac.in

U. S. Dixit et al. (eds.), *Current Progress in Engineering Sciences*,
https://doi.org/10.1007/978-981-96-8753-4_7

1 Introduction

In applications such as Micro Air Vehicles (MAVs) and small-scale wind turbines, airfoils often operate at comparatively low Reynolds numbers ($Re < 5 \times 10^5$) [1]. In this *Re* range, airfoils are prone to experiencing laminar boundary-layer separation on their suction surfaces, which can significantly impact their performance. When this separation happens, the flow field over the airfoil is predominantly affected by the formation of a separated shear layer. At higher angles of attack (*AoA*) or lower *Re*, this shear layer frequently fails to reattach to the airfoil surface, resulting in the creation of coherent structures in the wake of the airfoil [2, 3]. Conversely, at lower *AoA* or higher *Re*, the shear layer may reattach to the airfoil surface, resulting in the development of a separation bubble. The formation of coherent structures in the wake of an airfoil at low *Re* is of significant interest because it impacts airfoil performance and influences the flow field over downstream bodies. The properties of these structures are largely influenced by boundary layer behavior, including factors like the length of the separation bubble, the location of reattachment, and the transition to turbulence. However, the boundary layer over the airfoil is highly sensitive to changes in *AoA* and *Re*. These sensitivities can lead to significant variations in wake characteristics, which in turn affect aerodynamic forces and flow stability.

The flow field of the NACA 0009 airfoil exhibited three distinct forms depending on the *AoA* and *Re*: stable laminar flow, orderly vortex street, and unstable vortex street [4]. In the first phase, the wake remains laminar and stable for several chord lengths behind the airfoil. During the second phase, the wake transforms into a well-ordered vortex street. In the third phase, this vortex street becomes unstable, leading to the formation of a separation bubble on the suction side of the airfoil. As the angle of attack increases, the flow pattern transitions through these phases. Huang and Lin [3] investigated the flow field around the NACA0012 airfoil and distinguished four distinct modes of vortex shedding in the wake near the trailing edge. These modes: laminar, subcritical, transitional, and supercritical, are intricately linked to the behavior of the boundary layer on the airfoil's suction surface. At low *AoA*, the generation of vortex shedding behind the airfoil is significantly influenced by shear-layer instabilities. As the *AoA* increases, this shedding behavior evolves to include both low-frequency shedding and the superposition of high-frequency shear-layer instability waves. In the near-wake region, wake vortices were generated and shed alternately from the upper and lower sides of the turbulent wake. The presence of a separation bubble on the upper surface significantly reduced the coherence and length scale of these vortices. Yarusevych et al. [5] explored the coherent structures developed in the separated shear layer and wake of a NACA 0025 airfoil at *Re* range of 5.5×10^4–2.1×10^5, considering three *AoA*: 0°, 5°, and 10°. In each case, laminar boundary layer separation occurred on the airfoil's suction surface, and two distinct flow dynamics were examined: (i) shear layer separation with no reattachment and (ii) separation bubble development on the upper surface. Each of these flow regimes resulted in distinctly different characteristics in the formation and behavior of the roll-up and wake.

The wake behavior behind an airfoil exhibited characteristics similar to those of a bluff body at high *AoA* beyond stall [6]. Conversely, for lower *AoA*, even in cases where the separated shear layer did not reattach, the development of the airfoil wake was observed to diverge from that typically observed for bluff bodies. Apart from this, most previous studies have focused on analyzing vortex structures in the near-wake region of various bluff bodies, such as circular cylinders [7, 8], circular disks [9], and spheres [10]. In these studies, the power spectral density (PSD) of the velocity components is frequently utilized to identify structures and frequencies within the separated shear layer and wake region. The vortex shedding frequency (f) and its corresponding Strouhal number are crucial parameters for studying the periodic behavior in unsteady flow fields. For bluff body wakes, the Strouhal number remains relatively consistent across a broad range of *Re*. For instance, values of approximately 0.21 for a circular cylinder and 0.14 for a flat plate have been commonly reported [11]. In the wake of bluff bodies, the Strouhal number was found to be nearly constant with variation *Re*. At lower *Re*, when the separated shear layer does not reattach to the airfoil's upper surface, the wake vortex shedding structures of the airfoils such as NACA0012 [3], NACA0018 [12], and NACA0025 [5] are similar to those observed in bluff-body wakes. However, the Strouhal number shows greater variation with *Re* than is typically seen in bluff bodies at low *AoA* (<10°). At higher *AoA* (>10°), the Strouhal number variation remains constant with changes in *Re*.

According to the available literature, most researchers have concentrated on investigating the coherent structure characteristics in the wakes of symmetric airfoils and bluff bodies across a wide range of *Re*. These studies have significantly contributed to our understanding of vortex shedding, flow separation, and wake dynamics, providing valuable insights into the aerodynamic performance and stability of these airfoils and bluff bodies. Despite this extensive research, there remains a need to explore the vortex shedding behavior of thin-cambered low speed airfoils, especially for MAVs and wind turbine applications. Such airfoils have shown different aerodynamic behavior in terms of lift and drag compared to symmetric airfoils. Therefore, further research is essential to understand their unique wake dynamics and to optimize their design for improved efficiency and performance in these specific applications.

2 Experimental Setup

The E214 airfoil was selected for the fabrication of the wing model. This airfoil, categorized as a low Reynolds number airfoil, has been specifically chosen for its capability to generate higher lift at low angles of attack [13]. Such a characteristic is particularly beneficial for MAVs and small-scale wind turbines, which typically operate in a low Reynolds number regime where achieving sufficient lift can be challenging. The maximum thickness of the E214 airfoil is $0.111c$, located at $0.331c$ from the leading edge (LE), while the maximum camber is $0.037c$, positioned at $0.569c$ from the LE. Airfoil geometry is shown in Fig. 1. The chord length of this airfoil is 0.150 m, and the span size is 0.60 m. The model is made from Polylactic

Acid (PLA) and produced with a 3D printer employing fused deposition modeling (FDM) technology [**Model:** Pro2, **Make:** RAISE3D]. To minimize surface roughness, several key printing parameters were carefully selected. The layer height was set to 0.1 mm, and a 0.4 mm nozzle diameter was used to ensure finer details. The print speed was maintained at 35 mm/s while the print temperature was set to 210 °C and the bed temperature to 60 °C to achieve optimal adhesion and layer bonding. An infill density of 40% was chosen to balance strength and smoothness, and the cooling fan speed was set to 100% to enhance layer cooling and reduce warping. Additionally, the number of shells was chosen to be 3 to improve the strength of the model. These parameters collectively contribute to achieving a smoother surface finish and higher precision in the printed model. The wing model is mounted on a vertical strut using a mounting plate, with the other side of the strut fixed to a rectangular plate, as shown in Fig. 2. The *AoA* of the model is adjusted manually and measured using a digital spirit level. The span size of the wing covers the entire width of the test section, with a minimal gap between the tunnel side wall and the wing tip. Hence, air leakage through the wingtip can be neglected, allowing the flow to be assumed as two-dimensional (2D).

The experiments are conducted in a subsonic open-channel wind tunnel established in the Department of Mechanical Engineering at IIT Guwahati. This tunnel has a test section of 2 m × 0.6 m × 0.6 m and can achieve a maximum speed of up to 50 m/s. The tunnel speed is measured using a pitot-static tube connected

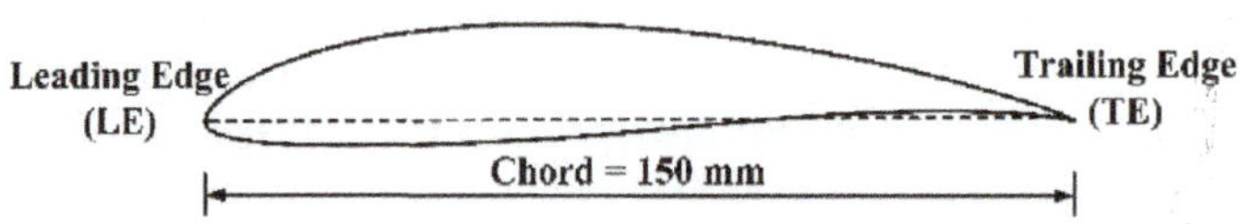

Fig. 1 E214 airfoil geometry

Fig. 2 Wing model mounted in test section

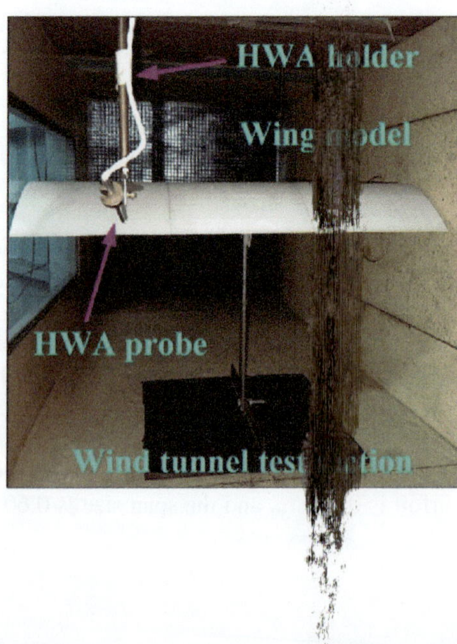

to an electronic manometer. The complete experimental setup for wake velocity measurement is shown in Fig. 3. The freestream turbulence level is found to be approximately 0.35% within the operating speed range. The velocity data in the wake region were obtained using constant temperature anemometers [**Model:** HWCTA-AMB717, **Make:** Sunshine Measurements]. The probe used in the anemometers was made of platinum-coated tungsten wire, measuring 3 mm in length and 10 μm in diameter. It was mounted on a holder connected to a remotely controlled 3D-traverse mechanism, allowing for precise movement of the probe in three directions with an uncertainty of less than 0.1 mm: along to the freestream (x), normal to the freestream (y), and along the spanwise (z). The traverse system is placed on top of the tunnel, enabling detailed mapping of the wake flow characteristics. The non-dimensional forms of these directions are represented as x/c, y/c, and z/c, where c is the chord length of the wing. Analog signals were captured using a National Instruments-built DAQ (Data Acquisition) system, specifically the NI-USB-6363 model. Data is obtained over a 10-s sampling period at a rate of 10 kHz and is stored in the computer (PC). The acquired signals were post-processed using the NI-LabVIEW program. The power spectrum of the acquired signals was represented by the mean square amplitude (MSA). To evaluate the error in the frequency measurement, the standard deviation of the samples and the corresponding standard error of the mean is estimated [14]. The maximum standard errors of the mean are found to be less than ± 2% in the frequency estimation.

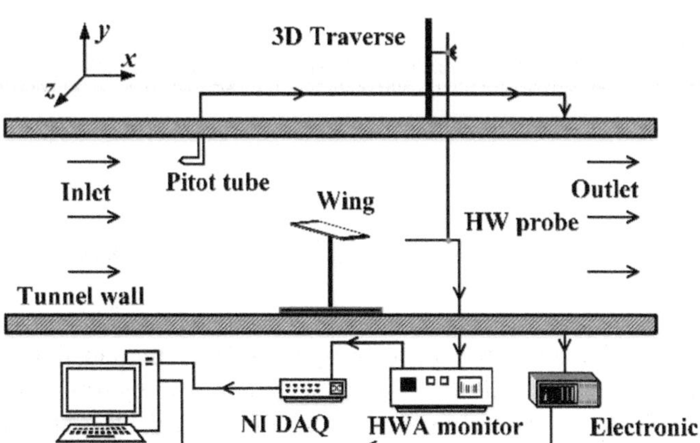

Fig. 3 Experimental setup

3 Results and Discussion

The wake shedding frequency analysis is carried out for the various post-stall angles of attack of the E214 airfoil at the *Re* range of 3×10^4–7×10^4. It has been reported for various symmetric airfoils that their wake behaves similarly to bluff body wakes near stall or post-stall *AoA*. Therefore, it is particularly interesting to explore the wake behavior of this cambered E214 airfoil in the post-stall region. The stall angle for the E214 airfoil was observed to be between 11° and 12° for the present *Re* range [13]. This study investigates the behavior of vortex shedding with the variation of three major factors: probe locations, *AoA*, and *Re*.

To measure wake frequency, the hot-wire probe is positioned at various locations downstream of the airfoil, including along the streamwise (x/c), spanwise (z/c), and orthogonal (y/c) directions. The specific wake domain tested is as follows: in the streamwise direction, the probe is placed within the range of $1.5 \leq x/c \leq 3.0$; in the vertical direction, it is moved from below the LE of the airfoil up to $y/c = 0.30$, and above the LE up to $y/c = 0.25$. For spanwise measurements, the probe is adjusted from the quarter ($z/c \geq 1.0$) to the mid-span length ($z/c \leq 2.0$). The origin of the x, y, and z coordinate system was positioned at the LE when the model was at a zero *AoA*. Examining the frequency response in the wake along the streamwise direction, the probe is positioned at four specific locations: $x/c = 1.5$, 2.0, 2.5 and 3.0 while keeping the vertical position fixed at $y/c = 0.15$ and the spanwise position at $z/c = 1.5$, as shown in Fig. 4. The acquired signal from the hot-wire probe undergoes signal processing using Power Spectral Density (PSD), which involves computing the power based on the amplitude of the Fourier-transformed data. In general, the squared amplitude of the Fourier transform is proportional to the power spectrum [15]. The power spectrum illustrates the distribution of power across various frequencies, helping to identify the dominant frequency within the signal. This analysis is typically expressed as Mean Squared Amplitude (MSA), providing a clear view of the signal's frequency content. The power spectrum analysis of instantaneous data in the wake region at different locations is presented for an *AoA* of 12° and a *Re* of 5×10^4 (Fig. 4).

The dominant frequency represents the most significant energy-carrying frequency within the wake, which is crucial for understanding the flow dynamics and the aerodynamic behavior of the airfoil in the post-stall regime. The dominant frequency is found to be 19.6 Hz at a probe location of $x/c = 1.5$, which is almost identical to the frequency response at a probe location of $x/c = 2.0$. It is observed that for probe locations from $x/c = 1.5$–2.0, a single dominant frequency is present in the spectrum and remains nearly constant despite variations in the probe location. This suggests a stable and periodic wake flow structure in this region. However, when the probe moved further downstream from $x/c = 2.0$–3.0, multiple frequencies appeared in the power spectra, indicating more complex flow dynamics. The presence of multiple frequencies suggests interactions between different flow structures or turbulence in the wake region. Beyond $x/c = 3.0$, the hot-wire signal becomes

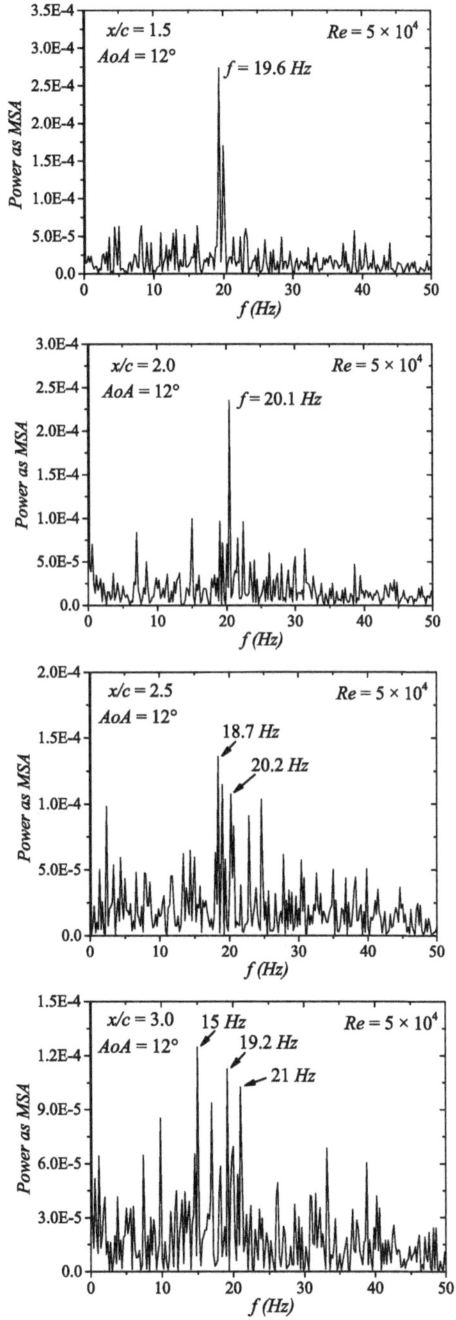

Fig. 4 Frequency response at different locations along the streamwise direction in the wake region

irregular, resulting in the absence of a dominant frequency peak in the power spectrum at the tested *Re* and *AoA*. Additionally, the amplitude corresponding to the dominant frequency decreases as the probe moves from near the trailing edge ($x/c = 1.5$) to further downstream ($x/c = 3.0$). This decrease in amplitude signifies a reduction in the strength and coherence of the wake vortices, highlighting the transition from a more organized wake flow to a turbulent wake with a more dispersed energy distribution. Similar tests are also conducted with the probe positioned at various points normal to the freestream within the range of $-0.3 \leq y/c \leq 0.25$, while keeping the other parameters constant. The results reveal no significant changes in vortex shedding frequency across these positions.

The frequency domain diagram at various *AoA* is shown in Fig. 5 to examine the effect of *AoA* on the wake characteristics of the E214 airfoil. The results are provided for an *AoA* range from 12° to 24° at a *Re* of 5×10^4, with a fixed probe location at $x/c = 2.0$, $y/c = 1.5$, and $z/c = 1.5$. When the *AoA* increases from 12° to 24°, the dominant frequency decreases from 20.1 to 12.3 Hz. This decline is likely due to the increased separation of the flow from the airfoil surface at higher *AoA*, which leads to larger, slower-moving vortices and, consequently lower wake shedding frequencies. A similar effect of *AoA* on the shedding frequency is observed for symmetric airfoils [3]. Additionally, the amplitude of the signal increases with an increase in *AoA*. This is due to the higher energy associated with the larger separated flow regions, which results in stronger vortex shedding.

This study is also extended to investigate the effect of *Re* on the vortex shedding frequency, as shown in Fig. 6. The *Re* chosen for this investigation ranges from 3×10^4 to 7×10^4, and power spectra are presented at an *AoA* of 12°, with a fixed probe location at $x/c = 2.0$, $y/c = 1.5$, and $z/c = 1.5$. It is observed that as *Re* increases, the dominant frequency of vortex shedding also increases. This is because higher *Re* corresponds to higher flow velocities, which result in increased energy in the flow. The additional energy can result in faster vortex formation and shedding, leading to an increase in dominant frequencies in the wake. Similar behavior of f_s as a function of *Re* for symmetric airfoils has been reported in previous studies [3, 12].

The Strouhal number (*St*) corresponding to the shedding frequency (f) is determined for various *Re* and *AoA* of the E214 airfoil, as illustrated in Fig. 7. The findings indicate that for a given *AoA*, as *Re* increases, the f also increases, but the *St* remains nearly constant. This is because the Strouhal number is defined as $St = fc/U$, where c is the chord length of the airfoil, and U is the freestream velocity. When *Re* increases, U also increases, which results in a proportional increase in the f. As a result, the ratio of f/U stays nearly constant, keeping the Strouhal number consistent. This observation agrees with reported studies for NACA0018 and NACA0025 airfoils [5, 12]. In contrast, for a given *Re*, *St* decreases with increasing *AoA*. The Strouhal number is estimated as 0.60 at $AoA = 12°$ with *Re* of 5×10^4. When *AoA* increases to 24°, this value decreases to 0.36 for the same *Re*. For bluff bodies such as cylinders, the Strouhal number is typically found to be around 0.21 [11], which is lower compared to streamlined bodies like airfoils. The higher Strouhal number for airfoils at lower *AoA* suggests more organized and regular vortex shedding, indicative of a stable wake. As the *AoA* increases, the flow separation becomes more pronounced, leading

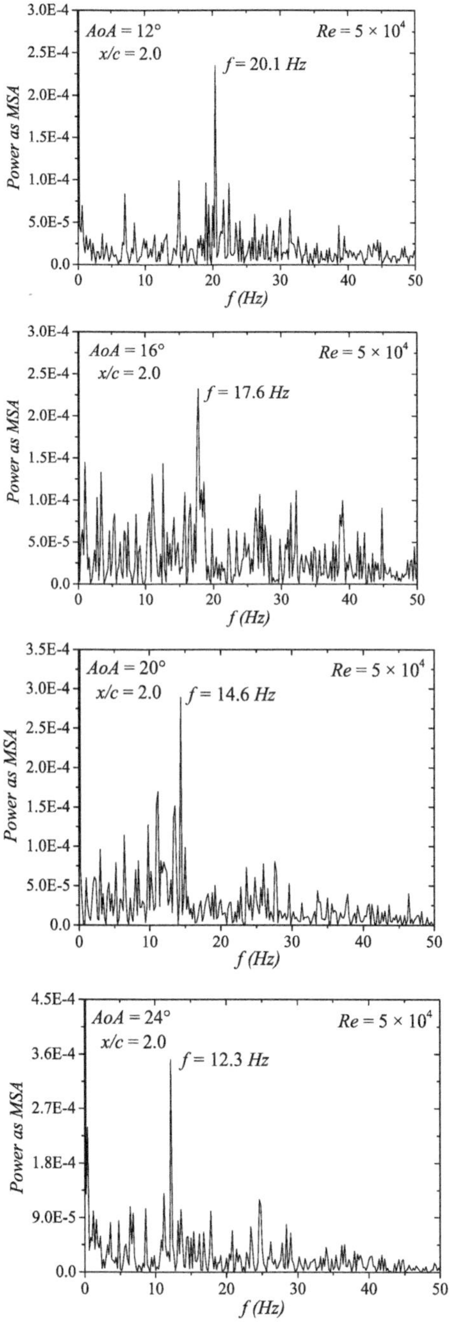

Fig. 5 Frequency response in the wake region of an airfoil at various *AoA*

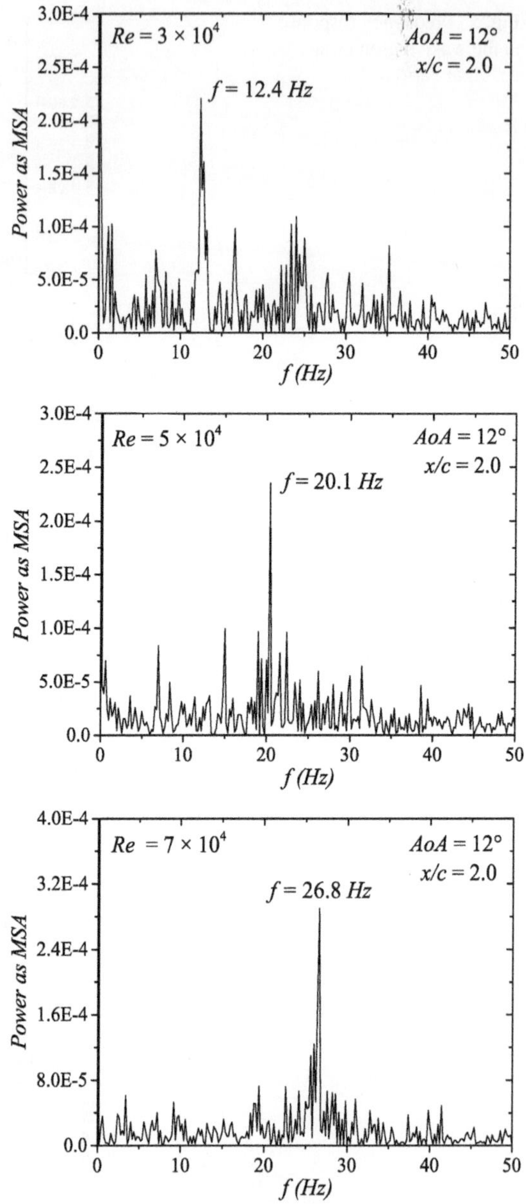

Fig. 6 Frequency response in the wake region of an airfoil at various *Re*

to a decrease in the Strouhal number. This lower Strouhal number at higher *AoA* indicates a more complex and irregular wake flow, resembling the wake behavior observed in bluff bodies.

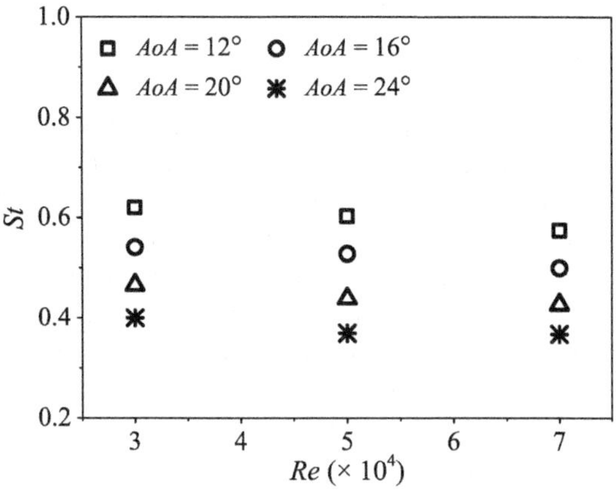

Fig. 7 Variation of Strouhal number with *Re* at various *AoA*

4 Conclusion

The wake characteristics of the low-speed E214 airfoil were investigated experimentally within the *Re* range relevant to MAVs and small-scale wind turbine applications. The experiments were carried out in an open-circuit wind tunnel with *Re* values ranging from 3×10^4 to 7×10^4. This study examined the effects of three major parameters on wake behavior: probe location, *Re*, and *AoA*. The results indicated that within the probe locations streamwise from $x/c = 1.5$–2.0, a single dominant frequency was consistently observed, suggesting a stable wake structure in this region. Beyond $x/c = 2.0$, multiple frequencies appeared in the power spectra, highlighting the increased complexity and interactions of vortices in the wake. Additionally, the amplitude of the dominant frequency decreased with increasing distance from the trailing edge, which is indicative of energy dissipation as vortices move downstream. The shedding frequency was found to increase with *Re* at a given *AoA*. Conversely, it decreased with increasing *AoA* at a fixed *Re*. The analysis of the Strouhal number showed that for a fixed *AoA*, the Strouhal number remained nearly constant with increasing *Re*. This stability is due to the proportional relationship between the shedding frequency and freestream velocity. Conversely, for a fixed *Re*, Strouhal number decreased as *AoA* increased. This decrease is attributed to enhanced flow separation and the formation of larger, slower-moving vortices at higher *AoA*, which results in a lower Strouhal number. These findings are consistent with existing literature on similar airfoils and provide important insights into the wake dynamics of the E214 airfoil. Understanding these characteristics is crucial for optimizing the design and performance of MAVs and small-scale wind turbines, as it impacts their efficiency and operational stability.

References

1. Carmichael, B.H.: Low Reynolds number airfoil survey Vol. 1. NASA Contractor Report 165803 (1981).
2. Yarusevych, S., Sullivan, P.E., Kawall, J.G.: Coherent structures in an airfoil boundary layer and wake at low Reynolds numbers. Phys. Fluids. 18, (2006). https://doi.org/10.1063/1.2187069
3. Huang, R.F., Lin, C.L.: Vortex Shedding and Shear-Layer Instability of Wing at Low-Reynolds Numbers. AIAA J. 33, 1398–1403 (1995). https://doi.org/10.2514/3.12561
4. Elimelech, Y., Arieli, R., Iosilevskii, G.: Flow over NACA-0009 and Eppler-61 Airfoils at Reynolds numbers 5000 to 60,000. AIAA J. 45, 2414–2421 (2007). https://doi.org/10.2514/1.22376
5. Yarusevych, S., Sullivan, P.E., Kawall, J.G.: On vortex shedding from an airfoil in low-Reynolds-number flows. J. Fluid Mech. 632, 245–271 (2009). https://doi.org/10.1017/S0022112009007058
6. Huang, R.F., Lee, H.W.: Turbulence effect on frequency characteristics of unsteady motions in wake of wing. AIAA J. 38, 87–94 (2000). https://doi.org/10.2514/3.14382
7. Kravchenko, A.G., Moin, P.: Numerical studies of flow over a circular cylinder at Re_D=3900. Phys. Fluids. 12, 403–417 (2000). https://doi.org/10.1063/1.870318
8. Rai, M.M.: A computational investigation of the instability of the detached shear layers in the wake of a circular cylinder. J. Fluid Mech. 659, 375–404 (2010). https://doi.org/10.1017/S002211201000251X
9. Shenoy, A.R., Kleinstreuer, C.: Flow over a thin circular disk at low to moderate Reynolds numbers. J. Fluid Mech. 605, 253–262 (2008). https://doi.org/10.1017/S0022112008001626
10. Rodriguez, I., Borell, R., Lehmkuhl, O., Perez Segarra, C.D., Oliva, A.: Direct numerical simulation of the flow over a sphere at Re = 3700. J. Fluid Mech. 679, 263–287 (2011). https://doi.org/10.1017/jfm.2011.136
11. Roshko, A.: On the drag and shedding frequency of two-dimensional bluff bodies. NACA Technical Note 3169 (1954).
12. Yarusevych, S., Boutilier, M.S.H.: Vortex shedding of an airfoil at low reynolds numbers. AIAA J. 49, 2221–2227 (2011). https://doi.org/10.2514/1.J051028
13. Selig, M.S., Donovan, J.F., Fraser, D.B.: Airfoils at Low Speeds. Soartect 8, SoarTech Publications, 1504 North Horseshoe Circle Virginia Beach, Virginia (1989).
14. Moffat, R.J.: Describing the uncertainties in experimental results. Exp. Therm. Fluid Sci. 1, 3–17 (1988). https://doi.org/10.1016/0894-1777(88)90043-X
15. Schneider, P., Xhafa, F.: Data stream processing: models and methods. In: Anomaly Detection and Complex Event Processing over IoT Data Streams. pp. 29–47. Elsevier (2022).

Comparative Analysis of Mechanical Properties of AA6061-T6 Fabricated by Single- and Multi-Track Friction Stir Additive Manufacturing

Ankan Das, Himangshu Kalita, Pankaj Biswas, and Sajan Kapil

Abstract Friction Stir Additive Manufacturing (FSAM) is an emerging technique in the field of additive manufacturing that applies the principle of friction stir welding for adding sheets as layer-by-layer fashion. This study aims to conduct a comparative analysis of the mechanical properties of AA6061-T6 components fabricated using single-track and multi-track FSAM. Single-track FSAM involves the deposition of material along a single path, while multi-track FSAM incorporates multiple overlapping tracks to build up the material. The primary focus is on evaluating how these two fabrication approaches influence the microstructure, hardness, tensile strength, and fractography analysis of the resultant AA6061-T6 specimens. By understanding the differences in mechanical properties between single-track and multi-track FSAM-fabricated AA6061-T6, this research seeks to provide insights into optimizing process parameters for enhanced performance. The findings will contribute to the broader knowledge base of additive manufacturing techniques, potentially guiding the development of more efficient and reliable manufacturing processes for high-performance aluminum alloys.

Keywords Friction stir additive manufacturing · Aluminum · Single track · Multi-track · Temperature · Tensile strength · Hardness

A. Das (✉) · H. Kalita · P. Biswas · S. Kapil
Department of Mechanical Engineering, Indian Institute of Technology Guwahati, Assam, India
e-mail: a.das@iitg.ac.in

H. Kalita
e-mail: k.himangshu@iitg.ac.in

P. Biswas
e-mail: pankaj.biswas@iitg.ac.in

S. Kapil
e-mail: sajan.kapil@iitg.ac.in

U. S. Dixit et al. (eds.), *Current Progress in Engineering Sciences*,
https://doi.org/10.1007/978-981-96-8753-4_8

1 Introduction

Friction Stir Additive Manufacturing (FSAM) is a solid-state process that works below the melting temperature of the substrate material. It counters the phase transformation-related solidification defects, such as cracking and porosity, which are prominent in fusion-based additive manufacturing processes [1]. Basically, FSAM is a type of sheet lamination technology where additive manufacturing and machining are combined synergistically to form a near-net shape [2]. In this process, the first two sheets are held together one on another, and then a rotating Friction Stir Welding (FSW) tool moves over the top sheet. The tool's probe is inserted into both of the sheets to facilitate the stirring process. Due to the rubbing action between the tool shoulder and the top sheet, frictional heat is generated, which is sufficient to deform the material plastically [3]. It is an energy-efficient and cost-effective process for joining two layers. A similar process is repeated for all the layers until the built height is achieved. The previous research done on FSAM showed the potential applications in various engineering applications such as airplane stiffeners and stringers, hydrogen tanks, pressure vessels, material-based functionally graded materials, electronically embedded structures [4]. Most of the research reported on FSAM is based on alloys of aluminum 2xxx, 5xxx, and 7xxx series, magnesium, and a few steel grades. The focus of the studies revolved around the single-track FSAM, where each layer was joined in one welding pass. The mechanical properties of the fabricated blocks' strength, hardness, and toughness are enhanced due to parametric optimization, grain refinement, and the absence of any solidification-related defects. The most advantageous part of this process is the manipulation freedom of microstructure along the building height [5–7]. The re-stirring phenomenon is the key to such a manipulation mechanism. Notably, the presence of a hooking defect in the thermo-mechanically affected zone (TMAZ) is reported in the works of literature. Its presence is very detrimental to the mechanical strength of the fabricated part. Eliminating such defects and increasing the joint area overlapping toolpaths may be very effective [8]. Henceforth, this study is motivated to investigate the differences between the single-track and multi-track FSAM in terms of peak temperature rise, mechanical properties, and microstructural appearance of the stirred zone.

2 Experimental Procedure

In this study, Al–Mg-Si alloy (AA6061-T6) is used as a substrate material. In single-track experiment sheets, the dimensions are 100 mm × 180 mm × 4 mm. Multi-track fabrication utilizes the sheets having dimensions of 150 mm × 180 mm × 4 mm. An FSW tool made of H13 tool steel is utilized during the experiments. The tool probe profile is cylindrical taper left-hand threaded, as shown in Fig. 1c. Tool rotational and traverse speeds varied from 875 to 1230 rpm and 41 to 82 mm/min, respectively. Other process parameters, such as plunge depth and tool tilt angle, were maintained

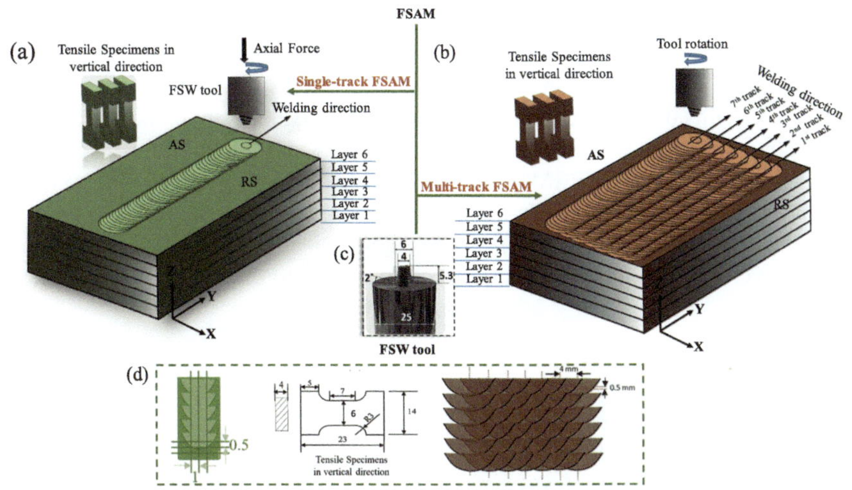

Fig. 1 **a** Schematic of single-track FSAM, **b** schematic of multi-track FSAM, **c** tool design, **d** dimensions of tensile specimens and hardness estimation patterns

constant at 0.2 mm and 2.5°, respectively. The effective tool pin overlap between adjacent tracks was maintained at 50% in the case of multi-track FSAM. Mechanical testing samples and metallurgical samples were cut out from the fabricated blocks. Vertical tensile specimens were taken from the building direction and dimensions shown in Fig. 1d. The crosshead speed of the tensile testing machine was 1 mm/min. The planning of micro-hardness estimation on the cross sections is shown in Fig. 1d. The load exerted by the diamond indenter on the specimen during the hardness estimation was 100 kgf. To ensure the accuracy and reliability of the results, each test was conducted three separate times for every parameter under consideration. An IR camera recorded the rise of peak temperature during FSAM. Fractography analysis was done using a field emission scanning electron microscope (FESEM) on the fracture surfaces.

3 Results and Discussion

Mechanical and microstructural properties of the fabricated blocks are directly influenced by the frictional heat generated in the process. Tempered AA6061-T6 is susceptible to the precipitation-strengthening mechanism. At higher exposure temperatures, secondary precipitate dissolution occurs in the alloy, which reduces the mechanical strength compared to the as-received material [9]. The rise in temperature, which varies with process parameters and operating positions, is plotted in Fig. 2. It shows that a rotational speed of 1230 rpm causes a rise in temperature irrespective of tool traverse speed. The number of tracks increases on the advancing side (AS), and the surface temperature also increases. But in both cases, the temperature reduces as the built height increases. Minimum and maximum temperatures recorded in single- and multi-track FSAM are 343 and 427, 340 and 452 °C, respectively. This temperature variation results in an altercation of microstructure development in the stirred zones.

Figure 3 illustrates the microstructural features and stress–strain responses of single- and multi-track FSAM under varying parametric conditions. In the micrograph corresponding to a single-track sample produced with the 875/82 parameter set, a hooking defect is observed on the AS, along with the presence of a micro-crack at the layer interface. The hooking defect arises due to the tendency of plastically

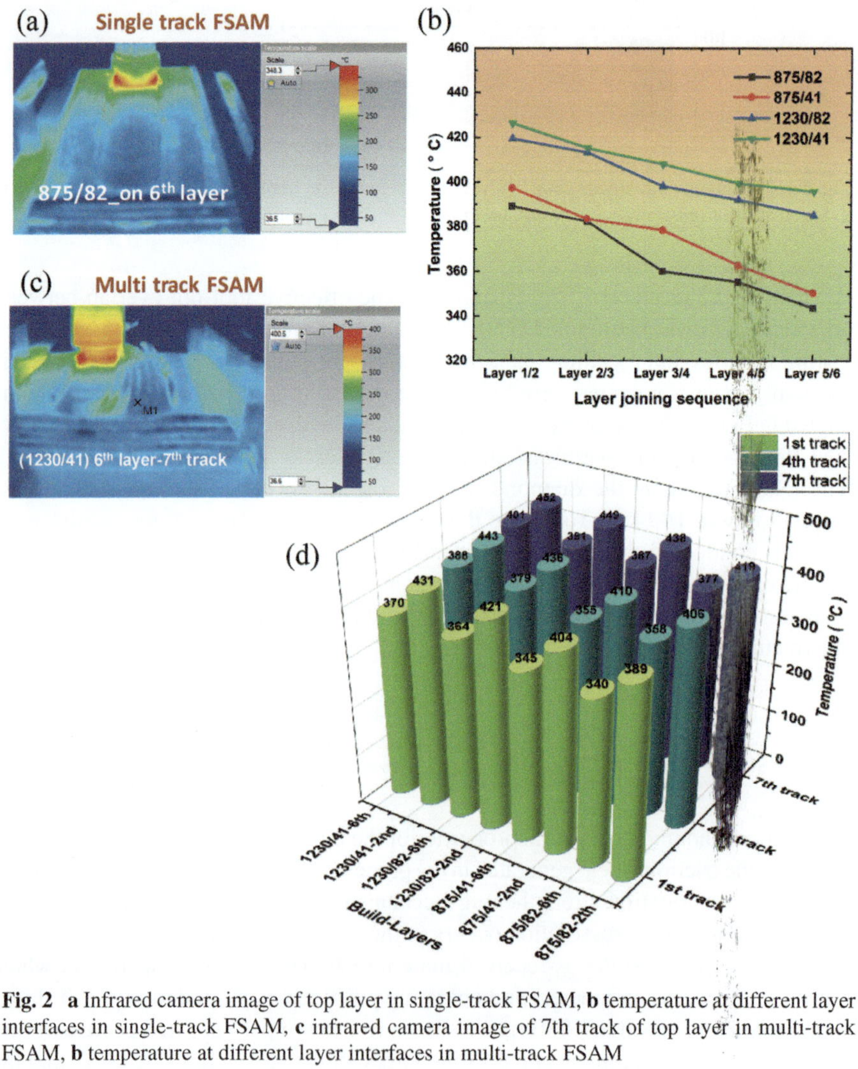

Fig. 2 **a** Infrared camera image of top layer in single-track FSAM, **b** temperature at different layer interfaces in single-track FSAM, **c** infrared camera image of 7th track of top layer in multi-track FSAM, **b** temperature at different layer interfaces in multi-track FSAM

deformed material to flow outward from a two-layer interface along the boundary of the stirred zones. Conversely, in multi-track FSAM, these defects are absent, and the layer interfaces are well fused. Also, a dip at 2% strain has been observed in Fig. 3d for 82 mm/min samples; this dip can be attributed to the material undergoing a structural rearrangement or phase transformation, such as twinning, dislocation movement, or a shift in crystal lattice structure, which temporarily reduces stress. The dip could represent the initiation of damage, such as micro-crack formation, void initiation, or localized yielding, leading to a momentary drop in stress before further strain hardening or failure [8]. The re-stirring action over 50% of the stirred zone on the advancing side mitigates the formation of hooking that would otherwise occur after the initial pass. The use of multiple tracks expands the bonded area, effectively eliminating or displacing the hooking defect away from critical sections. The detrimental effects of these defects are reflected in the reduced tensile strength of single-track FSAM samples, as shown in Fig. 3d. In contrast, the enhanced tensile strength of multi-track FSAM specimens can be attributed to the densely bonded adjacent layers and the resulting refined grain structures.

The highest ultimate tensile strength (UTS) achieved was 246 ± 6 MPa for single-track FSAM and 398 ± 5 MPa for multi-track FSAM. Specimens fabricated using the 875/82 parametric combination exhibited lower tensile strength in both configurations, likely due to the relatively high tool traverse speed of 82 mm/min. This higher traverse speed reduces heat generation per unit area and time, leading to insufficient heat and axial force, which are critical for ensuring dense bonding between layers. The yield strength (YS) of single-track specimens was approximately 141 ± 6 MPa, as illustrated in Fig. 4a. In contrast, the yield strength in multi-track FSAM specimens was higher, as shown in Fig. 4b. A traverse speed of 41 mm/min resulted in improved mechanical properties in both single- and multi-track FSAM. An elongation of approximately 18% was achieved in the 875/41 specimen for multi-track FSAM. Figure 4c, d presents the Vickers micro-hardness distribution across the stirred zones. In single-track specimens, hardness values range between 75 and 85 $HV_{0.1}$, while the maximum hardness of 119 $HV_{0.1}$ is achieved in the 875/41 specimen for multi-track FSAM. The average micro-hardness within the stirred zones is higher in multi-track FSAM compared to single-track specimens. Additionally, hardness increases with build height, consistent with the Hall–Petch relationship, which links smaller grain diameters to increased hardness. The re-stirring effect caused by overlapping toolpaths in multi-track FSAM results in further grain refinement within the stirred zones. It is important to note that the average hardness is reduced compared to the base material, likely due to the dissolution of precipitates during the process.

Fracture analysis of tensile samples provides insights into the material's behavior under stress and the mechanisms that lead to failure. The overall appearance of the fracture surface is examined to categorize the fracture mode. Figure 5 shows the high-resolution images of the fracture surface are taken to study the fracture morphology.

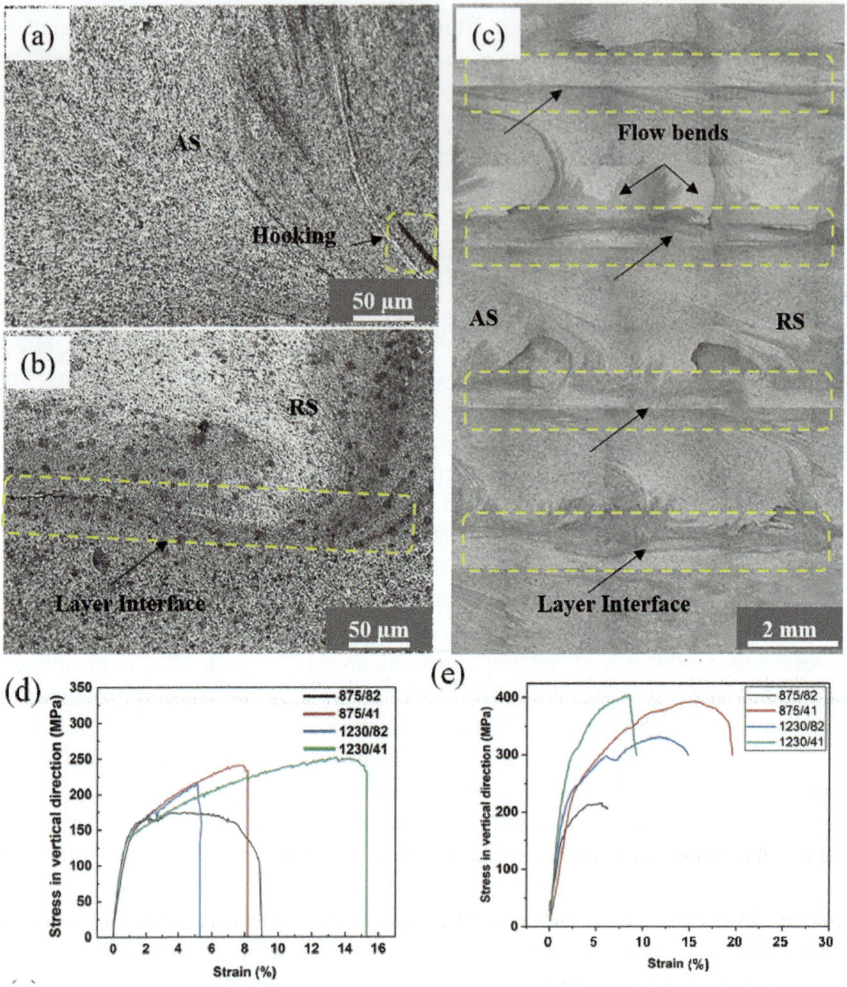

Fig. 3 Showing the microstructure and stress–strain graph of **a**, **b**, **d** single-track FSAM, **c**, **e** multi-track FSAM

In the case of the single-track specimen, the fracture surface exhibits inconsistent deformation, as evidenced by the inhomogeneous distribution of dimples. Additionally, the presence of cleavage facets indicates regions of brittle fracture, with little to no plastic deformation. The fracture surface also shows both primary and secondary cracks, which further suggests the occurrence of mixed-mode failure, combining characteristics of both ductile and brittle fracture mechanisms. In the case of the multi-track specimen, the fracture surface shows concordant deformation, characterized by homogeneous and deep dimples, which are indicative of a uniform plastic deformation process. Deep dimples typically signify higher toughness and lower propensity for crack propagation. The presence of tear ridges further suggests significant plastic

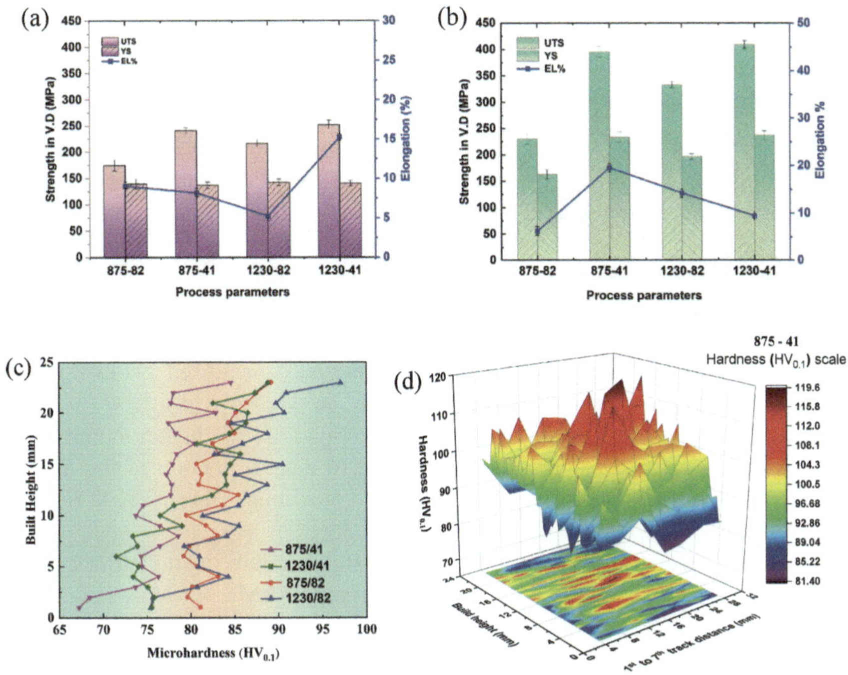

Fig. 4 Showing strength in vertical direction and hardness distribution, respectively, for **a**, **c** single-track FSAM **b**, **d** multi-track FSAM

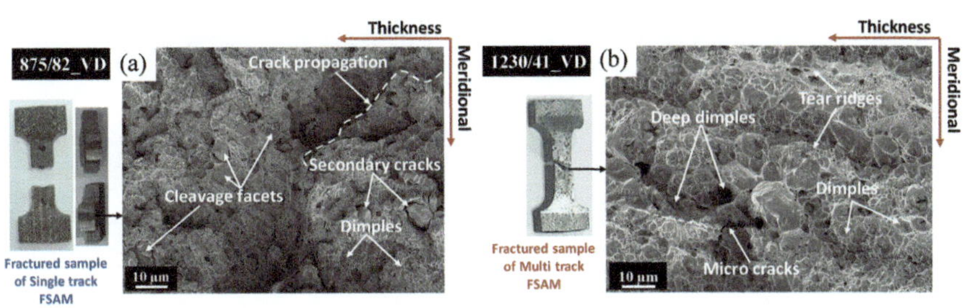

Fig. 5 Showing FESEM images of fracture surfaces for **a** single-track (875/82) FSAM and **b** multi-track (1230/41) FSAM

deformation preceding the fracture. Although micro-cracks are observed on the fracture surface, the overall mode of failure is predominantly ductile, as evidenced by the material's ability to undergo substantial plastic deformation before rupture.

4 Conclusions

The following conclusions can be drawn from the above comparative study:

- As the build height increases, the peak temperature decreases in both cases, with the bottom layers acting as a heat sink. Multi-track FSAM retains more heat within the build compared to single-track FSAM.
- Hardness gradually increases with build height in both cases.
- Overall, the hardness is reduced after additive manufacturing compared to the base material, which can be attributed to the dissolution of strengthening precipitates.
- However, the average micro-hardness is higher in multi-track FSAM, owing to the refined grain structure, with the highest hardness measured at 119 HV0.1 in the 875/41 multi-track specimen.
- The ultimate tensile strength in the build direction is 58.4% higher in multi-track FSAM, ranging from 246 to 398 MPa, due to the elimination of the hooking defect in the middle of the stirred zones. The maximum yield strength recorded in multi-track FSAM is 230 ± 8 MPa.
- Fractography analysis indicates a mixed mode of failure in single-track FSAM and predominantly ductile fracture in multi-track FSAM.

Acknowledgements The authors would like to acknowledge the contribution of the Central Workshop members and Advance Welding and Fabrication Lab research group. We also acknowledge Technology Innovation Hub (TIH), Indian Institute of Technology Guwahati, India, which has provided experimental facilities.

References

1. Galy, C., Le Guen, E., Lacoste, E., and Arvieu, C., Main Defects Observed in Aluminum Alloy Parts Produced by SLM: From Causes to Consequences, Addit. Manuf., **22**, pp. 165–175. (2018).
2. Das, A., Medhi, T., Kapil, S., & Biswas, P. Different build strategies and computer-aided process planning for fabricating a functional component through hybrid-friction stir additive manufacturing. *International Journal of Computer Integrated Manufacturing*, *37*(3), 350-371. (2024).
3. Das, A., Medhi, T., Kapil, S., & Biswas, P, Multi-track multi-layer friction stir additive manufacturing of alloy. *Progress in Additive Manufacturing*, 1–21 (2023).
4. Mishra, R. S., Haridas, R. S., & Agrawal, P. Friction stir-based additive manufacturing. *Science and Technology of Welding and Joining*, *27*(3), 141-165. (2022).
5. Palanivel, S., Nelaturu, P., Glass, B., and Mishra, R. S. Friction Stir Additive Manufacturing for High Structural Performance through Microstructural Control in an Mg Based WE43 Alloy, Mater. Des., **65**, pp. 934–952. (2015).
6. Yuqing, M., Liming, K., Chunping, H., Fencheng, L., and Qiang, L. Formation Characteristic, Microstructure, and Mechanical Performances of Aluminum-Based Components by Friction Stir Additive Manufacturing," Int. J. Adv. Manuf. Technol., **83**(9–12), pp. 1637–1647 (2016).
7. He, C., Li, Y., Zhang, Z., Wei, J., and Zhao, X., Investigation on Microstructural Evolution and Property Variation along Building Direction in Friction Stir Additive Manufactured Al–Zn–Mg Alloy, Mater. Sci. Eng. A, **777**, p. 139035 (2020).

8. Medhi, T., Das, A., Pankaj, P., Kapil, S., and Biswas, P. Multi-Pass Friction Stir Lap Welding of AA 6061-T6: Implication of Tool Pin Overlapping on Microstructure and Mechanical Properties of Joints, Soldag. e Insp., **27**. (2022).
9. Engler, O., Marioara, C. D., Aruga, Y., Kozuka, M., and Myhr, O. R. Effect of Natural Ageing or Pre-Ageing on the Evolution of Precipitate Structure and Strength during Age Hardening of Al–Mg–Si Alloy AA 6016, Mater. Sci. Eng. A, **759**, pp. 520–529 (2019).

A Novel Method of Calculating the Degrees of Freedom of Planar Linkages Using the Concepts of Linear Algebra

Anurag Patil, Aratrick Mondal, and Sajan Kapil

Abstract The calculation of DoF for any given mechanism is essential for its kinematic and dynamic analysis, and hence, a reliable and fast method to determine the DoF is of paramount importance. The conventional methods for DoF calculation are found to be inaccurate when dealing with various useful mechanisms. Therefore in this work, a novel matrix-based method is proposed that accurately calculates the DoF of every possible planar mechanism. A strong foundation is provided in the physical derivation of the method, and the mathematical premise is explained by the idea of vector space and linear transformation using the Fundamental Theorem of Linear Algebra. Several postulates have been formulated for a clear stepwise approach to apply this method. A few well-studied examples are elaborated as case studies to demonstrate the effectiveness and advantages of this proposed matrix-based method. A comprehensive analysis of the four-bar mechanism is undertaken to elucidate the profound implications of this method and its capacity to expand the horizons of mechanism analysis through this matrix-based technique. This endeavor serves to highlight the practical significance and broader applicability of this approach.

Keywords Planar mechanisms · Kinematic analysis · Degrees of freedom (DoF) · Linear transformation · Nullity

A. Patil · A. Mondal (✉) · S. Kapil
Department of Mechanical Engineering, Indian Institute of Technology Guwahati, Guwahati, Assam, India
e-mail: aratrickmondal02@gmail.com

A. Patil
e-mail: anup103@yahoo.com

S. Kapil
e-mail: sajan.kapil@iitg.ac.in

U. S. Dixit et al. (eds.), *Current Progress in Engineering Sciences*,
https://doi.org/10.1007/978-981-96-8753-4_9

1 Introduction

The design of a mechanical system necessitates the kinematic and dynamic analyses of the system components. This is where the need for DoF comes and it is basis of kinematic analysis that ascertains the possible range of motions of the system components. As per International Federation for the Promotion of Mechanism and Machine Science (IFToMM), DoF is the minimum number of independent variables needed to define the configuration of a kinematic chain or mechanism. It offers a quick analysis of the overall configuration of the mechanical system and gives an insight into the system's potential range of motions and the positions its components can attain. It is an indispensable tool for kinematic analysis, which is pivotal in various applications, including robotics and machinery design. With a clear understanding of the system's DoF, engineers and designers can tailor their designs to meet specific requirements and constraints, optimizing their designs for enhanced performance. Looking at it the other way around, the knowledge of DoF indirectly helps one to synthesize his/her own mechanisms according to the required output. Another critical advantage of assessing DoF is the prevention of overconstraining a mechanism. Overconstraints can lead to undue stress on system components, resulting in inefficiency and a higher probability of mechanical failure. Thus, understanding DoF serves as a safeguard against such issues.

1.1 History

The investigation into the calculation of DoF commenced during the mid-nineteenth century, featuring significant contributions from luminaries such as Chebyshev, Sylvester, Grübler, Somov–Malyshev, Kutzbach, and several other distinguished researchers. In recent years, a plethora of advanced formulations and techniques have been developed to surmount the constraints associated with conventional methodologies. This section offers a concise overview of both the traditional as well as modern DoF calculating methodologies, underscoring the progressive evolution of DoF calculation techniques over time and the pivotal role they play in simplifying complex kinematic analysis.

1.1.1 Chebyshev

He was likely the first person to propose the relation for computing the DoF of linkages. In the year 1854, Chebyshev published his first paper on linkages, followed by several extensions. One of the most important works includes *The theory of the mechanisms known under the name "parallelograms", 1854*. The word "parallelogram" denotes a specific mechanism, namely, Watt's parallelogram, and its improvements which were initially analyzed by Chebyshev for developing his first formulation. Chebyshev [1] derived an equation that holds true for any linkage with one DoF:

$$3n - 2(l + p) = 1 \tag{1}$$

where n is the number of bars (kinematic links), l and p are the numbers of revolute joints which are adjacent, and non-adjacent to the fixed link, respectively.

1.1.2 Grübler

After a few years, Grübler presented a structural condition for one DoF planar mechanisms and later extended this condition to one DoF spatial complex mechanisms with only helical joints as represented by Eqs. (2) and (3), respectively:

$$3n - 2p - 4 = 0 \tag{2}$$

$$5h - 6n - 4 = 0 \tag{3}$$

where n is the total number of links in the mechanism including the fixed link, p is the number of revolute joints in the mechanism and h is the number of helical joints in the mechanism.

In the year 1883, Grübler proposed a more generalized version that included both lower and higher pair joints. This is valid for many different planar mechanisms. It was proposed as follows:

$$DoF = 3(n - 1) - 2p_1 - p_2 \tag{4}$$

where n number of links, p_1 number of lower pairs and p_2 number of higher pairs. This is one of the most famous DoF calculating formulas used till now. It indeed opened the door for more research into the mobility calculation for more complex mechanisms.

1.1.3 Somov–Malyshev

Though Grübler's formula presented a significant finding in the area of mobility, it was limited to kinematic pairs with 2 DoF. Subsequently, in 1923, Somov and Malyshev developed a new formula that is valid for spatial mechanisms and is given as follows:

$$DoF = 6(l - 1) - \sum_{k=1}^{5} k \cdot p_k \tag{5}$$

where l represents the number of independent loops and p_i denotes the DoF of kinematic pairs.

One can look at the works of Ruzinov [2] to get acquainted with the modifications of the Somov–Malyshev formula to tackle redundant mechanisms.

1.1.4 Grübler–Kutzbach

Now moving on to the Grübler–Kutzbach formula (G–K formula) presented in the year 1917, which was first derived by Grübler and later modified by Kutzbach. This classical formula [3, 4] used to calculate the DoF of a mechanism is arrived at by defining all possible DoF of the mechanism and then reducing it by the constraints imposed by the joints or kinematic pairs. The formulation is as follows:

$$DoF = \lambda(n - 1 - j) + \sum_{i=1}^{j} f_i \tag{6}$$

where $\lambda = 3$ for planar mechanism and $\lambda = 6$ for spatial mechanism with n links (including frame), j joints, and f_i joint freedom corresponding to the i-th joint, where, f is 1 for revolute and slider type of joints and 2 for cam-follower and pin-in-a-slot type of joints. If there are any passive member/s involved that reduce/s the theoretical DoF of the mechanism, then it is modified as follows:

$$DoF = \lambda(n - 1 - j) + \sum_{i=1}^{j} f_i - f_p \tag{7}$$

where f_p represents the passive DoF. One of the main limitations of this formula can be seen from here itself, i.e., the passive DoF has to be found out manually otherwise directly applying the Grübler–Kutzbach would lead to an incorrect result.

1.1.5 Freudenstein–Alizade

In the year 1975, Freudenstein and Alizade [5] collectively presented the mobility equation for mechanisms that contain mixed independent loops with variable general constraints. It basically focuses on the synthesis and analysis of robot manipulators. The formulation is as follows:

$$DoF = \sum_{i=1}^{M} p_i - \sum_{j=1}^{L} \lambda_i \tag{8}$$

where p_i represents the DoF of kinematic pairs, λ represents the loop motion parameters, L represents the number of independent loops and M represents the number of joints.

The formulations discussed so far represent traditional methodologies that are widely employed for analyzing mechanisms. While effective for many applications, these classical approaches exhibit limitations when dealing with mechanisms involving configuration-dependent mobility, overconstraints, or redundant DoF. Such cases highlight the inadequacy of traditional formulas in accurately capturing the true mobility of these mechanisms.

For instance, consider the augmented 4-bar mechanism illustrated in Fig. 5. The DoF of this mechanism varies depending on its configuration. In one scenario, where all the links connected to the frame are parallel to each other, the mechanism moves freely. However, in another scenario where these three links are not parallel, the mechanism becomes immobile. In this case, the three joints on the upper (horizontal) link move in such a way that the points fail to remain collinear, potentially causing the upper link to break. This illustrates that the DoF of the mechanism depends significantly on its configuration.

- For Grübler formula: Number of links, $n = 5$, Number of lower pairs, $p_1 = 6$ (revolute joint), Number of higher pairs, $p_2 = 0$

$$DoF = 3 \cdot (5 - 1) - 2 \cdot 6 = 0$$

- For Grübler–Kutzbach formula: Number of links, $n = 5$, Number of joints, $j = 6$ and Joint freedom, $f_i = 1 \, \forall \, 1 \le i \le 6$

$$DoF = 3 \cdot (5 - 1 - 6) + \sum_{i=1}^{6} 1 = 0$$

These calculations suggest a DoF of zero, but as demonstrated earlier, the augmented 4-bar mechanism has a configuration-dependent DoF, meaning its mobility changes with the arrangement of its links. This inconsistency highlights the inability of traditional methods to capture such nuances.

Another example is the PPPP linkage depicted in Fig. 12, which consists of three links (including the frame) that can slide relative to one another. Intuitively, it is evident that the links can move or slide freely while maintaining contact. Applying the traditional formulas to this mechanism produces the following results:

- For Grübler formula Number of links, $n = 3$, Number of lower pairs, $p_1 = 3$ (prismatic joint), Number of higher pairs, $p_2 = 0$

$$DoF = 3 \cdot (3 - 1) - 2 \cdot 3 = 0$$

- For Grübler–Kutzbach formula: Number of links, $n = 3$, Number of joints, $j = 3$ and Joint freedom, $f_i = 1\, \forall\, 1 \leq i \leq 3$

$$DoF = 3 \cdot (3 - 1 - 3) + \sum_{i=1}^{3} 1 = 0$$

Despite these results, the mechanism clearly exhibits mobility, further demonstrating the limitations of these traditional formulas.

Finally, consider the cam-follower mechanism depicted in Fig. 7, which contains a redundant DoF due to the rotation of the cam about its own center. The correct DoF for this mechanism should be 1, but the traditional formulas yield the following results:

- From Grübler formula: Number of links, $n = 4$ (including frame), Number of lower pairs, $p_1 = 3$ (revolute joint), Number of higher pairs $p_2 = 1$ (between cam and follower)

$$DoF = 3 \cdot (4 - 1) - 2 \cdot 3 - 1 = 2$$

- For Grübler–Kutzbach formula: Number of links, $n = 4$, Number of joints, $j = 4$ and Joint freedom, $f_i = 1\, \forall\, 1 \leq i \leq 3,\ f_4 = 2$

$$DoF = 3 \cdot (4 - 1 - 4) + \sum_{i=1}^{3} 1 + 2 = 2$$

In this case, the formulas fail to account for the redundant DoF, resulting in an incorrect DoF value of two.

These three examples illustrate the limitations of traditional methods in accurately analyzing mechanisms with configuration-dependent mobility, overconstraints, or redundancies. Numerous other mechanisms also exist where these methods fail to provide accurate results. These shortcomings underscore the need for new methodologies capable of addressing these challenges without requiring manual intervention during the calculation process.

1.2 Recent Formulations

Several adaptations and refinements of the traditional DoF calculation techniques have been proposed; nevertheless, they remain reliant on manual intervention and necessitate the employment of human intellect and considerable time when executing calculations. Since the beginning of the twenty-first century, new methodologies have emerged, built on a variety of innovative concepts, introducing a more systematic approach to DoF computations.

1.2.1 Weidong–Jingjun

In 2013, Weidong and Jingjun [6] proposed a new method wherein they devised their own concepts of real-constraint higher pairs, redundant-constraint higher pairs, full-constraint lower pairs, half-constraint lower pairs, and additional condition numbers. It was proposed as follows:

$$DoF = 3n - 2p_{LA} - p_{LH} - p_H - C \tag{9}$$

where n is the number of movable components, p_{LA} is the fully constrained low number, p_{LH} is the semi-constrained low number p_H is the real constrained high number and C is the motion additional condition number. The high and low numbers used in this method is analogous to the number of higher and lower pairs in Grübler's formula. Though this method accurately overcomes the limitations caused by over-constrained/redundant members, the condition number that satisfies the motion in each mechanism is different and thus needs to be analyzed separately which makes this method time-consuming.

1.2.2 Huang–Liu–Zeng

Huang et al. [7] proposed a method based on screw theory to find the mobility of mechanisms. They basically modified the Grübler–Kutzbach criteria to address all the complications associated with complicated mechanisms especially pertaining to modern parallel mechanisms listed by Gogu [20]. The formulation is as follows:

$$DoF = d(n - g - 1) + \sum_{i=1}^{g} f_i + \nu \tag{10}$$

where n is the number of links, g is the number of kinematic joints, f_i the number of freedoms of the i-th joint, ν is the number of parallel-redundant-constraint factors and $d = 6 - \lambda$ is the order of the mechanism where λ is the common constraint of the mechanism. As one can intuitively guess, the determination of the common and parallel-redundant-constraint factors makes this method mathematically intensive and difficult for common engineering applications.

1.2.3 Osgouie–Gard

One more recent formulation involves the Jacobian Method proposed by Osgouie and Gard [8]. This method involves using the Jacobian Matrix to predict the mobility of a mechanism. This method incorporates writing n different loop equations μ_j $(j = 1, 2 \ldots n)$ in terms of m independent parameters q_k $(k = 1, 2 \ldots m)$. To avoid the trivial case, it is ensured that the Jacobian is non-zero [9]. The Jacobian matrix would look like:

$$J\left(\frac{\mu_1, \dots \mu_n}{q_1, \dots q_m}\right) = \begin{bmatrix} \frac{\partial \mu_1}{\partial q_1} & \frac{\partial \mu_1}{\partial q_2} & \dots\dots & \frac{\partial \mu_1}{\partial q_m} \\ \frac{\partial \mu_2}{\partial q_1} & \frac{\partial \mu_2}{\partial q_2} & \dots\dots & \frac{\partial \mu_2}{\partial q_m} \\ \vdots & \vdots & \vdots \; \vdots & \vdots \\ \frac{\partial \mu_n}{\partial q_1} & \frac{\partial \mu_n}{\partial q_2} & \dots\dots & \frac{\partial \mu_n}{\partial q_m} \end{bmatrix} \tag{11}$$

Then, the degree of freedom of the system is given as

$$DoF = m - p \tag{12}$$

where p is either n or r (rank of the Jacobian matrix) depending on certain conditions. Though the Jacobian method gives correct results for all cases, it is very time-consuming and requires higher computational costs as it involves calculating the derivatives of the loop equations with respect to each parameter. One more limitation of this method is that it is only applicable to closed-loop linkages.

1.2.4 Rajashekhar–Debasish

The Zebra Crossing method is another latest method by Rajashekhar and Debasish [10] that calculates the DoF quickly from a zebra crossing diagram similar to the graph theory used in [11]. This method involves the calculation of loops and identifying the black, white, and gray patches that represent joints, links, and combinations of both. The whole method has been subdivided into three cases, i.e., if it is an open loop or closed loop, if the mechanism can traced out in a lane or not, and finally if black, white, and gray patches are present or not. Their work has successfully covered both the problem of open-loop linkage as well as virtual constraints.

From the survey of existing methodologies, it becomes apparent that recent approaches, while addressing the challenge of overconstrained or redundant linkages in mechanisms, have introduced a trade-off in terms of increased computational complexity and time consumption. Consequently, the primary imperative is to develop a technique that effectively bridges the gap between traditional and contemporary methods, facilitating the swift and accurate calculation of DoF for a given instantaneous configuration. Thus a novel and efficient method rooted in matrix algebra, grounded on the principles of Vector Space, is proposed in this work which offers a rapid and precise DoF determination for the specified configuration. This method entails the formulation of a set of equations, as elucidated in the subsequent section, transcribed in matrix notation, and subsequently subjected to requisite matrix operations to yield the final result. An additional advantage intrinsic to this approach lies in the matrix's potential to furnish supplementary insights into the system's configuration.

2 Methodology

Consider a linkage with $l + 1$ rigid links including the frame. A joint is defined at the connection of two links where there is relative motion between them. At this point, a concern might arise regarding the preservation of the structural rigidity of the terminal link, which remains unconnected to any other links within the context of open linkages. A comprehensive resolution to this issue is addressed in this section later. The movement (or displacement) of each joint can be resolved in two directions (assuming only 2D motion is allowed). One refers to these joints as *nodal points*. Thus every i-th nodal point can be represented by its possible displacement as U_{iX} and U_{iY} in the X and Y directions, respectively.

Postulate 1 *The nodal point located at the joint between two links connected to the base or frame may be excluded from consideration, as its displacement is zero along any direction.*

Postulate 2 *For any link with n nodal points, the relative displacement between any two nodes on the link along the line joining them must be zero. This is to ensure the rigidity of the link. A similar concept has been used by Wang et al. [12, 13] to propose the DoF of planar mechanisms based on the constraint normal line. The constraint equation, in accordance with Fig. 1a will be:*

$$\boxed{U_k^X \cos\theta + U_k^Y \sin\theta - U_j^X \cos\theta - U_j^Y \sin\theta = 0} \tag{13}$$

Postulate 3 *If any link has n nodal points, out of which m nodal points are collinear (refer Fig. 1b, c and d), then some of the previous constraints become redundant, while few other constraints are required to maintain collinearity.*

(i) Along the common line, it is necessary and sufficient to have a constraint between adjacent points, since a constraint on two points encompassing another point does not guarantee that the interior parts will not expand/contract while maintaining a sense of rigidity among the outer two points. For m collinear points, $m - 1$ equations are obtained to satisfy this condition.

(ii) To maintain collinearity between the m points after the displacement is given, the resultant displacements of the m points perpendicular to the common line must satisfy the following constraints:

$$\boxed{l_{jk} V_i^N - (l_{ij} + l_{jk}) V_j^N + l_{ij} V_k^N = 0} \tag{14}$$

where $0 < i < j < k \leq m$, $V_i^N = U_i^Y \cos\theta - U_i^X \sin\theta l_{ij}$ = distance between i-th and j-th collinear nodal point.

(iii) It is necessary to prove the collinearity of any 3 points simultaneously, so this constraint results in $\binom{m}{3}$ equations.

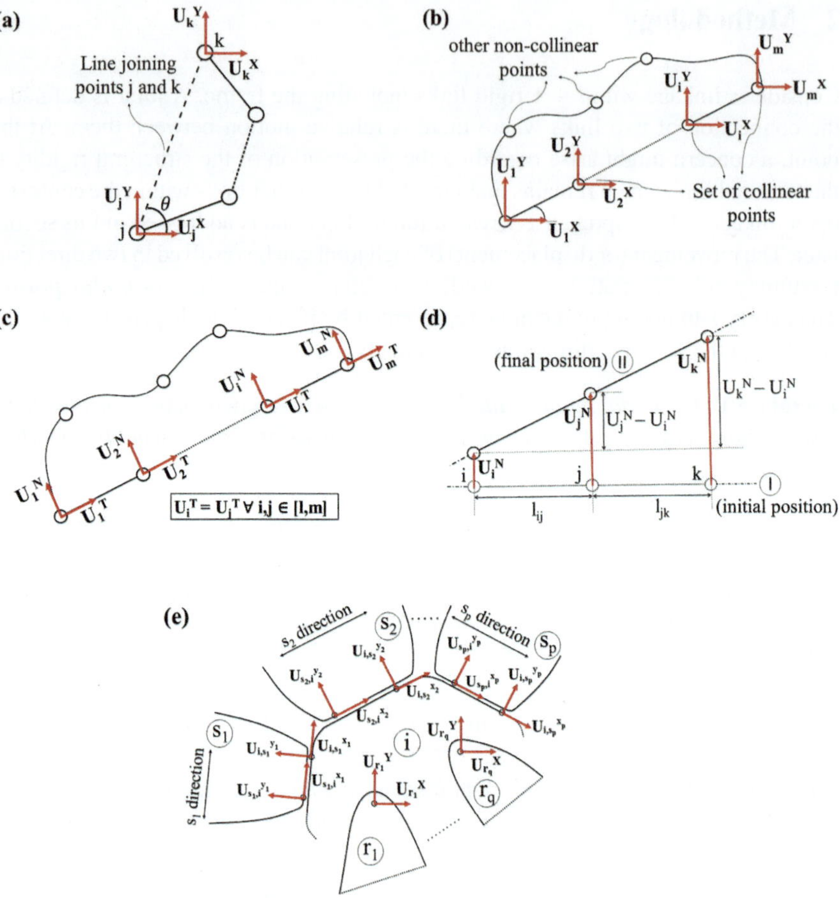

Fig. 1 **a** Displacement of two arbitrary nodal points. **b** A set of *m* collinear nodal points out of *n* number of total nodal points. **c** Resolved components of nodal displacement vectors along the line joining nodal points and perpendicular to it. **d** Formation of similar triangle by the nodal displacement components implying the collinearity condition. **e** A generalized body having *p* sliding joints and *q* revolute joints

Postulate 4 *Consider a body having revolute joints with q bodies and sliding joint with p bodies as shown in Fig. 1e. According to the definition of nodal points, it will have q nodal points at the revolute joints and infinite nodal points at each of the p sliding joints. The sliding nodal points will be present on both the bodies which are undergoing relative sliding motion. There will be two sets of infinite sliding nodal points. It is found that these infinite nodal points are redundant and can be replaced by only one nodal point which is at the periphery of the sliding contact of the respective bodies. The position is irrelevant and can be chosen as per the user's discretion. This results in two nodal points on the two bodies with sliding joint between them. The constraints required are as follows:*

(*i*) *For each sliding joint, the two nodal points as described above must have zero relative displacement in the direction perpendicular to the sliding motion to ensure contact between the two bodies at all times.*

$$\boxed{U_{rel}^{y_i} = 0} \quad (i = 1, 2, \ldots p) \tag{15}$$

where y_i refers to the generalized coordinate perpendicular to the i-th sliding direction.

(*ii*) *On a particular body, all nodal points should have the same displacement component as the sliding nodal point in the direction of sliding.*

$$\boxed{U_{s_1}^{i} = U_{s_2}^{i} = \cdots = U_{s_p}^{i} = U_{r_1}^{i} = U_{r_2}^{i} = \cdots = U_{r_q}^{i}} \quad (i = 1, 2, \ldots p) \tag{16}$$

where i denotes the i-th sliding direction, s refers to the sliding nodal point and r refers to the revolute nodal point.

(*iii*) *One can assume that the slider link does not undergo any rotation about its own axis, which implies that each sliding nodal point displacement is equal to one another, and can be completely replaced by a constraint in the direction perpendicular to sliding motion. Thus for a slider link, one only needs to take into consideration the p revolute joints and use constraints for rigidity of the link as described above, and a constraint that the resultant displacement perpendicular to sliding direction is zero.*

Postulate 5 *For any open link, that has an extension in its body through which output is extracted, then the definition of a nodal point can be adjusted and a nodal point is added at the point of extraction of output so that the rigidity constraint of the link is fulfilled.*

3 Mathematical Foundation and Physical Significance

If one applies all the postulates as described in the previous section to any mechanism whose nodal points have been identified, one can obtain a set of *linear equations* which represent the constrained displacement equations of each *nodal point*. The same set of equations can be written in the matrix form as shown below:

$$\underbrace{\begin{bmatrix} A_{11} & A_{12} & \cdots\cdots & A_{1(2n)} \\ A_{21} & A_{22} & \cdots\cdots & A_{2(2n)} \\ \vdots & \vdots & \vdots\ \ \vdots & \vdots \\ A_{l1} & A_{l2} & \cdots\cdots & A_{l(2n)} \end{bmatrix}}_{\text{Constraint Matrix}} \underbrace{\begin{bmatrix} U_1^X \\ U_1^Y \\ \vdots \\ U_n^X \\ U_n^Y \end{bmatrix}}_{\text{Displacement Vector}} = \underbrace{\begin{bmatrix} 0 \\ 0 \\ 0 \\ \vdots \\ 0 \end{bmatrix}}_{\text{Zero Vector}}$$

From the above matrix notation, one can infer the following points:

(i) The left-hand side represents a *linear transformation* $\mathcal{T} : \mathbb{R}^{2n} \rightarrow \mathbb{R}^{l}$ (where $\mathbb{R}^{2n}$ and $\mathbb{R}^{l}$ represents the $2n$ and l-dimensional real space, respectively), and its matrix representation is a $l \times 2n$ matrix such that $\mathcal{T}(\mathbf{u}) = A\mathbf{u}\,\forall\,\mathbf{u} \in \mathbb{R}^{2n}$, where A is a matrix of order $l \times 2n$ and $\mathbf{u}$ is a column vector.

(ii) The right-hand side is a zero vector of order l and it represents the transformed version of the displacement vector, $\mathbf{u}$.

(iii) The matrix that transforms the displacement vector from $\mathbb{R}^{2n}$ to $\mathbb{R}^{l}$ is termed as the *Constraint Matrix*, represented by A. It is a *rectangular matrix* with l rows and $2n$ columns and if $l = 2n$, then A becomes a *square matrix*.

(iv) The *nullspace* ($\mathcal{N}$) of the linear transformation $\mathcal{T}$ is defined as

$$\mathcal{N}(\mathcal{T}) = \mathbf{u} \in \mathbb{R}^{2n} \;\mid\; \mathcal{T}(\mathbf{u}) = \mathbf{0}_l$$

(v) Let the range of a matrix be represented as $\mathcal{R}$. Since A is the matrix representation of $\mathcal{T}$, it follows that $\mathcal{N}(\mathcal{T}) = \mathcal{N}(A)$ and $\mathcal{R}(\mathcal{T}) = \mathcal{R}(A)$. After performing row operations on the Constraint Matrix, one gets the Reduced Row Echelon Form (RREF) of A which is represented as RREF(A).

(vi) From the concepts of Linear Algebra, it can be easily shown that the $\mathcal{N}(A)$ is same as the $\mathcal{N}(RREF(A))$. Thus, $\mathcal{N}(\mathcal{T}) = \mathcal{N}(A) = \mathcal{N}(RREF(A)))$. One more important point to note over here is that $\mathcal{N}(RREF(A))$ forms a subspace of $\mathbb{R}^{2n}$.

$\mathcal{N}(\mathcal{T})$ forms a vector space, $\mathbb{V}$ which contains all the possible solutions of the equation $A\mathbf{u} = \mathbf{0}$. From the last section, since $\mathcal{N}(\mathcal{T}) = \mathcal{N}(RREF(A))$, the nullspace of RREF($A$) consists of all vectors $\mathbf{u}$ such that $A\mathbf{u} = \mathbf{0}$. This implies that any displacement pertaining to the mechanism will be included in the span of the nullspace. Thus, one needs to find the set of *linearly independent vectors* which *spans* the nullspace. As one might be aware, this set of linearly independent vectors is known as the *basis vectors*. In the very beginning, while defining DoF, it has been already mentioned that it is the minimum number of *independent* variables required to locate the position of each component of a rigid body system. Hence the word *independent* is important over here. The number of basis vectors thus represents the minimum number of independent variables required to define the displacement of the whole mechanism. In other words, it denotes the *degrees of freedom* of the space, more precisely to say the nullspace. This means, *the dimension of the nullspace, often referred to as the nullity, of the constraint matrix represents the DoF of the mechanism.*

From the essence of Linear Algebra, one can treat a linear transformation as a row picture (intersection of planes) or as a column picture (combination of columns). Here, the column picture is used as a foundation to derive the final result. The column space of A, represented by col(A) consists of all linear combinations of the columns of A and one seeks to find such combinations that produce the zero column vector. Before proceeding further, one needs to compute the RREF(A). Its significance will be clear in a few minutes. The RREF would look like the one shown below:

$$\boldsymbol{RREF(A)} = \begin{bmatrix} ① 0 * 0 \dots \dots 0 \\ 0 ① * 0 \dots \dots 0 \\ 0\ 0\ 0 ① \dots \dots 0 \\ \vdots \vdots \vdots \vdots \vdots \vdots ① \\ 0\ 0\ 0\ 0\ 0\ 0\ 0 \end{bmatrix}$$

$$\underbrace{\qquad\qquad}_{\textit{contains r pivots}}$$

The first non-zero entries of the rows, i.e., the *pivots* have been encircled in the above matrix representation. Since it is already known that $\mathcal{N}(\mathcal{T}) = \mathcal{N}(RREF(A))$ so instead of looking at $A\mathbf{u} = \mathbf{0}$ one can look at RREF(A)$\mathbf{u} = \mathbf{0}$. An important implication can be found from this equation if one expresses this in matrix representation.

$$\underbrace{\begin{bmatrix} \mathbf{1\ 0 * 0} \dots \dots \mathbf{0} \\ \mathbf{0\ 0 * 0} \dots \dots \mathbf{0} \\ \mathbf{0\ 0}\ 0\ \mathbf{1} \dots \dots \mathbf{0} \\ \vdots \vdots \vdots \vdots \vdots \vdots \mathbf{1} \\ \mathbf{0\ 0\ 0\ 0}\ 0\ 0\ 0 \end{bmatrix}}_{\textit{contains r pivots}} \begin{Bmatrix} U_{1X} \\ U_{1Y} \\ \vdots \\ U_{nX} \\ U_{nY} \end{Bmatrix} = \begin{Bmatrix} 0 \\ 0 \\ 0 \\ \vdots \\ 0 \end{Bmatrix}$$

One can easily inspect that there are r columns that contain the pivot variables (mentioned in bold). This is nothing but the dimension of $\mathcal{R}(A)$, which is also known as rank of A. The other set of columns that do not contain the pivot variables consists of free variables. They are linearly independent and their combination can be used to span the whole nullspace. The number of such columns consisting of free variables can be easily found by subtracting the columns with pivots from the total number of columns. The number of such independent columns is nothing but the *nullity* of the *constraint matrix.*

$$\boxed{dim(\mathcal{N}) = 2n - r} \tag{17}$$

This is supported by the Fundamental Theorem of Linear Algebra [14], which states that the nullity of a matrix is given by the difference between the number of columns of the matrix and the rank of the matrix.

$$dim(\mathcal{N}(T)) = 2n - dim(\mathcal{R}(T)) \tag{18}$$

$$dim(\mathcal{N}(A)) = 2n - dim(\mathcal{R}(A)) \tag{19}$$

$$dim(\mathcal{N}(RREF(\text{A})) = 2n - dim(\mathcal{R}(RREF(\text{A}))) \tag{20}$$

$$\boxed{dim(\mathcal{N}(RREF(\text{A})) = 2n - r} \tag{21}$$

There is an alternative way to understand the whole concept that makes more physical sense. The displacement vector, $\mathbf{u}$ initially lies in a $2n$-dimensional space. The dimension of this space, i.e., $2n$ represents the possible DoF of the mechanism without constraints. This basically represents the stage where the creation of the nodal points takes place without the presence of any links. As one introduces the links, the displacement vector is transformed from this $2n$-dimensional space to l-dimensional space, and some of this DoF gets restricted due to the constraints imposed by the links. The number of such constraints is given by r. So the actual DoF is given by the dimension of the initial space where the displacement vector lies, i.e., $2n$ minus the rank of the constraint matrix, r.

$$\text{Actual DoF} = \underbrace{\text{DoF possible of all nodal points without links}}_{2n} - \underbrace{\text{Number of constraints due to links}}_{r \text{ or } \mathcal{R}}$$

4 Results and Discussion

To validate the method introduced in this study and emphasize its efficiency in DoF calculations, nine distinct cases are examined, spanning from the basic four-bar mechanism to overconstrained and redundant linkages [15–17]. For the sake of simplicity, the paper does not illustrate the stepwise reduction of the constraint matrix to its RREF. The comprehensive evaluation of these cases underscores the robustness and practical applicability of the proposed method in accurately determining DoF, offering a valuable tool for the analysis of complex mechanical systems. Prior to delving into the case studies, it is imperative to elucidate the steps employed for determining the DoF of the mechanisms. The steps are outlined herein for the benefit of the readers.

Steps involved in the calculation of the DoF are as follows:

(i) Initially, it is essential to identify the nodal points of the mechanism based on the definition, taking into account Postulate 1.
(ii) One should then apply the remaining postulates accordingly to obtain a system of linear equations. This set of linear equations can then be written in matrix form as mentioned in Sect. 3.
(iii) At this point, one should convert the constraint matrix, A into its RREF. The rank of the RREF, $\mathcal{R}(\mathbf{RREF}(A))$ can then be calculated which is nothing but the number of non-zero rows of the RREF.
(iv) Finally, one can calculate the DoF of the mechanism by taking the difference between the number of columns and the $\mathcal{R}(\mathbf{RREF}(A))$.

By adhering to this straightforward four-step procedure, one can easily ascertain the DoF of planar mechanisms with ease. To maintain clarity and streamline the presentation of the case studies, the transformation of the constraint matrix into its RREF has been omitted.

4.1 4-Bar Planar Linkage

A four-bar planar linkage, more commonly known as a four-bar mechanism, consists of three movable links and one fixed link, all connected through revolute joints. It consists of three movable links and one fixed link all connected through revolute joints. Figure 2 depicts a four-bar linkage where link 1 can be assumed as the driver/input link which rotates about the revolute joint and transmits the motion to the driven/output link 3 with the help of coupler link 2. This mechanism is commonly employed for the conversion of revolute motion into reciprocating motion and vice versa. It also finds extensive application in path generation tasks, where precision in defining the trajectory of an end effector is of paramount importance.

Now, moving over to the formulation of displacement equations for each joint corresponding to every link, one can observe:

(i) **Link 1**: $U_1^X \cos\theta_1 + U_1^Y \sin\theta_1 = 0$
(ii) **Link 2**: $U_2^X \cos\theta_2 + U_2^Y \sin\theta_2 - U_1^X \cos\theta_2 - U_1^Y \sin\theta_2 = 0$
(iii) **Link 3**: $U_2^X \cos\theta_3 + U_2^Y \sin\theta_3 = 0$.

Now, writing the system of equations in Matrix form, one gets

$$\begin{bmatrix} \cos\theta_1 & \sin\theta_1 & 0 & 0 \\ -\cos\theta_2 & -\sin\theta_2 & \cos\theta_2 & \sin\theta_2 \\ 0 & 0 & \cos\theta_3 & \sin\theta_3 \end{bmatrix} \begin{Bmatrix} U_1^X \\ U_1^Y \\ U_2^X \\ U_2^Y \end{Bmatrix} = \begin{Bmatrix} 0 \\ 0 \\ 0 \end{Bmatrix} \tag{22}$$

After performing row operations on the Constraint Matrix, the RREF obtained is as follows:

$$\begin{bmatrix} 1\,0\,0 & \frac{\sin\theta_1 \sin(\theta_3-\theta_2)}{\cos\theta_3 \sin(\theta_1-\theta_2)} \\ 0\,1\,0 & \frac{\cos\theta_1 \sin(\theta_2-\theta_3)}{\cos\theta_3 \sin(\theta_1-\theta_2)} \\ 0\,0\,1 & \tan\theta_3 \end{bmatrix} \tag{23}$$

For any value of θ_1, θ_2 and θ_3 one finds that the nullity of the matrix is 1, which means the DoF of the mechanism is 1.

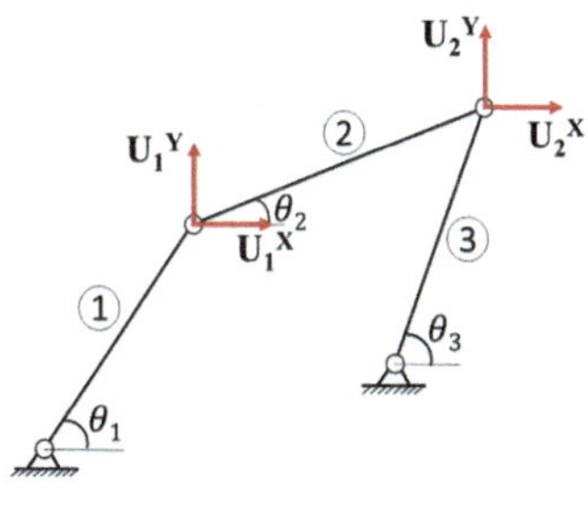

Fig. 2 Four-bar mechanism

Some important implications from the Constraint Matrix and its RREF for the four-bar mechanism are as follows:

(i) One can observe that $\mathcal{T}(\mathbf{u}) = A\mathbf{u}$ represents the transformation $\mathcal{T} : \mathbb{R}^4 \rightarrow \mathbb{R}^3$ thereby reducing the dimension of the vector space. This means, within the four-dimensional space of all possible vectors $\mathbf{u}$, the solutions to $A\mathbf{u} = 0$ form a three-dimensional subspace which is nothing but the *nullspace* of A.
(ii) There are 3 pivots in the RREF and they correspond to columns 1, 2, and 3. This implies column 4 is a dependent column, whereas columns 1, 2, and 3 are independent columns and column 4 lies in the plane generated by these three columns.
(iii) One can prominently observe that among all the columns, only one column, i.e., column 4 is influenced by the input variables (in this case only θ_3. Hence, the DoF aligns with the count of such "meaningful" or "significant" columns.
(iv) Since the dependent column does not have any say over the nullity, the DoF of this mechanism does not change whatever be the value of $\tan\theta_3$ or θ_3.
(v) The DoF does not depend at all on the dimensions of the link lengths as one can observe from the matrix form or its RREF.
(vi) There is at least one pair of elements in each row from which one can extract information about the orientation of the corresponding link. For example, row 2 contains cosine and sine values of the angle that link 2 makes with the horizontal.

One can extend similar implications to other mechanisms. The observations outlined above provide valuable insights into the information that can be derived from the constraint matrix and its RREF. An important general implication arises from this analysis: if the number of columns in the constraint matrix exceeds the number of rows, it cannot be a superstructure, indicating that it can be classified as either a mechanism or a structure. This conclusion can be easily verified from the definition of the nullity, which is the difference between the number of columns and the rank of the constraint matrix, and the fact that the rank is always less than or equal to the number of rows. As a result, the difference between the number of columns and the rank will consistently yield a non-negative value, underscoring the system's inherent nature as either a mechanism or a structure. While the insights presented may hold limited applicability within the domain of mechanical engineering, they bear noteworthy significance for civil engineers engaged in the design and assessment of diverse superstructures.

4.2 5-Bar Planar Linkage

In the pursuit of incorporating a 2 DoF mechanism into the mechanical system, the most straightforward and easy approach involves the integration of a five-bar linkage. The five-bar linkage is composed of five links, each connected to its adjacent counterparts through revolute joints, as depicted in Fig. 3. This mechanism finds applications in various engineering domains, including the domain of 3D printing, where it plays

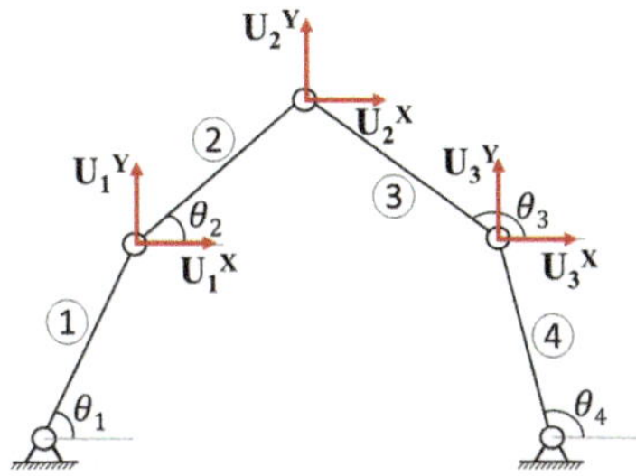

Fig. 3 Five-bar mechanism

a pivotal role in the design of 5-bar Parallel Scara Robots. The augmentation of an additional degree of freedom, in contrast to the four-bar linkage, imparts the system with heightened versatility and freedom of movement within its operational space.

One innovative application of the five-bar mechanism is in the development of prosthetic knee joints, as demonstrated by Yuanxi Sun et al. in their paper entitled "Design and Analysis of a Geared Five-Bar Mechanism for Prosthetic Knee Joints" [18]. This utilization underscores the adaptability and effectiveness of the five-bar linkage in addressing real-world challenges.

Now, shifting the focus toward the formulation of displacement equations for each joint, corresponding to every link, it becomes evident that:

(i) **Link 1**: $U_1^X \cos\theta_1 + U_1^Y \sin\theta_1 = 0$
(ii) **Link 2**: $U_2^X \cos\theta_2 + U_2^Y \sin\theta_2 - U_1^X \cos\theta_2 - U_1^Y \sin\theta_2 = 0$
(iii) **Link 3**: $-U_3^X \cos\theta_3 - U_3^Y \sin\theta_3 + U_2^X \cos\theta_3 + U_2^Y \sin\theta_3 = 0$
(iv) **Link 4**: $U_3^X \cos\theta_4 + U_3^Y \sin\theta_4 = 0$.

Now, writing the system of equations in Matrix form, one gets

$$\begin{bmatrix} \cos\theta_1 & \sin\theta_1 & 0 & 0 & 0 & 0 \\ -\cos\theta_2 & -\sin\theta_2 & \cos\theta_2 & \sin\theta_2 & 0 & 0 \\ 0 & 0 & \cos\theta_3 & \sin\theta_3 & -\cos\theta_3 & -\sin\theta_3 \\ 0 & 0 & 0 & 0 & \cos\theta_4 & \sin\theta_4 \end{bmatrix} \begin{Bmatrix} U_1^X \\ U_1^Y \\ U_2^X \\ U_2^Y \\ U_3^X \\ U_3^Y \end{Bmatrix} = \begin{Bmatrix} 0 \\ 0 \\ 0 \\ 0 \end{Bmatrix} \tag{24}$$

After performing row operations on the Constraint Matrix, the RREF obtained is as follows:

$$\begin{bmatrix} 1\,0\,0 & \frac{\sin\theta_1 \sin(\theta_2-\theta_3)}{\cos\theta_3 \sin(\theta_2-\theta_1)} & 0 & \frac{\sin\theta_1 \cos\theta_2 \sin(\theta_3-\theta_4)}{\cos\theta_3 \cos\theta_4 \sin(\theta_2-\theta_1)} \\ 0\,1\,0 & -\frac{\cos\theta_1 \sin(\theta_2-\theta_3)}{\cos\theta_3 \sin(\theta_2-\theta_1)} & 0 & -\frac{\cos\theta_1 \cos\theta_2 \sin(\theta_3-\theta_4)}{\cos\theta_3 \cos\theta_4 \sin(\theta_2-\theta_1)} \\ 0\,0\,1 & \tan\theta_3 & 0 & \tan\theta_3 - \tan\theta_4 \\ 0\,0\,0 & 0 & 1 & \tan\theta_4 \end{bmatrix} \tag{25}$$

For any value of $\theta_1, \theta_2, \theta_3$ and θ_4 one finds that the nullity of the matrix is 2, which means the DoF of the mechanism is 2.

4.3 Single Slider–Crank Mechanism

The slider–crank mechanism is one of the most commonly used mechanisms in both industrial and everyday applications. This mechanism serves the crucial purpose of converting revolute motion into translatory motion or vice versa, playing a pivotal role in a diverse array of mechanical systems, from hand pumps to internal combustion engines.

In this mechanism, as depicted in Fig. 4, link 3 is the slider, whereas link 2 effectively transmits motion between link 1 and the slider, also known as the coupler link. In the context of the slider–crank mechanism, it is noteworthy that the designation of the driver and driven links can vary, contingent upon the specific application of this mechanical system. Some noteworthy inversions of the single slider–crank mechanism are the Whitworth quick return mechanism and the Oscillating cylinder engine.

Now, proceeding with the formulation of displacement equations for each joint corresponding to every link, one can observe:

(i) **Link 1**: $U_1^X \cos\theta_1 + U_1^Y \sin\theta_1 = 0$
(ii) **Link 2**: $U_1^X \cos\theta_2 + U_1^Y \sin\theta_2 - U_2^X \cos\theta_2 - U_2^Y \sin\theta_2 = 0$
(iii) **Link 3**: $U_2^Y = 0$.

Now, writing the system of equations in Matrix form, one gets

$$\begin{bmatrix} \cos\theta_1 & \sin\theta_1 & 0 & 0 \\ \cos\theta_2 & \sin\theta_2 & -\cos\theta_2 & -\sin\theta_2 \\ 0 & 0 & 0 & 1 \end{bmatrix} \begin{Bmatrix} U_1^X \\ U_1^Y \\ U_2^X \\ U_2^Y \end{Bmatrix} = \begin{Bmatrix} 0 \\ 0 \\ 0 \end{Bmatrix} \tag{26}$$

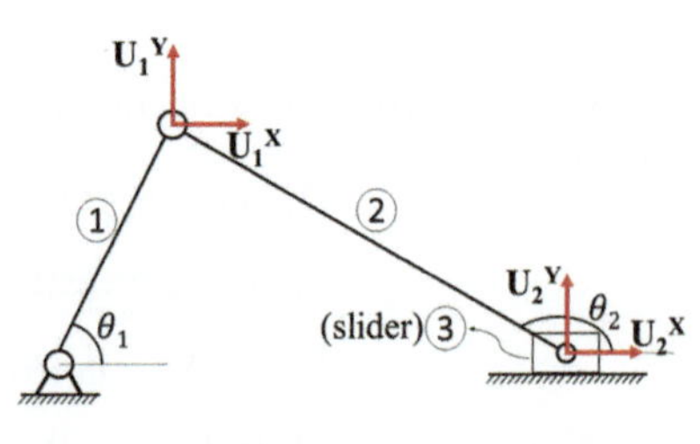

Fig. 4 Slider–crank mechanism

After performing row operations on the Constraint Matrix, the RREF obtained is as follows:

$$\begin{bmatrix} 1 & 0 & \frac{\sin\theta_1 \cos\theta_2}{\sin(\theta_2-\theta_1)} & 0 \\ 0 & 1 & -\frac{\cos\theta_1 \cos\theta_2}{\sin(\theta_2-\theta_1)} & 0 \\ 0 & 0 & 0 & 1 \end{bmatrix} \tag{27}$$

For any value of θ_1 and θ_2 one finds that the nullity of the matrix is 1, which means the DoF of the mechanism is 1.

4.4 Augmented 4-Bar Mechanism

In Fig. 5, a classic example of Grübler's Paradox is illustrated. The figure depicts a four-bar mechanism with the introduction of an intermediary link that connects from the coupler link to the frame, effectively transforming it into a five-bar linkage. This specific modification is typically employed to enhance the structural rigidity of the original four-bar mechanism, and it is a common practice in the design of parallel manipulators. What is particularly intriguing about this arrangement is that the outcome remains unchanged even if additional bars connecting the coupler link and the frame are incorporated into the mechanism, provided that all the links, apart from the coupler and the frame, maintain parallel alignment. The application of Grübler's formula leads to an unexpected result in these kinds of symmetrically oriented mechanisms.

The formulation of displacement equations for each joint corresponding to every link is as follows:

(i) **Link 1**: $U_1^X \cos\theta_1 + U_1^Y \sin\theta_1 = 0$
(ii) **Link 2**: $U_2^X \cos\theta_2 + U_2^Y \sin\theta_2 = 0$
(iii) **Link 3**: $U_3^X \cos\theta_3 + U_3^Y \sin\theta_3 = 0$
(iv) **Link 4.1**: $U_2^X - U_1^X = 0$
(v) **Link 4.2**: $U_3^X - U_2^X = 0$
(vi) **Collinearity Condition**: $bU_1^Y - (a+b)U_2^Y + aU_3^Y = 0$.

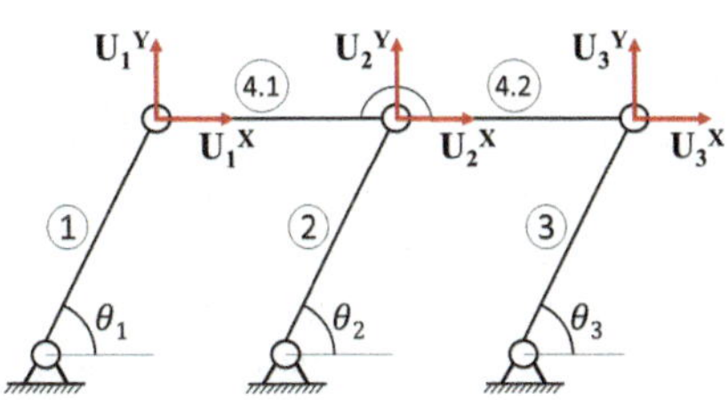

Fig. 5 Augmented 4-bar mechanism

Now, writing the system of equations in Matrix form, one gets

$$\begin{bmatrix} \cos\theta_1 & \sin\theta_1 & 0 & 0 & 0 & 0 \\ 0 & 0 & \cos\theta_2 & \sin\theta_2 & 0 & 0 \\ 0 & 0 & 0 & 0 & \cos\theta_3 & \sin\theta_3 \\ -1 & 0 & 1 & 0 & 0 & 0 \\ 0 & 0 & -1 & 0 & 1 & 0 \\ 0 & b & 0 & -(a+b) & 0 & a \end{bmatrix} \begin{Bmatrix} U_1^X \\ U_1^Y \\ U_2^X \\ U_2^Y \\ U_3^X \\ U_3^Y \end{Bmatrix} = \begin{Bmatrix} 0 \\ 0 \\ 0 \\ 0 \\ 0 \\ 0 \end{Bmatrix} \tag{28}$$

Now, there can be two cases:

(i) $\theta_1 = \theta_2 = \theta_3 = \theta$ (All the bars are parallel)
After performing row operations on the Constraint Matrix, the RREF obtained is as follows:

$$\begin{bmatrix} 1\,0\,0\,0\,0 & \tan\theta \\ 0\,1\,0\,0\,0 & -1 \\ 0\,0\,1\,0\,0 & \tan\theta \\ 0\,0\,0\,1\,0 & -1 \\ 0\,0\,0\,0\,1 & \tan\theta \\ 0\,0\,0\,0\,0 & 0 \end{bmatrix} \tag{29}$$

For any general case, the nullity comes out to be 1, meaning the DoF of the mechanism is 1.

(ii) $\theta_1 \neq \theta_2 \neq \theta_3$
After performing row operations on the Constraint Matrix, the RREF obtained is as follows:

$$\begin{bmatrix} 1\,0\,0\,0\,0\,0 \\ 0\,1\,0\,0\,0\,0 \\ 0\,0\,1\,0\,0\,0 \\ 0\,0\,0\,1\,0\,0 \\ 0\,0\,0\,0\,1\,0 \\ 0\,0\,0\,0\,0\,1 \end{bmatrix} \tag{30}$$

For any general case, the nullity comes out to be 0, meaning the DoF of the mechanism is 0.

Note: *If one apply the Grubler ($L = 5$, $P_1 = 6$, $P_2 = 0$) and Kutzbach ($N = 5$, $j = 6$, $\sum_{i=1}^{j} f_i = 6$) formula here, both will give DoF as 0. But one can see that depending upon the configuration the DoF of this system can be either 1 or 0. Thus the traditional formulas typically fail over here.*

4.5 Quick Return Mechanism

In Sect. 4.3, the analysis focused on the single slider–crank mechanism. This section now extends the examination to an inversion of the double slider–crank mechanism, as exemplified in Fig. 6, which embodies the quick return mechanism. This specific mechanism, employed in shaper and press machines, facilitates efficient operation by delivering reciprocating motion at its output (represented by link 5) while being actuated by a revolute input through link 1.

Now, proceeding with the formulation of displacement equations for each joint corresponding to every link, one can observe:

(i) **Link 1**: $U_1^X \cos\theta_1 + U_1^Y \sin\theta_1 = 0$
(ii) **Link 2**: $U_2^X \cos\theta_2 + U_2^Y \sin\theta_2 = 0$
(iii) **Link 3**: $U_1^X \sin\theta_2 - U_1^Y \cos\theta_2 = 0$
(iv) **Link 4**: $-U_2^X \cos\theta_3 + U_2^Y \sin\theta_3 + U_3^X \cos\theta_3 - U_3^Y \sin\theta_3$
(v) **Link 5**: $U_3^y = 0$.

Now, writing the system of equations in Matrix form, one gets

$$\begin{bmatrix} \cos\theta_1 & \sin\theta_1 & 0 & 0 & 0 & 0 \\ 0 & 0 & \cos\theta_2 & \sin\theta_2 & 0 & 0 \\ \sin\theta_2 & -\cos\theta_2 & 0 & 0 & 0 & 0 \\ 0 & 0 & -\cos\theta_3 & \sin\theta_3 & \cos\theta_3 & -\sin\theta_3 \\ 0 & 0 & 0 & 0 & 0 & 1 \end{bmatrix} \begin{Bmatrix} U_1^X \\ U_1^Y \\ U_2^X \\ U_2^Y \\ U_3^X \\ U_3^Y \end{Bmatrix} = \begin{Bmatrix} 0 \\ 0 \\ 0 \\ 0 \\ 0 \end{Bmatrix} \tag{31}$$

After performing row operations on the Constraint Matrix, the RREF obtained is as follows:

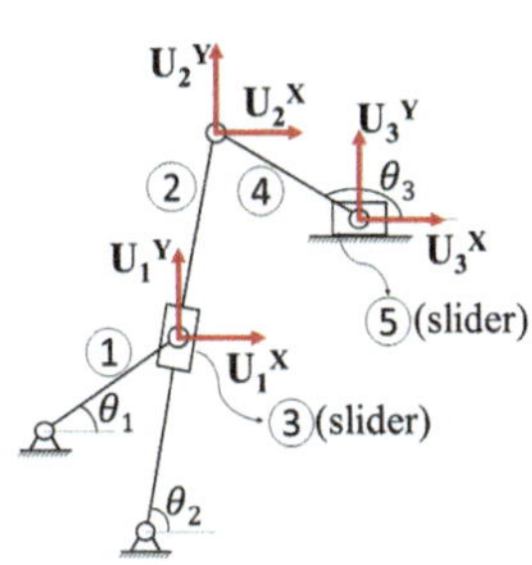

Fig. 6 Quick return mechanism

$$\begin{bmatrix} 1\,0\,0\,0 & 0 & 0 \\ 0\,1\,0\,0 & 0 & 0 \\ 0\,0\,1\,0 & \frac{-\cos\theta_3 \sin\theta_2}{\sin(\theta_3+\theta_2)} & 0 \\ 0\,0\,0\,1 & \frac{\cos\theta_3 \cos\theta_2}{\sin(\theta_3+\theta_2)} & 0 \\ 0\,0\,0\,0 & 0 & 1 \end{bmatrix} \tag{32}$$

For any value of θ_1, θ_2 and θ_3 one finds the nullity of the matrix is 1, which means the DoF of the mechanism is 1.

4.6 Cam Follower Mechanism

The inception of the simplified version of the Cam Follower mechanism can be traced back to as early as the third century. This mechanism fundamentally comprises three integral components: Firstly, the cam, which may undergo either revolute or reciprocating motion, serves as the driving component. Secondly, the follower which is capable of revolute, reciprocating, or oscillating motion, assumes the role of the driven member. And lastly, the stationary frame, which not only supports the cam but also provides the necessary guidance to facilitate the motion of the follower. As shown in Fig. 7, the cam (link 3), driven by circular motion, maintains continuous contact with the follower (link 2). The precise profile and orientation of the cam dictate the resulting motion exhibited by the follower. The most important thing to note here is that the cam and follower mechanism involves the higher pair of contact between the cam and the follower, i.e., the contact occurs at isolated points. This mechanism is primarily employed for the purpose of generating irregular motion patterns. Its primary applications include the operation of the inlet and exhaust valves of IC engines, retractable landing gear in aircraft, printing operations, etc.

Now, proceeding with the formulation of displacement equations for each joint corresponding to every link, one can observe:

(i) **Link 1**: $U_1^X \cos\theta_1 + U_1^Y \sin\theta_1 = 0$
(ii) **Link 2**: $-U_1^X \cos\theta_1 + U_1^Y \sin\theta_1 - U_2^X \cos\theta_2 - U_2^Y \sin\theta_2 = 0$
(iii) **Link 3**: $U_2^X \sin\theta_2 + U_2^Y \cos\theta_2 = 0$.

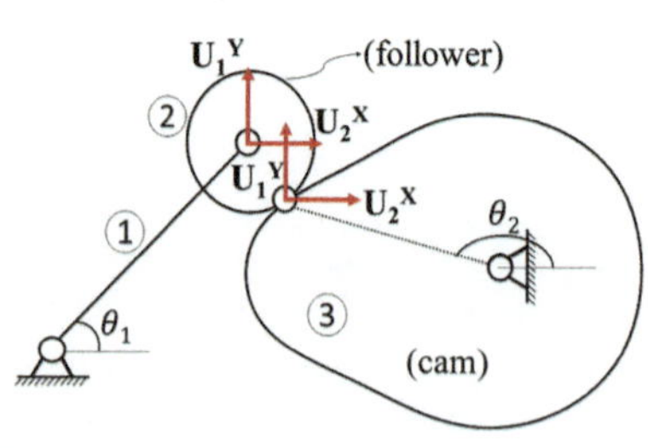

Fig. 7 Cam-follower mechanism

Now, writing the system of equations in Matrix form, one gets

$$\begin{bmatrix} \cos\theta_1 & \sin\theta_1 & 0 & 0 \\ -\cos\theta_1 & \sin\theta_1 & -\cos\theta_2 & -\sin\theta_2 \\ 0 & 0 & \sin\theta_2 & \cos\theta_2 \end{bmatrix} \begin{Bmatrix} U_1^X \\ U_1^Y \\ U_2^X \\ U_2^Y \end{Bmatrix} = \begin{Bmatrix} 0 \\ 0 \\ 0 \end{Bmatrix} \tag{33}$$

After performing row operations on the Constraint Matrix, the RREF obtained is as follows:

$$\begin{bmatrix} 1\,0\,0 & \frac{-\cos 2\theta_2}{2\sin\theta_1\cos\theta_1} \\ 0\,1\,0 & \frac{\cos 2\theta_2}{2\sin\theta_1\sin\theta_2} \\ 0\,0\,1 & \frac{\cos\theta_2}{\sin\theta_2} \end{bmatrix} \tag{34}$$

For any value of θ_1 and θ_2 one finds the nullity of the matrix is 1, which means the DoF of the mechanism is 1.

Note: *If one apply the Grubler* ($L = 4$, $P_1 = 3$, $P_2 = 1$) *and Kutzbach* ($N = 4$, $j = 4$, $\sum_{i=1}^{j} f_i = 5$) *formula here, both will give DoF as 2. But actually, the DoF of this system is 1. Here, the traditional formulas fail to detect the presence of redundant DoF. Thus the traditional formulas fail over here.*

4.7 Four-Bar Planar End-Effector

Recent research in DoF calculations has frequently relied on methodologies involving closed loops, exemplified by the Jacobian method by Osgouie et al. [8]. However, in the context of open-loop systems, commonly found in robotic manipulators, these approaches encounter challenges due to the inability to identify independent loop equations. The work included in this paper overcomes these limitations. A simple example of an open-loop system with 4 links and 3 revolute joints connecting them is presented in Fig. 8, illustrating the formulation of displacement equations for each joint, in accordance with Postulate 4, which includes the creation of an additional nodal point to ensure link rigidity.

The displacement equations are as follows:

(i) **Link 1**: $U_1^X \cos\theta_1 + U_1^Y \sin\theta_1 = 0$
(ii) **Link 2**: $U_1^X \cos\theta_2 + U_1^Y \sin\theta_2 - U_2^X \cos\theta_2 - U_2^Y \sin\theta_2 = 0$
(iii) **Link 3**: $U_2^X \cos\theta_3 + U_2^Y \sin\theta_3 - U_3^X \cos\theta_3 - U_3^Y \sin\theta_3 = 0$.

Fig. 8 Linkage having open-loop configuration

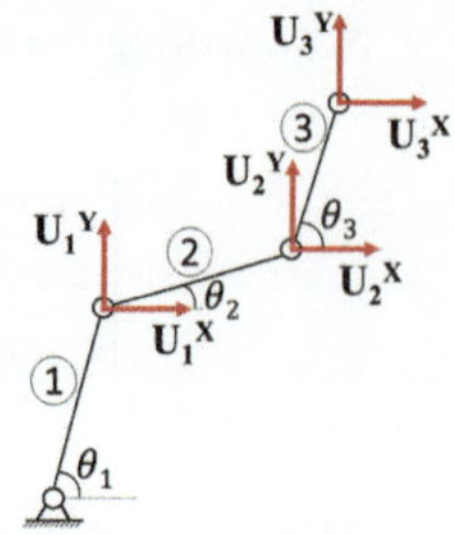

Now, writing the system of equations in Matrix form, one gets

$$\begin{bmatrix} \cos\theta_1 & \sin\theta_1 & 0 & 0 & 0 & 0 \\ \cos\theta_2 & \sin\theta_2 & -\cos\theta_2 & -\sin\theta_2 & 0 & 0 \\ 0 & 0 & \cos\theta_3 & \sin\theta_3 & -\cos\theta_3 & -\sin\theta_3 \end{bmatrix} \begin{Bmatrix} U_1^X \\ U_1^Y \\ U_2^X \\ U_2^Y \\ U_3^X \\ U_3^Y \end{Bmatrix} = \begin{Bmatrix} 0 \\ 0 \\ 0 \end{Bmatrix} \tag{35}$$

After performing row operations on the Constraint Matrix, the RREF obtained is as follows:

$$\begin{bmatrix} 1\,0\,0 & \frac{\sin\theta_1 \sin(\theta_2-\theta_3)}{\cos\theta_3 \sin(\theta_2-\theta_1)} & \frac{\cos\theta_2 \sin\theta_1}{\sin(\theta_2-\theta_1)} & \frac{\sin\theta_1 \cos\theta_2 \tan\theta_3}{\sin(\theta_2-\theta_1)} \\ 0\,1\,0 & \frac{\cos\theta_1 \sin(\theta_3-\theta_2)}{\cos\theta_3 \sin(\theta_2-\theta_1)} & \frac{-\cos\theta_1 \cos\theta_2}{\sin(\theta_2-\theta_1)} & \frac{-\cos\theta_1 \cos\theta_2 \tan\theta_3}{\sin(\theta_2-\theta_1)} \\ 0\,0\,1 & \tan\theta_3 & -1 & -\tan\theta_3 \end{bmatrix} \begin{bmatrix} U_1^X \\ U_1^Y \\ U_2^X \\ U_2^Y \end{bmatrix} = \begin{bmatrix} 0 \\ 0 \\ 0 \end{bmatrix} \tag{36}$$

For any value of θ_1, θ_2 and θ_3 one finds the nullity of the matrix is 3, which means the DoF of the mechanism is 3.

4.8 5-Link Structure

The structure depicted in Fig. 9 comprises five links, including the frame, interconnected by revolute joints, resulting in zero DoF. While this configuration lacks practical relevance in kinematic applications, it commonly finds utility in civil engineering. This analysis serves as an instructive illustration of the method's efficacy in discerning zero DoF mechanisms.

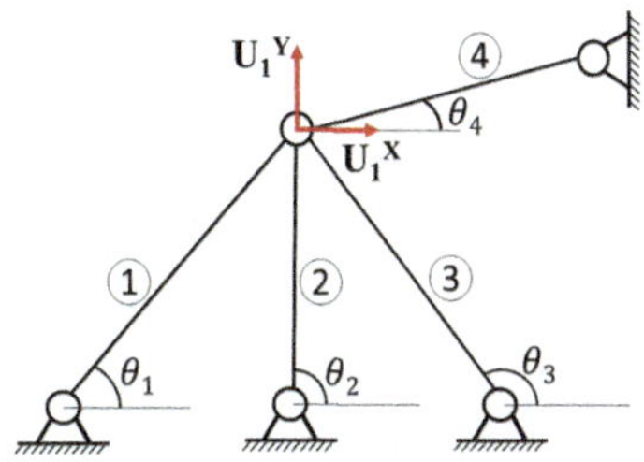

Fig. 9 A structure consisting of five links

The formulation of displacement equations for each joint corresponding to every link is as follows:

(i) **Link 1**: $U_1^X \cos\theta_1 + U_1^Y \sin\theta_1 = 0$
(ii) **Link 2**: $U_1^X \cos\theta_2 + U_1^Y \sin\theta_2 = 0$
(iii) **Link 3**: $U_1^X \cos\theta_3 + U_1^Y \sin\theta_3 = 0$
(iv) **Link 4**: $U_1^X \cos\theta_4 + U_1^Y \sin\theta_4 = 0$.

Now, writing the system of equations in Matrix form, one gets

$$\begin{bmatrix} \cos\theta_1 & \sin\theta_1 \\ \cos\theta_2 & \sin\theta_2 \\ \cos\theta_3 & \sin\theta_3 \\ \cos\theta_4 & \sin\theta_4 \end{bmatrix} \begin{Bmatrix} U_1^X \\ U_1^Y \end{Bmatrix} = \begin{Bmatrix} 0 \\ 0 \\ 0 \\ 0 \end{Bmatrix} \tag{37}$$

After performing row operations on the Constraint Matrix, the RREF obtained is as follows:

$$\begin{bmatrix} 1 & 0 \\ 0 & 1 \\ 0 & 0 \\ 0 & 0 \end{bmatrix} \tag{38}$$

For any value of θ_1, θ_2, θ_3 and θ_4 one finds that the nullity of the matrix is 0, which means the DoF of the mechanism is 0.

Note: *If one apply the Grubler* ($L = 5$, $P_1 = 7$, $P_2 = 0$) *and Kutzbach* ($N = 5$, $j = 7$, $\sum_{i=1}^{j} f_i = 7$) *formula here, both will give DoF as* -2. *But actually, the DoF of this system is 0. Thus the traditional formulas fail to correctly predict the DoF.*

4.9 RPPR Linkage

In Fig. 10, an unconventional zero DoF system is introduced, presenting a variation of the classic 4-bar linkage. This configuration deviates from the conventional 4-bar linkage by incorporating two slider joints and two revolute joints (thus called RPPR

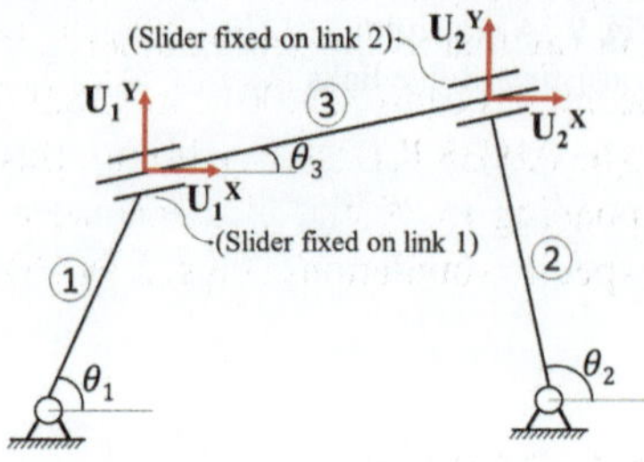

Fig. 10 RPPR linkage containing redundant link

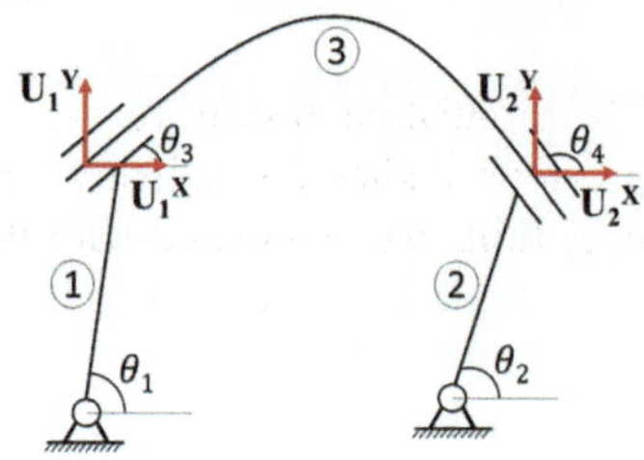

Fig. 11 Modified RPPR linkage with curved link

linkage, R stands for revolute and P stands for prismatic/slider joint), in contrast to the typical four revolute joints. Intuition suggests that motion is improbable within this mechanism unless a curved body is introduced to link 3, as illustrated in Fig. 11.

Now, shifting the focus toward the formulation of displacement equations for each joint, corresponding to every link, it becomes evident that:

(i) **Link 1**: $U_1^X \cos\theta_1 + U_1^Y \sin\theta_1 = 0$
(ii) **Link 2**: $U_2^X \cos\theta_2 + U_2^Y \sin\theta_2 = 0$
(iii) **Link 3**: $U_2^X \cos\theta_3 + U_2^Y \sin\theta_3 - U_1^X \cos\theta_3 - U_2^X \sin\theta_3 = 0$
(iv) **Slider Condition 1**: $U_1^X \sin\theta_3 - U_1^Y \cos\theta_3 = 0$
(v) **Slider Condition 2**: $U_2^X \sin\theta_3 - U_2^Y \cos\theta_3 = 0.$

Now, writing the system of equations in Matrix form, one gets

$$\begin{bmatrix} \cos\theta_1 & \sin\theta_1 & 0 & 0 \\ 0 & 0 & \cos\theta_2 & \sin\theta_2 \\ \sin\theta_3 & -\cos\theta_3 & 0 & 0 \\ 0 & 0 & \cos\theta_3 & -\sin\theta_3 \\ -\cos\theta_3 & -\sin\theta_3 & \sin\theta_3 & \cos\theta_3 \end{bmatrix} \begin{Bmatrix} U_1^X \\ U_1^Y \\ U_2^X \\ U_2^Y \end{Bmatrix} = \begin{Bmatrix} 0 \\ 0 \\ 0 \\ 0 \\ 0 \end{Bmatrix} \quad (39)$$

After performing row operations on the Constraint Matrix, the RREF obtained is as follows:

$$\begin{bmatrix} 1\,0\,0\,0 \\ 0\,1\,0\,0 \\ 0\,0\,1\,0 \\ 0\,0\,0\,1 \\ 0\,0\,0\,0 \end{bmatrix} \quad (40)$$

For any value of θ_1, θ_2 and θ_3 one finds that the nullity of the matrix is 0, which means the DoF of the mechanism is 0.

Note: *If one apply the Grubler* $(L = 4,\ P_1 = 4,\ P_2 = 0)$ *and Kutzbach* $(N = 4,\ j = 4,\ \sum_{i=1}^{j} f_i = 4)$ *formula here, both will give DoF as 1. But actually, the DoF of this system is 0. Thus the traditional formulas fail over here.*

4.10 PPPP Linkage

In some cases, there might be two slider joints on the same body. One such case is shown over here in Fig. 12. This linkage is called a PPPP linkage since it has four prismatic/slider joints. From intuition, one can easily that movement is possible between the two bodies while maintaining the sliding contact at all times. One can also infer that this movement is possible only because of the particular shape of the two bodies due to which both these bodies can move over each other. The two bodies are shown separately in Figs. 13 and 14.

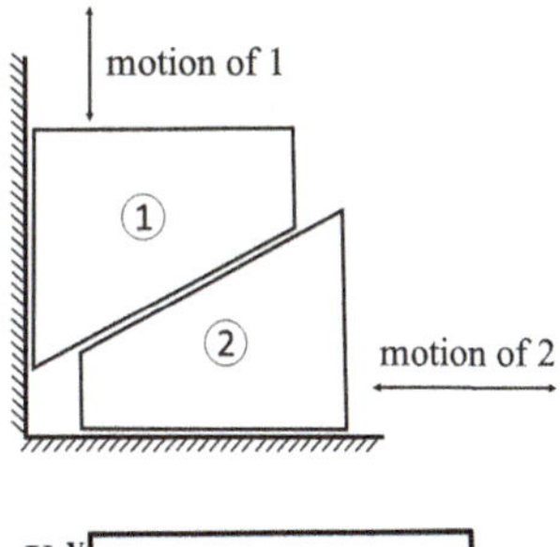

Fig. 12 PPPP linkage

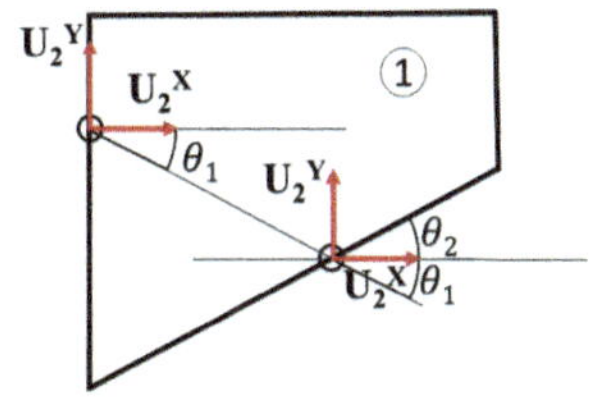

Fig. 13 Link 1 of PPPP linkage (Fig. 12)

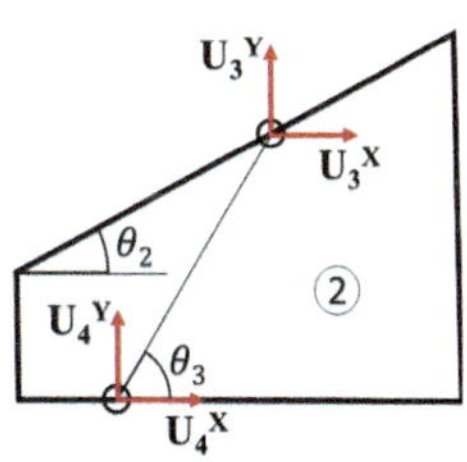

Fig. 14 Link 2 of PPPP linkage (Fig. 12)

The corresponding displacement equations are as follows:

(i) **Link 1 Rigidity**: $U_2^X \cos\theta_1 - U_2^Y \sin\theta_1 - U_1^X \cos\theta_1 + U_1^Y \sin\theta_1 = 0$
(ii) **Perpendicular to Sliding Direction 1**: $U_1^X = 0$
(iii) **Sliding Direction 1 (Link 1)**: $U_1^Y - U_2^Y = 0$
(iv) **Perpendicular to Sliding Direction 2**: $U_2^Y \cos\theta_2 - U_2^X \sin\theta_2 - U_3^Y \cos\theta_2 + U_3^X \sin\theta_2 = 0$
(v) **Sliding Direction 2 (Link 1)**: $U_2^X \sin\theta_2 + U_2^X \cos\theta_2 - U_1^X \sin\theta_2 - U_1^Y \cos\theta_2 = 0$
(vi) **Link 2 Rigidity**: $U_3^X \cos\theta_3 + U_3^Y \sin\theta_3 - U_4^X \cos\theta_3 - U_4^Y \sin\theta_3 = 0$
(vii) **Sliding Direction 2 (Link 2)**: $U_3^X \cos\theta_2 + U_3^Y \sin\theta_2 - U_4^X \cos\theta_2 - U_4^Y \sin\theta_2 = 0$
(viii) **Sliding Direction 3 (Link 2)**: $U_3^X - U_4^X = 0$
(ix) **Perpendicular to Sliding Direction 3**: $U_4^Y = 0$.

Now, writing the system of equations in Matrix form, one gets

$$\begin{bmatrix} -\cos\theta_1 & \sin\theta_1 & \cos\theta_1 & -\sin\theta_1 & 0 & 0 & 0 & 0 \\ 1 & 0 & 0 & 0 & 0 & 0 & 0 & 0 \\ 0 & 1 & 0 & -1 & 0 & 0 & 0 & 0 \\ 0 & 0 & -\sin\theta_2 & \cos\theta_2 & \sin\theta_2 & -\cos\theta_2 & 0 & 0 \\ -\sin\theta_2 & -\cos\theta_2 & \sin\theta_2 & \cos\theta_2 & 0 & 0 & 0 & 0 \\ 0 & 0 & 0 & 0 & \cos\theta_3 & \sin\theta_3 & -\cos\theta_3 & -\sin\theta_3 \\ 0 & 0 & 0 & 0 & \cos\theta_2 & \sin\theta_2 & -\cos\theta_2 & -\sin\theta_2 \\ 0 & 0 & 0 & 0 & 1 & 0 & -1 & 0 \\ 0 & 0 & 0 & 0 & 0 & 0 & 0 & 1 \end{bmatrix} \begin{Bmatrix} U_1^X \\ U_1^Y \\ U_2^X \\ U_2^Y \\ U_3^X \\ U_3^Y \\ U_4^X \\ U_4^Y \end{Bmatrix} = \begin{Bmatrix} 0 \\ 0 \\ 0 \\ 0 \\ 0 \\ 0 \\ 0 \\ 0 \\ 0 \end{Bmatrix} \tag{41}$$

After performing row operations on the Constraint Matrix, the RREF obtained is as follows:

$$\begin{bmatrix} 1 & -\tan\theta_1 & 1 & \tan\theta_1 & 0 & 0 & 0 & 0 \\ 0 & 1 & \cot\theta_1 & -1 & 0 & 0 & 0 & 0 \\ 0 & 0 & -\cot\theta_1 & 0 & 0 & 0 & 0 & 0 \\ 0 & 0 & 0 & 1 & \tan\theta_2 & -1 & 0 & 0 \\ 0 & 0 & 0 & 0 & 1 & \tan\theta_3 & -1 & -\tan\theta_3 \\ 0 & 0 & 0 & 0 & 0 & \frac{\sin(\theta_2-\theta_3)}{\cos\theta_3} & 0 & -\frac{\sin(\theta_2-\theta_3)}{\cos\theta_3} \\ 0 & 0 & 0 & 0 & 0 & 0 & 0 & 1 \\ 0 & 0 & 0 & 0 & 0 & 0 & 0 & 0 \\ 0 & 0 & 0 & 0 & 0 & 0 & 0 & 0 \end{bmatrix} \tag{42}$$

For any value of θ_1, θ_2 and θ_3 one finds that the nullity of the matrix is 1, which means the DoF of the mechanism is 1.

Note: *If one apply the Grubler* ($L = 3$, $P_1 = 3$, $P_2 = 0$) *and Kutzbach* ($N = 3$, $j = 3$, $\sum_{i=1}^{j} f_i = 3$) *formula here, both will give DoF as 0. But actually, the DoF of this system is 1. Thus the traditional formulas fail over here.*

4.11 Pratt Truss

In some mechanical systems, the introduction of additional linkages for the purpose of augmenting strength and rigidity can lead to localized constraints, resulting in an incorrect DoF when traditional formulas are applied. These methods require identification and removal of locally constrained sections to calculate the system's DoF accurately.

In contrast, the proposed matrix method is able to compute the correct DoF for the entire mechanism. The number of constraint equations are reduced to only independent constraints, as dependent equations vanish in the RREF form of constraint matrix. We consider a simple Pratt truss to showcase this. Trusses are structures made out of straight links, connected in multiple triangle loops which are designed to efficiently support loads by distributing them along the members. They are required to have 0 DoF, since they are used in bridges, roofs, towers, etc., and so must not have any free DoF. Consider Fig. 15, having 10 links including the roller support at one end.

Breaking links 2 and 9 according to postulate 2, and following the other steps accordingly, the formulation of displacement equations for each joint corresponding to every link is as follows:

(i) **Link 1**: $U_1^X \cos\theta_1 + U_1^Y \sin\theta_1 = 0$
(ii) **Sublink 2.1**: $U_2^X - U_1^X = 0$
(iii) **Sublink 2.2**: $U_3^X - U_2^X = 0$
(iv) **Collinearlity Condition (2.1 and 2.2)**: $U_1^Y - 2U_2^Y + U_3^Y = 0$
(v) **Link 3**: $U_1^X \cos\theta_2 + U_1^Y \sin\theta_2 - U_4^X \cos\theta_2 - U_4^Y \sin\theta_2 = 0$
(vi) **Link 4**: $U_1^Y \sin\theta_3 - U_1^X \cos\theta_3 - U_5^Y \sin\theta_3 + U_5^Y \cos\theta_3 = 0$
(vii) **Link 5**: $U_2^X \cos\theta_2 + U_2^Y \sin\theta_2 - U_5^X \cos\theta_2 - U_5^Y \sin\theta_2 = 0$
(viii) **Link 6**: $U_3^X \cos\theta_1 + U_3^Y \sin\theta_1 - U_5^X \cos\theta_1 - U_5^Y \sin\theta_1 = 0$
(ix) **Link 7**: $U_3^X \cos\theta_2 + U_3^Y \sin\theta_2 - U_6^X \cos\theta_2 - U_6^Y \sin\theta_2 = 0$
(x) **Link 8**: $U_3^Y \sin\theta_3 - U_3^X \cos\theta_3 - U_7^X \sin\theta_3 + U_7^X \cos\theta_3 = 0$
(xi) **Sublinkk 9.1**: $U_4^X = 0$
(xii) **Sublink 9.2**: $U_5^X - U_4^X = 0$
(xiii) **Sublinkk 9.3**: $U_6^X - U_5^X = 0$
(xiv) **Sublink 9.4**: $U_7^X - U_6^X = 0$
(xv) **Collinearlity Condition (9.1 and 9.2)**: $-2U_4^Y + U_5^Y = 0$

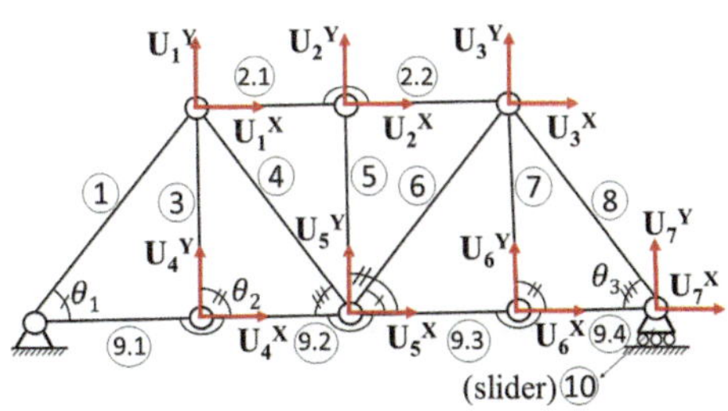

Fig. 15 Pratt Truss

(xvi) **Collinearlity Condition (9.2 and 9.3)**: $U_4^Y - 2U_5^Y + U_6^Y = 0$
(xvii) **Collinearlity Condition (9.3 and 9.4)**: $U_5^Y - 2U_6^Y + U_7^Y = 0$
(xviii) **Link 10**: $U_7^Y = 0$.

The Constraint Matrix, A for this case will be as follows: (for simplicity the whole equation is not shown in matrix form)

$$\begin{bmatrix}
\cos\theta_1 & \sin\theta_1 & 0 & 0 & 0 & 0 & 0 & 0 & 0 & 0 & 0 & 0 & 0 & 0 \\
-1 & 0 & 1 & 0 & 0 & 0 & 0 & 0 & 0 & 0 & 0 & 0 & 0 & 0 \\
0 & 0 & -1 & 0 & 1 & 0 & 0 & 0 & 0 & 0 & 0 & 0 & 0 & 0 \\
0 & 1 & 0 & -2 & 0 & 1 & 0 & 0 & 0 & 0 & 0 & 0 & 0 & 0 \\
\cos\theta_2 & \sin\theta_2 & 0 & 0 & 0 & 0 & -\cos\theta_2 & -\sin\theta_2 & 0 & 0 & 0 & 0 & 0 & 0 \\
\sin\theta_3 & -\cos\theta_3 & 0 & 0 & 0 & 0 & 0 & 0 & -\sin\theta_3 & \cos\theta_3 & 0 & 0 & 0 & 0 \\
0 & 0 & \cos\theta_2 & \sin\theta_2 & 0 & 0 & 0 & 0 & -\cos\theta_2 & \sin\theta_2 & 0 & 0 & 0 & 0 \\
0 & 0 & 0 & 0 & \cos\theta_1 & \sin\theta_1 & 0 & 0 & -\cos\theta_1 & -\sin\theta_1 & 0 & 0 & 0 & 0 \\
0 & 0 & 0 & 0 & \cos\theta_2 & \sin\theta_2 & 0 & 0 & 0 & 0 & -\cos\theta_2 & -\sin\theta_2 & 0 & 0 \\
0 & 0 & 0 & 0 & -\cos\theta_3 & \sin\theta_3 & 0 & 0 & 0 & 0 & 0 & 0 & \cos\theta_3 & -\sin\theta_3 \\
0 & 0 & 0 & 0 & 0 & 0 & 1 & 0 & 0 & 0 & 0 & 0 & 0 & 0 \\
0 & 0 & 0 & 0 & 0 & 0 & -1 & 0 & 1 & 0 & 0 & 0 & 0 & 0 \\
0 & 0 & 0 & 0 & 0 & 0 & 0 & 0 & -1 & 0 & 1 & 0 & 0 & 0 \\
0 & 0 & 0 & 0 & 0 & 0 & 0 & 0 & 0 & 0 & -1 & 0 & 1 & 0 \\
0 & 0 & 0 & 0 & 0 & 0 & 0 & -2 & 0 & 1 & 0 & 0 & 1 & 0 \\
0 & 0 & 0 & 0 & 0 & 0 & 0 & 1 & 0 & -2 & 0 & 1 & 0 & 0 \\
0 & 0 & 0 & 0 & 0 & 0 & 0 & 0 & 0 & 1 & 0 & -2 & 0 & 1 \\
0 & 0 & 0 & 0 & 0 & 0 & 0 & 0 & 0 & 0 & 0 & 0 & 0 & 1
\end{bmatrix}$$

When the matrix mentioned above is transformed into its RREF, the resulting rank of the matrix is established as 14.

$$\begin{bmatrix}
1&0&0&0&0&0&0&0&0&0&0&0&0&0 \\
0&1&0&0&0&0&0&0&0&0&0&0&0&0 \\
0&0&1&0&0&0&0&0&0&0&0&0&0&0 \\
0&0&0&1&0&0&0&0&0&0&0&0&0&0 \\
0&0&0&0&1&0&0&0&0&0&0&0&0&0 \\
0&0&0&0&0&1&0&0&0&0&0&0&0&0 \\
0&0&0&0&0&0&1&0&0&0&0&0&0&0 \\
0&0&0&0&0&0&0&1&0&0&0&0&0&0 \\
0&0&0&0&0&0&0&0&1&0&0&0&0&0 \\
0&0&0&0&0&0&0&0&0&1&0&0&0&0 \\
0&0&0&0&0&0&0&0&0&0&1&0&0&0 \\
0&0&0&0&0&0&0&0&0&0&0&1&0&0 \\
0&0&0&0&0&0&0&0&0&0&0&0&1&0 \\
0&0&0&0&0&0&0&0&0&0&0&0&0&1 \\
0&0&0&0&0&0&0&0&0&0&0&0&0&0 \\
0&0&0&0&0&0&0&0&0&0&0&0&0&0 \\
0&0&0&0&0&0&0&0&0&0&0&0&0&0 \\
0&0&0&0&0&0&0&0&0&0&0&0&0&0
\end{bmatrix}$$

Thus, one can find that for this configuration:

$$DoF = 14\ (number\ of\ columns) - 14\ (rank) = 0.$$

In the case of a structure/super structure, the number of pivots (rank) always comes out to be equal to column space (2n), thus arriving at a DoF of 0.

An innovative approach for the computation of DoF for planar mechanisms, employing a matrix manipulation technique, has been introduced. The effectiveness of this method is presented through a comprehensive examination of various scenarios, encompassing standard cases as well as complicated planar mechanisms. Starting with the fundamental four-bar mechanism and highlighting some crucial implications of the RREF matrix, it is demonstrated how this matrix-based approach proves highly valuable. Particularly, it is illustrated how this technique can accurately determine the DoF of mechanisms encompassing overconstrained and redundant links. Notably, in the case of the augmented four-bar mechanism with redundant links, one can identify a crucial observation: when the constraint matrix corresponding to the mechanism is reduced to its RREF, it becomes evident that two rows are equivalent. This equivalence directly implies that the corresponding displacement equations are linearly dependent, thereby highlighting the interplay of redundancy within the mechanism.

Another significant achievement of this method is that it correctly predicts the DoF of the cam-follower mechanism to be 1, indicating the ability for a link to move independently without inducing any movement in the remaining mechanism. This paper also introduces the analysis of a PPPP (4 planar joints) linkage, illustrating the critical discrepancy in the calculation of its DoF when employing traditional formulas. In this work, some cases with 0 DoF have also been demonstrated, popularly known as structures, thereby emphasizing the broader applicability of this novel matrix-based method. Importantly, one can find the lengths of individual links do not significantly impact the DoF unless they introduce alterations in the orientation of the links. Instead, the constraint matrix emerges as the focal point of this analysis. The constraint matrix assumes paramount significance in this work's approach, offering several invaluable insights. First and foremost, it enables the determination of DoF through the calculation of its nullity, providing a robust quantitative measure of the mechanism's mobility. Additionally, the constraint matrix facilitates the inference of specific displacements when others are known. This capability holds significant promise in predicting the deformation of lattice points within crystal structures, thereby underscoring the broader applicability and relevance of the proposed methodology.

However, when applied to high-mobility systems such as parallel manipulators or robotic end-effectors, certain computational challenges emerge. These systems often involve a number of constraints and higher degrees of mobility, leading to large and sparse matrices, which require significant computational resources for accurate analysis. Similar to the Pratt truss, cases where there are a large number of nodal points face increased dimensionality, which can result in higher computational time and memory usage. Although the method can robustly deal with redundant constraints owing to its foundation in the transformation of basis vectors, the increasing computational complexity for advanced systems must be acknowledged. To address these limitations, future work could explore the integration of efficient numerical solvers within the formulation. Algorithms for efficient sparse matrix factorization, such as those discussed in [19], could be incorporated to deal with the large sparse systems of linear equations obtained in these cases. Additionally, tailoring the method to incorporate specific structural symmetries or leveraging machine learning for the preconditioning of matrices could significantly enhance its scalability and applicability to complex mechanisms and applicability for real-time applications.

5 Conclusions

The proposed matrix-based method offers a rapid and efficient means of computing the DoF for planar mechanisms, particularly when displacement equations are expressed in matrix form. Through various case studies and empirical observations, it becomes apparent that this novel methodology effectively rectifies previous limitations inherent in DoF analysis. This approach addresses and overcomes limitations inherent in conventional methods such as Grubler's and Kutzbach's, which often fail to yield accurate results in scenarios involving redundant and overconstrained linkages. This fundamental challenge provided the impetus for the development of this innovative methodology. Notably, while the Jacobian method and other recent formulations necessitate heavy computation as well as advanced knowledge in different fields like screw theory, incurring substantial computational costs and requiring a good theoretical background, this method requires merely the calculation of the nullity of the constraint matrix, which can be accomplished very quickly and easily. In this work, the vector algebra approach and the matrix formulation have been presented, laying the foundation for its implementation.

In conclusion, the findings of this study emphasize the pivotal role of the constraint matrix in advancing the understanding of the mobility of planar mechanisms. The matrix-based method for calculating the degrees of freedom in planar mechanisms proves to be a highly effective and versatile tool, with the ability to accurately handle overconstrained and redundant mechanisms, and predict mobility across a range of scenarios. Despite the computational challenges associated with high-mobility systems, such as parallel manipulators and robotic end-effectors, the potential for further development of this approach—particularly with advanced numerical solvers and machine learning techniques—makes it a promising avenue for future research in the kinematics of complex mechanisms.

Acknowledgements The authors have received no financial support for the research, authorship, and/or publication of this article.

Disclosure of Interests The authors have no competing interests to declare that are relevant to the content of this article.

References

1. P. L. Chebyshev, Works(French), edited by A. A. Markov, and N. Sonin. *Œuvres de P. L. Tchebychef.* Commissionaires de l'Academie imperiale des sciences, 1899–1907.
2. L.D. Ruzinov. *Design of mechanisms by geometric transformations (translated from the Russian into English).* Iliffe Books Ltd., London, 1968.
3. Martin Grübler. Eine theorie des zwanglaufes und der ebenen mechanismen. *Verlag von Julius Springer, Berlin*, 1917.
4. Karl Kutzbach. Mechanische leitungsverzweigung, ihre gesetze und anwendungen. *Maschinenbau*, 8(21):710–716, 1929.
5. F Freudenstein and Rasim Alizade. On the degree of freedom of mechanisms with variable general constraint. In *Proceedings Fourth World Congress on the Theory of Machines and Mechanisms*, pages 51–56, 1975.
6. GUO Weidong and YU Jingjun. A new method of mobility calculation for planar mechanisms. *Journal of Mechanical Engineering*, 49(7):7, 2013.
7. Zhen Huang, JingFang Liu, and DaXing Zeng. A general methodology for mobility analysis of mechanisms based on constraint screw theory. *Science in China Series E: Technological Sciences*, 52(5):1337–1347, 2009.
8. A. Gupta, G. Karypis, and V. Kumar. Highly scalable parallel algorithms for sparse matrix factorization. *IEEE Transactions on Parallel and Distributed Systems*, 8(5):502–520, 1997.
9. Kambiz Ghaemi Osgouie and Bahman Gard. Using the matrix method to compute the degrees of freedom of mechanisms. *Journal of Applied and Computational Mechanics*, 3(3):158–170, 2017.
10. Sheryl A. Cole. *The Jacobian Matrix.* PhD thesis, The Kansas State Teachers College of Emporia, 1970.
11. VS Rajashekhar and G Debasish. Degrees of freedom analysis of mechanisms using the new zebra crossing method. *CoRR*, 2022.
12. Lung-Wen Tsai. *Mechanism design: enumeration of kinematic structures according to function.* CRC Press, 2000.
13. Guojun Luo, Xiaohui Wang, and Xianguo Yan. Geometric theorems and application of the dof analysis of the workpiece based on the constraint normal line. *Advances in Materials Science and Engineering*, 2021.
14. Guojun Luo, Xiaohui Wang, Xianguo Yan, et al. Dof calculation method of planar mechanisms based on constraint normal line. *Mathematical Problems in Engineering*, 2023.
15. Gilbert Strang. The fundamental theorem of linear algebra. *The American Mathematical Monthly*, 100(9):848–855, 1993.
16. Sarjit S Rattan. *Theory of machines.* Tata McGraw-Hill Education, 2014.
17. John J Uicker Jr, Gordon R Pennock, and Joseph E Shigley. *Theory of machines and mechanisms.* Cambridge University Press, 2023.
18. G.L.Kinzel K.J.Wladron. *Kinematics, Dynamics and Design of Machinery.* Wiley, 2004.
19. Yuanxi Sun, Wenjie Ge, Jia Zheng, and Dianbiao Dong. Design and evaluation of a prosthetic knee joint using the geared five-bar mechanism. *IEEE Transactions on Neural Systems and Rehabilitation Engineering*, 23(6):1031–1038, 2015.
20. Grigore Gogu. Chebychev–grübler–kutzbach's criterion for mobility calculation of multi-loop mechanisms revisited via theory of linear transformations. *European Journal of Mechanics-A/Solids*, 24(3):427–441, 2005.

Enhancing Aerodynamic Efficiency of Flat-Plate Wings at Low Reynolds Number with Inverted Dimples

Bastav Borah, Anand Verma, Vinayak N. Kulkarni, and Ujjwal K. Saha

Abstract This study investigates the effectiveness of using inverted dimples as passive sub-boundary-layer vortex generators to enhance aerodynamic performance in low Reynolds number conditions. Inverted dimples were strategically positioned along the span at multiple points on the suction side of a flat plate with an aspect ratio of 2. The primary objective was to initiate early transition to turbulent flow, thereby destabilizing the predominantly laminar flow, which is more prone to separation under these conditions. The dimples, each with a diameter of approximately 1.0 mm and a height of about 0.5 mm, were designed to induce localized turbulence, providing energy to the separated layer and promoting reattachment to the surface. The study demonstrated that the use of inverted dimples effectively triggered the transition from laminar to turbulent flow, reducing flow separation and enhancing aerodynamic efficiency. Flow visualization images confirmed the impact of the dimples, showing smooth laminar flow upstream and turbulent reattachment downstream. These findings indicate that applying inverted dimples on wings, especially in ground effect scenarios, can significantly improve performance by optimizing airflow and reducing drag. This research contributes to the broader understanding of passive flow control techniques and their potential applications in various aerodynamic contexts. By leveraging the principles of fluid dynamics and biomimicry, the use of inverted dimples presents a promising approach to enhancing the efficiency of wings and other aerodynamic surfaces, particularly in low Reynolds number environments where flow separation is a critical issue. Further studies could explore the scalability and applicability of this technique in more complex aerodynamic systems.

B. Borah (✉) · A. Verma · V. N. Kulkarni · U. K. Saha
Department of Mechanical Engineering, Indian Institute of Technology Guwahati, Guwahati, India
e-mail: bastav.borah@iitg.ac.in

A. Verma
e-mail: anand2016@iitg.ac.in

V. N. Kulkarni
e-mail: vinayak@iitg.ac.in

U. K. Saha
e-mail: saha@iitg.ac.in

U. S. Dixit et al. (eds.), *Current Progress in Engineering Sciences*,
https://doi.org/10.1007/978-981-96-8753-4_10

Keywords Flat-plate wings · Inverted dimples · Low Reynolds number

1 Introduction

The flow of air around wings at low Reynolds numbers is more susceptible to separation due to the lower speeds involved. This susceptibility arises because, at low speeds, the boundary layer is primarily laminar and lacks the energy needed to stay attached to the surface, making it more prone to separation. Flow separation can lead to increased drag and reduced lift, negatively impacting aerodynamic performance. Consequently, controlling and delaying flow separation in low Reynolds number conditions is essential for enhancing the efficiency and effectiveness of wings and other aerodynamic surfaces. The use of dimples as passive, sub-boundary-layer vortex generators has proven successful in improving heat exchanger design by enhancing thermal performance and fluid mixing [1–5].

Numerous studies have investigated the benefits of passive vortex generation to improve the aerodynamic efficiency of aerofoils [6–11]. These studies consistently confirm that boundary-layer protruding vortex generators enhance shear stress and turbulence transport near the wall by producing vortices. These augmentations help to delay flow separation, reduce drag, and improve overall aerodynamic performance, making passive vortex generators a valuable tool in aerofoil design and optimization. The implementation of low-profile vortex generators relies on the technique of 'minimal near-wall protuberance' to generate vortices that are sufficiently strong to counteract baseline flow separation [12, 13]. While this approach is effective, its success is highly dependent on the precise placement of the vortex generators relative to the flow separation point. Properly positioned, these generators can energize the boundary layer, delaying flow separation and reducing drag. However, incorrect placement can lead to diminished effectiveness, highlighting the critical need for careful optimization in their design and application.

Keeping these phenomena in mind, this paper explores the use of inverted dimples as passive, sub-boundary-layer vortex generators to enhance aerodynamic performance. While the majority of studies have focused on conventional dimples for flow control, the current study investigates flat-plate wings due to their design and fabrication simplicity, making inverted dimples an ideal choice. Inverted dimples can effectively generate localized turbulence, delay flow separation, and improve overall aerodynamic efficiency in these configurations. Hence, for flows at low Reynolds numbers, flat-plate wings modified with inverted dimples can play a significant role in a variety of applications. These applications include unmanned aerial vehicles (UAVs), micro-air vehicles (MAVs), and other low-speed aircraft, where maintaining aerodynamic efficiency is crucial. By introducing localized turbulence and enhancing boundary layer stability, inverted dimples can effectively delay flow separation and reduce drag, leading to improved performance and fuel efficiency. This approach can also be beneficial in automotive aerodynamics and wind turbine blade design, where controlling airflow at low Reynolds numbers can optimize overall system efficiency and performance.

2 Experimental Methodology

The experiments are conducted in the subsonic wind tunnel facility at IIT Guwahati, India. Figure 1 illustrates the schematic of the wind tunnel and the apparatus used. This facility features an open-circuit suction-type wind tunnel with a maximum wind speed capacity of 50 m/s. The wind tunnel has a square test section measuring 600×600 mm^2 and 2000 mm in length. Given the small size of the test model relative to the tunnel test section, no blockage corrections are considered for the present measurements. The wind tunnel is calibrated by varying the RPM of the suction fan and measuring the free stream velocity of air in the test section using a digital manometer. Figure 2 displays the calibration chart of the wind tunnel. The turbulence intensity in the test section is approximately 0.25% for the velocity range investigated. The wind tunnel also provides mounting provisions for the test model on the bottom wall and a 3D traverse on the top wall.

The dimensions and coordinates of the flat-plate model are detailed in Fig. 3. The test model, a flat plate, has a cross-section of 150×5 mm2 and a span of 300 mm. Constructed from plexiglass, the plate underwent milling and polishing operations to achieve a high-quality surface finish. The test model is mounted spanwise across the test section, secured between the side walls. It is attached to a pivoting mechanism allowing for an angle of attack (AoA) range from $-10°$ to $20°$, with an accuracy within $\pm0.5°$.

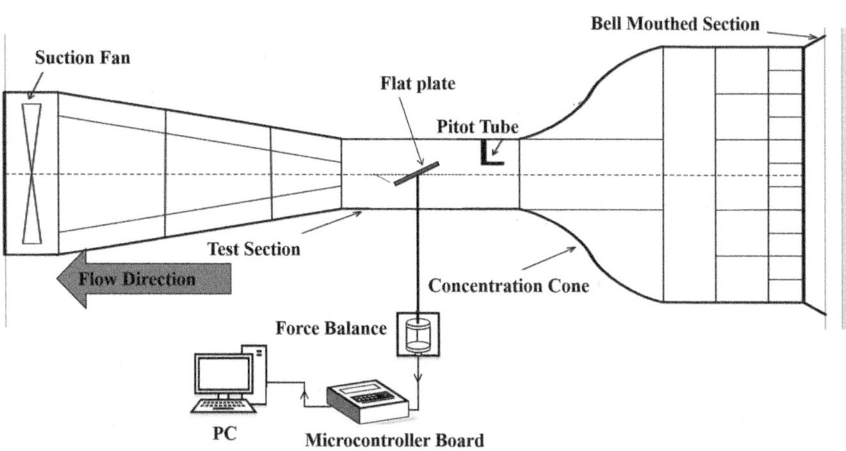

Fig. 1 Schematic of wind tunnel along

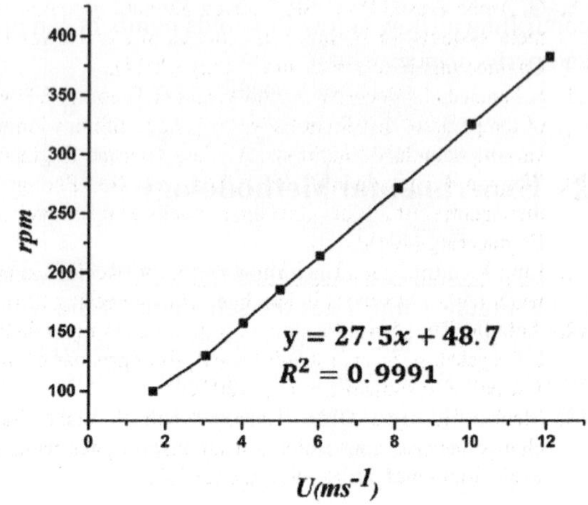

Fig. 2 Wind tunnel calibration

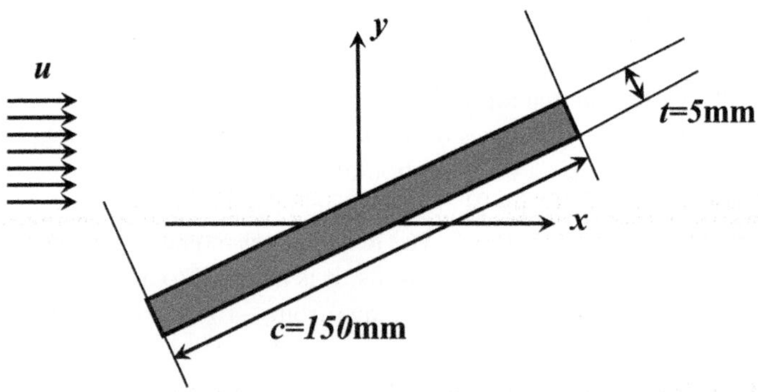

Fig. 3 Flat-plate Wing

An array of inverted dimples is strategically positioned along the span at multiple locations on the suction side of the flat-plate wing with an aspect ratio (AR) of 2 to initiate an early transition to turbulent flow.

The inverted dimples have a diameter of approximately 1.0 mm and a height of around 0.5 mm. The suction side of the flat plate is dotted with 15 dimples arranged along the span at three chordwise locations: 0.25c, 0.5c, and 0.75c. These dimples are distributed in the following combinations to study their effect on optimizing flow control and facilitating early transition to turbulent flow:

(a) Single row at location 0.25c from LE.
(b) Single row at location 0.5c from LE.
(c) Single row at location 0.75c from LE.
(d) Double rows at location 0.25c and 0.5c from LE.
(e) Double rows at location 0.5c and 0.75c from LE.

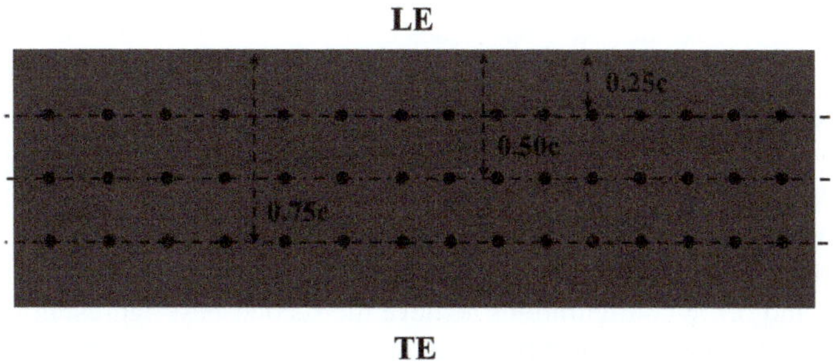

Fig. 4 Location of inverted dimples on the suction side of wing

Fig. 5 Flat-plate wing with inverted dimples at 0.25c location

Each of these combinations is tested, and the aerodynamic coefficients are analyzed and compared to determine any resulting improvements. The arrangement of the inverted dimples is illustrated in Figs. 4 and 5, with Fig. 4 showing the schematics and Fig. 5 presenting a real image.

3 Results and Discussion

For all the configurations of the modified wing, the lift-to-drag (L/D) ratio is measured and compared with an unmodified wing. The lift and drag forces in these configurations are measured, and the L/D is calculated. The lift-to-drag ratio (L/D) is a critical metric in aerodynamics, representing the efficiency of an aerofoil or wing. A higher L/D ratio indicates that the wing produces more lift for a given amount of drag, which is desirable for optimizing performance, fuel efficiency, and overall aerodynamic effectiveness.

The L/D plots at different Re are shown in Fig. 6a–d. From these plots, it can be observed that the 0.75c configuration consistently exhibits a better L/D ratio compared to the unmodified wing, especially at higher AoA. This improvement is particularly noticeable at the higher Re values of 13×10^4 and 15×10^4, where the 0.75c configuration consistently outperforms the unmodified wing at AoAs of 14°,

16°, 18°, and 20°. At the lower Re values of 8×10^4 and 10×10^4, the 0.75c configuration performs similarly to the unmodified wing, showing superior performance at some AoAs and slightly inferior performance at others. This indicates that while the 0.75c configuration is highly effective at higher Reynolds numbers, its benefits are less pronounced but still present at lower Reynolds numbers. These percentage improvement figures are listed in Table 1. No other configuration is able to achieve better performance compared to the unmodified wing at any other AoAs or Re. The 0.50c and 0.75c configurations achieve the second-best figures; however, their performance still lags significantly behind that of the unmodified wing.

This trend suggests that the 0.75c location of the dimples is particularly advantageous, enhancing the aerodynamic performance at critical AoAs. The substantial lead of the 0.75c configuration over others suggests a specific advantage in its ability to delay flow separation and reduce drag more effectively. There is a possibility that at these specific flow speeds and AoAs, the flow is sufficiently energized to reattach to the boundary layer due to the intermixing of fluid layers initiated by the inverted dimples. The strategically placed inverted dimples create localized turbulence, which

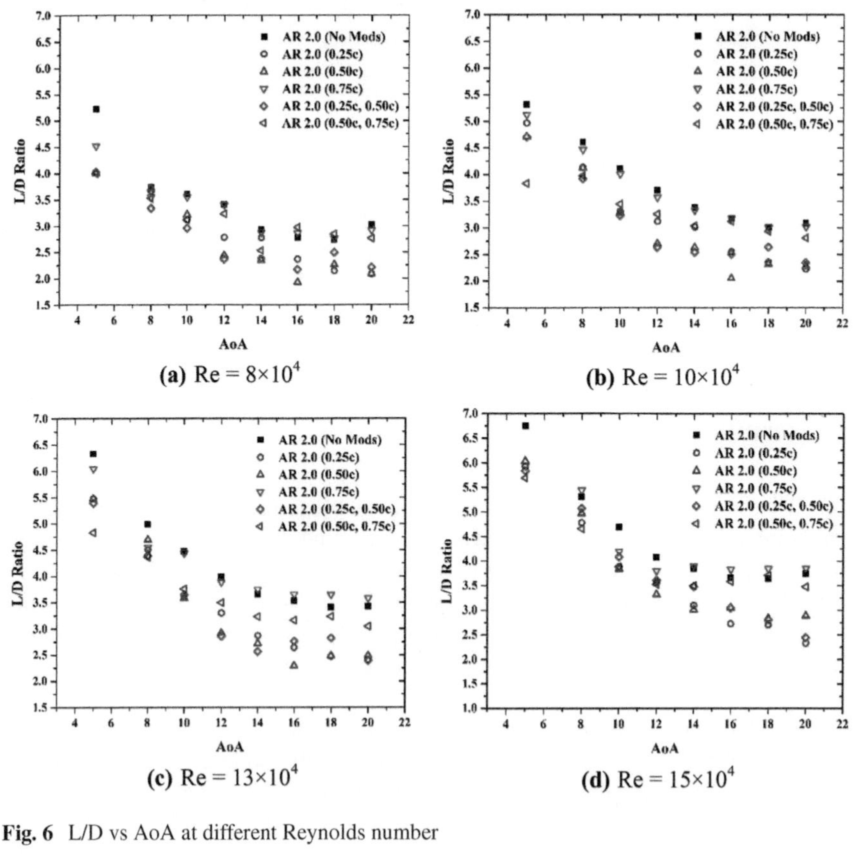

Fig. 6 L/D vs AoA at different Reynolds number

Table 1 Percentage improvement of L/D

AoA	Reynolds number	L/D		Improvement of 0.75c over unmodified wing (%)
		Unmodified wing	0.75c config. wing	
14°	15×10^4	3.85	3.90	1.42
	13×10^4	3.66	3.75	2.44
16°	15×10^4	3.65	3.83	4.83
	13×10^4	3.53	3.65	3.46
18°	15×10^4	3.64	3.84	5.49
	13×10^4	3.40	3.65	7.25
20°	15×10^4	3.74	3.85	2.90
	13×10^4	3.42	3.58	4.64

enhances the momentum exchange within the boundary layer. This increased turbulence helps to energize the flow, preventing premature separation and allowing the boundary layer to remain attached to the wing surface for longer periods.

At higher Re and AoAs, this effect becomes more pronounced, as the inverted dimples effectively disrupt the laminar flow and induce early transition to turbulence. This transition increases the boundary layer's energy, reducing the adverse pressure gradient's impact and promoting reattachment. Consequently, the wing experiences reduced drag and improved lift characteristics, leading to better overall aerodynamic performance.

The ability of inverted dimples to promote boundary layer reattachment highlights their potential in optimizing aerodynamic designs, particularly in applications where maintaining attached flow is critical. Further investigation into the precise mechanisms of this flow reattachment and the role of dimple geometry and placement could lead to even more effective aerodynamic enhancements. By understanding how these factors influence flow behavior, engineers can design more efficient wings and other aerodynamic surfaces, leveraging the benefits of inverted dimples to achieve superior performance across a range of conditions. These findings not only highlight the importance of precise placement but also suggest that the 0.75c configuration could be further optimized and explored for various aerodynamic applications, potentially leading to significant advancements in wing design and performance.

4 Conclusion

The use of inverted dimples as passive, sub-boundary layer vortex generators has demonstrated potential in improving aerodynamic performance, particularly at higher Reynolds numbers and angles of attack. The conclusion drawn from the current study can be summarized in the following points:

- They enhance aerodynamic performance, especially at higher Reynolds numbers and angles of attack.
- The dimples energize airflow and promote boundary layer reattachment, reducing drag and increasing lift.
- The 0.75c configuration shows superior performance over other configurations and the unmodified wing.

These findings highlight the importance of precise dimple placement and suggest further exploration and optimization of this technique for various aerodynamic applications.

Acknowledgements We would like to express our sincere gratitude to the faculty and staff of the Department of Mechanical Engineering at IIT Guwahati for their invaluable support and guidance throughout this research.

References

1. Ligrani, P.M. Flow visualization and flow tracking as applied to turbine components in gas turbine engines, *Meas Sci Tech*, 11, 992–1006 (2000).
2. Isaev, S.A. and Leont'ev, A.I. Analysis of the effect of viscosity on the vortex dynamics at laminar separated flow past a dimple on a plane with allowance for its asymmetry, *J Engineering Physics and Thermophysics*, 74(2), 339–346 (2001).
3. Bunker, R.S. and Donnellan, K.F. Heat transfer and friction factors for flows inside circular tubes with concavity surfaces, *Proc. ASME Turbo Expo*, Atlanta, Georgia, US (2003).
4. Khalatov, A.A. and Byerley, A. Flow characteristics within and downstream of spherical and cylindrical dimple configurations on a flat plate under laminar flow conditions, *Proc. of the ASME Turbo Expo*, Vienna, Austria, ASME Paper No. GT2004-53656 (2004).
5. Ligrani, P.M. and Park. J. Numerical predictions of heat transfer and fluid flow characteristics for seven different dimpled surfaces in a channel, *Num Heat Transfer (Part A)*, 47, 209–232 (2005).
6. Klausmeyer, S.M. A flow physics study of vortex generators on a multi-element aerofoil, AIAA 34th Aerospace Sciences Meeting and Exhibit, Reno (1996).
7. Isaev, S.A., Sudakov, A.G., and Baranov, P.A. Effect of supercirculation in a flow around a thick aerofoil with vortex cells, *Doklady Physics*, 46(3), 199–201, (2001).
8. Rae, A.J., Galpin, S.A. and Fulker, J. Investigation into the scale effect on the performance of sub boundary-layer on civil aircraft high- lift devices, AIAA 2nd Flow Control Conference (2002).
9. Lee, H-T. and Kroo, I.M. Computational Investigation of wings with miniature trailing edge control surfaces, AIAA 2nd Flow Control Conference (2004).
10. Van Der Berg, J.W., Maseland, J.E.J. and Brandsma, F.J. Low speed maximum lift and flow control, *Aerospace Science and Technology*, 8, 389–400 (2004).
11. Verma, A. and Kulkarni, V. Experimental Study of Self-Adjustable Flap for Fixed Wing MAV. *In Gas Turbine India Conference* (Vol. 83532, p. V002T08A009). American Society of Mechanical Engineers (2019).
12. Lin, J.C. Review of research on low profile vortex generators to control boundary-layer separation, *Progress in Aerospace Sciences*, 38, 389–420 (2002).
13. Verma, A., Borah, B. and Kulkarni, V. Cavity for Flow Control and Performance Improvement of Airfoil. *In Gas Turbine India Conference* (Vol. 85536, p. V001T05A004). American Society of Mechanical Engineers (2021).

Influence of Texture Geometry on the Performance of Textured Journal Bearing Considering Fluid Inertia Effect and Cavitation

Debajit Das and Sashindra Kumar Kakoty

Abstract This study compares the influence of three texturing geometries—spherical, cylindrical, and square on the performance of hydrodynamic journal bearings—while considering the mass-conserving boundary conditions and the effects of lubricant inertia. A modified Reynolds equation was developed to incorporate both fluid inertia and cavitation effects. The Jakobsson–Floberg–Olsson (JFO) boundary conditions, which address cavitation, were implemented using Elrod's algorithm. Considering protrusion texturing, the modified Reynolds equation was solved using the Gauss–Seidel iterative approach. The findings indicate that as inertia increases, square texturing reduces friction variables and enhances load-carrying capacity. However, the influence of lubricant inertia on the flow coefficient is minimal.

Keyword Fluid inertia · JFO boundary conditions · Textured journal bearing · Cavitation

1 Introduction

Surface texturing of bearing surfaces has turned out to be a useful method of improving hydrodynamic performance. Surface texture in bearings refers to deliberately changing the bearing surface by adding micro-scale patterns or features [1]. The performance characteristics, however, may vary depending on the shape of the texture. Singh and Awasthi [2] examined the performance characteristics of a journal bearing system with four distinct texturing geometries. They found that the shape of the textures has a substantial influence on total bearing performance. The same research group also evaluated the effect of dimple depth and position [3] and reported that textures formed in the upstream zone of 126°–286° had a considerable effect

D. Das (✉) · S. K. Kakoty
Indian Institute of Technology, Guwahati, Assam, India
e-mail: debajit.das@iitg.ac.in

S. K. Kakoty
e-mail: sashin@iitg.ac.in

U. S. Dixit et al. (eds.), *Current Progress in Engineering Sciences*,
https://doi.org/10.1007/978-981-96-8753-4_11

on performance. Fesanghary and Khonsari [4] proposed a modified switch function for the cavitation algorithm proposed by Elrod [5]. Kango et al. [6] evaluated the performance of spherically textured and grooved hydrodynamic journal bearings. They implemented the Jakobsson–Floberg–Olsson (JFO) boundary conditions and found that the grooved bearing yielded better results than the spherical textured bearing. Syed and Sarangi investigated the influence of texture shape on parallel slider bearings [7]. Syed and Kakoty [8] investigated the spherically protruded and dimple-textured bearings under JFO boundary circumstances. They observed that texturing the bearing's second half yielded superior results. Manser et al. [9] investigated the effect of texture shape on journal bearings under the impact of JFO boundary conditions, utilizing eight different texture shapes. Ma et al. [10] investigated the effects of depth-to-diameter ratio on the average pressure for textured surfaces implementing JFO theory. They suggested critical values of depth-to-diameter ratio, which are twice the minimum film thickness, and the values increase with the increase in the minimum film thickness. Sharma et al. [11] investigated the influence of texturing on journal bearing lubricated with couple-stress fluid using the JFO boundary conditions. Their study showed that the combined effects of couple-stress fluids and texturing lower the performance of journal bearings. The effects of slip and cavitation on dynamically loaded sleeve bearings were analyzed by Wang et al. [12].

In addition to surface texturing, the fluid inertia effect must be taken into account while analyzing journal bearing performance parameters. Kakoty and Majumdar [13–16] conducted various studies to demonstrate the significance of taking the lubricant inertia effect into account when designing journal bearings. The fluid inertia effects in the case of a multilobe conical bearing were evaluated by Walicka et al. [17]. They used a couple of stress fluids as lubricants and were able to obtain increased clearance pressure with the increase in couple stress. Das and Kakoty [18] studied the effect of fluid inertia for textured journal bearings using Reynolds boundary conditions. Javorova and Alexandrov evaluated the combined effects of bearing flexibility and fluid inertia [19]. They demonstrated the need to take into account lubricant inertia when analyzing journal bearing performance characteristics.

The referenced literature accentuates that the geometry of texturing is crucial in determining the performance characteristics of textured journal bearings. Additionally, it highlights the need to address the lubricant inertia effect, particularly when working with low-viscosity lubricants. It also highlights the significance of implementing an effective cavitation boundary condition. However, the combined effects of cavitation and lubricant inertia for textured journal bearings with different texture geometries have not been explored. Therefore, this study aims to conduct a comparative analysis of textured journal bearings with three distinct texture geometries—spherical, cylindrical, and square—while incorporating the fluid inertia effect and applying the Jakobsson-Floberg-Olson cavitation boundary conditions for enhanced accuracy. Additionally, the research employs the numerically efficient progressive mesh densification (PMD) approach to optimize computational performance. [18].

2 Theory

2.1 Governing Equation

The modified Reynolds equation for textured journal bearings considering the lubricant inertia effect is given as [12]

$$\frac{\partial}{\partial\theta}\left(\overline{h}^3\frac{\partial\overline{p}}{\partial\theta}\right)+\frac{\partial}{\partial\overline{z}}\left(\overline{h}^3\frac{\partial\overline{p}}{\partial\overline{z}}\right)=\frac{1}{\delta^2}\frac{\partial\overline{h}}{\partial\theta}-\frac{\text{Re}*}{3\delta^2}\left(\frac{\partial\left(\overline{h}I_\theta\right)}{\partial\theta}+\frac{\partial\left(\overline{h}I_{\overline{z}}\right)}{\partial\overline{z}}\right), \tag{1}$$

The modified form of the governing equation, incorporating the JFO boundary conditions, has been derived as follows.

$$\frac{\partial}{\partial\theta}\left(g\overline{\beta h}^3\frac{\partial\xi}{\partial\theta}\right)+\left(\frac{D}{L}\right)^2\frac{\partial}{\partial\overline{z}}\left(g\overline{\beta h}^3\frac{\partial\xi}{\partial\overline{z}}\right)= 6\frac{\partial\left(\overline{h}\xi\right)}{\partial\theta} - 2\text{Re}^*\left[\frac{\partial\left(\overline{h}I_\theta\xi\right)}{\partial\theta}+\left(\frac{D}{L}\right)\frac{\partial\left(\overline{h}I_{\overline{z}}\xi\right)}{\partial\overline{z}}\right], \tag{2}$$

where g represents the switch function introduced by Elrod [5] and ξ is the fractional film content [8]. The derivation of Eq. (2) has been presented in Ref. [20]. The switch function algorithm is given as follows.

$$g=\begin{cases}0 & \xi<1\\ 1 & \xi\geq 1\end{cases}, \tag{3}$$

The pressure developed due to cavitation can be estimated by

$$\overline{p}=\overline{p}_c+\overline{\beta}\ \ln\xi, \tag{4}$$

where $\overline{\beta}$ is the non-dimensional bulk modulus and $\overline{p}_c$ is the cavitation pressure.

2.2 Film Thickness

Figure 1 presents the schematics of surface texturing in journal bearings. The fluid film thickness for the three different texturing geometries is shown in Fig. 2. For the protrusion textured journal bearing, the non-dimensional film thickness ($\overline{h}$) is given by

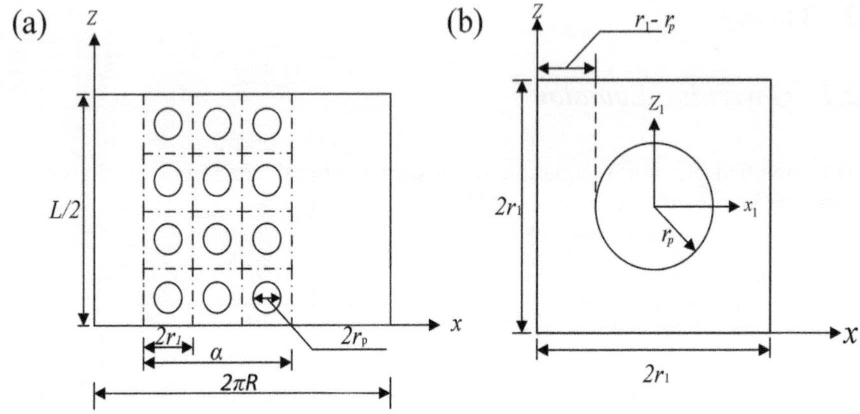

Fig. 1 Schematics of surface texture

$$\overline{h} = 1 + \varepsilon * \cos(\theta) - \overline{h}_p, \tag{5}$$

where $\overline{h}_p$ is the texture depth.

For Spherical texture,

$$\overline{h}_p = \begin{cases} -\left(\frac{1}{8\psi\delta} - \frac{\psi}{2\delta}\right) + \sqrt{\left(\frac{1}{8\psi\delta} + \frac{\psi}{2\delta}\right)^2 - \frac{1}{4\delta^2}\left((\overline{x}_1)^2 + (\overline{z}_1)^2\right)} & \text{when } \overline{x}_1^2 + \overline{z}_1^2 \le 1 \\ 0 & \text{when } \overline{x}_1^2 + \overline{z}_1^2 > 1 \end{cases}, \tag{6}$$

For Cylindrical texture,

$$\overline{h}_p = \begin{cases} \psi & \text{when } \overline{x}_1^2 + \overline{z}_1^2 \le 1 \\ 0 & \text{when } \overline{x}_1^2 + \overline{z}_1^2 > 1 \end{cases}, \tag{7}$$

For Square texture,

$$\overline{h}_p = \begin{cases} \psi & \text{when } |\overline{x}|_1 + |\overline{z}|_1 \le 1 \\ 0 & \text{when } |\overline{x}|_1 + |\overline{z}|_1 > 1 \end{cases}. \tag{8}$$

2.3 Numerical Solution

Equation (2) is discretized and solved by finite difference and Gauss–Seidel iterative methods, respectively. Initially, an inertia-less solution is obtained by setting the modified Reynolds number (Re^*) as 0. This calculated pressure is then utilized as

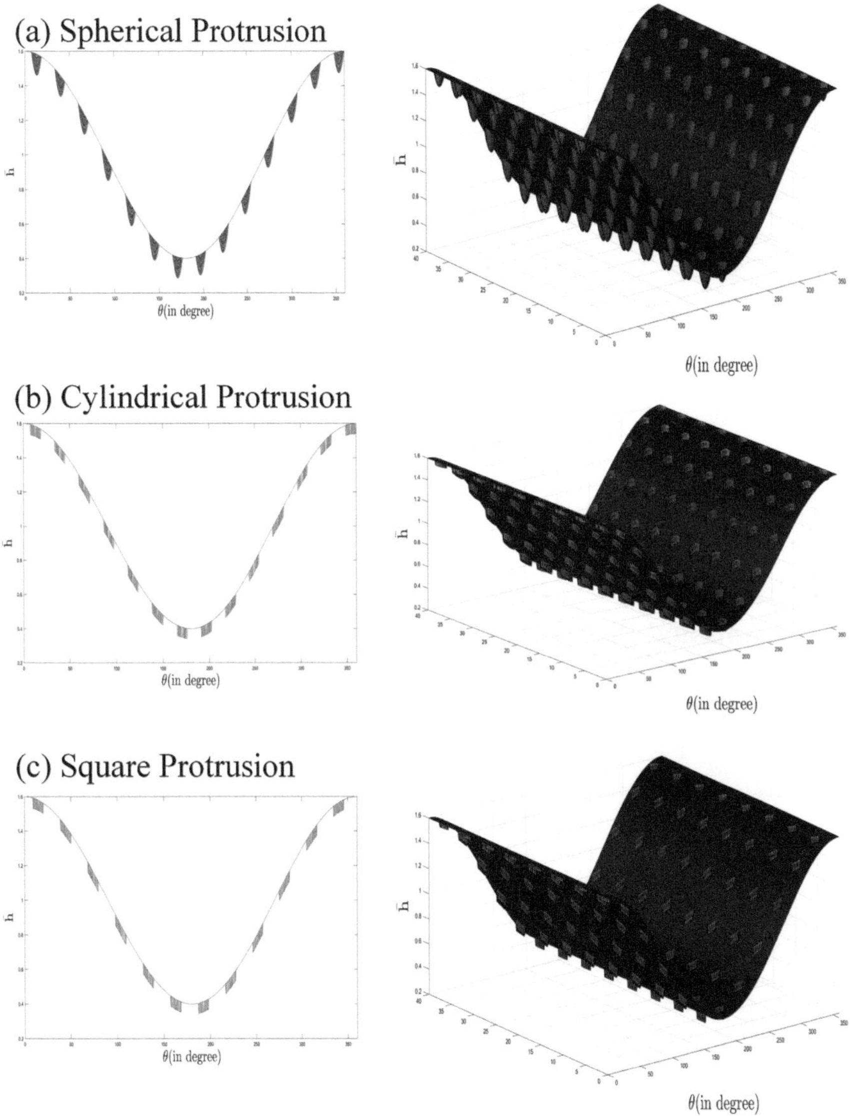

Fig. 2 Fluid film thickness with different texture geometries

input for the following iterations, incorporating finite values of the modified Reynolds number (Re^*). The convergence criteria for level 1 of the PMD method are set as 10^{-3}, which is decreased to 10^{-4} and 10^{-5} in level 2 and level 3 of PMD, respectively. After estimating the fractional film content, the pressure generated within the bearing is calculated using Eq. (3). The non-dimensional load-carrying capacity ($\overline{W}$), the friction variable (μ), and the flow coefficient ($\overline{q}_z$) can be estimated as follows:

$$\left.\begin{aligned}\overline{W}_{\overline{z}} &= \int_0^1 \int_0^{2\pi} \overline{p} \sin\theta \, d\theta \, d\overline{z} \\ \overline{W}_{\theta} &= -\int_0^1 \int_0^{2\pi} \overline{p} \cos\theta \, d\theta \, d\overline{z} \\ \overline{W} &= \sqrt{(\overline{W}_{\overline{z}})^2 + (\overline{W}_{\theta})^2}\end{aligned}\right\}, \tag{9}$$

$$\overline{q}_z = -\frac{1}{3} \int_0^{2\pi} \frac{\overline{h}^3}{2} \left(\frac{D}{L}\right) \frac{\partial \overline{p}}{\partial \overline{z}} \, d\theta, \tag{10}$$

$$\mu = \frac{1}{\overline{W}} \int_0^1 \int_0^{2\pi} \left(\frac{\overline{h}^2}{2} \frac{\partial \overline{p}}{\partial \theta} + \frac{1}{\overline{h}}\right) d\theta \, d\overline{z}. \tag{11}$$

3 Results and Discussion

3.1 Validation

To analyze the performance of textured journal bearings, a MATLAB program has been developed utilizing the PMD method, incorporating JFO boundary conditions and lubricant inertia. This analysis includes surfaces with spherical, cylindrical, and square textures, taking into account the combined influence of fluid inertia and cavitation.

The present simulation results are matched with the results of Elrod [5], as presented in Fig. 3. The findings from the present simulation demonstrate strong agreement with previously published results, and hence further analyses are carried out using the developed code. The progressively densified FDM grid sizes for the PMD method are 88 × 14, 176 × 28, and 352 × 56, respectively. The texturing and cavitation parameters are given in Table 1.

3.2 Steady State Analysis

The variation of $\overline{W}$ with respect to increasing ε and increasing Re^* is shown in Fig. 4. The non-dimensional load-carrying capacity of the bearing increases with increasing ε for all three texturing geometries. The texturing geometries, however, have less influence up to an eccentricity ratio of 0.7. Beyond this, square texturing produces

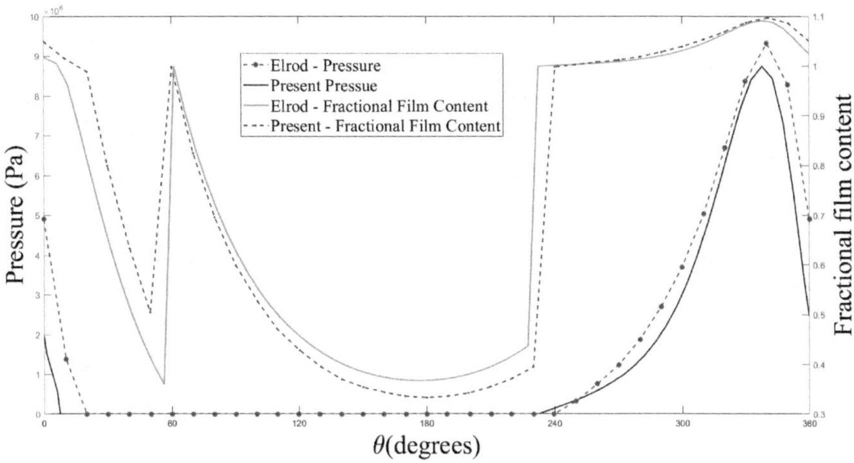

Fig. 3 Validation of current results with Elrod [5]

Table 1 Input parameters

Parameters	Values
Dimple aspect ratio (ψ)	0.06
Dimple area density (S_p)	0.15
Non-dimensional clearance (δ)	0.65
Eccentricity ratio (ε)	0.1–0.9
Texture portion (α)	1
Non-dimensional bulk modulus (β)	60
Modified Reynolds Number (Re^*)	0.5, 1.0, 1.5

a higher load-carrying capacity, followed by cylindrical and spherical texturing. Further, $\overline{W}$ decreases with the increase in modified Reynolds numbers for all the texturing configurations.

The variation of flow coefficients ($\overline{q}_z$) with increasing ε and Re^* is shown in Fig. 5. For all the cases of texturing $\overline{q}_z$ increases with increasing ε. Square texturing exhibits higher flow coefficients when Re^* is 1.0 and 1.5, particularly for $\varepsilon < 0.4$. The effect of lubricant inertia is, however, not significant.

Figure 6 presents the friction variable (μ) with respect to increasing ε and Re^* for spherically textured, cylindrically textured, and square textured journal bearings. μ decreases with increasing ε. The texturing geometry has little influence when $Re^* = 0.0$. With the increase in Re^*, the square texturing exhibits lesser friction variables when $\varepsilon \leq 0.3$. Further, with the increase in Re^*, μ is found to be decreasing for all cases of texturing geometry.

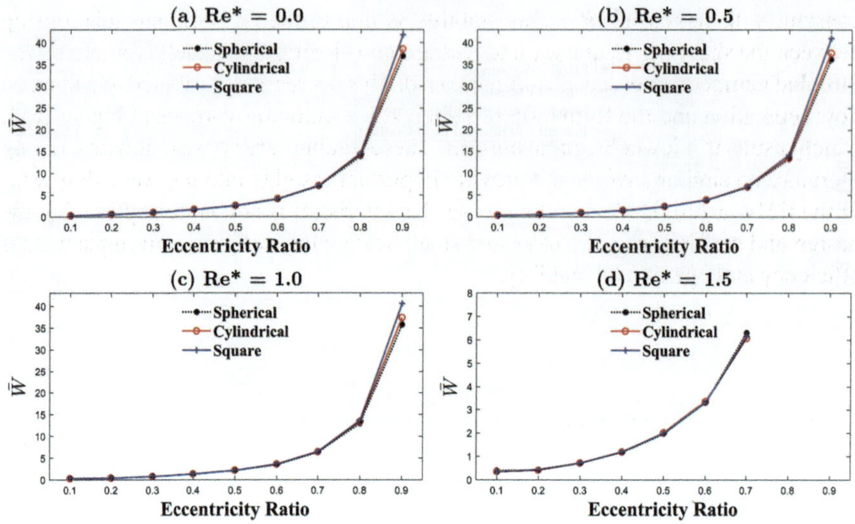

Fig. 4 Non-dimensional load-carrying capacity versus eccentricity ratio

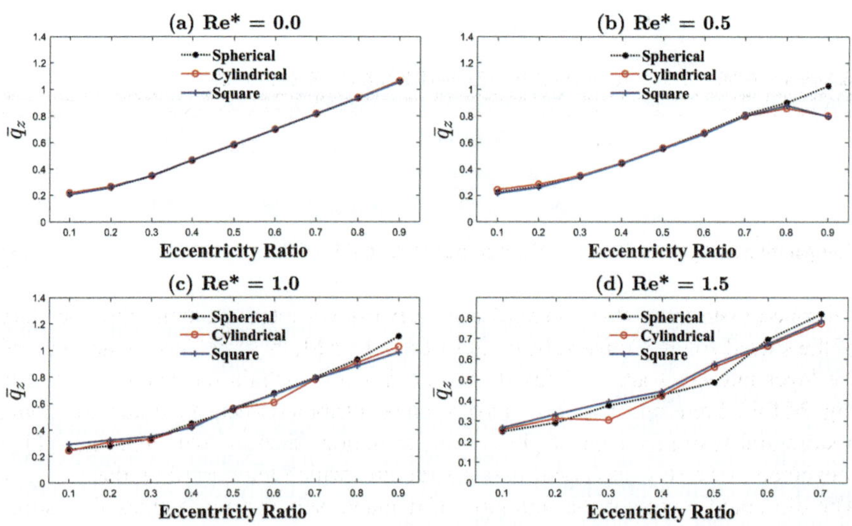

Fig. 5 Flow coefficients versus eccentricity ratio

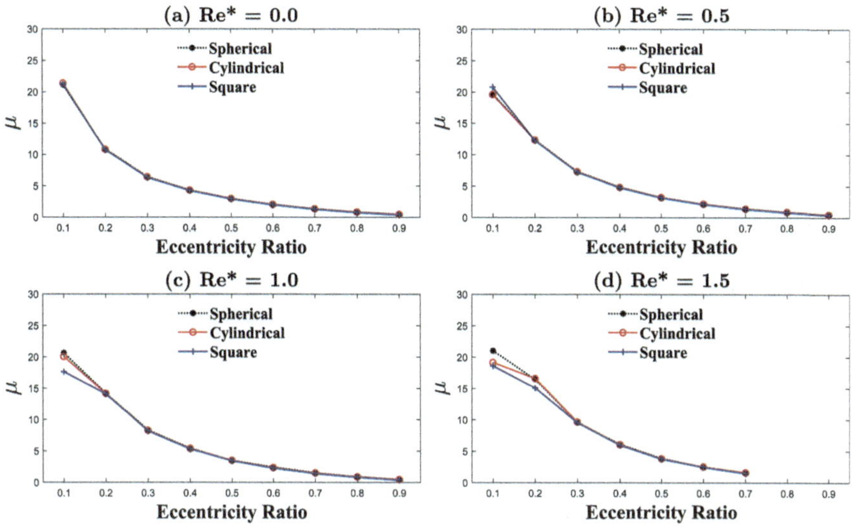

Fig. 6 Friction variable versus eccentricity ratio

4 Conclusion

This study compares the performance of journal bearings with three different texturing geometries, viz. spherical, cylindrical, and square protrusions texturing, considering the effects of lubricant inertia and using JFO boundary conditions. The conclusions derived from the above study are as follows:

- Square protrusion textured bearings produce higher load-carrying capacity, followed by cylindrical and spherical textured bearings for heavily loaded bearings and at high Re^*.
- Square textured bearings produce more flow coefficients for high Re^*. The lubricant inertia effect isn't very effective for the flow coefficient.
- Friction variables are reduced for square protrusion textured bearings when $\varepsilon \leq 0.3$. Increase in Re^* results in further reduction in friction variables for all three texturing geometries.

References

1. Gropper, D., Wang, L., Harvey, T.J.: Hydrodynamic lubrication of textured surfaces: A review of modeling techniques and key findings. Tribol Int. 94, 509–529 (2016). https://doi.org/10.1016/j.triboint.2015.10.009.
2. Singh, N., Awasthi, R.K.: Influence of texture geometries on the performance parameters of hydrodynamic journal bearing. Proceedings of the Institution of Mechanical Engineers, Part

J: Journal of Engineering Tribology. 235, 2056–2072 (2021). https://doi.org/10.1177/1350650120982691.
3. Singh, N., Awasthi, R.K.: Influence of dimple location and depth on the performance characteristics of the hydrodynamic journal bearing system. Proceedings of the Institution of Mechanical Engineers, Part J: Journal of Engineering Tribology. 234, 1500–1513 (2020). https://doi.org/10.1177/1350650120901979.
4. Fesanghary, M., Khonsari, M.M.: A modification of the switch function in the Elrod cavitation algorithm. J Tribol. 133 (2011). https://doi.org/10.1115/1.4003484.
5. Elrod, H.G.: A Cavitation Algorithm. J. of Lubrication Tech. 103, 350–354 (1981). https://doi.org/10.1115/1.3251669.
6. Kango, S., Sharma, R. K., Pandey, R. K.: Comparative analysis of textured and grooved hydrodynamic journal bearing. Proceedings of the Institution of Mechanical Engineers, Part J: Journal of Engineering Tribology. 228, 82–95 (2014). https://doi.org/10.1177/1350650113499742.
7. Syed, I., Sarangi, M.: Hydrodynamic lubrication with deterministic micro textures considering fluid inertia effect. Tribol Int. 69, 30–38 (2014). https://doi.org/10.1016/j.triboint.2013.08.011.
8. Syed, N.R., Kakoty, S.K.: Influence of spherical protruded and dimple texture on the journal bearing performance: A comparative theoretical analysis implementing JFO boundary conditions. Industrial Lubrication and Tribology. 74, 788–795 (2022). https://doi.org/10.1108/ILT-04-2022-0115.
9. Manser, B., Belaidi, I., Hamrani, A., Khelladi, S., Bakir, F.: Texture shape effects on hydrodynamic journal bearing performances using mass-conserving numerical approach. Tribology - Materials, Surfaces and Interfaces. 14, 33–50 (2020). https://doi.org/10.1080/17515831.2019.1666232.
10. Ma, C., Sun, J., Wang, Y., Yu, B., Yu, Q., Tu, Q.: On the optimum dimple depth-over-diameter ratio for textured surfaces. Advances in Mechanical Engineering. 9, 1–8 (2017). https://doi.org/10.1177/1687814017720085.
11. Sharma, N., Kango, S., Tayal, A., Sharma, R.K., Sunil, R.K.: Investigations on the Influence of Surface Texturing on a Couple Stress Fluid–Based Journal Bearing by Using JFO Boundary Conditions. Tribology Transactions. 59, 579–584 (2016). https://doi.org/10.1080/10402004.2015.1094840.
12. Wang, L.L., Zeng, Q.L., Zhang, X.: Cavitation and slip performance analysis of dynamically loaded sleeve bearing. Advances in Mechanical Engineering. 9 (2017). https://doi.org/10.1177/1687814017745439.
13. Kakoty, S.K., Majumdar, B.C.: Effect of fluid inertia on stability of flexibly supported oil journal bearings: linear perturbation analysis. Tribology International. 32, 217–228 (1999). https://doi.org/10.1016/S0301-679X(99)00036-5.
14. Kakoty, S.K., Majumdar, B.C.: Effect of Fluid Inertia on Stability of Oil Journal Bearings. Journal of Tribology. 122, 741–745 (2000). https://doi.org/10.1115/1.1288590.
15. Kakoty, S.K., Majumdar, B.C.: Effect of fluid inertia on the dynamic coefficients and stability of journal bearings. Proceedings of the Institution of Mechanical Engineers, Part J: Journal of Engineering Tribology. 214, 229–240 (2000). https://doi.org/10.1243/1350650001543133.
16. Kakoty, S.K., Majumdar, B.C.: Effect of fluid film inertia on stability of flexibly supported oil journal bearings: a non-linear transient analysis. Tribology Transactions. 45, 253–257 (2002). https://doi.org/10.1080/10402000208982548.
17. Walicka, A., Walicki, E., Jurczak, P.: Inertia effects in a multilobe conical bearing lubricated with a couple stress fluid. International Journal of Applied Mechanics and Engineering. 24, 439–451 (2019). https://doi.org/10.2478/ijame-2019-0027.
18. Das, D., Kakoty, S.K.: Effect of lubricant inertia on steady-state characteristics of textured hydrodynamic journal bearing. Industrial Lubrication and Tribology. 75, 333–342 (2023). https://doi.org/10.1108/ILT-09-2022-0289.
19. Javorova, J., Alexandrov, V.: Effects of fluid inertia and bearing flexibility on the performance of finite length journal bearing. In: IOP Conference Series: Materials Science and Engineering. Institute of Physics Publishing (2017). https://doi.org/10.1088/1757-899X/174/1/012039.

20. Das, D., Kakoty, S.K.: Effect of lubricant inertia on textured journal bearing implementing mass conserving (JFO) boundary conditions. 76, 1186–1196 (2024) https://doi.org/10.1108/ILT-07-2024-0276.

Design, Fabrication, and Flow Analysis of an Open Surface Recirculating Water Tunnel

Deepak Kumar Rathour and Atul Kumar Soti

Abstract A water tunnel is an instrument that provides a uniform water stream for testing flow around bluff bodies. Water tunnels are used for various applications, such as visualizing flow behavior over the bluff bodies and measuring fluid forces on the submerged structure. Water tunnels are used to conduct flow visualization around models. Water tunnels are preferred over wind tunnels for flow measurement, like particle image velocimetry (PIV), because compared to air, the kinematic viscosity of water is smaller by a factor of about 15, yielding higher Reynolds numbers at lower velocities. Simultaneously, the time scales of flow structures are increased, thus leading to an improved determination of transient effects with more ease and accuracy (Greenblatt et al. 2016); Schmidt et al. 2017). The water tunnels can also be used to perform the experimental study of vortex-induced vibration. In this paper, a discussion is made about the design, fabrication, and flow analysis of an open surface recirculating water tunnel having a test section size width $\times$ height $\times$ length: $0.2 \times 0.2 \times 0.7$ m. A Crompton Make MIQ32, a three-phase centrifugal pump of capacity 3 Hp, is used to generate the range of flow velocities from 0.04 to 0.6 m/s. The change in flow speed is achieved by changing the rotational speed of the pump, which can be done with the help of 5 Hp Mistura variable frequency driver. The CAD models of the water tunnel components have been designed in SolidWorks, and flow behavior analysis has been performed in Ansys Fluent. Structural analysis of Water components is performed in Ansys Fluent to provide structural stability. Turbulent intensity of the water tunnel flow is measured with Particle Image Velocimetry and Constant Temperature Anemometer. Turbulent intensity in the test section of the water tunnel is below 2%. The relation between the flow speed and the frequency of VFD has also been established by performing finding the velocities at different VFD frequencies from PIV analysis of the flow in the test section. The relation between the flow speed and the VFD frequency is represented by $U = 0.011682x + 0.000550$.

D. K. Rathour (✉) · A. K. Soti
Department of Mechanical Engineering, Indian Institute of Technology Guwahati, Guwahati, Assam, India
e-mail: r.deepak@iitg.ac.in

A. K. Soti
e-mail: atulsoti@iitg.ac.in

U. S. Dixit et al. (eds.), *Current Progress in Engineering Sciences*,
https://doi.org/10.1007/978-981-96-8753-4_12

Keywords Water tunnel · Particle image velocimetry · Constant temperature anemometer · Wind tunnel

Abbreviations

VFD	Variable frequency drive
CTA	Constant temperature anemometer
PIV	Particle image velocimetry
GI	Galvanized iron
PVC	Polyvinyl chloride
U_{std}	Standard deviation of velocity
U_{mean}	Mean velocity

1 Introduction

A water tunnel is an instrument that provides a uniform flow stream for testing the flow around different bodies. Water tunnels are used to study the hydrodynamic behavior of submerged objects. Water tunnels are similar to Wind tunnels except for the fluid medium. In the case of water as a fluid medium, a higher Reynolds number can be achieved at low flow velocity as the density of the water is nearly 831 times that of air at 20 °C. Water tunnels are used to calculate the lift and drag forces on the different bluff bodies. Mainly, the water tunnels are used to perform the flow visualization around the models. In the case of water as a fluid medium, the time scale of the flow characteristics is longer in comparison to air. The fabrication cost of the water tunnel is much lower than the wind tunnels. Water tunnels are used to perform Particle Image Velocimetry for the flow over different objects. The interaction between the fluid flow and the objects can be visualized to optimize their shapes according to their intended application.

The present water tunnel has been designed to visualize the flow behavior over the different bluff bodies and to perform the experimental studies on vortex-induced vibration of different-shaped bluff bodies in uniform and steady flow. Uniform flow means that at a point, flow velocity should not change with respect to time, while steady means that flow velocity at a point should not change with respect to time. It will also help in visualizing the vortical structures from the VIV of different bluff bodies.

2 Water Tunnel Survey

A water tunnel survey is an important step to get some idea regarding the sizes and designs of the existing water tunnels. From the survey, we can get some idea about the size of the test section and the range of flow speeds. We can also get some idea about the contraction ratio for the design of the contraction section. The water tunnel designed by Zahari [9] has a test section of size (width × height × length): 0.5 m × 0.5 m × 1.3 m. The maximum flow speed in that water tunnel was 0.5 m/s. He used the water tunnel to perform the experimental studies on VIV of the circular cylinder. There was no contraction section in that water tunnel. The water tunnel designed and fabricated by Saeed et al. [6] has a test section of size 0.4 m × 0.5 m × 2.0 m. They have used a contraction ratio of 5 to design the contraction section. In the water, the flow speed in the test section can be varied from 0 m/s to 0.5 m/s. They have kept the length-to-width ratio of the contraction section at 1.16. They have used a 10 hp centrifugal pump having a maximum flow rate of 70 m^3/s to generate the flow in the test section. The water tunnel designed by Kalgutkar et al. [3] has a test section of size 0.1 m × 0.15 m × 0.7 m. The flow speed in the water tunnel can be varied from 0 to 0.21 m/s. They have used a 2D type of contraction section with a contraction ratio of 6 and the length-to-width ratio of 1.06. The water channel used by Soti et al. [8] has a test section of size 0.6 m × 0.8 m × 4.0 m and a 2D type of contraction section.

The design of the present water tunnel has been finalized. We have selected the width-to-length ratio = 1 in designing the contraction section. We have provided the 2.5 D type of contraction to reduce the length of the contraction section. The contraction ratio in the contraction section is 6. A contraction ratio of 3 is provided in the width direction, and 2 in the depth direction. The size of the test section is 0.2 × 0.2 × 0.7 m. The range of flow speeds in the test section is 0.04–0.6 m/s.

3 Water Tunnel Design

This section is focused on the design of the different components of the water tunnel. The design of the water tunnel depends on the application and the space availability. A CAD model of the water tunnel has been designed in SolidWorks. We have used the help of the existing water tunnel designs to design different components of the water tunnel. We have designed all the components of the water tunnel based on the test section size and flow velocities. The designed water tunnel has a test section of 0.2 × 0.2 × 0.7 m. The flow velocities in the test section can range from 0.04 to 0.6 m/s. The CAD model of the final design of the Water tunnel is shown in Fig. 1.

A water tunnel consists of the following components:

1. Inlet Plenum
2. Flow conditioning system
3. Contraction section

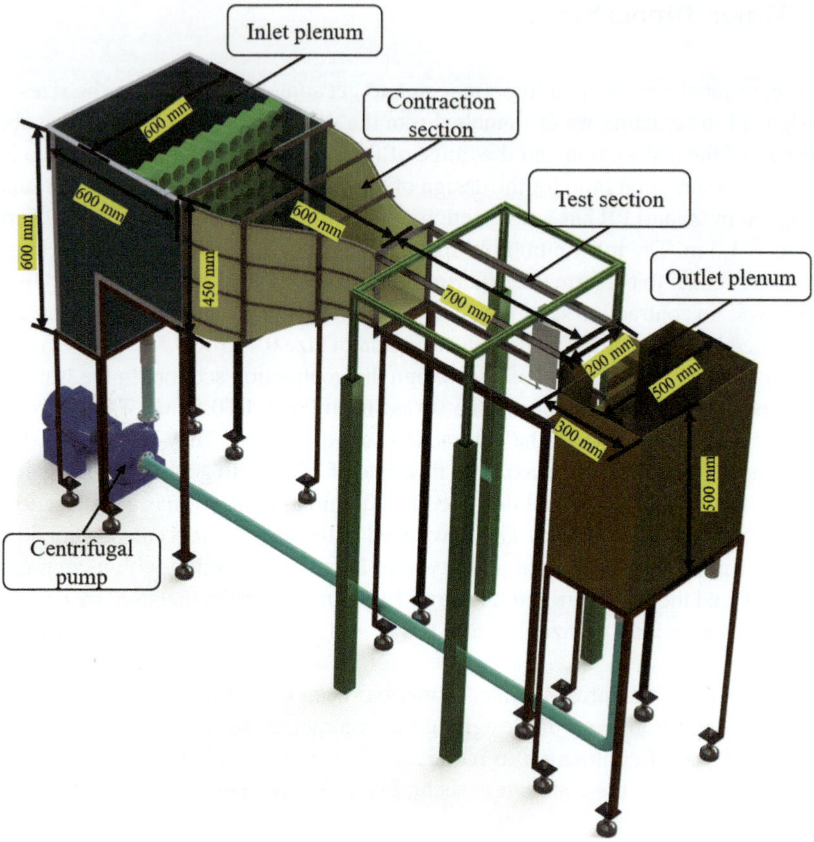

Fig. 1 CAD model of the water tunnel

4. Test section
5. Outlet plenum
6. Pump and piping system

3.1 Inlet Plenum

The Inlet plenum is the first component of the water tunnel, which connects the pump outlet. The flow coming from the pump has high speed and turbulence. To reduce the flow speed and turbulence, the inlet plenum must be large and wide. Because of the large size of the inlet plenum, the flow velocity will be reduced. The primary purpose of reducing the flow velocity is to reduce the disturbances and friction losses (as friction losses are proportional to V^2). The width and height of the inlet plenum can be decided according to the dimensions of the test section, the contraction ratio, and the type of contraction in the contraction section. The contraction ratio depends

Fig. 2 The inlet plenum of the water tunnel

on the speed required in the test section and the flow rate in the test section. Currently, we have a pump with a flow rate of 1130 – 500 l/min. We have decided to make a test cross-section 20 cm × 20 cm. The type of contraction we are providing is 2.5 D with a contraction ratio of 3 in the width direction and 2 in the depth direction. In the present water tunnel, the speed of the water coming from the centrifugal pump is 4 m/s. The CAD model of the Inlet Plenum is shown in Fig. 1, and the prototype with perforated sheet, meshes, and honeycomb is shown in Fig. 2.

By considering the above parameters, the size of the inlet plenum is determined to be 600 mm in width, 450 mm in depth, and 600 mm in length. The water depth in the plenum is up to 400 mm. All the flow conditioning media, such as honeycomb, mesh, and perforated sheets, are placed in this section.

3.2 Flow Conditioning System

A flow conditioning system is used to reduce the turbulence generated by the pumping system and distribute the flow evenly across the inlet of the contraction section. It consists of the following components:

Honeycomb: The purpose of using the flow conditioning system is to reduce the turbulence and disturbances present in the flow. Honeycomb reduces the transverse component of velocity fluctuation. Many researchers have conducted many experiments by changing the configuration and aspect ratio (length-to-diameter ratio) in the honeycomb panels and their effect on the flow quality. According to Vinayak

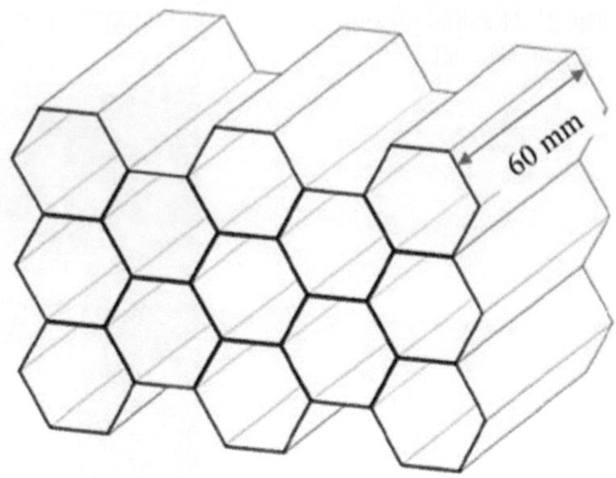

Fig. 3 Schematic view of the honeycomb used in the Water tunnel [3]

et al. [3], if we use the honeycomb with a length-to-diameter ratio of 8–12, then it is most effective in suppressing free-stream turbulence. If honeycombs of the optimal length are used, then the turbulence reduction coefficient is practically independent of the hydrodynamic drag coefficient, and the effectiveness of honeycomb turbulence suppression is almost independent of the oncoming flow velocity [5]. We have used the hexagonal cell honeycomb with a cell size of 6.4 mm and a thickness of 60 mm, hence the length-to-diameter ratio of 9.375. The dimension of the honeycomb panel is 600 mm ×450 mm. The schematic view of the honeycomb is shown in Fig. 3.

Screens: Mehta [4] had conducted the experiments with different material screens and of different wire diameters to examine the effects on the flow behavior. He observed that the plastic screens were different from the metal screens because of their different weaving properties and elastic fiber modules. Karnik and Tavouris [2] found that the large-spacing screens are more effective in reducing the mean flow non-uniformities than small-spacing screens of the same solidity and shape. Since flow quality depends on the wire diameter of screens, the selection of wire diameter becomes an important parameter along with the hole size of the screen. The wire diameter of the screen is chosen based on the Reynolds number, and for the selected screen, it is 50. We have used an SS 304 square cell screen of size M12 after the honeycomb. M represents the number of holes per inch^2 area. Some researchers found that higher opening screens should be used before the honeycomb, and small opening screens should be used after the honeycomb. One SS 304 perforated sheet, and a square hole screen of size M 10 have been used before the honeycomb.

3.3 Contraction Section

An iterative design for a contraction section installed for a low-speed small wind tunnel. The contraction section was designed with the help of 5th degree polynomial, which is mentioned in Eq. 1. The variables are shown in Fig. 4.

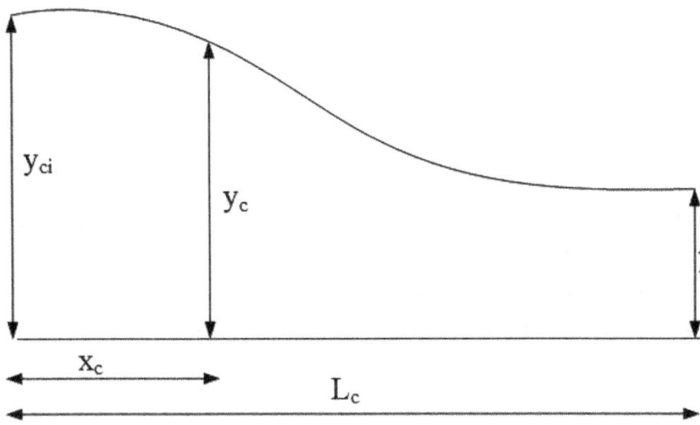

Fig. 4 Schematic of 5th degree polynomial

$$Y_c = Y_{c1} - (Y_{c1} - Y_{c0})\left[6\left(\frac{X_c}{L_c}\right)^5 - 15\left(\frac{X_c}{L_c}\right)^4 + 10\left(\frac{X_c}{Y_c}\right)^3\right] \quad (1)$$

We have used the contraction section of a 2.5D type contraction. We have provided a contraction ratio of 3 in the width direction and a contraction ratio of 2 in the depth direction. In the width direction $Y_{c1} = 0.3$ cm, $Y_{c0} = 0.1$ cm due to symmetry. And $Y_{c1} = 0.4$ cm, $Y_{c0} = 0.2$ cm in the vertical direction. Hence, we have used a contraction ratio of 6. The inlet cross section of the contraction section is 600 mm $\times$450 mm, and the outlet cross-section is 200 mm. A 50 mm extra height is provided to stop the splashing of water. Many researchers have suggested using the W/L ratio of the contraction section $\leq$1 in their research journals. The length of the contraction section is equal to the inlet width of the contraction section, which is 600 mm. A prototype of the contraction section is shown in Fig. 9.

Numerical simulation of the contraction section

The CFD analysis of the contraction section was carried out in Ansys Fluent software. In addition to simulating the streamlined flow across the section caused by the 5th degree polynomial curve, the primary goals of flow simulation are to determine the velocity contours obtained after the flow passes through the contraction section and to ensure that a straight streamline flow is obtained at the other end. The simulation results are shown in Fig. 5. From the simulation results, it can be seen that the flow at the exit of the contraction section is uniform, and all the streamlines are straight.

The flow analysis in the contraction section has been performed by applying the setup conditions mentioned in Table 1 in the Fluent solver setup. We have used the SST K-ω turbulence model because it is a combination of the k-omega model, which is suitable for near-wall flows, and the k-epsilon model, which is more suitable for flow away from the wall.

Fig. 5 Flow simulation results for the contraction section

Table 1 Fluent solver setup conditions for simulating the flow in Ansys Fluent

Parameters	Conditions
Analysis type	3D steady
Turbulence model	SST, K-ω standard
Solver	Pressure based
Material	Water
Inlet boundary condition	Velocity at inlet = 0.0833 m/s
Outlet boundary condition	Gauge pressure = 0
Outer walls	Symmetry
Walls	No-slip
Residual criteria	10^{-4}

The steady and incompressible form of continuity and Navier-Stokes equations has been solved in the flow simulation in the contraction section, outlet plenum, and in the complete water tunnel. These equations are as follows (2), (3).

Steady and incompressible continuity equation

$$\nabla \cdot \overrightarrow{V} = 0 \tag{2}$$

Steady and incompressible Navier-Stokes equations

$$\rho\left((\overrightarrow{V} \cdot \nabla)\overrightarrow{V}\right) = -\nabla p + \rho\overrightarrow{g} + \mu\nabla^2\overrightarrow{V} \tag{3}$$

where:

- ∇p: It represents the pressure gradient.
- $\rho\overrightarrow{g}$: It represents the body force.

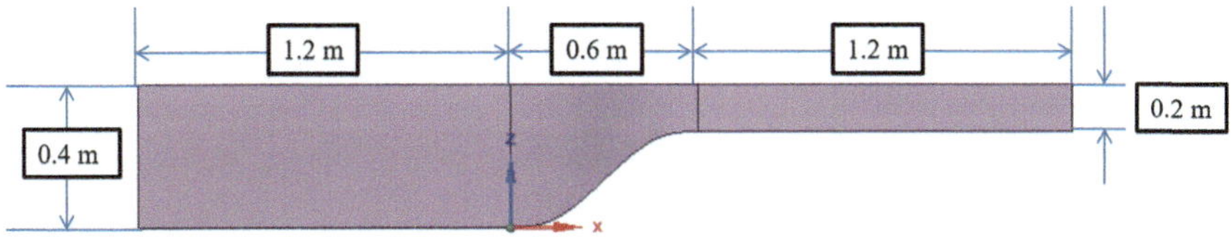

Fig. 6 The mathematical model of the contraction section with the dimensions in m

Table 2 No of nodes and elements in different no of divisions of the edge of the contraction section

Mesh type	No of divisions	No of nodes	No of elements
Mesh-1	120	213,900	193,952
Mesh-2	140	222,640	201,894
Mesh-3	160	230,920	209,418

- $\mu \nabla^2 \overrightarrow{V}$: It represents the diffusion term.

3.4 Mesh Independence Studies

The mathematical model of the contraction section is shown in Fig. 6.

A mesh independence study has been performed by taking three different types of meshes. In the first case flow has been simulated in the contraction section by applying the boundary conditions mentioned in Sect. 1 by starting from the coarse mesh (dividing the edge of the contraction section into 120 divisions). In the second case, we tried to refine the mesh by dividing the edge of the contraction section into 140 divisions. At last, we divided the edge of the contraction section into 160 divisions. The no of nodes and the no of elements present in three cases are shown in Table 2. The u-velocity at the outlet of the contraction section has been calculated from simulations for all three types of meshes, and it has been plotted along the depth of the contraction section. From Fig. 7 we can see that there is no variation in the velocity at the outlet of the contraction section. So we have used mesh 3 (160 divisions of the edge of the contraction section) in studying the flow behavior in the contraction section.

3.5 Test Section

We have designed the Square cross-sectional test section to get equal pressure intensity and equal velocity distribution. The length of the test section should be sufficiently long to observe the effects of vortices and wakes of the test objects. The water tunnel has a test section of size 200 mm × 200 mm and 700 mm long. The minimum and maximum achievable flow velocity in the test section are 0.04–0.6 m/s. The test

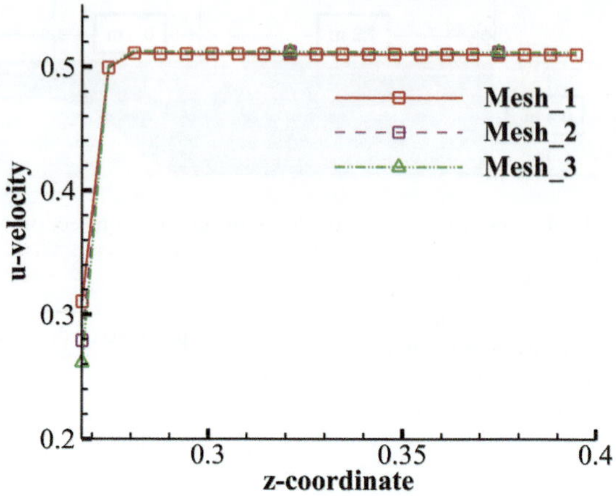

Fig. 7 The variation of u-velocity at the exit of the contraction section along the z-direction (along the depth of contraction section) for different mess sizes mentioned in Table 2

section of the water tunnel consists of the skeleton made from MS L angle, laser-cut flanges, and three glass panels. The side walls of the test section are made of clear toughened glass to avoid any scratches and to provide good transparency. The bottom glass panel has been glued with silicon to the bottom faces of the L angle. The side panels were glued with silicon above the bottom glass panel, and they were also glued to the side faces of the top L angles to provide a uniform cross section.

3.6 Outlet Plenum

The outlet plenum is the fourth component of the water tunnel. It is connected to the test section. There are two holes in the test section. The location and size of these two holes were found by the simulation of the outlet plenum in Ansys Fluent by applying the conditions mentioned in Table 1, except the inlet boundary condition as a velocity inlet of magnitude 0.5 m/s. The holes are located at the place where the vortex formation occurs. In the water tunnel, holes are located 200 mm from the inlet of the outlet plenum. The dimensions of the outlet plenum are 500 mm in depth and 300 mm in width. The sizes of both holes are the same, 80 mm. There is a bottom hole in the outlet plenum for the water flow recirculation in the water tunnel. A simulation of the outlet plenum has been done to find the location of holes for the discharge of water. The simulation results of the outlet plenum are shown in Fig. 8. The prototype of the Outlet plenum is shown in Fig. 9.

After performing a lot of iterations with the distances between the holes and their location from the start of the outlet, we finalized the distance between the holes and their locations from the start of the outlet plenum.

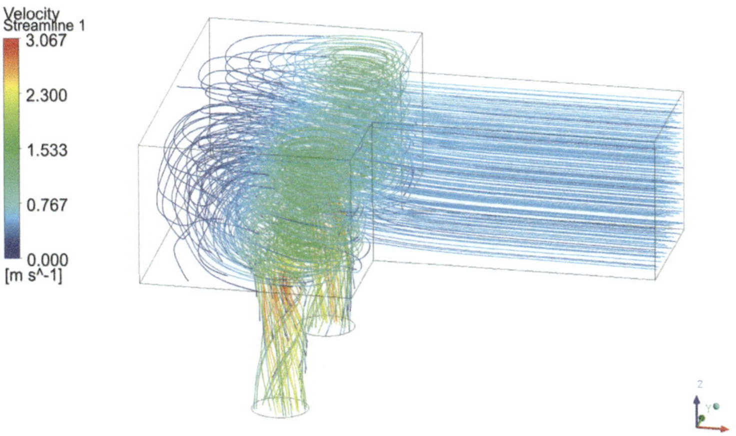

Fig. 8 Velocity streamlines in outlet plenum obtained from Ansys Fluent CFD simulation. The streamlines, colored from blue (low velocity) to red (high velocity), illustrate flow recirculation, swirling, and directional movement

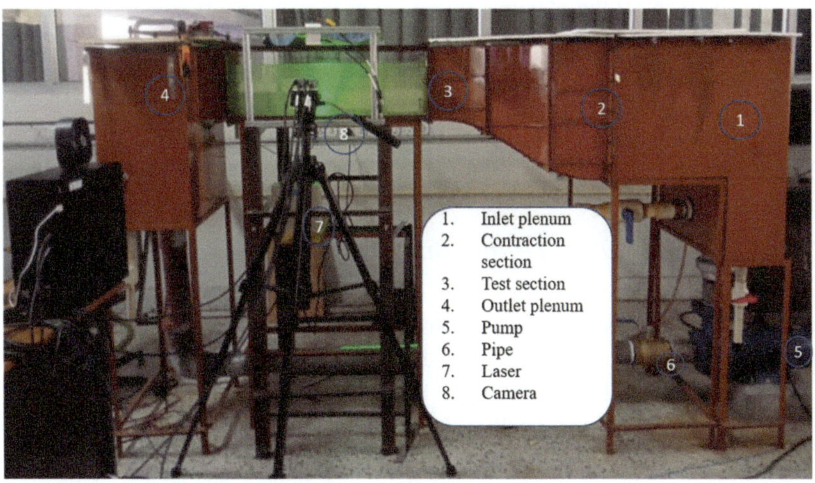

Fig. 9 Fabricated water tunnel setup at IITG workshop with labelled components

3.7 *Pump and Plumbing*

A 3 hp Crompton monoblock pump is used to recirculate the water in the water tunnel. The rotational speed of the pump is controlled by the variable frequency drive (VFD). The pump suctioned the water from the outlet plenum and delivered it to the inlet plenum. A 3″ ball valve is installed on the delivery side to regulate the flow rate in the test section. 3″ GI pipe has been used to connect the pump with the inlet and outlet plenum of the water. Flow speed in the test section can be varied by varying the rotational speed of the pump, which can be varied by varying the VFD frequency through Modbus software from a computer.

3.8 Supporting Frames

All the components of the water tunnel need to be supported above a frame for connection with the pump and piping system. To provide strong support to all the components of the water tunnel, a structural analysis has been performed for all the components and their support structure. Structural analysis has been done in Ansys Static Structure by importing the step file of the 3D CAD model from solid. In Static structural analysis, we can define the material of the components, fixed or supported components, and the load applied on the different sections of the components. After finalizing the designs of the components of the water tunnel and its support structure from a strength point of view, supporting frames were fabricated. All the frames are made of mild steel L angles of size 20 mm $\times$20 mm and 3 mm thick. After fabrication, the supporting structures were painted with red oxide primer and paint to avoid rusting. All the plenums of the water tunnel have been connected to each other with a 3 mm thick neoprene gasket and M6 MS nuts and bolts. After assembling all the components, clear silicon was used to make the tunnel leak-proof. Supporting frames are shown in Fig. 9. To level the water tunnel, the leveling screws have been provided at the bottom of the legs of the supporting structures.

3.9 Complete Water Tunnel Simulation

Numerical simulations of the CAD model of the full water tunnel without support structures have been done to verify the geometric dimensions and the shape of different plenums of the water tunnel. We simulated the flow in the complete water tunnel consisting of Inlet Plenum, Contraction section, Test section, and Outlet Plenum, by considering one honeycomb, one screen, and one perforated sheet in the Inlet Plenum. All the conditions mentioned in Table 1, except the inlet boundary condition, have been used in the Fluent Solver setup. The inlet boundary is given as a velocity inlet of 4 m/s. Inertial and viscous resistance have been defined for screens and honeycomb. We have kept the inertial resistance as 800 for both screens and the honeycomb. The simulation results are shown in Fig. 10. From the velocity contour at the midplane of the test section, it is clear that the velocity in the test section is uniform.

We have plotted the variation of the u and v components of velocity along the depth of the test section. The variation in the u component of velocity along the depth of the test section is shown in Fig. 11a. The variation in the v component of velocity along the depth of the test section is shown in Fig. 11b.

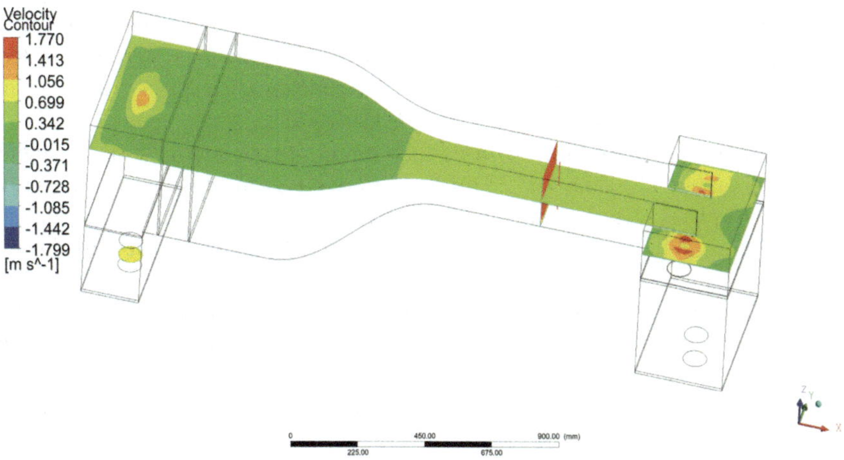

Fig. 10 Velocity contour at the midplane of the test section obtained from CFD simulation

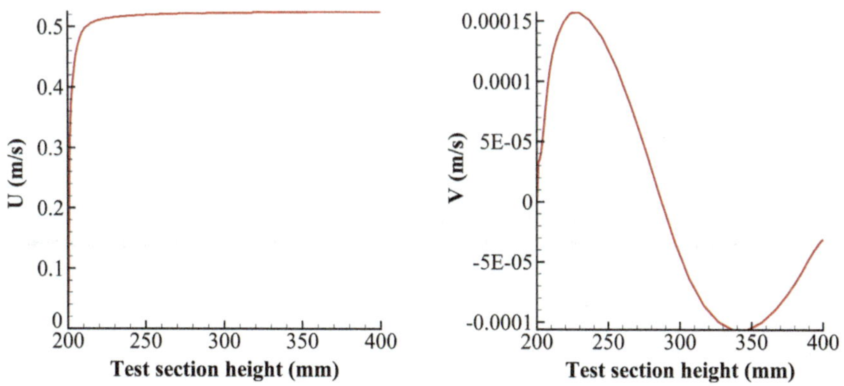

Fig. 11 **a** The variation of the u component of velocity along the depth of the test section, **b** the variation of the v component of velocity along the depth of the test section

4 Experimental Methodology

Turbulent intensity defines the quality of flow in the water tunnel. Turbulent intensity is defined as the ratio of the standard deviation of the velocity to the mean flow velocity. It can be found with intrusive and non-intrusive methods. In the intrusive method, we have to keep the measuring instrument inside the flow, such as a CTA and a Pitot tube. In a non-intrusive method, we don't need to do anything inside the flow. In the non-intrusive method, we can use Particle image velocimetry and laser Doppler. In this section discussion will be made about the description of the experimental setup required to measure the turbulent intensity of the water tunnel. Turbulent intensity can be measured by two methods:

1. Particle Image Velocimetry
2. Hot Wire Anemometer.

4.1 Particle Image Velocimetry

PIV is a non-intrusive method of determining the flow velocity. In non-intrusive measurement, the flow PIV particles are seeded in the flow field, and then the particles are tracked in a 2D plane by illuminating the plane with a monochromatic laser.

Experimental setup for PIV

The experimental set for PIV consists of the water tunnel, high-speed camera, laser optics, and seeding particles. To measure the flow velocity, we filled the water tunnel. For preparing the PIV particle solution, we put the 5 gms of the hollow glass spheres made of Borosilicate glass of mean diameter 9–13 μ m into one liter of water and then put them on a magnetic stirrer for 3–4 h. The density of the PIV particles is 1.10 ± 0.05 g/cm^3. We mix the PIV particle solution in the water tunnel while the flow is happening there. We have used the 200 m watt laser of 532 nm wavelength and 6 mm rod lens to generate the laser sheet. The PIV particles were illuminated by the laser sheet. The camera was positioned parallel to the laser sheet, and the images of the particles were captured as they moved along with the flow. We have used the Baselar camera with model no a2A 1920–160 μm Pro and a 16 mm focal length lens. This is a monochrome camera that has a Sony IMX392 CMOS sensor. The size of the sensor and the maximum resolution are 6.6×4.1 mm and 1920×1200 pixels (2.3 Mega Pixels). The maximum frame rate at full resolution is limited to 160 fps. This camera can be controlled by Pylon viewer software with USB 3.0 cables. These cameras do not require an external power supply. The experimental setup for the PIV is shown in Fig. 9. The images were captured at different frame rates according to the particle displacement. The frame rate was varied to capture the particle displacement from 8 to 10 pixels in the test window. The exposure and gain were also varied according to the flow speed. A 5 Hp VFD was used to control the speed of the three-phase 3 Hp Crompton make centrifugal pump. The flow speed in the water tunnel was varied by varying the frequency in the variable frequency drive. The frequency of the VFD was varied with the help of Modbus software through RS485 to USB converter. The data has been recorded by varying the VFD frequency from 4 to 50 Hz at an interval of 2 Hz. The recorded images were processed using the PIVlab software in MATLAB. The PIV results were analyzed to find the flow velocity components and it's variation at different locations of the test section. After finding the mean and standard deviation of the u-velocity, the turbulent intensity can be found by using Eq. 4.

$$TI = \frac{U_{\text{std}}}{U_{\text{mean}}} \times 100 \tag{4}$$

4.2 Constant Temperature Anemometer

A constant temperature anemometer is used to measure the velocity of fluids and the quick velocity fluctuations. It is an analogue type of instrument. The advantage of CTA over the other methods of measurement is that no data is lost, as the output is an analogue type, which means no information is lost. A constant temperature anemometer works on the basics of convective heat transfer, primarily related to fluid flow. The CTA has a very fine-diameter sensor wire, which is placed in the fluid flow. By making use of very fine sensor wire, electronics, and servo amplifiers, fast velocity fluctuations can be measured. CTA was calibrated from the PIV results. To measure the turbulent intensity in the water tunnel at different locations of the water, the CTA probe needs to be moved. The CTA probe was mounted on the Aluminum rod with the help of an NACA 0012 Aerofoil to avoid the flow perturbations. The CTA probe can be moved along the depth of the test section by using the different lengths of the Aerofoils. The velocity data at different VFD frequencies is recorded, and the turbulent intensity at different flow speeds can be calculated from the Streamware basics software.

5 Results and Discussion

The turbulent intensity measurement results from both methods will be discussed in this section.

5.1 PIV Results

PIV analysis of the flow at a location in the test section of the water tunnel shows that the u-velocity variation is very less with respect to time. It also shows that the u-velocity variation along the depth of the test section is very less. The mean and the standard deviation of the u-velocity at different pump frequencies were calculated from the PIV results. To establish a relation between the VFD frequency and the flow speed, a curve was fitted to the mean flow speed calculated from PIV at different VFD frequencies. Eq. 5 can represent the mean flow velocity in the test section at different VFD frequencies.

$$U = 0.011682x + 0.000550 \quad (5)$$

where x is the VFD frequency and U is the mean flow speed. Table 3 represents the Turbulent intensities in the test section flow calculated from PIV results at different flow speeds.

Table 3 Turbulent intensity from PIV at different VFD frequencies

VFD frequency (Hz)	ū (m/s)	Turbulent intensity (%)
10	0.11737	0.8261
16	0.1874	0.9759
20	0.234	0.7570
26	0.3042	0.3953
30	0.351	0.5614
36	0.421	0.7153
40	0.467	0.7995
46	0.537	0.8353

From Table 3, it can be seen that the Turbulent intensity is below 2% at VFD frequencies ranging from 4 to 40 Hz. In the test section at a point, there is negligible or no variation in the x-component of the mean flow velocity (u-velocity) with respect to time. The existence of the v component of the velocity is almost negligible.

The PIV analysis at 10 Hz VFD frequency and at different depths of the test section has been done to establish the velocity profile. The u-velocities at different depths of the test section have been plotted in Fig. 12.

The velocity profile at a point along the depth of the test section is uniform after the boundary layer thickness. It can be said that the flow in the test is uniform.

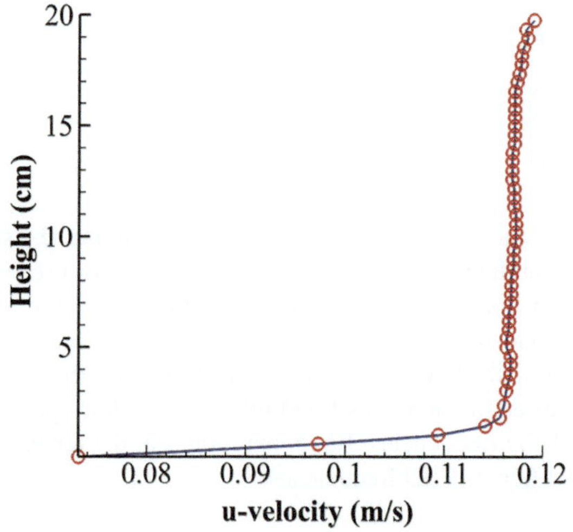

Fig. 12 The variation of the u component of velocity along the depth of the test section

Table 4 Turbulent intensity measured with CTA at different pump frequencies

Pump frequency (Hz)	$\bar{u}$ (m/s)	Turbulent intensity %
4	0.047	1.168
8	0.094	0.720
12	0.140	0.845
16	0.187	1.327
20	0.234	1.136
24	0.280	0.893
28	0.327	0.938
32	0.374	0.615
36	0.421	1.188
40	0.467	0.889

5.2 CTA Results

From Table 4, it can be seen that the turbulent intensity measured with CTA is a little bit higher in comparison to the turbulent intensity measured with PIV. This is due to the effect of the placement of the probe in the flow. The upcoming flow may get disturbed due to the probe. The turbulent intensity at different VFD frequencies (=pump frequencies) is below 2%. So, it represents that the flow in the test section is steady and uniform. This water tunnel can be used to study the flow visualization across the different bluff bodies.

6 Conclusions and Recommendations

The water tunnel, having a test section size of 200 mm × 200 mm and 700 mm in length, has been designed and fabricated. The water tunnel is completely leak-proof. The water tunnel consists of five components: (a) Inlet plenum, (b) Contraction section, (c) Test section, (d) Outlet plenum, (e) Pump and piping system. The flow speed in the water tunnel can be varied from 0.04 to 0.6 m/s by varying the VFD frequency. The quality of flow has been analyzed through performing PIV in the test section at different locations both along the length of and depth. From the PIV results, the turbulent intensity in the water tunnel is below 1%. The velocity profile at a point along the test section depth is parabolic in nature, which is shown in the figure. We have also checked the quality of flow with the help of a Constant temperature Anemometer. The turbulent intensity measured along the length and depth of the test section is below 2%. The flow quality analysis with both methods shows that the water tunnel has low turbulent intensity. This water tunnel can be used to perform the flow visualization around the different bluff bodies. It can also be used to perform experimental studies on the flow-induced vibration of different geometric shapes.

The water tunnels must be fabricated with stainless steel sheets, as mild steel will start rusting even after the painting. To connect the different sections of the water tunnel, neoprene gaskets of a suitable thickness should be used. To make the water tunnel leak-proof, clear silicone can be used. The test section of the water tunnel must be fabricated with the help of toughened clear glass. The material of construction of the pump impeller must be Stainless steel. All the pipes and fittings can be of PVC or stainless steel. The rail system should be designed to mount the laser and camera, so that the movement along and across the length of the test section can be easy and accurate. It will also benefit in terms of the alignment of the camera perpendicular to the laser plane. The CTA probe must be mounted on the rail support so that it can be moved along and across the test section to measure the turbulent intensity at different locations.

Acknowledgements The authors would like to acknowledge the financial support from Grant No. ICPSISPTIDF00228xSKD101A (R), provided by Technology Innovation Hub IIT Guwahati. The authors are grateful to the Mechanical Engineering Department at IIT Guwahati for funding the fabrication of a water tunnel for fluid mechanics at IIT Guwahati. Further, we are also grateful to the IIT Guwahati Central Workshop technicians, as they have helped fabricate the water tunnel and experimental setup.

References

1. Greenblatt, D., Mueller-Vahl, H., Strangfeld, C., Ol, M.V., Granlund, K.O.: High advance-ratio airfoil streamwise oscillations: Wind tunnel vs. water tunnel. In: 54th AIAA aerospace sciences meeting. p. 1356 (2016).
2. Karnik, U., Tavoularis, S.: Generation and manipulation of uniform shear with the use of screens. Experiments in Fluids 5(4), 247–254 (1987).
3. Kulkarni, V., Sahoo, N., Chavan, S.D.: Simulation of honeycomb–screen combinations for turbulence management in a subsonic wind tunnel. Journal of Wind Engineering and Industrial Aerodynamics 99(1), 37–45 (2011).
4. Mehta, R.D.: Turbulent boundary layer perturbed by a screen. AIAA journal 23(9), 1335–1342 (1985).
5. Mikhailova, N., Repik, E., Sosedko, Y.P.: Optimal control of free-stream turbulence intensity by means of honeycombs. Fluid dynamics 29(3), 429–437 (1994)
6. Saeed, M., Uddin, E., Mubashar, A., Zahir, S.U., Zaidi, A.A.: Design and development of low-speed water tunnel. In: 2018 15th International Bhurban Conference on Applied Sciences and Technology (IBCAST). pp. 614–619. IEEE (2018).
7. Schmidt, H.J., Woszidlo, R., Nayeri, C., Paschereit, C.: Separation control with fluidic oscillators in water. Experiments in Fluids 58, 1–17 (2017).
8. Soti, A.K., Zhao, J., Thompson, M.C., Sheridan, J., Bhardwaj, R.: Damping effects on vortex-induced vibration of a circular cylinder and implications for power extraction. Journal of Fluids and Structures 81, 289–308 (2018).
9. Zahari, M., Dol, S.: Design and development of low-cost water tunnel for educational purpose. In: IOP Conference Series: Materials Science and Engineering. vol. 78, p. 012040. IOP Publishing (2015).

Numerical Analysis of the Airfoil's Maximum Blade Thickness Position for Variable-Pitch Vertical-Axis Wind Turbine

Dhanjita Medhi and Atul Kumar Soti

Abstract For a long time, people have tried a variety of methods to capture wind energy, and out of these, one of the most promising devices is the vertical-axis wind turbine (VAWT), in particular, the Darrieus turbine. A lot of research has been done to improve the Darrieus vertical-axis turbine's performance, and focusing on the same, the current study presents a pitch angle control mechanism of turbine blades having NACA 6-series profiles, which is called the variable-pitch mechanism. For this variable-pitch vertical-axis turbine, a 2-D computational work has been done with two NACA 6-series blades, i.e., NACA 64,221 and NACA 66,021, to see the impact on VAWT performance. The study's goal is to evaluate the two distinct 6-series profiles in terms of the turbine's torque coefficient and power coefficient. These two profiles differ by their maximum blade thickness position, which is represented as a proportion of the chord line of the airfoil. The numerical simulations are carried out by using the unsteady Reynolds-averaged Navier–Stokes (URANS) equation and SST k–ω turbulence model. The findings show that employing NACA 64,221 over NACA 66,021 leads to a power coefficient improvement of 64% for the optimal tip speed ratio (TSR). Moreover, variable-pitch turbines with 6-series cambered blade profiles are able to produce more torque for all tip speed ratios compared to 6-series symmetric blade profiles.

Keywords NACA 6-series blade · Variable-Pitch VAWT · CFD

Nomenclature

N Blade numbers

D. Medhi (✉) · A. K. Soti
Department of Mechanical Engineering, Indian Institute of Technology Guwahati, Guwahati, Assam, India
e-mail: m.dhanjita@iitg.ac.in

A. K. Soti
e-mail: atulsoti@iitg.ac.in

U. S. Dixit et al. (eds.), *Current Progress in Engineering Sciences*,
https://doi.org/10.1007/978-981-96-8753-4_13

C	Chord length of blade (m)
R	Radius of turbine rotor (m)
λ	Tip speed ratio
Ω	Rotational speed of turbine (rad/s)
V_∞	Flow velocity at inlet (m/s)
σ	Solidity ratio
β	Blade pitch angle (degree)
α	Angle of attack (degree)
A_0	Blade pitch angle amplitude (degree)
Re	Reynolds number
w	Relative flow velocity (m/s)
C_P	Power coefficient
C_m	Torque coefficient
P	Power produced by turbine (watt)
T	Torque generated by turbine rotor (Nm)
ρ	Density of fluid (kg/m^3)
A	Rotor swept area (m^2)
θ	Rotor rotational/azimuthal angle (degree)
k	Turbulence kinetic energy (m^2/s^2)
υ	Kinematic viscosity (m^2/s)
ω	Specific dissipation rate (1/s)
ε	Energy dissipation rate (m^2/s^3)
μ	Dynamic viscosity ($N\text{-}s/m^2$)

1 Introduction

The emissions from burning fossil fuels that cause global warming have altered the environment in many ways since the Industrial Revolution. The quantity of fossil fuels is likewise finite and is running out daily. Therefore, in order to provide sustainable power production in the future, it is imperative to use various alternative energy sources, and in that sense, renewable energy sources are crucial. One of the renewable energy technologies that is developing the fastest is wind energy, which is a limitless resource. Globally, this kind of technology is expanding rapidly, and the wind industry has increased its total installed capacity in many countries. The wind turbine is the primary tool used to harvest wind energy, and hence, research has concentrated on vertical-axis wind turbines as a potential energy extraction method. Vertical-axis wind turbines (VAWT) and horizontal-axis wind turbines (HAWT) are the two main varieties of wind turbines, and each has pros and cons of its own. The major advantage of vertical-axis types is that to can function in more irregular wind patterns and catch wind coming from any direction without yawing [1]. Although HAWTs are more popular because of their superior efficiency and aerodynamics, VAWTs are more practical for on-shore uses [2]. The typical examples of VAWTs include the

Savonius and Darrieus wind turbines, a drag-type and lift-type device, respectively. Airfoil-shaped blades are used in lift-type turbines, namely the H-rotor and Darrieus rotor, to harness the lift force produced by fluid motion around an airfoil at an angle to the blade chord (referred to as the angle of attack, α). In H-rotor or Darrieus rotor, the blades are positioned around the rotor axis in either a fixed-pitch or variable-pitch configuration.

The stationary or fixed-pitch turbine's minor disadvantages are low efficiency, poor self-starting, and a brief effective life due to large swings [3]. On the other hand, the variable-pitch turbines, by changing their blade pitch angle (angle formed by the blade chord line and the tangential flow velocity) with the rotational angle of the turbine, can significantly improve starting torque, which causes the turbine to self-start. Furthermore, a fluctuating pitch mechanism of VAWT can produce torque with minimal torque ripple and high efficiency [4]. Abdalrahman et al. [5] used a 2-D unsteady numerical model, and findings demonstrated that an H-type VAWT's power generation performance can be enhanced by the use of a blade pitch control system. Kirke and Paillard [6] employed two distinct models, DMS (Double-Multiple Stream tube) and 2-D RANS (Reynolds Average Navier Stokes). They claimed that the two main shortcomings of fixed-pitch vertical-axis wind turbines, like their incapacity to self-start and overspeed, can be addressed by variable-pitch wind turbines. Sagharichi et al. [7] demonstrated a CFD study at various solidities to determine the ideal solidity efficiency value for variable-pitch VAWT. The findings demonstrated that for getting the self-starting torque, high solidities are chosen for variable-pitch blades. Rezaeiha et al. [8] examined the Darrieus wind turbine's aerodynamic behavior at different blade pitch angles between -7° to $+3^{\circ}$. The findings demonstrated that -2° Pitch angle at *TSR* (tip speed ratio) 4 resulted in a 6.6% improvement in power coefficient. Majority of the researches cited above has focused on improvement of VAWT performance by implementation of variable-pitch blades. Therefore, the current study has selected variable-pitch VAWT to improve its performance by utilizing two different NACA 6-series airfoils.

For vertical-axis turbines, the airfoil is the most crucial part, and the performance of the turbine is greatly influenced by different airfoils. Mohamed [9] examined 20 distinct symmetry and asymmetry airfoils and demonstrated that when compared to symmetrical NACA airfoils, the S-1046 airfoil increased the maximum power coefficient by 26.73%. Under identical flow and kinematic conditions, [10] used a cambered airfoil to see the impact on aerodynamic performance by implementing an asymmetric NACA 4412 airfoil and a symmetric NACA 0012 airfoil. The findings showed that when operating at low tip speed ratios, Darrieus-type vertical-axis turbines perform better when equipped with cambered airfoils. Additionally, [11] performed a 2-D numerical simulation by using NACA 0012 and compared the v2-f model with the k–ω SST model with low Reynolds number correlation. They discovered that the RANS method worked well for quickly designing or studying low Reynolds number airfoils. Raciti Castelli et al. [12] presented a CFD model for evaluating the energy performance of a straight-bladed Darrieus VAWT and, after validating it with experimental data, proposed a complete simulation for a traditional NACA 0021 three-bladed rotor. Sun and Huang [13] have used NACA-021 series

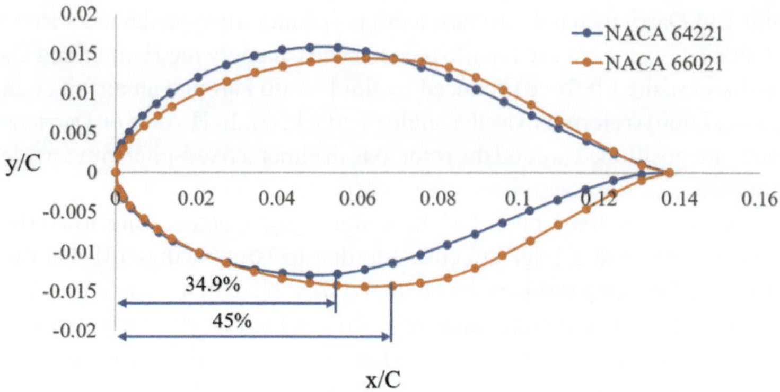

Fig. 1 Airfoils NACA 66,021 and NACA 64,221

airfoils to numerically analyze the impact of boundary layer suction position on airfoil and VAWT performance enhancement. They mentioned that the performance of airfoils and VAWT can be enhanced by using appropriate multi-suction slots while maintaining a consistent total suction momentum. The impact of blade thickness, Reynolds number, and speed ratio was examined by [14] using NACA 0012, NACA 0015, NACA 0018, and NACA 0021 airfoil blades. Similarly, [15] investigated how airfoil thickness affected VAWT performance. Although many studies have been done on a variety of airfoils, NACA 6-series airfoils for variable-pitch VAWT were not the subject of the studies mentioned above.

In the current study, research has focused to do numerical work on a 3-bladed variable-pitch vertical-axis turbine with NACA 64221 and NACA 66021 blade profiles. This study's primary goal is to compare the data from these two profiles in terms of the turbine's aerodynamic performance. Fig. 1 represents the airfoil configurations of NACA 64221 and NACA 66021.

The significance of the NACA 64221 airfoil is that the maximum blade thickness is 21% at the position of 34.9% of the chord line, and the maximum camber is 1.1% at 50% chord line. Similarly, NACA 66021 profile signifies that the maximum blade thickness is 21% at 45% chord length and 0% camber at 0% chord line.

2 CFD Methodology

2.1 *Turbine Design*

The turbine geometry for the current 2-D simulation has been selected on the basis of [7], which is tabulated in Table 1.

Table 1 Geometric details of turbine model

Parameters	Value
Turbine diameter, D	1.03 m
Solidity ratio, σ	0.80
Chord length of blade, C	0.1373 m
Blade numbers, N	3
Blade profile	NACA 64,221, NACA 66,021
Velocity of flow at inlet, V_∞	9 m/s
Tip speed ratio, λ	0.5, 1, 1.5, 2, 2.5, 3, 3.5
Rotational angles, θ	0°–360°
Pitch angle Amplitude, $A°$	10°

2.2 Design Parameters and Aerodynamic Parameters

Solidity Ratio (σ): It is defined by,

$$\sigma = NC/R \tag{1}$$

Tip Speed Ratio (λ): The tip speed ratio (TSR) is defined by,

$$\lambda = \Omega R/V_\infty \tag{2}$$

Coefficient of Power (Cp): The coefficient of power definition is given by,

$$Cp = P/(0.5\rho AV_\infty^3) \tag{3}$$

Coefficient of Moment (C_t): The coefficient of moment definition is provided by,

$$C_t = T/(0.5\rho AV_\infty^2 R) \tag{4}$$

2.3 Variable-Pitch Technique

Implementation of blade pitching helps to decrease the angle of attack, α (the angle between the relative flow velocity, w, and the blade chord line), through which the stalling of the blade and vortex shedding can be prevented. It improves the power and torque coefficients by adjusting the angle of attack to a roughly optimal value. As stated by authors [16–18], the angle of attack, α is related with blade pitch angle, β and azimuthal angle, θ as:

$$\alpha = \tan^{-1}(\sin\theta/(\lambda + \cos\theta)) - \beta \tag{5}$$

For a stationary or fixed-pitch turbine, the blade pitch angle $\boldsymbol{\beta} = 0$, while in the case of a variable-pitch turbine, the blade pitch angle $\boldsymbol{\beta}$ is varying with $\boldsymbol{\theta}$. In the present study, $\boldsymbol{\beta}$ is varying sinusoidally with $\boldsymbol{\theta}$ which is expressed as follows:

$$\beta = A_0.\sin\theta \tag{6}$$

where, A_0 is the blade pitch amplitude, $\theta = \Omega.t =$ Turbine rotational movement, Ω is the rotating speed of the rotor zone, and t is the time of motion.

2.4 Computational Domain and Boundary Conditions

Considering the simulation on the variable-pitch mechanism, the whole computational domain is divided into 3 parts: (a) the fixed domain for allowing the full development of the wake, (b) the rotating domain for implementing the rotational motion of the turbine rotor, and (c) the blade sub-domain for giving the pitching motion to blades. Figures 2 and 3 show the computational domain with boundary conditions. By referring to the literature [4, 7], the length of the computational domain measured from the rotor axis of rotation to the inner wall is 7D, from the rotor axis to the outer domain is 14D, and the domain width is 14D, where D is the diameter of the turbine. The rotating domain size is 1.65D, and the blade sub-domain size is 0.34D.

A velocity intake with a constant flow velocity of 9 m/s is considered for the inlet boundary condition, and a pressure outlet equal to atmospheric pressure is considered

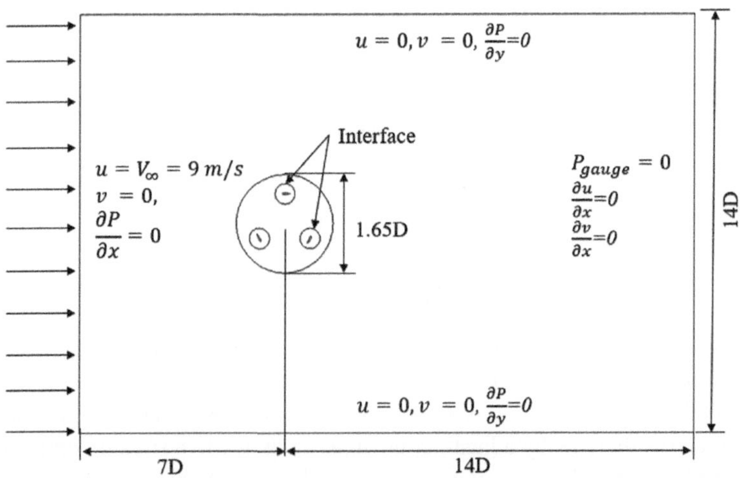

Fig. 2 A diagram of the whole computational domain along with the boundary conditions

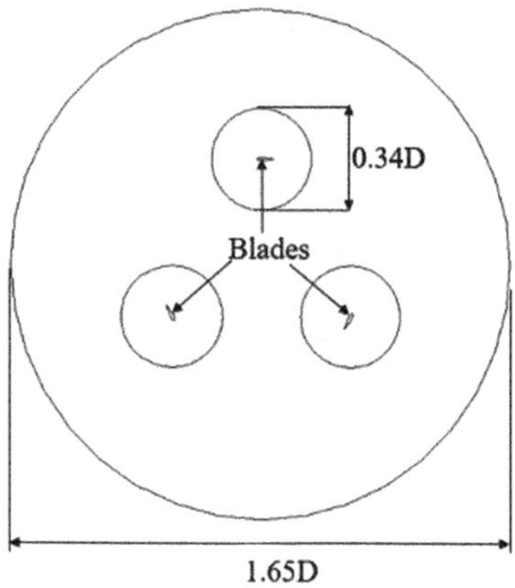

Fig. 3 Rotating domain with blade sub-domains

for the outlet boundary condition. The blade surface is set up as a moving wall with a no-slip boundary condition, while the walls of the fixed domain are described as a "symmetry" boundary condition. The blade sub-domains are tilted with a pitch angle that is related to the azimuthal angle, causing them to rotate independently around the axis that passes through 25% of their chord line. The computation is initialized with constant velocity ($\boldsymbol{u} = \boldsymbol{V}_{\infty} = 9\boldsymbol{m}/\boldsymbol{s}$, $\boldsymbol{v} = 0$) and pressure $\boldsymbol{p} = 0$.

For all simulations at different TSRs within the range of 0.5–3.5, a pitch angle amplitude of 10° is chosen. The 10° pitch angle amplitude has been chosen in accordance with the literature [4] as the highest self-starting performance was achieved by a sinusoidal pitch with 10° amplitude. The TSRs are changing from 0.5 to 3.5 while varying the turbine rotor's rotational speed by maintaining the constant inlet flow velocity. The rotating speed values corresponding to the various TSRs are described in Table 2.

Table 2 Angular velocity of the turbine rotor in relation to TSR

TSR, λ	Angular velocity, Ω (rad/s)
0.5	8.73
1	17.47
1.5	26.21
2	34.95
2.5	43.68
3	52.43
3.5	61.16

2.5 Meshing and Solver Setup

In order to create a mesh for both the rotating and fixed domains, quadrilateral components have been utilized. A face size of 9 mm is used to refine the rotating domain, while a body size of 1.5 mm is utilized to refine the blade sub-domains for all the cases. The boundary layers close to the blades are resolved by capturing a good y^+ value using inflation layers and maximum boundary layer thickness. A closed view of the grid arrangement at the blade wall is displayed in Fig. 4a and b.

In the context of numerical analysis for vertical-axis turbines, the fluid is considered as turbulent and incompressible. Reynolds number (***Re***) is an important parameter for predicting the flow behavior, which can be defined based on the blade chord length is:

$$Re = \frac{wC}{\upsilon} \tag{7}$$

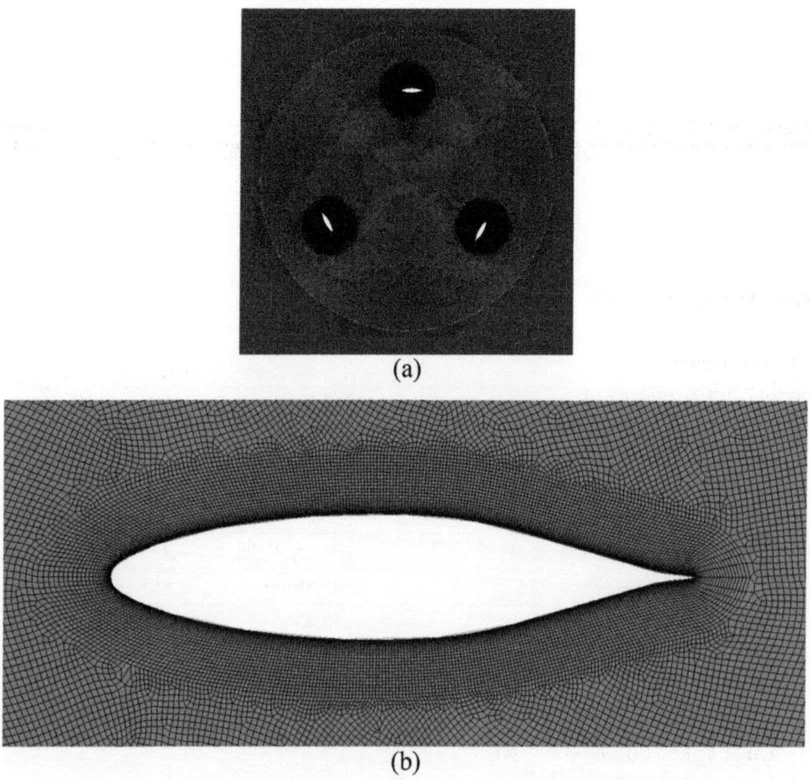

Fig. 4 Details of mesh: a rotating domain, b mesh near blades

The symbols meaning are described in the nomenclature section. Using a mathematical approximation, to calculate the Reynolds number, $\boldsymbol{w}$ can be replaced by $\boldsymbol{\Omega R}$, which is related to $\boldsymbol{\lambda}$ (Eq. 2). According to [19], $\boldsymbol{\Omega R}$ can be written as $\boldsymbol{\lambda_{Cpmax}} V_{\infty}$. Then Eq. 7 becomes

$$Re = \frac{\lambda_{\mathrm{Cpmax}} V_{\infty} C}{\upsilon} \tag{8}$$

Here, $\boldsymbol{\lambda_{\mathrm{Cpmax}}}$ is the tip speed ratio at which maximum power coefficient ($\boldsymbol{Cp}$) is achieved. Table 3 summarizes the Reynolds parameters related to the present model.

Considering the flow as turbulent, an unsteady Reynolds-averaged Navier–Stokes equation is employed to numerically simulate the uneven flow around the turbine. The associated governing equations for this process are described as:

$$\frac{\partial \overline{u}_j}{\partial x_j} = 0 \tag{9}$$

$$\frac{\partial \overline{u}_i}{\partial t} + \frac{\partial \overline{u}_j \overline{u}_i}{\partial x_j} = -\frac{1}{\rho}\frac{\partial \overline{p}}{\partial x_i} + \frac{1}{\rho}\frac{\partial}{\partial x_j}\left(\mu\left(\frac{\partial \overline{u}_i}{\partial x_j} + \frac{\partial \overline{u}_j}{\partial x_i}\right) - \rho\overline{u_i\prime u_j\prime}\right) + \overline{f}_i \tag{10}$$

The fluid density ρ, velocity u, time t, and Cartesian coordinate components x_i and x_j are represented in the equation above. Pressure, kinematic viscosity, and body force per unit volume are represented by the variables p, υ, and $\overline{f}_i$, respectively. It is necessary to model the additional turbulent or Reynolds stress tensor, referred to as $\rho\overline{u_i\prime u_j\prime}$, using traditional turbulence models.

Choosing a turbulent model among the many intricate flow-induced motion simulations is a major step forward, given the turbulent flow around the Darrieus rotor. Thus, most have chosen the SST $k - \omega$ turbulence model because it performs better for boundary layers that have a free shear layer, a negative pressure gradient, or no pressure gradient at all. As mentioned in the above literature, [4, 7, 20, 21], etc. have used this turbulence model in their numerical work, which uses two-equation models, one for the turbulent kinetic energy equation of k in conjunction with an unfavorable time scale equation for ω. The traditional $k - \varepsilon$ model is applied to the shear layers which are away from the surfaces, whereas the $k - \omega$ model is applied to the boundary layer's sub-layer. Hence, the SST $k - \omega$ turbulence model is employed for the present simulation as it can accurately predict flow characteristics in the near-wall regions and distinct flow areas [3]. The formulation of the SST $k - \omega$ model [22] is as follows:

Table 3 Reynolds number parameters

Blade chord length, C (m)	Inlet flow velocity, V_{∞} (m/s)	λ_{Cpmax}	Kinematic viscosity of air (m^2/s)	Reynolds number (Re)
0.1373	9	2.5	14.6×10^{-6}	2.1×10^5

$$\frac{\partial(\rho k)}{\partial t} + \frac{\partial(\rho u_j k)}{\partial x_j} = P_k - \rho\beta^* k\omega + \frac{\partial}{\partial x_j}\left[\left(\mu + \frac{\mu_t}{\sigma_k^*}\right)\frac{\partial\omega}{\partial x_j}\right] \quad (11)$$

$$\frac{\partial(\rho\omega)}{\partial t} + \frac{\partial(\rho\overline{u}_j\omega)}{\partial x_j} = \alpha\frac{\omega}{k}P_k - \rho\beta\omega^2 + \frac{\partial}{\partial x_j}\left[\left(\mu + \frac{\mu_t}{\sigma_\omega^*}\right)\frac{\partial\omega}{\partial x_j}\right] \quad (12)$$

The eddy viscosity, μ_t in the model can be expressed as:

$$\mu_t = \rho\frac{k}{\omega} \quad (13)$$

The coefficients of this model, as follows by [22], are:

$$\alpha = \frac{5}{9}, \beta = 0.075, \beta^* = 0.09, \sigma_k^* = \sigma_\omega^* = 2 \quad (14)$$

The working fluid is set as wind, and FVM (finite volume method) is used to discretize the governing equations. The second-order upwind technique is used to discretize the turbulence, momentum, and transport terms of the Navier–Stokes equations. For the transient formulation, a first-order implicit scheme is employed, and the temporal step size of 1° is taken into consideration in the unsteady simulation, for which a time independence study is conducted in the sections below. The total simulation is conducted for 10 revolutions, considering 50 iterations per time step. After 4–5 complete revolutions, simulations showed that the solution has converged, and the outcomes of the last rotation are chosen for further analysis. The semi-implicit method for pressure-linked equation (SIMPLE) algorithm is used to solve the pressure–velocity coupling. The convergence conditions are specified so that the continuity, x-momentum, y-momentum, k and ω residuals are less than 10^{-4} [23]. A low residual threshold ensures that the numerical solution closely satisfies these governing equations, minimizing numerical errors. For VAWT, due to flow unsteadiness and complex turbulence, a lower residual criterion is often necessary to capture accurate flow patterns. Convergence in aerodynamic forces or moments ensures accurate prediction of load distributions, which are critical for a turbine design perspective. Table 4 summarizes the parametric setup for the current study.

2.6 Grid Independence

A grid independence study is a critical step in numerical simulations, which gives accuracy in results. Therefore, in the present analysis, three unstructured grids comprising coarse, medium, and fine are carried out for the profile NACA 64,221 by using 221,946, 268,566, and 334,648 nodes. Table 5 describes the grid details which has been used for the mesh independence test.

Table 4 Parametric settings for the present numerical study

Parameters	Conditions
Analysis type	2-D transient
Turbulence model	k-ω SST
Material	Air
Scheme	Simple
Gradient	Green Gauss node based
Spatial discretization: Momentum, turbulent kinetic energy, and specific dissipation rate	Second order upwind
Transient formulation	First order implicit
Residual criteria	10^{-4}

Table 5 The grid details along the NACA 64,221 blade profile

Grid	Division count near blade	Boundary layers	Number of nodes	Maximum y^+
Coarse	200	25	221,946	4.87
Medium	250	35	268,566	0.96
Fine	300	45	334,648	0.24

For the three grid resolutions, the total torque coefficient versus azimuth angle at optimal TSR of 2.5 is displayed in Fig. 5 for two revolutions of turbine rotation. The figure shows that the total torque coefficients (all 3 blades) difference between medium and finer grids is considerably less than that of the coarse and finer grids. The coarse mesh shows more negative torque coefficient values comparatively medium and finer ones. Hence, the medium size of grid resolution has been selected for further simulation in the present work in order to reduce computational time and achieve more accuracy in results.

2.7 *Time Independence*

Time independence analysis is an essential stage for numerical simulation since the time step size for transient simulations significantly affects the results. This study ensures the accuracy of the simulation while avoiding excessive computational costs associated with unnecessarily small time steps. Hence, for the time independence test, the simulation is conducted for 3 different time step sizes by using the NACA 64,221 airfoil at an optimal TSR of 2.5. Time step sizes are calculated based on rotational angular increments of 0.25°, 0.50° and 1° which corresponds to 0.00009975, 0.0001995, and 0.000399 s, respectively. Figure 6 shows the variation in total torque coefficient with respect to azimuthal angle.

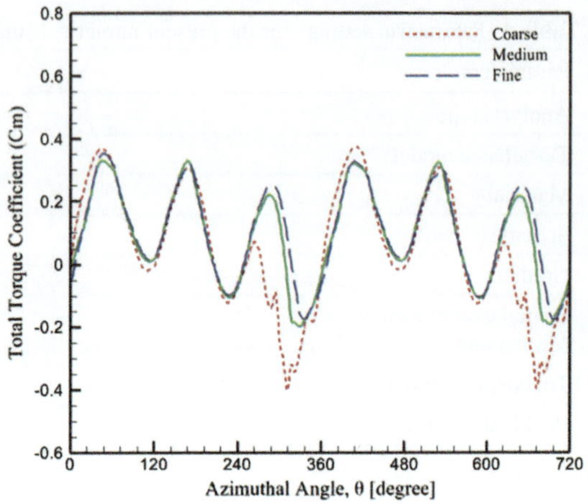

Fig. 5 Total torque coefficient versus azimuthal angle in terms of mesh independency at optimum TSR = 2.5

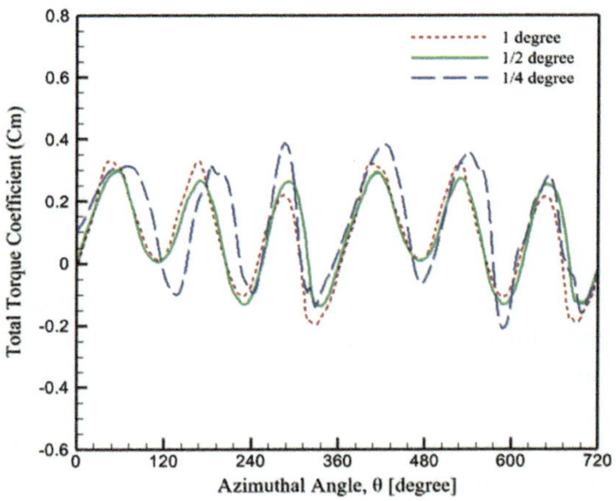

Fig. 6 Total torque coefficient versus azimuthal angle in terms of time independent at optimum TSR = 2.5

The above figure predicts that for time step sizes of 0.50° and 1°, there is very less difference in the torque coefficient between the two, while in the case of 0.25° the difference is comparatively higher. In light of this finding, 1° time step size is selected for all other simulations in order to reduce computational time and ensure accurate results.

2.8 Validation

To validate the numerical simulation, the present results are compared with the experimental results, as provided by [12]. The turbine diameter is 1030 mm, and the height

Table 6 The turbine's specifications, as tested in [12]

Parameters	Value
Turbine diameter, D	1030 mm
Solidity ratio, σ	0.50
Chord length of blade, C	85.8 mm
Blade numbers, N	3
Blade profile	NACA 0021
Velocity of flow at inlet, V_∞	9 m/s
Tip speed ratio, λ	0.5–3.5
Rotational angles, θ	0°–360°

is 1456.4 mm. The blade chord length is 85.8 mm. The dimensions and specifications of the Darrieus wind turbine and CFD model tested by [12] are displayed in Table 6.

Figure 7 presents the comparison of the present numerical results with the experimental data by plotting *Cp* versus λ curve as a validation graph. The computational results are in good agreement with the experimental data, and they predict the power coefficients trend for a wide range of TSRs. However, the current numerical power coefficients are higher than the experimental ones for TSRs larger than or equal to 3.0, and a similar kind of observation has been mentioned by [23, 24]. The discrepancy between the experimental and numerical findings may be the consequence of utilizing a 3-D model for the experiment and a 2-D CFD simulation. Additionally, the current analysis does not include the connections between the rotor hub and blades. Because of this limitation in 2-D numerical analysis, the difference in *Cp* may arise for higher TSRs.

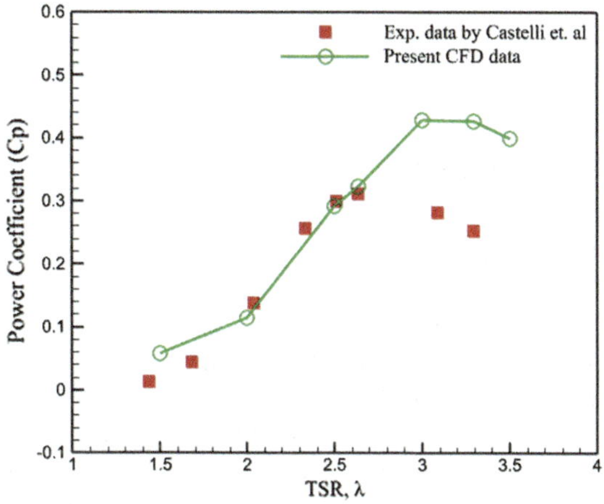

Fig. 7 Validation of the present numerical model with experimental results of [12]

3 Results and Discussion

The results of 2-D computational work for a 3-bladed variable-pitch vertical-axis turbine have been discussed in this chapter. As the goal of this study is to compare the numerical data of NACA 64,221 and NACA 66,021 blade profiles in terms of performance, the results of both profiles are compared in the section below.

At the TSR of 2.5, NACA 64,221 and NACA 66,021 results are compared in terms of single-blade moment coefficient as shown in Fig. 8.

The results are compared for one revolution cycle of turbine rotation, i.e., from 0° to 360° rotational angle. Both profiles exhibit periodic behavior as θ varies from 0° to 360°, reflecting the unsteady aerodynamic forces experienced by a single blade during a full rotation. The moment coefficient value peaks in between $\theta = 60°–90°$ and exhibit negative C_m in certain regions. The graph shows that NACA 64,221 gives higher single-blade moment coefficient values compared to the NACA 66,021 profile, approximately over the entire range. The higher peak C_m of NACA 64,221 makes it more effective in generating torque in the azimuthal phase, where lift dominates. This comparison highlights the importance of airfoil geometry on blade aerodynamics. Further analysis, like integrating C_m over a full rotation, i.e., total average moment coefficient of all blades, and examining power coefficients would provide a more comprehensive understanding. This comparison is conducted in the current study and summarized in the upcoming sections.

Figure 9 displays the average coefficient of moment (C_m) values of the simulated model over the TSR range of 0.5–3.5.

The average $\boldsymbol{C_m}$ values are calculated from the combined moment coefficient values of all three blades, and from these coefficient of moment ($\boldsymbol{C_m}$) values, the average torque of the wind turbine can be found. **The maximum moment coefficient value of 0.144 is obtained at TSR 2 for the profile of NACA 64,221,** and 0.077 is obtained for the NACA 66,021 profile. It implies that there is a 45% improvement in the moment coefficient value by using NACA 64,221 over NACA 66,021.

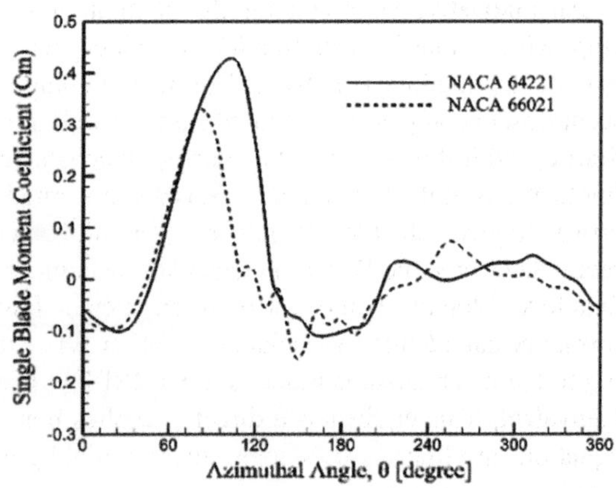

Fig. 8 Single blade moment coefficient comparison of NACA 64,221 and NACA 66,021 blade profiles at $TSR = 2.5$

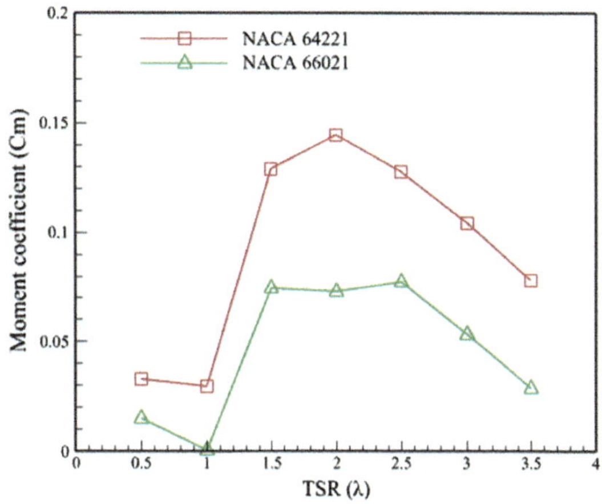

Fig. 9 Average moment coefficient comparison of NACA 64,221 and NACA 66,021 blade profiles

Similarly, the performance of the turbine is determined by the coefficient of power (C_p) curve, which is displayed in Fig. 10.

The graph compares the power coefficient (C_p) versus tip speed ratio (TSR) for the airfoils NACA 64,221 and NACA 66,021. The power coefficient is a key metric in VAWT performance, representing the efficiency of the turbine. It can be seen from the diagram that the modeled airfoil's optimum efficiency appeared around a TSR of 2.5. A maximum C_p of 0.35 at a TSR = 2.5 is achieved by NACA 64,221. Both the airfoils show a low power coefficient at lower TSR, where blades experience unfavorable aerodynamic conditions. The profile NACA 64,221 shows better performance compared to NACA 66,021 at all TSR values. This is consistent with the earlier observation that NACA 64,221 exhibits higher and smoother C_m, which contributes to better torque generation and overall power. Hence, it can be recommended to prioritize the NACA 64,221 profile for maximizing energy output.

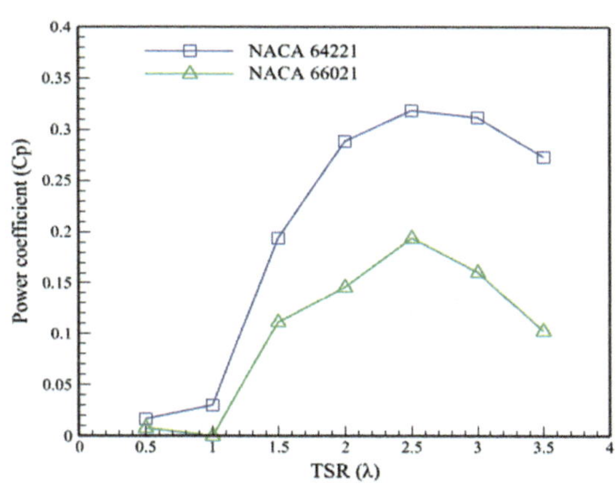

Fig. 10 Power coefficient comparison of NACA 64,221 and NACA 66,021 blade profiles

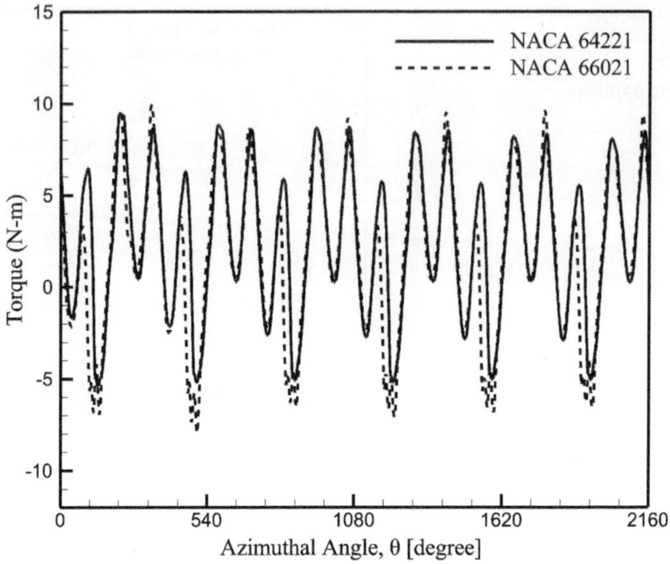

Fig. 11 Average torque comparison of NACA 64,221 and NACA 66,021 blade profiles at $TSR = 2.5$

In a similar manner, the average torque values for both the profiles are compared at TSR = 2.5 in the below Fig. 11. The torque values are obtained from moment coefficient values of all three blades by using Eq. 4.

From the figure, it is observed that both profiles exhibit a highly periodic torque pattern, reflecting the cyclic aerodynamic loading on the blades during rotation. The torque curve trend for both profiles is similar, but the profile NACA 64,221 consistently produces a higher peak torque compared to NACA 66,021. This correlates with the higher moment coefficient (C_m) and power coefficient (C_p) observed in the previous data. The amplitude difference indicates that NACA 64,221 extracts more aerodynamic energy from the wind, resulting in greater torque generation. Although both airfoils exhibit phases of negative torque, NACA 66,021 shows relatively higher negative torque than that of NACA 64,221. Also, the torque fluctuation of NACA 66,021 is more than that of NACA 64,221, which can impact the total torque value or the starting torque values of the variable-pitch turbine. The torque graph confirms the superior performance of NACA 64,221 over NACA 66,021. NACA 64,221 not only generates higher torque but also demonstrates better stability, reduced negative contributions, making it a more efficient airfoil choice for variable-pitch VAWTs.

4 Conclusion

In this paper, the 2-D numerical work has been conducted on a 3-bladed variable-pitch vertical-axis turbine with two 6-series NACA airfoil-shaped blades. The fluid is assumed to be completely turbulent and incompressible. The study employs unsteady Reynolds-averaged Navier–Stokes equations for numerical simulations, indicating a reliance on these equations to model the fluid behavior accurately. The SST k–ω turbulence model is selected due to its effectiveness in handling boundary layers with free shear layers and varying pressure gradients. The medium mesh and time step size of 1° are selected for the numerical simulation. Considering the variable-pitch mechanism, the simulation has been done with a blade pitch angle amplitude of 10°. . For each scenario, the moment coefficient values were calculated for various TSRs (0.5, 1, 1.5, 2, 2.5, 3, 3.5). Additionally, the power coefficient change through various tip speed ratio values was used to create the wind turbine map. Based on the results observed, it is recommended to use the NACA 64,221 blade profile for future study of a variable-pitch vertical-axis turbine, as it gives superior performance compared to the other profiles.

Future research will involve detailed comparisons of various airfoil profiles, particularly NACA 6-series airfoils, to determine which configurations yield superior aerodynamic performance for VAWTs. Investigations can be carried out for the blade pitch control system as they can significantly improve the starting torque and overall efficiency of VAWTs. Also, research can be extended for various blade pitch angle amplitudes with different incoming flow velocities by fixing the rotational speed of the turbine.

Acknowledgements The authors would like to acknowledge the financial support from Grant no. ICPSISPTIDF00228xSKD101A (R), provided by Technology Innovation Hub IIT Guwahati.

References

1. M. R. Hasan, M. R. Islam, G. M. H. Shahariar, and M. Mashud, "Numerical analysis of vertical axis wind turbine," *2014 9th Int. Forum Strateg. Technol. IFOST 2014*, no. October, pp. 318–321, 2014, https://doi.org/10.1109/IFOST.2014.6991130.
2. D. Ragni, C. Simão Ferreira, and M. Barone, "Experimental and numerical investigation of an optimized airfoil for vertical axis wind turbines," *32nd ASME Wind Energy Symp.*, no. November, 2014, https://doi.org/10.2514/6.2014-0171.
3. M. Zamani, M. J. Maghrebi, and S. R. Varedi, "Starting torque improvement using J-shaped straight-bladed Darrieus vertical axis wind turbine by means of numerical simulation," *Renew. Energy*, vol. 95, pp. 109–126, 2016, https://doi.org/10.1016/j.renene.2016.03.069.
4. A. Sagharichi, M. J. Maghrebi, and A. Arabgolarcheh, "Variable pitch blades: An approach for improving performance of Darrieus wind turbine," *J. Renew. Sustain. Energy*, vol. 8, no. 5, 2016, https://doi.org/10.1063/1.4964310.

5. G. Abdalrahman, W. Melek, and F. S. Lien, "Pitch angle control for a small-scale Darrieus vertical axis wind turbine with straight blades (H-Type VAWT)," *Renew. Energy*, vol. 114, pp. 1353–1362, 2017, https://doi.org/10.1016/j.renene.2017.07.068.
6. B. K. Kirke and B. Paillard, "Predicted and measured performance of a vertical axis wind turbine with passive variable pitch compared to fixed pitch," *Wind Eng.*, vol. 41, no. 1, pp. 74–90, 2017, https://doi.org/10.1177/0309524X16677884.
7. A. Sagharichi, M. Zamani, and A. Ghasemi, "Effect of solidity on the performance of variable-pitch vertical axis wind turbine," *Energy*, vol. 161, pp. 753–775, 2018, https://doi.org/10.1016/j.energy.2018.07.160.
8. A. Rezaeiha, I. Kalkman, and B. Blocken, "Effect of pitch angle on power performance and aerodynamics of a vertical axis wind turbine," *Appl. Energy*, vol. 197, pp. 132–150, 2017, https://doi.org/10.1016/j.apenergy.2017.03.128.
9. M. H. Mohamed, "Performance investigation of H-rotor Darrieus turbine with new airfoil shapes," *Energy*, vol. 47, no. 1, pp. 522–530, 2012, https://doi.org/10.1016/j.energy.2012.08.044.
10. P. Ouro, T. Stoesser, and L. Ramírez, "Effect of Blade Cambering on Dynamic Stall in View of Designing Vertical Axis Turbines," *J. Fluids Eng. Trans. ASME*, vol. 140, no. 6, 2018, https://doi.org/10.1115/1.4039235.
11. S. Wang, D. B. Ingham, L. Ma, M. Pourkashanian, and Z. Tao, "Numerical investigations on dynamic stall of low Reynolds number flow around oscillating airfoils," *Comput. Fluids*, vol. 39, no. 9, pp. 1529–1541, 2010, https://doi.org/10.1016/j.compfluid.2010.05.004.
12. M. Raciti Castelli, A. Englaro, and E. Benini, "The Darrieus wind turbine: Proposal for a new performance prediction model based on CFD," *Energy*, vol. 36, no. 8, pp. 4919–4934, 2011, https://doi.org/10.1016/j.energy.2011.05.036.
13. J. Sun and D. Huang, "Numerical investigation of boundary layer suction control positions on airfoils for vertical-axis wind turbine," *J. Mech. Sci. Technol.*, vol. 35, no. 7, pp. 2903–2914, 2021, https://doi.org/10.1007/s12206-021-0614-0.
14. M. C. Claessens, "The design and testing of airfoils for application in small vertical axis wind turbines," *Masters Thesis*, pp. 1–137, 2006.
15. M. Elkhoury, T. Kiwata, and E. Aoun, "Experimental and numerical investigation of a three-dimensional vertical-axis wind turbine with variable-pitch," *J. Wind Eng. Ind. Aerodyn.*, vol. 139, pp. 111–123, 2015, https://doi.org/10.1016/j.jweia.2015.01.004.
16. R. FIRDAUS, T. KIWATA, T. KONO, and K. NAGAO, "Numerical and experimental studies of a small vertical-axis wind turbine with variable-pitch straight blades," *J. Fluid Sci. Technol.*, vol. 10, no. 1, pp. JFST0001–JFST0001, 2015, https://doi.org/10.1299/jfst.2015jfst0001.
17. C. Li, Y. Xiao, Y. lin Xu, Y. xin Peng, G. Hu, and S. Zhu, "Optimization of blade pitch in H-rotor vertical axis wind turbines through computational fluid dynamics simulations," *Appl. Energy*, vol. 212, no. July 2017, pp. 1107–1125, 2018, https://doi.org/10.1016/j.apenergy.2017.12.035.
18. Z. Zhao *et al.*, "Variable pitch approach for performance improving of straight-bladed VAWT at rated tip speed ratio," *Appl. Sci.*, vol. 8, no. 6, 2018, https://doi.org/10.3390/app8060957.
19. S. Brusca, R. Lanzafame, and M. Messina, "Design of a vertical-axis wind turbine: how the aspect ratio affects the turbine's performance," *Int. J. Energy Environ. Eng.*, vol. 5, no. 4, pp. 333–340, 2014, https://doi.org/10.1007/s40095-014-0129-x.
20. S. Wang, Z. Tao, S. Wang, L. Ma, D. B. Ingham, and M. Pourkashanian, "Notice of Retraction: Numerical investigations on dynamic stall associated with low reynolds number flows over airfoils," *ICCET 2010 - 2010 Int. Conf. Comput. Eng. Technol. Proc.*, vol. 5, 2010, https://doi.org/10.1109/ICCET.2010.5486034.
21. G. Abdalrahman, M. A. Daoud, W. W. Melek, F. Lien, and E. Yee, "Design and Implementation of an Intelligent Blade Pitch Control System and Stability Analysis for a Small Darrieus Vertical-Axis Wind Turbine," 2022.
22. J. E. Bardina, P. G. Huang, and T. J. Coakley, "Turbulence modeling validation," *28th Fluid Dyn. Conf.*, no. May, 1997, https://doi.org/10.2514/6.1997-2121.

23. R. Nobile, M. Vahdati, J. Barlow, and A. Mewburn-Crook, "Dynamic Stall for a Vertical Axis Wind Turbine in a Two-Dimensional Study," *Proc. World Renew. Energy Congr. – Sweden, 8–13 May, 2011, Linköping, Sweden*, vol. 57, pp. 4225–4232, 2011, https://doi.org/10.3384/ecp110574225.
24. M. H. Mohamed, A. M. Ali, and A. A. Hafiz, "CFD analysis for H-rotor Darrieus turbine as a low speed wind energy converter," *Eng. Sci. Technol. an Int. J.*, vol. 18, no. 1, pp. 1–13, 2015, https://doi.org/10.1016/j.jestch.2014.08.002.

Exergy Analysis of a Three-Stage Liquid Piston Hydrogen Compressor for Large-Scale Energy Storage of Hydrogen Gas

Digvijay Kumar Gupta and Karuna Kalita

Abstract Long-duration energy storage using hydrogen can play a major role during India's transition from a fossil fuel-based economy to a green energy economy and change India into a self-sustained energy export economy from a major energy import economy. We need a long-duration energy storage system to improve the use and reliability of intermittent renewable energy. Due to the large gravimetric energy density and very small volumetric energy density, it is required to compress hydrogen and store it in a tank or underground cavern for further use in the transportation and industrial sectors. However, hydrogen can be stored in gas, liquid, and solid states. Hydrogen in gaseous form stored at very high pressure, i.e., 25, 35, and 70 MPa, is the most promising solution for transportation applications. This paper presents a comprehensive exergy analysis of multistage hydrogen compression and storage. Several parameters, including isentropic efficiencies of the compressors, pressure ratio, pressure in the storage vessel, preset pressure inside the storage vessel, etc., are considered in this analysis. This analysis has yielded significant insights pertaining to exergy destruction, exergy efficiency, and loss in power happening to the irreversibility of various parts/components of the system. The obtained results demonstrate that the amount of exergy destroyed and, thus, the compressor's exergy efficiency, is significantly affected by the working pressure ratio and the compressor's isentropic efficiency. The analysis found that the increase in storage pressure usually causes an improvement in the exergy efficiency of all three compressors. When the compressor is operating at low isentropic efficiency, η is equal to 0.70, the improvement becomes more apparent. For the low value of preset pressure inside the vessel, i.e., 10 MPa, more exergy is stored during storage of the hydrogen gas at max pressure, i.e., 70 MPa. For the storage process, it has been observed that the exergy destruction is mainly affected by the preset pressure inside the storage tank, and it

D. K. Gupta (✉)
SART, Indian Institute of Technology, Guwahati, India
e-mail: g.digvijay@iitg.ac.in

K. Kalita
Mechanical Engineering Department, Indian Institute of Technology, Guwahati, India
e-mail: karuna.kalita@iitg.ac.in

U. S. Dixit et al. (eds.), *Current Progress in Engineering Sciences*,
https://doi.org/10.1007/978-981-96-8753-4_14

can be minimized by using an isobaric filling process. The mechanical exergy stored in the hydrogen gas during isochoric storage at 700 bar is 380.49 MJ/m^3.

Keywords Exergy · Hydrogen energy · Isobaric storage · Compressor · Long-duration energy

1 Introduction

The adoption of the Paris Agreement in 2016, aimed at significantly decreasing the dangers of climate change and global warming, is proof that understanding of the concurrent environmental repercussions caused by greenhouse gas emissions has increased. The common Indian population will have access to clean fuel thanks to the national hydrogen program being implemented. Both reliance on fossil fuels and the import of crude oil will decline as a result. Another goal is for our nation to become a hub for the export of green ammonia and hydrogen [1]. Due to the intermittent character of most renewable energy sources such as wind, solar, and tidal, it has inherent problems for their exploitation on a large scale. In order to improve the dependability and dispatchability of power production facilities powered by renewable sources, Energy Storage Systems have emerged as a crucial component. By storing extra electricity generated at times of low demand or cheap production costs to be used during the peak periods at a higher price, the integration of Energy Storage Systems delivers additional economic benefits, expediting the transition to a flexible, affordable, and electrified future.

Energy Storage systems can be classified based on discharge time as

(a) Very Short-term storage (<5 min)
(b) Short-term storage (5 min–4 h)
(c) Medium-term storage (4–200 h)
(d) Long-term storage (>200 h).

Very short-term storage includes flywheels and supercapacitors. Short-term storage comprises electrochemical batteries. Medium-term storage includes thermo-mechanical solutions like CAES and pumped hydro storage. Long-term storage includes hydrogen, ammonia, and biogas [2]. Large-scale energy storage technologies, such as Underground Hydrogen Storage (UHS), and medium-scale energy storage systems such as Compressed Air Energy Storage (CAES), play critical roles in stabilizing energy grids and supporting renewable energy integration. Both technologies utilize underground formations such as salt caverns, aquifers, or reservoirs to store energy and exergy (available energy), requiring specific geological conditions for optimal implementation. Under pressure, hydrogen gas and air are injected into subterranean structures in geological storage into UHS and CAES, respectively, where they can later be retrieved as needed. There are currently two large-scale CAES systems in operation: the Huntorf CAES plant was built in Germany in 1978, and the McIntosh CAES plant was built in Alabama, USA, in 1991 [3]. Hypster project,

Table 1 The energy density of hydrogen

The energy density of hydrogen	HHV	LHV
Gravimetric energy density (MJ/kg)	141.86	120.0
Volumetric energy density (MJ/m^3)	12.75	10.8

which was officially launched in Jan 2021, first large-scale Green H_2 underground storage demonstrator at the site Etrez, France, which has a capacity of 3 tons of green hydrogen storage capacity in a salt cavern below 500–1500 m from the earth's surface [4].

Due to the large gravimetric energy density and low volumetric energy density, which is mentioned in Table 1, it is required to compress hydrogen, having a molecular weight of Hydrogen is 2.01594 g/mole and a density of 0.08988 g/l at 0 °C, for use in transportation, storage, and industry. Different hydrogen compression technologies are available and are mainly divided into mechanical and non-mechanical compression of hydrogen. Mechanical compressors include reciprocating piston compressors, Diaphragm compressors, linear compressors, and liquid piston compressors. Mechanical compressors include reciprocating piston compressors, Diaphragm compressors, linear compressors, and liquid piston compressors [5]. A diaphragm compressor is suitable for microscale applications due to its compactness and high efficiency. Liquid compressors are positive displacement devices using liquids to directly compress a gas, working in the absence of a mechanical piston. These are considered quasi-isothermal compression. The heat generated during compression is mostly absorbed by the liquid, and the nearby cylinder wall makes the process nearly isothermal. It has far more efficiency than other mechanical compressors, with approximately 83% [6]. After applying various heat transfer enhancement technologies during compression through the liquid piston compressors, Isothermal efficiency can be achieved from 64 to 95% [5]. Using various technologies, numerous researchers have tried over the years to lower energy loss and enhance the CAES plant's performance [7–12], but very little literature is available on UHS systems.

Exergy analysis is a conventional analysis technique based on the second law of thermodynamics, which discloses the law of energy loss in the transfer process. It can assist us in determining which system components lose a significant amount of energy and in examining how various elements affect system performance. Exergy analysis of a high-pressure four-stage hydrogen gas storage system has been carried out for mixed mass flow rate with varying inlet and storage pressure of hydrogen with predefined mechanical efficiency. The result showed that increasing the inlet pressure of hydrogen gas decreases the energy consumption for the compression and storage process while increasing exergy efficiency [13]. Exergy analysis of the hydrogen supply chain, including underground storage, has been carried out from the concept of exergy-return on exergy-investment (ERoEI), to investigate the exergetic efficiency. The result showed that the overall exergetic efficiency of the electricity-to-hydrogen-to-electricity conversion process can reach up to 25%. The analysis further has indicated that the exergetic efficiency of underground storage components ranges from 72 to 92%. A substantial fraction of the exergy is lost during compression and injection of the stored hydrogen [14]. The effects of multiple parameters on a proposed novel system's exergy efficiency and destruction are discussed. The system's environmental and economic aspects are assessed. The findings indicate that the system's exergy efficiency can be somewhat increased by raising the ambient and electrolyzer operating temperatures. The system's and fuel cell-organic rankine cycle's exergy efficiency quickly rose to 51.91% and 58.33%, respectively, as the fuel cell operating temperature rose. The system exergy efficiency rises by 6.87% when the hydrogen storage pressure is raised [15].

In this study, a novel three-stage liquid piston compression with intercooling and underground storage of green hydrogen produced by an electrolyzer has been considered. Computation methods have been applied for exergy analysis of these systems. A commercial proton exchange membrane water electrolyzer (PEMEL) has been considered in this system, which delivers hydrogen at a pressure of 3–4 MPa [16]. For the Exergy analysis of this proposed system, the variation of the isentropic efficiency of the liquid piston compressor is taken from 70 to 90%. To compress the hydrogen from 3 to 70 MPa, the three-stage liquid piston hydrogen compression with a constant compression ratio has been considered in this exergy analysis. An in-depth analysis of mechanical exergy stored during the isochoric filling process has been considered for predefined pressure in storage vessels. This paper approaches an innovative way to introduce some main concepts and parameters about the exergy evaluation for the three-stage hydrogen underground gas storage subsystem.

2 System Description

The electrolyzer provides hydrogen at 3 MPa, which will be fed to the hydrogen compressor with a compression ratio of 2.86 to increase the volumetric energy density. For minimum power with perfect intercooling between stages, there is a theoretically best relation used for the calculation of the compression ratio of each stage to compress hydrogen from 3 to 70 MPa [17]. A heat exchanger has been employed between the three-stage compressor to improve the isothermal compression efficiency of the storage system. The compression process is to be continued till the pressure at the final stage of the compressor reaches its preset storage pressure. A schematic figure of a three-stage liquid piston hydrogen compressor with a large-scale storage system is depicted in Fig. 1. During exergy analysis of this system, the kinetic and potential energies, as well as the small pressure losses, are ignored in the process, and it is assumed that the filling process in the storage system is steady state. Atmospheric temperature is 298 K and the final pressure inside the storage system is 70 MPa.

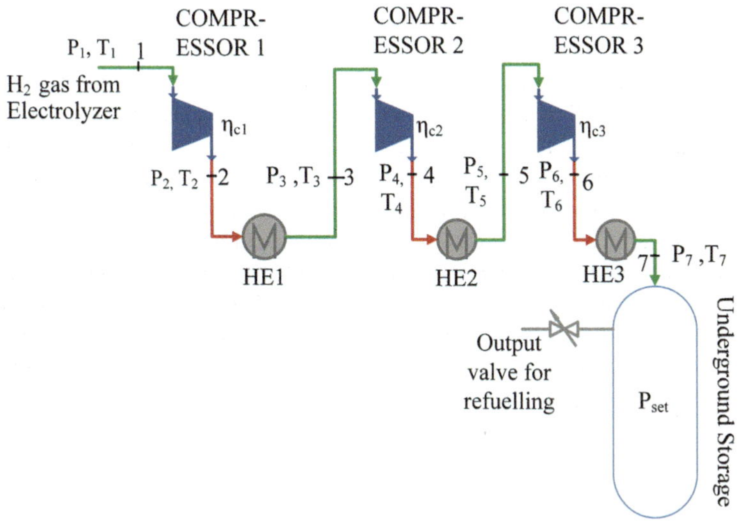

Fig. 1 A schematic of a three-stage hydrogen compression with a large-scale storage system

3 Exergy Analysis

3.1 Compressor

For the compression process, the exergy destruction equation [18] is given below,

$$I_c = \dot{E}_c^c - \dot{E}_c^h \tag{1}$$

I_C= the rate of exergy destruction, i.e., irreversibility,

$\dot{E}_C^c$ = exergy transfer rate to the compressor during compression work.

$\dot{E}_C^h$= exergy transfer rate to hydrogen gas by compressor.

The rate of exergy transfer to the compressor during compressor work is the same as the rate of work given to the compressor ($\dot{W}_c$) for compression of gas

$$\dot{E}_c^c = \dot{W}_c \tag{2}$$

The rate of change in exergy of a process, i.e., the rate of exergy transfer to hydrogen gas, is defined as

$$\dot{E}_c^h = \dot{m}_h\left[\left(h_f - h_i\right) - T_0\left(s_f - s_i\right)\right] \tag{3}$$

where change in enthalpy of the process will be

$$h_f - h_i = c_p(T_f - T_i) \quad (4)$$

And, the change in enthalpy of the process will be

$$(s_f - s_i) = C_p \ln \frac{T_f}{T_i} - R \ln \frac{P_f}{P_i} \quad (5)$$

In Eq. (5), subscript *f* denotes to the post-compression state, subscript *i* refers to the initial state before compression, and subscript *0* denotes the atmospheric state. Also, different parameters $\dot{W}_C$, $\dot{m}_h$, h, s, T, and P are the rate of compression work, hydrogen mass transfer rate, enthalpy, entropy, temperature, and pressure, respectively [7]. In addition, R denotes the specific gas constant, and C_P denotes the constant pressure specific heat of hydrogen. In the above equation, we can find the value of T_f using the following relations:

$$T_f = T_i + \frac{T_{fs} - T_i}{\eta_c}, \quad T_{fs} = T_1 \left(\frac{P_f}{P_i}\right)^{\frac{k-1}{k}} \quad (6)$$

In Eq. (6), T_{fs} indicates the isentropic hydrogen temperature following compression. The rate of exergy transfer to hydrogen gas can be subdivided into two components, referred to as mechanical exergy and thermal exergy, as follows

$$\dot{E}_c^h = \dot{E}_{CM}^h + \dot{E}_{CT}^h \quad (7)$$

where

$\dot{E}_{CM}^h = \dot{m}_h \left[RT_0 ln \frac{P_f}{P_i}\right]$, mechanical exergy

$\dot{E}_{CT}^h = \dot{m}_h [C_P ((T_f - T_i) - T_0 ln \frac{T_f}{T_i})]$, thermal exergy stored by hydrogen gas.

Also, the compressor's exergy efficiency is defined by

$$\eta_{ex} = \frac{\dot{E}_c^h}{\dot{E}_c^c} \quad (8)$$

where

$$E_c^c = W_c = \frac{W_{c,s}}{\eta_c} \quad (9)$$

$$\dot{E}_c^h = \dot{m}_h \left[RT_0 \ln \frac{P_f}{P_i} + C_p \left\{(T_f - T_i) - T_0 \ln \frac{T_f}{T_i}\right\}\right] \quad (10)$$

$$W_{c,s} = M_h \left(\frac{k}{k-1} RT_i \left[\left(\frac{p_f}{p_i}\right)^{\frac{k-1}{k}} - 1\right]\right) \quad (11)$$

3.2 Storage Vessels

The heat exchanger is employed before the entry of the storage system from compressor 3, as shown in Fig. 1. Furthermore, during the isochoric hydrogen filling process, the rate of exergy transfer to the storage vessel, which has constant volume considering only mechanical exergy, can be obtained.

The mechanical exergy transfer rate to the storage vessel by the hydrogen gas during filling is given by

$$\dot{E}_v^h = \dot{m}_h \left[RT_0 \ln \frac{P_f}{P_0} \right] \tag{12}$$

When pressurized hydrogen is delivered to the vessel, the total mechanical energy transferred to the storage vessel is given by

$$\dot{E}_v^h = M_h \left[RT_0 \ln \frac{P_f}{P_0} \right] \tag{13}$$

Moreover, the mechanical exergy transfer rate to the storage vessel during the hydrogen filling process can be measured using

$$\dot{E}_v^h = \dot{m}_h \left[RT_0 \ln \frac{P}{P_0} \right] \tag{14}$$

In Eq. (14), P is the tank pressure continuously increasing during the filling process from P_{set} to P_{max}. P_{set} is the initial condition of storage vessel pressure at the start of the hydrogen filling operation. The total mechanical exergy stored in the vessel when storing hydrogen inside the storage vessel is given by

$$E_v^h = \int \dot{E}_v^h.dt = \int RT_0 \ln \frac{P}{P_0}.dt = VP_{\max} \left[\ln \frac{P_{\max}}{P_0} - \frac{P_{set}}{P_{\max}} \ln \frac{P_{set}}{P_0} + \frac{P_{set}}{P_{\max}} - 1 \right] \tag{15}$$

4 Results and Discussion

In this section, the results calculated during exergy analysis of the three-stage compression with a storage vessel are shown. In Fig. 2, variations of the exergy delivered to the compressor during compression work are depicted versus the range of output pressure of three different compressors. According to the figure, input exergy is to be increased when a high value of compression pressure is required. For instance, when compressor 1 compresses hydrogen from 3 MPa to an output pressure of 8.58 MPa, and if the isentropic efficiency is increased from 0.70 to 0.90,

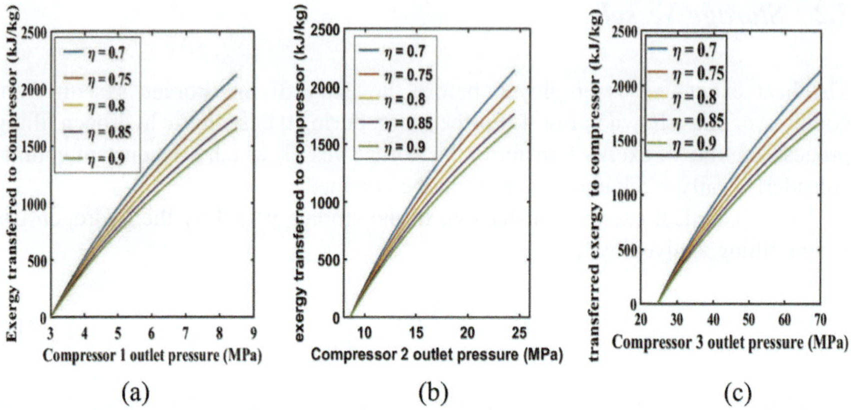

Fig. 2 Exergy transfer versus compressor pressure for different isentropic efficiencies **a** compressor 1, **b** compressor 2, **c** compressor 3

the compressor's input exergy needs are reduced from 2.129 to 1.656 MJ/kg. The depicted figure shows that the low isentropic efficiency causes high total exergy input to all three compressors. Similar trends are being followed in the exergy transfer to the hydrogen gas by the compressor in all three compressors, as shown in Fig. 3. Furthermore, the effects of the compressor's isentropic efficiency during compression are clearly noticeable. For hydrogen gas compression in which the hydrogen is pressurized from 3 MPa to the storage pressure (70 MPa), the system's high exergy input is caused by the use of an efficient compressor. It goes without saying that the system will be more efficient due to less exergy destruction. Variations of the exergy destruction in all three compressors working at different efficiencies are depicted in Fig. 4 in the given range of hydrogen pressure. It is depicted that for low compression efficiency, exergy destruction will be more.

Figure 5 depicts the variations of the exergy efficiency of all three compressors for different ranges of storage pressure and isentropic efficiencies. This figure illustrates how the exergy efficiency of all three compressors typically increases as storage pressure rises. When the compressor is operating at a low isentropic efficiency, this improvement is more apparent, i.e., η is equal to 0.70. For instance, by considering a compressor operating with the isentropic efficiency of 0.70, if the storage vessel pressure changes from 3 to 8.58 MPa, the exergy efficiency is increased from 0.70 to 0.787.

Figure 5 illustrates the total exergy stored in the hydrogen gas during the storage period in the vessel at different preset values of hydrogen pressure inside the vessel. It has been clearly visible from this figure that for the low value of preset pressure inside the vessel, i.e., 10 MPa, more exergy is stored during storage of the hydrogen gas at max pressure, i.e., 70 MPa. In other terms, a low preset value of hydrogen gas inside the vessel requires a high exergy to store hydrogen at the desired maximum pressure, i.e.,70 MPa. This is happening due to exergy destruction due to the high irreversibility. It is also clearly visible that increasing the preset value causes less

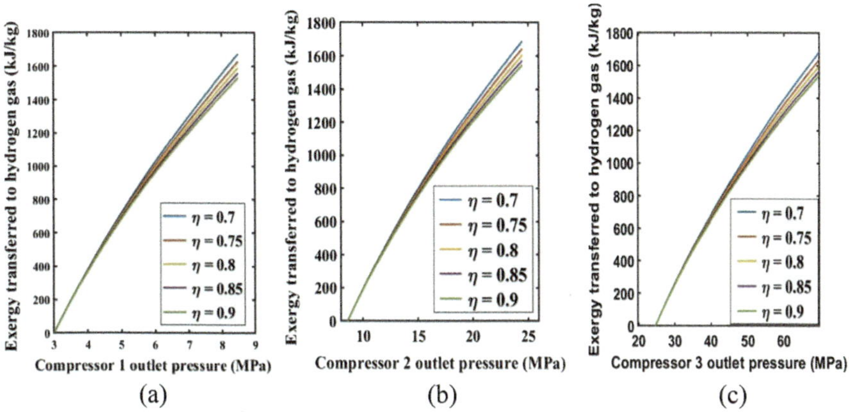

Fig. 3 Exergy transfer to hydrogen gas versus output pressure **a** compressor 1, **b** compressor 2, **c** compressor 3

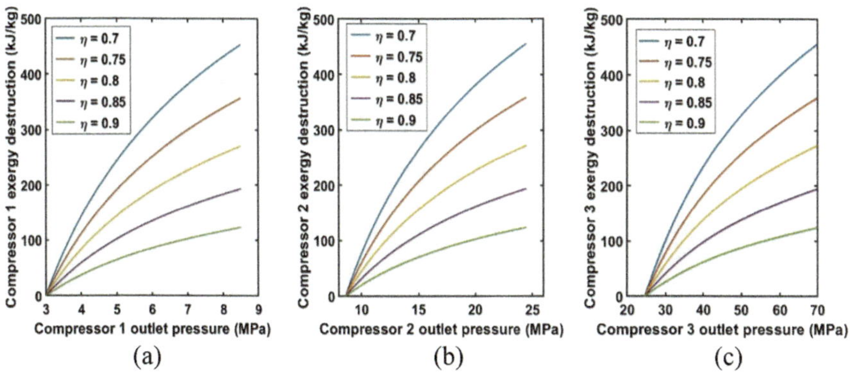

Fig. 4 Exergy destructions versus output pressure **a** compressor 1, **b** compressor 2, **c** compressor 3

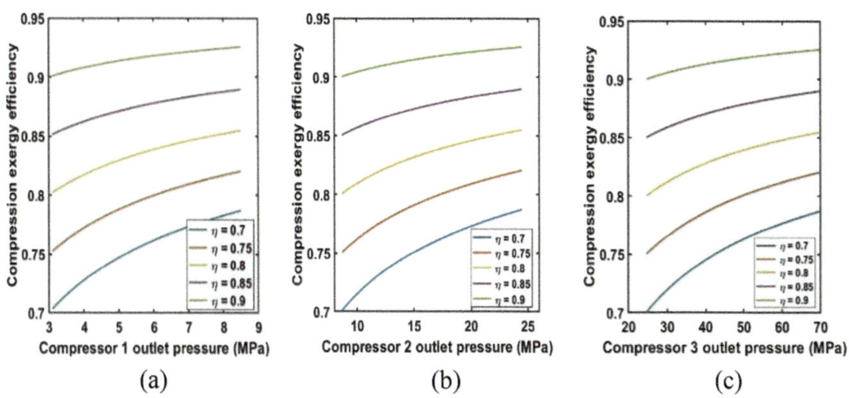

Fig. 5 Exergy efficiency versus output pressure **a** compressor 1, **b** compressor 2, **c** compressor 3

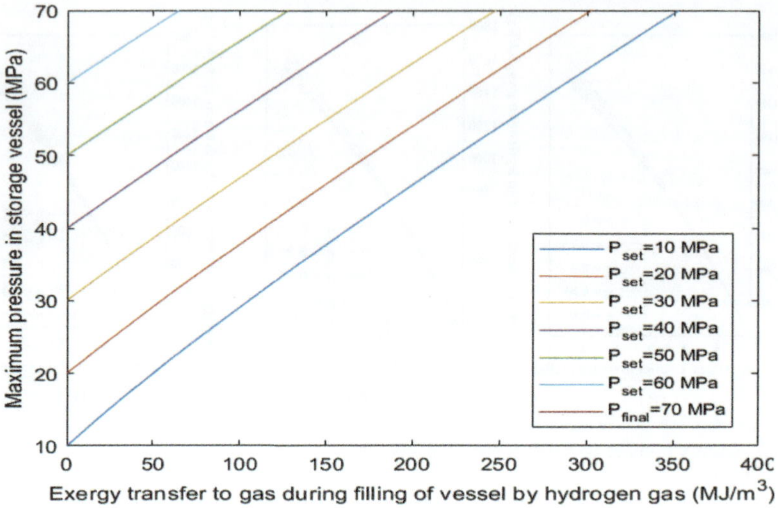

Fig. 6 Exergy transferred to hydrogen gas during filling of the storage vessel versus the maximum pressure of the storage vessel under predefined pressure in the vessel

exergy required to store hydrogen gas, which means that if the preset value is about to the maximum desired pressure, no exergy destruction happens. This can be easily achieved by storing hydrogen at isobaric conditions, which is shown in Fig. 6. The mechanical exergy stored in the hydrogen gas during isochoric storage for 700 bar is found to be 380.49 MJ/m^3.

5 Conclusions

In this paper, the performance of different components of a three-stage liquid piston hydrogen compressor with large-scale storage was evaluated based on exergy analysis. A wide range of parameters, such as isentropic efficiencies of the compressors, pressure ratio, storage pressure of hydrogen gas in the vessel, preset (initial) pressure inside the storage vessel, etc., are considered in this analysis. Applying this parameter, the effects of the output pressure of three compressors on the exergy of the compressors and compressed hydrogen gas are investigated at different isentropic efficiencies of these compressors. The analysis found that the increase in storage pressure usually causes an improvement in the exergy efficiency of all three compressors. When the compressor is operating at a low isentropic efficiency, η is equal to 0.70, the improvement becomes more apparent. The obtained results reveal that the working pressure ratio and the compressor's isentropic efficiency have a significant influence on the value of the exergy destroyed and, consequently, the exergy efficiency of the system. For the storage process, it has been observed that exergy destruction is mainly affected by the preset pressure inside the storage tank.

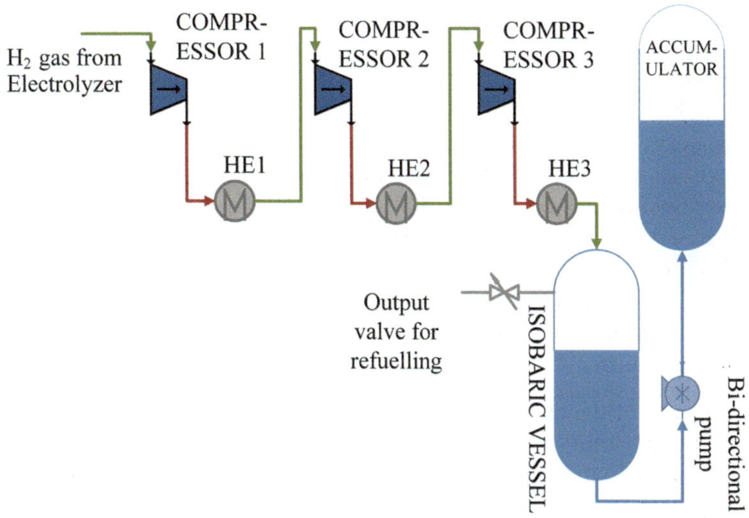

Fig. 7 Three-stage hydrogen gas compressor with isobaric storage of hydrogen

This destruction happens due to free expansion, i.e., irreversibility during gas expansion in vessels. With less preset pressure, there will be more exergy destruction. It can be minimized by using an isobaric filling process, which is illustrated in Fig. 7. During isobaric storage of hydrogen gas, losses due to irreversibility are minimized. Comparison of the exergy stored and destroyed during isochoric and isobaric storage can be explored further. For the low value of preset pressure inside the vessel, i.e., 10 MPa, more exergy is stored during storage of the hydrogen gas at max pressure, i.e., 70 MPa. A new and innovative storage system has been proposed to improve the efficiency of the overall hydrogen storage system. This study has considered hydrogen gas as an ideal gas, so in future work, this exergy analysis can be done considering a real gas. Exergy transfer rate can also be calculated in future work. The comparative analysis of isobaric and isochoric storage of hydrogen gas can also be done.

References

1. Prakash Jha, Braj N Singh, and Tripti Ratn, INDIA'S GREEN HYDROGEN REVOLUTION-An Ambitious Approach. MINISTRY OF NEW AND RENEWABLE ENERGY, 2024.
2. B. Cárdenas, L. Swinfen-styles, J. Rouse, and S. D. Garvey, "Short-, medium-, and long-duration energy storage in a 100% renewable electricity grid: A UK case study," *Energies (Basel)*, vol. 14, no. 24, Dec. 2021, https://doi.org/10.3390/en14248524.
3. A. M. Rabi, J. Radulovic, and J. M. Buick, "Comprehensive Review of Compressed Air Energy Storage (CAES) Technologies," Mar. 01, 2023, *Multidisciplinary Digital Publishing Institute (MDPI)*. https://doi.org/10.3390/thermo3010008.
4. HyPSTER and Clean Hydrogen Partnership, "HyPSTER Project, Hydrogen Pilot Storage for large Ecosystem Replication," Jan. 2022.

5. G. Sdanghi, G. Maranzana, A. Celzard, and V. Fierro, "Review of the current technologies and performances of hydrogen compression for stationary and automotive applications," Mar. 01, 2019, *Elsevier Ltd*. https://doi.org/10.1016/j.rser.2018.11.028.
6. J. D. Van de Ven and P. Y. Li, "Liquid piston gas compression," *Appl Energy*, vol. 86, no. 10, pp. 2183–2191, 2009, https://doi.org/10.1016/j.apenergy.2008.12.001.
7. H. Mozayeni, X. Wang, and M. Negnevitsky, "Exergy analysis of a one-stage adiabatic compressed air energy storage system," *Energy Procedia*, vol. 160, no. 2018, pp. 260–267, 2019, https://doi.org/10.1016/j.egypro.2019.02.145.
8. G. Grazzini and A. Milazzo, "Thermodynamic analysis of CAES/TES systems for renewable energy plants," *Renew Energy*, vol. 33, no. 9, pp. 1998–2006, Sep. 2008, https://doi.org/10.1016/j.renene.2007.12.003.
9. G. Grazzini and A. Milazzo, "A thermodynamic analysis of multistage adiabatic CAES," in *Proceedings of the IEEE*, Institute of Electrical and Electronics Engineers Inc., 2012, pp. 461–472. https://doi.org/10.1109/JPROC.2011.2163049.
10. H. Mozayeni, M. Negnevitsky, X. Wang, F. Cao, and X. Peng, "Performance Study of an Advanced Adiabatic Compressed Air Energy Storage System," in *Energy Procedia*, Elsevier Ltd, 2017, pp. 71–76. https://doi.org/10.1016/j.egypro.2017.03.108.
11. Z. Zhang, M. Wang, T. Zheng, N. Zhang, Y. Zhang, and S. Mei, "Performance Analysis of Advanced Adiabatic Compressed Air Energy Storage System with Constant-volume and Adiabatic Gas storage Model under Multi-cycle Operation," in *iSPEC 2020 - Proceedings: IEEE Sustainable Power and Energy Conference: Energy Transition and Energy Internet*, Institute of Electrical and Electronics Engineers Inc., Nov. 2020, pp. 1425–1430. https://doi.org/10.1109/iSPEC50848.2020.9351278.
12. J. and F. X. and H. Y. and S. J. and W. L. and D. Z. and X. G. and W. H. Liang Lixiao and Hou, "Thermodynamic Analysis of Multi-stage Compression Adiabatic Compressed Air Energy Storage System," in *Advances in Heat Transfer and Thermal Engineering*, Y. Wen Chuang and Yan, Ed., Singapore: Springer Singapore, 2021, pp. 863–868.
13. M. Ozsaban, A. Midilli, and I. Dincer, "Exergy analysis of a high pressure multistage hydrogen gas storage system," *Int J Hydrogen Energy*, vol. 36, no. 17, pp. 11440–11450, Aug. 2011, https://doi.org/10.1016/j.ijhydene.2011.03.087.
14. B. Baghirov, D. Voskov, and R. Farajzadeh, "Exergetic efficiency and CO2 intensity of hydrogen supply chain including underground storage," *Energy Conversion and Management: X*, vol. 24, Oct. 2024, https://doi.org/10.1016/j.ecmx.2024.100695.
15. Z. Meng, K. Wang, J. Di, Z. Lang, and Q. He, "Energy analysis and exergy analysis study of a novel high-efficiency wind-hydrogen storage and power generation polygeneration system," *Int J Hydrogen Energy*, vol. 57, pp. 338–355, Feb. 2024, https://doi.org/10.1016/j.ijhydene.2024.01.013.
16. R. Hancke, T. Holm, and Ø. Ulleberg, "The case for high-pressure PEM water electrolysis," *Energy Convers Manag*, vol. 261, no. December 2021, 2022, https://doi.org/10.1016/j.enconman.2022.115642.
17. "Reciprocating Process Compressor Design Overview," in *A Practical Guide to Compressor Technology*, John Wiley & Sons, Ltd, 2006, ch. 2, pp. 29–62. https://doi.org/10.1002/9780471929789.ch2.
18. M. J. Moran and H. N. Shapiro, "Fundamentals of Engineering Thermodynamics," John Wiley & Sons, Inc.

Strength Analysis and Multi-response Optimization of Friction Stir Welding for Al-Cu Metals Through GRA

Suman Pandipati, Kelli Durgaprasad, Dowluru Sreeramulu, Putti Venkata Siva Teja, and Srihari Palli

Abstract Friction stir welding (FSW) is an energy-efficient, solid-state joining technique with a wide range of applications. It was particularly effective for combining incompatible metals, such as copper (Cu) and Aluminium (Al). The purpose of this research was to design and build tools for FSW in order to ascertain the best welding and rotational speeds for Al–Cu butt joint welding. The study investigates the effect of process parameters such as tilt angle, pin offset, tool design, welding speed, and rotational speed on the quality of FSW joint. Mechanical properties are assessed using tensile, hardness, and impact testing. A nearly flawless joints was observed at a rotational speed of 1100 rpm. Furthermore, multi-response optimizations are carried out using a Taguchi L9 orthogonal array to ascertain the ideal set of FSW process parameters for optimized tensile strength and hardness. The optimal mechanical properties, encompassing ultimate tensile strength (132.609 N/mm^2), hardness, and impact strength, were attained with a cylindrical tool pin profile at a rotational speed of 1100 rpm and a traverse speed of 50 mm/min. Mechanical characteristics were decreased due to excessive or insufficient tool rotational speed to traverse speed ratios leading to overheating or insufficient material stirring. These results provide valuable insights into the relationship between process parameters and joint quality, and observed the rotational speed was having a major impact on the joint properties.

Keywords Friction stir welding · Al–Cu joint · Aluminium alloy · Copper · GRA optimisation · Mechanical properties

---

S. Pandipati · K. Durgaprasad (✉) · D. Sreeramulu · S. Palli
Department of Mechanical Engineering, Aditya Institute of Technology and Management, Tekkali, India
e-mail: durgaprasadnita@gmail.com; kdurgaprasad.mech@adityatekkali.edu.in

S. Pandipati
e-mail: suman.mech@adityatekkali.edu.in

S. Palli
e-mail: psrihari.mech@adityatekkali.edu.in

P. V. S. Teja
Department of Mechanical Engineering, Dhanekula Institute of Engineering and Technology, Ganguru, Andhra Pradesh, India

U. S. Dixit et al. (eds.), *Current Progress in Engineering Sciences*,
https://doi.org/10.1007/978-981-96-8753-4_15

1 Introduction

The welding institute, UK has invented and patented the friction stir welding (FSW) process in 1991 [1]. It was highly advantageous in welding of commercial aluminum (Al) and copper (Cu) alloys to achieve better weld properties [2]. It was a solid-state welding process, which was best suited for producing welds of high thermal conductivity materials with no distortion and porosity. Currently, the researchers were more focused on joining of dissimilar materials. Effective welding of dissimilar materials was crucial for applications in the sectors such as chemical, aerospace, and power industries. In the power sector for the high current and voltage distribution lines, aluminum and copper alloys were extensively used due to better electrical and thermal conductivities [3]. However, the joining of these alloys using conventional arc welding methods results in various defects such as porosity, formation of inter metallic compounds (IMCs), and lack of fusion. To overcome these difficulties, the solid-state joining process, FSW was used by many industries.

Various studies were carried out on the dissimilar FSW of Al–Cu alloys to achieve better weld properties such as higher ultimate tensile strength (UTS), and reduction in formation of IMCs, along with reduced distortions in the weld. Khajeh, R., et al., conducted studies on FSW of Al–Cu alloy to achieve a defect free joint with minimum amount of IMCs in the weld zone [4]. Also, the microstructural and mechanical properties were highly dependent on the tool rotational speed (W) and tool traverse speed (V). Optimum process parameters were found out to obtain higher tensile strength. Weld quality also depends on the type of tool profile used such as cylindrical, tapered and square profiles. By using a cylindrical tool profile, a sound weld joint was achieved by Celik et al. [3]. Weld ruptures were generally observed in the Al heat affected zone (HAZ) and weld zones. In the weld zone, due to the formation of Al–Cu layer and IMCs higher hardness were observed in the fusion zone of the weld. The smoothest surface was achieved when the rotational speed was at its highest, specifically 1400 rpm and welding speed was set at 20 mm/min [5]. UTS of 113 MPa was achieved at a traverse speed of 80 mm/min with 70.62% of joint efficiency. At the optimum values of heat input and traverse speeds at 80 mm/min, material was plasticized by allowing the tool and work piece interaction enabling to flow and achieve a better weld joint [6]. An offset in the tool and fixing the Cu plate on advancing side and Al plate on retreating side resulted in defect free weld with a thin and uniform IMCs along the butt joint. And at higher values of W, a stacking layered structure was developed leading to easy failure of the joint [7]. It was well presented that the process parameters such as W, V, tool pin profiles, and offset of the tool pin were having the major influence of the quality of weld. While several studies have focused on dissimilar FSW of AA6082 and pure Cu, the uniqueness of this study lies in the investigation of the combined effects of key process parameters such as rotational speed, traverse speed, and tool pin profiles on the strength and quality of welds for 6 mm thick plates. This systematic approach provides deeper insights into optimizing the process for improved mechanical properties. Therefore, a systematic study has been carried out on the Al–Cu alloy with 6 mm thick at

(a)

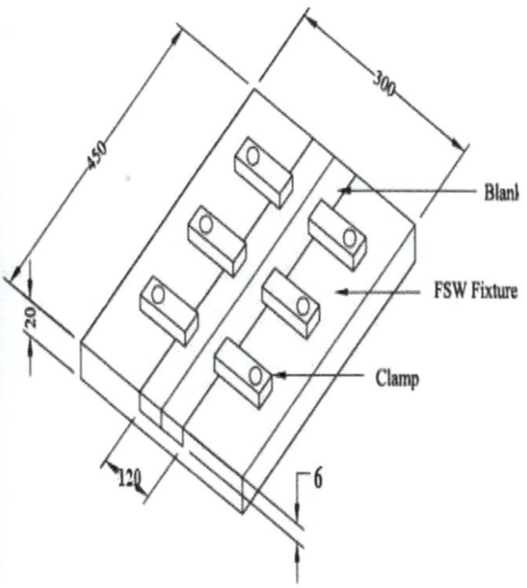

(b)

Fig. 1 Friction stir welding: **a** machine setup, and **b** workpiece fixture

three different rotational and traverse speeds in combination with three different tool profiles (tapered, cylindrical, and square). And a detailed analysis was conducted to analyses the strength, hardness and multi-response optimization was done using Grey relation analysis (GRA).

2 Materials and Methodology

2.1 Experimental Setup

A vertical Numerical control milling machine acts as an FSW machine to perform the welding process. The machine comprised of various auxiliary components including work table, motor system, and clamping mechanisms. The workpieces to be welded were securely clamped to the worktable. The tool was fastened to the tool post and the motors controlled all the rotational and translational movements. The experimental setup and workpiece fixture were shown in Fig. 1.

2.2 Fabrication of Types of Tool Pin Profiles

Geometrical characteristics such as the pin's form, length, and diameter are critical in the FSW process. In the current study, experiments are carried out with cylindrical,

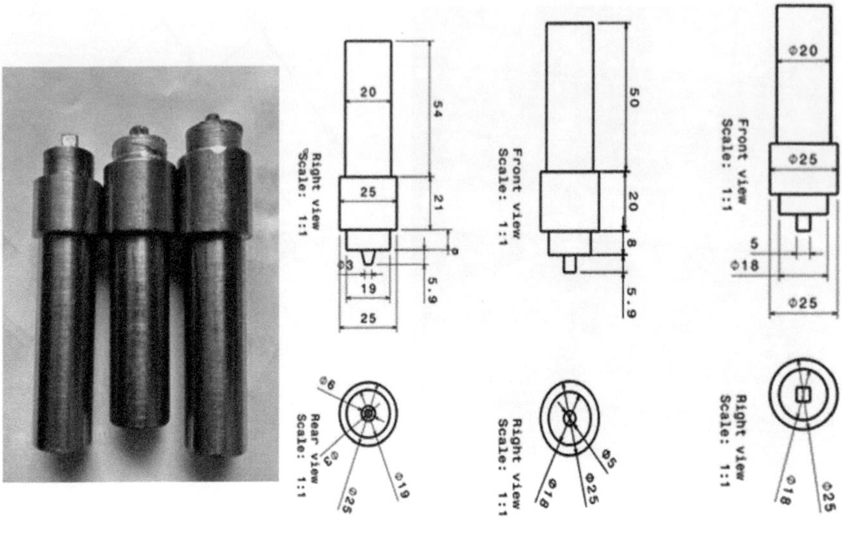

(a) (b) (c) (d)

Fig. 2 Tool pin profiles, **a** fabricated tools, **b** tapered pin, **c** cylindrical pin, and **d** square pin

tapered, and square tool pin profiles (TPP) to explore the influence of TPPs on weld strength. In this study, TPPs are chosen based on the thickness of the work material, ease of manufacturing, and efficacy in generating weld connections. Figure 2 depicts the three different types of FSW tools and their pin profiles.

H13 tool steel was used to manufacture the tool pins with 80 mm shoulder height, 5.8 mm pin length, and a shoulder diameter of 18 mm. It was made on a lathe machine. Figure 2 depicts the different TPPs, such as tapered pin, cylindrical pin, and a square pin. Tempering was conducted on the tool pins to increase its hardness for ensuring optimal performance of the tool.

2.3 Experimental Procedure

The materials AA6082 and Copper of thickness 6 mm were selected for the experiments. The weld plate of 150 mm × 75 mm × 6 mm has been extracted from the standard size flat.

The joint was fabricated in one pass welding, tool moves at right angles to the plate rolling direction, placing AA6082 and Copper. Experiments were carried out at various rotational speeds (900, 1100, and 1400 rpm), and traverse speeds (25, 50, and 75 mm/min), using cylindrical, taper, and square tools fabricated of H13 tool steel. The combination of process parameters was achieved using L9 orthogonal array. The aluminum plates of 150 mm × 75 mm × 6 mm were perfectly clamped

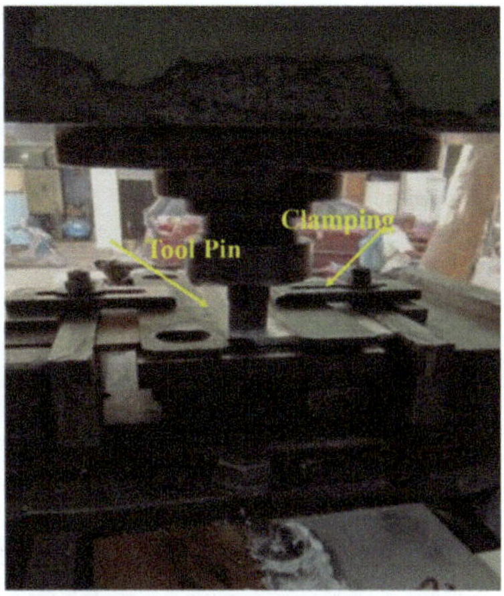

Fig. 3 FSW experimentation

in the FSW machine, with a work table with a backing plate shown in Fig. 3. In the downward direction, the tool is penetrated into the butting edges of the plate. Due to higher frictional heating, they result in a more intense mixing of the material. The specifications of FSW machine were presented and the numerical controlled FSW machine built up by R.V Machine Tool, Model No: SPM104, as shown in Fig. 1. To prevent separation of the attached butting edges, AA6082 and Copper plates with a thickness of 6 mm are positioned and tightly secured using backing plates. The tool was placed in the tool holder at a 0° tilt angle, and the tool pin was thrown to a specified depth at the edges of the butting surface of the plates to be joined. The tool was moved transversely forward after residing at the end, and the joint was produced with a single pass. After the weld is completed, the tool is removed, leaving an aperture at the end. Both the plates were comfortably coupled together to form a butt joint configuration.

2.4 *Tensile Test*

The tensile behavior of FSW joints was evaluated using the computerized universal testing machine (UTM) at a cross-head velocity of 5 mm/min, as illustrated in Fig. 4a. Tension testing revealed the details of the ductility and strength of the material under axial tensile stress. Under specific conditions, this information can help compare materials, alloy development, quality control, and design. The test samples were cut from the fabricated joints and machined as per the standard ASTM as shown in Fig. 4b. Three identical specimens were evaluated to get the average ultimate tensile strength value.

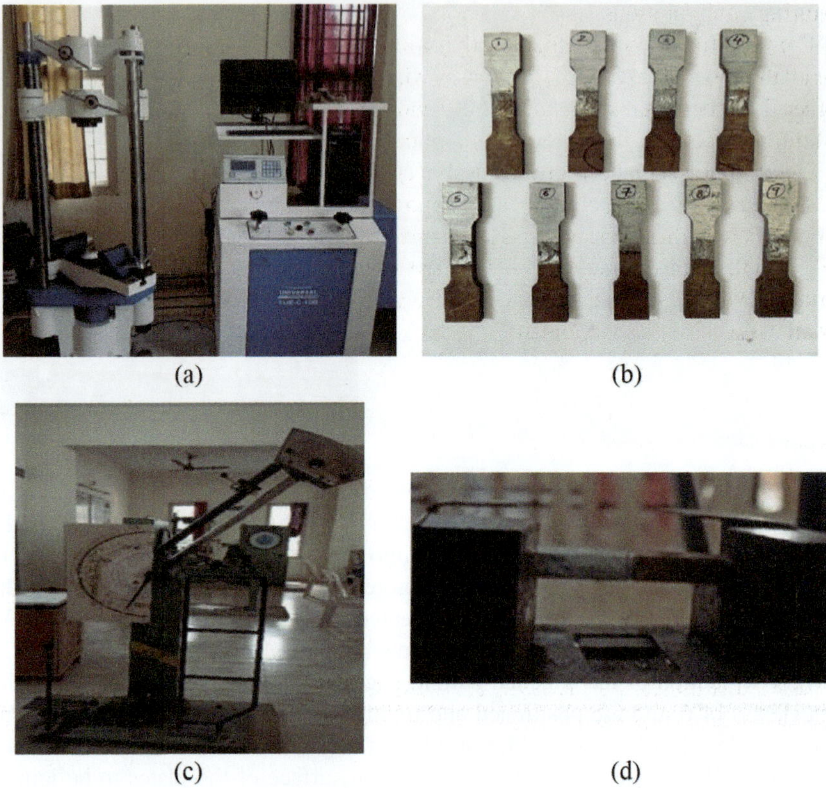

Fig. 4 **a** Computerized UTM, **b** tensile specimens, **c** Charpy machine, and **d** Charpy test specimen

2.5 Charpy Impact Test

Impact testing was designed to compute the resistance to a suddenly applied force onto a material. This test measures the energy absorbed before the fracture. The Charpy test was more popularly used to examine various materials' relative toughness or impact toughness. The sample will engross energy until it yields when the wedge hammer strikes the specimen. At this point, the specimen will initiate to endure plastic distortion at the groove. Three samples were extracted from FSW joints and machined as per ASTM E23 standard as shown in Fig. 4c, and d.

2.6 Hardness Test

Rockwell hardness test was carried out according to the ASTM E-384 standard. Common scales include Rockwell B (HRB) for softer materials and Rockwell C

(HRC) for harder materials. Applied a preliminary minor load to the specimen to ensure proper contact between the indenter and the surface followed by the major load. Measured the depth of indentation using a dial gauge to calculate the Rockwell hardness value based on the depth of penetration or the position of the indenter, considering the scale, load, and indenter type used.

3 Results and Discussion

3.1 *Effect of Process Parameters on Tensile Strength*

The tensile behavior of the weld joints plays an essential contribution in determining weld standards, and hence the configuration of various FSW regions plays a pivotal role. The process parameters and the obtained results of ultimate tensile strength including yield strength were presented in Table 1.

The UTS of FSW for Al–Cu joints was measured concerning the values of W, V and TPP. At lower values of W, the UTS of FSW joints is lower and increases correspondingly with an increased rotational speed up to a certain level as shown in Fig. 5. It is true irrespective of the tool pin shape. Also, the UTS is inversely proportional to the tool traverse speed. As the traverse speed increases at a constant tool rotational speed, the ultimate tensile strength is reducing as shown in Fig. 5. This is due to higher traverse velocities of the tool, there is insufficient generation of heat resulting in improper stirring of the metal in the stir zone. This leads to a porosity in the weld, causing the weld to break at lower tensile loads as shown in Fig. 5b. Figure 5b represents the fractured specimens of experiments 2, and 3 where the traverse velocities are 50 and 75 mm/min. The specimen at 75 mm/min having lack of material in the weld zone as represented in the Fig. 5b. The UTS is higher

Table 1 Process parameters and tensile properties of all weld samples

Exp. No.	Rotational speed (W), rpm	Welding speed (V), mm/min	Tool pin profile	Yield stress, N/mm^2	Ultimate tensile strength (UTS), N/mm^2
1	900	25	Cylinder	23.683	79.552
2	1100	50	Cylinder	105.359	132.609
3	1400	75	Cylinder	28.583	76.582
4	1100	75	Taper	1.52	79.552
5	1400	25	Taper	30.672	83.245
6	900	50	Taper	27.722	76.325
7	1400	50	Square	30.338	79.941
8	900	75	Square	75.372	70.365
9	1100	25	Square	30.413	80.645

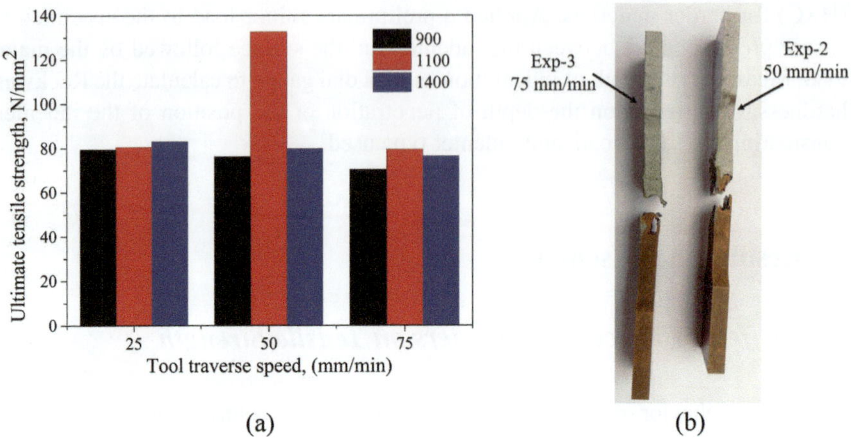

Fig. 5 Tensile strength at different combinations: **a** tool traverse speed versus UTS at various rotational speeds, and **b** fractured tensile specimens at 50 and 75 mm/min

at the cylindrical pin profile than the tapered cylindrical and square pin profile. The cylindrical tool pin has higher mating edges than other TPPs; then, the material is stirred well in these conditions. At low rotational speeds, the generated frictional heat is insufficient to adequately soften and plasticize the material, resulting in deficient joint strength. At higher W, high temperatures are generated, resulting in reduction in UTS. At low transverse speeds, the tool travel time is increased throughout the weld zone. Then in these conditions, high temperatures are generated resulting in coarsening of the grains, thereby reducing the strength of the weld [8]. At medium values of W and V, the ratio of W/V is sufficient enough to generate the heat and stir the material to obtain a better weld joint strength as indicated in Table 1 for experiment 2.

3.2 *Hardness of Al–Cu FSW Weld Bead*

Hardness of the welded joints plays a significant contribution in determining weld standards. Hence, the configuration of various FSW nugget zone regions plays a pivotal role. The hardness of FSW Al–Cu joints has been measured concerning W, V, and TPP. At lower values of W, the hardness of the FSW nugget zone is low and increases correspondingly with an increased W. Because, at higher values of W, the aluminum material melts faster as compared to copper. Due to higher thermal conductivity of Al, the heat from the molten metal was transferred to Cu in addition to heat generated from the tool shoulder resulting in earlier formation of molten pool, which in turn causes the better stirring of the metal in the stir zone. Also, grain refinement is critical in material strengthening because the size of grain in the weld is substantially finer than that of parent material [9]. As the rotating speed increases

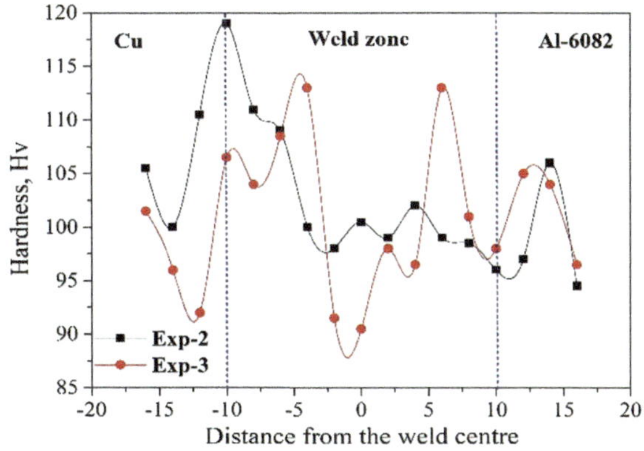

Fig. 6 Effect of tool rotational speed on the hardness

beyond a certain limit, reduction in the hardness was observed. The low hardness is the result of the material becoming softer due to the intense heat generation in the process. At low transverse speeds, the tool travel time is grown throughout the weld zone. Due to these conditions, high temperatures are generated. At high transverse speeds, the tool travel time is decreased throughout the weld zone. Lower temperatures are caused at high transverse speeds, and the material will not stir properly, resulting in lower hardness. A moderate values of W/V ratio was good enough to produce better welds resulting in achieving better mechanical properties. Hardness values were plotted between two process parameters (Exp-2, and Exp-3), which are having moderate W/V ratio (Exp-2) as compared to lower W/V ratio (Exp-3). A uniform hardness values were achieved for medium values of W, and V as compared to other as shown in Fig. 6. The hardness was higher at the cylindrical pin profile than the tapered and square pin profile. The cylindrical TPP having higher hardness due to higher mating edges than other TPPs; then, the material was stirred well in these conditions.

The hardness values have significant differences in the weld nugget and HAZ compared to the base material. The nugget zone's average hardness is significantly lower than that of the base material. This is due to strong plastic deformation at higher temperatures, which results in a fine grain structure in the weld zone. A soft zone is also generated, nearly symmetrically, on both sides of the weld joint line. The lateral surface of the weld nugget has a higher hardness as a result of material deformation.

3.3 *Impact Strength of Al–Cu Weld Specimen*

The effect of rotational speed on the impact strength of the Al–Cu weld beads at different traverse speeds was shown in Fig. 7. It was observed that the impact strength of the weld beads was reduced at lower values of W, because the frictional heat generation was less at 900 rpm leading to reduced plastic flow of material at stir zone

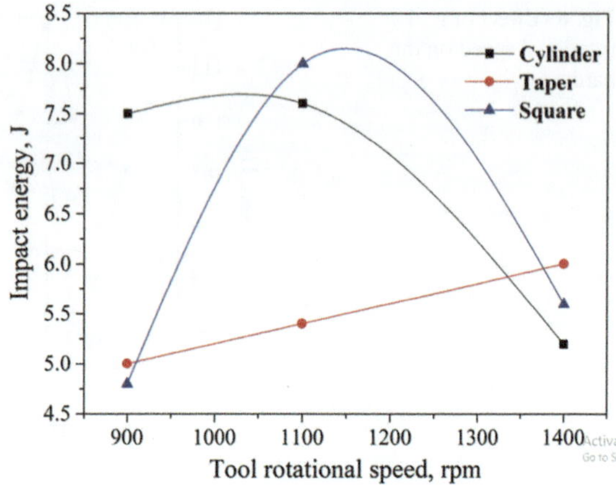

Fig. 7 Effect of tool rotational speed on the impact energy

resulting from the lower value of impact strength and the enhanced values of W, the flowability of plasticized material improved leading to enhanced impact strength. The impact strength was better at 1100 rpm and 50 mm/min. The less impact strength value was obtained at a traverse speed of 75 mm/min. This was due to less time of interaction between tool and work piece leading to poor plasticizing state. This in turn results in the defect with porosity and lack of fusion in the stir zone. As the values of V were enhanced from 25 to 50 mm/min, heat input reductions caused relatively refined grains, thus rising impact strength. It was also identified that joints produced by tapered TPPs have less impact strength at any value of traverse speeds. Whereas, other TPPs produced the joints were having the highest impact strength at 25–50 mm/min. The highest impact strength was obtained at a traverse speed of 50 mm/min.

3.4 Grey Relation Analysis

Taguchi method was used to find the optimal parameters of a process with a single performance characteristic. For multi performance characteristics, grey relational analysis along with Taguchi was used. The GRA, which can be perceived as overall quality of the weld joint was calculated as per the methodology [8] followed by Pandav et al., The following equations were used to calculate the normalization and deviation data followed by grey relation coefficient.

$$xs_i(j) = \frac{|x_i(j) - \min x_i(j)|}{\max x_i(j) - \min x_i(j)} \tag{1}$$

$$\Delta x_i(j) = |x_0(j) - xs_i(j)| \tag{2}$$

$$\xi_i(j) = \frac{\Delta\min + p\Delta\max}{\Delta x_i(j) + p\Delta\max} \tag{3}$$

where, $x_i(j)$ is the reference value of the jth enabler of ith alternative and $x(j)$ is the objective of ith enabler. Once the normalization is conducted, deviation $\Delta x_i(j)$ of the reference $x_0(j)$ is calculated using the Eq. (2) were presented in the Tables 2 and 3. The coefficient of the grey relation was obtained using the Eq. (3) and was presented in Table 4. For every performance characteristic variable, Δmin and Δmax are the minimum and maximum deviations. The identification factor (p) varies between 0 and 1. It was determined that 0.5 was the characteristic coefficient required to calculate the grey relation grade (GRG).

The overall weld quality of each metric at the specified level was selected, and grey relation analysis was used to calculate their mean as represented in Table 4, and the optimized values were obtained for experiment 2.

Table 2 Normalization of data

Exp. No.	W, rpm	V, mm/min	Tool pin profile	UTS (N/mm^2)	Impact energy, J	Hardness of weld zone
1	900	25	Cylinder	0	0.84375	0.630837359
2	1100	50	Cylinder	1	0.875	0.646537842
3	1400	75	Cylinder	0.099837728	0.125	0.033816425
4	1100	75	Taper	0.165116242	0.1875	1
5	1400	25	Taper	0.164055847	0.375	0.082930757
6	900	50	Taper	0.095708617	0.0625	0.798711755
7	1400	50	Square	0.153805369	0.25	0.988888889
8	900	75	Square	0.4013271	0	0
9	1100	25	Square	0.14755547	1	0.100644122

Table 3 Deviation sequence of data

Exp. No.	W, rpm	V, mm/min	Tool pin profile	UTS (N/mm^2)	Impact energy, J	Hardness of weld zone
1	900	25	Cylinder	1	0.15625	0.369163
2	1100	50	Cylinder	0	0.125	0.353462
3	1400	75	Cylinder	0.900162	0.875	0.966184
4	1100	75	Taper	0.834884	0.8125	0
5	1400	25	Taper	0.835944	0.625	0.917069
6	900	50	Taper	0.904291	0.9375	0.201288
7	1400	50	Square	0.846195	0.75	0.011111
8	900	75	Square	0.598673	1	1
9	1100	25	Square	0.852445	0	0.899356

Table 4 GRG and rank

Exp. No.	W, rpm	V, mm/min	Tool pin profile	UTS (N/mm^2)	Impact energy, J	Hardness of weld zone	GRG	Rank
1	900	25	Cylinder	0.333	0.76190	0.369163	0.51478	5
2	1100	50	Cylinder	1	0.8	0.353462	0.84646	1
3	1400	75	Cylinder	0.357	0.36363	0.966184	0.36629	9
4	1100	75	Taper	0.374	0.38095	0	0.68887	2
5	1400	25	Taper	0.374	0.44444	0.917069	0.39540	8
6	900	50	Taper	0.356	0.34782	0.201288	0.4543	6
7	1400	50	Square	0.371	0.4	0.011111	0.53967	3
8	900	75	Square	0.455	0.33333	1	0.43891	7
9	1100	25	Square	0.369	1	0.899356	0.53406	4

Table 5 Confirmation experiment results

Exp. No.	W, rpm	V, mm/min	Tool pin profile	UTS (N/mm^2)	Impact energy, J	Hardness of weld zone
2	1100	50	Cylinder	132.609	7.6	56.56
Confirmation test						
2	1100	50	Cylinder	130.85	7.54	55

After identifying the ideal parameter values, the last phase of the GRA study is to predict and confirm the enhancement of GRG, in this work. To validate the findings in the analysis, a confirmation test was conducted on the experiment 2. Three runs were conducted and the average values of the confirmation test were presented in Table 5 indicating a difference of less than 1.5%.

4 Conclusions

This study investigated the effects of process parameters, including rotational speed, traverse speed, and tool pin profile, on the mechanical properties of Friction Stir Welded (FSW) joints between aluminum (AA6082) and copper plates of 6 mm thickness. The best mechanical properties, including ultimate tensile strength (132.609 N/mm^2), hardness, and impact strength, were achieved using a cylindrical tool pin profile at a rotational speed of 1100 rpm and a traverse speed of 50 mm/min. This parameter set was validated through confirmation tests, with less than 1.5% variation in results. Cylindrical tool pin profiles demonstrated superior tensile strength and hardness, whereas square profiles exhibited better impact strength. These differences underscore the importance of tool geometry in FSW. A moderate W/V ratio produced optimal results, ensuring sufficient heat generation and material mixing. Excessive or insufficient W/V ratios led to reduced mechanical properties due to overheating or inadequate material stirring, respectively. The integration of Taguchi's method and Grey Relational Analysis (GRA) enabled effective multi-response optimization, highlighting the most significant factors and their optimal levels.

References

1. Pandipati, S., Potnuru, S., & Sreeramulu D. Multi-response optimization of friction stir lap welding of aluminum plates through GRA-PCA. Eng Res Express 2023;4.
2. Chodhuri, B., Durgaprasad, K., Debbarma, S., Majumder, A., & Saha SC. Parametric optimization and cost estimation of AA6061 weldments in friction stir welding. Int J Mech Prod Eng 20124;2:01–8.
3. Celik S, Cakir R. Effect of friction stir welding parameters on the mechanical and microstructure properties of the Al-Cu butt joint. Metals (Basel) 2016;6. https://doi.org/10.3390/met6060133.
4. Khajeh R, Jafarian HR, Seyedein SH, Jabraeili R, Eivani AR, Park N, et al. Microstructure, mechanical and electrical properties of dissimilar friction stir welded 2024 aluminum alloy and copper joints. J Mater Res Technol 2021;14:1945–57. https://doi.org/10.1016/j.jmrt.2021.07.058.
5. Bhushan RK, Sharma D. Investigation of mechanical properties and surface roughness of friction stir welded 2020.
6. Muthu MFX, Jayabalan V. Tool travel speed effects on the microstructure of friction stir welded aluminum-copper joints. J Mater Process Technol 2015;217:105–13. https://doi.org/10.1016/j.jmatprotec.2014.11.007.
7. Xue P, Ni DR, Wang D, Xiao BL, Ma ZY. Effect of friction stir welding parameters on the microstructure and mechanical properties of the dissimilar Al-Cu joints. Mater Sci Eng A 2011;528:4683–9. https://doi.org/10.1016/j.msea.2011.02.067.
8. Pandav AN, Patil RA, Pandipati S. Process parameters settings of friction stir welding using multi-response optimization for aluminum alloys. Eng Res Express 2024;6. https://doi.org/10.1088/2631-8695/ad4511.
9. Elmetwally HT, SaadAllah HN, Abd-Elhady MS, Abdel-Magied RK. Optimum combination of rotational and welding speeds for welding of Al/Cu-butt joint by friction stir welding. Int J Adv Manuf Technol 2020;110:163–75. https://doi.org/10.1007/s00170-020-05815-8.

Experimental Studies on Direct Ink Writing of Alumina Using Methyl Cellulose as Gelling Agent

Gaurav Prakash, Sukhomay Pal, and P. S. Robi

Abstract 3D printing is an advanced manufacturing technique that enables rapid manufacturing of complex, intricate geometries, which are often challenging through conventional processing routes. Ceramic structures are often processed using 3D printing due to benefits such as the elimination of mold requirements, design flexibility, and economic viability. In this study, Direct Ink Writing (DIW), a material extrusion-based 3D printing process, is used to fabricate alumina structures. The printable ceramic suspension was prepared using Alumina (57 wt.%), methylcellulose (2.7 wt.%), and distilled water, which was decided through repeated trials. The key rheological properties, such as yield stress, viscosity, and storage/loss modulus of the suspension, were characterized. The printable suspension was 3D printed using DIW, and the printed parts were subjected to drying, debinding, and sintering to obtain dense parts. Microstructural characterization and physical properties of the parts were analyzed to assess the quality of the printed parts. It is shown that the prepared suspension displays shear-thinning behavior, and the printed parts exhibited good extrudability and shape-retention capability. The results show that the sintering temperature of 1500 °C at a holding time of 6 h is a suitable thermal condition to impart density to the final parts. The microstructure shows strong mechanical bonding among the grains at the prescribed sintering profile, having an average grain size of 1.64 μm.

Keywords 3D printing · DIW · Rheology · Alumina

G. Prakash · S. Pal (✉) · P. S. Robi
Department of Mechanical Engineering, Indian Institute of Technology Guwahati, Guwahati, Assam, India
e-mail: spal@iitg.ac.in

G. Prakash
e-mail: gaurav.prakash@iitg.ac.in

P. S. Robi
e-mail: psr@iitg.ac.in

U. S. Dixit et al. (eds.), *Current Progress in Engineering Sciences*,
https://doi.org/10.1007/978-981-96-8753-4_16

1 Introduction

Ceramics are now widely used in a broad range of applications, including the chemical industry, automotive, biomedical devices, refractories, and aerospace, due to their excellent combination of mechanical, chemical, and thermal properties [1]. These advanced applications require high-quality control in processing techniques. Traditional ceramic processing methods, such as slip casting, tape casting, gel casting, die pressing, and injection molding, require additional post-processing steps (drying, debinding, and sintering) to densify the product. However, these techniques face challenges such as high processing times and high costs. Furthermore, the high melting point, brittleness, and low ductility of ceramics pose challenges for conventional manufacturing techniques, as machining is often involved. Complex geometries are also difficult to fabricate due to molding and casting limitations [2, 3]. Additive manufacturing (AM) has rapidly evolved in recent years as a promising solution for ceramics manufacturing. AM involves the layer-by-layer stacking of materials, enabling parts to be built directly from CAD models, in contrast to traditional subtractive manufacturing. This approach addresses many of the challenges associated with ceramic manufacturing and is still being actively explored. Several AM techniques are currently in use for ceramic processing, including stereolithography, selective laser sintering, binder jetting, and material extrusion. Among these, Direct Ink Writing (DIW), an extrusion-based process, shows immense potential for fabricating a wide range of complex ceramic components. DIW is an extrusion-based additive manufacturing technique where a ceramic suspension, consisting of a mixture of powder and binder, is extruded through a nozzle following a computer-generated CAD design to create three-dimensional structures. The printed green body undergoes post-processing treatments, including drying, debinding, and sintering, to enhance its quality and strength. The DIW process is highly sensitive to the rheological behavior of the suspension. The suspension must exhibit shear-thinning behavior with respect to shear rate and sufficient yield stress to solidify rapidly after deposition [4]. DIW printing of ceramics has been extensively studied by researchers. Zhu et al. [5] investigated the influence of printing parameters on the performance of Al_2O_3 refractory products prepared using DIW. They found that interlayer gaps could be minimized by reducing the layer height to 1 mm (50% of the nozzle diameter). Rueschhoff et al. [6] fabricated near-net-shaped dense ceramic components and studied the effect of solid loading on the rheology, density, microstructure, and mechanical properties of the components. They reported no significant differences in density, grain size, or mechanical properties with varying solid loading. Baltazar et al. [7] examined the effects of infill orientation and sintering temperature on the mechanical properties of robocasted dense alumina components. They achieved a relative density of 99% and a hardness of 17 GPa when sintered at 1600 °C for 2 h. Carloni et al. [8] fabricated transparent Al_2O_3 ceramics and compared the results with the traditional cold isostatic pressing technique. A two-step vacuum sintering process resulted in higher density and reduced grain size compared to a single-step sintering profile. Several additives have been employed in DIW to improve the printability of suspension, such as PEG, PVA, ethylene glycol, etc. [9]. In this study, alumina was processed using the DIW technique, and gelation agent methyl cellulose was employed to optimize the rheological properties and improve the printability. This research focuses on obtaining complex high-precision alumina samples printed using the DIW process. The printed green body was sintered at 1500 °C for a holding time of 6 h. The microstructural characterization of the sintered specimen was analyzed. Since this method has high efficiency and low cost, it can be of great significance for industrial applications involving complex shapes.

2 Experimental

2.1 Preparation of Alumina Slurries

Alumina powder was chosen as the raw material to process alumina ceramics. Methyl cellulose was used as a gelling agent. Distilled water was used as a solvent to prepare the ceramic suspension. The alumina powders were ball-milled for 4 h to reduce the particle size. The constituents were weighed accurately and stirred for 1 h to obtain a uniform suspension. The whole mixture was ball-milled for 2 h to achieve a uniform suspension. The suspension was prepared with solid loading of 57 and 2.7 wt.% methyl cellulose.

2.2 Processing of Ceramic Parts by DIW

The CAD model was prepared using SolidWorks software, and Cura was used to slice the parts and generate the G-code for printing. An extrusion-based DIW method was employed to print the ceramic samples. The schematic of the DIW setup is illustrated in Fig. 1. In DIW, the alumina-based suspension is extruded through the nozzle at room temperature. The alumina green body was obtained using the design of suspension design mentioned above. The green bodies were dried at room temperature for 24 h to remove the residual moisture and subsequently sintered at 1500 °C for 6 h to densify the sample.

2.3 Characterization

The morphology and elemental composition of the starting and ball-milled powders were examined using Field Emission Scanning Electron Microscopy (FESEM) equipped with Energy Dispersive Spectroscopy (EDX). Particle size was measured using ImageJ software. X-ray Diffraction (XRD) was used to analyze the phase

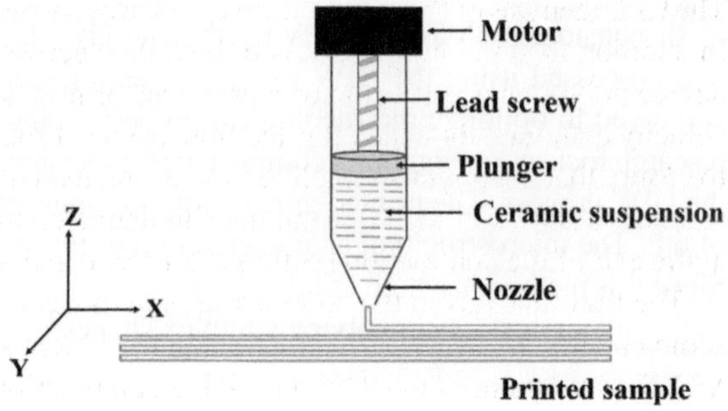

Fig. 1 Schematic representation of DIW setup

composition of the starting powders. The rheological behavior of the suspension was evaluated to determine the optimal composition for DIW. The shear rate was varied from 0 to 100 s^{-1}. Viscoelastic analysis was conducted at a constant frequency of 10 Hz. Thermogravimetric analysis (TGA) was performed to determine the debinding temperature. The microstructure of the sintered sample was analyzed using FESEM.

3 Results and Discussion

The FESEM micrographs of as-received and ball-milled alumina powder are shown in Fig. 2a and b. From the micrograph, it is shown that agglomerations are present in the starting powder, which is reduced after ball milling. The particle size of the starting and ball-milled powder is 54 and 5.365 μm, respectively, as shown in Fig. 2c and d.

XRD of starting powder and ball-milled powder is shown in Fig. 3. The XRD patterns were recorded within the range of 10°–100°, and the major peaks correspond to (311), (222), (400), (511), (440), and (444) planes. The peaks confirm that it is evident that XRD profiles of ball-milled powder are similar to the starting powder, indicating no change in the phase structure of powders. The XRD results are in agreement with JCPDS card number 00-001-1303.

The rheological results are shown in Fig. 4. For the rheological analysis, a parallel plate configuration (diameter = 25 mm) is used with a gap between the stationary and rotating plates set to 2.8 mm. It can be inferred from the curve that the suspension with 57 wt.% alumina and 2.7 wt.% methyl cellulose. The shear rate was varied from 0 to 100 s^{-1}.

Shear thinning behavior was observed for the designed suspension for the smooth extrudability of the suspension through the nozzle. Viscosity of the suspension decreases as the shear rate is increased, which is due to the weakening of alumina—binder interactions. The flow behavior can be represented by the Herschel-Bulkley model:

$$\tau = \tau_0 + K * \gamma^n$$

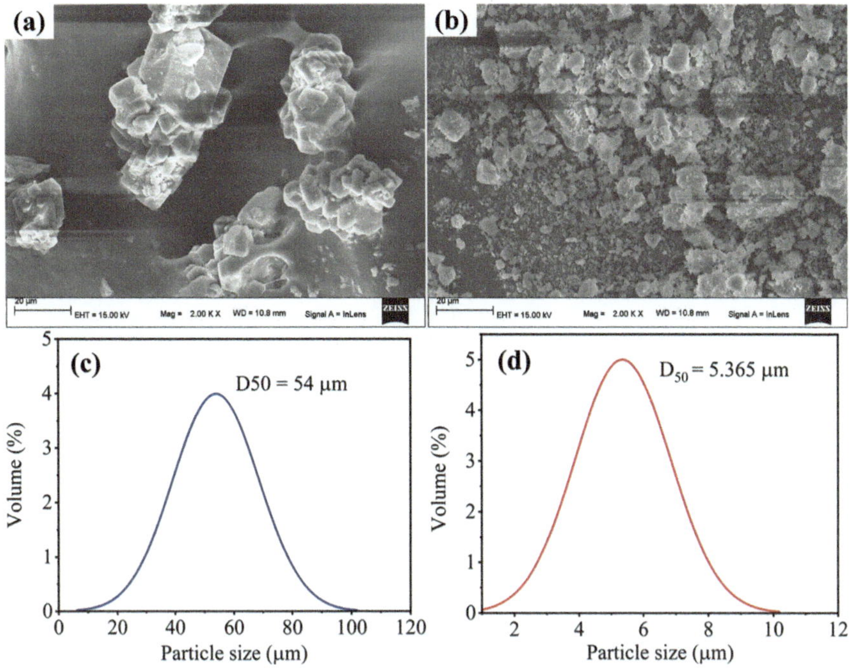

Fig. 2 Morphology of **a** starting alumina powder, **b** ball-milled alumina powder for 4 h, **c** particle size of starting alumina powder, **d** particle size of ball-milled powder

Fig. 3 XRD of as-received and ball-milled powder

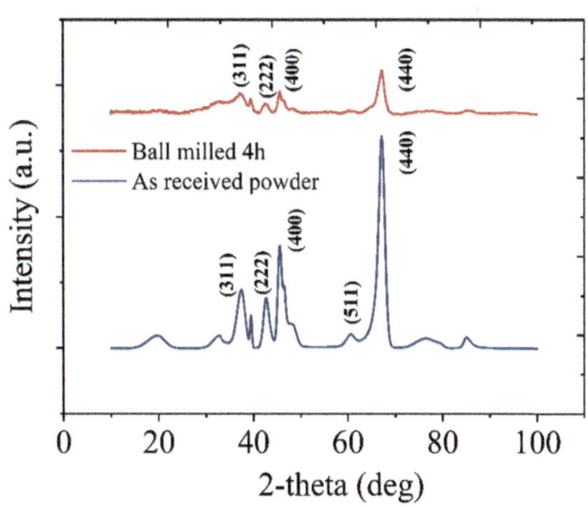

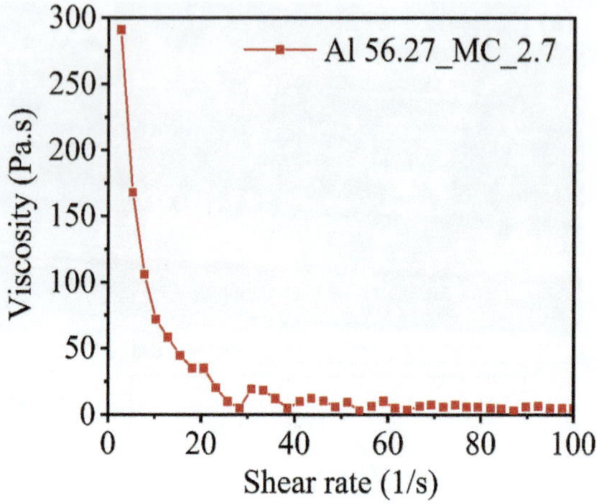

Fig. 4 Flow curve for 57 wt.% alumina and 2.7 wt.% methyl cellulose

Table 1 Rheological parameters

K (Pa. s^n)	τ_0 (Pa)	n
434.09	183.04	0.661

where τ is the shear stress (Pa), τ_o is the yield stress (Pa), K is the flow consistency parameter, γ is shear rate (s^{-1}), and n is the flow behavior index.

The parameters are summarized in Table 1.

Figure 5 shows the viscoelastic results of the suspension containing 57 wt.% alumina and 2.7 wt.% methyl cellulose. The dependence of $G\prime$ and G'' is evaluated at varying strain rates at a constant frequency of 1 Hz.

From the graph, it can be observed that the $G\prime$ is constant and higher than $G^{''}$.. This region is the Linear Viscoelastic Region (LVR). Point 1 is the crossover point, which is the transition from solid to liquid like behavior. After the crossover point, the

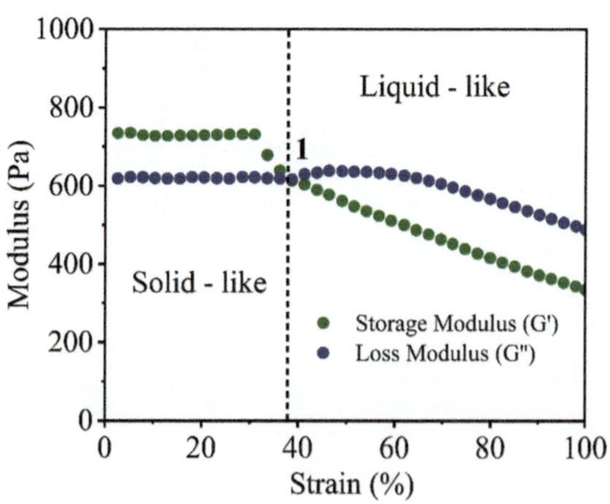

Fig. 5 Storage ($G\prime$) and loss modulus ($G^{''}$) curve vs strain rate for 57 wt.% alumina and 2.7 wt.% methyl cellulose

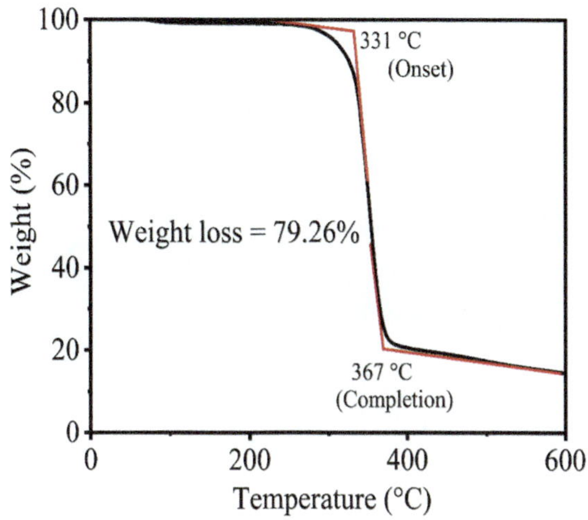

Fig. 6 Thermogravimetric analysis of methyl cellulose binder

viscous behavior dominates over elastic behavior, which corresponds to the flow of suspension through the nozzle. The elastic modulus value at the LVR region is 750 Pa, which is sufficient to retain the structure of the printed object. A thermogravimetric study of methyl cellulose was conducted to determine the debinding temperature of the printed sample, as shown in Fig. 6. It can be observed that major weight loss (80%) occurred when the sample was heated from 331 to 367 °C.

3.1 DIW Printing and Characterization

The above prepared suspension was 3D printed using the DIW process. The printing parameters used for printing are listed in Table 2. The layer height was taken as 2 mm, and the nozzle diameter was 3 mm. The sample was printed with 100% infill density. A 2D concentric infill pattern was used for printing, closely resembling the contours of the object's walls.

The 3D printed sample, along with the macrograph, is shown in Fig. 7. It can be observed that the printed sample displays good self-supporting and shape-retention capability. The printed sample shows the capability of DIW to print complex shapes. The macroview of the part exhibits a fine finish along the layer.

The dimensional accuracy was determined from observing the macroview of the printed samples, as shown in Fig. 8. The average infill distance is 2.883 mm. The

Table 2 Printing parameters for DIW

Nozzle diameter (mm)	Layer height (mm)	Extrusion width (mm)	Printing speed (mm/s)	Infill %
3	2	5	25	100

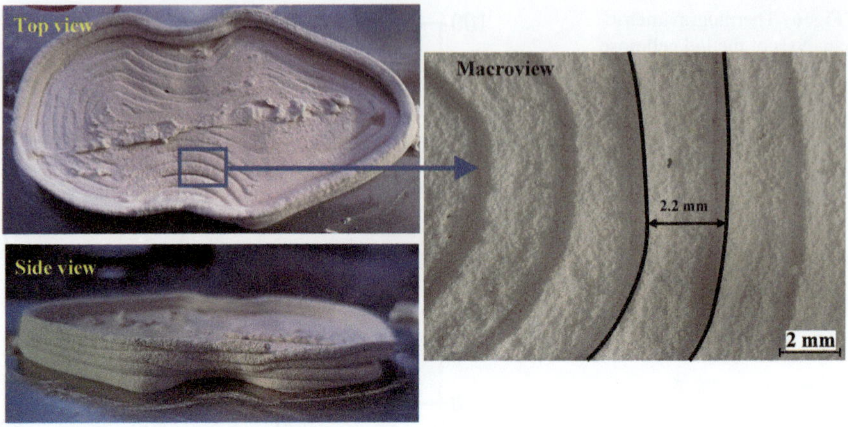

Fig. 7 3D printed sample and its macroview

deviation from the actual measurement is 3.9% which shows a fine printing using the DIW technique.

The post-processing schedule is illustrated in Fig. 9. The sample was dried at room temperature for 24 h for moisture removal. The sample was debinded in a muffle furnace at 400 °C for 3 h to remove the binder, which was determined through thermogravimetric analysis. It is then heated to 800 °C at a heating rate of 5 °C/min with a holding time of 30 min. The sample is finally heated to 1500 °C at a heating rate of 5 °C/min and held for 6 h followed by furnace cooling.

Fig. 8 Macroview of the printed part

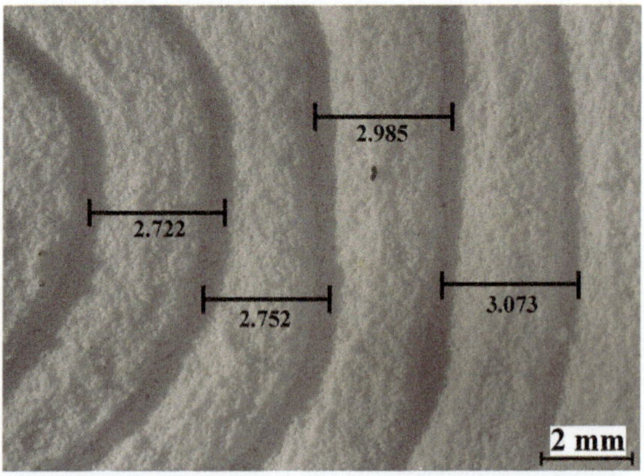

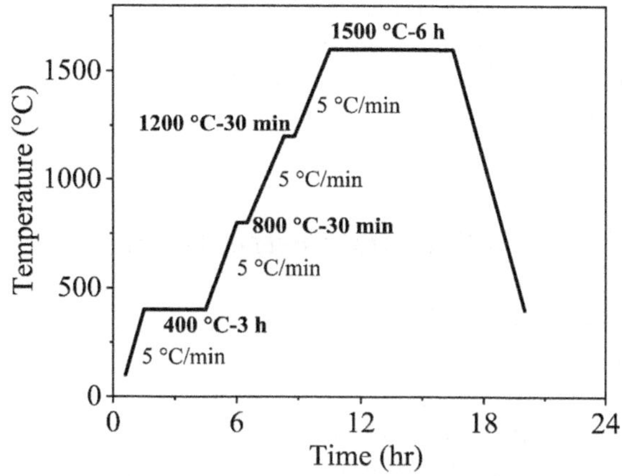

Fig. 9 Post-processing treatment of DIW printed specimen

The sintered microstructure is shown in Fig. 10a, which reveals that neck formation takes place at 1500 °C. Sintering results in the strong mechanical bonding between the grains. Some pores are also present in the microstructure. The average grain size after sintering is 1.64 μm, as shown in Fig. 10a.

It can be observed that grain coarsening is inhibited during sintering, which results in a fine microstructure. There is a significant change in grain size, which may be due to a different sintering mechanism which may be operative during sintering. EDX analysis of the sintered sample depicted in Fig. 10b shows that alumina and oxygen are the major elements in the atomic ratio present in the microstructure. The elemental mapping of the sintered sample is shown in Fig. 10c–f. It can be observed that Alumina and oxygen are distributed homogeneously in the matrix with some traces of Magnesium.

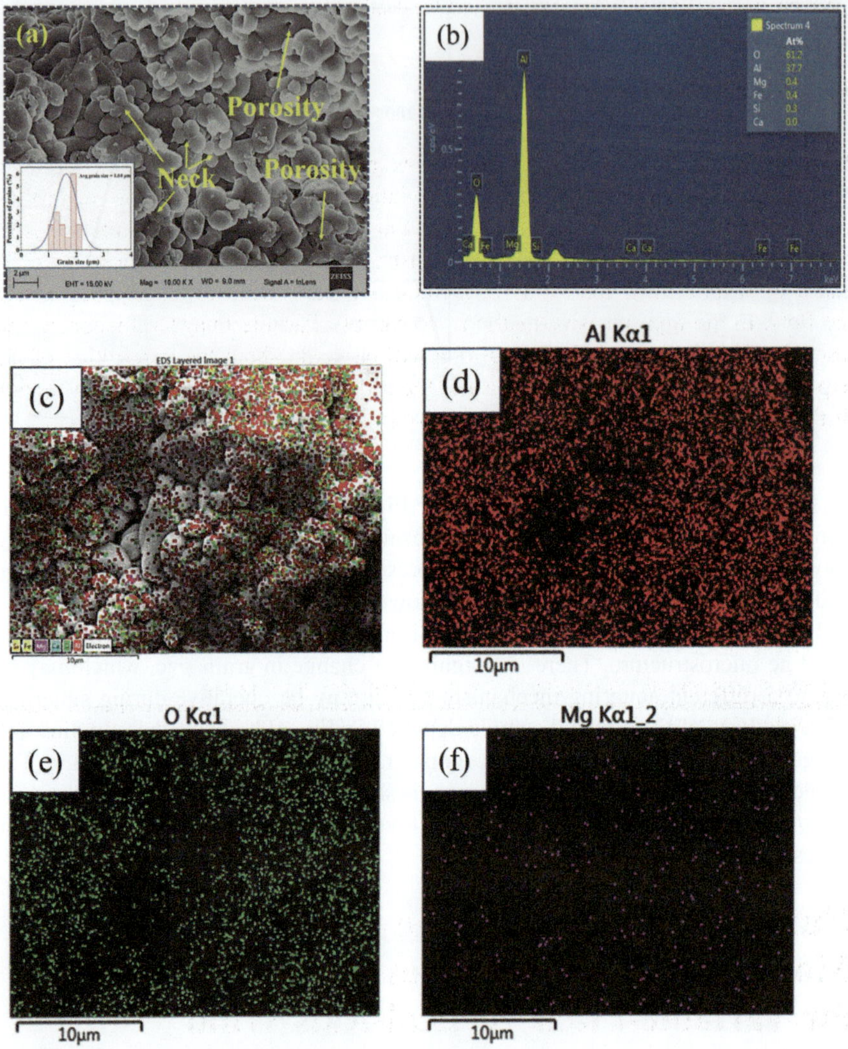

Fig. 10 FESEM micrograph of **a** sintered sample, **b** EDX point scan of sintered sample, **c–f** Elemental mapping of sintered sample

4 Conclusion

In this study, Al_2O_3 ceramic is processed using the DIW technique using a printable suspension composition, i.e., Alumina (57 wt.%), methyl cellulose (2.7 wt.%). The complex shape was printed, which possesses good consistency between layers and has sufficient yield stress to prevent slumping and sustain the weight of deposited layers. The additive used (methyl cellulose) was decomposed and volatilized during the debinding process. The post-processing was undertaken to impart strength and quality to the part. The sintered microstructure shows strong bonding between the grains having average grain size of 1.64 μm. The low cost and simplicity of the DIW method make it a compelling approach for producing geometrically complex ceramic products.

References

1. Z. Chen *et al.*, "3D printing of ceramics: A review," *Journal of the European Ceramic Society*, vol. 39, no. 4. Elsevier Ltd, pp. 661–687, Apr. 01, 2019. https://doi.org/10.1016/j.jeurceramsoc.2018.11.013.
2. S. M. Olhero, P. M. C. Torres, J. Mesquita-Guimarães, J. Baltazar, J. Pinho-da-Cruz, and S. Gouveia, "Conventional versus additive manufacturing in the structural performance of dense alumina-zirconia ceramics: 20 years of research, challenges and future perspectives," *J. Manuf. Process.*, vol. 77, no. July 2021, pp. 838–879, 2022, https://doi.org/10.1016/j.jmapro.2022.02.041.
3. S. Jang, S. Park, and C. Jun Bae, "Development of ceramic additive manufacturing: process and materials technology," *Biomed. Eng. Lett.*, vol. 10, no. 4, pp. 493–503, 2020, https://doi.org/10.1007/s13534-020-00175-4.
4. Y. Wang, Y. Bu, and X. Wang, "Advances in 3D printing of structural and functional ceramics: Technologies, properties, and applications," *J. Eur. Ceram. Soc.*, vol. 44, no. 14, p. 116653, 2024, https://doi.org/10.1016/j.jeurceramsoc.2024.05.075.
5. K. Zhu *et al.*, "Additive manufacturing of SiO2–Al2O3 refractory products via Direct Ink Writing," *Ceram. Int.*, vol. 46, no. 17, pp. 27254–27261, 2020, https://doi.org/10.1016/j.ceramint.2020.07.210.
6. L. Rueschhoff, W. Costakis, M. Michie, J. Youngblood, and R. Trice, "Additive Manufacturing of Dense Ceramic Parts via Direct Ink Writing of Aqueous Alumina Suspensions," *Int. J. Appl. Ceram. Technol.*, vol. 13, no. 5, pp. 821–830, 2016, https://doi.org/10.1111/ijac.12557.
7. J. Baltazar, P. M. C. Torres, J. Dias-de-Oliveira, J. Pinho-da-Cruz, S. Gouveia, and S. Olhero, "Influence of filament patterning in structural properties of dense alumina ceramics printed by robocasting," *J. Manuf. Process.*, vol. 68, no. PA, pp. 569–582, 2021, https://doi.org/10.1016/j.jmapro.2021.05.043.
8. D. Carloni, G. Zhang, and Y. Wu, "Transparent alumina ceramics fabricated by 3D printing and vacuum sintering," *J. Eur. Ceram. Soc.*, vol. 41, no. 1, pp. 781–791, 2021, https://doi.org/10.1016/j.jeurceramsoc.2020.07.051.
9. S. Yang *et al.*, "Rapid prototyping of ceramic lattices for hard tissue scaffolds," *Mater. Des.*, vol. 29, no. 9, pp. 1802–1809, 2008, https://doi.org/10.1016/j.matdes.2008.03.024.

Natural Sealants for Siddha Medicine System: Earthworm Burrow Clay Characterization

Himakshi Baishya, G. Pugazhenthi, and Senthilvelan Selvaraj

Abstract This research project investigates earthworm burrow clay as a potential sealing material for Siddha medicine vessels, focusing on comprehensive characterization. Siddha medicine, a traditional system in India, uses earthen vessels for medicine preparation and storage. Traditionally, medicines are sealed in earthen vessels using *seelaimann*, a termite cavern soil that adheres well due to its fine particles and good bonding characteristics. Earthworm burrow clay appears to be similar to termite cavern soil. However, there is a notable lack of scientific studies on this type of clay. Hence, this study aims to bridge this gap by comprehensively characterizing earthworm burrow clay through advanced analytical techniques. Understanding its physical, chemical, morphological, and mechanical properties will enhance its application in traditional practices. This research will integrate traditional knowledge with modern science, improving traditional medicine preparation methods and developing more effective sealing materials for Siddha medicine vessels.

Keywords Earthworm burrows clay · Siddha medicine · Sealant · Clay characterization

---

H. Baishya (✉) · S. Selvaraj
Department of Mechanical Engineering, Indian Institute of Technology Guwahati, Guwahati, Assam, India
e-mail: b.himakshi@iitg.ac.in

S. Selvaraj
e-mail: ssvelan@iitg.ac.in

G. Pugazhenthi
Department of Chemical Engineering, Indian Institute of Technology Guwahati, Guwahati, Assam, India
e-mail: pugal@iitg.ac.in

U. S. Dixit et al. (eds.), *Current Progress in Engineering Sciences*,
https://doi.org/10.1007/978-981-96-8753-4_17

1 Introduction

This research project focuses on exploring the potential of earthworm burrow clay as a sealing material for earthen vessels used in Siddha medicine manufacturing, encompassing a comprehensive characterization of its properties and suitability. Siddha medicine is a traditional system of medicine in India that relies on traditional practices and the use of earthen vessels for the preparation and storage of medicines. Research in traditional medicine is robust and expanding, with a focus on several key areas. Efforts are also being made to standardize and document treatment guidelines, conduct preclinical evaluations and toxicity testing, and assess pharmacological activity to ensure the safety and efficacy of these medicines. However, there is a lack of attention to the traditional earthen vessels and equipment that are used to prepare these medicines. In the traditional Siddha medicine system, the raw materials for medicine are placed in the bottom closure, which is then covered with the top closure and sealed using a traditional sealant made from a cloth coated with water-wetted termite cavern soil known as *seelaimann* [1]. The sealed earthen vessel called an *agal* is subsequently dried in sunlight. These agals are then fired using traditional fuels such as cow dung cake, which leads to calcination of the raw materials and removal of impurities. Earthworm burrow clay is similar to termite cavern soil as it adheres well due to its fine particles and binding nature when mixed with water [2–4].

The key distinction lies in the habitat adaptation, as termites are exceptional soil engineers in hot, dry regions, but it is difficult to find them in other regions, whereas earthworms' natural habitat is moist environments, particularly in clay soils [5]. Earthworms inhabit and feed on soil, interacting with microbes and significantly contributing to the mixing and decomposition of soil organic matter. Their activities enhance soil quality by increasing organic carbon content and forming stable macro aggregates, which improve soil structure, protect organic matter from microbial attack, and enhance soil porosity and stability. Earthworm feeding and burrowing are essential for carbon cycling and soil aggregation, as these stable aggregates protect soil organic carbon from decomposition [6]. This enhanced soil structure may be beneficial for creating effective sealants, such as those used in traditional practices like sealing earthen vessels. However, there is a notable lack of scientific literature and characterization studies on the specific types of clay, such as earthworm burrow clay, used in these traditional practices. Clay is a natural, fine-grained material made up primarily of these minerals, known for its plasticity when wet and its ability to harden upon drying or firing. Clay minerals, including kaolinite, smectite, vermiculite, mica, chlorite, palygorskite, and sepiolite groups, are some of the most important and valuable raw materials worldwide [7]. Each type of clay mineral exhibits unique properties that make it suitable for various applications. For example, smectite clay minerals are ideal for sealing nuclear waste repositories due to their low water permeability, plasticity during deformation, and high swelling capacity. These characteristics prevent water infiltration and allow the material to self-heal potential cracks, maintaining the repository's integrity [8]. Hence, a comprehensive analysis would help to understand the different properties of clay and enhance its application in traditional practices, and bridge the gap by thoroughly characterizing the earthworm burrow clay to be used as a potential sealant in the Siddha medicine manufacturing system through advanced analytical techniques.

2 Materials and Methods

2.1 Materials

The material considered for this study was earthworm clay, obtained from North Guwahati. Assam, India (situated at 91° 44′ E longitude and 26° 10′ N). The clay was subjected to various preparation steps, which included overnight drying at 45°C, grinding using a mortar pestle, and sieving with a standard sieve size of 125 microns to ensure a uniform particle size distribution and remove sand and plant residue. The prepared clay samples were then characterized using a series of analytical techniques to determine their structural, compositional, and thermal properties (Fig. 1).

Fig. 1 **a** Location of earthworm burrow clay sample collection, **b** samples after overnight drying, **c**, **d** optical microscope images of the dried sample showing surface morphology and pores

2.2 Characterization Techniques

Field Emission Scanning Electron Microscopy (FESEM) was employed to examine the surface morphology of the earthworm clay. The samples were analyzed using a FESEM (Make: Zeiss, Model: Sigma 300) after gold coating to study the particle size, shape, and surface texture. Energy Dispersive X-ray Analysis (EDAX) was used to determine the elemental composition of the earthworm clay. EDAX was performed using FESEM (Make: Zeiss, Model: Sigma 300). The samples were prepared in the same way for FESEM. The data obtained provided quantitative and qualitative analysis of the elemental distribution within the clay samples. Inductively Coupled Plasma Mass Spectrometry (ICP-MS) was conducted to quantify trace elements present in the earthworm clay. Approximately 10 mg of the clay was digested using a mixture of concentrated 7 ml nitric acid and 3 ml hydrofluoric acid. The digested samples were then analyzed using an ICP mass spectrometer (Make: Agilent, Model: 7850 ICP-MS). This technique allowed for the detection and quantification of the elements with high sensitivity and precision. X-ray diffraction (XRD) was used to identify the crystalline phases present in the earthworm clay. Powdered samples were prepared and analyzed using a 9KW Powder X-ray diffraction System (Make: Rigaku Technologies, JAPAN, Model: Smartlab) with Cu-Kα radiation. The data were collected over a 2θ range of 5°–85° with a step size of 0.02. The obtained diffraction patterns were compared with standard reference patterns to identify the mineralogical composition of the clay. Fourier Transform Infrared Spectroscopy (FTIR) was performed to identify the functional groups and molecular bonding in the earthworm clay. The spectra were recorded using an FTIR Spectrophotometer (Make: PerkinElmer Model: Spectrum Two), over a range of 4000–400 wavenumbers. The characteristic absorption bands were analyzed to determine the presence of specific functional groups. Thermogravimetric Analysis—Derivative thermogravimetry (TGA-DTG) and Differential Scanning Calorimetry-) : TGA-DTG and DSC were used to assess the thermal properties and stability of the earthworm clay. Approximately 10 mg of the clay was subjected to a controlled heating program from 25 to 1000 °C at a heating rate of 10.0 K/min using a High-Temperature Differential Scanning Calorimetry (DSC)/Thermo Gravimetric (TG) System (Make: Netzsch, Model: STA449F3A00). The pH of the earthworm clay was measured to determine its acidity or alkalinity. A slurry was prepared according to ASTM D4972-13 by mixing 5 g of clay with 10 ml of distilled water. The mixture was stirred and allowed to equilibrate. The pH was then measured using a calibrated pH meter.

3 Result and Discussion

3.1 pH

The earthworm burrowed clay sample was found to be neutral with a pH of 7.2. A neutral pH (around 7) is generally less aggressive than alkaline or acidic conditions, which can lead to the degradation of clay minerals. For instance, high pH levels (above 11) can cause certain clays, like bentonite, to degrade, whereas neutral pH helps maintain their structural integrity over time [9].

3.2 X-ray Diffraction and Phase Identification

The X-ray Diffraction (XRD) analysis of the clay sample provides a comprehensive diffraction pattern, revealing several distinct peaks corresponding to various crystalline phases. It revealed significant insights into its mineralogical composition, which is crucial for evaluating its potential use as a natural sealant material. The minerals include Biotite, Kaolin, Quartz, and Despujolsite, which are identified based on their Joint Committee on Powder Diffraction Standards (JCPDS) card numbers. The identified peaks, along with their 2θ values, d-spacing, Miller indices (hkl), and associated minerals, are summarized in Table 1.

Kaolinite ($Al_2Si_2O_5(OH)_4$):. Kaolinite is a layered silicate mineral with a 1:1 structure, consisting of a tetrahedral silica (SiO4) layer and an octahedral alumina [AlO2(OH)4] layer. The layers are bonded by van der Waals forces and hydrogen bonds, forming a dioctahedral structure. This regular sheet structure gives kaolin good plasticity and improves the adhesive performance, making it an important material in ceramics. [10]. Kaolinite can be identified by JCPDS Card No.: 00-029-1488.

Quartz (SiO_2) is one of the most common and widely distributed minerals in the Earth's crust. It is composed of silicon and oxygen atoms in a continuous framework of SiO_4(silicon-oxygen tetrahedra). It is stable at high temperatures, so it can provide strength and stability for the sealant material [11]. It has been identified by JCPDS Card No.: 01-079-1906.

Biotite ($K_2(Fe_{2.786}Mg_{2.321}Ti_{0.550})(Al_{2.413}Si_{5.587}O_{20})(OH)_4$) is a monoclinic 2:1 phyllosilicate mineral from the mica group, with a complex chemical composition that includes potassium, iron, magnesium, aluminum, silicon, and hydroxide ions. Plays a crucial role in soil chemistry through its weathering process. This process alters the biotite lattice, releasing potassium (K) into the soil solution and transforming it into secondary clay minerals such as vermiculite, smectite, and kaolinite [12]. It can be identified by JCPDS Card No.: 01-083-1366.

Despujolsite ($Ca_3Mn(SO_4)_2(OH)_6(H_2O)_3$) contains calcium, manganese sulfate hydroxide, which is a rare mineral transparent to translucent mineral with lemon-yellow color and found in hydrothermal manganese deposits [13]. Despujolsite has been identified by JCPDS Card No.: 01-072-0388 (Fig. 2).

Table 1 XRD data of earthworm burrow clay for phase identification

2 theta (°)	d-spacing [Å]	h	k	l	Phase identification
8.84	10.053	0	0	2	Biotite
12.3	7.235	0	0	1	Kaolin
19.84	4.47683	1	1	0	Kaolin
20.78	4.255	1	0	0	Quartz
24.82	3.588	0	0	2	Kaolin
25.48	3.493	2	0	1	Despujolsite
26.6	3.356	1	1	2	Despujolsite
27.42	3.259	1	1	4	Biotite
27.88	3.196	1	0	3	Despujolsite
34.14	2.625	1	1	6	Biotite
34.86	2.574	2	0	3	Despujolsite
36.46	2.463	1	1	0	Quartz
38.52	2.343	−1	1	3	Kaolin
39.42	2.287	1	0	2	Quartz
40.24	2.241	2	2	3	Biotite
41.7	2.175	−2	2	4	Biotite
42.36	2.133	2	0	0	Quartz
45.74	1.984	2	2	1	Kaolin
50.08	1.820	3	0	4	Despujolsite
54.8	1.675	2	2	4	Despujolsite
55.3	1.662	0	1	3	Quartz
59.9	1.491	1	1	4	Kaolin
62.26	1.456	−2	2	11	Biotite
63.94	1.432	3	3	0	Kaolin
67.68	1.385	1	2	2	Quartz
68.24	1.375	2	2	6	Despujolsite
75.56	1.257	3	0	2	Quartz
77.58	1.230	2	2	0	Quartz
79.84	1.200	1	2	3	Quartz
81.38	1.184	3	1	0	Quartz

3.3 *Surface Morphology and Microstructure*

The FESEM image taken at a magnification of 15 and 25 K of the earthworm burrow clay provides insights into its microstructural characteristics. The clay particle morphology exhibits a flaky and irregular shape, which might be advantageous for sealant applications. Smaller particles could aid in filling the gaps between larger

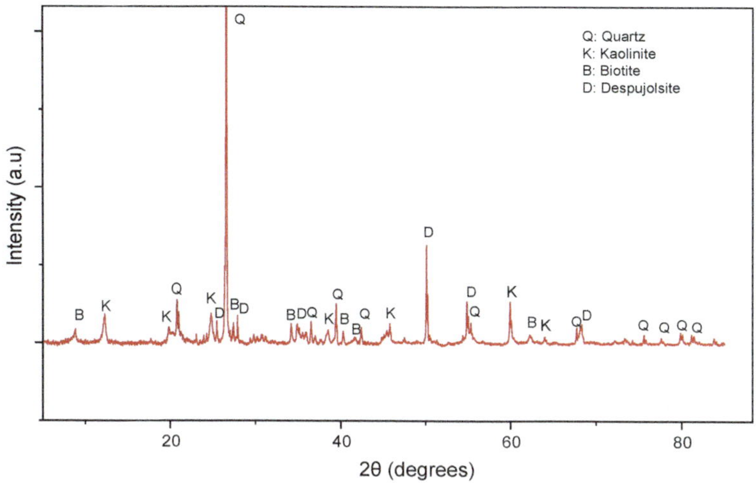

Fig. 2 XRD patterns of earthworm burrow clay

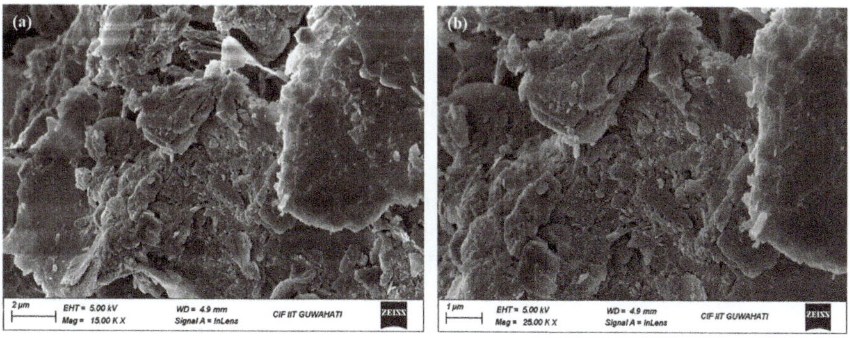

Fig. 3 Fesem images of the earthworm burrow, **a** 15 K magnification, **b** 25 K magnification

ones, to achieve stronger bonding, whereas the irregular shapes could contribute to mechanical interlocking, enhancing the shear strength of the sealant [14] (Fig. 3).

3.4 Qualitative Compositional Analysis Using EDAX

EDAX was employed to find the different elements present in the clay sample, and the result indicated the presence of Silicon, Aluminum, Iron, Calcium, and Potassium as major constituents, while other oxides such as Magnesium, Sulfur, Sodium, Manganese, and Titanium in trace amounts. The EDAX data also suggests this clay has carbon content, which might not be beneficial for sealant application as it will burn at elevated temperatures, leading to an increase in porosity (Fig. 4).

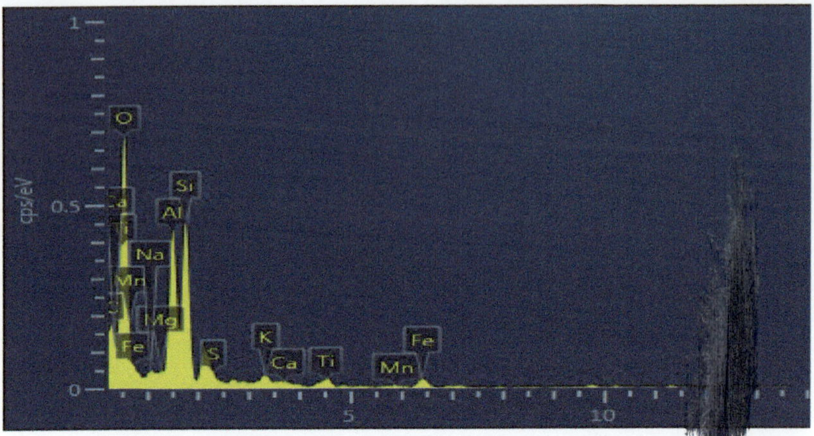

Fig. 4 EDAX graph showing the presence of different elements in the earthworm burrow clay

3.5 Quantitative Compositional Analysis Using ICP-MS

ICP-MS was done to quantify the elements present in the clay sample, which revealed significant concentrations of several trace elements. Aluminum (Al) was found in high concentrations (3867.707 ppb), and Ferrous (1658.46 ppb), which is consistent with the known abundance of Al in clay minerals. Notably, elements such as Potassium also showed elevated levels, which further confirms the presence of biotite. These findings could provide insights into the geochemical processes occurring within earthworm burrows and their potential impact on the clay composition and suitability for sealant application. The ICP-MS data is represented in Table 2.

Table 2 ICP-MS of earthworm burrow clay

Element	Mass	Conc	CPS
Na	23	388.323	1,088,686
Mg	24	196.53	265,474.5
Al	27	3867.707	1,969,211
K	39	1591.281	1,911,564
Ca	43	43.843	1503.51
Mn	55	44.577	430,025.3
Fe	56	1658.46	22,296,666

Fig. 5 FTIR spectra for the OH stretching region

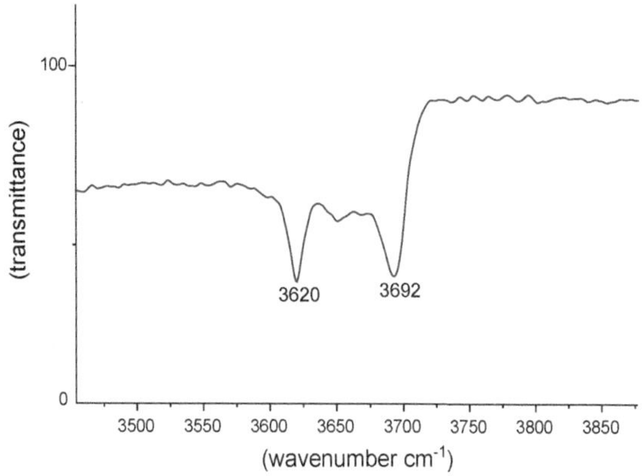

3.6 Fourier Transform Infrared Spectroscopy (FTIR)

The FTIR spectrum of the natural sealant material reveals several characteristic absorption bands corresponding to key functional groups in clay minerals. The OH stretching vibrations at 3620 and 3692 cm^{-1} are indicative of the presence of kaolinite as one of the primary components of the earthworm burrow clay used. [15, 16] The band located at 909 cm^{-1} is attributed to the deformation vibration of Al–OH. These bands suggest confirming the presence of kaolinite, which may contribute to the adhesive properties of the sealant [10]. The presence of a doublet at 797 and 756 cm^{-1} and a single located at 694 cm^{-1} in the figure is attributed to the quartz vibrations. The Si–O stretching bands around 531, 464, and 1002 cm^{-1} further confirm the presence of silicate minerals, which are crucial for the structural integrity of the sealant [17]. The analysis confirmed the presence of minerals such as Kaolinite and Quartz in the soil sample. These minerals contribute to the soil's plasticity and mechanical strength, which may enhance the performance of sealants in various applications (Figs. 5 and 6).

3.7 Thermogravimetric Analysis (TGA) and Derivative Thermogravimetry (DTG)

The TGA and DTG analysis provides valuable information on the thermal stability and decomposition behavior of the clay, which can be used to predict and tailor the material properties for sealant application. Based on the DTG and TGA curves, the initial mass loss up to 200 °C can be attributed to the removal of physically adsorbed water, which was due to the presence of the OH group and other volatile components commonly found in clays. This dehydration process is typical and expected in

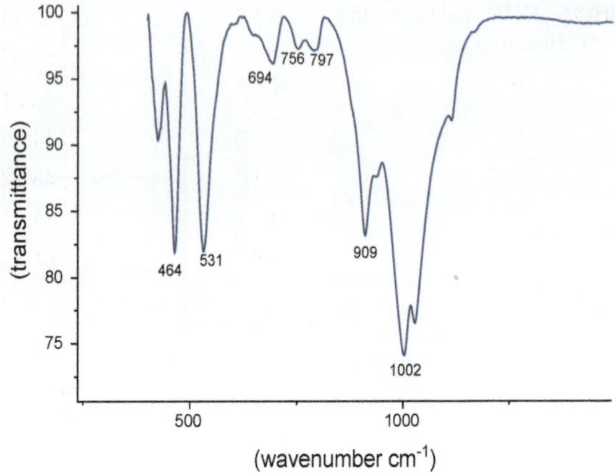

Fig. 6 FTIR spectra for Al–OH and SiO bands

clays [18]. There are significant changes in the DTG curve with multiple peaks, indicating different decomposition events or phase transformations. DTG curve around 400–600 °C, approximately 12% mass loss is observed, which can be attributed to the dehydroxylation of kaolinite, which releases chemically bound water. At around 1000 °C, the remaining mass is about 78% of the original, indicating that approximately 22% of the sample's mass was lost during heating. This suggests that the clay matrix and minerals have undergone significant thermal decomposition, but a substantial portion remains stable up to 1000 °C, which may be due to the presence of quartz, which may contribute to the sealant material's thermal stability since quartz remains stable at high temperatures (up to 1200 °C) [7] (Fig. 7).

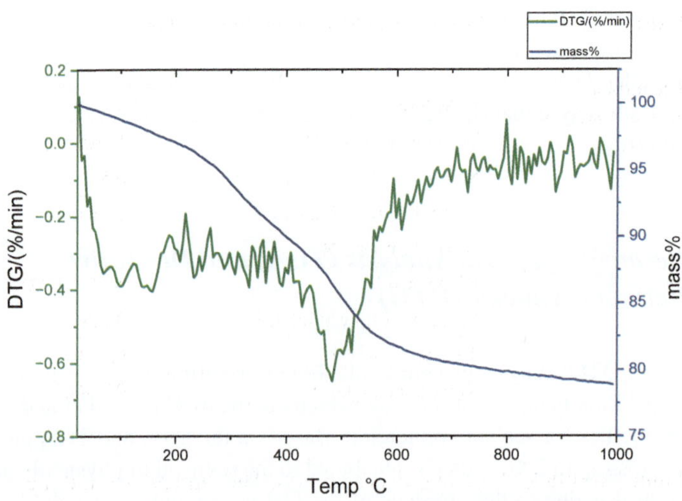

Fig. 7 TGA-DTG curves of earthworm burrow clay up to 1000 °C

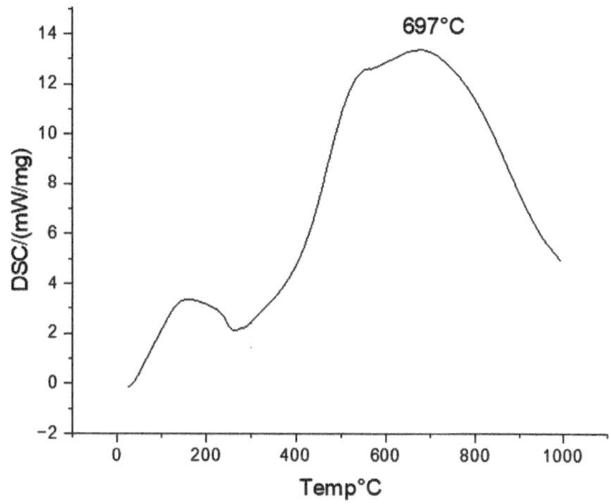

Fig. 8 DSC curve of earthworm burrow clay up to 1000 °C

3.8 Differential Scanning Calorimetry (DSC)

Differential Scanning Calorimetry (DSC) provides crucial insights into its thermal stability and phase transitions. These properties are especially relevant for its potential use as a natural sealant material. An exothermic peak at 697 °C can be attributed to the pyrolysis of organic matter [18]. The DSC curve shows a broad endothermic peak or stabilization in the 900–1000 °C temperature range, indicating the onset of sintering where particle bonding and densification occur. Sintering enhances the mechanical properties and longevity of the sealant (Fig. 8).

4 Conclusion

The analysis of clay samples from earthworm burrows aims to explore their potential as a natural sealant material. In the absence of termite clay, commonly found in dry regions, earthworm clay from moist areas may serve as an alternative. The clay's neutral pH of 7.2 indicates stability over time. XRD analysis identified key minerals such as Kaolinite, Quartz, Biotite, and Despujolsite, which may contribute to its structural integrity and plasticity. FESEM and EDAX analysis highlighted its favorable microstructural characteristics and elemental composition, while ICP-MS confirmed significant concentrations of essential trace elements. FTIR analysis confirmed the presence of functional groups that could be crucial for adhesive properties. Thermal analysis through TGA, DTA, and DSC indicated that the clay exhibits thermal stability up to 1000 °C, with significant phase transformations and sintering onset, potentially enhancing its mechanical properties. Further investigations to quantify carbon content and assess adhesiveness are required to confirm its suitability as a sealant material. Overall, the comprehensive characterization of the earthworm burrow clay indicates its promising potential for use as a natural sealant material.

References

1. A. S. G. Kumar and S. Selvaraj, "Preliminary investigation of temperature and pressure profile of pudam in the preparation of Siddha medicine Padigara parpam," *J Ayurveda Integr Med*, vol. 14, no. 5, Sep. 2023, https://doi.org/10.1016/j.jaim.2023.100762.
2. S. Wen, M. Shao, and J. Wang, "Earthworm burrowing activity and its effects on soil hydraulic properties under different soil moisture conditions from the loess plateau, china," *Sustainability (Switzerland)*, vol. 12, no. 21, pp. 1–13, Nov. 2020, https://doi.org/10.3390/su12219303.
3. P. Jouquet, M. Lepage, and B. Velde, "Termite soil preferences and particle selections: Strategies related to ecological requirements," *Insectes Sociaux - INSECTES SOC*, vol. 49, pp. 1–7, Mar. 2002, https://doi.org/10.1007/s00040-002-8269-z.
4. "Soil Quality Indicators Earthworms."
5. *Soil Ecology*. Dordrecht: Kluwer Academic Publishers, 2003. https://doi.org/10.1007/0-306-48162-6.
6. A. Józefowska *et al.*, "Stability of Aggregates Made by Earthworms in Soils with Organic Additives," 2021. https://doi.org/10.3390/agronomy11030421.
7. P. Ptáček, F. Šoukal, T. Opravil, J. Havlica, and J. Brandštetr, "The kinetic analysis of the thermal decomposition of kaolinite by DTG technique," *Powder Technol*, vol. 208, no. 1, pp. 20–25, Mar. 2011, https://doi.org/10.1016/J.POWTEC.2010.11.035.
8. J. Cuadros, "Clay as sealing material in nuclear waste repositories," *Geology Today*, vol. 24, no. 3, pp. 99–103, May 2008, https://doi.org/10.1111/j.1365-2451.2008.00667.x.
9. M. Lado, M. Ben-Hur, and I. Shainberg, "Clay Mineralogy, Ionic Composition, and pH Effects on Hydraulic Properties of Depositional Seals," *Soil Science Society of America Journal - SSSAJ*, vol. 71, Mar. 2007, https://doi.org/10.2136/sssaj2006.0169.
10. B. Zhang, Z. Chang, J. Li, X. Li, Y. Kan, and Z. Gao, "Effect of kaolin content on the performances of kaolin-hybridized soybean meal-based adhesives for wood composites," *Compos B Eng*, vol. 173, p. 106919, Sep. 2019, https://doi.org/10.1016/J.COMPOSITESB.2019.106919.
11. S. Das, K. Dana, N. Singh, and R. Sarkar, "Shrinkage and strength behaviour of quartzitic and kaolinitic clays in wall tile compositions," *Appl Clay Sci*, vol. 29, pp. 137–143, Apr. 2005, https://doi.org/10.1016/j.clay.2004.10.002.
12. C. Srinivasarao *et al.*, "Soil potassium fertility and management strategies in South Asian agriculture," *Advances in Agronomy*, vol. 177, pp. 51–124, Jan. 2023, https://doi.org/10.1016/BS.AGRON.2022.11.001.
13. A. El-Shater, A. M. Mansour, M. R. Osman, A. A. A. El Ghany, and A. A. El-Samee, "Evolution and significance of clay minerals in the Esna Shale Formation at Dababiya area, Luxor, Egypt," *Egyptian Journal of Petroleum*, vol. 30, no. 2, pp. 9–16, Jun. 2021, https://doi.org/10.1016/J.EJPE.2021.03.001.
14. X.-Z. Wang, Y.-L. Weng, X. Wang, and W.-J. Chen, "Interlocking mechanism of calcareous soil," *Yantu Lixue/Rock and Soil Mechanics*, vol. 39, pp. 3113–3121, Sep. 2018, https://doi.org/10.16285/j.rsm.2017.2426.
15. S. Kaufhold, M. Hein, R. Dohrmann, and K. Ufer, "Quantification of the mineralogical composition of clays using FTIR spectroscopy," *Vib Spectrosc*, vol. 59, pp. 29–39, Mar. 2012, https://doi.org/10.1016/J.VIBSPEC.2011.12.012.
16. J. Madejová, "FTIR techniques in clay mineral studies," *Vib Spectrosc*, vol. 31, no. 1, pp. 1–10, Jan. 2003, https://doi.org/10.1016/S0924-2031(02)00065-6.
17. S. Louati, S. Baklouti, and B. Samet, "Geopolymers Based on Phosphoric Acid and Illito-Kaolinitic Clay," *Advances in Materials Science and Engineering*, vol. 2016, pp. 1–7, Aug. 2016, https://doi.org/10.1155/2016/2359759.

18. G. Lo Dico, L. Lisuzzo, V. Carcelén, G. Cavallaro, and M. Haranczyk, "Thermogravimetric Analysis of Moisture in Natural and Thermally Treated Clay Materials," *Materials*, vol. 17, no. 10, 2024, https://doi.org/10.3390/ma17102231.

Pineapple Leaf Fibers: Physical, Mechanical, and Morphological Characterization for Its Application in Sustainable Polymer Composites

Parthapratim Barman, Paladugu Rakesh, Somaiah Chowdary Mallampati, and Ujendra Kumar Komal

Abstract Natural fibers have been identified as reinforcing materials in polymer composites to fabricate sustainable materials. Natural fibers have low density compared to synthetic fibers, resulting in high specific strength and stiffness, making them useful as lightweight reinforcing materials and composites. In the current study, the properties of locally available pineapple leaf fibers (PALFs) have been investigated for their potential as a reinforcement in sustainable polymer composites. The fibers were extracted using the water retting method. Alkali treatment has been performed using 5 wt.% of NaOH solution to make the fibers compatible with the polymeric matrix phase. The diameter of the fibers is measured as 234 ± 33.50 μm and 190 ± 34.01 μm before and after treatment, respectively. The FTIR and EDX results show the removal of hemicellulose, lignin, and impurities present on the fiber surface. The tensile strength of raw and treated fibers is measured as 252.14 ± 12.35 MPa and 296.62 ± 14.21 MPa. Atomic force microscopy (AFM) and scanning electron microscopy (SEM) are utilized to examine the surface topography of the treated and untreated fibers.

Keywords Natural fiber · Bio-composites · Surface morphology · Alkali treatment · Tensile strength

P. Barman · P. Rakesh · S. C. Mallampati · U. K. Komal (✉)
Department of Mechanical Engineering, Indian Institute of Technology Guwahati, Guwahati, Assam, India
e-mail: ukk@iitg.ac.in

P. Barman
e-mail: b.parthapratim@iitg.ac.in

P. Rakesh
e-mail: r.paladugu@iitg.ac.in

S. C. Mallampati
e-mail: somuchow23@iitg.ac.in

U. S. Dixit et al. (eds.), *Current Progress in Engineering Sciences*,
https://doi.org/10.1007/978-981-96-8753-4_18

Abbreviations

PALF	Pineapple Leaf Fiber
NF	Natural Fiber
SEM	Scanning Electron Microscopy
AFM	Atomic Force Microscopy
EDX	Energy Dispersive X-ray Spectroscopy
FTIR	Fourier Transform Infrared Spectroscopy
R_a	Average Roughness
R_q	Root-mean-square (RMS) Roughness
R_t	Maximum Peak-to-valley Height
R_z	10-Point Average Absolute Height Roughness
R_{sk}	Skewness
R_{ku}	Kurtosis

1 Introduction

The usage of ecologically sound, sustainable, renewable resources is becoming more significant due to the depletion of the world's petroleum reserves, increasing greenhouse gas emissions, global warming, and carbon footprint; all of these create severe risks to coming generations [1–3]. Increasing environmental concerns, strict laws, and government regulations have limited the use of non-biodegradable materials. In engineering applications, composite materials have been made by strengthening polymers with durable synthetic fibers such as glass, carbon, aramid, etc. These artificial materials find widespread use in contemporary technology, including electronics, sports equipment, automobile parts, aviation and naval structures, hydrogen storage tanks, and building supplies [4]. The application of synthetic fibers as reinforcement in polymer composite materials has been restricted due to their non-biodegradability, non-recyclability, disposal, health risks, etc. Looking for substitute bio-based products made from naturally replenishable resources is an excellent way to slow down the deterioration of the environment. Therefore, the research fraternity has explored an alternative solution for synthetic fibers as reinforcing materials. Since natural fibers have unique qualities and are better for the environment than synthetic fibers, using cellulose-based fibers from sustainable plant sources has drawn much interest from various industries. Natural fibers (NFs) such as jute, flax, and hemp have been proposed as alternatives to synthetic fibers because of their excellent mechanical qualities, affordability, biodegradability, low carbon footprint, and adaptability [5, 6]. In recent years, interest has risen in natural fiber (NF) based polymer composites due to their excellent mechanical, biodegradable, lightweight, and economical qualities [7].

Many plant-derived fibers, such as jute, kenaf, bamboo, flax, sisal, and pineapple leaf fiber (PALF), in addition to lignocellulosic agricultural residues like rice husk, coir, and sugar bagasse, are appropriate sources for the application of circular economy strategies. These fibers can be used as reinforcements in a polymer resin to improve its mechanical properties, thermal stability, degradation, etc. NFs are generally hydrophilic, and polymers are hydrophobic, so the compatibility of these two is challenging. To enhance compatibility, the surface of NF is usually treated using various chemical treatments, such as alkali, silane, isocyanate, peroxide, benzoylation, enzymatic treatment, microwave treatment, etc. [8, 9]. Several parameters influence the physical, mechanical, and thermal characteristics of NFs, such as harvesting stage, plant growth, local climate, fiber location in the plant, species of plant, crop location, fiber extraction methods, age of fiber, transportation condition, and storage condition [10, 11]. NFs consist of chemical constituents such as cellulose, hemicellulose, lignin, pectin, wax materials, and moisture content, and the percentage of the chemical composition of these groups varies for different NFs. As a result, it impacts natural fibers' physical, mechanical, thermal, and degradation properties [12]. Therefore, investigation of the characteristics of NF is essential before its use as reinforcement in polymeric composites. Barman et al. [13] determined raw and alkali-treated bamboo fibers' physical, mechanical, and morphological characteristics. The experimental results showed that after 12 wt% of alkali treatment, there was a 40% improvement in tensile strength. SEM images showed an improvement in surface morphologies after alkali treatment, which improves the interfacial interlocking between fiber and resin in the composites. In this case, mercerization significantly improved the properties of the fiber for reinforcement. Akbar et al. [14] experimentally investigated the effect of 5 wt.% of NaOH treatment on PALF. SEM images showed that NaOH treatment enhanced surface roughness by removing surface contaminants and partially dissolving hemicellulose, lignin, and wax in the fiber bundle. According to the authors' findings, alkali-treated samples exhibited a higher density and fewer pores on the surface of fibers.

PALF is considered a good solution to act as reinforcement for polymer composite materials. PALF is mainly composed of cellulose, hemicellulose, lignin, and pectin. The pineapple plant is said to contain the largest cellulose content of any NF, which increases its strength and utility in products. The properties and, consequently, applications of fibers are directly impacted by their chemical composition. Compared to other common NFs, PALFs have a higher cellulose content, which demonstrates their strong mechanical strength [15]. Table 1 shows the wt.% of chemical constituents of some common NFs.

In the present study, PALFs were treated with NaOH solution, and the effect of alkali treatment on PALF was determined experimentally. The influence of chemical treatment on PALF has been investigated in terms of physical, mechanical, chemical, and surface characteristics. Figure 1 illustrates the flowchart of the present work in detail.

Table 1 Wt.% of chemical constituents of different NFs [16, 17]

Natural fiber	Cellulose (wt.%)	Hemicellulose (%)	Lignin (%)	Wax (%)	Moisture (%)
PALF	70–83	–	5–12.7	–	11.8
Coir	32–43	0.15–0.25	40–45	–	–
Hemp	68	15	10	0.8	6.2
Jute	64.4	12	11.8	0.7	1.1
Banana	56–63	20–25	7–9	3	–
Bagasse	55.2	16.8	25.3	–	–
Flax	64.1	16.7	2.0	3.3	3.9
Agave	68.42	4.85	4.85	0.26	7.69
Oil palm	65	10.12	17.5	4	–
Kenaf	45–57	8–13	21.5	0.8	6–12
Bamboo	73.83	12.49	10.15	3.16	–
Sisal	60–78	10.0–14.2	8.0–14	2.0	10–22
Abaca	56–63	20–25	7–9	3	–

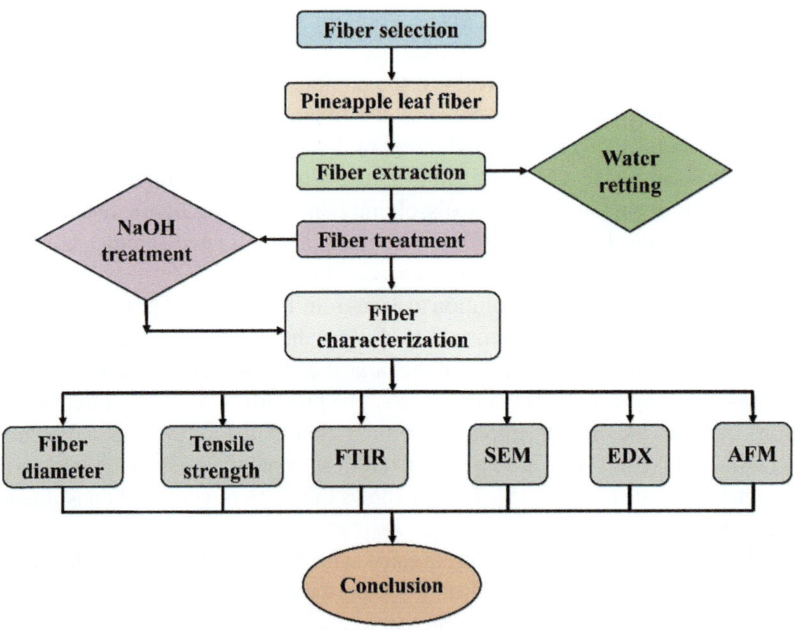

Fig. 1 Flowchart of methodology

Fig. 2 **a** Pineapple leaves, **b** water-retted dried raw fibers, **c** untreated, and **d** alkali-treated fibers

2 Materials and Methods

2.1 Fiber Extraction

Pineapple leaves were collected from the local village of Assam, India (Fig. 2a). The Water retting method was used to extract the fibers. PALFs were soaked in water for 5 days, and then, by using the manual extraction method, fibers were removed from the stem as shown in Fig. 2b. The extracted fibers were dried under ambient conditions for 24 h and then kept inside the hot air oven for 6 h at 60 °C. The dried fibers were kept under sealed packaging to avoid moisture absorption from the air and prevent contamination from dirt, pollutants, etc.

2.2 Alkali Treatment

NaOH pellets were acquired from Sigma-Aldrich, India. Fibers were submerged in 5 wt.% of NaOH solution with a fiber-to-alkali solution ratio of 1:10 for 5 h at room temperature. The fibers were then extensively rinsed with distilled water to eliminate alkali substances from the fiber surface. After cleaning, fibers were kept in a hot air oven for 12 h at 60 °C. The final dried samples were kept in a plastic-sealed packet. Figure 2c and d show the final dried samples of untreated and treated fibers. Equation 1 represents the chemical reaction that takes place during mercerization between the alkali solution and the fiber surface [18, 19].

$$\text{Fiber} - OH + NaOH \rightarrow \text{Fiber} - O - Na + H_2O + \text{Surface impurities} \quad (1)$$

2.3 Fiber Diameter

A stereo optical microscope (make: Nikon) was used to measure the diameter of fibers at different locations. Considering the variation in fiber diameter along the length, at least 5 specimens were employed, and the diameter was measured at 7 different locations of each fiber along its length, and the average value has been reported.

2.4 Tensile Properties

The tensile strength of untreated and alkali-treated PALFs was determined as per ASTM D3379 standard using a 5 kN Electromechanical Universal Testing Machine (make: Zwick Roell, Z005TN). All the tests were performed at a strain rate of 2 mm/min and an ambient temperature of 25 °C with a relative humidity of 65%. PALF samples with a gauge length of 50 mm were fixed with glue on an emery paper to determine tensile strength, as shown in Fig. 4a. At least 5 specimens of each fiber type were tested, and average values of tensile strength, Young's modulus, and elongation at break were reported.

2.5 FTIR Analysis

FTIR spectroscopy was used to determine the functional groups of PALF before and after treatment. FTIR apparatus (make: PerkinElmer, Singapore; model: Spectrum 2) was used for the analysis. PALFs were converted to powder form using ball milling. FTIR spectra were recorded from 4000 to 400 cm^{-1} with a resolution of 2 cm^{-1}.

2.6 Surface Morphology

The surface morphology of PALF was analyzed using SEM (model: Gemini 500 FE-SEM). This device is additionally connected to an Energy Dispersive X-ray Spectrometer (EDX) for elemental characterization. At different magnifications, the PALFs' surface and diameter were examined. Fiber samples were sputter-coated with gold before SEM examination to prevent charging and improve the quality of SEM images.

2.7 Atomic Force Microscopy (AFM)

The surface topography of fibers was recorded using tapping mode at 3 locations. A minimum of three samples were utilized for each fiber type. The surface roughness was measured using AFM and characterized in terms of R_a, R_q, R_t, R_z, R_{sk}, and R_{ku} parameters.

3 Results and Discussion

3.1 Fiber Diameter

Figure 3a, b shows the fiber diameter measured using an stereo optical microscope. The average fiber diameter of untreated and treated PALF is measured as 234 ± 33.50 μm and 190 ± 34.01 μm, respectively. After alkali treatment, the diameter reduces due to removing hemicellulose, lignin, wax, surface impurities, etc. [20]. The aspect ratio (L/d) is enhanced by alkali treatment as the fiber diameter is reduced while the length remains essentially unchanged. An improved aspect ratio is advantageous for fiber reinforcement in composites as it enhances the surface area available for stress transfer from the polymer to the fibers [21].

3.2 Tensile Strength

After alkali treatment, the PALF showed an increase in tensile strength. The percentage increment in tensile strength after alkali treatment was 17.56%, as shown in Fig. 4b. This is caused by the partial removal of amorphous hemicellulose and non-cellulosic materials. The abundance of these non-cellulosic materials in PALF generally does not account for their strength. Instead, the partial removal of hemicellulose strengthens the fiber by promoting the creation of hydrogen bonding between the microfibrils, which enhances the fibrils' close packing. The homogeneity of the fibers is increased, and the interfibrillar regions' ability to support load is enhanced by this refinement in the formation of bonds between the cellulose fibrils [22]. Figure 4c shows the average tensile strength of alkali treated and untreated PALF, along with their standard deviation.

3.3 Fourier Transform Infrared Spectroscopy (FTIR)

Spectra show the characteristic bands of absorption linked to the lignin, hemicellulose, and cellulose fractions, suggesting that these constituents are present in the

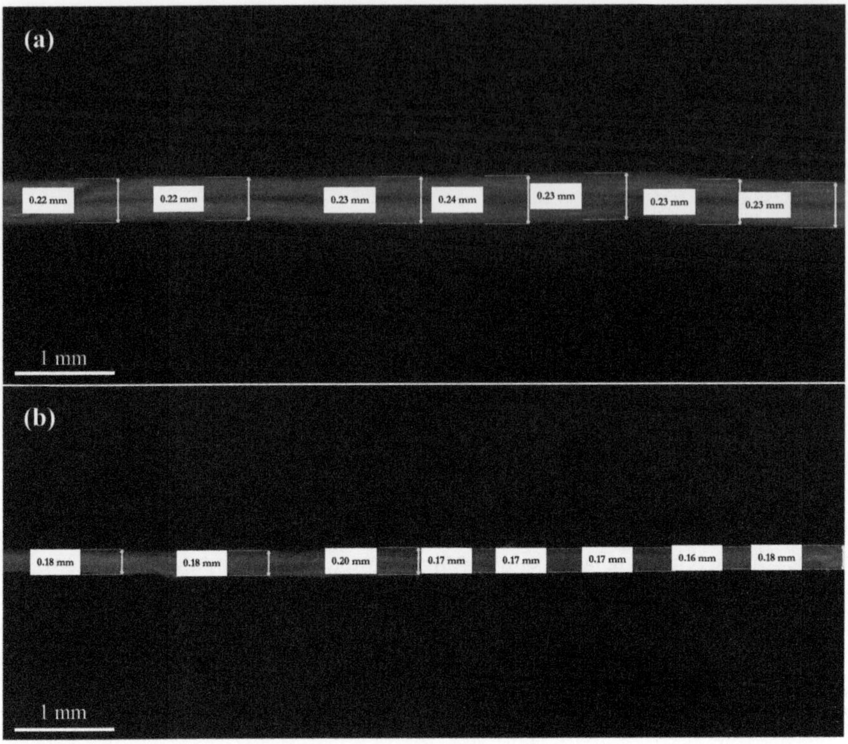

Fig. 3 Diameter of **a** untreated, **b** alkali treatment PALFs

fibers. Figure 5b shows a decrease in the peak after alkali treatment. It indicates that the treatment has affected specific functional groups or bonds. Alkali treatment can remove certain impurities or surface contaminants, decreasing peaks associated with these impurities. An FTIR peak will drop in intensity when the concentration of a specific functional group or compound within the sample is lowered. Figure 5b shows that after alkali treatment, peaks corresponding to -OH stretching vibrations shift (within -OH stretching region, 3200–3500 cm^{-1}) due to changes in hydrogen bonding. Fibers become hydrophilic because they contain hydroxyl groups. As PALF is a lignocellulose fiber, it interacts with water through hydrogen bonding on a large scale. It is noted that -OH groups present in the amorphous phase can easily form hydrogen bonds with water, but very little water interaction with the hydroxyl groups present on the surface of the crystalline phase. The FTIR spectra of raw and alkali-treated samples are relatively comparable, indicating that no additional functional groups are introduced after alkali treatment. Table 2 contains an overview of every additional peak seen on the FTIR spectra.

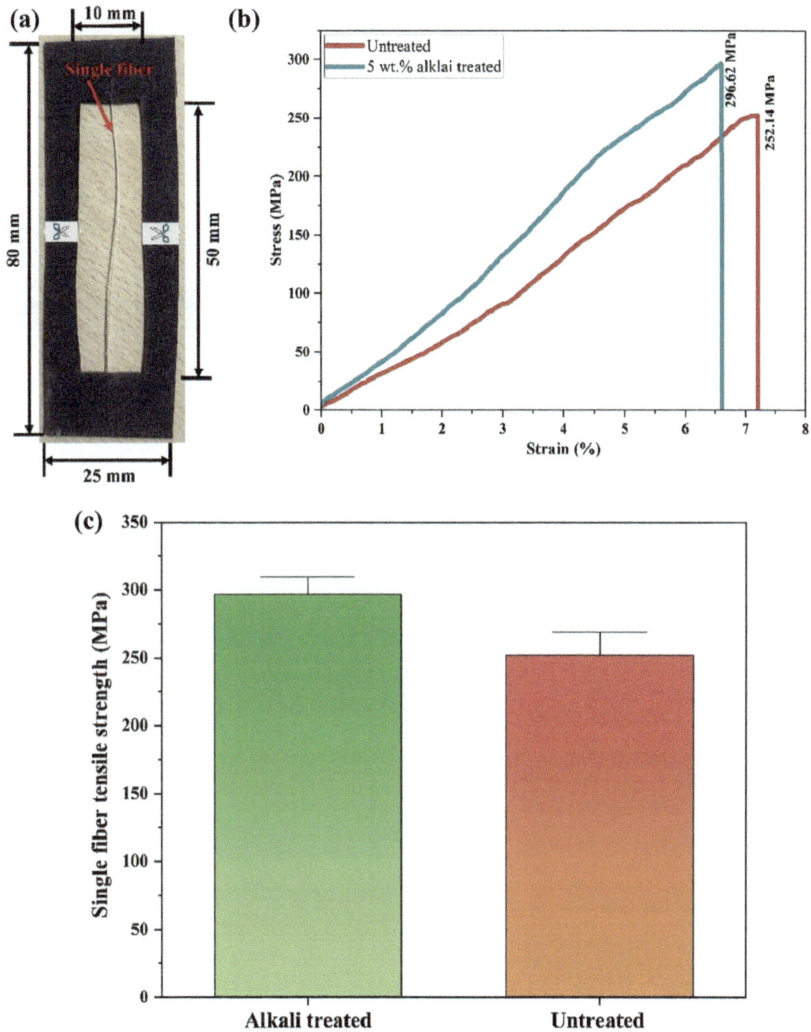

Fig. 4 **a** Single fiber test specimen, **b** tensile stress–strain curve, **c** single fiber tensile strength

3.4 Scanning Electron Microscopy (SEM)

SEM is an excellent method for analyzing the surface topography of both treated and untreated fibers. Figures 6a, b and 7a, b show the SEM pictures of raw and alkali-treated PALF, respectively. Figure 6 has a relatively smooth surface covered by a waxy layer and impurities, which protects the fibers but hinders the interfacial interaction between fibers and polymers. On the surface of the untreated fiber, parenchyma cells, waxes, and polysaccharides such as pectin, lignin, and hemicellulose were observed [30, 31]. The presence of cracks on the surface suggests natural

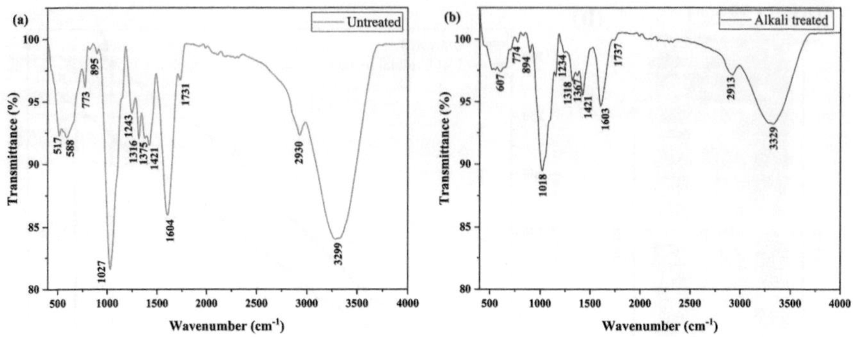

Fig. 5 Transmittance band spectrum of (a) raw and (b) NaOH-treated PALF

Table 2 Transmittance band assignment in FTIR spectrum

Functional group	Wavenumber (cm^{-1})		References
	Raw	5 wt.% alkali treated	
-OH groups stretching vibration	3299	3329	[23, 24]
Cellulose and hemicellulose -CH stretching	2930	2913	[23, 24]
Acetyl and uranic ester groups, and C=O bonds indicate the structure of hemicellulose	1731	1737	[25, 26]
The absorption band of lignin, indicating aromatic vibration of C=C in-plane	1604	1603	[25, 26]
-OH in-plane bending of cellulose structure	1421	1421	[27]
Assigned to CH_3 bending	1375	1367	[28]
C-H deformation of lignocellulose	1316	1318	[29]
C-O stretching of the acetylene groups present in hemicellulose	1243	1234	[29]
Stretching vibration C-O/C=C	1027	1018	[27]
Vibration of β-glycosidic linkage between monosaccharides	895	894	[23]

imperfections and potential weaknesses within the fiber. These cracks might be due to environmental exposure or inherent structural defects. After alkali treatment, the fiber surface is seen to be free from amorphous materials and all impurities (Fig. 7b). The alkali-treated fiber shows grooves along its length (Fig. 7a). These grooves result from removing the waxy layer and other surface impurities during the alkali treatment. The surface also shows micro-cracks, which can occur due to the harshness of the alkali treatment. These micro-cracks can increase the surface area and improve the mechanical interlocking with the polymer matrix. Overall, the treated

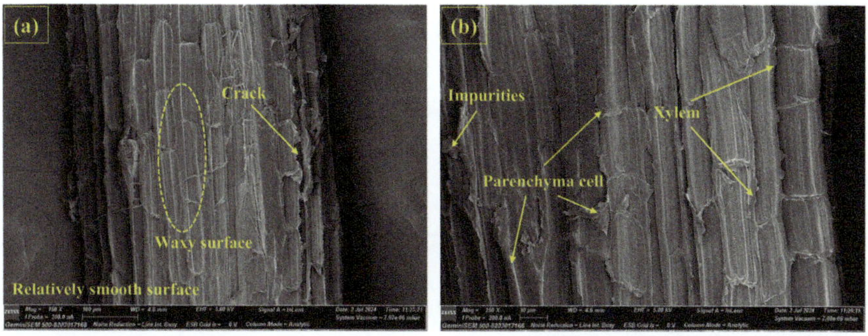

Fig. 6 **a** and **b** SEM micrograph of untreated PALF at different magnifications

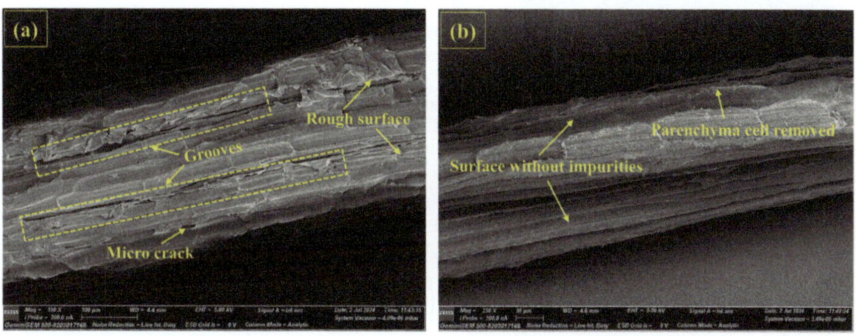

Fig. 7 **a** and **b** SEM micrograph of alkali-treated PALF at different magnifications

fiber has a rougher surface than the untreated fiber. This roughness is beneficial for enhancing the mechanical bonding between the fiber and the polymer matrix in composite materials. Exposure of hydroxyl groups (-OH) after treatment facilitates reaction sites for polymer and improves mechanical interlocking, thereby increasing fiber/matrix interfacial adhesion [32]. Alkali treatment is necessary for NFs to have better mechanical and interfacial characteristics. By exposing crystalline cellulose and eliminating amorphous substances like lignin and hemicellulose, it improves the fiber's mechanical strength, thermal stability, and compatibility with the matrix.

3.5 Energy Dispersive X-Ray Spectroscopy (EDX)

Figure 8a, b depicts the EDX location and patterns of untreated and alkali-treated fibers. Higher amounts of carbon and oxygen are present on both treated and untreated fiber surfaces, indicating the higher concentration of cellulose present in untreated and treated fibers [33]. An increase in carbon content from 34.2 to 37.60% was observed, which supports the literature that alkali treatment enhances cellulose

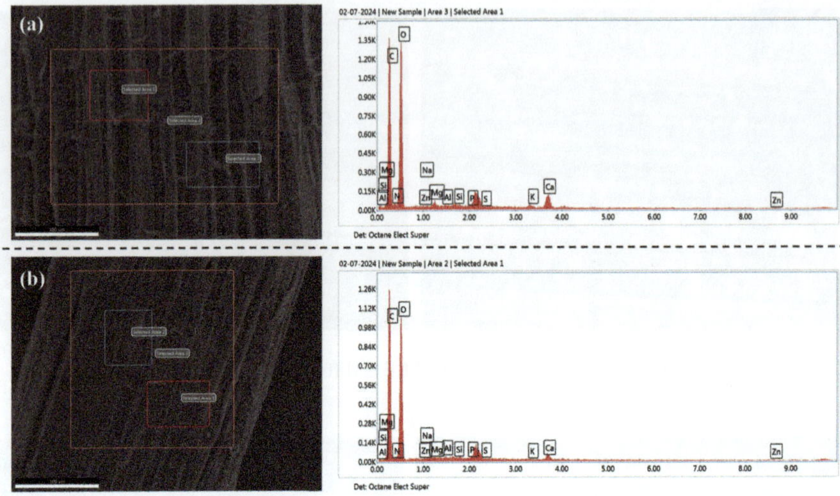

Fig. 8 Surface element distribution of **a** untreated and **b** alkali-treated PALF

content in the fiber [34]. The presence of aluminum (Al), silicon (Si), and calcium (Ca) is attributed to the existence of impurities and other substances like lignin, wax, etc. [33, 35]. However, after treatment, Al, Si, and Ca content decreased due to the partial removal of lignin, hemicellulose, and waxy substances from the fiber surface. Na is absent in the fibers. The literature shows that after alkali treatment, the % of Na rises due to the deposition of Na on the fiber surface from NaOH [34, 36]. However, the present study shows no evidence of Na after alkali treatment. This indicates that the fibers were properly neutralized after the alkali treatment. Table 3 shows the different wt. % and atomic percentages of elements distributed on the surface of PALF.

3.6 Atomic Force Microscopy (AFM)

Figure 9a and d depicts the surface topography of untreated and alkali-treated PALFs. Considerable modification can be seen on the surface of the fibers after being treated with NaOH solution. Alkali treatment has a more pronounced effect on fiber roughness. It was also clear from SEM micrographs that the alkali treatment developed more noticeable hills and valleys as well as roughness. Higher roughness is advantageous for improved fiber-matrix interaction. From the AFM image, two regions can be easily identified: a shiny region and a dull region. The shiny region and dull region correspond to crystalline and amorphous parts, respectively [37]. The R_a value of untreated fiber is 178.20 nm, which indicates that the surface is covered with hemicellulose, lignin, wax, and other impurities [38]. SEM also confirmed the same. Surface skewness (R_{sk}) of treated fibers is predominantly peaks, and untreated fibers

Table 3 Amount of element distribution in weight % and atomic % on the surface of fibers

Elements	Raw PALF		Alkali-treated PALF	
	Weight (%)	Atomic (%)	Weight (%)	Atomic (%)
C	34.2	42.36	37.60	45.50
N	2.80	2.99	2.60	2.70
O	55.23	51.43	54.86	49.76
Na	0	0	0	0
Mg	0.73	0.43	0.56	0.30
Al	0.33	0.20	0.23	0.11
Si	0.40	0.20	0.36	0.20
P	0.73	0.36	0.60	0.30
S	0.60	0.30	0.60	0.26
K	0.63	0.23	0.26	0.10
Ca	3.3	1.26	1.20	0.46
Zn	1	0.23	0.83	0.16

have more valleys than peaks. Surfaces with more peaks may have better adhesion due to more contact points between fibers and resin. R_{sk} for alkali-treated fibers is − 0.2365, which indicates porous fiber surface [39]. After alkali treatment, there is a significant improvement in RMS values (113.57%) and 10 points average absolute height roughness (225%). Table 4 contains different roughness parameters of both fibers. Therefore, it can be concluded that fiber mercerization improves the fiber surface roughness, essential for improved mechanical interlocking of PALFs with polymeric matrices.

4 Conclusions

In the present work, characterizing PALFs before and after alkali treatment has provided significant insights into their structural, chemical, morphological, and mechanical properties. Fibers from the pineapple leaf are extracted using water retting methods, and different characteristics have been experimentally studied. The following findings have been reached:

- After treatment due to the removal of surface materials (hemicellulose lignin, wax, impurities) present on the fiber surface, the diameter of the fiber decreases. Optical microscopy has shown a reduction in the diameter from 234 ± 33.50 μm to 190 ± 34.01 μm of the fibers post-treatment.
- The strength of the PALF after alkali treatment is improved. 5 wt.% NaOH-treated fiber (296.62 ± 14.21 MPa) has better mechanical properties than untreated PALF (252.14 ± 12.35 MPa).

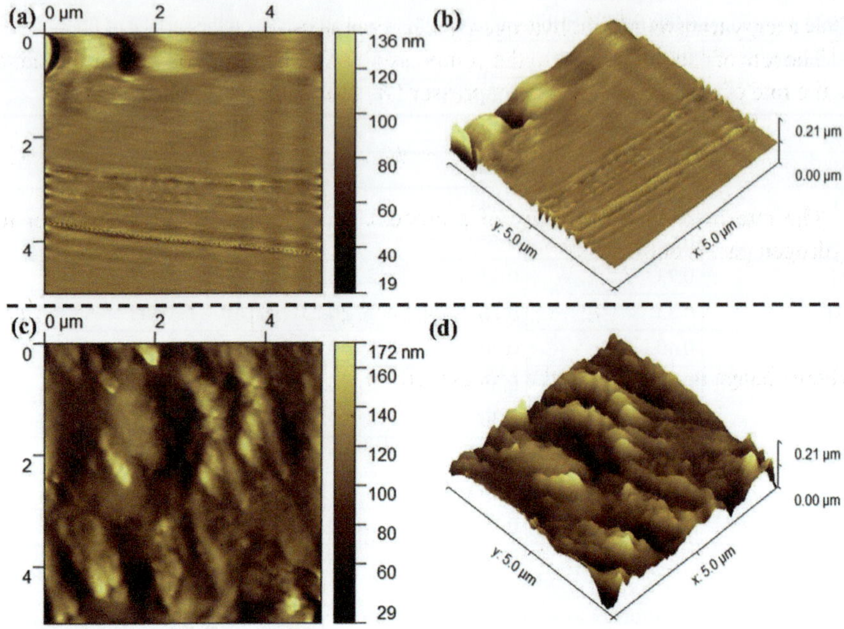

Fig. 9 2D and 3D AFM topography of **a**, **b** raw and **c**, **d** alkali-treated PALF

Table 4 Different surface roughness parameters of PALF

Fibers	R_a (nm)	R_q (nm)	R_z (nm)	R_{sk} (nm)	R_{ku} (nm)
Untreated	178.20	34.62	16.31	1.692	15.90
Alkali treated	448.60	73.94	52.66	−0.2365	3.041

- SEM and AFM analysis provide a detailed surface topology of the fibers, showing increased surface roughness post-treatment, essential for enhancing the interfacial bonding between fiber and matrix.
- FTIR spectral differences provide evidence of the decrease of the peaks associated with amorphous content (hemicellulose, lignin, wax, and pectin) as well as exposure and reorganization of cellulose in the fibers, enhancing their interfacial interaction with epoxy in composites.
- EDX analysis also confirms the removal of hemicellulose, lignin, wax, and other surface impurities. The reduction in non-cellulosic constituents enhances the cellulose content, which is primarily responsible for improving the mechanical properties of fibers.

The findings of the current research suggest that the alkali treatment of PALFs improved the physical, mechanical, chemical, and surface characteristics that are essential for NFs, which have immense potential to be used as reinforcement as an alternative to synthetic fiber in polymeric composites for various structural and non-structural applications.

Acknowledgements The authors gratefully acknowledge the infrastructural and financial support provided by IIT Guwahati under Startup Grant (Project No: xxMESUGIITG01391xUKK001.xls) and infrastructural support provided by the North East Centre for Biological Sciences and Healthcare Engineering (NECBH), IIT Guwahati, and the Department of Biotechnology (DBT), Govt. of India, under Project No. BT/NER/143/SP44675/2023. The authors would also like to thank the Centre for the Environment, IIT Guwahati, for providing the FTIR facility.

References

1. J. O. Ighalo *et al.*, "Biochar from coconut residues: An overview of production, properties, and applications," *Ind. Crops Prod.*, vol. 204, no. August, 2023, https://doi.org/10.1016/j.indcrop.2023.117300.
2. A. Melelli *et al.*, "Anticipating global warming effects: A comprehensive study of drought impact of both flax plants and fibres," *Ind. Crops Prod.*, vol. 184, no. April, 2022, https://doi.org/10.1016/j.indcrop.2022.115011.
3. S. Ovalı, O. Eryılmaz, and S. Uyanık, "Exploring the potential of sustainable natural cellulosic fiber from Sorghumbicolor (Sorghumvulgare var. technicus) stem for textile and composite applications," *Cellulose*, vol. 31, no. 5, pp. 3289–3302, 2024, https://doi.org/10.1007/s10570-024-05800-4.
4. O. Eryilmaz, "Revalorization of cellulosic fiber extracted from the waste stem of Brassica oleracea var. botrytis L. (cauliflower) by characterizing for potential composite applications," *Int. J. Biol. Macromol.*, vol. 266, no. March, 2024, https://doi.org/10.1016/j.ijbiomac.2024.131086.
5. B. A. Amel, M. T. Paridah, R. Sudin, U. M. K. Anwar, and A. S. Hussein, "Effect of fiber extraction methods on some properties of kenaf bast fiber," *Ind. Crops Prod.*, vol. 46, pp. 117–123, 2013, https://doi.org/10.1016/j.indcrop.2012.12.015.
6. G. Diouf and P. B. Gning, "Influence of the Extraction Process on the Mechanical and Morphological Properties of Technical Fibers from Typha domingensis," *J. Nat. Fibers*, vol. 21, no. 1, 2024, https://doi.org/10.1080/15440478.2024.2357744.
7. C. Baley, M. Gomina, J. Breard, A. Bourmaud, and P. Davies, "Variability of mechanical properties of flax fibres for composite reinforcement. A review," *Ind. Crops Prod.*, vol. 145, no. September 2019, 2020, https://doi.org/10.1016/j.indcrop.2019.111984.
8. Pankaj, C. S. Jawalkar, and S. Kant, "Critical Review on Chemical Treatment of Natural Fibers to Enhance Mechanical Properties of Bio Composites," *Silicon*, vol. 14, no. 10, pp. 5103–5124, 2022, https://doi.org/10.1007/s12633-021-01194-1.
9. T. P. Naik, S. Gairola, I. Singh, and A. K. Sharma, "Microwave-assisted alkali treatment of sisal fiber for fabricating composite as non-structural building materials," *Constr. Build. Mater.*, vol. 411, no. November 2023, 2024, https://doi.org/10.1016/j.conbuildmat.2023.134651.
10. H. H. Chiu and W. Bin Young, "The Longitudinal and Transverse Tensile Properties of Unidirectional and Bidirectional Bamboo Fiber Reinforced Composites," *Fibers Polym.*, vol. 21, no. 12, pp. 2938–2948, 2020, https://doi.org/10.1007/s12221-020-0109-0.
11. M. Asim *et al.*, "A review on pineapple leaves fibre and its composites," *Int. J. Polym. Sci.*, vol. 2015, 2015, https://doi.org/10.1155/2015/950567.
12. S. Chokshi, V. Parmar, P. Gohil, and V. Chaudhary, "Chemical Composition and Mechanical Properties of Natural Fibers," *J. Nat. Fibers*, vol. 19, no. 10, pp. 3942–3953, 2022, https://doi.org/10.1080/15440478.2020.1848738.

13. P. Barman, P. P. Dutta, M. Bardalai, and P. P. Dutta, "Experimental investigation on bamboo fiber reinforced epoxy polymer composite materials developed through two different techniques," *Proc. Inst. Mech. Eng. Part C J. Mech. Eng. Sci.*, vol. 238, no. 6, pp. 2185–2204, Mar. 2024, https://doi.org/10.1177/09544062231195474.
14. A. Akbar and R. Gnanamoorthy, "Characterization of Treated and Untreated Pineapple Leaf Fiber for Engineering Applications," *Mater. Circ. Econ.*, vol. 6, no. 1, 2024, https://doi.org/10.1007/s42824-024-00116-x.
15. J. Jain and S. Sinha, "Pineapple Leaf Fiber Polymer Composites as a Promising Tool for Sustainable, Eco-friendly Composite Material: Review," *J. Nat. Fibers*, vol. 19, no. 15, pp. 10031–10052, Nov. 2022, https://doi.org/10.1080/15440478.2021.1993478.
16. O. Faruk, A. K. Bledzki, H. Fink, and M. Sain, "Progress in Polymer Science Biocomposites reinforced with natural fibers : 2000–2010," *Prog. Polym. Sci.*, vol. 37, no. 11, pp. 1552–1596, 2012, https://doi.org/10.1016/j.progpolymsci.2012.04.003.
17. L. Mohammed, M. N. M. Ansari, G. Pua, M. Jawaid, and M. S. Islam, "A Review on Natural Fiber Reinforced Polymer Composite and Its Applications," *Int. J. Polym. Sci.*, vol. 2015, 2015, https://doi.org/10.1155/2015/243947.
18. N. H. Sari, I. N. G. Wardana, Y. S. Irawan, and E. Siswanto, "The effect of sodium hydroxide on chemical and mechanical properties of corn husk fiber," *Orient. J. Chem.*, vol. 33, no. 6, pp. 3037–3042, 2017, https://doi.org/10.13005/ojc/330642.
19. K. Seisa, V. Chinnasamy, and A. U. Ude, "Surface Treatments of Natural Fibres in Fibre Reinforced Composites: A Review," *Fibres Text. East. Eur.*, vol. 30, no. 2, pp. 82–89, 2022, https://doi.org/10.2478/ftee-2022-0011.
20. M. A. S. Oliveira, K. L. Pickering, T. Sunny, and R. J. T. Lin, "Treatment of hemp fibres for use in rotational moulding," *J. Polym. Res.*, vol. 28, no. 2, pp. 1–9, 2021, https://doi.org/10.1007/s10965-021-02414-3.
21. A. E. Bekele, H. G. Lemu, and M. G. Jiru, "Study of the Effects of Alkali Treatment and Fiber Orientation on Mechanical Properties of Enset/Sisal Polymer Hybrid Composite," *J. Compos. Sci.*, vol. 7, no. 1, pp. 1–11, 2023, https://doi.org/10.3390/jcs7010037.
22. U. K. Komal and I. Singh, "Sustainable Treatments of Pineapple Leaf Fibers for Polylactic Acid Based Biocomposites," *J. Nat. Fibers*, vol. 19, no. 16, pp. 13438–13456, 2022, https://doi.org/10.1080/15440478.2022.2095550.
23. M. Maache, A. Bezazi, S. Amroune, F. Scarpa, and A. Dufresne, "Characterization of a novel natural cellulosic fiber from Juncus effusus L.," *Carbohydr. Polym.*, vol. 171, pp. 163–172, 2017, https://doi.org/10.1016/j.carbpol.2017.04.096.
24. S. Tanpichai and S. Witayakran, "All-cellulose composite laminates prepared from pineapple leaf fibers treated with steam explosion and alkaline treatment," *J. Reinf. Plast. Compos.*, vol. 36, no. 16, pp. 1146–1155, Aug. 2017, https://doi.org/10.1177/0731684417704923.
25. Y. Song, X. Li, G. Han, Y. Zhang, and W. Jiang, "Comparison of Different Methods to Produce Pineapple Leaf Fibers with Steam Explosion," *J. Nat. Fibers*, vol. 18, no. 1, pp. 12–20, 2021, https://doi.org/10.1080/15440478.2019.1612301.
26. D. M. Chaves, J. C. Araújo, C. V. Gomes, S. P. Gonçalves, R. Fangueiro, and D. P. Ferreira, "Extraction, characterization and properties evaluation of pineapple leaf fibers from Azores pineapple," *Heliyon*, vol. 10, no. 4, 2024, https://doi.org/10.1016/j.heliyon.2024.e26698.
27. V. Johny, A. Kuriakose Mani, S. Palanisamy, V. K. Rajan, M. Palaniappan, and C. Santulli, "Extraction and Physico-Chemical Characterization of Pineapple Crown Leaf Fibers (PCLF)," *Fibers*, vol. 11, no. 1, pp. 1–15, 2023, https://doi.org/10.3390/fib11010005.
28. J. Jaafar, J. P. Siregar, A. N. Oumer, M. H. M. Hamdan, C. Tezara, and M. S. Salit, "Experimental investigation on performance of short pineapple leaf fiber reinforced tapioca biopolymer composites," *BioResources*, vol. 13, no. 3, pp. 6341–6355, 2019, https://doi.org/10.15376/biores.13.3.6341-6355.
29. H. Rahman, S. Rana, A. Das, and R. Alagirusamy, "Studies on interfacial shear strength of pineapple leaf fibre from agro-waste reinforced polypropylene composites: Influence of fibre length and carding parameters," *J. Thermoplast. Compos. Mater.*, vol. 37, no. 5, pp. 1748–1773, 2024, https://doi.org/10.1177/08927057231200008.

30. P. Senthamaraikannan and M. Kathiresan, "Characterization of raw and alkali treated new natural cellulosic fiber from Coccinia grandis.L," *Carbohydr. Polym.*, vol. 186, no. October 2017, pp. 332–343, 2018, https://doi.org/10.1016/j.carbpol.2018.01.072.
31. S. Sahu *et al.*, "Characterization of natural fiber extracted from Bauhinia vahlii bast subjected to different surface treatments: A potential reinforcement in polymer composite," *J. Nat. Fibers*, vol. 20, no. 1, 2023, https://doi.org/10.1080/15440478.2022.2162185.
32. L. Yan, N. Chouw, and X. Yuan, "Improving the mechanical properties of natural fibre fabric reinforced epoxy composites by alkali treatment," *J. Reinf. Plast. Compos.*, vol. 31, no. 6, pp. 425–437, 2012, https://doi.org/10.1177/0731684412439494.
33. A. Venugopal and S. K. Boominathan, "Physico-chemical, Thermal and Tensile Properties of Alkali-treated Acacia Concinna Fiber," *J. Nat. Fibers*, vol. 19, no. 8, pp. 3093–3108, 2022, https://doi.org/10.1080/15440478.2020.1838998.
34. J. S. N. Raju, M. V. Depoures, and P. Kumaran, "Comprehensive characterization of raw and alkali (NaOH) treated natural fibers from Symphirema involucratum stem," *Int. J. Biol. Macromol.*, vol. 186, no. July, pp. 886–896, 2021, https://doi.org/10.1016/j.ijbiomac.2021.07.061.
35. J. S. N. Raju, M. V. Depoures, J. A. B. I. H. U. L. L. A. Shariff, and S. U. D. H. I. R. Chakravarthy, "Characterization of Natural Cellulosic Fibers from Stem of Symphirema Involucratum Plant," *J. Nat. Fibers*, vol. 19, no. 13, pp. 5355–5370, 2022, https://doi.org/10.1080/15440478.2021.1875376.
36. Y. Bulut and A. Aksit, "A comparative study on chemical treatment of jute fiber: Potassium dichromate, potassium permanganate and sodium perborate trihydrate," *Cellulose*, vol. 20, no. 6, pp. 3155–3164, 2013, https://doi.org/10.1007/s10570-013-0049-6.
37. L. Ravindran, S. M.S., and S. Thomas, "Novel processing parameters for the extraction of cellulose nanofibres (CNF) from environmentally benign pineapple leaf fibres (PALF): Structure-property relationships," *Int. J. Biol. Macromol.*, vol. 131, pp. 858–870, Jun. 2019, https://doi.org/10.1016/j.ijbiomac.2019.03.134.
38. X. Fang, J. Xu, H. Guo, and Y. Liu, "The Effect of Alkali Treatment on the Crystallinity, Thermal Stability, and Surface Roughness of Bamboo Fibers," *Fibers Polym.*, vol. 24, no. 2, pp. 505–514, 2023, https://doi.org/10.1007/s12221-023-00093-z.
39. M. Kathirselvam, A. Kumaravel, V. P. Arthanarieswaran, and S. S. Saravanakumar, "Assessment of cellulose in bark fibers of Thespesia populnea: Influence of stem maturity on fiber characterization," *Carbohydr. Polym.*, vol. 212, no. February, pp. 439–449, 2019, https://doi.org/10.1016/j.carbpol.2019.02.072.

Comparative Aerodynamic Study Between Fuselage-Wing Combination of Tube-Wing Design and BlendedWing Design Surveillance UAV in Subsonic Speed Regime

Reni Varghese and Santosha Kumar Dwivedy

Abstract In this paper, a small category surveillance-type fuselage-wing combination of a Blended Wing Body (BWB) UAV and a conventional Tube-wing UAV were designed. The designs were created using MATLAB code and modeled in Open VSP and Solid Works. The Aerodynamic analysis was carried out (excluding the Propellers) using Open VSP and ANSYS Fluent. The fuselage-wing configuration of UAVs was studied to determine the impact of the integrated role of the fuselage, wing attachment, and shape on Aerodynamic performance. Then, the results of the two UAV models obtained from Open VSP and Ansys Fluent were compared. Findings from this study indicate that the coefficient of lift at various Angle of attack (AOA) is higher for the Tube-wing UAV, even when the tail was excluded. However, when the Aerodynamic efficiency is considered, the BWB UAV outperforms its counterpart. Furthermore, the moment coefficient of the BWB UAV has a negative slope, indicating stability. In contrast, the Tube-wing UAV has a positive slope (unstable) for the moment coefficient, which necessitates the attachment of the tail section for stability. Thus, further decreasing the Aerodynamic efficiency of the Tube-wing UAV. Additionally, it is also found that at lower AOA, the Open VSP and CFD results show close agreement. Thus, it can be concluded that the BWB UAV design remains more Aerodynamically efficient than the conventional Tube-wing Design despite the tail being excluded.

Keywords Blended wing · Unmanned aerial vehicle · Configuration layout · Aerodynamic study · Surveillance · UAV

R. Varghese
Department of Design, Indian Institute of Technology Guwahati, Guwahati, Assam, India
e-mail: vreni@iitg.ac.in

S. K. Dwivedy (✉)
Department of Mechanical, Indian Institute of Technology Guwahati, Guwahati, Assam, India
e-mail: dwivedy@iitg.ac.in

U. S. Dixit et al. (eds.), *Current Progress in Engineering Sciences*,
https://doi.org/10.1007/978-981-96-8753-4_19

1 Introduction

The market of the unmanned aerial vehicle (UAV) in the twenty-first century will triple [1]. One of the most efficient configurations is the Blended-Wing Body (BWB). It has numerous advantages: less fuel consumption, large internal volume, high lift, almost elliptic lift distribution, and lower operating expenses than conventional Tube-wing Design [2]. BWB is a revolutionary platform. The BWB has a future-proof design because it has a huge fuselage volume, which allows it to accommodate future fuel-hydrogen easily [3]. There have been a variety of advances in BWB, like variable Camber technology [4]. The review paper can be referred for more details on the advances in the BWB [5]. BWB does not have a tail. It usually consists of an exterior section (Wing) and a middle segment (center body or fuselage), with the area in between serving as the blending area where the center body smoothly blends into the wing [6]. In this paper, a small category surveillance type fuselage-wing combination BWB UAV and a Tube-wing UAV are designed. Later, aerodynamic analysis (CFD) is done by excluding the propellers, and then they are compared with each other on critical aerodynamic parameters. The mission requirements for designing this small UAV are considered from the existing UAVs in the market (like Rotla mini UAV, SpyLite, Orbiter 2, etc.), which can be used for multiple applications like precision agriculture, surveying, surveillance, etc. A high wing was selected for the Surveillance UAV to minimize obstruction, especially when imaging targets are directly below the UAV. Additionally, it provides stability to the platform during low-speed flight and when flying in gusty environments.

The specification of it is:

(a) Wingspan is around 3 m.
(b) Maximum Takeoff Weight to be within 10 kg.
(c) Endurance to be a minimum of 20 min.
(d) Speed range is 20–30 m/s.
(e) Payload = 1 kg for the surveillance mission.

2 Method

As per requirements and using standard equations in [7, 8] MATLAB code was written. The key parameters for designing the UAV were found from the Code. The details of the MATLAB code are explained in the next section. Then, the key parameters were used to make the 3D model initially in Open VSP and Solid Works software. The first step in modeling was to divide it into three sections, i.e., the fuselage or center section, the blending section, and the outer Wing section. As per the definition of BWB, it includes both a wing and a fuselage. A combination of wing and fuselage (center body in BWB) is considered as a platform for the BWB UAV.

The important features of the aerofoil profile selected for both models are detailed in the next section. Due to stability reasons, a BWB-type platform requires a reflex camber airfoil profile for the center section [7, 9]. The center section should have a comparatively large thickness so that it has enough space to store the batteries, payloads, etc. [7, 9]. For the tip section, an airfoil was selected, which has high aerodynamic performance at the Operating Reynolds number of the UAV.

For the blending section, the airfoil profile selected was such that the root smoothly transitioned into the tip profile. Later, a CFD analysis was carried out at 20 m/s and 30 m/s speeds using Open VSP and Ansys. Around the model an unstructured grid was generated. On the walls of the model, 10 inflation layers were created so that the boundary layer phenomena could be accurately modeled (Fig. 1). Finally, the Airfoil with the below mentioned important properties was selected from the Airfoil tools website [10].

In the Ansys CFD analysis, to ensure high accuracy, the Reynolds-Averaged-Navier-Stokes (RANS) equations were solved, coupled with a low-Reynolds turbulence model. The Spalart-Allmaras and k-omega SST turbulence models were used as they are reliable for external flow conditions [9, 11, 12]. More details of settings are mentioned below and in [9, 11, 12]. A steady-state method was used to perform this comparative study.

3 Results and Discussions

3.1 BWB Model Details

For both configurations, the same baseline UAV was used. Initially assuming not much of a major difference is there, and later, through analysis, the difference is shown. A MATLAB program was used for calculating the dimensions of the model as per the mission requirements in an iterative process (for efficient flying at low speed, 20 m/s speed was considered in the design calculation). In the MATLAB code, initially, a similar mission baseline UAV weight and ratios of structure and propulsion were considered. Later, the battery weight as per the new mission was estimated and added to the previous step. The new UAV weight was estimated using iteration such that the difference between the initial and the new weight is less than 10e-10. The Detailed program steps are based on the formulas and steps described in references [7, 9, 13]. The minimum fuselage (center body in BWB) volume required was fixed as per the existing similar UAVs in the market (like Rotla mini UAV, SpyLite, Orbiter 2, etc.), such that it can include the battery and the payload [9]. A similar process was followed for the fuselage design of Tube-wing UAV. In both the models, the wing planform area and its properties are kept constant for comparison (i.e, fuselage and center body excluded). The payloads are flight controllers, servos, pitot tubes, cameras with gimbals, telemetry, antennas, etc., which come up to around 1 kg in weight. Figures 2 and 3 show two sections: the center section and the wing section. The Blending section is not separately shown (as it lies between these two sections). For the Tube-wing UAV, the wing section extends to the center line of the fuselage to

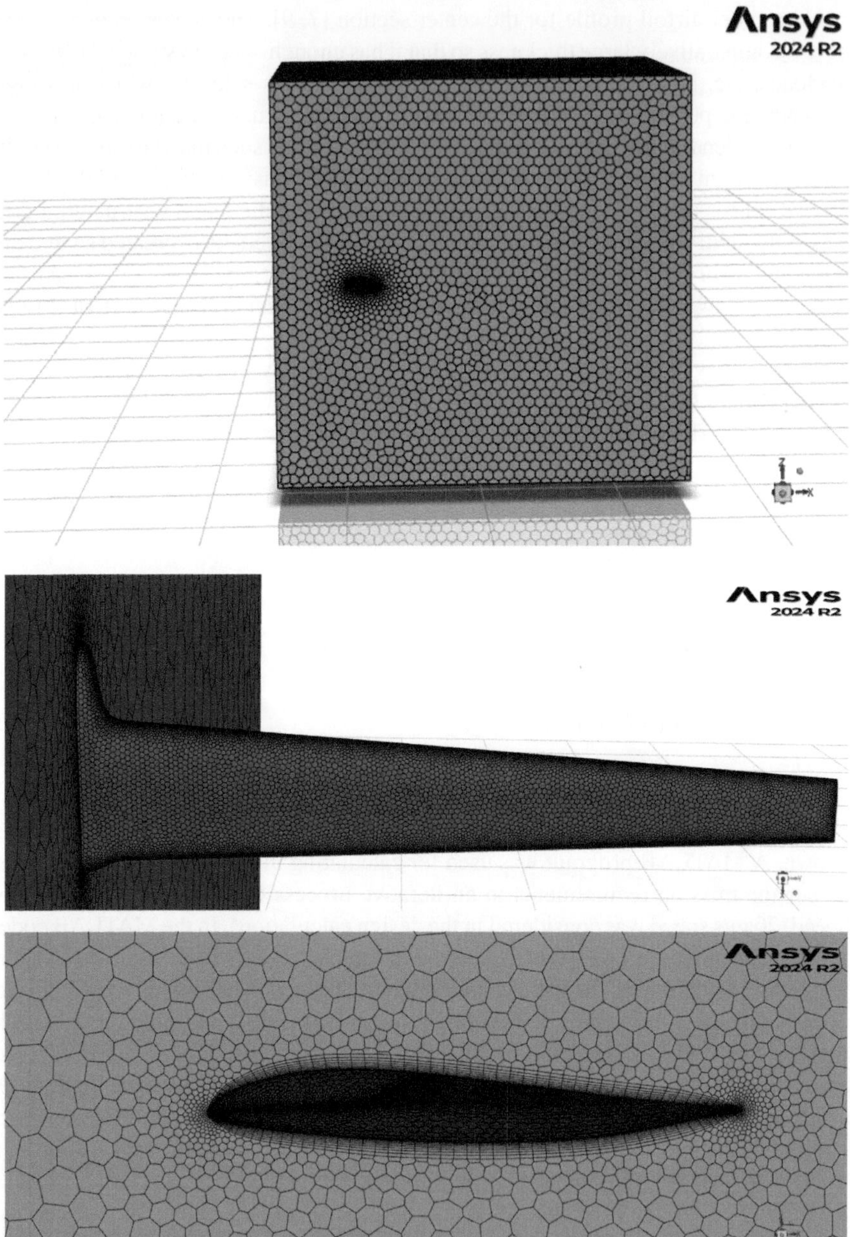

Fig. 1 Computational mesh in Ansys 2024 on the (**a**) Tube-wing, (**b**) BWB UAV-wing view, (**c**) BWB UAV-Airfoil view

form the center section. The Wing section properties are kept the same in both UAV models for comparison purposes, except for the Airfoils.

The major Design here is the wing. In the Wing design process, 18 parameters are found [7, 8]. They are as follows:

a. the wing reference area is S
b. number of wings is 1
c. vertical position relative to the fuselage is high-wing
d. Horizontal position relative to the fuselage is such that the CG of the UAV has to be Close to the Mean Aerodynamic center of the wing.
e. cross-section mentioned below
f. aspect ratio (AR);
g. taper ratio (λ);
h. tip chord (Ct);
i. root chord (Cr);
j. Mean aerodynamic chord (MAC);
k. span (b);
l. twist angle is 2 deg and −2 deg for the root and tip of the wing
m. The sweep angle is very less, as it is for low speed.
n. dihedral Angle is zero
o. the incidence angle of the wing is 2 deg
p. high-lifting devices are not used in the present case, as these increase drag
q. aileron is not considered.
r. Other wing accessories are not considered.

The reasons for the above selections are as per the mission requirements of the current UAV. The Airfoil should satisfy the below-mentioned criteria efficiently in the operating Reynolds number range of the UAV specially during Cruise [6, 9, 13]:

a. should have the highest maximum lift coefficient
b. should have a proper high-design lift coefficient
c. should have the lowest minimum drag coefficient
d. should have the highest lift-to-drag ratio
e. should have the highest lift curve slope
f. should have the lowest (closest to zero; negative or positive) pitching moment coefficient
g. should have proper stall quality in the stall region (the variation must be gentle, not sharp)
h. The center body airfoil should have enough thickness so as to accommodate the battery and payload and be aerodynamically efficient.

Finally, another program was run in a loop in MATLAB by varying the values of Aspect ratio, lift-to-drag ratio, and Wing loading, and the most efficient parameters were finalized as below:

General-BWB

(a) Maximum takeoff mass = 7.5 kg

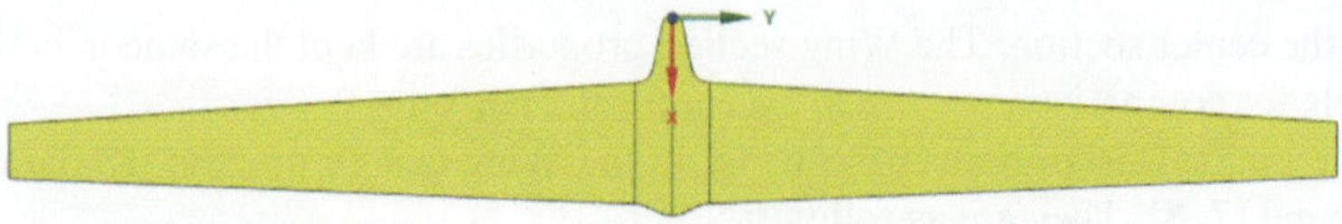

Fig. 2 BWB UAV model

(b) Wing Loading = 10 kg/m^2
(c) Aspect Ratio = 15
(d) Lift to drag ratio = 20
(e) Span = 3.33167 m

Wing dimensions (Fig. 2)

(f) Area of the Wing (S) = 0.717 m^2
(g) Average chord of the wing = 0.22 m
(h) Coefficient of lift for the Airfoil to be used in wings is around 0.3955
(i) The coefficient of drag for the Airfoil to be used in wings is around 0.019778
(j) Wing root chord = 0.3048 m
(k) Wing tip chord = 0.1378 m

Centre section dimensions

(l) Centre body root chord = 0.5 m
(m) Center body tip chord = 0.3048 m
(n) Span of the center body only = 0.2 m
(o) The blending strength used is gentle with a value = 0.5 (Fig. 2)

Airfoils selected for BWB

(p) For Centre body–Eppler 325 flying Wing airfoil
Max thickness 12.6% at 34.3% chord
Max camber 1.8% at 16.3% chord
(q) For Wings – Eppler 331 flying Wing airfoil
Max thickness 11.2% at 28.3% chord
Max camber 2.7% at 23.6% chord
(r) Other general Characteristics are – a high lift-to-drag ratio, smooth stall, high coefficient of lift.

3.2 Tube-Wing Model Details

General

(a) Maximum takeoff mass = 7.5 kg
(b) Wing Loading = 10 kg/m^2
(c) Aspect Ratio = 15
(d) Lift to drag ratio = 20

(e) Span = 3.33167

Wing dimensions

(f) Area of the Wing (S) = 0.7174 m^2
(g) Average chord of wing = 0.22 m
(h) The coefficient of lift for the Airfoil to be used in wings is around 0.3955
(i) The coefficient of drag for the Airfoil to be used in wings is around 0.019778
(j) Wing root chord = 0.3048 m
(k) Wing tip chord = 0.1378 m

Fuselage dimensions

It's a circular section cylinder with a length of 2.5 m and a diameter = 0.31 m with zero radius at the ends (Fig. 3).

Airfoil selected for Tube-Wing

(l) For the Centre body – no airfoil is required
(m) For Wings – Eppler 550 airfoil
    Max thickness 18.2% at 34% chord
    Max camber 2.2% at 29.4% chord
(n) Other general Characteristics are – a high lift-to-drag ratio, smooth stall, high coefficient of lift.

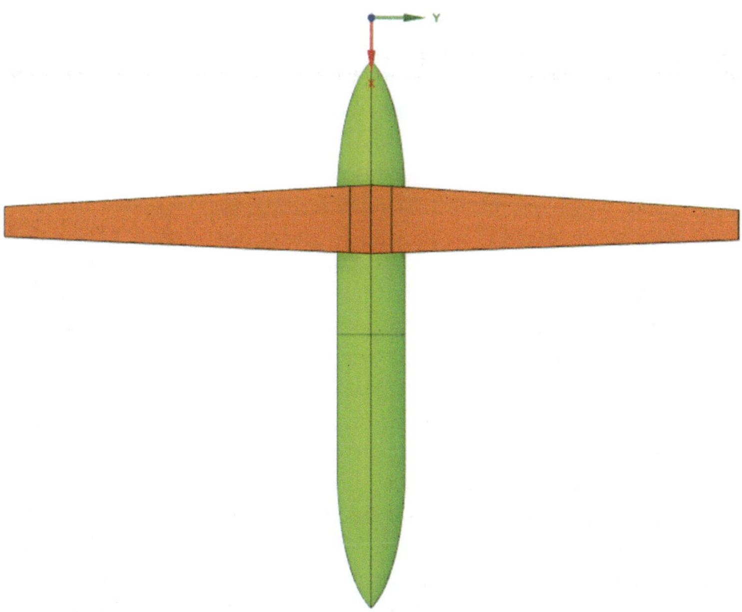

Fig. 3 Tube-Wing UAV model

3.3 Aerodynamic Analysis of the Model

The Open VSP and Ansys were used for analysis with the following settings:

Reference chord got from the Open VSP is used for Analysis. The remaining Ansys settings is mentioned below. The analysis was done at 20 m/s and 30 m/s, but the graph below shows only for 30 m/s, as this shows higher forces acting when compared to 20 m/s. The second-order upwind formulation is used for Turbulent Kinetic energy. Steady state simulation is performed using the pressure-based coupled solver. The k-omega SST model is selected for the Viscous formulation. The Backflow Turbulent Intensity (%) = 1 and Backflow Turbulent Viscosity Ratio = 2 is considered. For the analysis Planform area is considered. The Planform area values for BWB is 0.796 m^2 and for Tube-wing is 0.777 m^2. This difference in area is due to the bigger center section of the BWB.

3.3.1 BWB UAV and Tube-Wing UAV Aerodynamic Properties

The below graphs are drawn between critical Aerodynamic parameters such as Lift to drag ratio VS AOA etc. See Fig. 4.

1. **Lift Coefficient versus Drag Coefficient (CL vs. CD)**

Implications:

- This graph helps in evaluating aerodynamic efficiency, particularly how much lift an aircraft generates for a given amount of drag.
- A higher CL per CD value suggests better efficiency, meaning more lift is generated for less drag.
- For BWBs, the absence of a distinct fuselage and being equipped with a lifting fuselage (i.e, which produces lift) reduces drag and increases lift. However, BWB UAVs have less CL value due to the reflex airfoil, which decreases the CL per CD value compared to a conventional UAV without the tail section, as in this case.
- Conventional Tube-and-wing UAVs, often exhibit higher parasite drag due to fuselage-wing interactions and higher CL due to high lifting airfoil. The absence of a tail section and the presence of a non-lifting fuselage in the Tube-wing UAV have led to a higher CL and higher CD value than the BWB UAV.

Applications:

- Used in performance calculations to determine glide ratios, cruise efficiency, and endurance.
- Helps in drag reduction strategies for both UAVs and conventional aircraft.
- Crucial for optimizing wing designs and selecting airfoils.

2. **Lift Coefficient versus Angle of Attack (CL vs. AOA)**

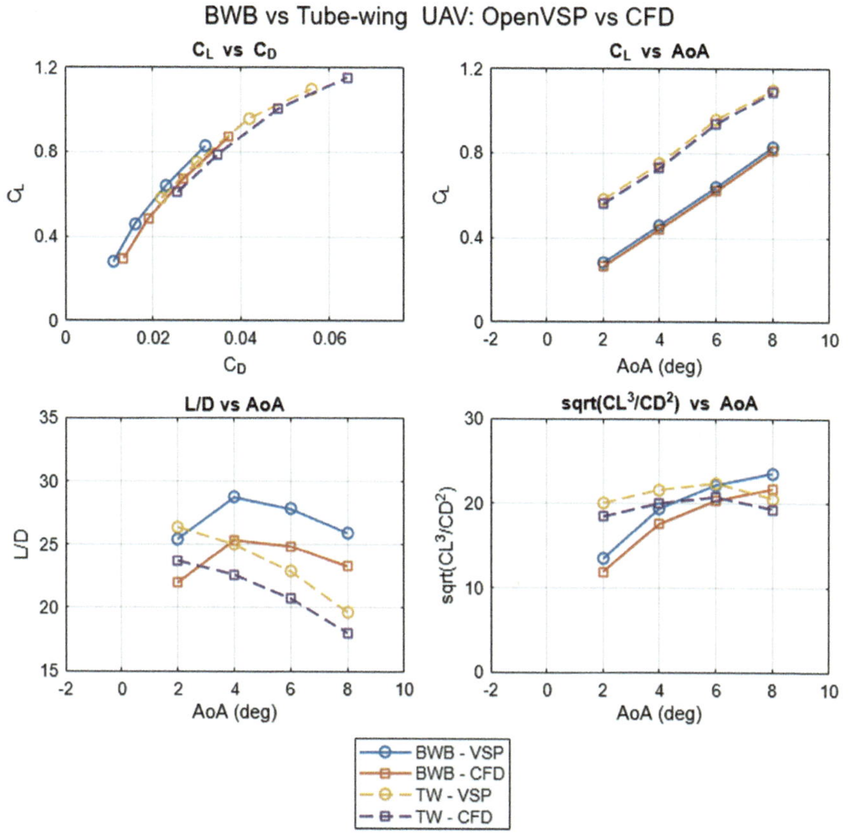

Fig. 4 Aerodynamic analysis plots of Tube-Wing and BWB UAV from Open VSP and Ansys Fluent Software

Implications:

- This graph shows how the lift coefficient changes with the angle of attack, indicating stall characteristics.
- BWBs, due to their integrated lifting surface, generally have a smoother stall behavior and higher lift efficiency compared to traditional designs. In the above plot, the BWB has lower CL value comparatively due to the use of reflex airfoils.
- Conventional UAV (tube-wing) rely on their high-lift airfoils but tend to experience a more abrupt stall due to wing-fuselage interactions.

Applications:

- Essential for stability and control analysis in UAVs.
- It helps optimize wing twist, camber, and leading-edge devices.
- Determines maximum AOA before stall, which is crucial for takeoff and landing performance.

3. **Moment Coefficient Slope Sign**

Implications:

- The sign of the slope value illustrates the stability characteristics of the UAV by showing how the pitching moment changes with AOA.
- BWBs, in this case, have a negative slope, which means BWB UAV is stable. The tailless design of BWB needs a more complex pitch stability profile, requiring careful control surface adjustments. The value of the moment coefficient is less comparatively in BWB as it uses reflex airfoils.
- Thus cause the Tube-wing UAV has a positive slope (i.e, unstable due to fuselage and wing) and needs a tail section to provide stability, further decreasing aerodynamic efficiency.

Applications:

- Used for control system design in UAVs to ensure pitch stability.
- Helps in autopilot tuning for stable flight in both UAVs and manned aircraft.
- Determines trim conditions to minimize control effort and drag.

4. **Lift-to-Drag Ratio versus Angle of Attack (CL/CD vs. AOA)**

Implications:

- This graph is crucial in understanding the Aerodynamic efficiency of an aircraft at different AOAs.
- BWBs often show higher L/D ratios due to reduced parasitic drag and increased lift due to the lift producing fuselage.
- Conventional UAVs have less L/D comparatively, as their efficiency is impacted by non-lift producing fuselage drag and its attachment to the wing.

Applications:

- Determines the most fuel-efficient cruise conditions.
- Helps in designing UAVs for endurance and range optimization.
- Useful in analyzing energy efficiency in electric propulsion aircraft.

5. **$\surd$(CL^3/CD^2 versus Angle of Attack (AOA)**

Implications:

- It is helpful for UAVs, as it indicates optimal conditions for maximizing lift production while minimizing drag.
- BWB UAVs can have a higher $\surd$(CL^3/CD^2, at higher AOA due to better aerodynamic integration.
- Conventional UAV have lower $\surd$(CL^3/CD^2, at higher AOA as it suffer more from fuselage and interference drag.

Applications:

- Used in endurance and range calculations for UAVs and commercial aircraft.

- Helps in optimizing aerodynamic efficiency for long-duration flights.
- Essential in comparing innovative designs against traditional aircraft.

4 Conclusion

The fuselage-wing configuration of UAVs was studied to determine the impact of the integrated role of the fuselage, wing attachment, and airfoil shape on Aerodynamic performance. For a **BWB UAV**, the aerodynamic advantages are evident in improved efficiency (higher L/D and better drag reduction) and stability. However, pitch stability and control require advanced design strategies. In contrast, a **conventional Tube-wing UAV** has stability only when tail section is included. Due to tail section, it suffers from further higher drag and less efficiency during flight. Understanding these aerodynamic graphs is essential for optimizing UAVs and conventional aircraft designs. These graphs enable Engineers to make informed decisions when designing future air vehicles based on the integrated contribution of fuselage, wing and airfoil.

A comparative aerodynamic study was conducted between a fuselage-wing combination of BWB and Tube-wing Surveillance UAV in a low subsonic speed regime. The models were designed s using MATLAB code, and Airfoil tools were used to select airfoils. Open VSP and Ansys Fluent was used for Analysis. The study revealed that the lift coefficient with AOA is higher for the Tube-wing UAV. However, the Aerodynamic efficiency is higher for the BWB UAV. Additionally, the BWB UAV moment coefficient has negative slope (indicating stability). In contrast, the Tube-wing UAV needs a tail to make its moment coefficient slope negative. The higher performance of the BWB UAV is attributed to the center body's ability to generate lift in BWB, the smooth blending of wings into the center body, and the selection of low-moment aerofoils. It was also found that at a lower AOA of attack, the Open VSP values are very close to the results obtained from Ansys Fluent. It can be concluded that the Aerodynamic performance of the BWB design outperforms that of the Tube-wing design, in the low-speed regime (even when excluding tail surface). This higher Aerodynamic efficiency of the BWB UAV will further increase comparatively when the tail surfaces are added to the Tube-wing UAV model considered in the study. The results of the study follow the trends as mentioned in the literature. In the future, complete UAV models of BWB design and Tube-wing design, along with their control surfaces, can be compared.

References

1. S.J. Zaloga, D. Rockwell, P. Finnegan (2015), "World Unmanned Aerial Vehicle Systems Study – 2015 market profile and forecast", Teal Group Corporation.

2. R. Liebeck (2004), Design of the blended wing body subsonic transport, J. Aircraft. 41 (1), pg.- 10–25.
3. Adler, Eytan J.; Martins, Joaquim R. R. A. (2024), Blended Wing Body Configuration for Hydrogen-Powered Aviation Journal of Aircraft, 61 (3), pp. 887–901.
4. Wang, Yutong Lan, Qingsheng; Zhou, Zhu; Yang, Tihao; Song, Chao, (2024), Design and aerodynamic analysis of blended wing body with variable camber technology, Journal of Beijing University of Aeronautics and Astronautics, 50 (4), pp. 1292–1307.
5. Okonkwo, P. and Smith, H., (2016), "Review of evolving trends in blended wing body aircraft design," Progress in Aerospace Sciences, Vol. 82, pp. 123.
6. Daniel Raymer (2012) Aircraft Design a Conceptual Approach, AIAA Series.
7. Mohammad H. Sadrey (2013), Aircraft Design, Aerospace series, Wiley.
8. P. Panagiotou, P. Kaparos, C. Salpingidou, K. Yakinthos., (2016), "Aerodynamic Design of a MALE UAV", Aerospace Sci. Technol. 50, pp. 127–138.
9. Qin, N., Vavalle, A., Le Moigne, A., Laban, M., Hackett, K., and Weinerfelt, P., (2004), "Aerodynamic considerations of blended wing body aircraft," *Progress in Aerospace Sciences*, Vol. 40, pp. 321–343.
10. "Airfoil Tools" Available: http://airfoiltools.com/.
11. P.R. Spalart, S.R. Allmaras, (1992), "A one-equation turbulence model for aerodynamic flows"", in: 30th Aerospace Sciences Meeting & Exhibit, American Institute of Aeronautics and Astronautics, Reno, NV, pp. 5–21.
12. P.R. Spalart, C.L. Rumsey., (2007), "Effective inflow conditions for turbulence models in aerodynamic calculations", AIAA J. 45, pp. 2544–2553.
13. Anderson, J. D., (2011), "Fundamentals of aerodynamics", New York: McGraw-Hill.

Kinematic Analysis of a Mechanism of an Integrated Inter- and Intra-Row Paddy Field Weeder Using SolidWorks

Sapunii Sebastian and Karuna Kalita

Abstract This paper presents the kinematic analysis of a paddy field weeder designed for inter- and intra-row weeding operations. There are numerous types of paddy field weeders in use all over the world, for both wet and dry land. However, most of them are designed for use in inter-row weeding. Weeds in an intra-row zone can significantly reduce yield. Hence, their removal is necessary to achieve a good yield. This study aims to analyze the kinematics of a newly designed paddy field weeder mechanism for both inter- and intra-row weeding. Various components and linkages necessary for the desired motion of the weeder have been designed using SolidWorks software. The components were assembled, and motion simulations were conducted. The kinematic analysis is based on the slider-crank mechanism. This mechanism is later used for intra-row weeding. The rotor blade for inter-row weeding was made to rotate at 200 rpm. The analysis reveals that the intra-row weeding unit traverses between the paddy plants following a sinusoidal path without disturbing them. This could eliminate weeds in the intra-row zone and possibly increase the crop yield.

Keywords Intra-row · Kinematic analysis · Slider crank mechanism · SolidWorks · Weeds

1 Introduction

Many weeders for inter-row cultivation have been designed and fabricated. However, a weeder designed for intra-row cultivation in wetland paddy fields is not known [1]. Inter-row paddy field weeders extract weeds only in the inter-row zone (the zone along the rows), and the weeds between the plants (intra-row zone) remain unharmed.

S. Sebastian (✉)
School of Agro and Rural Technology, IIT Guwahati, Guwahati, India
e-mail: s.sebastian@iitg.ac.in

K. Kalita
Mechanical Engineering Department, IIT Guwahati, Guwahati, India
e-mail: karuna.kalita@iitg.ac.in

U. S. Dixit et al. (eds.), *Current Progress in Engineering Sciences*,
https://doi.org/10.1007/978-981-96-8753-4_20

Apart from inter-row weeds, intra-row weeds have the potential to significantly lower yields in the 18–76% range [2–5], and therefore their eradication is also critical.

The slider-crank mechanism exhibits both linear and rotational motion simultaneously and is applied in various mechanical devices. This particular mechanism transforms the circular motion of the disk (acting as a crank) into the translational motion of the lateral plates via a trapezoidal plate (acting as a slider). The study uses SolidWorks software, which is a comprehensive product lifecycle visualization tool for computer-aided design and engineering. Its many capabilities make it simple for users to develop, design, and model the products [6]. This study aims to analyze the motion of various components necessary for the weeding units and trace a desired path suitable for intra-row weeding units.

2 The Working Mechanism of the Weeder

The circular disc is driven by a chain sprocket system using an IC engine as the prime mover's source. A rectangular link connects the circular disc to the trapezoidal plate. Two lateral plates are attached to the trapezoidal plate via a linear motion slider. As the trapezoidal plate reciprocates, the lateral plates move outward and inward in a lateral direction, which slides on the round bar. For intra-row weeding, the two lateral plates extend into a shank with spikes at the tip. The power for the inter-row weeding unit is derived from the circular disc to the rotor shaft fitted with an L-type blade via chain and sprocket systems.

3 Creation of Individual Components and Assembly

Figure 1 shows a labeled 3-D model of the designed inter-cum intra-row paddy field weeder. Each of the individual components required for the functioning of the weeder was designed and saved. Using the *insert components* command, the individual components were inserted into the *Assembly* section. The desired components were then selected or browsed from the saved folder for assembly. Based on the row-to-row and plant-to-plant spacing of the paddy (here taken as 200 mm for both spacings), the dimensions of the major components (trapezoidal plate, length of connecting bar, diameter of the circular disc) were decided. The length of the connecting bar was decided based on a trial and error method using the SolidWorks software so that free reciprocating motion could be played between the circular disc and trapezoidal plate without interfering with each other. The lateral extension (linear displacement) of the intra-row weeding unit ranged from 29 to 386 mm. However, for ideal weeding conditions at 200 mm $\times$ 200 mm plant spacing, the minimum extension should be kept at 200 mm for each intra-row weeding unit so that it could traverse around the plant without disturbing it. Furthermore, the amount of lateral

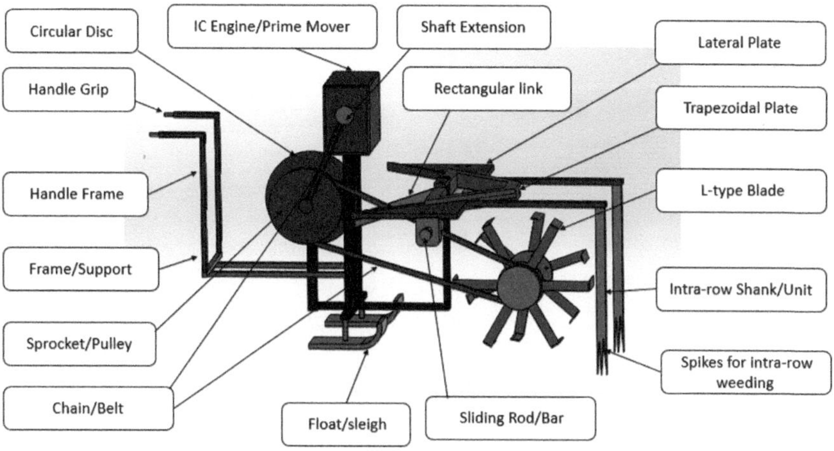

Fig. 1 A 3-D model of the weeder

extension can be varied as required by changing the length of the crank, as the movement of lateral motion depends on it. The height of the intra-row weeding unit was kept higher than the height of the paddy plant to avoid damage during the weeding operation. Thus, the overall dimensions (height, length, and width) of the machine were 600, 1450, and 800 mm, respectively.

3.1 *The Stages of Motion Study*

The motion study's phases, which comprise the following, were carried out in accordance with Nedelcu et al. [7]. It includes activation of the SolidWorks Motion Module, specification of the *Rotary Motor, Force, Gravity, Motion mates, and* execution of the design study.

The *Rotary Motor* was activated to the face of the circular disc, and the *constant speed* was set to 125 rpm for operating the intra-row weeding unit. The operating rotational speed of the rotor blade for the paddy field weeder, however, varies from 180 to 250 rpm. The difference in the variation of the rpm can be adjusted by changing the gear ratio.

A *Linear Motor* was applied to the base of the frame to propel the machine in the forward direction. The speed of the *Linear Motor* was set at 416.67 mm/s to match the usual operating speed of the paddy field weeder (i.e., 1.5 kmph). The activation of the *Rotary* and *Linear Motor* for simulation is shown in Fig. 2a and b.

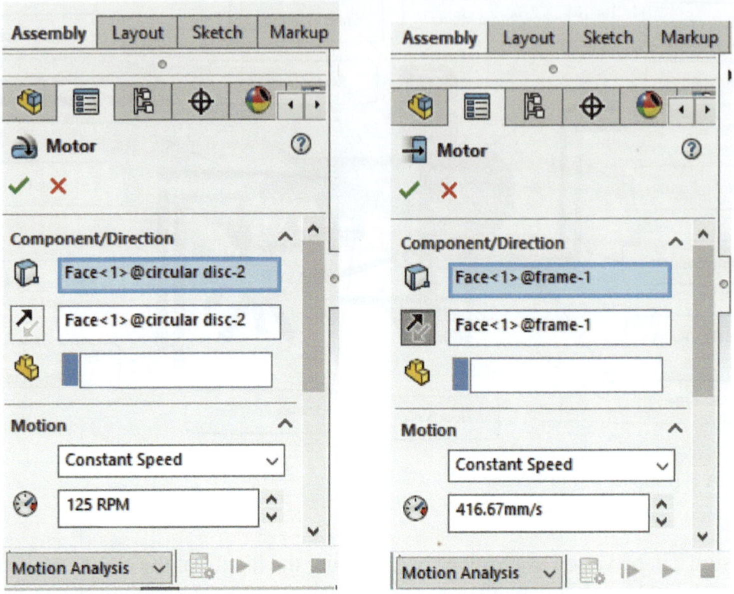

(a) Rotary motor (b) Linear motor

Fig. 2 Screenshots depicting the activation of the rotary and linear motor

3.2 Running the Design Study

The modeled design assembly was analyzed as follows [8].

Clicking the *Calculate* button initiates running the study. The *Motion Analysis Messages* window shows specific information during the computation period, and the following options will become accessible at the end of the computation period. There are various ways to interact with the animation: by pressing the button, the animation can start from the beginning, the button helps to play the animation starting from the current point, and the button stops the animation. The slider button enables us to set the animation to a current point. The button 1x was used to select the speed of operation. The button was used to save the animation file.

4 Position Analysis of the Weeding Units

The position analysis of the intra-row weeding unit is shown in Fig. 3. The power derived from the IC engine was made to rotate the circular disc in a clockwise direction. The circular disc was made to rotate, which provides reciprocating motion to the trapezoidal plate via a rectangular link. The two inclined sides of the trapezoidal

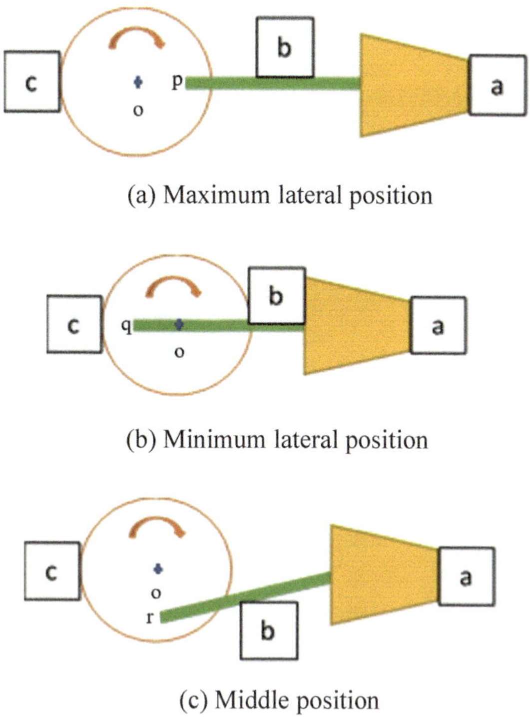

Fig. 3 A sketch depicting various positions of the trapezoidal plate (**a**: trapezoidal plate; **b**: rectangular link; **c**: circular disk)

plate, in turn, led to the lateral movement of the adjoining lateral plates when it reciprocated back and forth.

Figure 3a shows the maximum lateral position of the intra-row unit when the circular disc rotates in the clockwise direction. In other words, the lateral distance between the two intra-row units is farthest when the connecting point of the rectangular link and circular disc acting as the crank is at point p. The distance between the center point o and the points p, q, r represents the length of the crank.

Figure 3b shows the minimum lateral position of the intra-row unit. This occurs when the connecting point of the rectangular link and circular disc is at point q. At this point, the distance between the two intra-row units is minimal.

Figure 3c is the position of the intra-row unit when it is exactly in between the two plants. This occurs when the connecting point of the rectangular link and circular disc is at point r. As the inter-row unit operates in the middle row among the three rows, the intra-row unit operates in the two adjacent rows when it is in the maximum lateral extension position. During the minimum lateral position, it operated within the rows of the inter-row unit. The intra-row unit traverses in the sinusoidal path along with the linear forward motion of the weeder. The maximum and minimum lateral extension of the intra-row unit are presented in Fig. 4.

Fig. 4 A top-view depicting the simulation of the weeder

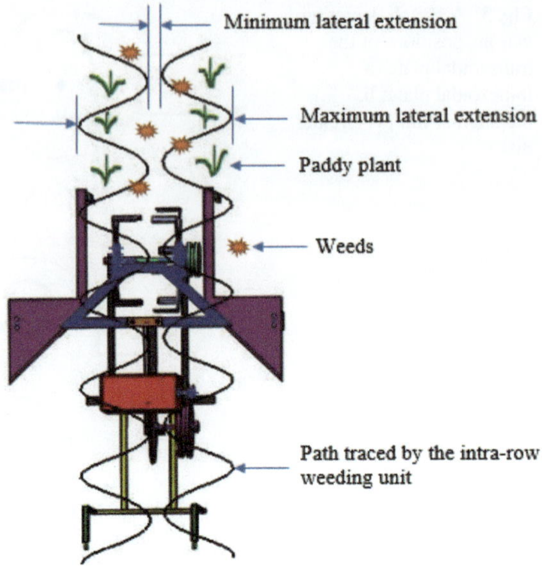

5 Simulation Results

The *Motion Analysis* button was activated for path tracing of the intra-row unit. The *Timeline* for the simulation was set to run for 8 s. Activation for path tracing of intra-row units is shown in Fig. 5.

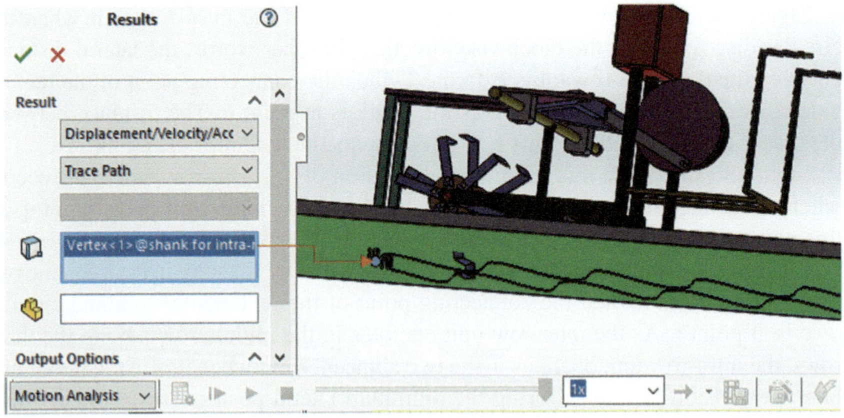

Fig. 5 Activation of the *Trace Path* for the intra-row unit

5.1 The Path Traced by the Intra-Row Unit

The button *Results and Plots* was clicked to view the path traced by the intra-row as shown in Fig. 4. The plot was defined by following the steps given below:

- *Displacement/Velocity/Acceleration*;
- *Trace path*;
- The tip of the intra-row weeding unit was selected.

The top view of the path traced by the intra-row unit is shown in Fig. 4. To avoid damage to the paddy plant during the weeding operation, the intra-row weeding unit was designed to traverse a sinusoidal path.

5.2 Plot of Linear Velocity Versus Time

To view the linear velocity curve as a function of time, the button *Results and Plots* was clicked. The plot was defined as follows:

- *Displacement/Velocity/Acceleration*;
- *Linear Velocity*;
- *Magnitude;*
- Select the face of one of the intra-row weeding units for the *simulation element*.

The maximal and minimal values of linear velocities of the intra-row weeding unit were 1109 and 407 mm/s, respectively, as shown in Fig. 6. The linear velocity parameters were generated based on the speed of machine travel, which is assumed at 1.5 km h^{-1}. However, when it comes to actual field conditions, the actual speed will depend upon the skill of the operator, too. To ensure precise and damage-free weeding operations, this value can be optimized by proper selection of gear ratio. Furthermore, considering the machine's travel speed of 1.5 km h^{-1}, the machine will take 0.48 s to cover a linear distance of 200 mm plant spacing. Thus, the intra-row unit should cover an approximate minimum distance of 400 mm (twice the plant spacing) at the same time (i.e., 0.48 s) for its outward and inward journey to maintain its synchronization, considering a minimum lateral movement of 200 mm outward and a minimum 200 mm inward. Hence, the speed of the intra-row weeding units should traverse at 833.34 mm/s to synchronize with the machine speed and hence avoid plant damage.

5.3 Linear Displacement as a Function of Time

To view the linear displacement as a function of time, the button *Results and Plots* was selected. The plot was defined by:

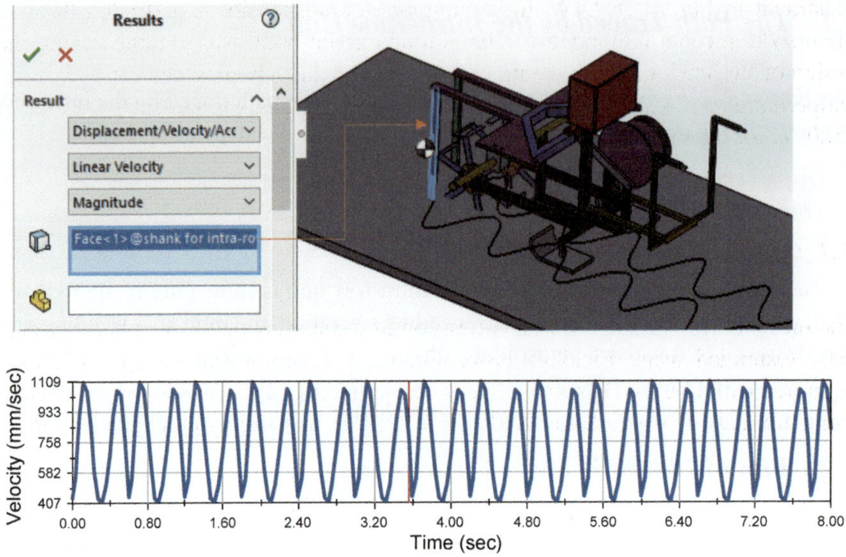

Fig. 6 Linear velocity of intra-row weeding unit versus time

- *Displacement/Velocity/Acceleration*;
- *Linear Displacement*;
- *Magnitude*.

The faces of both the intra-row weeding units were selected for the *simulation element*.

The value of linear displacement of the intra-row weeding units varies from 29 to 386 mm, as shown in Fig. 7. The amount of lateral extension (linear displacement of the intra-row unit) can be varied as required by changing the length of the crank, as the movement of lateral motion depends on it. The minimum amount of lateral extension was kept at 200 mm for 200 mm by 200 mm plant spacing to avoid plant damage.

6 Limitations and Further Scope

In this study, the row-to-row and hill-to-hill spacing of the paddy plant was assumed to be planted at 200 × 200 mm, maintaining an equal distance. In reality, it would be difficult to maintain a constant spacing of the paddy plant during planting. However, the machine was designed in such a way that it is flexible to handle some variability to some extent. The travel speed of the machine was assumed to be constant, as per the simulation, traveling at 1.5 km h^{-1}. In actual field conditions, it would be challenging to maintain a constant speed due to slippage. The tyne/spikes movements

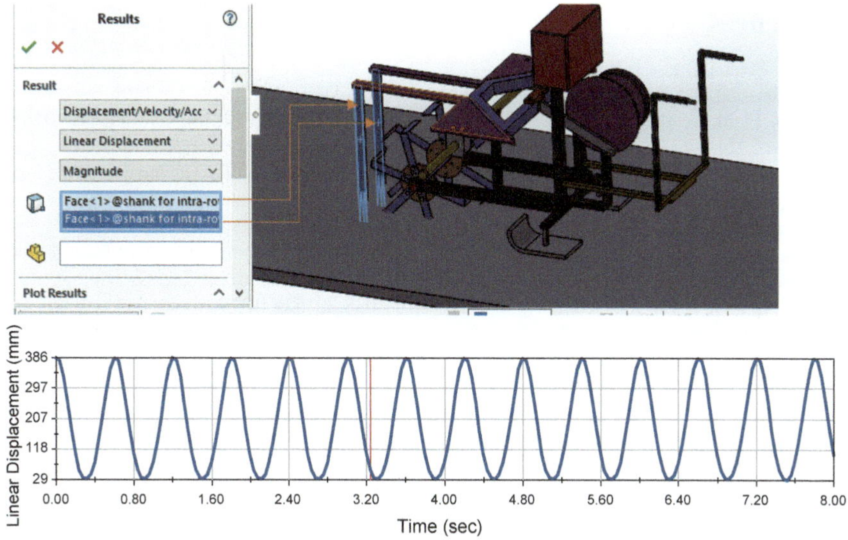

Fig. 7 Linear displacement of intra-row weeding units versus time

of the intra-row weeding unit were considered moving in a perfect sinusoidal curve without any obstacles such as rocks and tree roots. This condition made it easier to model. The operating depth of the weeder was assumed to be 50 mm.

Notwithstanding the previously mentioned restrictions, there is a significant chance that an inter-cum intra-row paddy field weeder based on the slider-crank mechanism will be introduced. The mechanism would be cost-effective for end users because it is made to function without the need for any sensors or hydraulic systems.

7 Conclusions

This paper presents the steps to analyze the working mechanism of inter-cum intra-row paddy field weeder in SolidWorks software. The mechanism to function various components was studied, which was based on the slider-crank mechanism. The analysis was focused on the movement of the intra-row weeding unit, and hence, the path to be traveled by the intra-row weeding unit has been traced. The motion of various components necessary for the operation was also studied. Position analysis of the intra-row weeding unit in coherence with associated components was also presented in the study. This kinematic analysis lays the groundwork for further refinement and development of efficient weeding mechanisms, contributing to improved agricultural practices.

References

1. S. Sebastian and K. Kalita, "Theoretical Analysis and Design of Inter Cum Intra-row Wetland Paddy Field Weeder," *Crop Prot.*, no. 10 April 2024, 106696, 2024, https://doi.org/10.1016/j.cropro.2024.106696.
2. Y. Gharde, P. K. Singh, R. P. Dubey, and P. K. Gupta, "Assessment of yield and economic losses in agriculture due to weeds in India," *Crop Prot.*, vol. 107, no. January, pp. 12–18, 2018, https://doi.org/10.1016/j.cropro.2018.01.007.
3. N. S. Chandel, H. Tripathi, and V. K. Tewari, "Evaluation and Adoption Scope of Rotary Power Weeder for Weed Management in Vegetable Crops," *Int. J. Bio-resource Stress Manag.*, vol. 6, no. 4, p. 534, 2015, https://doi.org/10.5958/0976-4038.2015.00076.7.
4. S. Sebastian and K. Kalita, "Study on Comparative Performances of Selected Paddy Weeders Developed in India," *J. Agric. Eng.*, vol. 59, no. 4, pp. 333–345, 2022, https://doi.org/10.52151/jae2022594.1786.
5. O. S. Alba, L. D. Syrovy, H. S. N. Duddu, and S. J. Shirtliffe, "Increased seeding rate and multiple methods of mechanical weed control reduce weed biomass in a poorly competitive organic crop," *F. Crop. Res.*, vol. 245, no. January 2020, p. 107648, 2020, https://doi.org/10.1016/j.fcr.2019.107648.
6. K. Vardaan and P. Kumar, "Design, analysis, and optimization of thresher machine flywheel using Solidworks simulation," *Mater. Today Proc.*, vol. 56, pp. 3651–3655, 2022, https://doi.org/10.1016/j.matpr.2021.12.348.
7. D. Nedelcu, G. R. Gillich, A. Bloju, and I. Padurean, "The kinematic and kinetostatic study of the shaker mechanism with SolidWorks Motion," *J. Phys. Conf. Ser.*, vol. 1426, no. 1, 2020, https://doi.org/10.1088/1742-6596/1426/1/012025.
8. D. Nedelcu, M. D. Nedeloni, and D. Daia, "The kinematic and dynamic analysis of the crank mechanism with SolidWorks Motion," *Recent Adv. Signal Process. Comput. Geom. Syst. Theory - ISCGAV'11, ISTASC'11*, no. January 2015, pp. 245–250, 2011.

A Terminal Nonlinear Sliding Mode Control-Based Rotor and Active Magnetic Bearing System

Shivani Raj and Rajiv Tiwari

Abstract To cater to the uncertainties in system dynamic parameters and disturbances, the sliding mode control (SMC) law is often used. The terminal SMC uses a nonlinear sliding surface (SS) to reduce the vibration caused by rotor movement over a specific time. A 1-degree-of-freedom (DOF) model is selected for a rotor and active magnetic bearing (AMB) system to observe the control action. It is found that the nonlinear term in the SS makes the control law effective in vibration reduction. Also, the control law performed better with system uncertainties and can be applied to low-speed motors and turbines.

Keywords Uncertainty · Sliding mode control · Rotor · Sliding manifold · Vibration · Active magnetic bearing

1 Introduction

Low-speed rotors cover a wide range of machinery operations. They find applications in motors used in permanent magnet [1], wind turbines [2], mines [3], ships [4], and ports [5]. During their operation, the rotors are supported by bearings, which can be of two types: conventional, where the bearing action is done by contact with the rotor, and magnetic, where the bearing action is done by no contact with the rotor using a magnetic field. The active magnetic bearing (AMB) is used to suppress the rotor vibration using controllers and actuators run by current coils acting as an electromagnet [6, 7]. A schematic diagram of an AMB with a rotor is shown in Fig. 1. It can be seen that the sensor gives the position signal to the controller. The controller sends the signal to the amplifier, and the amplified current signals are given

S. Raj (✉) · R. Tiwari
Guwahati, India
e-mail: rshivani@iitg.ac.in

R. Tiwari
e-mail: rtiwari@iitg.ac.in

U. S. Dixit et al. (eds.), *Current Progress in Engineering Sciences*,
https://doi.org/10.1007/978-981-96-8753-4_21

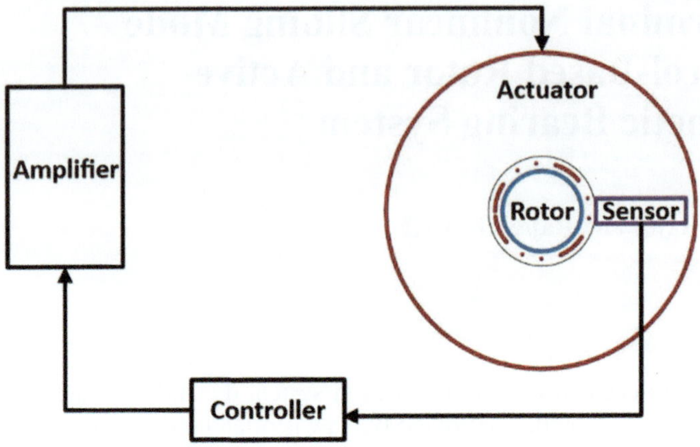

Fig. 1 A representation of the rotor and AMB system

to the coils in the actuator to produce the desired magnetic current [8]. Based on this current, the rotor is displaced to its designated position.

Different control methods can be chosen depending on the demands of the rotating machinery. A position-following controller is developed for the rotor used in permanent magnets to enhance the performance at low-speed operations [9]. A PID control-based AMB is used to estimate the imbalance in two rotors through the identification method, and the same control law is used to suppress the vibrations due to excitations and faults [10]. The robust control method is selected for the AMB, where the control law is implemented by decoupling it for high and low modes of frequency [11]. Wu et al. [12] proposed a 5-degree-of-freedom (DOF) model with foundation movement and applied PID and feedforward control laws to eliminate the vibration. In [13], the PD control law is derived by defining the desired eigenvalue regions in real and imaginary planes for a 4 DOF to reduce vibrations caused by unbalanced forces.

However, due to the unmodeled system dynamics, disturbances, and uncertainties in system parameters, sliding mode control (SMC) is used [14]. This control law provides a definite time system that states convergence to their desired values. A second-order SMC law is proposed in [15] to deal with the forces developed due to the unbalanced mass in the rotor. The radial basis function network (RBFN) estimates the unknown system parameters while designing an SMC law for a hybrid AMB [16]. A terminal SMC is proposed for a quick convergence in a definite time with better performance for a thrust AMB [17]. Here, the nonlinearity of the sliding surface (SS) is introduced to ignore the singularity due to the developed control law.

In [18], a nonlinear SS is proposed so that the surface adjusts during the settling time, and the convergence of states is done with minimal overshoot. In this paper, the nonlinear SS is selected with a terminal index to improve vibration suppression performance in a 1-degree-of-freedom (DOF) rotor and AMB system. This paper aims to design a control law based on terminal nonlinear SS to minimize the vibration and the states' transient response.

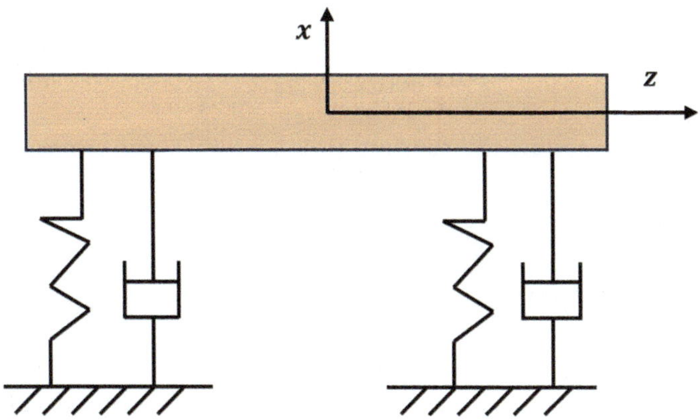

Fig. 2 A 1 DOF rotor bearing system

2 Mathematical Model

Figure 2 shows the schematic of a 1 DOF rotor and AMB system. The equation of motion is considered as [8],

$$m\ddot{x} + c\dot{x} + (k - k_\delta)x = F_{un} - k_i u \tag{1}$$

The unbalanced force due to eccentric mass is given by $F_{un} = me\omega^2 \sin\omega t$, ω is the rotational speed of the rotor, its mass is m, the damping coefficient of the bearing is c, and the stiffness coefficient of the bearing is k. Rotor movement is considered in the x direction. The AMB displacement coefficient and current stiffness are given by k_δ and k_i, respectively. The control current in the AMB is given by u.

The motivation of this work is to reduce the vibrational displacement x in the presence of an unbalanced force F_{un}. The input current u is designed, and it is fed back to the AMB to decrease the oscillations in the x direction. Due to the AMB stiffness k_δ, the system's overall stiffness can be manipulated to achieve the desired result. Moreover, Eq. 1 represents a closed-loop dynamic system for a control law design.

3 Control Law Design

The terminal nonlinear SMC (TNSMC) law is designed. For this, Eq. 1 can be written as,

$$\ddot{x} = m^{-1}(F_{un} - k_i U - (k - k_\delta)x - c\dot{x}) \tag{2}$$

Given the advantage of variable gain, an SS is chosen as,

$$s = \dot{x} + (Q_1 - \varnothing Q_2)x^{m/n} \tag{3}$$

where m and n are positive real numbers, Q_1 is a feedback gain obtained from pole placement, Q_2 is a positive definite matrix satisfying the relation,

$$Q_2(Q_1)^T + (Q_1)Q_2 = R \tag{4}$$

where $R > 0$ and,

$$\varnothing = -\alpha\exp(-\gamma x^2) \tag{5}$$

where $\alpha, \gamma > 0$. Equation 3 is differentiated to get,

$$\dot{s} = \ddot{x} + \frac{m}{n}(Q_1 - \varnothing Q_2)x^{\frac{m}{n}-1}\dot{x} - \dot{\varnothing}Q_2x^{\frac{m}{n}} \tag{6}$$

The following law is chosen for reaching the sliding surface,

$$\dot{s} = -\kappa s - \sigma sign(s) \tag{7}$$

such that,

$$\left|\left|m^{-1}F_{un}\right|\right| \le \sigma \tag{8}$$

The control law is selected as,

$$u = -k_i^{-1}\left((k - k_\delta)x + c\dot{x} + m\begin{pmatrix} -\kappa s - \sigma sign(s) + \dot{\varnothing}Q_2x^{\frac{m}{n}} \\ -\frac{m}{n}(Q_1 - \varnothing Q_2)x^{\frac{m}{n}-1}\dot{x} \end{pmatrix}\right) \tag{9}$$

3.1 Stability

The following Lyapunov function is chosen as,

$$V = \frac{1}{2}s^2 \tag{10}$$

Equation (10) is differentiated as,

$$\dot{V} = s\dot{s} \tag{11}$$

Equation (6) is substituted in Eq. (11) to get,

$$\dot{V} = s\left(\ddot{x} + \frac{m}{n}(Q_1 - \varnothing Q_2)x^{\frac{m}{n}-1}\dot{x} - \dot{\varnothing} Q_2 x^{\frac{m}{n}}\right) \tag{12}$$

Equation (2) is substituted in Eq. (12) to get,

$$\dot{V} = \left(m^{-1}(F_{un} - k_i U - (k - k_\delta)x - c\dot{x}) + \frac{m}{n}(Q_1 - \varnothing Q_2)x^{\frac{m}{n}-1}\dot{x} - \dot{\varnothing} Q_2 x^{\frac{m}{n}}\right) \tag{13}$$

Equation (9) is substituted in Eq. (13) to get,

$$\dot{V} = s\left(m^{-1}F_{un} - \kappa s - \sigma\, sign(s)\right) \tag{14}$$

From Eqs. (8) to (14), it can be deduced that $\dot{V} < 0$. This implies that the control law given in Eq. (9) is successful in converging the states to the origin. The time for reaching the SS is given by,

$$t = \frac{|x(0)|^{1-\frac{m}{n}}}{(Q_1 - \varnothing Q_2)\left(1 - \frac{m}{n}\right)} \tag{15}$$

where $x(0)$ is the initial state of the system.

4 Results

The control law is simulated in MATLAB. The system parameters are $m = 4$ kg, $c = 0.1$ Ns/m, $k = 20$ kN/m, $k_\delta = 72500$ N/m, $k_i = 29$ N/A, $e = 0.0002$ m, $\omega = 100$ rad/s. The controller parameters are $Q_1 = 2.9$, $\alpha = 0.03$, $\gamma = 0.01$, $Q_2 = 71$, $m = 3$, $n = 4$, $\kappa = 0.02$, $\sigma = 4$. The TNSMC is compared with the PD control law [8] given by,

$$u = K_P x + K_D \dot{x} \tag{16}$$

where $K_P = 3500$ and $K_D = 8$. The simulation is run for $t = 0.5$ s, and the responses are compared with TNSMC and PD control laws using an unbalanced force. Figure 3 shows the plots for rotor movement in the x direction and the control current applied by the AMB. It can be seen from Fig. 3 that the response obtained from TNSMC is reduced more than that of the PD response. Also, the control current is reduced during the steady state.

A parameter uncertainty of +20% is given to the rotor mass and rotor stiffness with unbalanced force, and the plots are obtained as shown in Fig. 4. The steady state response in the left plot shows decreased vibration displacements, and the right plot

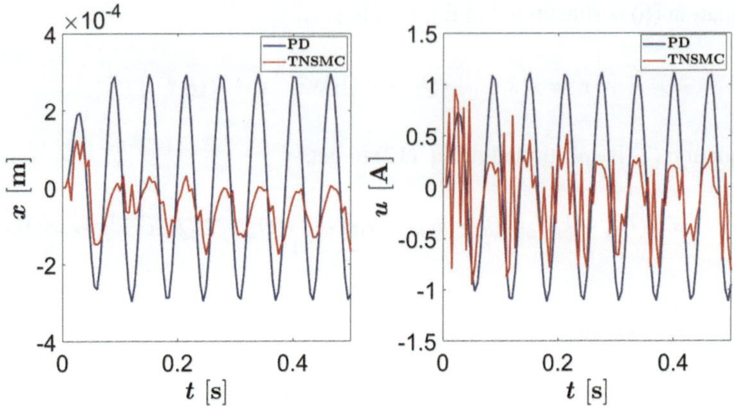

Fig. 3 The displacement (left) and control current (right) response for TNSMC and PD control laws

response shows decreased control current in steady state. It can be observed that the TNSMC performs better than PD control in the presence of uncertainty in system dynamic parameters.

Due to terminal parameters in TNSMC, the overshoot in the states can be manipulated by changing the terms m and n. However, due to the limited time convergence property of TNSMC, some amount of overshoot is observed before the settling time.

The rotor speed is increased to 150 rad/s, and the responses are obtained without any uncertainty but with an unbalanced disturbance to analyze the effect for different speed operations. It can be derived from Fig. 5 that TNSMC is effective in reducing the vibration in the positive x direction, however, there is vibration reduction similar to PD control in the negative x direction. Also, there is an increase in the control

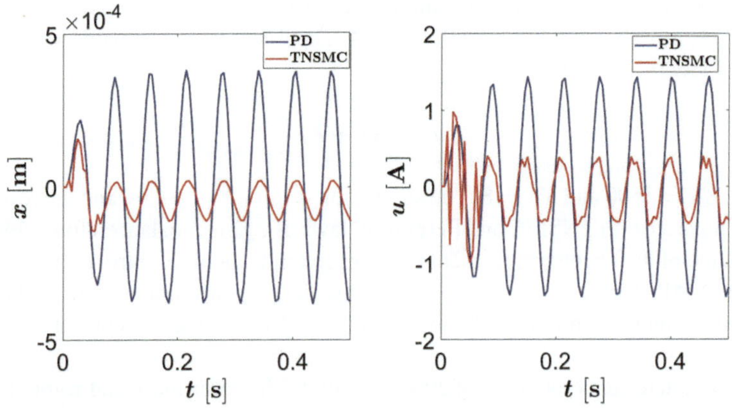

Fig. 4 The displacement (left) and control current (right) response for TNSMC and PD control laws with 20% uncertainty

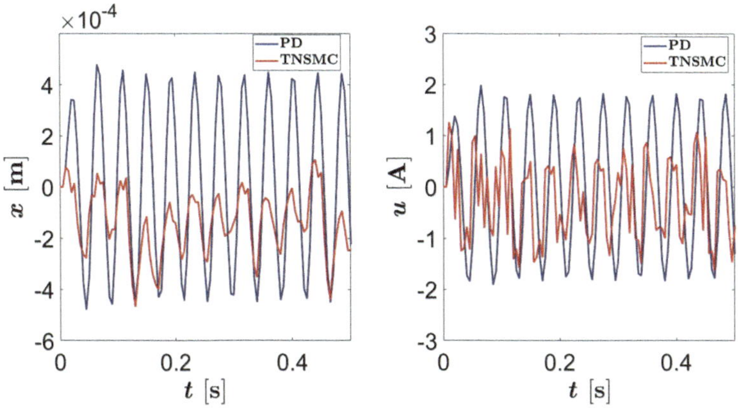

Fig. 5 The displacement (left) and control current (right) response for TNSMC and PD control laws with increased rotor speed of 150 rad/s

current as seen in Figs. 3 and 4 indicate that the control law works in a limited range of low-speed operations. Here, as the rotor speed increases, the unbalanced force $\boldsymbol{F}_{un}$ given in Eq. (1) also increases because $\boldsymbol{F}_{un}$ is a function of the square of the rotor speed. To suppress the vibrations induced due to the increased unbalanced force, the TNSMC applied more control current than it did at low rotor speed.

However, if the speed of the rotor is further increased, the control current further increases to suppress the vibrations caused due to the unbalanced force. However, there is a limitation of the magnetic material to produce the required magnetic flux [19], and it can't produce beyond a certain value. This phenomenon is called magnetic saturation. So even though at increased rotor speed, the controller applies increased control current, it will not be transmitted to the rotor because of the saturation effect. This will lead to reduced control performance in terms of vibration suppression because AMB is not providing the required magnetic force due to its saturation. Thus, the TNSMC will not be effective at high-speed operations.

One of the potential strategies that may address this limitation is to develop AMB with a nonlinear model where the effect of magnetic saturation is taken care of [19]. Another solution is to use robust control methods such as H_∞ or μ-synthesis methods in the design of the control law for AMB. Based on this limitation, the TNSMC can be developed considering the unbalanced force as parametric uncertainty, where the increased unbalanced force at the high-speed rotor is taken care of [20].

5 Conclusions

The TNSMC is effective in reducing the vibration due to the unbalanced mass and system uncertainties. A 1 DOF rotor and AMB model are selected, and the rotor is run at low speed; the derived control law can be applied to any low-speed machinery operations. When the control law is applied at increased speed, the control current becomes higher than that of the PD control, even though TNSMC effectively reduces vibration. Thus, the control law design can be further evolved for high-speed applications.

Acknowledgements The authors thank the Advanced Dynamics and Vibration Lab, Mechanical Department, Indian Institute of Technology, Guwahati, India, for the consistence support. The Ministry of Education, India, funds this work.

References

1. Bouheraoua, M., Wang, J., Atallah, K.: Speed control for a pseudo direct drive permanent-magnet machine with one position sensor on low-speed rotor. IEEE transactions on industry applications 50(6), 3825–3833 (2014).
2. Guerrero, J. M., Lumbreras, C., Reigosa, D., Fernandez, D., Briz, F., Charro, C.B.: Accurate rotor speed estimation for low-power wind turbines. IEEE Transactions on Power Electronics 35(1), 373–381 (2019).
3. Malafeev, S. I., Zakharov, A. V., Safronenkov, Y. A.: A new series of asynchronous frequency-controlled motors for mining excavators. Russian Electrical Engineering 90, 299–303 (2019).
4. Liang, L., Zhao, P., Zhang, S., Yuan, J., Wen, Y.: Simulation and analysis of Magnus rotating roll stabilizer at low speed. Ocean Engineering 142, 491–500 (2017).
5. Ahamad, N. B. B., Su, C. L., Zhaoxia, X., Vasquez, J. C., Guerrero, J. M.: Modeling and controls of flywheel energy storage systems for energy harvesting from harbor electrical cranes. In 2018 IEEE industry applications society annual meeting (IAS), IEEE, pp. 1–8.
6. Bleuler, H.: A survey of magnetic levitation and magnetic bearing types. JSME international journal. Ser. 3, Vibration, control engineering, engineering for industry 35(3), 335–342 (1992).
7. Srinivas, R. S., Tiwari, R., Kannababu, C.: Application of active magnetic bearings in flexible rotordynamic systems–A state-of-the-art review. Mechanical Systems and Signal Processing 106, 537–572 (2018).
8. Tiwari, R.: Rotor systems: analysis and identification. CRC press (2017).
9. Bu, F., Xuan, F., Yang, Z., Gao, Y., Pan, Z., Degano, M., Gerada, C.: Rotor position tracking control for low speed operation of direct-drive PMSM servo system. IEEE/ASME Transactions on Mechatronics 26(2), 1129–1139 (2020).
10. Ranjan, G., Tiwari, R., Nemade, H.B.: Robust dynamic balancing of dual rotor-active magnetic bearing system through virtual trial unbalances as low and high frequency magnetic excitation. ASCE-ASME Journal of Risk and Uncertainty in Engineering Systems, Part B: Mechanical Engineering 9(1), 011203 (2023).
11. Sahinkaya, A., Sawicki, J. T.: Computationally efficient implementation of robust controllers in active magnetic bearing systems. Mechanical Systems and Signal Processing 144, 106902 (2020).
12. Wu, H., Zhang, L., Zhou, J., Hu, Y.: Dynamic analysis and vibration control of a rotor-Active magnetic bearings system with base motion. Journal of Vibration and Control 30(11–12), 2697–2708 (2024).
13. Zhou, T., Zhu, C.: Robust proportional-differential control via eigenstructure assignment for active magnetic bearings-rigid rotor systems. IEEE Transactions on Industrial Electronics 69(7), 6572–6585 (2021).
14. Utkin, V.I., Vadim, I.: Sliding mode control. Variable structure systems: from principles to implementation. vol. 66, pp. 1 (2004).

15. Kandil, M. S., Dubois, M. R., Bakay, L. S., Trovao, J. P. F.: Application of second-order sliding-mode concepts to active magnetic bearings. IEEE Transactions on Industrial Electronics 65(1), 855–864 (2017).
16. Zad, H. S., Khan, T. I., Lazoglu, I.: Design and adaptive sliding-mode control of hybrid magnetic bearings. IEEE Transactions on Industrial Electronics 65(3), 2537–2547 (2017).
17. Chen, S. Y., Lin, F. J.: Robust nonsingular terminal sliding-mode control for nonlinear magnetic bearing system. IEEE Transactions on Control Systems Technology 19(3), 636–643 (2010).
18. Raj, S., Reddy, B. S.: Emergency Braking Control Design for a Dual Cart-Trolley Overhead Crane. In 3rd International Conference on Electrical, Computer, Communications and Mechatronics Engineering (ICECCME), pp. 1–6. IEEE, Spain (2023).
19. Wang, D., Wang, N., Ye, C. and Chen, K.: Research on analytical bearing capacity model of active magnetic bearings based on magnetic saturation. IET Electric Power Applications, 11(9), 1548–1557 (2017).
20. Mystkowski, A., Pawluszewicz, E. and Dragašius, E.: Robust nonlinear position-flux zero-bias control for uncertain AMB system. International Journal of Control, 88(8), 1619–1629 (2015).

A Comprehensive Analysis of Energy-Efficient Swing Drive System in Excavator Machine Using Independent Metering Control

Shubhyanshu Singh and Ajit Kumar

Abstract This study investigates the energy savings achieved through the implementation of the Independent Metering Control (IMC) for precise angular position control in the Hydraulic Swing Drive (HSD) system of an excavator machine. The opening of the valve ports is regulated by a PID controller to control the angular position of the swing motor. This study involves a comparative analysis between the conventional and IMC-based HSD systems. A detailed simulation model of the HSD system is developed on the MATLAB/Simulink platform, incorporating the IMC control strategies. Under varying loading conditions, key performance factors such as energy consumption, efficiency, and response time are examined. The simulation results show that the IMC system significantly reduces the energy consumption during the swing operation with an energy-saving efficiency of 41.56% compared to the conventional system. Furthermore, the results also show that the IMC system enhances the precision of swing operations. The paper concludes that the IMC system promotes sustainable practices in heavy machinery operations by reducing energy consumption and operational costs.

1 Introduction

In recent years, there has been a growing emphasis on enhancing the energy efficiency of hydraulic systems in heavy machinery to address environmental concerns and reduce operating expenses. The versatility and robustness of hydraulic excavators

S. Singh · A. Kumar (✉)
Department of Mechanical Engineering, Indian Institute of Technology (Indian School of Mines), Dhanbad, Jharkhand, India
e-mail: ajit.ism185@gmail.com

S. Singh
e-mail: shubhyanshusingh1103@gmail.com

S. Singh
School of Civil and Mechanical Engineering, Curtin University, Bentley, WA, Australia

U. S. Dixit et al. (eds.), *Current Progress in Engineering Sciences*,
https://doi.org/10.1007/978-981-96-8753-4_22

make them indispensable in a variety of mining, construction, and demolition operations. A critical function of these machines is the swing operation, which involves rotating the excavator's upper structure to precisely position the arm, boom, and bucket. The conventional systems incorporate traditional proportional directional control valves (PDCV), which feature mechanically linked meter-in and meter-out orifices. This design ensures robustness and ease of control [1]; however, they are intrinsically inefficient. The design of conventional PDCVs inevitably maintains a pressure at the actuator's outlet, creating an undesirable load that the actuator must work against, thereby reducing the efficiency of the system [2]. To attain accurate regulation and substantial energy reduction, researchers have explored various types of metering valve-controlled systems, such as independent metering controlled systems [3] and programmable valve-controlled systems [4]. By introducing IMC, the physical connection between the meter-in and meter-out orifices is removed, enabling a range of valve activities to be performed using different control techniques. For independent metering systems, numerous researchers have already established control algorithms to achieve desired flow rates at different operating pressures across control valves for the linear actuators. Below are descriptions of a few research studies.

Huh et al. [5] describe the fundamental design principles for creating an Independent Metering Valve tailored for Medium Commercial Vehicles (MCV). The study highlights the critical role of IMVs in optimizing hydraulic flow control, enhancing system efficiency, and reducing energy consumption in commercial vehicles. Hu et al. [6] showed that energy savings and superior tracking performance may be achieved by controlling a single-rod cylinder using a programmable valve and a Hybrid-Control algorithm. Shenouda et al. [7] found that the systems with two cylinders have effectively used a two-layer controller technique for Synchronized-Motion-Control. Liu et al. [8] explored the SMISMO valve configuration for synchronized control of two cylinders. The proposed system employed linearization feedback technology to linearize the nonlinear characteristics of the SMISMO control system. This approach significantly improved trajectory tracking performance, surpassing the traditional PID control strategy. Wang et al. [9] focused on an electro-hydraulic meter-in and meter-out independent regulation system designed to enhance the smoothness of acceleration and deceleration in valve control systems with high inertia loads, particularly in hydraulic systems used in construction machines. This method aims to improve the performance and stability of such systems by providing more precise control over the hydraulic flow.

Choi et al. [10] investigated the application of independent metering valves in excavators to enhance energy efficiency. The research demonstrates that using independent metering valves significantly improves the energy-saving performance of excavators, contributing to more efficient hydraulic systems and reduced fuel consumption. Hi et al. [11] explored electro-hydraulic velocity and position control in mobile construction equipment using independent metering valve control. The research focuses on enhancing automation and precision in control systems for construction machinery, particularly through a position-pressure cooperative control strategy that improves the simultaneous control of position and pressure. Ge et al. [12]

investigated the performance of hydraulic excavators using a combined pump and valve separate meter-in and meter-out circuits. The research highlights the significant reduction in pressure losses through the valve, improvements in energy efficiency, and enhanced machine stability achieved by implementing this configuration system in hydraulic excavators. Ahamad et al. [13] focused on developing a new generation of IMV circuits designed for boom cylinder control. This innovative IMV circuit seeks to minimize the number of required valves by incorporating a directional control valve while maintaining control over velocity and pressure, potentially offering greater energy savings compared to conventional proportional valve control systems.

Jansson et al. [14] delve into innovative control strategies for mobile hydraulic systems, proposing the separation of meter-in and meter-out orifice controls. This approach allows for precise regulation of hydraulic fluid flow, resulting in significant improvements in system efficiency, stability, and response. By optimizing the flow control independently at both the input and output stages, the study demonstrates enhanced performance and energy savings in mobile applications like construction and agricultural machinery. Abuowda et al. [15] introduced micro-independent metering, which allows precise control of hydraulic fluid flow at a micro level, enhancing system responsiveness and efficiency. This novel approach improves the accuracy and performance of electro-hydraulic systems in applications such as robotics and industrial machinery, offering significant advancements in mechatronics. Tabor et al. [16] developed a technique that utilizes four valve independent metering coupled with load feedback to achieve precise control over actuator movements. The approach enhances the efficiency and responsiveness of hydraulic systems by optimizing fluid flow and adapting to varying loads in real-time. Singh et al. [17] controlled the speed of the valve-controlled hydromotor using fuzzy control. Thakur et al. [18] discussed different types of motor speed control, including the IMC technique.

This research aims to explore the potential of energy savings in the swing operation of hydraulic excavators through the application of IMC and incorporating a PID controller in the proposed HSD systems. By developing and analyzing detailed simulation models of the proposed HSD system in MATLAB/Simulink environment, the study seeks to quantify the improvements in energy efficiency and operational performance. The rest of this paper is structured as follows: Sect. 2 outlines both the conventional and proposed HSD systems. Section 3 focuses on the PID controller used for regulating the valves. Section 4 outlines the simulation model of the proposed HSD system. Section 5 presents the results, highlighting the energy savings and performance enhancements achieved with IMC. Finally, Sect. 6 concludes the paper, summarizing the key contributions and benefits of adopting IMC in hydraulic excavators.

2 System Description

2.1 Conventional HSD System

The swing drive system in hydraulic excavators is crucial for enabling the rotation of the upper structure, which includes the boom, arm, and bucket. This rotation allows for a wide range of operational movements without the need to reposition the entire machine, significantly enhancing operational flexibility. Figure 1 shows the schematic diagram of a conventional HSD system. The conventional swing drive system consists of a prime mover (1), which drives the swing pump (2). The swing pump is an axial piston pump that supplies the necessary pressurized hydraulic fluid to the swing motor (7) through a 4/3-way direction control (DC) valve (4). The DC valve regulates the flow and pressure of hydraulic fluid directed to the swing motor. This helps in providing the fine adjustments to the swing speed, position, and direction based on operator input. The relief valves (5b and 5c) control the acceleration and deceleration of the swing motor.

The ports of the swing motor are connected with the same DC valve; due to this, when controlling the incoming flow of the swing hydromotor, the outgoing flow from the hydromotor is also regulated because when the orifice area of port A is regulated by moving the spool, the orifice area of port B is also changed. This causes unwanted pressure at the output of the motor against which the swing pump has to work. This makes the conventional HSD system less energy efficient.

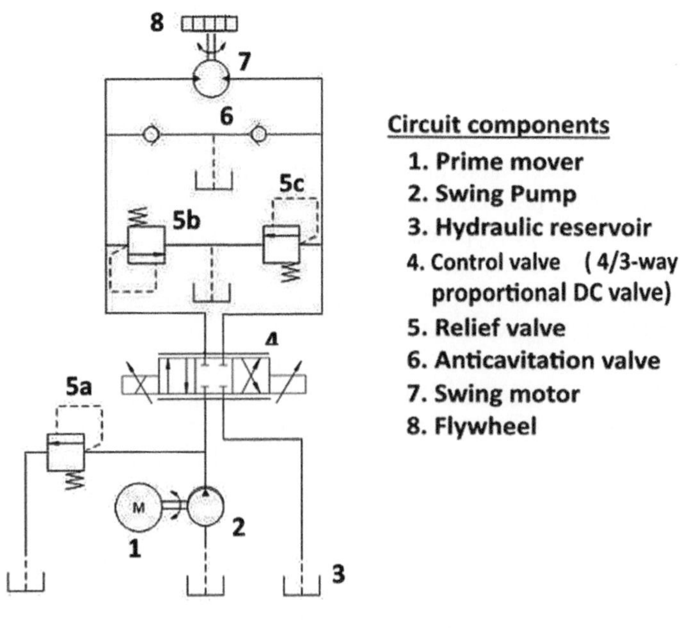

Fig. 1 Schematic diagram of a conventional HSD system [19]

2.2 *Proposed HSD System*

The proposed HSD system incorporates the IMC technique to overcome the limitations of the conventional HSD system, as mentioned in Sect. 2.1. The IMC system uses two control valves (4, 5) in which one of the control valves is connected to the inlet port of the swing motor while the other control valve is connected to the outlet port of the swing motor, as shown in Fig. 2. This configuration helps in independent metering of the inlet and outlet flow of the swing motor. The control valves are actuated by a PID controller for precise angular movement of the swing motor.

3 PID Controller for Conventional and Proposed HSD Systems

The PID controller is used to regulate the orifice port area of the control valve to control the angular position of the swing motor in both conventional and proposed HSD systems. Figure 3 shows the schematic of the PID-controlled HSD system. A PID controller is a control loop feedback mechanism widely used in industries. This algorithm comprises three distinct parameters: Proportional (P), Integral (I), and Derivative (D), which are used to maintain the desired set point by minimizing the error between the process variable and the set point. The PID controller's output $u(t)$ can be represented by Eq. (1).

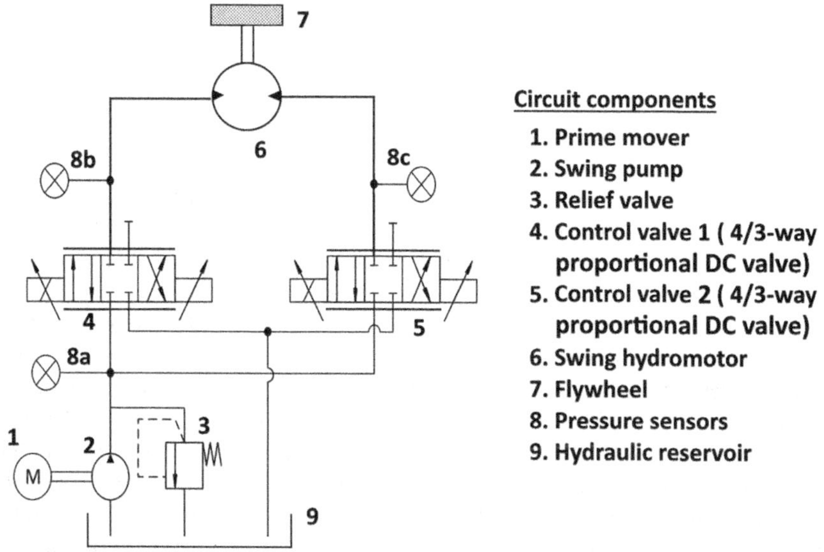

Fig. 2 Schematic diagram of the proposed HSD system based on the IMC technique

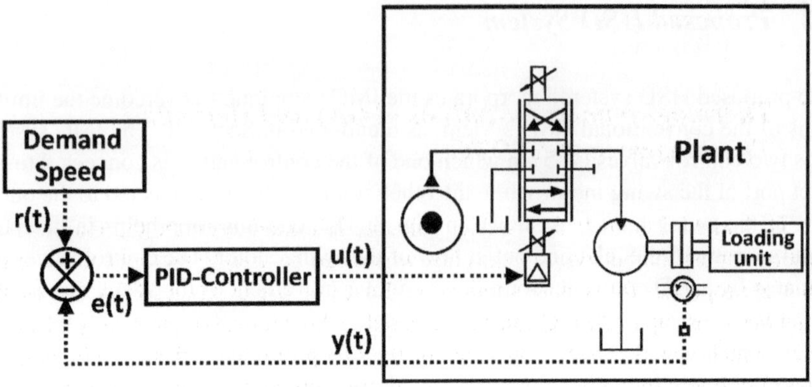

Fig. 3 Schematic diagram of PID controlled HSD system

$$u(t) = K_p \times e(t) + K_i \times \int_0^t (e(\tau) \times d\tau) + K_d \frac{de(t)}{dt} \quad (1)$$

where $e(t)$ is the error between the reference and the output variable at time t, K_p is the proportional gain, K_i is the integral gain, and K_d is the derivative gain. The proportional term $K_p e(t)$ provides a control output that is proportional to the error. The integral term $K_i \int_0^t e(\tau)d\tau$ sums the past errors to eliminate residual steady-state error. The derivative term $K_d \frac{de(t)}{dt}$ predicts future errors based on the rate of change, adding a damping effect to the system.

3.1 *Tuning of PID Parameters (Ziegler–Nichols Tuning Method)*

There are several methods for tuning PID parameters, such as the Ziegler–Nichols tuning method, the Cohen-Coon tuning method, the Lambda tuning (IMC tuning), the automatic tuning and adaptive control method, and the trial-and-error method. Among all the Ziegler–Nichols method was selected due to its practicality and effectiveness for systems with nonlinearities, such as the analyzed energy-saving swing system. It provides a systematic approach to obtain stable and responsive PID parameters with minimal computational effort. While advanced methods may offer slight improvements, the Ziegler–Nichols method ensures reproducibility, simplicity, and efficiency, aligning with the study's focus on evaluating energy regeneration performance without overcomplicating tuning. Ziegler–Nichols method provides a systematic way to determine K_p, K_i, and K_d based on the system's response. It involves setting the K_i, and K_d to zero and increasing the K_p until the system oscillates with

Table 1 PID parameters tuned by the Ziegler–Nichols tuning method for conventional and proposed HSD systems

HSD system	PID parameters (tuned by the Ziegler–Nichols tuning method)		
	$K_p = 0.6K_u$	$K_i = \frac{2K_p}{P_u}$	$K_d = \frac{K_p P_u}{8}$
Conventional HSD system	2.1	0.19	5.89
Proposed HSD system	1.5	0.11	4.9

a constant amplitude. The ultimate gain K_u and ultimate period P_u are then used to calculate the PID parameters. The value of tuned parameters for both conventional and proposed HSD systems are given in Table 1.

4 Simulation Model of Conventional and Proposed HSD Systems

Simulation models for both HSD systems are developed in MATLAB/Simulink environment. The parameters used for the Simulink block for the simulation are given in Table 2. Figure 4 depicts the Simulink model of the proposed HSD system. The model is developed by considering the following assumptions:

- Leakage losses through the control valves are negligible.
- Prime mover drives the swing pump with constant speed irrespective of the pressure at its outlet port.

Table 2 Block parameters used to develop the Simulink model of the proposed HSD system

S. No	Parameters	Values
1	Swing pump displacement	28 cc/rev
2	Swing pump shaft speed	1500 rpm
3	Volumetric efficiency of swing pump	85%
4	Mechanical efficiency of swing pump	95%
5	Swing motor displacement	14 cc/rev
6	Volumetric efficiency of swing pump	83%
7	Mechanical efficiency of swing pump	98%
8	Flywheel inertia	10 kgm^2
9	Setting the pressure of the relief valve (5b, 5c) in the conventional HSD system	52 bar
10	Setting the pressure of the main relief valve (3) in the proposed HSD system	100 bar
11	Maximum opening area of control valves	1×10^{-4} m^2

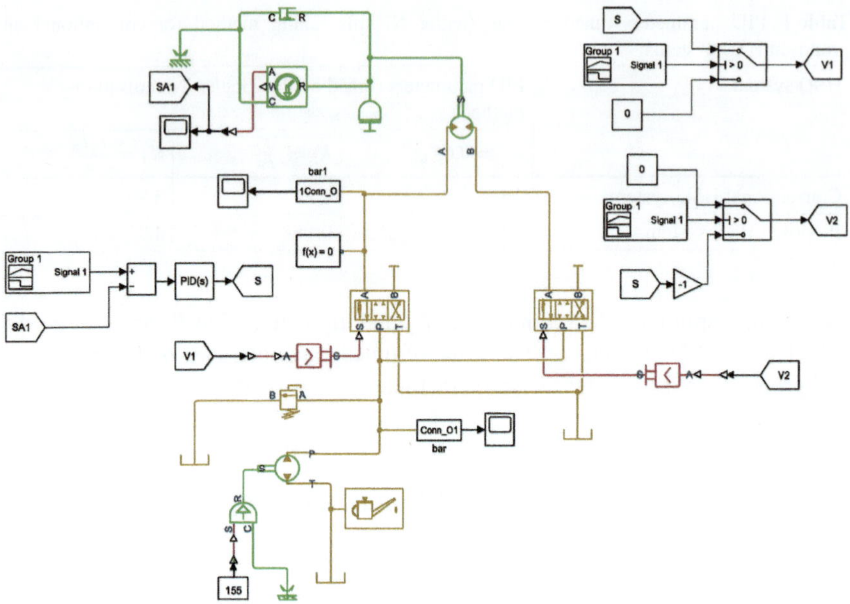

Fig. 4 Simulink model of the proposed HSD system

- The volumetric and mechanical efficiencies of the swing motor and pump are constant and independent of the load on it.
- The inertia of the moving parts of the swing motor is negligible.

The simulation parameters in Table 2 are based on assumptions chosen to represent typical conditions for hydraulic excavator systems. These assumptions were used to evaluate the performance of the proposed control technique rather than to model a specific real-world system.

While the assumed values may not precisely reflect real-world conditions, limiting accuracy and generalizability, they are sufficient to demonstrate the effectiveness and robustness of the control approach.

5 Result and Discussion

The upper structure of the excavator rotates 360° to load and dump the materials. It takes 4–6 s to complete one rotation during swing operation. Considering the demand of 1 revolution in 4 s, the demand signal will be a ramp signal. This ramp signal increases linearly from 0 to 360° in 4 s, then it remains at 360° for 4–16 s (refer to Fig. 5).

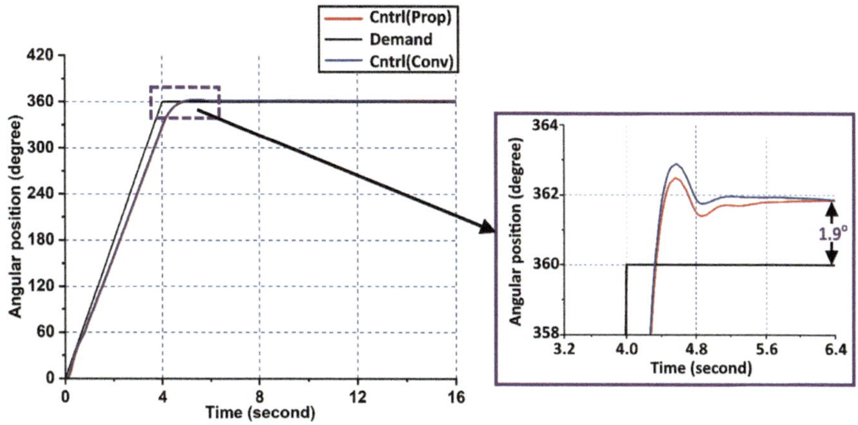

Fig. 5 Angular position variation of the flywheel attached to the swing motor shaft of the PID-controlled conventional and proposed HSD system with respect to time

Figure 5 shows the demand angular position and controlled position of the conventional and proposed PID-controlled HSD system. It can be observed from the inset curve of Fig. 5 that the maximum overshoot for the conventional system is more than the proposed system. But the conventional system settles more aggressively than the proposed system. However, both systems have a steady-state error of 0.53% (1.9°).

Figure 6 depicts the inlet pressure curve of the swing motor for conventional and proposed HSD systems. The inset curve of Fig. 6 shows that the maximum pressure overshoot for the conventional system (52 bar) is less than that of the proposed system (86 bar). This happens because the setting pressure of the relief valve in the case of a conventional system is less than that of the proposed system. However, in both systems, the pressure curve settles at the same time. It is also found that pressure has negative values during the fluctuation, which can cause cavitation in the system. Cavitation can significantly affect system performance by causing vibration, noise, and damage to components. It also impacts control stability. So, there is a need for an anti-cavitation valve to overcome the cavitation in the system.

Figure 7 shows the pressure variation at the outlet of the swing motor for both the conventional and proposed HSD systems. It can be seen that the maximum outlet pressure of the swing motor of a conventional system is 13.8 bar and settles at 1.1 bar. This outlet pressure occurs due to unwanted control of the port orifice of the control valve, which is connected to the outlet port of the swing motor. In the case of the proposed system, the pressure overshoot is 1.1 bar, which is much less than in the conventional system because the swing motor outlet port connected to the control valve port orifice is fully open.

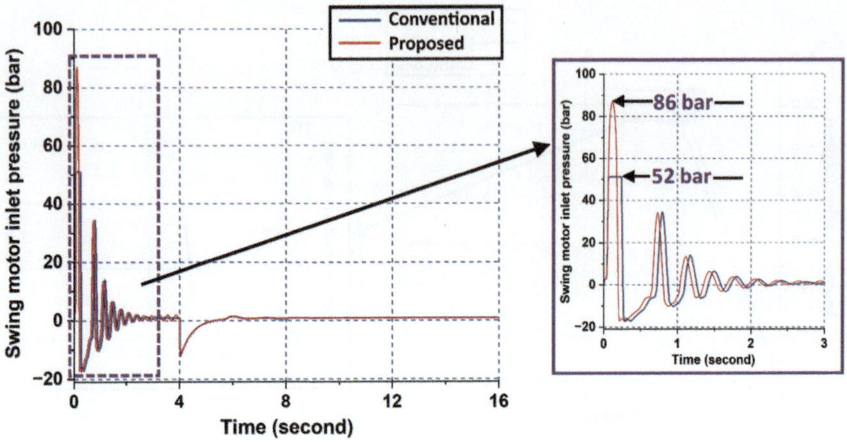

Fig. 6 Variation of inlet pressure of swing motor of conventional and proposed HSD system with respect to time

Fig. 7 Variation of outlet pressure of swing motor of conventional and proposed HSD system with respect to time

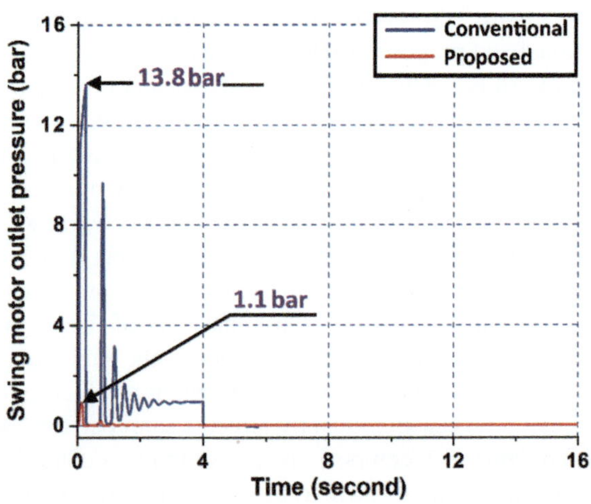

Figure 8 depicts the energy consumption by conventional and proposed HSD systems with respect to time. It is found that the energy consumption by conventional HSD systems is greater than that of the proposed HSD system. At the end of 16th second, the energy consumed by the conventional system is 493 J, and by the proposed system is 290 J. The extra 203 J energy consumed by the conventional system is against the motor outlet pressure as well as the energy wasted by the relief valve (5b). The energy saved by the proposed HSD system is given by the following equations:

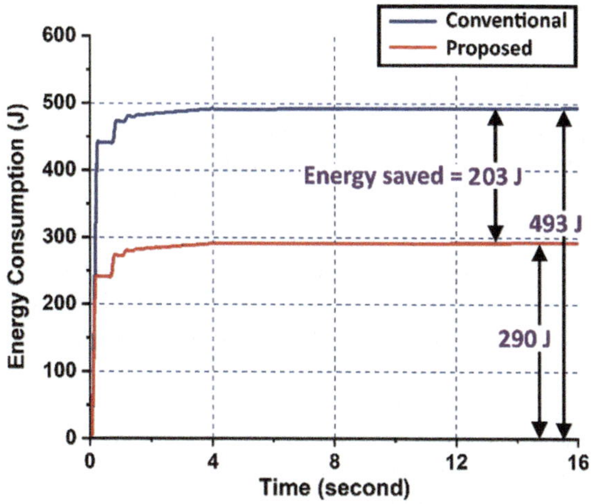

Fig. 8 Energy consumption of conventional and proposed HSD systems

$$\text{Energy saved} = \begin{bmatrix} \text{Energy consumed by conventional HSD system} \\ -\text{Energy consumed by proposed HSD system} \end{bmatrix} \tag{2}$$

$$\text{Energy saving \%} = \frac{\text{Energy saved}}{[\text{Energy consumed by conventional HSD system}]} \times 100 \tag{3}$$

$$\text{Energy saved} = 493\,\text{J} - 290\,\text{J} = 203\,\text{J}$$

$$\text{Energy saving \%} = \frac{203}{493} \times 100 = 41.18\%$$

6 Conclusion

This paper presents a novel energy-efficient swing drive (HSD) system based on an Independent Metering Control (IMC) strategy. The proposed HSD system integrates a PID controller to achieve precise positioning of the excavator's upper structure during swing operations. Compared to conventional systems, the proposed system exhibits a lower maximum overshoot in the position versus time plot. However, it also demonstrates a higher peak pressure demand. Notably, the results indicate that the proposed HSD system reduces energy consumption by 41.18% compared to conventional systems. Future research could explore optimizing the peak pressure demand, integrating advanced control algorithms, and extending the application of the proposed system to other types of heavy machinery to further enhance its performance and energy efficiency.

References

1. Eriksson, B., & Palmberg, J. O. (2011). Individual metering fluid power systems: challenges and opportunities. *Proceedings of the institution of mechanical engineers, part I: journal of systems and control engineering, 225*(2), 196–211.
2. Singh, S., Kumar, A., & Thakur, S. (2025). Sustainable energy solutions for hydraulic excavators: A comprehensive review and novel energy regeneration approach. *Proceedings of the Institution of Mechanical Engineers, Part E: Journal of Process Mechanical Engineering,*https://doi.org/10.1177/09544089251324561
3. Zhang, B., Zhong, Q., Niu, M., Hong, H., & Yang, H. (2018). *Design of Control System for Independent Metering Valve*. Universitätsbibliothek der RWTH Aachen.
4. Liu, S., & Yao, B. (2006). Automated onboard modeling of cartridge valve flow mapping. *IEEE/ASME transactions on mechatronics, 11*(4), 381–388.
5. Huh, J., & Jung, G. H. (2018). Basic Design for Development of IMV for MCV. *Journal of Drive and Control, 15*(3), 49–56.
6. Hu, H., & Zhang, Q. (2002). Realization of programmable control using a set of individually controlled electrohydraulic valves. *International Journal of Fluid Power*, *3*(2), 29–34.
7. Shenouda, A., & Book, W. (2008). Optimal mode switching for a hydraulic actuator controlled with four-valve independent metering configuration. *International journal of fluid power*, *9*(1), 35–43.
8. Liu, Y., Xu, B., Yang, H., & Zeng, D. (2009, July). Simulation of separate meter in and separate meter out valve arrangement used for synchronized control of two cylinders. In *2009 IEEE/ASME International Conference on Advanced Intelligent Mechatronics* (pp. 1665–1670). IEEE.
9. Wang, Q., Gu, L., & Lu, Y. (1999, November). Research on digital control of meter-in and meter-out independent regulating for high inertia load. In *ASME International Mechanical Engineering Congress and Exposition* (Vol. 16479, pp. 109–114). American Society of Mechanical Engineers.
10. Choi, K., Seo, J., Nam, Y., & Kim, K. U. (2015). Energy-saving in excavators with application of independent metering valve. *Journal of mechanical science and technology, 29*, 387–395.
11. Shi, J., Quan, L., Zhang, X., & Xiong, X. (2018). Electro-hydraulic velocity and position control based on independent metering valve control in mobile construction equipment. *Automation in Construction, 94*, 73–84.
12. Ge, L., Dong, Z., Huang, W., Quan, L., Yang, J., & Li, W. (2015, August). Research on the performance of hydraulic excavator with pump and valve combined separate meter in and meter out circuits. In *2015 international conference on fluid power and mechatronics (FPM)* (pp. 37–41). IEEE.
13. Ahamad, M., Dinh, Q. T., Nahian, S. A., & Ahn, K. K. (2015). Development of a New Generation of Independent Metering Valve Circuit for Hydraulic Boom Cylinder Control. *International Journal of Automation Technology, 9*(2), 143–152.
14. Jansson, A., & Palmberg, J. O. (1990). Separate controls of meter-in and meter-out orifices in mobile hyraulic systems. *SAE transactions*, 377–383.
15. Abuowda, K., Noroozi, S., Dupac, M., & Godfrey, P. (2019, March). Algorithm design for the novel mechatronics electro-hydraulic driving system: Micro-independent metering. In *2019 IEEE International Conference on Mechatronics (ICM)* (Vol. 1, pp. 7–12). IEEE.
16. Tabor, K. A. (2005). *A novel method of controlling a hydraulic actuator with four valve independent metering using load feedback* (No. 2005-01-3639). SAE Technical Paper.
17. Singh, A., Kushwaha, P., Ghoshal, S. K., & Singh, S. (2022, December). Dynamic Analysis of a Low-Speed, High-Torque Hydro-Motor Drive System to Maintain a Constant Drive Speed Using ANN-Based Fuzzy PI Controller. In *International Conference on Industrial Problems on Machines and Mechanism* (pp. 201–210). Singapore: Springer Nature Singapore.
18. Thakur, S., Kumar, A., & Singh, S. (2023). A Critical Review of the Various Energy-Saving Models Used in Hydrostatic Power Transmission Drives for Enhancing the Efficiency of Chairlift Transport System. *Arabian Journal for Science and Engineering*, 1–22.

19. Singh, S., Kumar, A., & Thakur, S. (2023, December). Development of an Accumulator-Based Energy Regenerative Technology (AERT) for the Hydraulic Swing Mechanism of an Excavator Machine. In Conference on Fluid Mechanics and Fluid Power (pp. 221–232). Singapore: Springer Nature Singapore.

Comparative Usability Evaluation of Three Different Gaming Controllers

Mouly Bhowmick, Krishna Chaitanya Mallampalli, Abhishek Shrivastava, and Sougata Karmakar

Abstract Gaming controllers are input devices that provide a tangible or intangible interfaces for the gamers to interact with the videogame. Regardless of age and gender, those controllers are increasing in demand in the growing market of the videogame industry. Manufacturers have created a wide range of controllers, but those are not equally suitable for every user population due to the shortcomings of the controllers, including challenging operations and anthropometric and biomechanical incompatibility that ultimately leads to an increase in physical and mental demand. While the design of the controllers is greatly influenced by the buttons/switches and joystick configurations, it can be very difficult for players, especially inexperienced ones, to determine which game controller is best for them. Therefore, the goal of this study is to identify the best gaming controller configuration for improved user performance. A comparative usability study was conducted with three different gaming control interfaces with the participation of the 20–40 age range of the student population. The testing methods included anthropometric and biomechanical compatibility study with the controllers considering the hand anthropometry and total range of motion data (5th, 50th, and 95th percentile data) respectively. Materials included the NASA-TLX to compare the perceived workload by the users, the System Usability Scale (SUS) to evaluate the ease of use, and heuristic evaluation to identify usability issues with three controllers. The results indicated that the anthropometric and biomechanical compatibility were reasonable with the selected controllers. The NASA-TLX and SUS scores showed variation in mean value with no significant

M. Bhowmick (✉) · A. Shrivastava · S. Karmakar
Center for Intelligent Cyber Physical Systems, Indian Institute of Technology Guwahati, Assam, India
e-mail: m.bhowmick@iitg.ac.in

A. Shrivastava
e-mail: shri@iitg.ac.in

S. Karmakar
e-mail: karmakar.sougata@iitg.ac.in

K. C. Mallampalli · A. Shrivastava · S. Karmakar
Department of Design, Indian Institute of Technology, Guwahati, Assam, India
e-mail: chaitanya555@rnd.iitg.ac.in

U. S. Dixit et al. (eds.), *Current Progress in Engineering Sciences*,
https://doi.org/10.1007/978-981-96-8753-4_23

difference after performing one-way ANOVA. Through heuristic evaluation, a wide range of usability issues have been identified. A total of 40 issues were listed for console 1, console 2, and console 3. Future study indicates the incorporation of more participants to strengthen the validity of the results obtained.

Keywords Ergonomic evaluation · Gaming control interface · Usability methods · Tangible control interface · Graphical user interface

1 Introduction

The gaming controller industry is experiencing rapid growth due to the continuous technological advancement and the extensive development of the gaming sector worldwide [1, 2, 4]. The number of gaming users has greatly increased due to the widespread availability of different gaming platforms, such as consoles, personal computers (PCs), and mobile devices [3, 4]. Different controller requirements for every platform create a diverse market for controllers that serve various ecosystems. To analyze the market report of gaming controllers, the following segments have been considered: product type, compatibility, connectivity, end-use, distribution channel, and geographical region. Based on the product type, the market is divided into joysticks, gamepads, steering wheels, trackballs, throttle quadrants, etc. [5, 6]. Moreover, according to industry experts, gaming controllers are likely to integrate augmented reality (AR) and virtual reality (VR) intensively to provide gaming players with even more immersive experiences [7, 8]

Despite advancements in user interfaces and new digital game genres, it has been found that not all people can play all the games and are also not comfortable with all types of gaming controllers [9, 10, 11]. From the younger generation to the older, the disabled population is involved in video game playing with different gaming controllers [10, 11]. The users have their own game choices and preferences for gaming controllers. Poorly designed or overly complex gaming controllers can affect the users. Therefore, developing other intuitive interfaces can contribute to key driving factors in expanding demographics [9, 10, 12].

It has been found that amongst all the controller segments in the gaming controllers' market, the 'gamepad' segment held the largest market share in 2022 as it provides a more user-friendly and intuitive control scheme [4, 5]. It can also be used across various platforms, such as consoles, PCs, and mobile devices. In action and adventure games, precision is crucial for performing tasks like controlling, shooting, aiming, etc. This precision in the gamepad is achieved by the features incorporated in it, like buttons, joysticks, triggers, or other input methods. The arrangement of these features in the gamepad is a very crucial part in gaming control designing process [13]. The ergonomic design allows players to hold and use the gaming controller comfortably [9]. Any uncomfortable experience playing the videogame with the controllers leads to difficulties in operation, increased physical [14] and mental workload [15], physical exertion, and pain in fingers and hands [16]. In such scenarios, the configuration of features in gaming controllers provides a valuable contribution to the design of the controllers. But from a usability perspective, it becomes a very challenging task to identify suitable features for gaming controller design and to provide a better gaming experience. Hence the purpose of the experimental study is to come up with the identification of a suitable configuration of gaming controller for better user performance.

2 Materials and Methods

An experiment was conducted with the 20-40 years of age range of the student population of a higher academic institute with three types of gaming controllers with different configurations. Those are Gaming Controller 1 (Zebronics ZEB-V260JP Gamepad -Physical controller), Gaming Controller 2 (X Box 360 Gamepad—Physical Controller), and Gaming Controller 3 (Touchscreen of the smartphone as gaming controller—Tactile controller). Figures 1, 2 and 3 show the experimental study design, three different gaming controllers and participant playing with controllers respectively. Various ergonomic and usability aspects were evaluated for the gaming control interfaces under current studies. These evaluations included (a) an anthropometric and biomechanical compatibility study of the gaming controller configurations (b) the NASA-TLX [17] method for measuring perceived workload, (c) SUS score evaluation [18] to compare the ease of performance, and (d) Heuristic evaluation to identify the usability issues in those interfaces. Anthropometric compatibility studies were conducted to evaluate whether the sizes/dimensions of buttons/ switches and joysticks were compatible with the hand/ figure anthropometry of the users. Likewise, biomechanical compatibility was performed to assess the range of motions (ROM) of the fingers for operating or accessing the buttons or joystick. In this experiment, the main focus was on the controllers' upper front side. The specific controllers' section included the action buttons (1/2/3/4 or A/B/X/Y or glyphs), D-pad or direction buttons, and multidimensional analog sticks or joysticks, or thumb sticks, based on the task designed for the experiment.

2.1 *Participants*

Anthropometric compatibility. 43 students (28 males, 15 females), including undergraduate and postgraduate students, with an average age of 30.3 $\pm$ 4.2 years, participated in the experiment. The mean body height was 168.5 $\pm$ 9.2 cm, and the mean body weight was 72.5 $\pm$ 15.0 kg.

Biomechanical compatibility. 43 students (28 males, 15 females), including undergraduate and postgraduate students, with an average age of 30.3 $\pm$ 4.2 years,

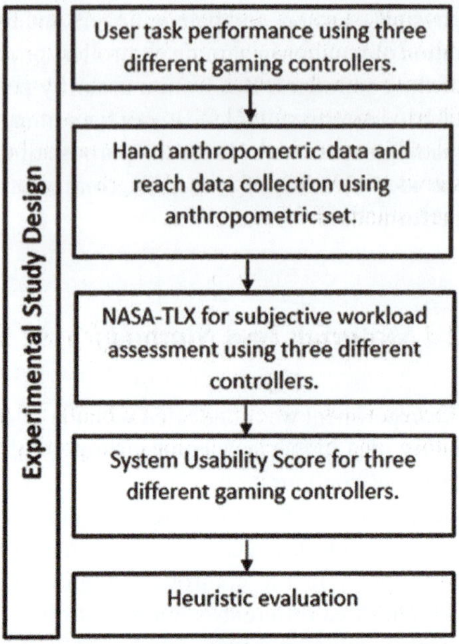

Fig. 1 Diagrammatic representation of the experimental study design

Fig. 2 Three different gaming controllers (Zebronics ZEB-V260JP Gamepad, XBox 360 Gamepad, and smartphone touchscreen as gaming control interface from left to right)

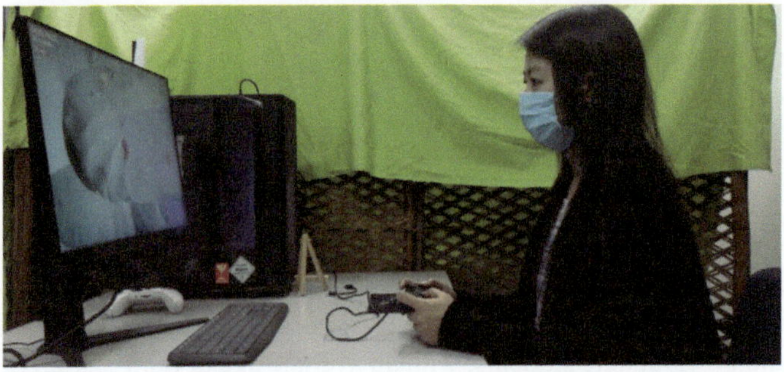

Fig. 3 Sample picture of a participant testing gaming controllers

participated in the experiment. The mean body height was 168.5 ± 9.2 cm, and the mean body weight was 72.5 ± 15.0 kg.

Perceived workload measurement. 15 students (8 males, 7 females) including both undergraduate and postgraduate with an average age of 30 ± 3.8 years, participated in the experiment. The mean body height was 170.4 ± 15.5 cm and the mean body weight was 75.5 ± 14.1 kg.

Usability (SUS Score) assessment. 15 students (8 males and 7 females), including undergraduate and postgraduate students, with an average age of 30 ± 3.8 years, participated in the experiment. The mean body height was 170.4 ± 15.5 cm and the mean body weight was 75.5 ± 14.1 kg.

Heuristic evaluation. Five experts in heuristic evaluation with an average age of 32 ± 3.5 years participated in the experiment. The mean body height was 168.4 ± 10.0 cm and the mean body weight was 72.2 ± 10.1 kg.

2.2 Data Collection and Analysis

Anthropometric compatibility. An anthropometric measuring set was used to take hand anthropometric dimensions, including length, width, span, depth, and circumference (Fig. 4).

The data was analyzed with SPSS V29.0 software. Shapiro–Wilk test was performed to see whether the data came from a normally distributed population. Statistical significance of $p < 0.05$ indicates the rejection of the null hypothesis. $P > 0.05$ indicates that the sample distribution is not significantly different from a normal distribution. The collected anthropometric data were calculated for 5th, 50th, and 95th percentile values.

Biomechanical compatibility. For the biomechanical study, the same participants were taken, and their total range of motion data for 26 dimensions, including wrist and forearm, index finger, and thumb, were taken by manual Goniometers (Figs. 4, 5).

The data was analyzed with SPSS V29.0 software. Shapiro-Wilk test was performed to see whether the data came from a normally distributed population. Statistical significance of $p < 0.05$ indicates the rejection of the null hypothesis.

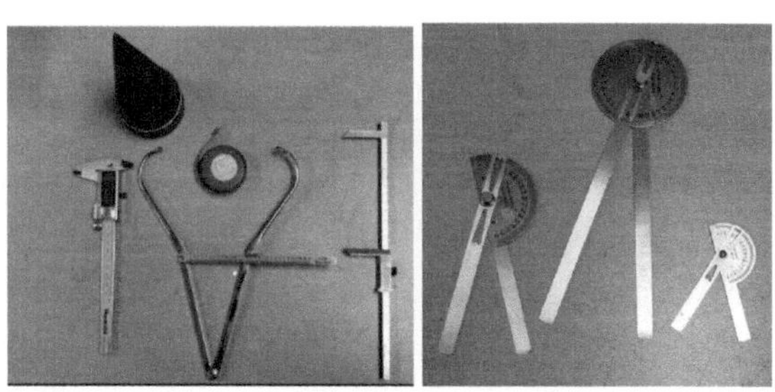

Fig. 4 The anthropometric set and goniometers that were used to measure hand dimensions and range of motion respectively

Fig. 5 Biomechanical compatibility study

$P > 0.05$ indicates that the sample distribution is not significantly different from the normal distribution.

Perceived workload measurement. In the present study, the NASA-TLX scores of the participants were evaluated to measure cognitive workload. Participants had to perform the given gaming task with three types of gaming controllers individually. After completing each task, they were supposed to fill out the questionnaire.

The data was analyzed with SPSS V29.0 software. Statistical significance taken as $p < 0.05$. A one-way ANOVA analysis was conducted to check whether there was any difference amongst the three controllers regarding participants' cognitive workload while using them for playing the game.

System Usability Scale (SUS). In the present study, SUS scores were evaluated to measure product usability. Participants had to perform the given gaming task with three types of gaming controllers individually. After completing each task, they were supposed to fill out the SUS questionnaire.

The data was analyzed with SPSS V29.0 software. Statistical significance taken as $p < 0.05$. A one-way ANOVA analysis was conducted to check whether there was any difference among the three controllers regarding SUS score while using them for playing the game.

Heuristic evaluation. In the present study, a heuristic evaluation of three controllers was conducted with five experts. Individual sessions were conducted with each expert. At the beginning of the session, three controller designs were shown to the experts, and they were asked to handle those. During this session, experts were free to ask questions about the controllers. All the queries raised by the experts were addressed during this session. This session also helped the experts to get familiar with the controllers' design before the start of the experiment. After exploring the controllers, participants were asked to use each console to play an online game. After completing the task, all the experts were asked to fill Nelson heuristic evaluation form. The experts rated each dimension of Nelson's heuristic on a scale of 1–10 (1—very good and 10 very bad). Following the Nelson heuristic evaluation, experts were asked to elicit recommendations/comments on controller redesign. For example, experts were asked to comment on

1. What design features of the consoller (1, 2, and 3) interface do you think is more user-friendly? What do you like about the consoller (1, 2, and 3)?
2. Do you feel any modifications/changes to the existing controller (1, 2, and 3) make the design more user-centric?

The session was audio-recorded and photographed.

3 Results

Table 1 shows the anthropometric compatibility. For anthropometric compatibility with the buttons, the button diameter should be in the 50th percentile of the width of the distal part of the finger by which it is operated. In this case, a good number of people whose finger width is the same or near the 50th percentile value will be able to operate the button. In the case of designing the gap between two buttons, it should be considered that the larger finger width (95th percentile) should not touch the adjacent button. The distance of the joysticks from the middle of the control interface handle (where the thumb base is placed should be small enough to reach it by the 5th percentile of the population.

Table 2 shows the biomechanical compatibility. The range of motion for different body joints particularly in hand and fingers may vary according to the task requirement or task context while operating various controllers. Those ROM for operating different types of controllers should be within the comfort range. Here in this experiment, the total range of motion of various joints in hands and fingers were measured and the 5th, 50th and 95th percentile values were calculated to see whether the participant is able to operate the gaming controller comfortably or not. The maximum movement angle should be like the person with 5th percentile ROM of any finger can easily operate the button or switch.

Table 3 shows the NASA-TLX cognitive workload in three controllers. The results showed that there is no significant difference between the three different controllers. The possible reason for this result can be attributed to the small sample size. However, the mean value of cognitive workload (NASA TLX score) of controller 2 was the highest (50.7) among the three controllers, followed by controller 3 (42.0) and

Table 1 Anthropometric compatibility study

Anthropometric measurements	5th percentile (cm)	50th percentile (cm)	95th percentile (cm)
Length of the thumb (reach)	5.1	6.6	7.6
Width of the distal thumb	1.5	2.0	2.3

Table 2 Biomechanical compatibility study

Range of motion	5th percentile (in degree)	50th percentile (in degree)	95th percentile (in degree)
Thumb flexion distal thumb (IP)	24	59.0	85.8
Thumb abduction	25.4	55.0	82.0

controller 1 (40.0). The implication of this study is that controller1 is preferred by the participants as it has less cognitive load.

The results showed no significant difference among the three controllers. The possible reason for this result can be attributed to the small sample size. However, the mean usability rating of controller 1 was the highest (72.5) amongst the three controllers, followed by controller 3 (68.1) and controller 2 (62.8). The implication of this study is that console 1 is preferred by the participants as its usability is better (Table 4).

Heuristic evaluation: Through heuristic evaluation, a wide range of usability issues have been identified. A total of 40 issues were listed for controller 1, 2, and 3. Table 5 shows the usability issues identified by experts for each gaming controller in the study. Three out of five experts have the opinion that Console 1 and Console 2 have more usability issues (Table 5).

Table 3 Comparison of NASA-TLX for gaming controller 1, 2 and 3

Gaming console control	N (sample size)	Minimum	Maximum	Mean	Std. deviation	Homogeneity variance based on means	Significance value for one-way ANOVA
1	15	6.7	96.3	40.0	24.7	1.000	**0.478**
2	15	5.0	94.3	50.8	24.8		
3	15	9.7	82.3	42.6	24.4		

Table 4 Comparison of SUS score for three gaming controllers 1, 2 and 3

Gaming console control	N (sample size)	Minimum	Maximum	Mean	Std. deviation	Homogeneity variance based on means	Significance value for one-way ANOVA
1	15	22.5	97.5	72.5	24.6	0.154	0.470
2	15	25.0	100.0	62.8	21.9		
3	15	25.0	1000	68.2	17.1		

Table 5 Usability issues identified by heuristic experts for three different gaming controllers

SL No.	Gaming controller	Issues identified separately
1	1	40
2	2	37
3	3	34

4 Discussion

After conducting the usability evaluation of all three controllers, it has been found that there is no such anthropometric and biomechanical incompatibility with respect to the gaming controllers, as all three controllers are standard ones in the market. Gaming controller 1 has the lowest perceived workload (NASA-TLX Score) and highest product usability score (SUS Score), followed by gaming controller 3 and gaming controller 2. On the other hand, in the case of the heuristic evaluation, the least usability issues were found in gaming controller3, followed by Gaming controller 2 and Gaming controller 1. The mismatch of results in heuristic evaluation with the perceived workload measurement and product usability score occurred probably due to gaming controller 3 being a touchscreen controller, as heuristic evaluation is specially designed to evaluate graphical user interfaces [20] but rarely for physical control interfaces. Because the NASA- TLX and SUS Score evaluation show the same result for evaluating those three gaming controllers, it can be concluded that gaming controller 1 is the best, followed by gaming controller 3 and gaming controller 2.

Previous studies indicate that a very important aspect of a gaming controller is that it should allow a player to move the user's hand and body in such a way that the digital experience mimics its real-life counterpart and gives the user a comfortable experience. On the other hand, the gaming controller's naturalness plays a vital role in gaming enjoyment. In car racing games, a steering wheel controller was perceived as more natural than a keypad, joystick, and gamepad controller [21], naturally contributing to the game's enjoyment. However, to have a comfortable gaming experience, the ergonomic design of a gaming controller is very much necessary [22]. The comfort of the controller mainly depends on the practical shape and button layout while playing a video game for a prolonged period of time without changing the grip [22]. Brown and Mackenze conducted a study to explore the relationship between gaming controller usability and hand size by comparing the Xbox 360 controller, Wii remote with Nunchuck attachment, and dual shock 3 controller [23]. The finding showed that people with larger hand sizes have to move their hands to reach more of the buttons without adjusting the hand grip, and in their experiment, the Xbox controller showed the highest correlation between hand size and hand movements, which was probably due to the slightly larger size of the controller. They also concluded that for the ease of use of the controller, it should be designed by focusing on the reachability of the buttons [24].

ratu analyzed their study's qualitative and quantitative differences between mouse and gaming controllers. They recorded the workload using the NASA-TLX questionnaire after completing every task given with each input device. The findings supported that the mouse was preferable for its ease of use and best suited for almost every task [23].

Seo and Kang experimented to determine the comfort level of smartphones and physical gaming controllers. The findings showed that direct touchscreen interaction was more effective for the two-handed input control task, and physical input was more effective for the one-handed input controlled task [25].

The present study was conducted only for a very few students (43 students for biomechanical and anthropometric compatibility study, 15 students for measuring perceived workload and ease of use, and five experts for heuristic evaluation), and it was conducted in a higher institute on undergraduate and postgraduate students only. The future study includes further incorporating more participants, including gender specificity and socio-economic cultural specificity, to conduct a more effective comparative usability study of the gaming controllers. The study was conducted only for three types of gaming controllers, and it should include incorporating more controllers with diverse specifications. The present study was conducted with the front-facing buttons of the controllers only. So in the next phase of the study, the user interaction efficiency of LB (Left Bumper), RB (Right Bumper), LT (Left Trigger) and RT (Right Trigger) within gameplay scenarios should be included. The usability evaluation study conducted here is very limited. Future studies will indicate the need for more usability studies by adding instruments like EMG and eye tracker to validate the result. The task specificity for the controllers should be more analyzed.

5 Conclusion

If a larger study with more controls and gamers could be executed, the ergonomics and usability evaluations presented in the current study would help decide the suitable configuration of the gaming controller for better user performance.

References

1. Akyaman, S., & Alppay, E. C. (2024). Effects of industrial design properties of game controllers on player experience. Entertainment Computing, 50, 100687.
2. Cui, Z., Jiang, J., & Zhu, Y. (2021, December). Research on Future Business Strategy Reform of Gaming Console Brands. In 2021 3rd International Conference on Economic Management and Cultural Industry (ICEMCI 2021) (pp. 1395-1401). Atlantis Press.
3. Poh, M. (2019). Evolution of video games user interface (UI).
4. Volckmann II, W. M. (2024). Consumer and critic reception of video game platforms: Trends from 2002–2022. Entertainment Computing, 48, 100599.
5. https://www.theinsightpartners.com/reports/gaming-controller-market267
6. https://www.bain.com/insights/gamer-survey-young-players-reshape-the-industry-gaming-report-2024/
7. Hameed, A., Perkis, A., & Möller, S. (2021, June). Evaluating hand-tracking interaction for performing motor tasks in VR learning environments. In 2021 13th International Conference on Quality of Multimedia Experience (QoMEX) (pp. 219-224). IEEE.
8. Jin, Y., Ma, M., & Zhu, Y. (2022). A comparison of natural user interface and graphical user interface for narrative in HMD-based augmented reality. Multimedia Tools and Applications, 81(4), 5795-5826
9. Glinert, E. M. (2008). The human controller: usability and accessibility in video game interfaces (Doctoral dissertation, Massachusetts Institute of Technology).

10. Boot, W. R., Andringa, R., Harrell, E. R., Dieciuc, M. A., & Roque, N. A. (2020). Older adults and video gaming for leisure: lessons from the Center for Research and Education on Aging and Technology Enhancement (CREATE). Gerontechnology, 19(2).
11. Martinez, J. J., Froehlich, J. E., & Fogarty, J. (2024, May). Playing on hard mode: accessibility, difficulty and joy in video game adoption for gamers with disabilities. In Proceedings of the 2024 CHI Conference on Human Factors in Computing Systems (pp. 1-17).
12. Malone, L. A., Padalabalanarayanan, S., McCroskey, J., & Thirumalai, M. (2017). Assessment of active video gaming using adapted controllers by individuals with physical disabilities: a protocol. JMIR Research Protocols, 6(6), e7621
13. Torok, L., Pelegrino, M., Lessa, J., Trevisan, D. G., Vasconcelos, C. N., Clua, E., & Montenegro, A. (2015, July). Evaluating and customizing user interaction in an adaptive game controller. In International Conference of Design, User Experience, and Usability (pp. 315-326). Cham: Springer International Publishing.
14. Rupp, M. A., Oppold, P., & McConnell, D. S. (2013, September). Comparing the performance, workload, and usability of a gamepad and joystick in a complex task. In Proceedings of the Human Factors and Ergonomics Society Annual Meeting (Vol. 57, No. 1, pp. 1775-1779). Sage CA: Los Angeles, CA: SAGE Publications.
15. Reinhardt, D., Haesler, S., Hurtienne, J., & Wienrich, C. (2019, March). Entropy of controller movements reflects mental workload in virtual reality. In 2019 IEEE Conference on Virtual Reality and 3D User Interfaces (VR) (pp. 802-808). IEEE.
16. Yesodha, K. K. R. K., Narasimhan, V., Li, Y., & Craig, B. (2017, June). Ergonomic evaluation of videogame controllers. In International Conference on Applied Human Factors and Ergonomics (pp. 384-391). Cham: Springer International Publishing.
17. Hart, S. G., & Staveland, L. E. (1988). Development of NASA-TLX (Task Load Index): Results of empirical and theoretical research. In Advances in psychology (Vol. 52, pp. 139-183). North-Holland.
18. Brooke, J. (1996). SUS-A quick and dirty usability scale. Usability Evaluation in Industry, 189(194), 4-7.
19. Nielsen, J. (1995). How to conduct a heuristic evaluation. Nielsen Norman Group, 1(1), 8.
20. Skalski, P., Tamborini, R., Shelton, A., Buncher, M., & Lindmark, P. (2011). Mapping the road to fun: natural video game controllers, presence, and game enjoyment. New Media & Society, 13(2), 224-242.
21. Gutierrez, A. M. J., Cadiz, L., Filoteo, N., Juan, I., & Leopando, B. (2019, October). Enhancement of gaming experience and performance through an ergonomically designed console chair. In International Conference on Industrial Engineering and Operations Management (pp. 31-41).
22. Hernández, H. A., Khan, A., Fay, L., Roy, J. S., & Biddiss, E. (2018). Force resistance training in hand grasp and arm therapy: feasibility of a low-cost videogame controller. Games for Health Journal, 7(4), 277-287.
23. Bratu, C. I. (2023). Which input method is more efficient: a comparison between mouse and controller (Bachelor's thesis, RWTH Aachen University).
24. Brown, M. A., & MacKenzie, I. S. (2013). Evaluating video game controller usability as related to user hand size. In Proceedings of the International Conference on Multimedia and Human Computer Interaction (pp. 1-9).
25. Seo, S. D., & Kang, S. (2019). A comparison study of the smartphone gaming control. Journal of Usability Studies, 14(4).
26. Mendonca, C. J., Malone, L. A., Mohanraj, S., & Thirumalai, M. (2023). The usability of a touchpad active video game controller for individuals with impaired mobility: observational study. JMIR Rehabilitation and Assistive Technologies, 10, e41993.

MLADS: Multi-layered Attribute-Based Access Control and Delegation System for Cloud-Enabled IoT Healthcare System

Mridul Kailashchandra Mishra and Udai Pratap Rao

Abstract Advancements in cloud computing have helped Internet of Things (IoT) systems to enable complex and highly scalable IoT solutions in healthcare, which has led to concerns about unauthorized access to the sensitive resources of cloud-enabled IoT (CE-IoT). Traditional access control methods often cannot handle complex resource allocation and permission delegation properly, resulting in inefficiencies in access control and delegation and vulnerabilities in security. To overcome these issues, we have proposed a novel Attribute-Based Access control (ABAC)-based framework for IoT-based healthcare, enabling secure access to resources layered by utilizing a centralized policy repository. Moreover, the framework facilitates the automatic delegation of permissions based on predefined policies and attributes. The proposed framework security has undergone a comprehensive informal security analysis, confirming its effectiveness in achieving the required security. The proposed framework takes significantly less computation and communication costs compared to existing schemes.

Keywords IoT · Access control · ABAC · CE-IoT

1 Introduction

Cloud computing-enabled Internet of Things (CE-IoT) has transformed many industries, including healthcare. CE-IoT allows real-time data collection, improved care delivery models, and remote patient monitoring [1]. Studies have indicated that CE-IoT-based healthcare systems can help reduce total healthcare costs while improving patient-care outcomes [2]. For instance, deploying CE-IoT in chronic disease management has demonstrably improved patient self-management capabilities and

M. K. Mishra (✉) · U. Pratap Rao
Department of Computer Science and Engineering, National Institute of Technology, Patna, Bihar, India
e-mail: mishrak.ph23.cs@nitp.ac.in

U. Pratap Rao
e-mail: udai.cs@nitp.ac.in

U. S. Dixit et al. (eds.), *Current Progress in Engineering Sciences*,
https://doi.org/10.1007/978-981-96-8753-4_24

compliance with medication [3]. However, leveraging the full potential of CE-IoT depends on robust security mechanisms to protect sensitive data.

The usage of IoT is ever-increasing due to the rise in the number of intelligent devices being connected to the web. IoT systems offer intelligent and convenient solutions but also put sensitive and personal information at security risk from cyber-attacks. Concerns have been raised [4] about possible security lapses in IP-based cameras. A manufacturer's IP-based cameras may have allowed attackers to obtain unauthorized access, steal confidential information, or even interfere with camera operations. According to an analysis by HP, all smartwatches have vulnerabilities that could allow for unauthorized access (in early 2023) [5]. A new Mirai variant hijacks IoT devices by exploiting vulnerabilities [6]. Unauthorized access and remote code execution are critical vulnerabilities in IoT devices [7], highlighting the importance of robust access control regulations to prevent unauthorized access.

CE-IoT in healthcare requires robust security measures to safeguard sensitive data, but traditional access control methods struggle to manage complex policy frameworks and delegate access rights effectively [8]. Due to this, ABAC emerges as a reasonable solution.

ABAC is an essential method for safe delegation and access control in CE- IoT, covering the rising resource management demand based on user and resource characteristics. By dynamically granting access based on attributes associated with persons, devices, data, and the surrounding environment, ABAC enables more granular access control than traditional role-based access control (RBAC) [9]. For example, nurses possessing specific attributes are granted access to read prescriptions and document patients' medication history. In contrast, qualified doctors can read and write patient vitals and prescriptions. ABAC also enables permission delegation. In situations where medical staff are unavailable, ABAC can automatically delegate required access to another doctor as per defined policies, ensuring secure patient care.

Studies show that ABAC facilitates secure and efficient access control in healthcare systems [10, 11]. ABAC can meet the complex and dynamic access control requirements in CE-IoT healthcare because it handles complex policy structures and dynamic delegation.

Delegation is an essential element of efficient access control. It involves transferring authorization or authority from one entity (such as a person, device, or service) to another [12]. In the context of the IoT, delegation grants specific access privileges or permissions in compliance with defined rules, allowing for dynamic and flexible access control in highly dynamic distributed contexts. Benefits of delegation include efficient resource management, granular access control, reduced administrative overhead, and improved compatibility.

By integrating access and delegation rules, we propose a novel framework for managing access control in CE-IoT that optimizes workflows and reduces device complexity through an automatic permission delegation system, layered access control for various resources (cloud, virtual, and physical), and a single policy repository. Additional benefits include reduced administrative effort, improved device-to-device communication due to a unified policy store, and generally more effective and flexible access control for intricate CE-IoT setups.

1.1 Research Contributions

Traditional approaches to access control in CE-IoT separate access control and delegation rules, leading to potential inefficiency and complexity. The framework proposes a novel approach that combines access control and delegation rules into a single, layered policy structure, aiming to overcome these limitations and address critical challenges in CE-IoT access control. The following contributions are included:

- **Development of a Layered Access Control Model**: The MLADS framework introduces layer-wise access control and delegation policies to manage each layer with policies appropriate to their specific needs and more granularity, thereby reducing unnecessary restrictions and improving overall system performance.
- **Isolation and Security Enhancement**: Layer-Specific Security Measure helps implement isolated access control policies for each layer to minimize the attack surface, significantly increasing the robustness and security of the healthcare IoT system and maintaining data integrity and confidentiality.
- **Scalability and Flexibility in Policy Management**: Our framework introduces a scalable access control framework with layered policy management, enabling independent updates and modifications within each access layer. This enhances scalability and flexibility in managing access policies across diverse IoT environments, supporting dynamic adaptation to evolving regulatory and operational requirements.

1.2 Organization

Section 2 discusses state-of-the-art methods and compares them against our proposed framework. Section 3 explains the proposed framework's system and security models and requirements. Section 4 highlights the preliminary requirements, and Sect. 5 discusses the details of the proposed framework. Sections 6 and 7 provide the security and performance analysis of the proposed framework. Section 8 discusses the conclusions and future work.

2 Background and Related Work

The access control model is designed to regulate who (subject/intended user) can do what (tasks/operation) and when on which resource. ABAC achieves this with the help of attributes of the user and the user. Unlike traditional access control policies, ABAC grants access to resources' services based on the attributes possessed [13] by the requester. ABAC uses user attributes (like department ID and project ID), system attributes, and environmental attributes (like location and time) to determine whether to grant access.

The prime focus of this research is to propose a fine-grained and efficient access control framework for CE-IoT healthcare system. The access model should support flexible delegation when needed. Our framework involves permission delegation for partial delegation. A study of related works in IoT access control is presented in this section.

2.1 Literature Review

This section studies the current state of research on IoT access control, presenting a review of the most relevant and recent scholarly literature.

Aghili et al. [14] introduced a Multi-level Security Attribute-Based Access Control (MLS-ABAC) scheme for IoT applications based on the NIST ABAC model and employs Attribute-Based Encryption (ABE) for fine-grained control over encrypted data. While the scheme meets NIST's ABAC requirements and offers efficient encryption and decryption, it faces limitations. Existing ABE schemes may need to fully align with NIST's criteria or offer sufficient computational efficiency for IoT environments. Additionally, the paper does not address scalability, practical implementation challenges, or potential security vulnerabilities, and the experimental results may not capture performance issues in large-scale deployments.

Choksy et al. [16] proposed an ABAC model tailored for IoT healthcare, featuring highly customizable delegation capabilities. This model provides precise access control over patient medical records on cloud servers, utilizing a Quality Factor (QF) to govern delegation and limit attribute transfers. Integration of delegation regulations into the ABAC framework could enhance performance.

Thomaz et al. [17] have proposed a cloud-based that performs complex processing tasks on IoT data while enforcing access control policies to ensure privacy and security. This framework uses memory enclaves for access policy enforcement but faces challenges such as the complexity of enclave development and performance overhead due to the limited resources of IoT devices.

Li et al. [18] proposed the SEWAC scheme for cloud-assisted Industrial IoT (IIoT) applications, which allows data owners to share sensitive data via the cloud under self-defined access structures. Concerns include scalability with a large number of attributes and users, as well as the complexity involved in managing weighted attributes and access structures.

Fugkeaw et al. [19] integrated fog computing, Ciphertext Policy-ABE, and blockchain technology to enable secure, fine-grained, and scalable sharing of Electronic Medical Records (EMRs) in a cloud environment. However, the computational cost of CP-ABE is high, which may be a challenge for resource-constrained IoT devices, and the cost of policy updates is significant.

In [20], Bagga, Palak, et al. have introduced a novel access control system utilizing user biometrics for video surveillance setups. This system includes mutual authentication between users and smart devices through cloud servers to establish secure communication session keys. However, the paper provides limited insight into scalability and performance aspects and does not adequately address potential vulnerabilities.

The RSA2C method presented in [21] aims to enhance access control in cloud-assisted IoT communications by improving resource allocation and request processing efficiency. However, the focus is more on ensuring service availability than providing a detailed access control mechanism.

The MSLShard framework [23] uses an adaptive network sharding scheme based on network distance, node credibility, and access frequency, along with a multidimensional subjective logic trust management system for blockchain-based access control in IoT environments. However, the practical implementation and real-world feasibility of MSLShard should be sufficiently discussed, potentially impacting its adoption. The study needs specific performance metrics for evaluating scalability and security, raising questions about its robustness. The research must also explore potential vulnerabilities or attack vectors to ensure the framework's security in IoT environments.

Alshehri et al. [24] proposed a simplified access control approach using a consortium blockchain, and a reliable consensus mechanism is proposed. Limitations include potential disruption of access control if the blockchain faces technical issues or attacks and the computational expense of revoking access for many devices.

The combo-chain scheme described in [25] combines hierarchical ABAC, smart contracts, and sharding for flexible access policy management. While sharding enhances efficiency, managing shard groups and ensuring consistent policies can be complex. Vulnerabilities in smart contracts might lead to access control bypass.

The C-SACS system in [26] uses cloud-based interference assessment to identify interference before node transmission, addressing hidden and exposed node issues. However, the methodology needs more details on how this approach handles individual device characteristics or user permissions.

Zhang et al. [27] proposed an access control scheme leveraging blockchain technology and ABE for precise control, swift attribute revocation, and transaction-level modifications. Blockchain reliance may introduce scalability and performance challenges, and frequent updates and revocations could lead to increased computational and storage overhead.

The scheme outlined in [28] uses ABAC to manage unauthorized access, data tampering, and non-compliance in real-time Industrial IoT settings. It incorporates timing and task logic into ABAC attributes and multi-level hash identity authentication but needs to address noncompliant behavior, which may increase complexity for resource-limited IoT devices. The efficiency of access revocation for numerous devices is also a concern.

The scheme introduced in [29] enhances CP-ABE key delegation with specific content and permissions restrictions. It includes limitations on delegation frequency, delegatee revocation, and tracing prohibited delegations. However, the efficiency of the revocation process and associated computational costs require further assessment.

Table 1 shows comparison of the proposed framework with the state-of-the art methods for IoT access control.

Table 1 Comparison with state-of-the-art methods

Schemes	Delegation assisted by third party	Control on delegation chain length	Control on further delegation	Consider trust score	Delegation of permission level	Delegation revocation	Delegation flexibility	Granularity of access control
Aghili et al. [14]	×	×	×	×	✓	×	Moderate	Fine-grained
Bhatt et al. [15]	×	×	×	×	×	×	Low	Moderate
Choksy et al. [16]	×	✓	✓	✓	×	CA	High	Moderate
Thomaz [17]	×	×	×	×	×	×	Moderate	Moderate
Li et al. [18]	×	✓	×	×	×	×	High	Moderate
Fugkeaw et al. [19]	×	✓	✓	×	×	×	High	Fine-Grained
Bagga et al. [20]	×	×	×	×	×	×	Low	Moderate
Kousalya et al. [21]	×	×	×	×	×	×	Moderate	Moderate
Sun et al. [22]	✓	×	×	×	✓	×	Moderate	Fine-Grained
Tian et al. [23]	×	×	×	×	×	×	Moderate	Moderate
Alshehri et al. [24]	✓	×	×	×	×	×	Low	Moderate
Bakhtiary et al. [25]	✓	×	×	×	×	×	High	Fine-Grained
Adamczyk et al. [26]	✓	×	×	×	×	×	Moderate	Moderate
Zhang et al. [27]	✓	×	×	×	✓	×	Moderate	Moderate
Wang et al. [28]	×	×	×	×	×	×	Moderate	Moderate
Porwal et al. [29]	✓	×	×	×	✓	×	High	Fine-Grained
Sifou et al. [30]	×	×	×	×	×	×	Moderate	Moderate
Bhatt et al. [31]	×	×	×	×	×	×	Moderate	Moderate
Proposed MLADS scheme	×	✓	✓	×	✓	Users & AAs	Very High	Highly Fine-Grained

2.2 *Findings from Literature*

The reviewed literature highlights several common issues in access control frameworks for healthcare IoT environments. Considering the complexity of safely generating, distributing, and maintaining cryptographic keys—which can result in high administrative costs and possible vulnerabilities—complex key management becomes an important issue. Another issue that affects system performance is high computational overhead in frameworks that use complex cryptographic approaches like Attribute-Based Encyption (ABE). Depending on centralized systems, such as servers or attribute authority can cause issues such as limited scalability and single points of failure. Furthermore, the possibility of attribute collusion among entities may lead to breach of data confidentiality and integrity, highlighting the significance of strong security measures to successfully reduce these risks. To deal with these issues, there is a need for developing robust and effective access control solutions for healthcare solutions in IoT.

So, the common issues in CE-IoT access control methods are:

1. **Scalability and Performance issues due to**:
   - Large numbers of devices, attributes, and access policies [14, 18, 19, 24, 27].
   - Complex cryptographic operations can introduce overhead and computational challenges [17, 19, 20, 28].
2. **Complexity and Manageability**:
   - Managing fine-grained access control policies with numerous attributes can be complex for administrators and users [16, 21, 25, 30].
   - Techniques like ABE or blockchain might increase the overall complexity of resource-limited devices [19, 27, 28].
3. **Revocation Challenges**:
   - Efficient revocation of access for users or devices can be difficult, especially when privacy preservation is a concern [16, 19, 27, 29].
   - Revoking access might require updates to access control policies and potentially re-encryption of data, which can be computationally expensive for large deployments [27, 29].

As observed in existing literature, utilizing ABAC for access control proves advantageous in dynamic IoT environments, offering fine-grained access control and the potential for attribute-based delegation. However, this approach may result in a complex policy structure. We have to accept a trade-off for the complexity of policy structure to enforce efficient and fine-grained access control policies. Blockchains seem promising for policy enforcement, yet the expenses associated with policy updates and blockchain availability challenges render them inefficient for extensive IoT deployments unless a decentralized IoT system is specifically required.

Based on the literature review, ABAC proves to be more effective than other models in dynamic environments. Choksy et al. [16] presented a more robust and customizable delegation capability tailored for IoT healthcare, effectively addressing specific access control requirements. MLADS builds upon this by integrating multi-layered attribute-based access control mechanisms, enhancing security and delegation flexibility in cloud-enabled IoT environments. We aim to develop an access control and delegation framework based on ABAC for CE-IoT that provides fine-grained access control, controlled access delegation and revocation, and combined ABAC-based access and delegation policies to improve manageability and efficiency.

3 System Model, Security Model, and Security Requirements

Here, we have defined an ABAC-based framework for CE-IoT healthcare scenarios, discussing the system's architecture. We demonstrate the system and define the security requirements that must be adhered to for secure resource access.

3.1 System Model

We present a system model for a healthcare system in which CE-IoT-enabled gadgets monitor patients. These devices continuously generate healthcare data and are safely saved in the cloud. Each patient registered with a healthcare provider has a cloud-based Electronic Health Record (EHR) that they uniquely identify. Attribute-based policies control access to EHRs, guaranteeing that only authorized users, under specified circumstances, can access patient data. According to patient consent and policy compliance, users from various sectors, such as medical students, insurance agents, specialist health practitioners, international healthcare professionals, and nurses, may need access to these EHRs. Figure 1 depicts the components mentioned below:

- **IoT Healthcare Provider (IHP)**: IoT-enabled devices such as blood pressure monitors, heart rate monitors, and temperature sensors continuously monitor vital signs. Data is securely transmitted to a centralized cloud infrastructure, ensuring the integrity and confidentiality of critical health metrics essential for real-time patient assessment.
- **Attribute Authority (AA)**: The AA within the MLADS framework is responsible for managing and assigning attributes to users, devices, and resources involved in the patient monitoring system. We consider AAs as trusted entities. IHP will be responsible for assigning environmental attributes (like location).
- **Users**: Healthcare providers (doctors, nurses), technicians for IoT devices, and administrators utilize IHP resources using specific attributes within the MLADS

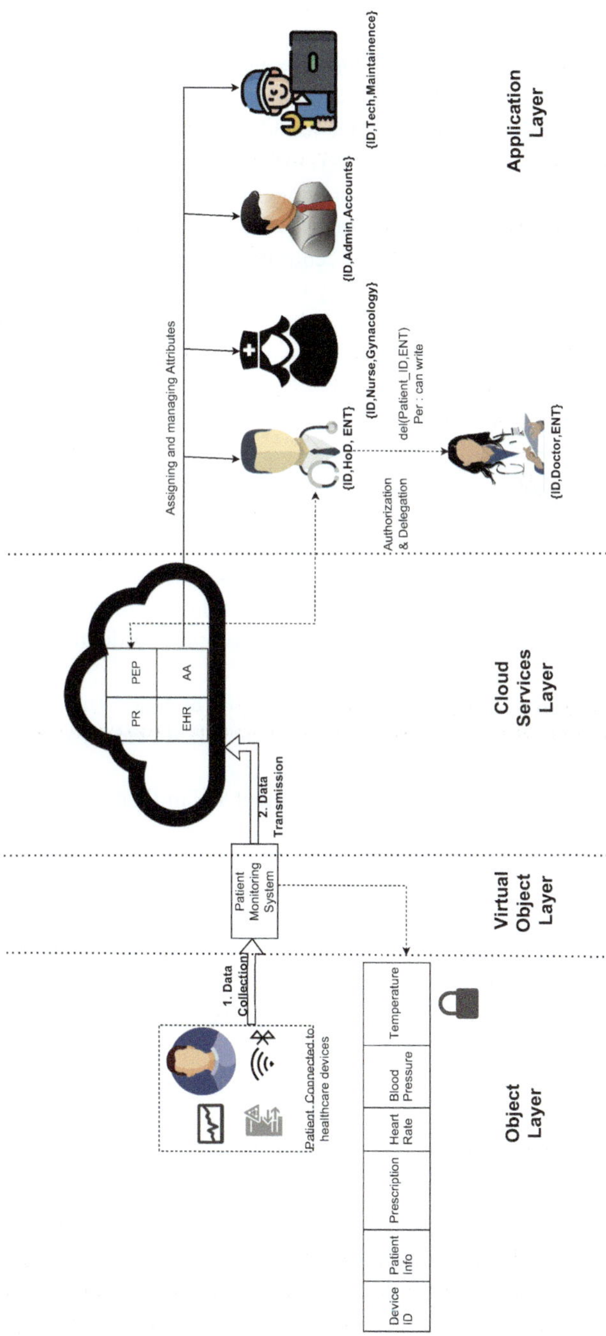

Fig. 1 System architecture

framework. Attributes managed include user roles, device types, patient identifiers, and contextual data (e.g., time, location), managed and assigned by the AAs.
- **Cloud Infrastructure**: In the proposed system framework, patients' personal information and sensitive data are collected by IHS and stored in a secure cloud such as EHS. The data is analyzed using this cloud architecture to get insightful information. Predefined policies are the foundation of the access control paradigm and are likewise kept in the cloud. Therefore, a secure cloud is necessary for our system model to guarantee patient data integrity and security and the efficacy of access control mechanisms. Within the MLADS system, the cloud infrastructure provides scalability, availability, and centralized policy management by hosting the Policy Decision Point (PDP), Policy Enforcement Point (PEP), and AAs.

3.2 Security Model

The security model of MLADS has following features and assumptions:

- **Trusted Secure Party**: IHP and AA are considered trusted and secure and tasked with managing and overseeing the system.
- **Data Encryption and Secure Communication**: Industry-standard- encryption (TLS/SSL for transit, AES-256 for rest) protects data from unauthorized access. Secure, lightweight protocols ensure data integrity and confidentiality during transmission.
- **Authentication and Authorization**: Strong authentication (multi-factor, digital certificates) verifies user and device identities. ABAC enforces access policies based on attributes for authorized data access.
- **Intrusion Detection and Prevention**: To detect and mitigate threats, with regular security audits and vulnerability assessments ensuring resilience.
- **Delegation and Revocation**: Defined policies allow authorized delegation of access permissions, with on-demand revocation ensuring control.
- **Preventing Unauthorized Access and Collusion**: Measures are in place to prevent users from accessing more data than permitted, forging attributes, performing replay attacks, and colluding for unauthorized access.
- **Regulatory Compliance**: Compliance with HIPAA, GDPR, and other standards is maintained through regular policy updates.

3.3 Security Requirements

Here, we have defined the essential security requirements that an access control framework for a CE-IoT-based healthcare system must fulfill as following:

- **Patient Privacy**: EHRs saved in the cloud contain highly sensitive and private data that must be shared with proper authorization. Access to EHRs should strictly

adhere to access policies based on the attributes possessed by users. Delegation of access rights should only be valid if authorized by the delegator and must not violate the access policies defined for the EHRs.
- **Robust Attribute Management**: AAs must maintain and manage attributes effectively, ensuring they accurately reflect the current state of users' permissions and preventing unauthorized access attempts.
- **Resistance to Attribute Collusion**: Resistance to Attribute Collusion Users must not collude to combine their attributes to gain unauthorized access. The system should detect and prevent attempts to exploit attribute combinations from multiple users to bypass access controls. Attribute collusion is considered invalid and should be effectively mitigated.
- **Controlled Access Delegation**: The delegator should be able to specify and control the delegation process. While delegation is an essential feature of ABAC, constant interaction with AAs should not be required for each delegation action to ensure operational efficiency.
- **On-Demand Attribute Revocation**: The framework must support attribute revocation mechanisms managed by AAs and authorized users, al- allowing immediate revocation of access when attributes expire or become invalid.

4 Preliminaries

4.1 User Attributes

In the proposed MLADS framework, attributes are crucial in defining and enforcing access control policies. Attributes can be categorized into several types based on their nature and usage within the system.

Static Attributes These attributes are typically fixed and do not change frequently. Examples include user roles (e.g., doctor, nurse, administrator), device types (e.g., blood pressure monitor, heart rate monitor), and patient identifiers (e.g., patient ID, medical record number). Static attributes provide foundational information used in access control decisions.

Dynamic Attributes These attributes may change over time or based on specific conditions. They often include contextual information such as time of access, location, and patient condition (e.g., critical status). Dynamic attributes enhance the granularity of access control by allowing policies to adapt to changing circumstances in real time.

Delegated Attributes Integral to the framework is the concept of delegated attributes, which allow users to transfer access permissions to other users temporarily. Delegated attributes enable flexible access management, facilitating scenarios where, for instance, a senior physician delegates access to patient records to a junior colleague for a defined period or purpose. This feature supports collaborative care while ensuring access remains controlled and compliant with organizational policies and regulatory requirements.

4.2 Structure of User Attributes

Here, the attributes assigned to users are presented as assignment and delegation tokens.

User Assignment Token As depicted in Fig. 2 the Assignment Token is a cryptographic token issued by the AA within the MLADS framework. It encapsulates the attributes assigned to a specific user, device, or resource, ensuring secure and verifiable authorization decisions. The token comprises a header, payload, and signature. The header includes a unique Token ID, the AA's identifier, issuance timestamp, and expiration timestamp. The payload contains the subject's unique identifier (user, device, or resource), a list of assigned attributes (roles, permissions, and contextual information like location and time), and a reference to applicable ABAC policies. The signature is a digital signature generated by the AA using its private key, ensuring the token's integrity and verifying its issuance by the trusted AA. Strong encryption protocols (e.g., TLS for data in transit and AES-256 for data at rest) and digital signatures (e.g., RSA or ECDSA) ensure the token's security.

User Delegation Token As represented in Fig. 3 the delegation Token facilitates the secure delegation of access rights and attributes within the MLADS framework. It enables a user (delegator) to delegate specific permissions to another user (delegatee) while ensuring the security and verifiability of the delegated rights. The token includes a header, payload, and signature. The header features a unique Token ID, the delegator's identifier, issuance timestamp, and expiration timestamp. The payload contains the unique identifiers of both the delegator and delegatee, a list of delegated attributes and permissions, conditions or constraints under which the permissions can be exercised (e.g., time-bound or location-specific), and a reference to govern-

Header					Assigned Attributes			Signature
ID	Issued By	Issued To	Issue Date	Expiry Date	User	Device	Contextual	

Fig. 2 Structure of user assignment token

Header					Attributes		Revocation Condition		Signature
ID	Issued By	Issued To	Issue Date	Expiry	User	Conditions	Revocation Policy	Immediate Revocation	

Fig. 3 Structure of user delegation token

ing ABAC policies. The digital signature is generated by the delegator using their private key, ensuring the token's integrity and authenticity. Strong encryption protocols for secure transmission and robust digital signatures are essential for maintaining the token's security and integrity.

4.3 Composition of Delegated Attributes

As described above, Attributes in the MLADS framework are categorized as user attributes, device attributes, and contextual environment.

Structure of Assigned Attributes Each attribute in the MLADS framework is represented by a tuple containing several elements:

- **Attribute Identifier (Attr_ID)**: A unique identifier for the attribute.
- **Attribute Type (Attr_Type)**: Specifies the attribute type (user, device, contextual).
- **Attribute Value (Attr_Value)**: The actual value of the attribute (e.g., "doctor" for a user role).
- **Issuer Identifier (Issuer_ID)**: Identifies the entity that issued the attribute.
- **Validity Period (Valid_From, Valid_To)**: Specifies the period during which the attribute is valid.
- **Signature (Sig_Issuer)**: A digital signature from the issuer to ensure the authenticity and integrity of the attribute.

The formal structure can be defined as

$$\begin{aligned}\texttt{Attribute} = (&\texttt{Attr\_ID}, \texttt{Attr\_Type},\\ &\texttt{Attr\_Value}, \texttt{Issuer\_ID},\\ &\texttt{Valid\_From}, \texttt{Valid\_To},\\ &\texttt{Sig})\end{aligned} \tag{1}$$

Delegation of Attributes Supervising and monitoring delegation ensures the delegation of attributes is secure and authorized. A delegation token is sent with every delegation and consists of:

- **Delegator Identifier (Delegator_ID)**: The identifier of the user who is delegating the attributes.
- **Delegatee Identifier (Delegatee_ID)**: The identifier of the user to whom the attributes are being delegated.
- **Delegated Attributes (Delegated_Attrs)**: The set of attributes being delegated.
- **Delegation Validity Period (Del_Valid_From, Del_Valid_To)**: The validity period for the delegated attributes.
- **Delegation Signature (Sig_Delegator)**: A digital signature from the delegator to ensure the authenticity and integrity of the delegation, which can be formalized as

$$\begin{aligned}&\texttt{Delegation\_Token = (Delegator\_ID, Delegatee\_ID,}\\&\texttt{Delegated\_Attrs,Attr\_Value, Del\_Valid\_From,}\\&\texttt{Del\_Valid\_To, Sig\_Delegator)}\end{aligned} \tag{2}$$

4.3.1 Aggregated Chains

Aggregated chains in the MLADS framework represent the sequence of attribute assignments and delegations, providing a traceable path of access control decisions. Consider a scenario as described below where a senior doctor delegates access rights to another doctor, who may further delegate some rights to a nurse. Each delegation step is captured in an aggregated chain, ensuring traceability and accountability:

1. **Initial Attribute Assignment**:
   - The AA assigns initial attributes to the senior doctor.
   - Let A_{SD} represent the set of attributes assigned to the senior doctor.
   - The assignment token T_{SD} is generated as

$$\texttt{T\_SD = Enc\_K\_SD(ID\_SD,A\_SD, t\_issue, t\_exp, Sig\_AA)} \tag{3}$$

2. **Delegation from Senior Doctor to Another Doctor**:
   - The senior doctor delegates a subset of their attributes to another doctor.
   - Let $A_D \subseteq A_{SD}$ represent the attributes delegated to the other doctor.
   - The delegation token T_D is generated as

$$\begin{aligned}&\texttt{T\_D = Enc\_K\_D (ID\_SD, ID\_D, A\_D,}\\&\texttt{t\_issue, t\_exp, Sig\_SD)}\end{aligned} \tag{4}$$

3. **Delegation from Doctor to Nurse**:
   - The doctor delegates a subset of their attributes to a nurse.
   - Let $A_N \subseteq A_D$ represent the attributes delegated to the nurse.
   - The delegation token T_N is generated as

$$\texttt{T\_N = Enc\_K\_N (ID\_D, ID\_N, A\_N, t\_issue, t\_exp, Sig\_D)} \tag{5}$$

Each token in the aggregated chain needs verification for the integrity and authenticity of delegation. Digital signatures and attribute validity must be verified along with their delegation.

The aggregated chain can be represented as a sequence of tokens:

$$\texttt{Chain = \{T\_SD, T\_D, T\_N\}} \tag{6}$$

Each token is connected through delegation, showing a full record of attributes and delegations.

The verification function γ for each token in the chain is

$$\gamma(T_i) = \begin{cases} 1 & \text{if } \mathrm{Sig}(i, d_i) \text{ is valid and } A_i \geq A(d_i) \\ 0 & \text{otherwise} \end{cases} \tag{7}$$

The entire chain is considered valid if

$$\gamma_{\text{chain}} = \prod_{i=1}^{n} \gamma(T_i) = 1 \tag{8}$$

indicating that all tokens in the aggregated chain are valid and correctly delegated.

4.4 Cryptographic Background

The MLADS framework relies on various cryptographic techniques to ensure healthcare data's security, integrity, and confidentiality. Symmetric encryption using AES-256 helps secure data at rest and during transmission [32]. Thus, the EHR and communication between cloud and IoT devices are secured. Asymmetric encryption, utilizing RSA with 2048-bit keys, provides secure key exchange and confidentiality [33] during attribute assignment and delegation, safeguarding communication between Attribute Authorities (AAs) and users. Digital signatures, generated through algorithms like DSA or RSA, assure the authenticity and integrity [34] of assignment and delegation tokens, ensuring that the source is trusted. Additionally, secure hash functions such as SHA-256 generate signatures for data [35], detecting any tampering during storage or transmission. Moreover, secure communication protocols like SSL and TLS encrypt data in transit [36], safeguarding attribute management and access control decisions. As a collective effect of these cryptographic methods, the MLADS framework protects sensitive healthcare data, maintaining confidentiality, security, and authorized access, vital for trust and compliance in healthcare settings.

4.5 Resources and Their Access Policy

4.5.1 Structure of Resources

We will primarily discuss the cloud-stored EHR within the MLADS framework. Sensitive and private patient healthcare data is stored in every EHR. The structure of these resources is designed to facilitate management, utilization, access, security, and efficiency. The resource structure's key elements are:

1. **Resource Identifier (RID)**: Every EHR is given a unique identification to help distinguish it uniquely from other records. Through the RID, access requests are accurately forwarded to the intended resource.
2. **Patient Information**: Personal information like name, age, gender, and phone number are included in this. This information is essential to identify patients and link their identity to their EHR.
3. **Medical Data**: This contains several health indicators, including blood pressure, heart rate, temperature, prescriptions, past medical history, test results, and diagnosis. This information is essential for medical personnel to evaluate the patient's condition and administer the necessary care.
4. **Access Control Attributes**: These attributes specify the EHR-related access policies. They provide details regarding certain access permissions, the patient's consent, and the circumstances in which the data is accessible.
5. **Audit Logs**: This maintains records of the EHR's updates and successful and attempted accesses. In order to verify compliance with security policies and regulatory requirements, these logs are essential for monitoring and auditing access.

4.6 Access Policies

Access policies represent the required criteria for access to a particular EHR object, identified as "EHRid.obj". Policies are defined for each "EHRid.obj" object. One or more policies can be defined per "EHRid.obj". An access control policy P is defined as

$$P = \{T, A\} \tag{9}$$

wherein **T** is a Boolean equation with attributes related to disjunction ($\vee$) and conjunction ($\wedge$). **A** represents the three operations that can be performed on the object: read, write, and edit. The user can access the data only when one of the set policies for that particular item is satisfied; otherwise, the request is denied.

Example Policies:

1. **Doctor Access to Patient EHRs**

$$\begin{aligned} P = \{&\text{Role} = \text{Doctor}\wedge \\ &\text{Specialization} = \text{Cardiology}\wedge \\ &\text{PatientID} \in \text{AssignedPatients}\} \end{aligned} \tag{10}$$

2. **Nurse Access to Medical Data During Working Hours**

$$P = \{\text{Role} = \text{Nurse} \wedge \text{DataSensitivity} = \text{Low} \wedge \text{AccessTime} \in \text{ShiftHours}\} \tag{11}$$

3. **Medical Student Access for Educational Purposes**

$$P = \{\text{Role} = \text{MedicalStudent} \wedge \\ \text{Purpose} = \text{Educational} \wedge \\ \text{DataAnonymized} = \text{True}\} \tag{12}$$

4. **Insurance Agent Access for Claims Processing**

$$P = \{\text{Role} = \text{InsuranceAgent} \wedge \\ \text{PatientConsent} = \text{True} \wedge \\ \text{DataForClaimsProcessing} = \text{True}\} \tag{13}$$

5. **Delegation of Access from Senior Doctor to Assistant Doctor**

$$P = \{\text{Role} = \text{SeniorDoctor} \wedge \\ \text{Delegation} = \text{True} \wedge \\ \text{DelegateeRole} = \text{AssistantDoctor}\} \tag{14}$$

Consider an example of the policy for a patient with id xyz and object oxygen level (OL), denoted as EHRxyz, OL. The access control policy P could be expressed as

$$P = \{F(A)\} \quad \text{where} \quad T = (\text{Role} = \text{Doctor} \wedge \text{Specialization} = \text{Pulmonology}) \\ \vee (\text{Role} = \text{Any}) \\ A = \{\text{'read'}, \text{'edit'}\} \tag{15}$$

5 Proposed Framework

As discussed in Sect. 3, the MLADS framework ensures secure and efficient access control in CE-IoT healthcare environments. Below is a detailed description of the framework, which incorporates various AAs and utilizes advanced cryptographic techniques to manage attribute assignments and delegations.

The cryptographic terms used in the MLADS framework are represented in Table 2, which provides a clear understanding of the symbols and operations involved in the scheme.

In the MLADS framework, AAs manage and distribute attributes based on their specific domains. Each AA performs the following tasks:

1. **Secret Key Initialization**: Each AA selects a secret key SK_{AA} and publishes a corresponding public key PK_{AA}.
2. **Attribute Secret Components**: For each attribute they manage, AAs generate a secret component SC_{attr}.

3. **PKI Certificates**: All AAs and users have pre-loaded PKI certificates, which include their public and private keys.

Users in the MLADS framework can become delegators, effectively acting as virtual AAs during the delegation phase. Each delegator initializes themselves with their public–private key pair and selects a secret element for each attribute they wish to delegate. The delegation process is as follows:

1. **Delegator Initialization**: A user (delegator) initializes their keys:
   - Private Key: SK_D
   - Public Key: PK_D
2. **Attribute Delegation**: The delegator chooses secret components for the attributes to be delegated and signs these attributes using their private key.

Attributes are assigned and delegated in the form of signed tokens. These tokens include assignment tokens and aggregated tokens:

1. **Assignment Tokens**: Signed by the AA, these tokens represent the initial attribute assignments to users.
   - Assignment Token: $\tau_{\text{assign}} = \text{Sign}_{SK_{AA}}(A_i)$
2. **Aggregated Tokens**: During delegation, attributes are combined and signed by the delegator, resulting in aggregated tokens.
   - Aggregated Token: $\tau_{\text{agg}} = \text{Sign}_{SK_j}(A_i \| \text{Delegatee Info})$

The MLADS framework supports on-demand attribute revocation by altering the secret component of a specific attribute. This revocation only affects the particular attribute without impacting the entire system.

The AA or delegator can update the secret component SC_{attr} to revoke the attribute. $SC_{\text{attr}} \rightarrow SC_{\text{attr}}^{\text{new}}$.

Users who wish to access EHR must request the PEP and authenticate with the AA using their PKI certificates. The process is as follows:

1. **Mutual Authentication**: Users authenticate using pre-loaded PKI certificates, ensuring communication between legitimate entities.
2. **Access Request**: The authenticated user submits an access request to the PEP, specifying the EHR object ID `EHR_id_obj` and the desired action (read, write, edit).

Table 2 Cryptographic notations for MLADS

Symbol	Description
SK_AA	Secret key of the AA
PK_AA	Public key of the AA
SC_attr	Secret component for a specific attribute managed by the AA
SK_p	Private key of the delegator
PK_D	Public key of the delegator
T_assign	Assignment token, signed by the AA (T_assign = Sign_SK_AA(A))
T_agg	Aggregated token, signed by the delegator (T_agg = Sign_SK_D(A ‖ Delegatee Info))
EHR_id_obj	Electronic Health Record object ID
PEP	Policy Enforcement Point
PR	Policy Repository
P	Access control policy
T	Boolean equation of attributes derived from policy
Sign_SK_AA	Digital signature using the AA's secret key
Sign_SK_D	Digital signature using the delegator's secret key
attr'	New secret component after revocation
PKI	Public Key Infrastructure

3. **Policy Retrieval and Verification**:
   (a) The PEP forwards the request to the Policy Repository (PR), which retrieves the relevant access policy `P`.
   (b) The PEP generates a Boolean equation `T` from the policy and provides it to the user.
   (c) The user selects a selection of attributes that fulfill the Boolean equation and sends the signed tokens to the PEP.
4. **Token Verification**:
   (a) **Primary Verification**: The PEP verifies each token to ensure its validity and integrity.
   (b) **Secondary Verification**: Tokens undergo additional verification to check for any revocations by the AA or delegator.
   (c) **Access Decision**: Based on the verification results, the PEP either grants or denies access to the requested EHR object.

5.1 Details of Proposed Framework

The ECC curve used in the proposed framework leverages ECC-based digital signatures for signing and verifying tokens. ECC provides strong security with smaller key sizes compared to other cryptographic algorithms. It is designed to offer robust and adaptable access control mechanisms, including attribute management, token formation, and validation processes, as outlined in the following subsections.

5.2 Initialization

In the proposed framework, we use three main types of attributes: identity attributes, role attributes, and context attributes. All of them are crucial in defining access control policies to ensure resource access only to authorized users. Identity attributes uniquely identify users, containing attributes like user IDs or biometric data. Role attributes assign specific roles to users, characterizing roles such as doctor, nurse, or patient. Context attributes provide contextual details necessary for allowing or denying access, e.g., time and location. The underlying structure of assigned attributes allows effective management and utilization in the framework. Each attribute's type, value, validity period, and unique identifier define it. For authentication and integrity, attributes are signed using ECC. The Service for User Profiles AA is in charge of managing user profiles and assigning and modifying characteristics. When users want access to resources, AAs provide them with the necessary attributes.

5.3 Token Generation

AAs issue assignment tokens, signed using the AA's private key to ensure authenticity and integrity. The tokens include attributes assigned to a user.

Assignment tokens issued by AAs, similarly, delegation tokens are generated when a user delegates access rights to another, containing the delegator's attributes, the delegatee's identity, and the delegation permissions, and are digitally signed to ensure authenticity.

5.4 Access Decisions

Access decisions in the proposed scheme are determined by evaluating a user's attributes against predefined access policies. Let U be a user and $A_U = \{a_1, a_2, \ldots, a_n\}$ be the set of attributes associated with U. Each Resource R_i has an associated access control policy P_i which can be expressed as a Boolean function over the set of attributes. Mathematically, let P_i be the access policy for resource R_i. Then, $P_i(A_U) = T$, here T is a Boolean equation with attributes related logical disjunction ($\vee$) and conjunction ($\wedge$). Here, T is of the form: $T = \bigwedge_{i=1}^{n}(A_{v_i} \vee A_{\neg v_i})$, where $\bigwedge$ and $\vee$ represent the logical AND and OR operations, respectively. A stands for the available actions (read, write, and edit) on the object. The user can access data only if one of the established policies for that specific object is met; if not, the request is rejected.

The user U is granted access to R_i if and only if $P_i(A_U) = \text{True}$. The access decision process can be formally described by the following equation:

$$\text{Access} = \begin{cases} \text{Granted} & \text{if } P_i(A_U) = \text{True} \\ \text{Denied} & \text{otherwise} \end{cases} \tag{16}$$

5.5 *Automatic Delegation*

Automatic delegation in our scheme allows a user (delegator) to delegate certain access rights to another user (delegatee). Let D represent the delegation process, PL be the permission level, and DC be the delegation count.

The delegation token DT can be represented as

$$DT = (PL, DC, EH_{R_{\text{id,obj}}}, T_p) \tag{17}$$

where T_p is the delegation timestamp.

The delegation process involves the following steps:

1. Validate the existing token.
2. Check if further delegation is allowed by evaluating DC.
3. Assign the new permission level PL' and update DC.
4. Generate the delegation token DT.

Mathematically, the delegation process can be expressed as

$$\begin{aligned} DC' &= DC - 1 \\ PL' &< PL \\ DT' &= (PL', DC', EH_{R_{\text{id,obj}}}, T_b) \end{aligned} \tag{18}$$

5.6 *On-Demand Attribute Revocation*

On-demand attribute revocation ensures that invalid or expired attributes are revoked, preventing unauthorized access. Let $A_u = \{a_1, a_2, \ldots, a_n\}$ be the set of attributes associated with user U, and R_A be the revoked attribute set.

The attribute revocation process can be described as follows:

1. Identify the attributes to be revoked.
2. Update the user's attribute set by removing the revoked attributes.
3. Ensure the updated attribute set satisfies the access policy.

Mathematically, let A'_u be the updated attribute set after revocation:

$$A'_u = A_u \setminus R_A \tag{19}$$

The revocation process can be expressed as

$$A'_u = \{a \in A_u \mid a \notin R_A\} \tag{20}$$

If $P(A'_u) = \text{True}$, access is granted; otherwise, access is denied. The access decision post-revocation is given by

$$\text{Access} = \begin{cases} \text{Granted} & \text{if } P(A'_u) = \text{True} \\ \text{Denied} & \text{otherwise} \end{cases} \tag{21}$$

5.7 *Validation*

Primary validation is performed by the Policy Enforcement Point (PEP) to verify the authenticity and integrity of the tokens submitted by the user. This involves checking the digital signatures and the validity periods of the tokens, represented as

$$PV = \text{Verify}\,(\text{Signature}_{AA}, AT) \wedge \text{Verify}\,(\text{Signature}_{\text{Delegator}}, DT) \tag{22}$$

Secondary validation ensures that the attributes within the tokens meet the access control policies defined for the requested resource. This involves evaluating the Boolean equations representing the policies represented as

$$\text{SV} = \text{Evaluate}\,(\text{Policy}, \text{Attributes}). \tag{23}$$

6 Security Analysis

In this section, we assess the suggested access control framework's security and show how our method can resist unauthorized access to cloud-hosted patient healthcare data. Here, we outline the hardness issue and underlying assumptions upon which the security of our framework rests. We investigate a the suggested strategy's security features:

Theorem 1 *The MLADS framework ensures data confidentiality and integrity through the use of industry-standard encryption protocols and secure communication channels.*

Proof MLADS secures data both in transit and at rest using industry-standard cryptographic techniques. For data in transit, TLS/SSL encrypts plaintext data M into ciphertext $E_{\text{TLS/SSL}}(M)$, ensuring that decryption without the corresponding keys is computationally infeasible. For data at rest, AES-256 encryption transforms plaintext M using key K into ciphertext $C_E = \text{AES-256}(K, M)$, similarly safeguarding against unauthorized decryption. Data integrity during transmission is maintained through cryptographic hash functions like SHA-256, where $H(M)$ represents the hash of message M. Any alteration in M results in a different $H(M)$, enabling tampering detection. Secure lightweight protocols, ensuring both data integrity and confidentiality, are represented as $P(M) = [E_{\text{TLS/SSL}}(M), H(M)]$

Theorem 2 *The MLADS framework ensures strong authentication and authorization through multi-factor authentication, digital certificates, and ABAC.*

Proof MLADS secures authentication using multi-factor authentication (MFA), where multiple independent factors $A_1, A_2, \ldots, A_n$ are required, significantly reducing the probability of a breach. Digital certificates, leveraging public key infrastructure (PKI), include a user's certificate $Cert_U(PK_U)$ with their identification ID_U and a Certificate Authority's signature $Sig_{CA}(Cert_U(PK_U))$, ensuring the authenticity of public keys. ABAC further enhances access control through policies expressed as Boolean expressions T, granting access if a user possesses attributes $a_1, a_2, \ldots, a_m$ that satisfy T, thereby minimizing unauthorized access. Together, MFA, digital certificates, and ABAC ensure robust authentication and authorization.

Theorem 3 *The MLADS framework provides effective delegation and revocation control, ensuring secure attribute delegation and timely revocation.*

Proof Delegation in MLADS is managed through roles where D is the delegator, d is the delegatee, and A is the attribute. Delegation is denoted as $D \rightarrow d : A$, with the delegation token T_{deleg} comprising $\{A, Sig_D(A), Exp\}$, where $Sig_D(A)$ is the delegator's signature on A and Exp is the expiration time. The delegatee d validates delegated access using T_{deleg}, verifying $Sig_D(A)$ and Exp. Revocation is managed by updating the attribute's secret component s to s', with the revocation represented as $Revoke(A, s')$, where $s' \neq s$; tokens using the outdated s become invalid. Immediate updating of s ensures minimal probability of using a revoked attribute, ensuring effective delegation and revocation control.

Theorem 4 *The MLADS framework prevents unauthorized access and attribute collusion, ensuring secure and exclusive access control.*

Proof The MLADS framework prevents unauthorized access through well-defined access control policies P_{auth}. Access is granted only if users possess attributes $\{a_1, a_2, \ldots, a_m\}$ that satisfy P_{auth}. If a user lacks the required attributes, P_{auth} evaluates to false, denying access. Attribute collusion is prevented by centralized control mechanisms: users U_1 and U_2 with distinct attribute sets $\{a_1, a_2\}$ and $\{a_3, a_4\}$, respectively, are prevented from colluding by ensuring that combined attributes $\{a_1, a_2, a_3, a_4\}$ are invalid due to centralized attribute management tied to individual user identities. Thus, MLADS effectively safeguards against unauthorized access and attribute collusion.

Theorem 5 *The MLADS framework prevents signature forgery.*

Proof ECC is used for digital signatures due to its strong security properties. The security against signature forgery is based on the difficulty of the Elliptic Curve Discrete Logarithm Problem (ECDLP). To generate a digital signature, the following process is used let S be the private key, let T be the public key attribute token, let H be a cryptographic hash function. The signature σ is computed as $\sigma = S \cdot H(T)$. For verification, the equation $P \cdot H(T) = \sigma$, where P is the public key. ECC ensures that without knowing S, it is computationally infeasible to generate a valid signature σ, making this scheme resistant to forgery.

Theorem 6 *The proposed framework resists impersonation and replay attacks.*

Proof Impersonation attacks are prevented through mutual authentication using preloaded PKI certificates. Let U be a user, pu be the user's public key, and su be the user's private key. During the authentication phase, the user proves their identity by generating a signature $\sigma_U = su \cdot H(M)$, where M is a challenge message. The verifier checks the validity by computing $Pu \cdot H(M) = \sigma_U$. Replay attacks are mitigated using nonces and timestamps within the authentication protocol. Each request includes a unique nonce N and a timestamp T_N, ensuring that the same message cannot be reused.

Theorem 7 *The proposed scheme guards against collusion of assignment tokens.*

Proof In MLADS, each assignment token is tied to specific attributes and is digitally signed by the issuer. Suppose users U_1 and U_2 attempt to collude by combining their tokens. Each token T_i is signed individually as $\sigma_i = s^A \cdot H(T_i)$. For Verification we verify $P_A \cdot H(T_i) = \sigma_i$. Without the issuer's private key s^A, users cannot create a valid aggregated token T_{agg} that combines attributes from U_1 and U_2. Thus, the scheme guards against collusion.

Theorem 8 *The proposed scheme provides control against the delegation of attributes.*

Proof In MLADS, attribute delegation is controlled by the delegator's policies. Suppose a delegator D wishes to delegate attribute A to delegatee E. The delegation token T_d1 includes $T_d1 = \{A, P_D, P_E, \sigma_D\}$, where $\sigma_D = s_D \cdot H(A \| P_E)$. Verification involves checking $P_D \cdot H(A \| P_E) = \sigma_D$. The control mechanism ensures that only authorized attributes are delegated, and any unauthorized delegation attempts will fail verification.

Theorem 9 *The proposed scheme is secure against unauthorized access of objects using delegation tokens.*

Proof Access to objects using delegation tokens is verified by the Policy Enforcement Point (PEP). Let O be the object, and T_d be the delegation token for E. Verification checks $P_D \cdot H(O \| P_E) = \sigma_D$. If T_d passes verification and matches the policy requirements, access is granted. Otherwise, access is denied, ensuring security against unauthorized access.

Theorem 10 *The proposed framework enforces patient privacy.*

Proof Patient privacy in MLADS is enforced through fine-grained access control policies and encrypted data storage. Access to EHR is governed by policies that require specific attributes. Let P_{privacy} be the privacy policy $P_{\text{privacy}} = (A_{\text{doctor}} \wedge A_{\text{specialization}}) \vee (A_{\text{nurse}} \wedge A_{\text{department}})$. Only users with valid attributes satisfying P_{privacy} can access the EHR, ensuring patient privacy is maintained.

Theorem 11 *The proposed framework provides security against malicious behavior of legitimate users.*

Proof MLADS monitors and logs all access requests and delegations. Each action is associated with the user's public key and signed tokens. In case of malicious behavior, the logs provide an audit trail. The auditing process verifies $\sigma = S_U \cdot H(\text{request})$ and $P_U \cdot H(\text{request}) = \sigma$. If malicious behavior is detected, the user's access can be revoked and their tokens invalidated, maintaining overall system security.

Theorem 12 *The MLADS framework ensures resistance against man-in-the-middle (MITM) attacks.*

Proof In MLADS, resistance against MITM attacks is ensured through a combination of cryptographic protocols and secure communication mechanisms. Each user U_i possesses a unique public–private key pair (PK_{Ui}, SK_{Ui}) and a digital certificate $Cert_{Ui}$ issued by a Certificate Authority (CA), which includes their public key PK_{Ui} to authenticate their identity. Mutual authentication between user U_i and server S is achieved through the exchange and verification of digital certificates. Subsequently, a session key K_{session} is established via the Diffie–Hellman key exchange mechanism, ensuring that no third party can eavesdrop on or modify messages [37]. During this process, both parties generate nonces (N_U and N_S) to verify each other's identity. Verified nonce signatures are exchanged using asymmetric encryption, where private keys sign the nonces to prevent unauthorized spoofing. The Diffie–Hellman key exchange mechanism involves transmitting "key generation" data pre-supplied by both parties before any personal data exchange, minimizing threats to data integrity. For every user attribute U_1 assigned, an assignment token $Token_{\text{attribute}}$ is generated (e.g., for the attribute "Organization," $Token_{\text{attribute}} =$ "XYZ"). Thus, the combination of PKI architecture, mutual authentication, nonce verification, and secure data transmission mechanisms within MLADS ensures robust resistance against MITM attacks.

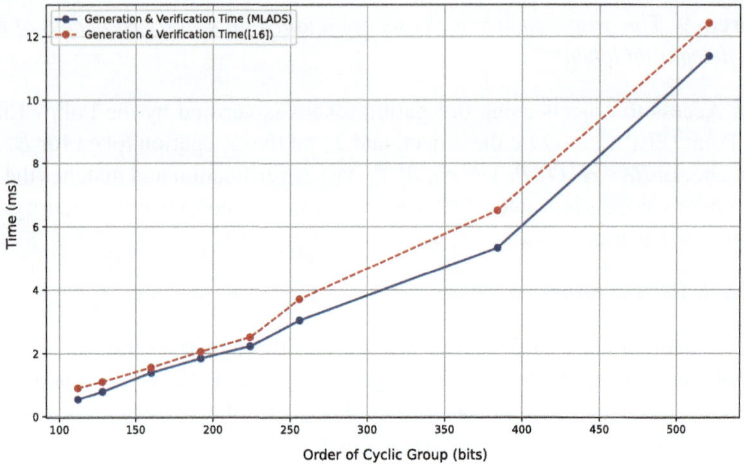

Fig. 4 Variation of computation time in relation to signed assignment token with group order

Table 3 Performance metrics for cryptographic operations at various bit lengths

Order of cyclic group (bits)	Generation time of signed token (MLADS) (ms)	Verification time of signed token (MLADS) (ms)	Generation & verification time of signed token [16] (ms)
112	0.15	0.4	0.9
128	0.2	0.6	1.1
160	0.3	1.1	1.55
192	0.35	1.5	2.05
224	0.43	1.8	2.50
256	0.62	2.42	3.70
384	0.9	4.43	6.50
521	2.40	8.97	12.40

7 Performance Analysis

This section presents a comparative performance analysis of the proposed MLADS framework against the controlled access delegation scheme in [16] based on computational cost. To address concerns about the complexity and feasibility of the proposed scheme, we provide a detailed explanation alongside results to demonstrate its efficiency and practicality. The computational costs of generating and verifying signed assignment tokens and aggregated tokens have been evaluated using simulations. The simulations were conducted on a machine with the following specifications: Processor: 11th Gen Intel(R) Core(TM) i5-11400H @ 2.70 GHz, Installed RAM: 8.00 GB, System Type: 64-bit operating system, x64-based processor.

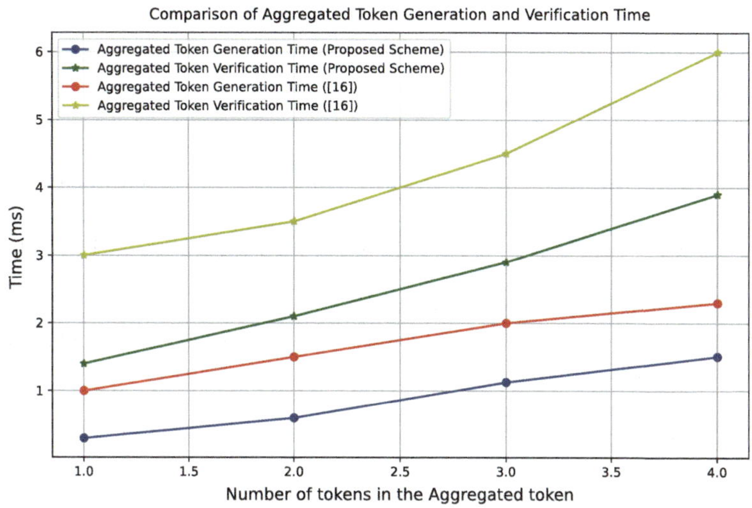

Fig. 5 Comparison of aggregated token generation and verification time

We used elliptic curve cryptography (ECC) with the elliptic curve defined as $y^2 = x^3 + ax + b$ over a 512-bit finite field. Each attribute token is assumed to be 2000 bits, which accommodates both the issuer's and the receiver's public keys, given the assumption of 512-bit ECC public keys. The x-axis in Fig. 4 represents the order of cyclic groups in bits, while the y-axis represents the average computational time of signed token in milliseconds. The simulation results show that the computational cost increases with the order of the cyclic group. The computation cost stays within acceptable limits in spite of this rise. For instance, the NIST special publication on key management states that the same level of security may be achieved with a configuration that uses an elliptic curve over a 512-bit finite field and a group order of 256 bits as with a 2048-bit RSA key [38]. Under this configuration, the average generation time for a signed assignment token is 0.66 ms, and the verification time is 2.65 ms, which is more efficient than the scheme proposed in [16], as shown in Table 3.

Figure 5 shows the variation in computational cost with the number of tokens in an aggregated token. The x-axis shows number of tokens, y-axis shows average computational time. Results suggest linear relation between tokens and computational cost. This is because addition and multiplication operations increase linearly with number of tokens in aggregated token. Specifically, if the number of tokens in an aggregated token is L, the average generation time is 0.35 ms, and the average verification time is 1.20 L ms. This performance is more efficient compared to the scheme in [16], as detailed in Table 4.

Table 4 Performance metrics for aggregated tokens

Number of tokens in the aggregated token	Proposed scheme		ABAC scheme in [16]	
	Generation time (ms)	Verification time (ms)	Generation time (ms)	Verification time (ms)
1	0.3	1.4	1.0	3.0
2	0.6	2.1	1.5	3.5
3	1.13	2.9	2.0	4.5
4	1.51	3.9	2.3	6.0

8 Conclusion and Future Work

MLADS integrates access control and delegation rules into a unified policy store, supporting both automated and on-demand delegation revocation while ensuring the security of token generation and verification through cryptographic techniques. Our rigorous performance analysis computational efficiency. Particularly noteworthy are the significant improvements MLADS offers over traditional ABAC systems with fully flexible delegation mechanisms in IoT healthcare contexts, highlighting its practical feasibility and robust security–performance balance. In future work, we intend to integrate advanced revocation mechanisms that minimize the impact on system performance and security. We also aim to investigate the use of predictive analytics to anticipate access control needs and adjust policies dynamically.

Disclosure of Interests The authors have no competing interests to declare relevant to this article's content.

Acknowledgments This work is supported by a research project funded by IHUB-NTIHAC Foundation, IIT Kanpur (Sanction Order No.: IHUB-NTIHAC/2021/01/8), under the aegis of the National Mission on Interdisciplinary Cyber-Physical System (NM-ICPS), DST, GoI.

References

1. Xu, H., Berres, A., Yoginath, S.B., Sorensen, H., Nugent, P.J., Severino, J., Tennille, S.A., Moore, A., Jones, W., Sanyal, J.: Smart mobility in the cloud: Enabling real-time situational awareness and cyber-physical control through a digital twin for traffic. IEEE Transactions on Intelligent Transportation Systems. 24(3), 3145–3156 (2023)
2. Aceto, G., Persico, V., Pescapé, A.: Industry 4.0 and health: Internet of things, big data, and cloud computing for healthcare 4.0. Journal of Industrial Information Integration. 18, 100129 (2020)
3. Kelly, J.T., Campbell, K.L., Gong, E., Scuffham, P.: The Internet of Things: Impact and implications for health care delivery. Journal of Medical Internet Research. 22(11), e20135 (2020)
4. SecurityWeek: Multiple Vulnerabilities Found in Popular IP Cameras. SecurityWeek, 2021. https://www.securityweek.com/multiple-vulnerabilities-found-popular-ip-cameras/

5. SecurityWeek: All Smartwatches Vulnerable to Attack: HP Study. SecurityWeek, 2023. https://www.securityweek.com/all-smartwatches-vulnerable-attack-hp-study/ (Accessed: 2024-04-04)
6. Arghire, I.: New Mirai Variant Leverages 10 Vulnerabilities to Hijack IoT Devices. SecurityWeek, 2021. https://www.securityweek.com/new-mirai-variant-leverages-10-vulnerabilities-hijack-iot-devices/
7. Iqal, Z.M., Selamat, A., Krejcar, O.: A Comprehensive Systematic Review of Access Control in IoT: Requirements, Technologies, and Evaluation Metrics. IEEE Access. (2023)
8. Jaiswal, K., Anand, V.: A survey on IoT-based healthcare system: potential applications, issues, and challenges. In: Advances in Biomedical Engineering and Technology: Select Proceedings of ICBEST 2018, pp. 459–471. Springer (2021)
9. Qian, H., Li, J., Zhang, Y., Han, J.: Privacy-preserving personal health record using multi-authority attribute-based encryption with revocation. International Journal of Information Security. 14, 487–497 (2015)
10. Zhang, Y., et al.: A hybrid access control model for attribute-based access control in cloud-based internet of things. IEEE Internet of Things Journal. 6(4), 7005–7015 (2019)
11. Xu, S., Ning, J., Li, Y., Zhang, Y., Xu, G., Huang, X., Deng, R.H.: Match in my way: Fine-grained bilateral access control for secure cloud-fog computing. IEEE Transactions on Dependable and Secure Computing. 19(2), 1064–1077 (2020)
12. Wang, Y., Ragothaman, K., Lawrence, M., et al.: Access Control for IoT: A Survey of Existing Research, Dynamic Policies and Future Directions. IEEE Communications Surveys and Tutorials. (2023)
13. Hu, V. C., Ferraiolo, D., Kuhn, R., Friedman, A. R., Lang, A. J., Cogdell, M. M., Schnitzer, A., Sandlin, K., Miller, R., Scarfone, K., et al.: Guide to Attribute Based Access Control (ABAC) Definition and Considerations (Draft). *NIST Special Publication*, **800**(162), 1-54 (2013).
14. Aghili, S.F., Sedaghat, M., Singelée, D., Gupta, M.: MLS-ABAC: Efficient multi-level security attribute-based access control scheme. Future Generation Computer Systems. 131, 75–90 (2022)
15. Bhatt, S., Sandhu, R.: Abac-cc: Attribute-based access control and communication control for internet of things. In: Proceedings of the 25th ACM Symposium on Access Control Models and Technologies, pp. 203–212. ACM (2020)
16. Choksy, P., Chaurasia, A., Rao, U.P., Kumar, S.: Attribute based access control (ABAC) scheme with a fully flexible delegation mechanism for IoT healthcare. Peer-to-Peer Networking and Applications. 16(3), 1445–1467 (2023)
17. Thomaz, G.A., Guerra, M.B., Sammarco, M., Detyniecki, M., Campista, M.E.M.: Tamper-proof access control for IoT clouds using enclaves. Ad Hoc Networks. 147, 103191 (2023)
18. Li, Q., Zhang, Q., Huang, H., Zhang, W., Chen, W., Wang, H.: Secure, efficient, and weighted access control for cloud-assisted industrial IoT. IEEE Internet of Things Journal. 9(18), 16917–16927 (2022)
19. Fugkeaw, S., Wirz, L., Hak, L.: Secure and Lightweight Blockchain-enabled Access Control for Fog-Assisted IoT Cloud based Electronic Medical Records Sharing. IEEE Access. 11, 48556–48568 (2023)
20. Bagga, P., Mitra, A., Das, A.K., Vijayakumar, P., Park, Y., Karuppiah, M.: Secure biometric-based access control scheme for future IoT-enabled cloud-assisted video surveillance system. Computer Communications. 195, 27–39 (2022)
21. Kousalya, A., Sakthidasan, K., Latha, A.: Reliable service availability and access control method for cloud assisted IOT communications. Wireless Networks. 27, 881–892 (2021)
22. Sun, J., Yuan, Y., Tang, M., Cheng, X., Nie, X., Aftab, M.U.: Privacy-preserving bilateral fine-grained access control for cloud-enabled industrial IoT healthcare. IEEE Transactions on Industrial Informatics. 18(9), 6483–6493 (2021)
23. Tian, J., Tian, J.F., Du, R.: MSLShard: An efficient sharding-based trust management framework for blockchain-empowered IoT access control. Journal of Parallel and Distributed Computing. 185, 104795 (2023)

24. Alshehri, S., Bamasaq, O., Alghazzawi, D., Jamjoom, A.: Dynamic secure access control and data sharing through trusted delegation and revocation in a blockchain-enabled cloud-IoT environment. IEEE Internet of Things Journal. 10(5), 4239–4256 (2022)
25. Bakhtiary, V., Mirabi, M., Salajegheh, A., Erfani, S.H.: Combo-Chain: Towards a hierarchical attribute-based access control system for IoT with smart contract and sharding technique. Internet of Things. 25, 101080 (2024)
26. Adamczyk, C., Samorzewski, A., Grzyb, M., Kliks, A.: Cloud-based Spectrum Access Control System for Dense IoT Networks. In: 2022 International Conference on Software, Telecommunications and Computer Networks (SoftCOM), pp. 1–6. IEEE (2022)
27. Zhang, L., Li, X., Wu, Q., Rezaeibagha, F.: Blockchain-aided anonymous traceable and revocable access control scheme with dynamic policy updating for the cloud IoT. IEEE Internet of Things Journal. (2023)
28. Wang, J., Zhu, M., Li, M., Sun, Y., Tian, Z.: An access control method against unauthorized and non-compliant behaviors of real-time data in industrial IoT. IEEE Internet of Things Journal. (2023)
29. Porwal, S., Mittal, S.: A fully flexible key delegation mechanism with efficient fine-grained access control in CP-ABE. Journal of Ambient Intelligence and Humanized Computing. 14(9), 12837–12856 (2023)
30. Sifou, F., AlShahwan, F., Hammoud, A., Marwan, M., Hammouch, A.: Implementing Policy Rules in Attributes Based Access Control with XACML within Cloud-Enabled IoT Environment. J. Commun. 15(1), 107–112 (2020)
31. Bhatt, S., Pham, T.K., Gupta, M., Benson, J., Park, J., Sandhu, R.: Attribute-Based Access Control for AWS Internet of Things and Secure Industries of the Future. IEEE Access 9, 107200–107223 (2021).
32. Mohammad, H.M., Abdullah, A.A.: Enhancement process of AES: a lightweight cryptography algorithm-AES for constrained devices. TELKOMNIKA (Telecommunication Computing Electronics and Control). 20(3), 551–560 (2022)
33. Gulen, U., Baktir, S.: Side-Channel Resistant 2048-Bit RSA Implementation for Wireless Sensor Networks and Internet of Things. IEEE Access. 11, 39531–39543 (2023)
34. Taleb, A.R., Vergnaud, D.: Speeding-up verification of digital signatures. Journal of Computer and System Sciences. 116, 22–39 (2021)
35. Sadeghi-Nasab, A., Rafe, V.: A comprehensive review of the security flaws of hashing algorithms. Journal of Computer Virology and Hacking Techniques 19(2), 287–302 (2023)
36. Nandakumar, K., Vinod, V., Akbar Batcha, S.M., et al.: Securing data in transit using data-in-transit defender architecture for cloud communication. Soft Computing 25(18), 12343–12356 (2021)
37. Pan, J., Qian, C., Ringerud, M.: Signed (group) diffie–hellman key exchange with tight security. Journal of Cryptology. 35(4), 26 (2022)
38. Barker, E., Dang, Q.: Draft NIST Special Publication 800-57 Part 3 Revision 1. Recommendation for Key Management, 2014.

Sub-terahertz On-Chip Filter Design for 6G Wireless Communication Systems

Ritika Negi, Ashish Kumar, Debabrata Sikdar, and Ashwini Sawant

Abstract In this paper, we propose a novel sub-terahertz on-chip bandpass filter utilizing substrate integrated plasmonic waveguide (SIPW) technology, compatible with the standard 0.13 μm SiGe process technology. Our design incorporates a Comb-like array on the topmost metal layer of the substrate integrated waveguide (SIW) to excite spoof surface plasmon polaritons (SSPPs) modes. The proposed filter is designed to operate within the passband of 249–297 GHz, achieving the maximum insertion loss of 2 dB and the reflection coefficients better than 10 dB. These results validate the effectiveness of the proposed filter, highlighting its potential for practical applications in sub-terahertz frequency bands for 6G communication systems.

Keywords Sub-terahertz · Substrate integrated waveguide · Spoof surface plasmon polaritons · Bandpass filter

1 Introduction

The advent of 6G communication signifies the dawn of a new age for mobile communication systems. This advancement demands higher data rates, lower latency, and seamless connectivity, which cannot be achieved solely by relying on existing networking and low-frequency wireless transmission technologies. To meet these goals, it is essential to incorporate the millimeter-wave frequency band into the com-

R. Negi (✉) · A. Kumar · D. Sikdar · A. Sawant
Department of Electronics and Electrical Engineering, Indian Institute of Technology Guwahati, Guwahati, Assam, India
e-mail: n.ritika@iitg.ac.in

A. Kumar
e-mail: ashish2003@iitg.ac.in

D. Sikdar
e-mail: deb.sikdar@iitg.ac.in

A. Sawant
e-mail: ashwinisawant@iitg.ac.in

U. S. Dixit et al. (eds.), *Current Progress in Engineering Sciences*,
https://doi.org/10.1007/978-981-96-8753-4_25

munication spectrum and explore new frequency bands in the sub-terahertz range (100–300 GHz). These higher frequencies offer wider bandwidths but present challenges in designing and integrating on-chip components like filters. Conventional planar transmission lines, like microstrip lines, are not suitable for millimeter-wave and sub-terahertz frequency bands due to their high transmission loss and coupling effects [1].

Surface Plasmon Polaritons (SPPs) and their derivatives, such as spoof surface plasmon polaritons (SSPPs), have emerged as promising solutions for addressing these challenges. SPPs are unique surface electromagnetic waves that travel along conductor/dielectric surfaces at optical frequencies, offering sub-wavelength confinement and high-density integration potential for optical circuits and components [2–4]. The concept of SSPPs, introduced in 2004, mimics the behavior of SPPs in the microwave and terahertz frequency ranges, enabling highly field-confined, flexible, and localizable electromagnetic modes in the sub-wavelength region [5]. SSPPs have been employed in various passive microwave and terahertz devices, including couplers, splitters, and antennas, demonstrating their versatility and effectiveness [6–10].

Despite these advancements, existing on-chip bandpass filters still face limitations in terms of bandwidth and insertion loss. Filter designs, such as those using substrate integrated waveguides (SIWs) and hybrid circuits combining SSPPs and traditional waveguide modes, have shown promising results but require large spaces or exhibit higher insertion losses [6, 11]. Furthermore, most of these devices are fabricated using conventional printed circuit board techniques, limiting their applicability in integrated circuit technology [12–16].

In this paper, we introduce a novel on-chip bandpass filter for the sub-terahertz range, achieving a passband of 249–297 GHz with an insertion loss of less than 2 dB. This design meets the high-performance requirements for filters in 6G wireless communication systems. Our work advances beyond previous designs by providing a more compact, efficient, and easily integrable solution, facilitating the development of advanced on-chip components for future communication technologies [17–20].

2 Design of Sub-terahertz On-Chip Filter

In this paper, we designed a sub-terahertz bandpass filter using 0.13 μm SiGe process technology. The structure consists of a 2 μm thick top metal layer over a silicon substrate, separated by 2.8 μm thick silicon dioxide (SiO_2) layer, with the top metal layer connected to the ground metal layer by metallic vias. Simulations were carried out with a 50 μm thick silicon and SiO_2 film having relative permittivities of 11.9 and 4.1, and loss tangents of 0.001 and 0.01, respectively. It is evident, from Fig. 1, that proposed SSPPs and groove SSPPs display almost identical dispersion characteristics, exhibiting a low-pass filter response onversely, in the SIW case, high-pass filter behavior is observed, which is influenced by the distance W_{sipw} between bilateral vias and via radius.

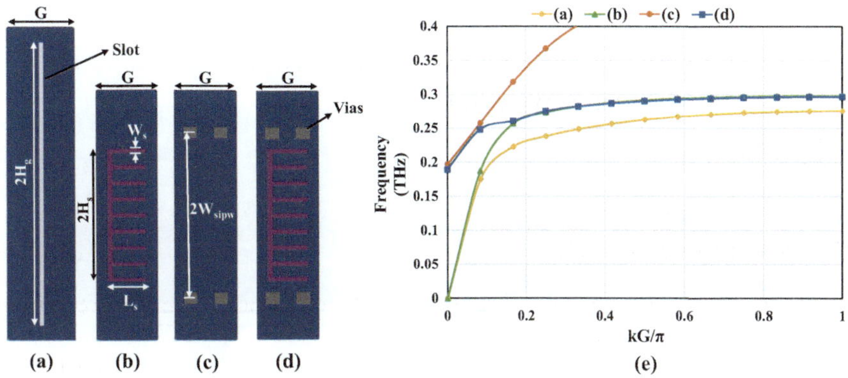

Fig. 1 **a** Schematic representation of the conventional groove SSPPs unit cell. **b** Schematic representation of the proposed SSPPs (comb-like) unit cell. **c** Schematic representation of the SIW unit cell. **d** Schematic representation of the proposed SIPW unit cell. **e** Dispersion curves for the unit cell. Dimensions are specified as follows: $H_g = 150\,\mu m$, $H_s = 90\,\mu m$, $G = 60\,\mu m$, $W_{sipw} = 100\,\mu m$, $L_s = 37\,\mu m$, with a via size of $15\,\mu m \times 15\,\mu m$

By combining SIW and SSPPs, a bandpass filter is created with low- and high-cutoff frequencies dictated by the SIW and SSPPs, respectively. The dispersion curve of the SIPW resembles that of the SIW at lower frequencies and the SSPPs at higher frequencies. After analysing the dispersion curves, we have proposed a design for Sub-terahertz bandpass filter.

The proposed filter topology, as illustrated in Fig. 2, includes three main components: the input/output grounded coplanar waveguide (GCPW) feeding section, the transition from GCPW-to-SIW, and the transition from SIW-to-SIPW, along with the SIPW repeating array section. A grounded CPW transmission line, which offers lower dispersion and improved heat dissipation compared to microstrip transmission lines, is employed as the feeding pad for the SIPW, making it ideal for sub-terahertz integrated circuits [11]. The short-circuit groove with a quarter-wavelength facilitates efficient mode conversion between GCPW and SIW [11]. Additionally, the SIW-to-SIPW transition features four linearly increasing comb-like arrays, which efficiently excite the SSPPs mode and facilitate proper momentum and mode matching between SIW and SSPPs. This transition ensures smooth signal propagation and minimal loss, optimizing performance across the sub-terahertz frequency range. The integration of these components results in a filter with enhanced performance characteristics, including improved signal integrity and frequency selectivity, crucial for advanced sub-terahertz applications.

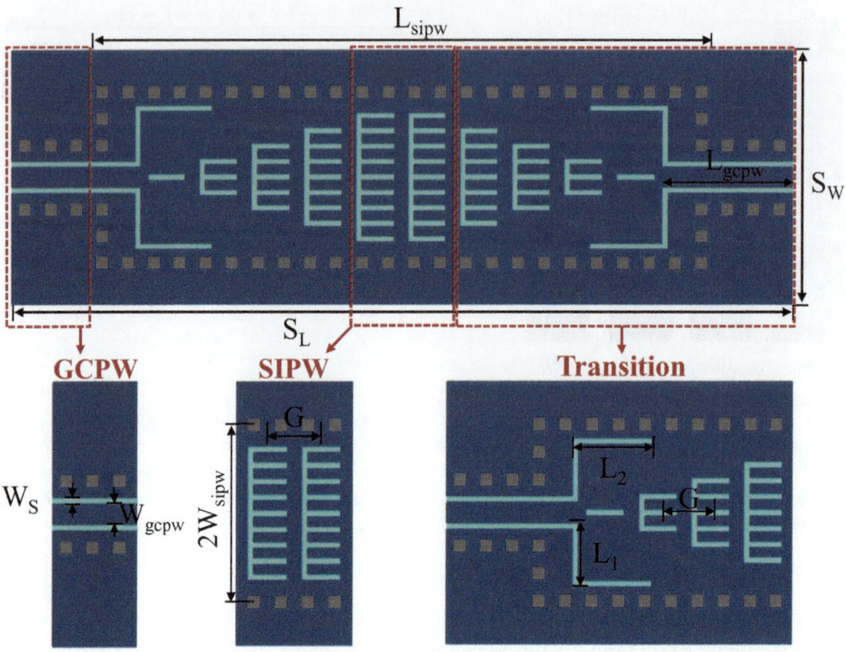

Fig. 2 Structure of the proposed sub-THz bandpass filter utilizing an SIPW unit cell, designed with 0.13 μm SiGe process technology. ($W_{gcpw} = 22\,\mu$m, $L_{gcpw} = 140\,\mu$m, $S_W = 285\,\mu$m, $S_L = 900\,\mu$m, $G = 60\,\mu$m, $W_s = 6\,\mu$m, $W_{sipw} = 96\,\mu$m, $L_{sipw} = 720\,\mu$m, $L_1 = 72\,\mu$m and $L_2 = 80\,\mu$m)

3 Results and Discussion

The simulated S-parameters, including reflection coefficients (S_{11}) and transmission coefficients (S_{21}) of the proposed sub-terahertz on-chip bandpass filter, are shown in Fig. 3. The results are simulated on the CST Microwave Studio.

The filter shows excellent bandpass characteristics and efficient signal propagation over a frequency range from 249 to 297 GHz, with (S_{21}) values less than 2 dB and in-band (S_{11}) values better than 10 dB. This performance indicates minimal insertion loss and strong return loss, ensuring high signal integrity. The filter's design effectively suppresses out-of-band interference, contributing to a sharp and well-defined passband.

To confirm the filter's passband adjustment capability, a parameter study was performed. The simulated transmission coefficients S_{21} with varying geometrical parameters are depicted in Figs. 4, 5, and 6.

It is evident that the SIPW band-pass filter's high cutoff frequency is influenced by the length (H_s) and width (L_s) of the SSPPs, whereas low cutoff frequency is determined by the width (W_{sipw}) of the SIW. The effect of H_s on S_{21} is illustrated in Fig. 4.

As H_s increases, the high-cutoff frequency moves to higher values (blue-shifts), the low-cutoff frequency remains constant, resulting in a broader bandwidth. Similarly, as shown in Fig. 5, increasing L_s moves the right edge of the S_{21} curve to lower frequencies. This occurs because the larger geometrical size of the SSPPs results in a higher propagation constant, which lowers the asymptotic frequency and consequently reduces the high-cutoff frequency.

The S_{21} curves for different SIW widths (W_{sipw}) are shown in Fig. 6. It is clear that W_{sipw} impacts only the low-cutoff frequency. As W_{sipw} increases, the low-cutoff frequency shifts to a higher frequency (blue-shifts). This is due to the fact that the cutoff frequency of the primary mode in the SIW is influenced by its width, W_{sipw}. Thus, the proposed band-pass filter's bandwidth can be easily adjusted by modifying the geometric parameters of the SSPPs and SIW.

The electric field distribution at 280 GHz within the passband is shown in Fig. 7. It is apparent that effective mode conversion is achieved, enabling terahertz waves to travel through the entire bandpass filter with minimal radiation loss.

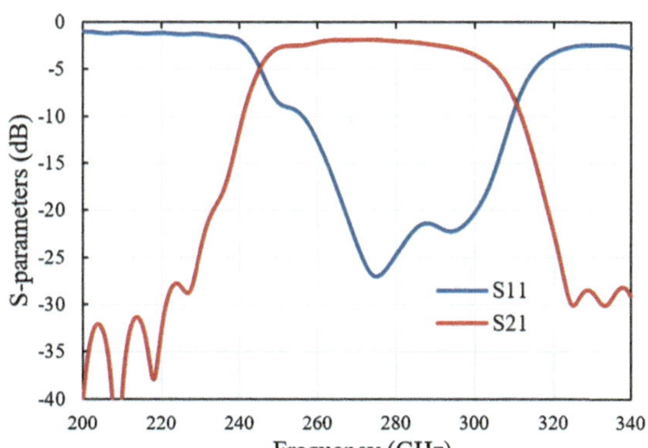

Fig. 3 The simulated S-parameters of the proposed sub-terahertz on-chip band-pass filter

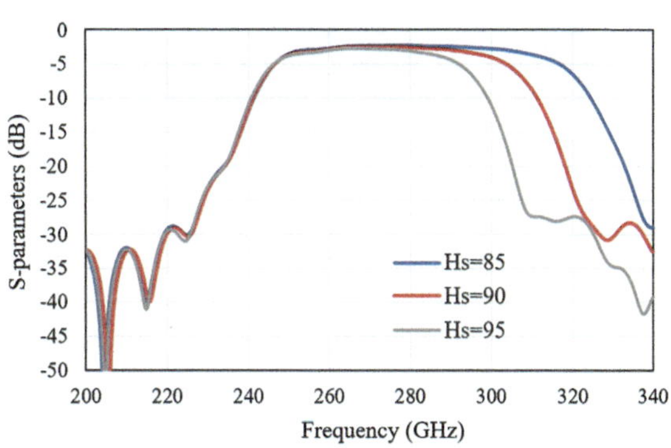

Fig. 4 The transmission coefficient S_{21} analysis graph of the proposed band-pass filter with different values of H_s

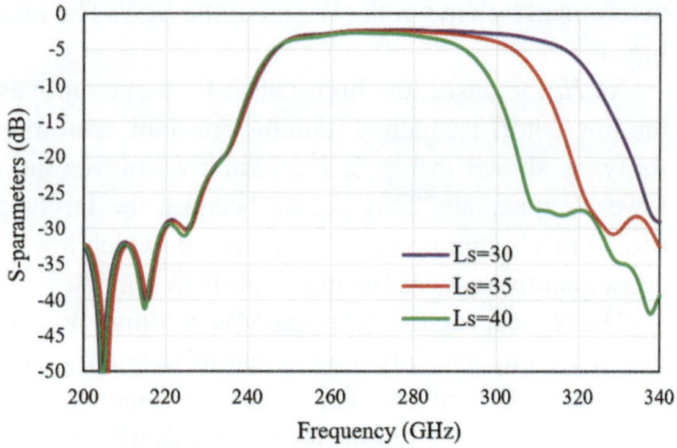

Fig. 5 The transmission coefficient S_{21} analysis graph of the proposed band-pass filter with different values of L_s

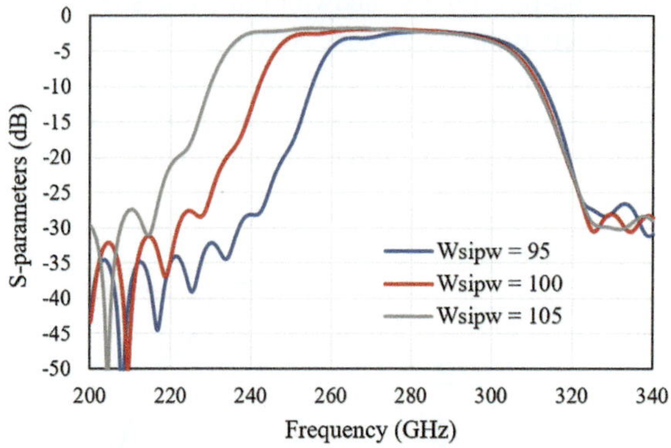

Fig. 6 The transmission coefficient S_{21} analysis graph of the proposed band-pass filter with different values of W_{sipw}

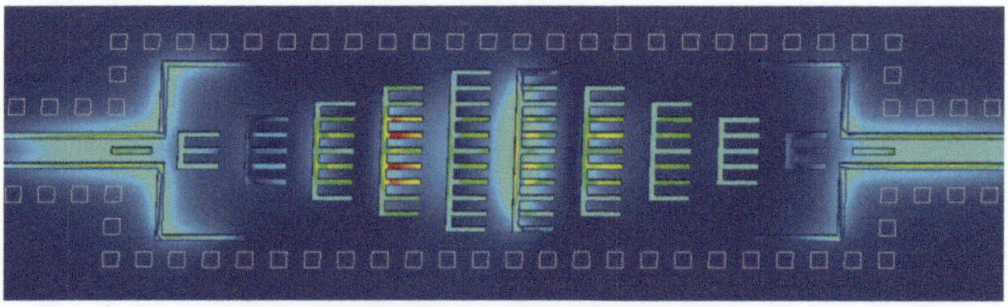

Fig. 7 The simulated normalised electric field distributions of proposed sub-terahertz bandpass filter's topmost metal at 280 GHz

4 Conclusion

This paper introduces a novel sub-terahertz on-chip bandpass filter using SIPW technology compatible with 0.13 μm SiGe process technology. By incorporating a Comb-like array on the topmost metal layer of the SIW, we facilitate support for SSPPs. Through CST Microwave Studio simulations, we characterize SSPPs and SIW unit cells, revealing distinct low pass and high pass filter responses. Operating between 249 and 297 GHz, the filter achieves a maximum insertion loss of 2 dB in the passband and an in-band reflection coefficient better than 10 dB. The geometric parameters of the filter can be tuned to flexibly engineer the bandwidth of the proposed filter. These advancements in miniaturization and integration capabilities position our design as promising for next-generation communication systems, particularly in 6G networks, highlighting its potential for practical deployment in sub-terahertz frequency bands.

References

1. Cao, X., Wang, C., Li, W., Cai, Q.: An on-chip bandpass filter using complementary slit-ring-resonator-loaded spoof surface plasmon polaritons with a flexible notch-band. Micromachines **14**(3), 607 (2023)
2. Barnes, W.L., Dereux, A., Ebbesen, T.W.: Surface plasmon subwavelength optics. Nature **424**(6950), 824–830 (2003)
3. Yin, L., Vlasko-Vlasov, V.K., Pearson, J., Hiller, J.M., Hua, J., Welp, U., Kimball, C.W.: Subwavelength focusing and guiding of surface plasmons. Nano Letters **5**(7), 1399–1402 (2005)
4. Xu, K.D., Guo, Y.J., Deng, X.: Terahertz broadband spoof surface plasmon polaritons using high-order mode developed from ultra-compact split-ring grooves. Optics Express **27**(4), 4354–4363 (2019)
5. Pendry, J.B., Martin-Moreno, L., Garcia-Vidal, F.J.: Mimicking surface plasmons with structured surfaces. Science **305**(5685), 847–848 (2004)
6. Zhang, Q., Zhang, H.C., Wu, H., Cui, T.J.: A hybrid circuit for spoof surface plasmons and spatial waveguide modes to reach controllable band-pass filters. Scientific Reports **5**(1), 16531 (2015)
7. Yi, H., Qu, S.W., Bai, X.: Antenna array excited by spoof planar plasmonic waveguide. IEEE Antennas and Wireless Propagation Letters **13**, 1227–1230 (2014)
8. Zhang, Q., Zhang, Q., Chen, Y.: Spoof surface plasmon polariton leaky-wave antennas using periodically loaded patches above PEC and AMC ground planes. IEEE Antennas and Wireless Propagation Letters **16**, 3014–3017 (2017)
9. Gao, X., Hui Shi, J., Shen, X., Feng Ma, H., Xiang Jiang, W., Li, L., Jun Cui, T.: Ultrathin dual-band surface plasmonic polariton waveguide and frequency splitter in microwave frequencies. Applied Physics Letters **102**(15) (2013)
10. Qi, C., Liao, S., Xue, Q.: Frequency splitter based on spoof surface plasmon polariton transmission lines. Applied Physics Letters **113**(16) (2018)
11. Zhu, H., Zhang, Y., Ye, L., Li, Y., Xu, Y., Xu, R.: On-chip terahertz bandpass filter based on substrate integrated plasmonic waveguide. Results in Physics **27**, 104553 (2021)
12. Guo, Y.J., Xu, K.D., Tang, X.: Spoof plasmonic waveguide developed from coplanar stripline for strongly confined terahertz propagation and its application in microwave filters. Optics Express **26**(8), 10589–10598 (2018)

13. Guo, Y.J., Xu, K.D., Deng, X., Cheng, X., Chen, Q.: Millimeter-wave on-chip bandpass filter based on spoof surface plasmon polaritons. IEEE Electron Device Letters **41**(8), 1165–1168 (2020)
14. Fernández-Domínguez, A.I., Moreno, E., Martín-Moreno, L., García-Vidal, F.J.: Terahertz wedge plasmon polaritons. Optics Letters **34**(13), 2063–2065 (2009)
15. Jiang, T., Shen, L., Wu, J.J., Yang, T.J., Ruan, Z., Ran, L.: Realization of tightly confined channel plasmon polaritons at low frequencies. Applied Physics Letters **99**(26) (2011)
16. Jin Zhou, Y., Jiang, Q., Jun Cui, T.: Bidirectional bending splitter of designer surface plasmons. Applied Physics Letters **99**(11) (2011)
17. Garcia-Vidal, F.J., Martin-Moreno, L., Pendry, J.B.: Surfaces with holes in them: new plasmonic metamaterials. Journal of Optics A: Pure and Applied Optics **7**(2), S97 (2005)
18. Guo, Y.J., Xu, K.D., Deng, X.: Tunable enhanced sensing of ferrite film using meander-shaped spoof surface plasmon polariton waveguide. Applied Physics Express **12**(11), 115502 (2019)
19. Ye, L., Chen, Y., Xu, K. D., Li, W., Liu, Q. H., Zhang, Y.: Substrate Integrated Plasmonic Waveguide for Microwave Bandpass Filter Applications. IEEE Access **7**, 75957–75964 (2019)
20. Zhang, B., et al.: Narrowband SIW-SSPP Hybrid Bandpass Filter With Compact Profile at Ka-Band. IEEE Access **11**, 98305–98314 (2023)

Analysis of Preventive Measures Against Buffer Overflow Attacks on an IoT Operating System

Dipendra Gurung, Sandeep Gurung, Mohan Pratap Pradhan, and Adarsh Sharma

Abstract The interconnected nature of devices today poses a major security challenge. A breach on Internet of Things (IoT) ecosystem can have far reaching drastic impacts on mainstream systems. Defensive mechanisms are in place at various levels. However, IoT devices are particularly vulnerable due to their resource constrained nature. Buffer overflow attacks utilize vulnerabilities in code to inject malicious code or hijack control. Although countermeasures are available, modern attacks are still able to bypass them. This paper presents an assessment of the built-in defense mechanisms against the code injection and *return-to-libc* attack on the Kali Linux and Raspbian operating systems. The tradeoff between performance and security has been a challenge for IoT devices. Together with available and new techniques, a balance between the two is needed.

Keywords IoT · Buffer overflow · DEP · Stack canaries · ASLR · ROP

1 Introduction

Securing computer systems has become of paramount importance in today's times. One of the reasons for this is because all kinds of devices are now connected to the internet. Devices on the Internet of Things (IoT) are particularly vulnerable to threats [1]. These devices are resource constrained, and purpose built. Hence, security is often an afterthought and a breach on these devices can potentially compromise

D. Gurung (✉)
Department of Computer Applications, Sikkim Manipal Institute of Technology, Sikkim, India
e-mail: dipendra.gu@smit.smu.edu.in

S. Gurung · A. Sharma
Department of Computer Science and Engineering, Sikkim Manipal Institute of Technology, Sikkim, India
e-mail: sandeep.gu@smit.smu.edu.in

M. P. Pradhan
Department of Computer Applications, Sikkim University, Sikkim, India
e-mail: mppradhan@cus.ac.in

U. S. Dixit et al. (eds.), *Current Progress in Engineering Sciences*,
https://doi.org/10.1007/978-981-96-8753-4_26

the entire ecosystem, as evident from the Mirai botnet attack which compromised IoT devices for launching DDoS attacks on mainstream systems [2]. Buffer overflow attack is a class of attack in which an attacker tries to overflow a memory buffer with the intention of injecting malicious code, redirecting control or crashing the application [3]. Various measures have been implemented to protect systems against a buffer overflow attack. These include Data Execution Prevention (DEP), stack canaries, and Address Space Layout Randomization (ASLR) [4]. Use of safe alternatives like *fgets()* instead of *gets()* or memory safe language like Rust has also been recommended [4]. However, the techniques available cannot be fully applied on IoT devices due to their resource constrained nature. Moreover, C is still the preferred language for writing code for IoT devices as a C program is compiled directly into machine code and thus it executes faster as compared to code written in other languages which require additional interpretation and checks [3, 5]. The overhead incurred due to extra checks and interpretation is not desirable in the case of IoT devices with limited resources. Also, the availability of legacy code in C that can be used has made the transition to other safe languages a bit more distant. This paper focuses on code injection attack on the stack using a buffer overflow vulnerability and the *return-to-libc* attack. An analysis on the effectiveness of the preventive measures mentioned above is made by considering the Kali Linux and Raspbian operating systems. The analysis considers the prevention provided by each technique individually as well as collectively.

Release 2024.2 of the Kali Linux OS was used on a virtual machine running on an Intel Core i5 processor with 4 GB of RAM. Raspbian OS version 10 running on Raspberry PI 3 with ARMv7 Quad-Core processor and 1 GB of RAM was used. Both the operating systems are equipped with the three security techniques under consideration.

2 Background

Buffer overflow attacks exploit weaknesses in a program's memory management, allowing attackers to manipulate the program's execution or corrupt its data [3]. These attacks typically target buffers that handle input data without adequate bounds checking, which is common in programming languages like C and C++ [6, 7]. A stack-based buffer overflow is a type of security vulnerability that occurs when a program writes more data into a buffer on the stack than it can hold. The stack is a region of memory used to store local variables and function call information. When a program fails to validate the size of input data before copying it into a buffer, an attacker can craft malicious input to overwrite adjacent memory locations.

By carefully overwriting specific parts of the stack, such as function return addresses, the attacker can gain control over the program's execution flow. This can lead to the execution of arbitrary code that the attacker injects into the stack, potentially enabling unauthorized access, privilege escalation, or the execution of malicious commands. To exploit a stack-based buffer overflow, an attacker typically crafts input data containing carefully constructed payloads, including shellcode, which is a small piece of assembly code designed to grant unauthorized access to the system. Once the vulnerable program processes the malicious input, it can inadvertently execute the injected code, leading to a complete compromise of the system [8].

Some of the recent instances of buffer overflow attack include CVE-2023-32,560, which was a stack-based buffer vulnerability in Ivanti Avalanche which could cause service disruption or allow attackers to execute arbitrary code [9]. CVE-2023-33,010 was a buffer overflow vulnerability in some Zyxel products, which allowed an unauthenticated attacker to perform remote code execution or impose DoS conditions [10]. Buffer overflow attacks have also been prominent across the IoT devices in recent years. CVE-2023-27,217, was a vulnerability on the Wemo Mini Smart Plug V2 which allowed remote code injection through exploitation of the feature that enables users to change the device name [11].

Recent instances of buffer overflow attacks depict that although various defense mechanisms have been in place, this class of attacks are one of the most prevalent ones. Data Execution Prevention (DEP) is a security mechanism that prevents the execution of code on the stack [6]. However, an attacker would still be able to find other executable regions in the memory and place a malicious code there and then redirect control by writing the malicious code's address on the stack. On the other hand, stack canaries use arbitrary values which are placed on the stack. When a function attempts to return, the canary value present before the return address is inspected for tampering [6]. If a change in the value is detected, a buffer overflow attempt is detected. Address Space Layout Randomization (ASLR) is another technique that randomizes memory locations of certain memory regions to make it difficult for an attacker to easily locate vulnerabilities [12]. With ASLR enabled, the stack and the libc can be loaded at different locations every time an executable is prepared for execution [4]. This makes any information about location of vulnerabilities collected by an attacker irrelevant for subsequent use.

3 Implementation

This section presents a discussion on the implementation of a code injection attack and the *return-to-libc* attack, both carried out using a buffer overflow vulnerability. During the first phase, the attacks were implemented by initially switching off all the three security mechanisms under consideration namely DEP, stack canaries and ASLR. Next, the attacks were launched by enabling one mechanism at a time and finally all the three mechanisms were enabled to analyze their collective effectiveness. It is to be noted that 32-bit variants of both the operating systems were used. This was done as 32-bit devices are more common in the IoT ecosystem, as they are less expensive and less resource intensive. Moreover, although the aforementioned security measures work well on 64-bit systems, the same cannot be said for 32-bit systems [6]. Hence, 32-bit variants of the operating systems were deliberately chosen for this work to analyze the effectiveness of the security measures and make a comparative study between the two operating systems.

3.1 Vulnerable Program and Design of Exploit

For the purpose of implementation of the attack, a C program with a known vulnerability was used. Figure 1 illustrates the vulnerable C program. The C library function *strcpy()* which copies the contents of *string* into *buffer* is the vulnerability in the program, as while copying it does not check the boundary limits of *buffer*. Hence, it would continue copying the contents of string even if the size of buffer is exceeded, thereby overwriting the contents of adjacent memory locations. This would overwrite the return address of the *vulnerable_function* leading to a crash when it attempts to return. However, instead of crashing the program, a suitable payload can be crafted to inject a shell code or redirect control. By doing this, the shell code would be executed instead of the function returning the control correctly.

Figure 2 illustrates the call stack before *strcpy()* executes. 100 bytes are allocated for *buffer* and the return address of *vulnerable_function* is also present on the stack.

Shell Code Injection Attack. The shell code used for the attack spawns a shell which can be used to execute any command on the target system. For the design of the payload, the buffer is filled with an attack string consisting of 100 A's. This is

Fig. 1 Vulnerable program

```
#include<stdio.h>
#include<string.h>
void vulnerable_function(char* string) {
    char buffer[100];
    strcpy(buffer, string);
}
int main(int argc, char** argv) {
   vulnerable_function(argv[1]);
   printf("Done!");
  return 0;
}
```

Fig. 2 Stack before *strcpy* executes

<argument>
<return address>
<old %ebp>
<100 bytes of
...
buffer>
<argument>
<address of buffer>

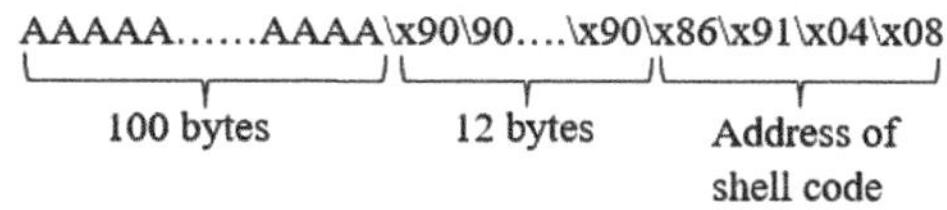

Fig. 3 Payload for shell code injection

followed by 12 NOP instructions to properly align the shell code in the stack. The address of the shell code follows this and overwrites the return address. The complete payload would appear as in Fig. 3.

Upon successful execution of the payload the stack would look as in Fig. 4.

Figure 5 shows the snapshot of the stack after attack taken from GDB on Kali Linux. Here, 0×41 represents the letter A, 0×90 the NOP instruction and 0× 08049186 the address of the shell code. For the implementation of this attack the program shown in Fig. 1. was compiled as *gcc -m32 -z execstack -fno-stack-protector -no-pie -g vuln_prog.c -o vuln_prog.*

Here, *-z execstack* flag indicates that the stack is executable, *-fno-stack-protector* indicates that stack canaries are not enabled and *-no-pie* indicates that Position Independent Executable (PIE) is disabled, meaning that the code is not relocatable. PIE is the ASLR equivalent for executables [13]. In other words, all the three security measures stack canaries, ASLR and DEP are disabled. The attack was successful in this case.

On the other hand, when DEP was enabled, the attack was unsuccessful and resulted in a segmentation fault. Similarly with stack canaries enabled, the code execution was terminated with an error message stating "stack smashing detected." Likewise, with PIE enabled, a segmentation fault was encountered. With all the

Fig. 4 Stack after attack

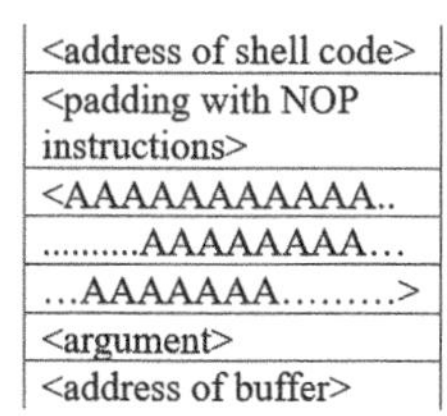

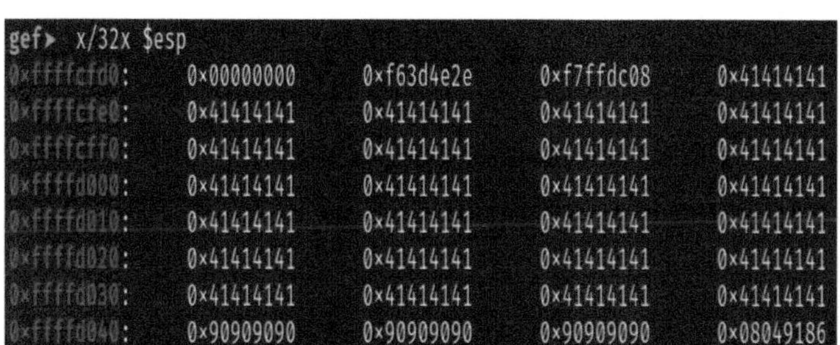

Fig. 5 GDB instance of stack after attack on Kali Linux

three security measures enabled, the execution terminated with a "stack smashing detected" error.

The attack was implemented similarly on the Raspbian OS. DEP, stack canaries and ASLR individually were effective against the attack. They were also collectively able to protect the system against the code injection attack.

Return-to-libc attack. This attack relies on using functions available in the *libc*, instead of injecting malicious code. Generally the *libc* function *system()* is used with "/bin/sh" to invoke a shell [8]. The addresses of *system()* and *"/bin/sh"* can be easily known by examining the executable with GDB. With the addresses known it is only a matter of modifying the stack to redirect control to execute *system()*. For this attack as well, the code shown in Fig. 1 was used. The payload was designed to fill the buffer with a string consisting of 100 A's, followed by 12 bytes of NOP instructions, the address of *system()*, an arbitrary return address indicating the return address of the call to *system()*. This was followed by the address of "/bin/sh". This is illustrated in Fig. 6.

The change that a successful execution of the payload makes to the stack is illustrated in Fig. 7.

Figure 8 shows the GDB snapshot of the stack after attack with the payload shown in Fig. 6. on Raspbian OS. Here, 0×76e8d9c8 is the address of *system()* and 0 × 76f80b6c is the address of */bin/sh*, which is passed as an argument to *system()*.

With DEP, stack canaries and ASLR disabled, the attack was successful. In the next attempt DEP was enabled. The attack was still successful with DEP enabled. This is because in this attack, no code is executed on the stack, instead the control is redirected to an existing function in *libc*. Hence, DEP is not able to detect the attack. In the third attempt, stack canaries were enabled. The attack was not successful, the error message "stack smashing detected" was generated. In the next attempt

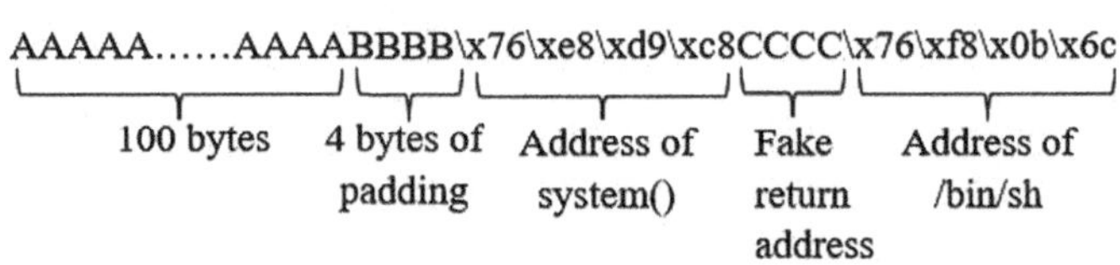

Fig. 6 Payload for return-to-libc attack

Fig. 7 Stack after return-to-libc attack

<address of /bin/sh>
<fake return address>
<address of system>
<padding>
<AAAAAAAAAAA..
..........AAAAAAAA...
...AAAAAAA.........>
<argument>
<address of buffer>

```
gef➤  x/32x $sp
0x7efff3f0:    0x76fffc70    0x7efff71a    0x00000005    0x41414141
0x7efff400:    0x41414141    0x41414141    0x41414141    0x41414141
0x7efff410:    0x41414141    0x41414141    0x41414141    0x41414141
0x7efff420:    0x41414141    0x41414141    0x41414141    0x41414141
0x7efff430:    0x41414141    0x41414141    0x41414141    0x41414141
0x7efff440:    0x41414141    0x41414141    0x41414141    0x41414141
0x7efff450:    0x41414141    0x41414141    0x41414141    0x41414141
0x7efff460:    0x42424242    0x76e8d9c8    0x43434343    0x76f80b6c
```

Fig. 8 GDB snapshot of stack after return-to-libc attack on Raspbian OS

with ASLR enabled, a segmentation fault occurred. In the final attempt, with all the measures enabled, the attack was prevented with a segmentation fault error.

Similar results were observed on the Kali Linux OS with DEP alone being ineffective against the attack while stack canaries and ASLR preventing the attack. All the three measures were also able to collectively prevent the attack.

4 Analysis

The implementation of the code injection and the *return-to-libc* attack on the Kali Linux and Raspbian OS provided an insight into the built-in defense mechanisms against these attacks.

Table 1 illustrates the summary of actions taken by DEP, stack canaries and ASLR against the attacks. It can be observed that the three preventive measures were individually effective against code injection attack using buffer overflow on both Kali Linux and Raspbian OS. The *return-to-libc* attack was successful with only DEP enabled on both the operating systems. However, stack canaries and ASLR were individually able to defend against the attacks on both the operating systems. It is important to note that the three measures complement each other and are together able to provide greater security. DEP is enabled by default on most Linux distributions including Raspbian OS [14]. However, it is hardware dependent and would work only if supported by the underlying architecture. Stack canaries on the other hand can be enabled by using the appropriate flags during compilation. The flags *"-fstack-protector"*, *"-fstack-protector-strong"*, and *"-fstack-protector-all"* provide basic, intermediate and advanced levels of protection respectively [15]. ASLR is also enabled by default on most Linux distributions with the default setting being full randomization [15]. Users also have the option of switching it off entirely or choosing the medium randomization setting. ALSR randomizes the *libc*, stack and other regions of memory. For randomizing the executable itself, PIE can be used by enabling the appropriate flag during compilation.

As observed from the implementation, the security measures are effective against the two attacks under consideration, but it is to be noted that these measures can be bypassed through different techniques. As evident DEP can be bypassed through the *return-to-libc* attack, as this attack does not execute any code on the stack. Stack

Table 1 Summary of action taken against attack by different measures

	Code injection attack			Return-to-libc attack		
	DEP	Stack canaries	ASLR	DEP	Stack canaries	ASLR
Kali linux	Segmentation fault	Stack smashing detected	Segmentation fault	Attack successful	Stack smashing detected	Segmentation fault
Raspbian OS	Segmentation fault	Stack smashing detected	Segmentation fault	Attack successful	Stack smashing detected	Segmentation fault

canaries can be bypassed through information leakage attacks, which leak the canary value placed on the stack. An attacker can design a payload to write the correct canary value, thus bypassing detection. Return-Oriented Programming (ROP) is a technique which can be used along with memory leakage attacks to bypass ASLR [16]. ROP works by chaining together small assembly instructions called gadgets to perform arbitrary operations like hijacking control [15, 17].

It is also to be noted that although the discussed measures provide security which cannot be undermined, they also incur additional overhead which may be in terms of more memory usage for storing additional information to support the protection mechanisms or greater execution time for resolution of dependencies.

Analysis of memory overhead on Kali Linux. Figure 9 presents an analysis of the memory taken during normal execution of the executable with all protections disabled and with each protection enabled on Kali Linux. The executable was executed 50 times each to prepare a dataset containing the used memory for each execution. In the figure, the y axis represents the memory used in kilo bytes and the x axis represents the number of executions. As observed, the memory used lies in between 1100 and 1500 kilobytes for all the scenarios. However, on careful observation the memory used by the executable with all protection disabled show a significantly less variation as compared to the other cases. This is also confirmed upon calculation as evident from Table 2.

Table 3 presents the memory overhead induced by the protection mechanisms on Kali Linux.

As observed ASLR incurs the highest overhead of 5.98% followed by DEP with 1.13% and canary with 0.97%. The high overhead in ASLR can be attributed to extra information that needs to be stored to support the randomization of memory regions.

Analysis of time overhead on Kali Linux. An analysis of the time taken during 50 runs of the executable is presented in Fig. 10. As illustrated in the figure four cases were considered. The execution time with all protections disabled lies in the range between 0.5 and 2 ms with only a few executions taking more than 1.5 ms. With stack canaries enabled, few spikes can be observed. However, the remaining execution times remain well below 1.5 ms. Similarly with DEP enabled, the execution time remains within 2 ms, with only one execution touching the 2 ms mark. On the other hand, with ASLR, the execution time crosses 2 ms in one instance.

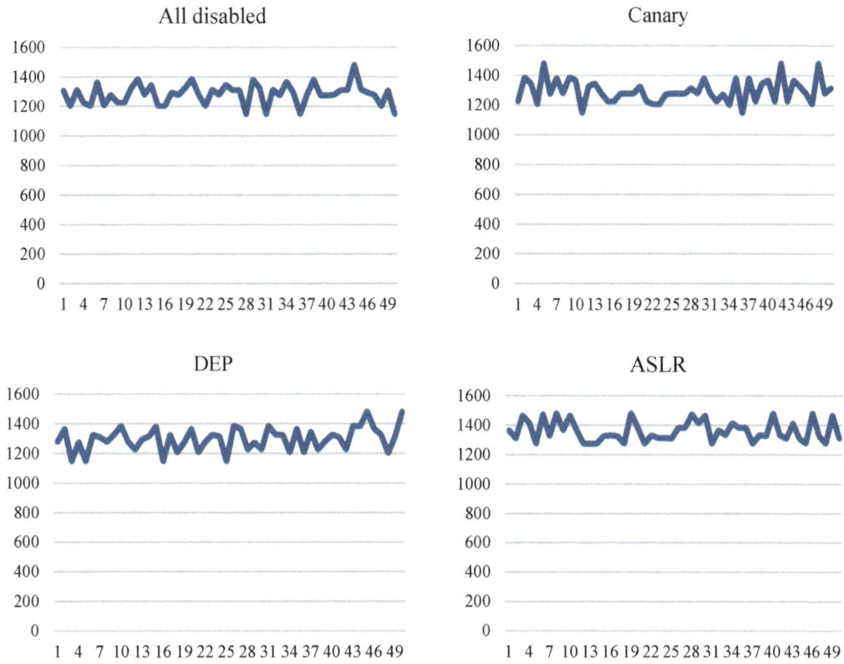

Fig. 9 Analysis of memory taken during normal execution with all protection disabled and with each protection enabled on Kali Linux *(Y axis represents memory in kilo bytes and X axis represents number of runs of executable)*

Table 2 Average memory used on Kali Linux

Protection mechanism used	Average memory used (KB)
All disabled	1284.88
Canary enabled	1297.44
DEP enabled	1299.36
ASLR enabled	1361.68

Table 3 Memory overhead on Kali Linux

Protection mechanism used	Overhead (%)
Canary enabled	0.97
DEP enabled	1.13
ASLR enabled	5.98

It can be noted from the analysis that the runtime overhead incurred by the protection mechanisms is minimal which could be attributed to the time taken for inspection of canary values or the resolution of randomized addresses in case of ASLR.

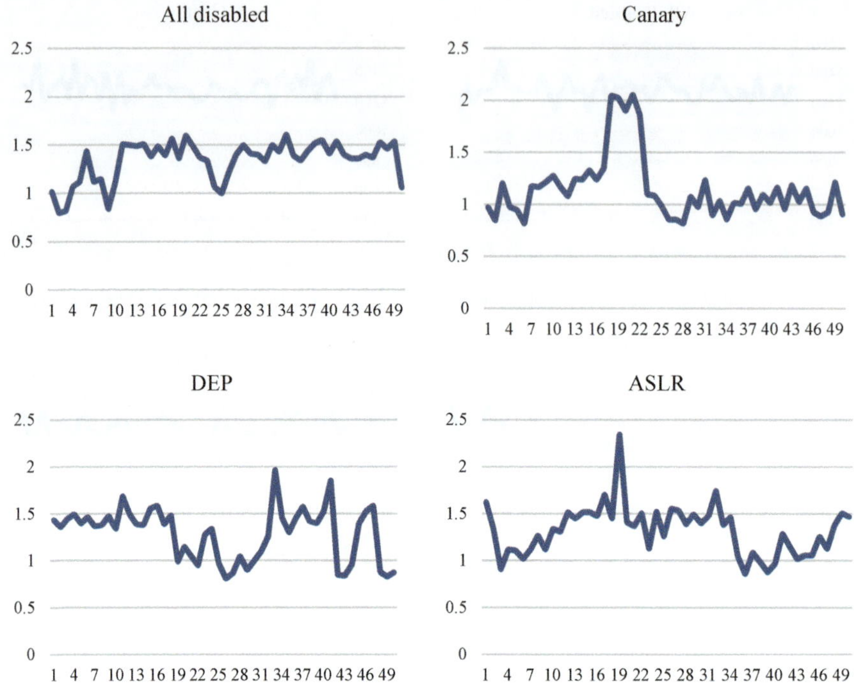

Fig. 10 Analysis of time taken during normal execution with all protection disabled and with each protection enabled on Kali Linux *(Y axis represents time in milliseconds and X axis represents number of runs of executable)*

Analysis of overhead on Raspbian OS. Figure 11 presents an analysis of the memory taken by the executable on Raspbian OS. The memory consumption lies in the range between 1000 and 1300 kilobytes. As with Kali Linux, ASLR incurs slightly more memory overhead as compared to the other two mechanisms. The average memory consumption is shown in Table 4.

The memory overhead incurred is present in Table 5.

As evident from Table 5 ASLR incurs a memory overhead of 1.44% which is slightly more than the other two mechanisms. DEP does not directly cause a memory overhead as it is implemented in hardware. The slight overhead could be due to factors such as memory fragmentation. Stack canaries would cause a slight overhead as a result of storing canary values. However, as observed it also seems to be negligible. The memory overhead of ASLR on Kali Linux (5.98%) is greater than that of Raspbian OS (1.44%). A possible reason for this could be a more robust implementation of ASLR on Kali Linux as compared to that in Raspbian OS which could be using a more lightweight equivalent.

The analysis of execution time for the Raspbian OS also showed a minimal overhead as was the case with Kali Linux. The protection offered by these mechanisms far outweighs the slight increase in execution time and a bit of memory overhead.

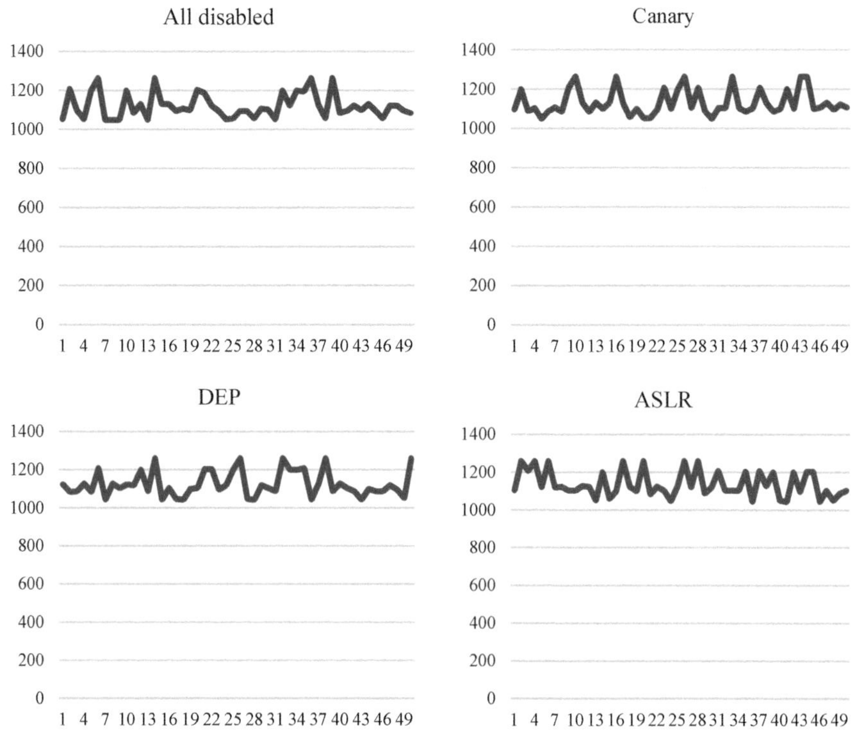

Fig. 11 Analysis of memory taken during normal execution with all protection disabled and with each protection enabled on Raspbian OS *(Y axis represents memory in kilo bytes and X axis represents number of runs of executable)*

Table 4 Average memory used on Raspbian OS

Protection mechanism used	Average memory used (KB)
All disabled	1121.92
Canary enabled	1132.32
DEP enabled	1122.776
ASLR enabled	1138.08

Table 5 Memory overhead on Raspbian OS

Protection mechanism used	Overhead (%)
Canary enabled	0.93
DEP enabled	0.08
ASLR enabled	1.44

Fig. 12 Memory used by executable on Kali Linux during an attack

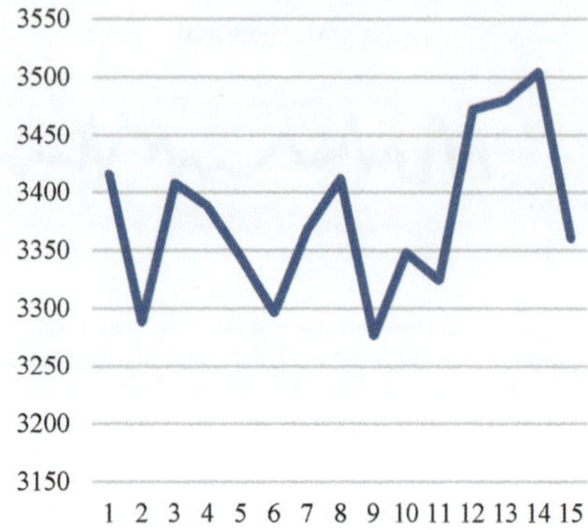

Fig. 13 Execution time on Kali Linux during an attack

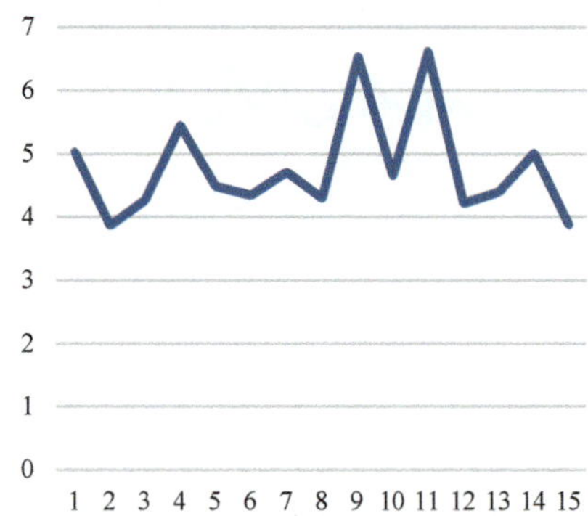

Analysis of memory and execution time could also be used to detect an attack. Figure 12 illustrates the memory usage of the same executable under an attack. As observed, in comparison to normal execution (Fig. 9), the executable under attack uses a significantly greater memory. For fifteen attack instances, the average memory used was 3378.933 kilobytes, a 162.9% increase over normal execution with all protections disabled.

Figure 13 presents the execution time of the executable under the attack for fifteen attack instances. The average execution time is 4.88 ms, an increase of 263.78% as compared to normal execution.

5 Conclusion

Buffer overflow attacks are still of major concern as evident from recent instances of attacks. With limited resources and minimal security considerations, IoT devices are particularly vulnerable to attacks. This paper has presented a comparison of the built-in security measures provided by the Kali Linux and Raspbian operating systems. As evident from the discussions, both the operating systems provided equal protection against the code injection and the *return-to-libc* attack. This could be attributed to the fact that the Raspbian operating system is based on Debian and thus inherits its security features. Moreover, the device used for the implementation with 1 GB of RAM and ARMv7 Quad-Core processor was not as resource constrained as IoT devices generally are.

The analysis of memory and execution time revealed a minimal overhead for both Kali Linux and Raspbian OS. The protection offered by the security measures outweighs the minimal overhead incurred. A similar analysis on more resource constrained IoT devices that support these measures would help in understanding their suitability for securing lightweight IoT infrastructures. Analysis during an attack depicted a drastic increase in memory and execution time. A real-time monitoring system based on this idea could be used to monitor abnormal resource usage and would be helpful in detecting attacks.

Advanced attacks that bypass security measures can also be studied to understand existing loopholes. Return-Oriented Programming is one such attack that can bypass ASLR and other measures [18]. The assessment can also be done for other more resource constrained IoT devices.

References

1. Abbasi, J. Wetzels, T. Holz and S. Etalle, "Challenges in Designing Exploit Mitigations for Deeply Embedded Systems," 2019 IEEE European Symposium on Security and Privacy (EuroS&P), Stockholm, Sweden, 2019, pp. 31–46, https://doi.org/10.1109/EuroSP.2019.00013.
2. Liu, C., Wu, YJ., Wu, JZ. et al. A buffer overflow detection and defense method based on RISC-V instruction set extension. Cybersecurity 6, 45 (2023). https://doi.org/10.1186/s42400-023-00164-x.
3. G. Mullen and L. Meany, "Assessment of Buffer Overflow Based Attacks On an IoT Operating System," 2019 Global IoT Summit (GIoTS), Aarhus, Denmark, 2019, pp. 1–6, https://doi.org/10.1109/GIOTS.2019.8766434.
4. Marco-Gisbert, H.; Ripoll Ripoll, I. Address Space Layout Randomization Next Generation. Appl. Sci. 2019, 9, 2928. https://doi.org/10.3390/app9142928.
5. Zili Shao, C. Xue, Q. Zhuge, M. Qiu, Bin Xiao and E. H. . -M. Sha, "Security protection and checking for embedded system integration against buffer overflow attacks via hardware/software," in IEEE Transactions on Computers, vol. 55, no. 4, pp. 443–453, April 2006, https://doi.org/10.1109/TC.2006.59.
6. H. M. Gisbert and I. Ripoll, "On the Effectiveness of NX, SSP, RenewSSP, and ASLR against Stack Buffer Overflows," 2014 IEEE 13th International Symposium on Network Computing and Applications, Cambridge, MA, USA, 2014, pp. 145–152, https://doi.org/10.1109/NCA.2014.28.
7. V. Iyer, A. Kanitkar, P. Dasgupta and R. Srinivasan, "Preventing Overflow Attacks by Memory Randomization," 2010 IEEE 21st International Symposium on Software Reliability Engineering, San Jose, CA, USA, 2010, pp. 339–347, https://doi.org/10.1109/ISSRE.2010.22.
8. https://hackmd.io/@minhkhoi1026/ret2libc.
9. https://www.bleepingcomputer.com/news/security/ivanti-avalanche-impacted-by-critical-pre-auth-stack-buffer-overflows/.

10. https://www.bleepingcomputer.com/news/security/netgear-warns-users-to-patch-recently-fixed-wifi-router-bug/.
11. https://sternumiot.com/iot-blog/mini-smart-plug-v2-vulnerability-buffer-overflow/.
12. PaX Team. (2001). Address Space Layout Randomization in Linux. Retrieved from http://pax.grsecurity.net/docs/aslr.txt.
13. Hovav Shacham, Matthew Page, Ben Pfaff, Eu-Jin Goh, Nagendra Modadugu, and Dan Boneh. 2004. On the effectiveness of address-space randomization. In Proceedings of the 11th ACM conference on Computer and communications security (CCS '04). Association for Computing Machinery, New York, NY, USA, 298–307. https://doi.org/10.1145/1030083.1030124.
14. https://www.devx.com/terms/data-execution-prevention/#:~:text=Is%20Data%20Execution%20Prevention%20enabled,configuration%20screens%20in%20most%20cases.
15. https://codearcana.com/posts/2013/05/28/introduction-to-return-oriented-programming-rop.html.
16. Kosolapov, Y.V. On Detecting Code Reuse Attacks. Aut. Control Comp. Sci. 54, 573–583 (2020). https://doi.org/10.3103/S0146411620070111.
17. Intel Corporation. (2019). Intel® Control-flow Enforcement Technology (Intel® CET): Enabling Hardware-enhanced Security for Control Flow Integrity. Intel White Paper.
18. Zhe Wang, Chenggang Wu, Jianjun Li, Yuanming Lai, Xiangyu Zhang, Wei-Chung Hsu, and Yueqiang Cheng. 2017. ReRanz: A Light-Weight Virtual Machine to Mitigate Memory Disclosure Attacks. In Proceedings of the 13th ACM SIGPLAN/SIGOPS International Conference on Virtual Execution Environments (VEE '17). Association for Computing Machinery, New York, NY, USA, 143–156. https://doi.org/10.1145/3050748.3050752.

WS_2 Nanosheet-Based Sensor Device for Iodine Detection in Rock Salt

Moti Kumar Jha, Supriya Khatoniar, Julaiba Tahsina Mazumder, and Ravindra Kumar Jha

Abstract To observe the presence of iodine ion in rock salt, here we proposed to use a simple two terminal sensor. The liquid-exfoliated WS_2 nanosheets were used for sensing the change in conductivity of the water in presence of different salts. The structural properties of the nanosheets are verified from X-ray diffraction (XRD) and field emission scanning electron microscopy (FESEM) images. A unique response observed in electrical conductivity in presence of rock salt when swept in the voltage range 0–2 V. The presence of iodine in rock salt could be responsible for such deviation as compared to other electrolyte solutions.

Keywords WS_2 · Salt solution · Iodine · Rock salt

1 Introduction

Iodine is a vital micronutrient essential for synthesizing thyroid hormones, which regulate metabolism and overall physiological functions. A deficiency or excess of iodine can lead to severe health conditions such as goiter and iodine deficiency disorders (IDDs) [1]. Monitoring iodine levels in commonly consumed salts is critical to maintaining the population's health, as dietary salt serves as the primary vehicle for iodine supplementation. Even if there is a chance of adulteration, accurate monitoring of iodine levels in salt is crucial. According to the report of Food Safety and Standard Authority of India (FSSAI) [2], there are reports on such cases where uncontrolled amounts of iodine intake may lead to the expansion of the thyroid gland, which is a typical cause of "endemic" goiter. In the National Family Health Survey IV report (NFHS IV (2015–2016)), the prevalence of self-reported goiter or thyroid disorder was 2.2% for women and 0.5% for men in the 15–49 age group; in the

M. K. Jha · S. Khatoniar · J. T. Mazumder · R. K. Jha (✉)
Department of Electronics and Electrical Engineering, Indian Institute of Technology Guwahati, Assam, India
e-mail: jha@iitg.ac.in

M. K. Jha
e-mail: jha@iitg.ac.in

U. S. Dixit et al. (eds.), *Current Progress in Engineering Sciences*,
https://doi.org/10.1007/978-981-96-8753-4_27

National Family Health Survey-V report (2019–2021), it was 2.7% for women and 0.5% for men in the same age group in India [3]. Traditional analytical techniques, such as ion chromatography and atomic absorption spectroscopy, are precise but expensive, complex, and unsuitable for widespread or routine applications. This gap has driven interest in developing cost-effective, portable, and user-friendly sensing devices. There are currently no publications reporting the development of electrical sensors specifically designed for iodine detection. This highlights a significant gap in research and technology development, underscoring the novelty and relevance of the proposed approach.

Tungsten disulfide (WS_2) [4] and other transition metal dichalcogenides (TMDs) [5] have shown great promise as materials for sensor applications because of their special blend of optical, chemical, and electrical characteristics. Covalently bound S-W-S monolayers held together by weak van der Waals forces make up the two-dimensional layered material known as WS_2. Two layers of sulfur atoms are positioned between a center layer of tungsten atoms organized hexagonally to form the monolayer structure. Because of its remarkable characteristics, such as its large surface area, adjustable bandgap, potent electrocatalytic activity, and improved charge transfer efficiency, WS_2 is a great option for iodine ion detection in liquid medium.

In this work, a sensor for iodine ion detection in salt solutions was created using exfoliated WS_2 nanosheets. The performance of the sensor was assessed by measuring the voltage–current (V–I) characteristics of a variety of salt solutions, such as tap water, daily-use salt, WS_2-treated rock salt, and WS_2-treated black salt, across a voltage range of 0–2.5 V. Five milliliters of deionized (DI) water were used to dissolve 0.5 g of each salt to create solutions. Because of the special characteristics of WS_2 nanosheets, the current response varied greatly among the samples, with WS_2-treated solutions displaying increased sensitivity and daily-use salt displaying the lowest current.

This work demonstrates the feasibility of using WS_2 nanosheets as a receptor material for developing cost-effective, efficient, and portable sensors [6] for iodine detection in salts. By addressing the critical gap in the availability of electrical sensors for iodine detection, this study bridges advanced material science with practical applications, tackling pressing public health concerns such as iodine deficiency and food safety compliance.

2 Experimental Details

The bulk WS_2 powder at a concentration of 6 mg ml^{-1} was mixed in an ethanol and water mixture (65:25). The solution was then dispersed for 6 hours in a bath sonicator. Then, the mixture was allowed to settle for 18 hours, followed by centrifugation for 45 minutes at 4500 rpm to separate the suspended nanosheets from the partly or unexfoliated WS_2 [7]. Finally, the supernatants were collected carefully to extract the nanosheets. To test the electrolytes, three different solutions of commonly used edible salts including rock salt, black salt, and normal salt were prepared by dissolving 0.5 g of each salt in 5 ml of DI water. The chemiresistive [8] device was formed by depositing few μL of WS_2 supernatants on an interdigitated electrode followed by drying. The response of the device in presence of electrolytes was observed by sweeping voltage from 0 to 2.5 V.

3 Results and Discussion

The formation of a monolayer of WS_2 is confirmed by X-ray diffraction (Make: Rigaku Technology, Model: Smartlab) data in Fig. 1a, exfoliated nanosheet XRD pattern reveals a single, strong peak at 14.3° that corresponds to diffraction from the (002) plane, the remaining peaks are either missing or noticeably weaker than the one strong peak, which is comparatively weaker than from bulk. The fact that no further peaks, such as those resulting from the vertical crystal planes symmetry (004), (006), and (008), are present indicates that the nanosheets have exfoliated along the (002) axis. Also, monolayer WS_2 structure is confirmed by WS_2 FESEM (Zeiss, Sigma 360) image in Fig. 1b.

The response of the sensor suggests that pristine WS_2 has low conductivity by nature, which might be caused by restricted charge mobility or surface imperfections. Deionized (DI) water is a polar solvent even though it has no dissolved ions. DI water may interact with the surface of materials like WS_2 because of its polarity, which is caused by the partial negative charge on the oxygen atom and the partial positive charge on the hydrogen atom. When DI water is added to WS_2 nanosheets, the polar water molecules engage in dipole interactions with the WS_2 surface. Even without free ions, the resultant structure promotes better charge transmission. Because the

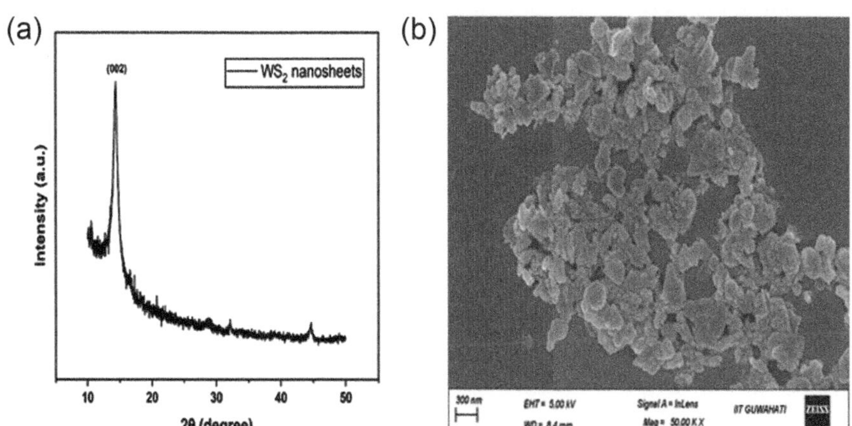

Fig. 1 **a** XRD spectra of exfoliated WS_2 nanosheets, **b** FESEM image of the WS_2 nanosheets

water molecules assist to arrange the charge distribution on the surface of WS_2, this effect increases the electrical conductivity of WS_2 in DI water.

Therefore, although DI water has a low concentration of ions, its polar characteristics allow for more effective interaction with WS_2, increasing conductivity by encouraging the flow of charge carriers. As a result of this interaction, an electric double layer forms at the water–WS_2 nanosheet interface, increasing the mobility of charge carriers close to the surface [9].

Reduced current results from the introduction of ions brought about by salts, which compete for adsorption sites and obstruct charge transfer. Because of its additions and contaminants, which further block the WS_2 surface and prevent conduction, table salt that is used regularly exhibits the lowest current among the salts.

The current–voltage (I–V) characteristics were observed on the bare WS_2-based device followed by the response observed after dropping different electrolytes such as DI water, and three different salt solutions mentioned earlier. The observed responses are presented in Fig. 2. It has been found that the device is responsive to each of the salt solutions after a voltage of above 0.5 V. If we observe, it can be seen that within the range of 0–1.75 V, the response of the device in the presence of rock salt is significantly different from other electrolytes. As iodine is a unique element that is present naturally in rock salt, the presence of this ion in the rock salt solutions could be responsible for the distinct modulation in the conductivity. The change in the V–I curve signifies that the sensor has a highly sensitive reaction kinetics to the salt solution. Therefore, the present device could be a promising alternative for the detection of iodine.

When compared to conventional ion sensors, the WS_2-based device's detection sensitivity was shown to be greater, providing a more dependable way to monitor iodine. The gadget also demonstrated outstanding sensor characteristics, such as quick reaction time, high repeatability, and sustained stability, which qualified it for useful, real-world uses. In order to increase conductivity and allow the sensor to function even in the absence of important ions [10], the contact between the polar

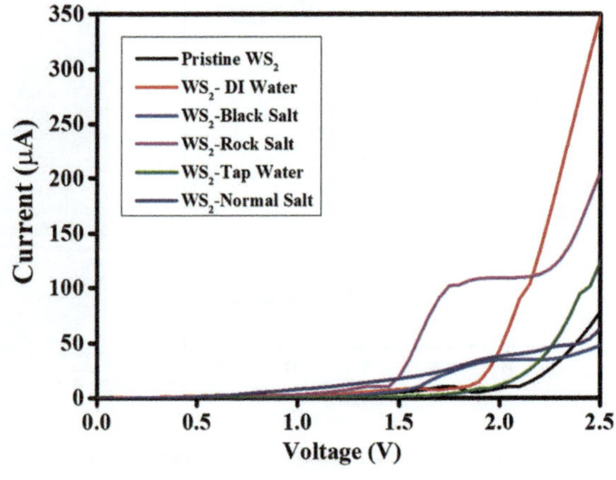

Fig. 2 I–V response curve in presence of different salt solutions

water molecules and the WS_2 surface is essential. The promise of WS_2 nanosheets as an affordable, user-friendly iodine detection solution is demonstrated by this combination of high sensitivity and advantageous sensor properties, providing a possible substitute for traditional methods for food safety and health monitoring.

4 Conclusion

The potential of WS_2 nanosheets as a sensing material for iodine ion detection in salt solutions was effectively proven in this work. The sensor was a potential option for iodine monitoring in daily life because of its high repeatability, quick reaction time, and outstanding sensitivity. Increased conductivity and better sensor performance resulted from the interaction between the polar water molecules and the WS_2 surface, which boosted the charge transfer efficiency. Notably, the WS_2-based gadget offered an affordable and easy-to-use substitute for conventional iodine detection techniques, which are sometimes intricate and costly. In addition to demonstrating its great accuracy in iodine ion detection, the WS_2 nanosheet-based sensor also showed that using 2D materials for environmental and public health monitoring is feasible. This gadget might be a dependable instrument for making sure salt's iodine levels are within acceptable bounds, helping to avoid iodine deficiency diseases (IDDs), thanks to its high sensitivity and advantageous sensor characteristics. Future research may concentrate on enhancing the sensor's scalability, stability, and incorporation into field-ready, portable devices for extensive application in food safety monitoring.

References

1. C. J. Eastman and M. B. Zimmermann, *The Iodine Deficiency Disorders*. MDText.com, Inc., South Dartmouth (MA), 2000. [Online]. Available: http://europepmc.org/books/NBK285556.
2. "Economic growth, rising incomes," 2017. [Online]. Available: www.tarina.tci.cornell.edu.
3. "GOVERNMENT OF INDIA MINISTRY OF HEALTH AND FAMILY WELFARE DEPARTMENT OF HEALTH AND FAMILY WELFARE."
4. W. Zhao *et al.*, "Evolution of Electronic Structure in Atomically Thin Sheets of WS $_2$ and WSe $_2$," *ACS Nano*, vol. 7, no. 1, pp. 791–797, Jan. 2013, https://doi.org/10.1021/nn305275h.
5. M. Chhowalla, Z. Liu, and H. Zhang, "Two-dimensional transition metal dichalcogenide (TMD) nanosheets," *Chem Soc Rev*, vol. 44, no. 9, pp. 2584–2586, 2015, https://doi.org/10.1039/C5CS90037A.
6. D. M. Wilson, S. Hoyt, J. Janata, K. Booksh, and L. Obando, "Chemical sensors for portable, handheld field instruments," *IEEE Sens J*, vol. 1, no. 4, pp. 256–274, 2001, https://doi.org/10.1109/7361.983465.
7. R. K. Jha and P. K. Guha, "Liquid exfoliated pristine WS2 nanosheets for ultrasensitive and highly stable chemiresistive humidity sensors," *Nanotechnology*, vol. 27, no. 47, Oct. 2016, https://doi.org/10.1088/0957-4484/27/47/475503.
8. G. Benedetto and K. A. Mirica, "Conductive Framework Materials for Chemiresistive Detection and Differentiation of Toxic Gases," *Acc Chem Res*, vol. 57, no. 19, pp. 2775–2789, Oct. 2024, https://doi.org/10.1021/acs.accounts.4c00319.

9. Y. Wang, T. Sohier, K. Watanabe, T. Taniguchi, M. J. Verstraete, and E. Tutuc, "Electron mobility in monolayer WS2 encapsulated in hexagonal boron-nitride," *Appl Phys Lett*, vol. 118, no. 10, Mar. 2021, https://doi.org/10.1063/5.0039766.
10. **C. Shahar *et al.*, "Surface Functionalization of WS $_2$ Fullerene-like Nanoparticles," *Langmuir*, vol. 26, no. 6, pp. 4409–4414, Mar. 2010,** https://doi.org/10.1021/la903459t.

Autoencoder-Based Centrality for Topic Representative Term Extraction in Short-Text Clustering

Tapan Mahata, Abhradeep Datta, Debanga Raj Neog, and Ashok Singh Sairam

Abstract Short-text clustering is crucial for extracting knowledge from online social media platforms. The task is challenging due to the limited text length and various forms of noise. Short texts about a specific topic often share common terms that are representative of a topic, known as topic-representative keywords. The goal is to identify these keywords by clustering at the word level. The short texts are represented as a word graph network where nodes represent words and edges represent co-occurrence or semantic relationships, capturing the context. We propose a novel centrality measure, Encoder Indexed Centrality (EIC), that is applied to the graph to identify the topic-representative keywords. The proposed metric is based on the latent representation of existing edge-weighted centrality measures. By measuring thematic closeness and semantic coherence, we discover semantically relevant word clusters around each keyword. This induced word cloud serves as the basis for short-text clustering. Our model, implemented on a short-text Twitter dataset, demonstrates superior performance compared to various baselines.

Keywords Short-text clustering · Keyword extraction · Graph centrality

T. Mahata (✉) · A. Datta · D. R. Neog · A. S. Sairam
Indian Institute of Technology Guwahati, Guwahati, Assam, India
e-mail: t.mahata@iitg.ac.in

A. Datta
e-mail: a.datta@iitg.ac.in

D. R. Neog
e-mail: dneog@iitg.ac.in

A. S. Sairam
e-mail: ashok@iitg.ac.in

U. S. Dixit et al. (eds.), *Current Progress in Engineering Sciences*,
https://doi.org/10.1007/978-981-96-8753-4_28

1 Introduction

Social media platforms have revolutionized communication among people in the form of shared interests and opinions about various social events. Text clustering, an unsupervised learning technique is primarily used to cluster the social media text and extract the topic-representative keywords. However, short-text clustering comes with its own set of challenges. Short texts create sparse feature vectors with TF-IDF (Term Frequency-Inverse Document Frequency) [13] or bag-of-words [8], complicating pattern recognition for clustering algorithms. Despite fewer words, a large vocabulary leads to high-dimensional feature spaces and the curse of dimensionality. Additionally, short texts are often ambiguous and noisy, lacking the context of longer documents, making it difficult to understand the meaning and to perform clustering.

Short-text clustering methods can be broadly categorized into three categories: distance-based, topic modeling, and deep learning-based. Distance-based clustering uses vector representations like bag-of-words and TF-IDF, applying distance metrics (e.g., Euclidean, Cosine, Manhattan, and Pearson) to group texts. Topic modeling clusters texts by identifying latent topics using techniques like LDA (Latent Dirichlet Allocation) [2] and LSA (Latent Semantic Analysis) [3]. NMF (Non-negative Matrix Factorization) [17] is a popular technique used in topic modeling and document clustering. Deep learning approaches involve word embeddings from language models and semantic composition to represent phrases or sentences, enabling more effective short-text grouping.

However, the above-mentioned clustering paradigms come with their own set of challenges. The high-dimensional feature space of the vector space model [14] fails to capture the rich sequential and contextual information of the data, especially short texts. Topic modeling lacks clear interpretability, complex parameter adjustment, and poor representation of short text. Topic models depend upon the rich co-occurrence of words in documents which is absent in short texts. Deep learning-based contextual embedding like BERT (Bidirectional Encoder Representations from Transformers) [4] solves the problem to some extent but comes with a high computational cost and is resource-intensive. The performance depends on the quality and diversity of the training data.

In this research, we overcome the challenge of short-text clustering by low-resource graph-based method to discover topic-representative keywords. The motivation is that short texts, being syntactically incomplete and less contextual, often share certain common terms that can effectively represent a topic. These inherent topics can be mapped to a cluster of short texts. We use the concept of graph centrality to mark the topic-representative keywords and also formulate a mechanism to discover semantically relevant word clusters around every topic-representative keyword. Our main contributions are as follows:

- We aim to solve the problem of short-text clustering at the word level by developing a word graph network where nodes represent unique words in the corpus and edges represent co-occurrence or semantic relationships between words, capturing the context.

- We use an autoencoder-based [7] dimension reduction technique to create a unique centrality measure called EIC (Encoder Indexed Centrality), which is based on the latent representation of existing edge-weighted centrality measures (Betweenness, Closeness, Degree, Eigenvector, Clustering Coefficient, and PageRank).
- EIC (Encoder Indexed Centrality) is applied on the word graph network with a window size of three to determine the topic-representative keywords.
- We discover closely bound semantically relevant word clusters around every topic-representative keyword by measuring thematic closeness and semantic coherence. This induced word cloud around each topic-representative keyword acts as a base for short-text clustering.

2 Related Works

The existing short-text clustering research can be reviewed in three categories: distance-based clustering, topic modeling, and deep learning-based clustering.

2.1 *Distance Based Clustering*

This method groups text vectors using distance metrics and clustering algorithms. The vector space model [14] represents documents as high-dimensional vectors, with TF-IDF highlighting term importance. To overcome the sparseness of using a vector space model in short texts, word embedding methods enhance feature representation. Clustering algorithms like k-means, hierarchical clustering, and DBSCAN are applied to these features. The choice of distance measure, such as Euclidean, Cosine, Manhattan, or Pearson, significantly impacts clustering performance, cluster shape, and result interpretation.

The high-dimensional feature space of the vector space model often fails to capture the sequential and contextual richness of short texts. This issue is mitigated by enhancing semantic representation using external knowledge bases like Wikipedia,[1] which address the sparsity of raw word features. For instance, Banerjee et al. [1] augmented TF-IDF representations with relevant Wikipedia concepts, while Zheng et al. [19] enriched short-text features by mapping documents to a hidden semantic space, adding virtual term frequencies of new words to improve clustering performance.

[1] https://www.wikipedia.org/.

2.2 Topic Modeling

Topic modeling-based clustering is a method that groups short texts by identifying underlying topics within a collection of documents. Unlike traditional clustering which directly groups documents based on similarity measures, topic modeling first identifies latent topics and then clusters documents based on these topics. LDA [2], LSA [3], and NMF [17] are some of the popular topic modeling techniques. However, they lack clear interpretability, complex parameter adjustment, and poor representation of short text. Also, the success of topic models depends upon the rich co-occurrence of words in documents which is absent in short texts.

Yin et al. [18] presented GSDMM for short-text clustering, using the Dirichlet Multinomial Mixture (DMM) model. GSDMM is a probabilistic generative model for the short-text corpus. It addresses the sparsity issue of short texts in clustering tasks by assuming that each short text is created by a single latent topic. The themes of brief texts are determined using Collapsed Gibbs sampling methods and utilized as cluster labels.

2.3 Deep Learning-Based Clustering

Deep learning has advanced short-text grouping through two main approaches: word embeddings and semantic composition at the sentence level. Word embeddings like Word2Vec [9], GloVe [11], and FastText [6] train language models to represent words as dense, low-dimensional vectors, forming semantic clusters. However, they treat words in isolation, neglecting context and sentence structure. Semantic composition models like BERT address these limitations by generating contextual embeddings, but they are computationally intensive and dependent on the quality of training data.

To tackle short-text clustering, we propose a low-resource method combining graph-based techniques with deep learning. Instead of clustering documents, we focus on identifying topic-representative keywords. We model this by creating a word graph network, where nodes are words and edges represent word co-occurrences. A novel centrality measure, Encoder Indexed Centrality (EIC), is introduced, integrating various edge-weighted centrality measures (Betweenness, Closeness, Degree, Eigenvector, Clustering Coefficient, and PageRank) using an autoencoder-based dimension reduction technique. EIC combines multiple centralities into a single index, providing a superior descriptor for keyword extraction.

3 Methodology

We first introduce the terminologies and notations in our proposed algorithms. A corpus D is a collection of m documents denoted as $D = \{d_1, d_2, d_3, \ldots, d_m\}$, where $|d_i|$ is the average number of words in each document d_i. The problem of short-text

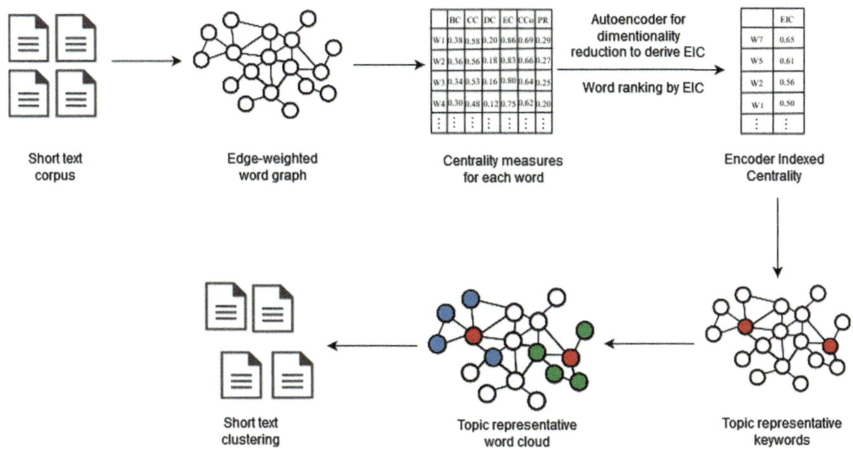

Fig. 1 Short-text clustering architecture using topic-representative keywords

clustering is to partition D into different groups, in which short texts in the same group are more similar to each other than those in other groups.

The computational framework of our proposed short-text clustering approach using topic-representative keywords is depicted in Fig. 1. First, we construct an edge-weighted word graph and compute various centrality measures for the nodes. These measures are integrated into a single metric, Enhanced Integrated Centrality (EIC), to identify the top words as topic-representative keywords. Clusters are then formed around these keywords by finding semantically relevant words. The resulting word clouds serve as the basis for clustering the short texts.

3.1 Edge Weighted Word Graph

The documents are tokenized into individual words, forming a vocabulary used to construct the word graph network $G = (V, E)$, where V consists of unique words and E represents corpus-level co-occurrence frequencies between two corresponding words within a window size of 3. A smaller window size is used to maintain precision, as words further apart contribute less to the document's semantics [15].

3.2 Node Ranking by Centrality Measures

We aim to identify the most important nodes in an edge-weighted word graph. Centrality measures rank the importance of vertices based on their capacity for influence. We use six centrality measures to rank the nodes: Degree, Betweenness, Closeness,

Eigenvector, Clustering coefficient, and PageRank. Each measure focuses on different network parameters, such as local information flow, network load, and regional connections. Below, we briefly discuss each of these measures.

- Degree centrality: The weighted degree centrality $C_D(v)$ is given by

$$C_D(v) = \sum_{u \in Ne(v)} w_{vu}$$

  where w_{vu} is the edge weight between nodes v and u, and $Ne(v)$ is the set of neighbors of v.
- Betweenness centrality: The weighted betweenness centrality $C_B(v)$ is given by

$$C_B(v) = \frac{\sum_{s \neq v \neq u} \frac{\sigma_{su}(v)}{\sigma_{su}}}{(N-1)(N-2)/2}$$

  where $\sigma_{su}(v)$ is the number of shortest paths between nodes s and u with node v acting as an intermediate node in the shortest path, σ_{su} is the total number of shortest paths between nodes s and u, N is the total number of nodes in the network.
- Closeness centrality: The weighted closeness centrality $C_C(v)$ is given by

$$C_C(v) = \frac{N-1}{\sum_{\forall u, u \neq v} d(u, v)}$$

  where $d(u, v)$ is the shortest path distance between nodes u and v, N is the total number of nodes in the network, $\sum_{u \neq v} d(u, v)$ is the total farness of node v, i.e., the sum of its distances to all other nodes.
- Eigenvector centrality: The eigenvector centrality $C_E(v)$ is given by

$$C_E(v) = \frac{1}{\lambda} \sum_{u \in Ne(v)} w_{vu} C_E(u)$$

  where $C_E(v)$ is the eigenvector centrality of the node v, $Ne(v)$ is the set of neighbors of node v. w_{vu} is the weight of the edge between nodes v and u, and λ is a constant (the largest eigenvalue of the adjacency matrix).
- Clustering Coefficient: The weighted clustering coefficient $C(v)$ is given by

$$C(v) = \frac{1}{d(v)(d(v)-1)} \sum_{s,u \in Ne(v), s \neq u} \frac{w_{vs} + w_{vu}}{2}$$

  where $d(v)$ is the degree of node v, $Ne(v)$ is the set of neighbors of node v, w_{vs} is the weight of the edge between node v and node s, w_{vu} is the weight of the edge between node v and node u the summation $\sum_{s,u \in Ne(v), s \neq u}$ is over all distinct pairs of neighbors s and u of node v.

– PageRank: The PageRank $PR(v)$ is given by

$$PR(v) = \frac{1-d}{N} + d \sum_{u \in M(v)} \frac{PR(u) \cdot w_{uv}}{\sum_{s \in N(u)} w_{us}}$$

where d is the damping factor ($0 \leq d \leq 1$), N is the total number of nodes in the network, $M(v)$ is the set of nodes that link to node v (in-neighbors of v), $N(u)$ is the set of nodes that node u links to (out-neighbors of u), w_{uv} is the weight of the edge from node u to node v, $\sum_{s \in N(u)} w_{us}$ is the sum of the weights of all outgoing edges from node u.

3.3 *Encoder Indexed Centrality (EIC): A Unified Measure*

Traditional centrality measures rank network nodes based on specific aspects such as Degree (local connectivity), Betweenness (bridging capability), or Eigenvector (influence through neighbors), often leading to conflicting results. No single measure comprehensively captures node importance.

The Encoder Indexed Centrality (EIC) provides a comprehensive solution for ranking network nodes by combining the strengths of multiple centrality measures into a single, unified index. By leveraging an autoencoder, EIC creates a latent representation that captures the essential features of various centrality measures, offering a holistic perspective on node importance. This approach enables EIC to generalize effectively across diverse datasets, as it focuses on the relationships between centrality measures rather than dataset-specific characteristics. Additionally, the dimensionality reduction inherent in the autoencoder architecture filters out noise and irrelevant variations, resulting in a robust and reliable representation of node importance.

EIC employs an autoencoder to reduce the six input centrality measures to a single latent representation, preserving essential information while minimizing reconstruction error. This unsupervised approach ensures that this new reduced representation generalizes effectively across datasets, as it captures the shared relationships between centrality measures rather than relying on dataset-specific features.

Figure 2 illustrates the architecture of the autoencoder used in the EIC model.

Encoder The encoder En maps the input $x \in \mathbb{R}^6$ to a one-dimensional latent representation $z \in \mathbb{R}$. The encoding process is represented as

$$h_l = R(W_l^e h_{l-1} + b_l^e), \quad l \in \{1, 2, \ldots, L_e\},$$

where:

- L_e denotes the total number of layers in the encoder,
- l is the index of the layer,

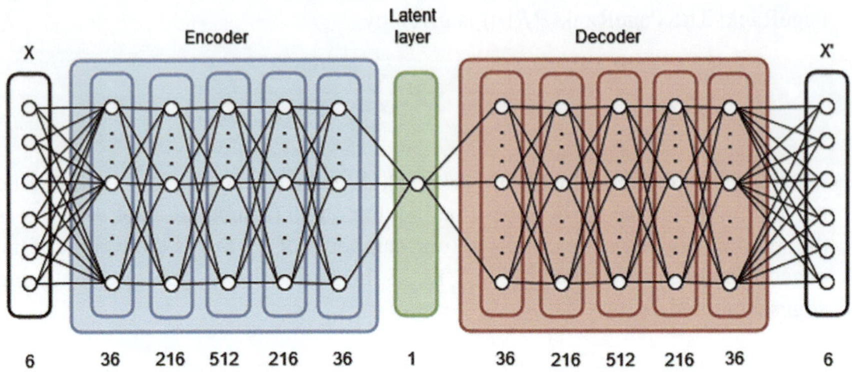

Fig. 2 Encoder indexed centrality

- $h_l \in \mathbb{R}^d$ is the output of the l-th layer, with d being the dimensionality of the layer output,
- $W_l^e \in \mathbb{R}^{d \times d'}$ and $b_l^e \in \mathbb{R}^d$ are the weight matrix and bias vector for the l-th layer,
- $R(x) = \max(0, x)$ is the ReLU activation function,
- $h_0 = x$ represents the input to the encoder.

The one-dimensional latent representation $z \in \mathbb{R}$ is computed at the final layer as

$$z = W_{L_e+1}^e h_{L_e} + b_{L_e+1}^e,$$

where $W_{L_e+1}^e \in \mathbb{R}^{1 \times d}$ and $b_{L_e+1}^e \in \mathbb{R}$.

Decoder The decoder *De* maps the one-dimensional latent representation $z \in \mathbb{R}$ back to the reconstructed output $x' \in \mathbb{R}^6$. The decoding process is represented as

$$g_l = R(W_l^d g_{l-1} + b_l^d), \quad l \in \{1, 2, \ldots, L_d\},$$

where:

- L_d denotes the total number of layers in the decoder,
- l is the index of the layer,
- $g_l \in \mathbb{R}^d$ is the output of the l-th layer,
- $W_l^d \in \mathbb{R}^{d' \times d}$ and $b_l^d \in \mathbb{R}^d$ are the weight matrix and bias vector for the l-th layer,
- $g_0 = z$ is the one-dimensional latent representation provided as input to the decoder.

The reconstructed output $x' \in \mathbb{R}^6$ is computed at the final layer as

$$x' = \sigma(W_{L_d+1}^d g_{L_d} + b_{L_d+1}^d),$$

where $W_{L_d+1}^d \in \mathbb{R}^{6 \times d}$, $b_{L_d+1}^d \in \mathbb{R}^6$, and $\sigma(x) = \frac{1}{1+e^{-x}}$ is the sigmoid activation function.

Autoencoder The autoencoder A is the composition of the encoder En and the decoder De. It is represented as

$$A(x) = De(En(x)).$$

Loss Function The autoencoder is trained to minimize the reconstruction error using the Mean Squared Error (MSE) loss function:

$$L(x, x') = \frac{1}{6}\|x - x'\|^2,$$

where $\| \cdot \|^2$ represents the squared Euclidean norm.

The six-dimensional input data is processed through the encoder's hidden layers, which learn complex relationships among the centrality features. The encoder compresses the data to a one-dimensional latent space, retaining only the essential features. The decoder then reconstructs the original six-dimensional data from this latent feature. Using ReLU activation functions, the autoencoder captures non-linear mappings and complex patterns in the data. After training, the encoder is used to transform the six-dimensional input into the one-dimensional latent feature called EIC. The top words ranked by EIC are termed as Encoder Indexed Keywords (EIK).

3.4 Topic Representative Word Cloud

After obtaining the EIK, we identify closely bound semantically relevant words. Algorithm 1 outlines the process of forming a topic-representative word cloud. The algorithm takes three inputs: the word graph (G), the EIK, and n, the predetermined number of topics to identify. The algorithm iterates through the EIK, examining each keyword. For any keyword that has not been visited, the $adjacent(word)$ function identifies its adjacent keywords. These adjacent keywords, along with the current keyword, are grouped into a word cloud (t), representing the word cloud for that keyword. To prevent redundancy, both the current keyword and its adjacent keywords are marked as visited. This process repeats until the word cloud set (T) contains at most n topics.

EIK acts as potential candidate keys representing the centroid of the word cloud and their corresponding one-hop neighbors with decreasing order of edge weight are the supporting keys of the word cloud. These terms are thematically close and semantically relevant to the topic-representative keywords (centroid).

Algorithm 1 Topic Representative Word Cloud Construction

```
Input: G = (V, E), EIK: Encoder Indexed Keywords, n: number of topics
Output: T: word cloud set
Init: T ← NULL, visited ← NULL
for each keyword ∈ EIK do
    if word ∉ visited then
        t ← adjacent(keyword) ∪ {keyword}
        T ← T.append(t)
        visited ← visited.append(t)
        if length(T) ≥ n then
            break
        end if
    end if
end for
return T
```

3.5 Short Text Clustering

The induced word cloud around each topic-representative keyword (centroid) acts as a base for short-text clustering. Each word cloud can be interpreted as a topic.

Let T is word cloud set represented as $T = \{t_1, t_2, \ldots, t_n\}$, where t_i is a word cloud containing at most k elements. We define the mapping between a short text and a word cloud by considering the length of their overlapped terms.

We assign each of the m short text $d_i \in D$, $1 \leq i \leq m$, to a word cloud t_j if d_i has the maximum number of common unigrams with the topic-representative word cloud t_j than any other word clouds. The following equation depicts the topic assignment for short texts:

$$\arg\max_{t_j \in T} |d_i \cap t_j|, \tag{1}$$

where $|d_i|$ indicates the number of words in document d_i, $|d_i \cap t_j|$ denotes the number of common words shared between d_i and t_j.

The time complexity of the proposed method is analyzed in two phases: preprocessing and assignment. In the preprocessing phase, each word cloud $t_j \in T$ is represented as a hash set containing k topic-representative keywords. Constructing hash sets for all n word clouds incurs a time complexity of $O(nk)$, where k is the number of keywords in each word cloud. This step is performed once and is independent of the document count. In the assignment phase, each document d_i is compared with all n word clouds to determine the one with the maximum overlap. For a single document, this operation has a time complexity of $O(n \cdot |d_i|)$, where $|d_i|$ is the average number of words in the document. For a dataset of m documents, the total complexity of the assignment phase is $O(m \cdot n \cdot |d_i|)$. Therefore, the overall time complexity of the method is $O(nk) + O(m \cdot n \cdot |d_i|)$. While the preprocessing step $O(nk)$ is executed only once, the assignment phase $O(m \cdot n \cdot |d_i|)$ scales with the dataset size. Given that $|d_i|$ is typically small for short texts, the method remains efficient in practical scenarios.

4 Experiment Details

In this section, we present a detailed analysis of the dataset used, the baseline methods to be compared, and evaluation metrics.

4.1 Dataset

The Twitter dataset used in our experiment is taken from [10] with 7.67 million tweets related to 6 different crisis events. However, 32,462 tweets are labeled with ground truth topics which include Sandy Hurricane, Alberta Floods, Boston Bombings, Oklahoma Tornado, Queensland Floods, and West Texas Explosion. The number of labeled tweets across the mentioned topics is given in Table 1.

4.2 Baseline Models

We compare our proposed approach with the below-mentioned methods for short-text clustering:

- LDA [2] is a generative probabilistic model used for topic modeling. It models documents as mixtures of topics, while topics are probability distributions over words. Each document has a topic distribution vector derived from word occurrence patterns. The k-means algorithm is used on the topic distribution for clustering tasks.
- NMF [17] in topic modeling is used to discover latent topics within a document collection. It is a dimension reduction technique that decomposes a document–term matrix into two lower-dimensional matrices representing the relationships between documents and topics, and topics and terms, respectively.
- GSDMM (Gibbs Sampling Dirichlet Multinomial Mixture) [18] is a model designed for clustering short texts. It is especially beneficial for short-text data that frequently lacks word co-occurrence information. The method works on the

Table 1 Dataset description

Topic	Number of labeled tweets
Sandy Hurricane	6138
Alberta Floods	5189
Boston Bombings	5648
Oklahoma Tornado	4827
Queensland Floods	5414
West Texas Explosion	5246

assumption that each short text is generated from a single latent topic and uses Gibbs sampling to get the topic index. Short texts related to the same latent topic are organized into clusters.

4.3 Evaluation Metrics

In our research, we use a short-text dataset labeled with ground truth topics. For evaluation, we automatically curate the set of ground truth keywords to better describe specific topics. We then evaluate how well the detected keywords from our topic model match these ground truth keywords. The automated addition of ground truth keywords is done using the KeyBERT model [5], parameterized with the respective Wikipedia article for each topic.

We evaluate our algorithm by comparing the ground truth topics with the detected topics. Ground truth topics are augmented with keywords from Wikipedia. Topic mapping is performed by matching keywords between each pair of ground truth and detected topics. After mapping, tweets are assigned to detected topics using Eq. (1). We use Precision, Recall, and F1-score to assess clustering accuracy, evaluating the relevance and accuracy of grouping tweets based on the discovered topics compared to the ground truth.

4.4 Result Analysis

This section explains the analysis of short-text clustering and the quality of topic-term discovery of short text.

Analysis of Short Text Clustering

4.5 Results and Discussion

Table 3 highlights the effectiveness of the proposed Encoder Indexed Centrality (EIC) compared to traditional edge-weighted centrality measures for short-text clustering. Among the conventional measures, PageRank and Closeness centrality perform relatively well, achieving F1-scores of 0.6536 and 0.6793, respectively. However, measures like Betweenness and Eigenvector centrality exhibit significantly lower effectiveness, with F1-scores of 0.3720 and 0.4055, respectively. EIC outperforms all traditional measures, achieving the highest Precision (0.7489), Recall (0.7564), and F1-score (0.7459). This superior performance is attributed to EIC's ability to

Table 2 Ground truth topic with keywords from Wikipedia

Topic	Representative Keyword Group
Sandy Hurricane	'hurricane', 'cyclone', 'sandy', 'landfall', 'storm', 'bahamas', 'caribbean', 'superstorm', 'cuba', 'tropical', 'jamaica', 'coast', 'coastal', 'atlantic', 'haiti'
Alberta Floods	'floods', 'flooding', 'alberta', 'flooded', 'disaster', 'evacuation', 'saskatchewan', 'wildfire', 'fort', 'rainfall', 'provincial', 'emergency', 'catastrophic', 'damages', 'canadian'
Boston Bombings	'tsarnaev', 'shootout', 'brothers', 'killed', 'suspects', 'terrorist', 'boston', 'marathon', 'bombs', 'bomb', 'bombing', 'shot', 'explosive', 'watertown', 'tamerlan'
Oklahoma Tornado	'tornado', 'tornadoes', 'storms', 'oklahoma', 'thunderstorm', 'windy', 'winds', 'moore', 'thunderstorms', 'storm', 'hurricane', 'meteorological', 'wind', 'ef5', 'weather'
Queensland Floods	'cyclone', 'bundaberg', 'evacuated', 'storms', 'flooding', 'queensland', 'rainfall', 'mundubbera', 'tropical', 'rescues', 'damage', 'affected', 'bay', 'impacted', 'burnett'
West Texas Explosion	'ammonium', 'explosion', 'exploded', 'fertilizer', 'explosives', 'nitrate', 'facility', 'waco', 'texas', 'west', 'emergency', 'killed', 'destroyed', 'damaged', 'firearms'

Table 3 Comparison of EIC with the centrality measures for clustering

Centrality	Precision	Recall	F1-score
Degree centrality	0.6582	0.6712	0.6646
Betweenness centrality	0.3846	0.3602	0.3720
Closeness centrality	0.6848	0.6739	0.6793
Eigenvector centrality	0.4011	0.4344	0.4055
Clustering coefficient	0.5023	0.5144	0.5082
PageRank	0.7047	0.6605	0.6536
EIC	**0.7489**	**0.7564**	**0.7459**

integrate multiple centrality measures into a unified latent representation using an autoencoder, providing a more comprehensive measure of node importance for clustering.

From Table 4, we observe that our proposed model performs the best against various baseline clustering algorithms in almost all metrics. Our model beats GSDMM, which is the highest-ranked clustering algorithm, by 6.8% in recall and 6.20% in F1-score. However, GSDMM exhibits a slightly higher precision (1.2%) score due to the fact that GSDMM explicitly models the distribution of topics within clusters and the distribution of words within topics. This dual-level modeling aids in the formation of more exact clusters by taking into account both document–topic and word–topic correlations. Our model beats NMF by a margin of 13% in precision, 6.23% in recall, 11.6% in F1-score. We observe that LDA is the least effective in

Table 4 Comparison of the proposed model with other clustering models

Model	Precision	Recall	F1-score
LDA [2]	0.4869	0.4715	0.4742
NMF [17]	0.6620	0.7120	0.6685
GSDMM [18]	**0.7579**	0.7084	0.7024
Our model	0.7489	**0.7564**	**0.7459**

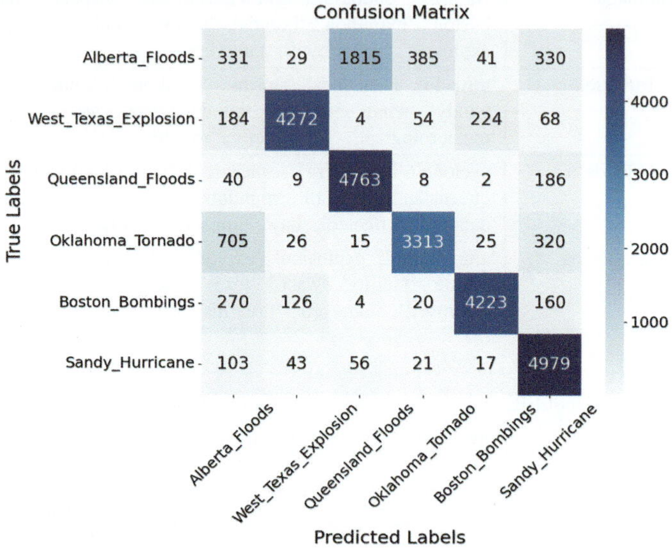

Fig. 3 Confusion matrix for EIC

short-text clustering as it requires long text where topics can be distributed over many words, and also decent word co-occurrence. The confusion matrix for calculation of the aforementioned metrics for EIC is represented in Fig. 3.

4.5.1 Analysis of Topic Term Discovery

In this section, we analyze the capacity to discover effective topic-representative keywords by LDA, NMF, and GSDMM as compared to our model. For each model, we represent the topics (Sandy Hurricane, Alberta Floods, Boston Bombings, Oklahoma Tornado, Queensland Floods, and West Texas Explosion) with the top-ranked generated keywords. Table 2 depicts the ground truth keywords for each topic. Tables 4, 5 and 6 show the resultant topic-representative keywords by LDA, NMF, and GSDMM.

We observe that EIC model (Table 8) can better uncover topic-relevant keywords as compared to the other models (Tables 5, 6 and 7).

Table 5 Topic representative word cloud from LDA

Topic	Representative keyword group
Sandy Hurricane	'hurricane', 'sandy', 'im', 'like', 'dont', 'help', 'get'
Alberta Floods	'flood', 'tornado', 'storm', 'australia', 'amazing', 'woman', 'oklahoma'
Boston Bombings	'boston', 'marathon', 'bombing', 'suspect', 'bomb', 'fbi', 'explosion'
Oklahoma Tornado	'oklahoma', 'prayer', 'thought', 'affected', 'go', 'everyone', 'boston'
Queensland Floods	'flood', 'queensland', 'crisis', 'australia', 'death', 'alberta', 'toll'
West Texas Explosion	'explosion', 'texas', 'plant', 'fertilizer', 'west', 'waco', 'video'

Table 6 Topic representative word cloud from NMF

Topic	Representative keyword group
Sandy Hurricane	'hurricane', 'sandy', 'im', 'like', 'name', 'safe', 'everyone'
Alberta Floods	'baby', 'led', 'stuffed', 'hoisted', 'powerful', 'bag', 'storm'
Boston Bombings	'boston', 'marathon', 'bombing', 'suspect', 'fbi', 'bomb', 'victim'
Oklahoma Tornado	'oklahoma', 'prayer', 'tornado', 'thought', 'go', 'affected', 'everyone'
Queensland Floods	'flood', 'queensland', 'australia', 'rise', 'water', 'crisis', 'photo'
West Texas Explosion	'explosion', 'texas', 'plant', 'fertilizer', 'west', 'waco', 'video'

Table 7 Topic representative word cloud from GSDMM

Topic	Representative keyword group
Sandy Hurricane	'hurricane', 'sandy', 'im', 'like', 'people', 'boston', 'get'
Alberta Floods	'flood', 'queensland', 'calgary', 'australia', 'alberta', 'water', 'flooding'
Boston Bombings	'boston', 'bombing', 'marathon', 'suspect', 'flood', 'queensland', 'bomb'
Oklahoma Tornado	'oklahoma', 'tornado', 'prayer', 'help', 'victim', 'boston', 'affected'
Queensland Floods	'flood', 'australia', 'queensland', 'crisis', 'rise', 'water', 'toll'
West Texas Explosion	'explosion', 'texas', 'plant', 'fertilizer', 'west', 'waco', 'boston'

Table 8 Topic representative word cloud from EIC model

Topic	Representative keyword group
Sandy Hurricane	'hurricane', 'sandy', 'like', 'affected', 'everyone', 'im', 'name'
Alberta Floods	'people', 'rainfall', 'killed', 'prayer', 'dead', 'affected', 'go'
Boston Bombings	'boston', 'marathon', 'bombing', 'suspect', 'explosion', 'bomb', 'victim'
Oklahoma Tornado	'tornado', 'oklahoma', 'victim', 'affected', 'help', 'relief', 'moore'
Queensland Floods	'flood', 'queensland', 'crisis', 'australia', 'water', 'victim', 'alberta'
West Texas Explosion	'fertilizer', 'plant', 'explosion', 'texas', 'camera', 'west', 'caught'

- Sandy Hurricane: The resultant keywords for the topic "Sandy Hurricane" are almost comparable across all models. The inherent nature of these models can identify the most salient terms in the corpus that distinguish the given topic from the rest.
- Alberta Floods: The topics "Alberta Floods" and "Queensland Floods" consist of tweets about floods and hence, have overlapping keywords. Except for EIC, all the other models capture certain incorrect keywords. For example, LDA wrongly represents "Alberta Floods" with "australia" and "oklahoma" wherein "australia" should be attached with "Queensland Floods" and "oklahoma" with "Oklahoma Tornado". Similarly, GSDMM wrongly interprets "Alberta Floods" by "queensland" and "australia" which should be mapped with "Queensland Floods". NMF has a very weak representation of the same topic. Compared to the other models, EIC prevents the misrepresentation of the topics in case of "Alberta Floods".
- Boston Bombings: GSDMM wrongly represents the topic "Boston Bombings" with "flood" and "queensland" keywords. EIC and other models have a comparable keyword representation for the aforementioned topic.
- Oklahoma Tornado: For the topic "Oklahoma Tornado," LDA and GSDMM misinterprets the topic with "boston" keyword which is related to "Boston Bombings". EIC presents the most meaningful set of keywords in the form of "tornado," "oklahoma," "victim," and "relief" for the concerned topic.
- Queensland Floods: EIC manages to prevent misrepresentation of topics in most cases except for the topic "Queensland Floods."

We conclude that in most cases EIC successfully represents the topics with meaningful keywords as compared to the other models. The quality of EIC-based keywords is comparable with the ground truth keyword group for most of the topics. EIC-indexed keywords represent the thematic centroid of the word graph and the corresponding one-hop neighboring words with decreasing order of edge weight further add up to topic-representative word cloud, enhancing semantic relevance among them.

5 Conclusion

Our research improves short-text clustering by identifying thematically enriched keyword groups that represent specific topics. Using an edge-weighted word graph network, words are ranked by Encoder Indexed Centrality (EIC) to identify key topics. Clusters form around these keywords based on thematic closeness and semantic coherence, creating a word cloud for clustering. Experiments on Twitter data show our method outperforms baseline models in clustering accuracy and topic-term discovery.

In the future, we will try to solve the problem of topic misinterpretation when the same keywords fall under multiple topics with the help of word embedding models. We would experiment with multiple short-text data of different topical domains to ensure the robustness of our model.

References

1. Banerjee, S., Ramanathan, K., Gupta, A.: Clustering short texts using wikipedia. In: Proceedings of the 30th annual international ACM SIGIR conference on Research and development in information retrieval. pp. 787–788 (2007)
2. Blei, D.M., Ng, A.Y., Jordan, M.I.: Latent dirichlet allocation. Journal of machine Learning research **3**(Jan), 993–1022 (2003)
3. Deerwester, S., Dumais, S.T., Furnas, G.W., Landauer, T.K., Harshman, R.: Indexing by latent semantic analysis. Journal of the American society for information science **41**(6), 391–407 (1990)
4. Devlin, J., Chang, M.W., Lee, K., Toutanova, K.: Bert: Pre-training of deep bidirectional transformers for language understanding. arXiv preprint arXiv:1810.04805 (2018)
5. Grootendorst, M.: Keybert: Minimal keyword extraction with Bert. (2020)
6. Joulin, A., Grave, E., Bojanowski, P., Douze, M., Jégou, H., Mikolov, T.: Fasttext. zip: Compressing text classification models. arXiv preprint arXiv:1612.03651 (2016)
7. Liou, C.Y., Cheng, W.C., Liou, J.W., Liou, D.R.: Autoencoder for words. Neurocomputing **139**, 84–96 (2014)
8. Manning, C., Schutze, H.: Foundations of statistical natural language processing. MIT Press (1999)
9. Mikolov, T., Chen, K., Corrado, G., Dean, J.: Efficient estimation of word representations in vector space. arXiv preprint arXiv:1301.3781 (2013)
10. Olteanu, A., Castillo, C., Diaz, F., Vieweg, S.: Crisislex: A lexicon for collecting and filtering microblogged communications in crises. In: Proceedings of the international AAAI conference on web and social media. vol. 8, pp. 376–385 (2014)
11. Pennington, J., Socher, R., Manning, C.D.: Glove: Global vectors for word representation. In: Proceedings of the 2014 conference on empirical methods in natural language processing (EMNLP). pp. 1532–1543 (2014)
12. Phan, X.H., Nguyen, L.M., Horiguchi, S.: Learning to classify short and sparse text & web with hidden topics from large-scale data collections. In: Proceedings of the 17th international conference on World Wide Web. pp. 91–100 (2008)
13. Robertson, S.: Understanding inverse document frequency: on theoretical arguments for IDF. Journal of documentation **60**(5), 503–520 (2004)
14. Salton, G., Wong, A., Yang, C.S.: A vector space model for automatic indexing. Communications of the ACM **18**(11), 613–620 (1975)

15. Vega-Oliveros, D.A., Gomes, P.S., Milios, E.E., Berton, L.: A multi-centrality index for graph-based keyword extraction. Information Processing & Management **56**(6), 102063 (2019)
16. Vo, D.T., Ock, C.Y.: Learning to classify short text from scientific documents using topic models with various types of knowledge. Expert Systems with Applications **42**(3), 1684–1698 (2015)
17. Xu, W., Liu, X., Gong, Y.: Document clustering based on non-negative matrix factorization. In: Proceedings of the 26th annual international ACM SIGIR conference on Research and development in information retrieval. pp. 267–273 (2003)
18. Yin, J., Wang, J.: A dirichlet multinomial mixture model-based approach for short text clustering. In: Proceedings of the 20th ACM SIGKDD international conference on Knowledge discovery and data mining. pp. 233–242 (2014)
19. Zheng, C.T., Liu, C., San Wong, H.: Corpus-based topic diffusion for short text clustering. Neurocomputing **275**, 2444–2458 (2018)

Designing and Development of Can-Sized Satellite (CANSAT): Learnings, Challenges, and Implementations

Aman Raj, Vivaswat Sinha, and Shweta Gautam

Abstract This research involves the development, design, implementation, and navigation control of an advanced comeback Can-Sized Satellite (CANSAT) that will be launched to an elevation of about 725 m using an amateur rocket from the ground surface of a Can-Sized Satellite (CANSAT). The Can-Sized Satellite (CANSAT) is a component of a project that was developed for the NASA CANSAT Competition 2021–22. We have used Problem-Based Learning (PBL) initiative allowing students with different engineering backgrounds to work together on a real-world project. Through this program which was 11 months long and includes the phase from the making of PDR, CDR to final launch where students were able to use their technical skills to design, construct and develop a working Can-Sized Satellite (CANSAT) system. This study also explores the transition from a traditional classroom environment to an online PBL environment using Can-Sized Satellite (CANSAT) kits the feedback, benefits, and challenges of the project are presented, and students find it interesting and rewarding due to its interdisciplinary nature. The study also highlights the challenges and difficulties encountered and the solutions developed by the students and facilitators to overcome them.

Keywords CANSAT · Satellite · Rocket · Launch · Navigation control · Problem-based learning · Online PBL · Electronic engineering · Aeronautical engineering · Interdisciplinary project · NASA CANSAT

A. Raj
Department of Computer Science and Engineering, Indian Institute of Technology Guwahati, Guwahati, Assam, India

V. Sinha
Department of Computer Science and Engineering, Netaji Subhas University of Technology, New Delhi, India

S. Gautam (✉)
Department of Electronics and Communication Engineering, Netaji Subhas University of Technology, New Delhi, India
e-mail: shweta.gautam@nsut.ac.in

U. S. Dixit et al. (eds.), *Current Progress in Engineering Sciences*,
https://doi.org/10.1007/978-981-96-8753-4_29

1 Introduction

PBL method is a successful educational approach used worldwide in various disciplines within higher education institutions. It emphasizes active student participation and encourages the teacher to act as a facilitator, allowing students to solve real-world case studies and assignments related to situations they may face in future practice [1].

A CANSAT design is regarded as a crucial teaching tool for satellite development and design. As a result, understanding and experiencing the entire CANSAT design, test, launch, and recovery process is critical. CANSAT is made up of several fields such as electrical circuit design, management, aerodynamics, atmospheric science, communication, programming, and so on. As illustrated in Fig. 1 the rocket and CANSAT container, a simple CANSAT comprises of a Microcontroller, Accelerometer, Temperature and Pressure Sensors, Cameras, Structure, and Parachute. It also displays the CANSAT inside an amateur rocket. The purpose of an advanced level CANSAT is to deploy and land the CANSAT at a preset target point. CANSAT will also transmit sensor information to the ground control station.

Telemetry data includes accelerations, angular velocities, GPS data, temperature and pressure values, and the state of the actuators. An ultra-light microprocessor, pressure and temperature sensors, an accelerometer, gyroscope, camera, GPS, IR distance measurement sensors, an RF communication module to interact with the ground station PC, and three actuators were utilized to control the flight and ground

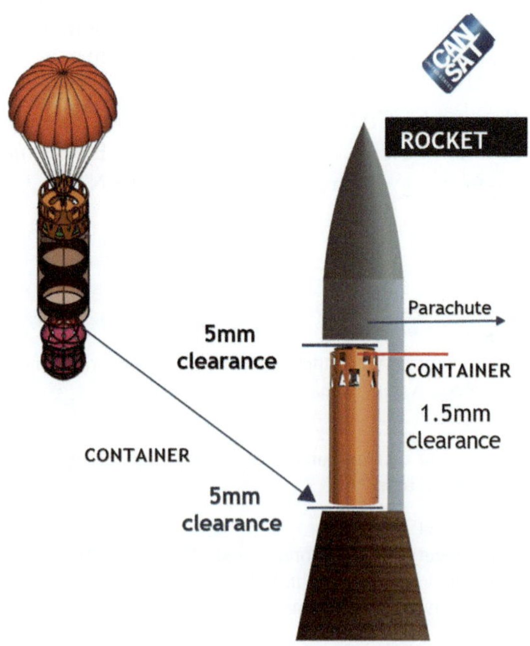

Fig. 1 The rocket and the CANSAT's container

parts, these are the motor driven p and rudder servos. PBL has been effectively utilized in different areas of higher education, various fields of study, including but not limited to, medicine, dentistry, health sciences, business, law, education, and engineering (such as aeronautical and electronics), have been included in the research [2–6].

2 Literature Review

In this section, relevant studies on making and designing of CANSAT with online learning, problem-based learning (PBL), and interdisciplinary learning are discussed. In the CLTP (CANSAT Leadership Training Program) at Wakayama University, a CANSAT has been developed and put into operation with the capability to return to its target. The CANSAT is equipped with a state-of-the-art MBED 32-bit microcontroller and features pressure, ultrasonic, GPS sensors, and a 2.4 GHz transmitter as its main components. The CANSAT's mission consists of two stages: first, it will fly back to its target with the help of a paraglider, and second, it will rove on the ground on wheels. The process will be monitored through GoogleEarth software at a ground station. This paper discusses the electronic and hardware design as well as the control algorithm of the CANSAT in depth [7].

Online learning refers to education that takes place through the internet, with both the teacher and the student having unique roles, responsibilities, and interactions within the online environment. Educators in online courses have diverse tasks to perform, such as pedagogical, instructional design, facilitation, social, administrative, and technological duties [8, 9]. The time and resources necessary to develop online material should not be underestimated, and can frequently surpass the time and resources required for traditional lecture-based courses [10, 11]. From a pedagogical standpoint, for example, the online educator must comprehend student involvement and promote active learning in a digital environment, including the creation of personalized material that engages and inspires students to develop autonomous learning [12].

The aim is to develop a descent control system for a CANSAT, which simulates a sensor payload's journey through the Martian atmosphere, collecting data on atmospheric composition during the flight. CANSAT serves as an inexpensive way to test space program designs and is an initial step in space exploration. The system design follows the guidelines of the 2016 International CANSAT Competition, which aims to imitate NASA's Mars glider project. The research presents a novel method for the descent control system, which was designed and implemented by cadets in accordance with the competition's requirements [13].

An online Problem-Based Learning (oPBL) technique is used in this research. Conventional PBL is a widely utilized and respected teaching paradigm in a variety of disciplines [14–17]. PBL began in the 6500 B.C–1500 B.C. in Indian Vedic Schools. It was created to address issues with using traditional teaching models to prepare students for various practical-based life learnings, some of the proofs can be found in the history books like Mahabharat and Ramayana where students follow the Gurukuls-based learnings and in today many medical students use it for clinical practice situations [18, 19], but it has evolved far beyond these beginnings to now promote student-centered, multidisciplinary education, and lifelong learning in professional practice across a variety of disciplines and faculties [19, 20].

CANSAT is a small satellite model that weighs between 350 and 1050 g and offers valuable training opportunities. These satellites are launched using a rocket and can reach an altitude of 150–2000 m. Once the rocket separates from the satellite, it performs tasks during free fall, which are similar to current space missions such as terraforming and airbag missions. The term CANSAT is a combination of the words "Can" and "Satellite" and originated from Japan and the United States, but undergraduate students in the researcher's country have been working on this system for the last 2 years. The satellite team, which was founded by undergraduate students, participated in international competitions and was awarded the MERIT Award for their research. This year, studies have been initiated in the Turkish Air Force Academy, marking the first step for those studying and interested in this field in their country's space activities [21].

The CANSAT platform has recently evolved from a development and interaction platform to an experimental platform for in-field sampling and data collecting at altitude. For example, [19] describes the design, data collection, analysis, and presentation of results for a series of air sampling experiments carried out using a CANSAT mini-satellite.

3 Methodology

The methodology for the development of the Can-Sized Satellite (CANSAT) focused on achieving a robust and efficient design, emphasizing practicality and scalability. The mission aimed to launch the CANSAT to an altitude of 670–725 m, ensuring controlled payload deployment and data acquisition throughout its descent. Key stages included the design of dual-stage parachute mechanisms for controlled descent, payload stabilization for accurate terrain imaging, and integration of essential sensors and communication modules for real-time telemetry.

A comprehensive simulation environment modeled atmospheric conditions, enabling iterative testing and refinement of the system. Power consumption and operational timing were meticulously calculated to optimize performance and ensure reliability during the mission. Active and passive stabilization strategies were employed to minimize swaying and maintain aerodynamic efficiency. The integration of advanced flight software and a ground control system facilitated seamless telemetry management and data visualization, ensuring the mission's success.

The primary motivation for this work stemmed from the need to create a cost-effective and scalable platform for satellite simulation and data collection. By leveraging the CANSAT, we aimed to provide a hands-on learning experience for students, bridging the gap between theoretical knowledge and practical applications in aerospace engineering. Additionally, this project sought to democratize access to satellite technology, making it more accessible for educational and experimental purposes.

Research Gap: While existing studies have focused on traditional satellite development, there is limited research on transitioning such projects into a low-cost, interdisciplinary, and modular framework. This work addresses this gap by presenting a practical model for designing and deploying a small-scale satellite that can simulate real-world missions. The integration of dual-parachute systems and remote telemetry further enhances its utility as a versatile educational tool. By filling this gap, the study opens new avenues for affordable satellite experimentation and skill-building in academic settings.

3.1 Base Design Requirements

The objective is to create a CANSAT that consists of a container and payload, which will be launched to an altitude between 670 and 725 m above the launch site and then deployed close to apogee. After deployment, the CANSAT should descend using a parachute at a speed of 15 m/s $\pm$ 5 m/s until it reaches an altitude of 400 m, where a larger parachute should be deployed to decrease the descent rate to 5 m/s $\pm$ 2 m/s. At a height of 300 m, the CANSAT should release a tethered payload, which should move 10 m away within 20 s while maintaining a video camera's orientation pointing toward the south at a 45-degree downward angle to capture the terrain in the video. All videos must be recorded and retrieved once the CANSAT has been recovered from the field.

At a certain altitude, the CANSAT should deploy a larger parachute and release a tethered payload to capture a video of the terrain. The payload should maintain a specific orientation during this time, and all recorded videos will be retrieved after recovery of the CANSAT. The size of the container and the dual parachute is given in Fig. 2.

3.2 Configurations of Container

The container is equipped with multiple sensors such as air pressure and temperature sensor, GPS sensor, two XBEEs, and two servos. The drogue parachute shroud lines will be fastened to the container mount, and the parachute will automatically deploy due to the air flow through the container's cuts. The primary parachute will be released using a rotatory spring mechanism at 400 m.

The design rationale of the payload configuration is as follows: a hybrid bullet shape is used to provide aerodynamic stability and increase drag. The use of vertical slots in the payload is intended to improve maneuverability and prevent swaying of the science payload. A motor–pulley release mechanism is employed to enable a

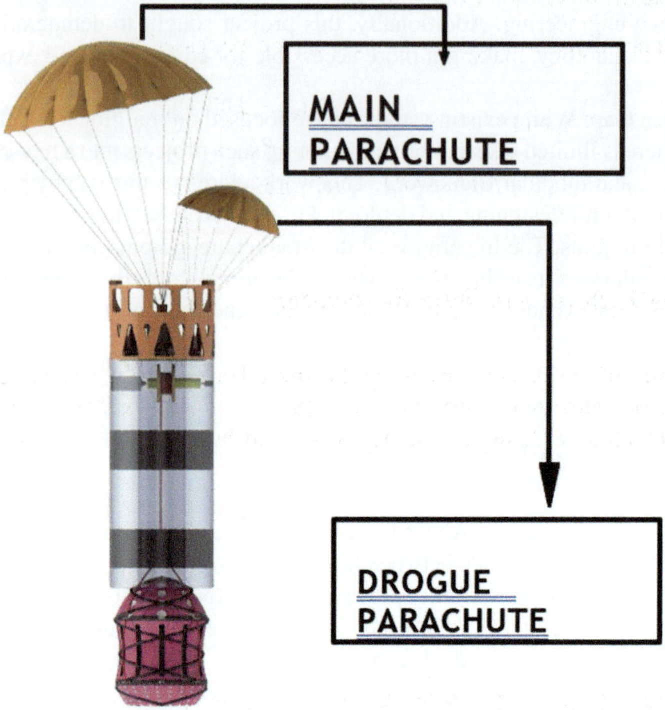

Fig. 2 The container with dual parachute

controlled release of the payload. In addition, both parachutes are of elliptical shape to ensure quick deployment. This design feature also provides a high drag coefficient while remaining lightweight and compact.

3.3 Payload Configuration

- An M5 insert is utilized to mount the fastener and securely attach both parts of the payload.
- M5 Countersunk fasteners are employed for mounting the printed circuit board (PCB).
- A battery compartment is provided for easy access.
- The base of the PCB serves as the attachment point for the vertical mounting.
- Four dedicated housings for fasteners are used to join the solid body and PCB enclosure.
- The camera is positioned to point at a 45° angle toward the Nadir.

3.4 Calculation Methodology

- Pre Launch Time (T_{Pre}) [Till the CANSAT Deployment] = 60 min
- Post Launch Time (T_{Post}) [Till CANSAT reaches ground] = Time taken by Container to reach 400 m + Time taken by container to reach 300 m where Payload deploys + Time taken by container to reach the ground

$$= 20.877\,\mathrm{s} + 18.382\,\mathrm{s} + 60.729\,\mathrm{s} = 99.988\,\mathrm{s} \approx 100\,\mathrm{s} \approx 1.67\ \mathrm{min}$$

- Recovery Stage Time (T_R) = 60 min
- Total Time (T_{Tot})

$$= T_{\mathrm{Pre}} + T_{\mathrm{Post}} + T_R = 121.67\ \mathrm{min} = 2.028\ \mathrm{hrs}$$

- Duty Cycle =

$$\left(\frac{\text{Time for which the component is active}}{\text{Total Time}}\right) \times 100\%$$

- Base Power = Current × Voltage
- Average Power Consumption = Duty Cycle × Base Power $\times T_{\mathrm{Tot}}$.

4 Project Design and Software Implementation

4.1 Project Design

I. **Design Strategies**

Passive Design Strategy: The payload design includes various tether attachment points to connect the payload with the tether. Hinge points are positioned in such a way that they can hold the payload uniformly and provide support in all directions to prevent swaying. The descent of the payload is regulated using a servo operated pulley, while different passive designs on the payload surface help prevent swaying shown in Fig. 3.

Active Design Strategy: To prevent swaying and wave drag during deployment, grid fins are attached to the payload as shown in Fig. 4. These fins are designed in a way that causes air to flow around them instead of through them, generating less drag. During the ascent to apogee, the fins are in a closed configuration, but they are opened during deployment using feedback from an altitude sensor and motor. The battery and servo motors controlling the fins are located at the bottom of the payload, which keeps the center of gravity downwards and maintains the NADIR direction.

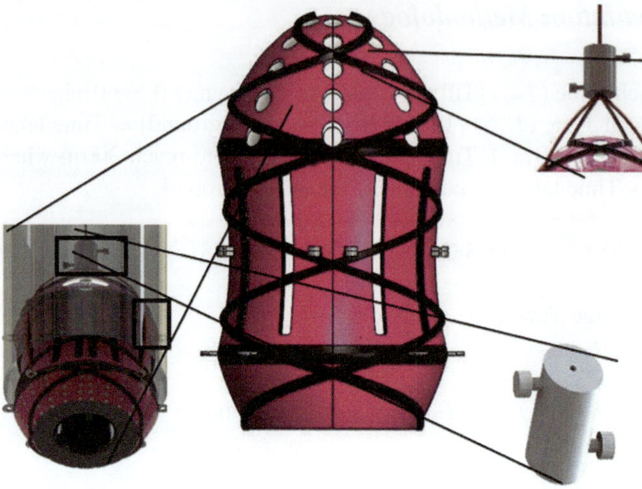

Fig. 3 Payload descent stability control overview

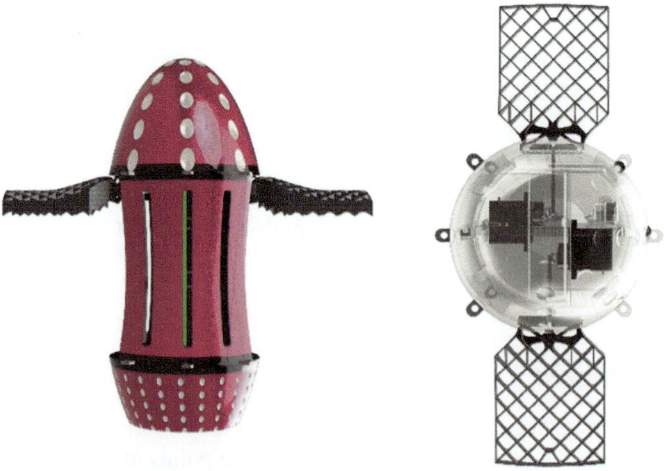

Fig. 4 Deployed configuration and servo controlled grid fins in payload

II. **Mechanical Part**

The mechanical design for the project includes several key structural components, such as the main cylindrical body, support structure for the science payload, parachute mount and chamber, payload deployment mechanism, spring mechanism for opening the main parachute lid, servo motor for shaft rotation, tether for connecting the payload with the pulley, and shaft for rotating the pulley. The container, parachute, and payload colors are orange and pink, respectively. Polycarbonate is chosen for the container and payload materials, while ripstop nylon is used for the parachute. The interface definitions include servo motor mountings and a rotatory disc mechanism for dual parachute deployments at various altitudes.

III. **Electrical Part**

This passage describes the various components involved in the design of an electrical system for a payload container. The design incorporates a microcontroller (either a Teensy 4.0 or SAM D21 Mini Breakout Board) to control and manipulate on-board data. Sensors are used to measure various types of data, including temperature, pressure, latitude, longitude, rotation, and the voltage of the battery. An internal real-time clock (RTC) is utilized to keep track of the date and time. Communication between the ground and the container, as well as between the container and payload, is facilitated through the use of an XBee Pro S2C module. An LED is used to indicate when the system is powered on, while a camera records video during the release and descent of the payload, as well as visuals from the payload's point of view. Data is stored on a micro SD card. To decrease voltage, LDOs (low-dropout regulators) are employed, with voltages reduced from 3.7 to 3.3 V, 7.4 to 5 V, and 6 V in the container, and 6–3.3 V and 6–5 V in the payload. Finally, an antenna is used to transmit and receive data between the two radios involved in the system.

IV. **Software Development Part**

A. **Container's Flight Software Design (FSW) and Ground Control System Design (GCS)**

Container will transmit its telemetry to the ground station throughout the flight at 1 Hz frequency and poll payloads telemetry at 4 Hz until it lands. This data is stored in its SD Card. In simulation mode the container pressure sensor is in standby mode. Once the payload is released from the container, they will relay their telemetry to the container at 4 Hz frequency. The payload telemetry is also stored in the SD Card. There is no change in simulation mode of payload FSW. We are using two variables Present Altitude and Previous Altitude to check altitude increasing/decreasing conditions.

Commands

ST command: Used to set the time in UTC.

CX ON/OFF command: Used to start/stop container telemetry to GCS.

PD command: Sent to payload to poll payload telemetry.

The Real-Time Plotting Software Design is used to read data from the serial port and plot real-time graphs for the container and payload in the Graphical User Interface as shown in Fig. 5. The development of the Ground Control Software involves using Commercial off-the-Shelf (COTS) software, including C/C++, Arduino, and XBEE Program Software. Calibration command has been used for the container to receive commands to start the calibration of sensors and payload. Various commands

	A	B	C	D
1	CMD	1077	SIMP	101325
2	CMD	1077	SIMP	101202.524
3	CMD	1077	SIMP	101078.969
4	CMD	1077	SIMP	100955.536
5	CMD	1077	SIMP	100834.618
6	CMD	1077	SIMP	100713.817
7	CMD	1077	SIMP	100594.328
8	CMD	1077	SIMP	100476.147
9	CMD	1077	SIMP	100359.271
10	CMD	1077	SIMP	100242.504
11	CMD	1077	SIMP	100128.228
12	CMD	1077	SIMP	100014.056
13	CMD	1077	SIMP	99901.178
14	CMD	1077	SIMP	99788.403
15	CMD	1077	SIMP	99678.102
16	CMD	1077	SIMP	99567.9
17	CMD	1077	SIMP	99458.98
18	CMD	1077	SIMP	99351.339
19	CMD	1077	SIMP	99244.974
20	CMD	1077	SIMP	99138.7
21	CMD	1077	SIMP	99033.698
22	CMD	1077	SIMP	98929.965
23	CMD	1077	SIMP	98827.497
24	CMD	1077	SIMP	98726.291
25	CMD	1077	SIMP	98625.169
26	CMD	1077	SIMP	98526.479
27	CMD	1077	SIMP	98427.87
28	CMD	1077	SIMP	98330.513
29	CMD	1077	SIMP	98233.234
30	CMD	1077	SIMP	98138.374
31	CMD	1077	SIMP	98043.589

Simulation Input

Fig. 5 Simulations

have been included in the interface as per the guidelines provided in this year's Mission Guide. In addition, telemetry data and media recording will be submitted to the judges after the completion of the flight, including the data for `.csv` files, screenshots of the interface, and media data from SD cards. The telemetry data is taken from Serial Ports and is diverted into two different files: `Flight_1077_C.csv` and `Flight_1077_T.csv`, which are first created and then stored with values. Overall, the article focuses on the technical aspects of data visualization and software design for real-time plotting, calibration, and telemetry data recording in the context of a specific mission.

Formula and Assumptions

$$\text{FSL} = \text{Gain (dB)} + \text{Power (dBm)} - \text{Cable Loss (dB)}$$

$$\textbf{Range} = 10^{\frac{FSL+20\log\lambda-21.98}{20}}$$

Cable Loss: 3 dB
Receiver Sensitivity: **−70 dBm**

Range has been calculated with respect to Receiver's antenna

B. **Simulation**

A profile of barometric pressure data that mimics the projected flight profile will be supplied at a rate of one sample per second by the team's Ground System for testing, pre-flight demonstration, and in a contingency when launch activities are not feasible. The flight software will use the received value instead of the real pressure sensor data after it has been received. Simulation output is shown in Fig. 6.

Key Points

- **FSW** changes from **FLIGHT MODE** to [•] SIMULATION MODE when both Simulation Enable and Active commands are sent to the Container Flight Software.
- Container Pressure Sensor is switched to Standby mode.
- A CSV with SIMP Commands is provided to GCS, which is read line-by-line by the GUI Backend and sent to the container at 1 Hz frequency.
- Only the container Pressure and Altitude data are simulated, and other sensor data are actual values; this data is sent to GCS in Telemetry format for plotting.
- When the simulated altitude value becomes constant as the pressure value becomes constant (i.e., the container has landed), the telemetry to GCS stops.

C. **GCS Overview**

Key Points

- The calibration, CX, ST, SIM, SIMP commands are transmitted from the ground station to the Container via Zigbee XBee Pro S2C Module, explorer module, and antenna.
- The calibration command will be transmitted from the ground station to the container.
- Telemetry data will be transmitted from the payload to the ground station via container at 300 m.
- During Simulation, for the connection using MQTT, the ground station acts as an MQTT Client that publishes telemetry data to the MQTT server on the topic named `teams/1077`.

This has been scaffolded with details and explanations of all the hardware components, programming tools, and additional reading material made accessible to each student group to design, build, and test a successful CANSAT system.

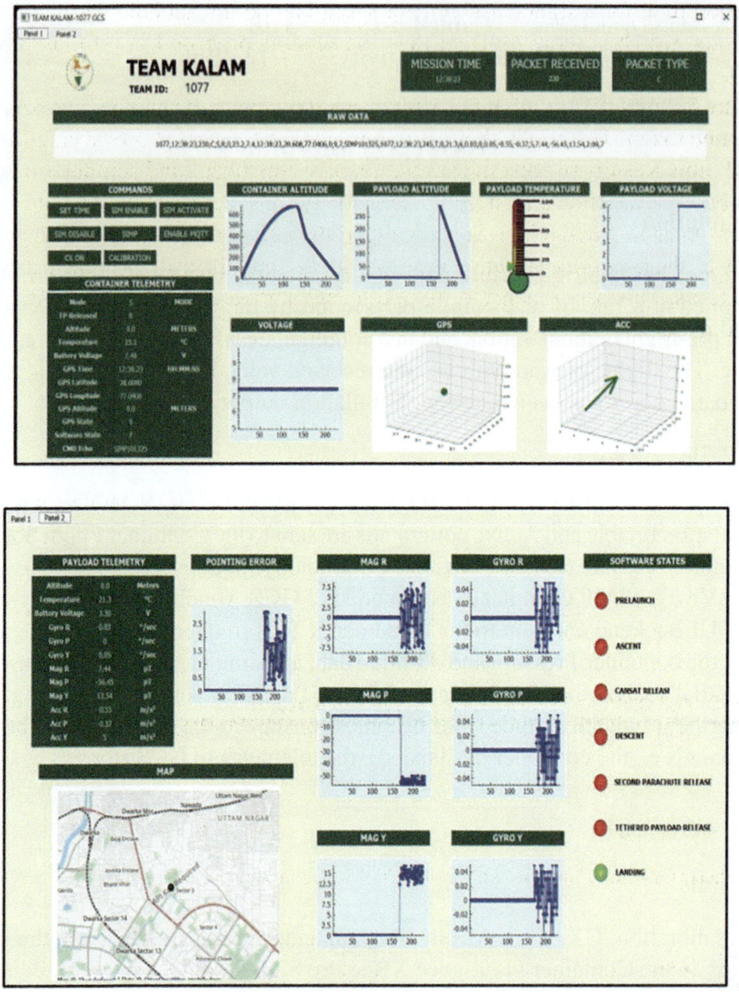

Simulation Output

Fig. 6 Simulations

GCS Specifications Include

- An external cooler with its own battery will be there, and to prevent direct exposure to sunlight, an umbrella will be installed.
- The auto-update feature will be disabled on Windows OS before the launch.
- The ground system software will also be downloaded as a backup.
- Battery backup will be greater than 3 h.

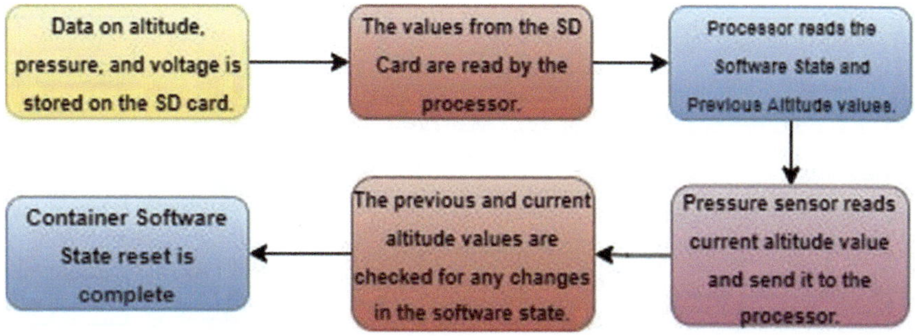

Fig. 7 Recovery methods

D. **Recovery Methods**

The software state is updated and stored on an SD card along with the altitude values. In order to recover our software in case of a reset, we have considered taking two variable approaches:

- One variable is **Previous Altitude**, which will store the latest value from the SD card.
- The **Current Altitude** variable will be fetched from the sensors.

These two variable values will help restore the software to its current state, as shown in Fig. 7.

V. **Project Implementation**

The mission described in the text involves the use of a **CANSAT**, a small satellite in the form of a can that can be launched into space. The mission is divided into five stages, as shown in Fig. 8.

Mission Stages

- **Stage 1: Preparation for Lift-off**
  The CANSAT was initialized, and the rocket armed with an electronic UWS (Universal Wiring System) module. This module ensured the separation of the rocket and the safe release of the CANSAT.
- **Stage 2: Lift-off**
  The ignition of the main rocket motor marked the launch beginning. The rocket's propulsion system was optimized for stable ascent, ensuring payload safety and precise trajectory alignment.
- **Stage 3: Reaching Apogee**
  At the apogee (670–725 m), the rocket transitioned to a horizontal position, activating the UWS module. This triggered the controlled release of the CANSAT.
- **Stage 4: Controlled Descent**
  The descent was executed with precision:

Fig. 8 Stages of CANSAT

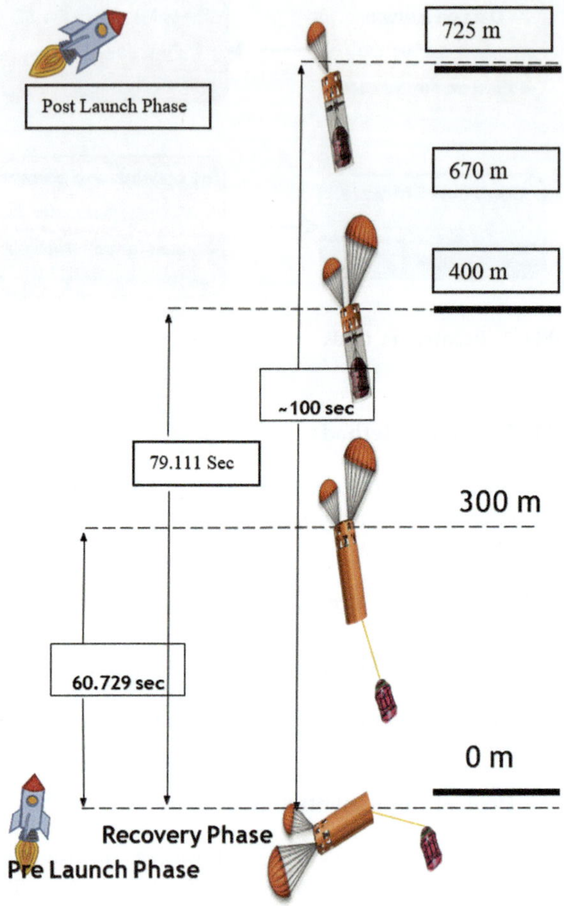

- A drogue parachute maintained stability and reduced speed to approximately 15 m/s ± 5 m/s until 400 m.
- At 400 m, a larger parachute deployed, reducing the descent rate to 5 m/s ± 2 m/s.

- **Stage 5: Landing and Recovery**
  The final stage involved the safe landing of both the rocket and the CANSAT, ensuring minimal impact damage and data retrieval.

Novelty and Targets Achieved

- **Dual-Parachute Deployment**: Ensured precise descent rates, reducing payload damage risks.
- **Payload Stabilization**: Integrated grid fins and a motor–pulley mechanism for minimal swaying and high-quality video capture.

- **Interdisciplinary Design**: Combined mechanical, electrical, and software engineering for satellite design.

Targets Achieved

- Successful telemetry data collection, including pressure, temperature, and GPS coordinates.
- Recovery of high-resolution terrain imagery.
- Validation of CANSAT's structural and functional integrity.

The mission not only achieved its technical objectives but also highlighted the potential of low-cost satellite models for educational and experimental purposes. This effort underscores the importance of interdisciplinary collaboration and innovative problem-solving in aerospace engineering. Overall, the mission involves using a CANSAT to collect data in space and safely returning it to the ground. The different stages of the mission involve preparation for lift-off, lift-off, reaching apogee, falling down at parachutes, and landing.

VI. **Impressions and Feedback**

To assess the project's efficacy, a variety of qualitative data was gathered, including student assessment scores, student comments, and facilitator observation and remark (s). Qualitative research was judged more acceptable for small samples since it provides a comprehensive description and analysis of a study issue without limiting the scope of the research or the nature of participant replies [22, 23].

The module evaluation form(s) that the students completed as part of each module were used to obtain student assessment scoring. Student input was obtained at the conclusion of the semester through student reflection(s) on the project. Feedback and comments from teachers/facilitators/judges were received directly from the teachers participated in the courses.

Overall, students were satisfied with the project, finding it interesting and useful in terms of its applicability to their respective subjects. "The project went well with our group, we finished every deliverable for the due date, and everything worked well in the organisation and in the execution of the CANSAT. We are excited to work on this project this semester and to adapt to an online version of the CANSAT project."

The CANSAT platform provided a perfect convergence between aeronautical and sensor/instrumentation technologies in electronics for students to use the theory and knowledge they had learned in the module(s): "Overall, I enjoyed the assignment. It's exciting for me as an international student to study a topic and then be able to use what I've learned." "The most valuable aspect was the experience acquired from working on a project .... It was a departure from past missions."

Because the project used a real-world and "genuine" topic, such as the development of the CANSAT, it appealed to the many skills-based disciplines involved (aeronautical and electronics), giving the students with transferable skills and knowledge [24, 25]. Because of the project's multidisciplinary character, students were able to collaborate with students from many fields and use their previous knowledge in new and unique ways.

VII. **Conclusion**

In conclusion, the CANSAT mission is a complex and multi-stage process that involves preparing the CANSAT and rocket for lift-off, ignition of the main rocket motor, reaching apogee, releasing the CANSAT, and landing safely on the ground. The use of the UWS module is a critical component in separating the rocket and releasing the CANSAT at the appropriate time. The design of the CANSAT includes the use of parachutes to ensure a safe landing, with the rocket and CANSAT landing separately. This mission requires careful planning and execution to ensure the success of each stage and the safe return of the CANSAT and rocket to the ground. Overall, the CANSAT mission provides a unique opportunity for students and researchers to learn about and experience the complexities of space technology and engineering in a hands-on way. It also highlights the importance of teamwork, careful planning, and attention to detail in the successful execution of space missions. The initiative shifted a real-world, timely, relevant, and useful problem from a traditional PBL environment to an online PBL context. Feedback, perceptions, rewards, and obstacles linked with the project are examined from the perspectives of both the student and the facilitator. While data collection and analysis for the case study are obviously restricted, feedback from all participants has been overwhelmingly favorable. The project has exceeded the organizers' expectations, and it is intended to continue implementing it in online or traditional PBL format for future iterations of the initiative.

References

1. Neville, A.J. and Norman, G.R.: PBL in the undergraduate MD program at McMaster University: three iterations in three decades. Academic Medicine **82**(4), 370–374 (2007)
2. Ahlfeldt, S., Mehta, S. and Sellnow, T.: Measurement and analysis of student engagement in university classes where varying levels of PBL methods of instruction are in use. Higher Education Research & Development **24**(1), 5–20 (2005)
3. Etchepareborda, P., Bierzychudek, M.E., Carducci, L., et al.: A project-based learning method applied to an introductory course in electronics engineering. In: 2018 IEEE World Engineering Education Conference (EDUNINE), pp. 1–4. IEEE (2018)
4. Mantri, A., Dutt, S., Gupta, J.P. and Chitkara, M.: Design and evaluation of a PBL-based course in analog electronics. IEEE Transactions on Education **51**(4), 432–438 (2008)
5. Brodeur, D.R., Young, P.W. and Blair, K.B.: Problem-based learning in aerospace engineering education. In: Proceedings of the 2002 American Society for Engineering Education Annual Conference and Exposition, Montreal, Canada, pp. 16–19 (2002)
6. Benedito, J.P., Perez-Alvarez, J. and Calzada, M.C.: PBL in the teaching of design in aeronautical engineering: Application and evolution of a consolidated methodology. The International Journal of Engineering Education **31**(1), 199–208 (2015)
7. Aydemir, M.E., Celebi, M., Ay, S., et al.: Design and implementation of a rover-back CANSAT. In: Proceedings of the 5th International Conference on Recent Advances in Space Technologies (RAST2011). https://doi.org/10.1109/RAST.2011.5966952
8. Baran, E., Correia, A.-P. and Thompson, A.: Transforming online teaching practice: Critical analysis of the literature on the roles and competencies of online teachers. Distance Education **32**(3), 421–439 (2011)

9. Evans, S., Knight, T., Walker, A. and Sutherland-Smith, W.: Facilitators' teaching and social presence in online asynchronous interprofessional education discussion. Journal of Interprofessional Care **34**(4), 435–443 (2020)
10. O'Donoghue, G., McMahon, S., Doody, C., et al.: Problem-based learning in professional entry-level therapy education: A review of controlled evaluation studies. Interdisciplinary Journal of Problem-Based Learning **5**(1), 54–73 (2011)
11. Serrano, D.R., Dea-Ayuela, M.A., Gonzalez-Burgos, E., et al.: Technology-enhanced learning in higher education: How to enhance student engagement through blended learning. European Journal of Education **54**(2), 273–286 (2019)
12. de Jong, N., Savin-Baden, M., Cunningham, A.M. and Verstegen, D.M.: Blended learning in health education: three case studies. Perspectives on Medical Education **3**(4), 278–288 (2014)
13. Kizilkaya, M.O., Oguz, A.E. and Soyer, S.: CANSAT descent control system design and implementation. In: 2017 8th International Conference on Recent Advances in Space Technologies (RAST). https://doi.org/10.1109/RAST.2017.8002947
14. Bijsmans, P. and Schakel, A.H.: The impact of attendance on first-year study success in problem-based learning. Higher Education **76**(5), 865–881 (2018)
15. Farmer, J.L. and Wilkinson, L.: Engineering success: Using problem-based learning to develop critical thinking and communication skills in a Chemical Engineering classroom. In: Proceedings of the Canadian Engineering Education Association (CEEA) (2018)
16. Greening, T.: Scaffolding for success in problem-based learning. Medical Education Online **3**(1), 4297 (1998)
17. McParland, M., Noble, L.M. and Livingston, G.: The effectiveness of problem-based learning compared to traditional teaching in undergraduate psychiatry. Medical Education **38**(8), 859–867 (2004)
18. Hmelo-Silver, C.E. and Barrows, H.S.: Problem-based learning: Goals for learning and strategies for facilitating. Essential Readings in Problem-Based Learning, 69–84 (2015)
19. Thorndahl, K.L. and Stentoft, D.: Thinking critically about critical thinking and problem-based learning in higher education: A scoping review. Interdisciplinary Journal of Problem-Based Learning **14**(1) (2020)
20. Boud, D. and Feletti, G.: Initiating problem-based learning at Harvard Medical School. In: The Challenge of Problem-Based Learning. Routledge, 81–88 (2013)
21. Soyer, S.: Small space can: CANSAT. In: Proceedings of the 5th International Conference on Recent Advances in Space Technologies (RAST2011). https://doi.org/10.1109/RAST.2011.5966950
22. Collis, J. and Hussey, R.: Business Research: A Practical Guide for Undergraduate and Postgraduate Students. Macmillan International Higher Education (2013)
23. Prasad, K.N.: Research methodology notes – data collection. Dynamic Tutorials and Services. Available at: https://www.dynamictutorialsandservices.org/2017/05/research-methodology-notes-data.html, last accessed 2022/10/26
24. Donnelly, R. and Fitzmaurice, M.: Collaborative project-based learning and problem-based learning in higher education: A consideration of tutor and student roles in learner-focused strategies. Emerging Issues in the Practice of University Learning and Teaching, 87–98 (2005)
25. Jurewitsch, B.: A meta-analytic and qualitative review of online versus face-to-face problem-based learning. Journal of Distance Education **26**(2), 1 (2012)
26. Ostaszewski, M., Dzierzek, K. and Magnuszewski, Ł.: Analysis of data collected while CANSAT mission. In: 2018 19th International Carpathian Control Conference (ICCC), pp. 1–4. IEEE (2018)

Extended Kalman Filter Implementation: A Comparative Study of Python-Based DSL and Vivado HLS for Target Tracking

Devansh Gupta, Rupak Kundu, Chiranjan Roy, N. Manikanta, Niraj N. Sharma, Mandar Datar, and Sachin B. Patkar

Abstract We have developed an Extended Kalman Filter (EKF) for target tracking, which estimates seven system states by observing four measurements. This was implemented using Xilinx Vivado High-Level Synthesis (HLS) and a Python-based Domain Specific Language (DSL). The Vivado HLS exported IP was integrated with a ZYNQ System on Chip (SoC) via an AXI interface and tested on the ZCU104 FPGA Board utilizing the PYNQ framework. We calculated and compared the execution times for both hardware and software implementations to estimate the speedup. In the Python DSL, the EKF was designed using custom-made floating-point units operating on IEEE754 single-precision floating-point numbers, also designed in the Python DSL. The EKF designed in Python DSL employs *SizedFIFOs* for external interfacing and *SimpleBRAMs* for internal data handling. Finally, the accuracy of the results from the Vivado HLS exported IP integrated with the ZYNQ SoC and the Python DSL-designed EKF were compared across four iterations of the EKF.

Keywords EKF · Python · HLS · DSL · RTL · FPGA · SoC

1 Introduction

The Kalman filter (KF) [1] is a versatile tool for estimating system states using measurements collected over time. It is widely employed in applications such as target state estimation and navigation. The fundamental principle behind the Kalman

D. Gupta (✉) · R. Kundu · C. Roy · N. Manikanta · N. N. Sharma · M. Datar · S. B. Patkar
Indian Institute of Technology, Bombay, India
e-mail: iitb.devansh19@gmail.com

S. B. Patkar
e-mail: patkar@ee.iitb.ac.in

U. S. Dixit et al. (eds.), *Current Progress in Engineering Sciences*,
https://doi.org/10.1007/978-981-96-8753-4_30

filter is its ability to estimate states by integrating information from two key models: the *process model* and the *measurement model*. Typically, these models are linear, but non-linear systems can also be addressed using techniques like the Extended Kalman Filter (EKF).

The KF operates as a recursive algorithm, meaning it continuously updates its estimates as new measurements become available. The process model represents how the system evolves over time, incorporating information about the system's dynamics and how it responds to inputs or disturbances. On the other hand, the measurement model outlines the relationship between the system's true states and the measurements obtained from sensors or instruments. While these models are ideally linear, they can be adapted for non-linear systems.

The primary objective of the KF is to minimize the estimation error by optimally weighting the information from the process and measurement models. It achieves this by computing a weighted average, known as the *Kalman Gain*, which takes into account the relative reliability of the process and measurement information. This approach ensures that the filter is robust in the presence of noise and uncertainty, providing accurate state estimates even with noisy measurements or imperfect knowledge of the system's behavior.

2 Background

2.1 Kalman Filtering Algorithm

The KF [3–5] is designed for linear dynamic systems. It assumes that both the process model and the measurement model are linear functions. If the system being estimated is truly linear, the KF can provide optimal estimates. Algorithm 1 [5] explains the KF algorithm.

$$\text{State Transition Model:} \quad x_k = F_{k-1} x_{k-1} + w_{k-1} \tag{1}$$

$$\text{Measurement Model:} \quad z_k = H_k x_k + v_k \tag{2}$$

Here, F_k and H_k are the linear functions of x.

Algorithm 1 Kalman Filter Algorithm

Require: Initial state estimate $\hat{x}_{0(+)}$ and covariance estimate $P_{0(+)}$
for $k = 1$ to N **do**
  Linear dynamic model:
  $x_k = F_{k-1}x_{k-1} + w_{k-1}, \quad w_k \sim \mathcal{N}(0, Q_k)$
  Linear measurement model:
  $z_k = H_k x_k + v_k, \quad v_k \sim \mathcal{N}(0, R_k)$
  Linear implementation equations:
    Computing the a priori estimate of x_k:
    $\hat{x}_{k(-)} = F_{k-1}\hat{x}_{k-1(+)}$
    Computing the predicted measurement:
    $\hat{z}_k = H_k \hat{x}_{k(-)}$
  Computing the priori covariance matrix:
  $P_{k(-)} = F_{k-1} P_{k-1(+)} F_{k-1}{}^T + Q_{k-1}$
  Computing Innovation covariance matrix:
  $S_k = H_k P_{k(-)} H_k{}^T + R_k$
  Computing the Kalman gain:
  $K_k = P_{k(-)} H_k{}^T (S_k)^{-1}$
  Conditioning the predicted estimate on the measurement:
  $\hat{x}_{k(+)} = \hat{x}_{k(-)} + K_k(z_k - \hat{z}_k)$
  Computing the a posteriori covariance matrix:
  $P_{k(+)} = (I - K_k H_k) P_{k(-)}$
end for

2.2 *Extended Kalman Filtering Algorithm*

The EKF [6] is specifically tailored to handle non-linear dynamic systems. It does this by linearizing the non-linear models at each estimation step. This linearization involves computing the Jacobian matrices of the process and measurement models. While the EKF can handle non-linear systems, it relies on these linearizations, which may introduce errors in highly non-linear systems. Algorithm 2 [6] explains the EKF algorithm.

$$\text{State Transition Model:} \quad x_k = f_{k-1}(x_{k-1}) + w_k \tag{3}$$

$$\text{Measurement Model:} \quad z_k = h_k(x_k) + v_k \tag{4}$$

Here f_{k-1} and h_k are the non-linear functions of x.

Algorithm 2 Extended Kalman Filter Algorithm

Require: Initial state estimate $\hat{x}_{0(+)}$ and covariance estimate $P_{0(+)}$

for $k = 1$ to N **do**

Nonlinear dynamic model:

$x_k = f_{k-1}(x_{k-1}) + w_{k-1}, \quad w_k \sim \mathcal{N}(0, Q_k)$

Nonlinear measurement model:

$z_k = h_k(x_k) + v_k, \quad v_k \sim \mathcal{N}(0, R_k)$

Nonlinear implementation equations:

Computing the predicted state estimate:

$\hat{x}_{k(-)} = f_{k-1}(\hat{x}_{k-1(+)})$

Computing the predicted measurement:

$\hat{z}_k = h_k(\hat{x}_{k(-)})$

Using Jacobian to linearize equations locally:

$F_{k-1} = \frac{\partial f_k}{\partial x}\Big|_{x=\hat{x}_{k-1(-)}}, \quad H_k = \frac{\partial h_k}{\partial x}\Big|_{x=\hat{x}_{k(-)}}$

Computing the priori covariance matrix:

$P_{k(-)} = F_{k-1} P_{k-1(+)} F_{k-1}^T + Q_{k-1}$

Computing Innovation covariance matrix:

$S_k = H_k P_{k(-)} H_k^T + R_k$

Computing the Kalman gain:

$K_k = P_{k(-)} H_k^T (S_k)^{-1}$

Conditioning the predicted estimate on the measurement:

$\hat{x}_{k(+)} = \hat{x}_{k(-)} + K_k(z_k - \hat{z}_k)$

Computing the posteriori covariance matrix:

$P_{k(+)} = (I - K_k H_k) P_{k(-)}$

end for

3 Methodology

In this study, we analyze a target tracking system where the seeker measures the range, closing velocity, and azimuth and elevation angles of a target. The primary objective is to accurately estimate the true positions and velocities of the target relative to the seeker [8]. The system model employs seven states, which are estimated using four measurements from the seeker. The states in the EKF are defined as follows:

- Relative distance to the target along the X, Y, and Z axes (in meters): 3 states
- Relative velocity of the target along the X, Y, and Z axes (in meters per second): 3 states
- Ballistic coefficient: 1 state.

The implementation of the EKF utilizes these seven states and four seeker measurements.

3.1 Xilinx Vivado High-Level Synthesis

The implementation of the EKF was carried out in Vivado HLS using the C programming language, with AXI-Lite and AXI-Stream interfaces. Given that the EKF designed for this study involves various matrix computations across matrices of different dimensions, specialized matrix multipliers were developed to handle the multiplication of rectangular matrices.

Furthermore, matrix inversion is a critical component of the EKF. To address this, two algorithms were compared: the traditional brute force method using the adjoint of the matrix, and the Gauss-elimination method [7]. The performance and efficiency of these algorithms were evaluated to determine the most suitable approach for matrix inversion within the EKF framework.

3.1.1 State Prediction Model (SPM)

SPM solves the differential equations of the system using Runge–Kutta method numerically to get prior state estimate $\hat{x}_{k(-)}$.

$$\frac{\partial x}{\partial t} = v_x \qquad \frac{\partial v_x}{\partial t} = kv_x - g \qquad \frac{\partial b}{\partial t} = 0$$
$$\frac{\partial y}{\partial t} = v_y \qquad \frac{\partial v_y}{\partial t} = kv_y \qquad \frac{\partial z}{\partial t} = v_z$$
$$\frac{\partial v_z}{\partial t} = kv_z$$

$$g = 9.81\ \text{m/s}^2 \tag{5}$$

$$v_m = \sqrt{v_x^2 + v_y^2 + v_z^2} \tag{6}$$

$$k = -0.18196 \cdot b \cdot e^{(-0.0001516584(x-11000.0))} \cdot v_m \tag{7}$$

where b is ballistic coefficient (seventh state of EKF).

3.1.2 H Matrix Module

It computes the H_k matrix. Let $\mathbf{x}$ be a vector of states, and $\mathbf{z}$ be a vector of measurements:

$$\mathbf{x} = \left[x, y, z, v_x, v_y, v_z, b\right]^T$$

$$\mathbf{z} = \left[R, \dot{R}, \theta_{ele}, \theta_{azi}\right]^T$$

Let

$$\begin{bmatrix} R \\ \dot{R} \\ \theta_{ele} \\ \theta_{azi} \end{bmatrix} = \begin{bmatrix} f_1(\mathbf{x}) \\ f_2(\mathbf{x}) \\ f_3(\mathbf{x}) \\ f_4(\mathbf{x}) \end{bmatrix}$$

Then the H matrix is given as

$$H = \begin{bmatrix} \frac{\partial f_1}{\partial x} & \frac{\partial f_1}{\partial y} & \frac{\partial f_1}{\partial z} & \frac{\partial f_1}{\partial vx} & \frac{\partial f_1}{\partial vy} & \frac{\partial f_1}{\partial vz} & 0.0 \\ \frac{\partial f_2}{\partial x} & \frac{\partial f_2}{\partial y} & \frac{\partial f_2}{\partial z} & \frac{\partial f_2}{\partial vx} & \frac{\partial f_2}{\partial vy} & \frac{\partial f_2}{\partial vz} & 0.0 \\ \frac{\partial f_3}{\partial x} & \frac{\partial f_3}{\partial y} & \frac{\partial f_3}{\partial z} & \frac{\partial f_3}{\partial vx} & \frac{\partial f_3}{\partial vy} & \frac{\partial f_3}{\partial vz} & 0.0 \\ \frac{\partial f_4}{\partial x} & \frac{\partial f_4}{\partial y} & \frac{\partial f_4}{\partial z} & \frac{\partial f_4}{\partial vx} & \frac{\partial f_4}{\partial vy} & \frac{\partial f_4}{\partial vz} & 0.0 \end{bmatrix}$$

where f_1, f_2, f_3, and f_4 are given by the following equations:

$$R = \sqrt{x^2 + y^2 + z^2} \tag{8}$$

$$\dot{R} = \frac{x \cdot v_x + y \cdot v_y + z \cdot v_z}{R} \tag{9}$$

$$\theta_{ele} = \tan^{-1}\left(\frac{x}{\sqrt{y^2 + z^2}}\right) \tag{10}$$

$$\theta_{azi} = \tan^{-1}\left(\frac{y}{z}\right) \tag{11}$$

In this context, x refers to the prior estimate of the state at time step k, denoted as $\hat{x}_{k(-)}$. The matrix H is a 4×7 matrix.

3.1.3 h Function Module

It computes the predicted measurements $\hat{z}_k = h_k(\hat{x}_{k(-)})$ using Eqs. 8–11.

3.1.4 State Transition Matrix Module

The state transition matrix F_{k-1} is computed by taking Jacobian of non-linear dynamic state equations, which gives us locally linearized system model. From Eqs. 5 to 7,

$$\rho_{\text{air}} = -0.18196e^{-0.0001516584(x-11000.0)} \tag{12}$$

$$v_m = \sqrt{v_x^2 + v_y^2 + v_z^2} \tag{13}$$

$$f_1 = \rho_{\text{air}} v_m v_x \tag{14}$$

$$f_2 = \rho_{\text{air}} v_m v_y \tag{15}$$

$$f_3 = \rho_{\text{air}} v_m v_z \tag{16}$$

$$F_{k-1} = \begin{bmatrix} 1.0 & 0.0 & 0.0 & \Delta t & 0.0 & 0.0 & 0.0 \\ 0.0 & 1.0 & 0.0 & 0.0 & \Delta t & 0.0 & 0.0 \\ 0.0 & 0.0 & 1.0 & 0.0 & 0.0 & \Delta t & 0.0 \\ 0.0 & 0.0 & 0.0 & 1.0 & 0.0 & 0.0 & f_1 \Delta t \\ 0.0 & 0.0 & 0.0 & 0.0 & 1.0 & 0.0 & f_2 \Delta t \\ 0.0 & 0.0 & 0.0 & 0.0 & 0.0 & 1.0 & f_3 \Delta t \\ 0.0 & 0.0 & 0.0 & 0.0 & 0.0 & 0.0 & 1.0 \end{bmatrix}$$

Fig. 1 Implementation of the EKF

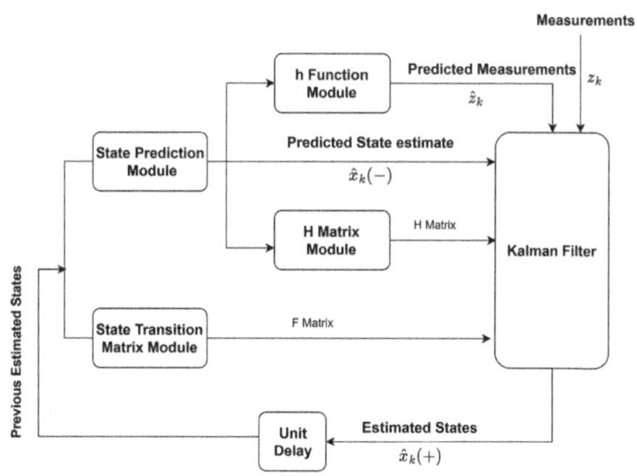

The various modules in EKF implementation are given in Fig. 1.

Table 1 shows the comparison of resource utilization for EKF using different matrix inversion algorithms. Specifically, the adjoint method demonstrates faster computation times compared to the Gaussian elimination method. However, this increase in speed comes at the cost of higher resource utilization. In contrast, the Gaussian elimination method, while slower, is more resource-efficient.

The EKF designed using Vivado HLS is exported as IP and integrated with Zynq Processing System and other modules such as AXI DMA and AXI Smart Connect and the generated bitstream was used to configure ZCU104 FPGA Board. The Resource utilization and Power breakup for integrated design is shown in Tables 2 and 3 respectively. In Table 3, the power utilization of the integrated design reveals that the SoC ARM processor constitutes almost 87% of the dynamic power consumption in both of the cases, while the rest of the designed hardware components, including

Table 1 Resource utilization of EKF for different matrix inversion algorithms used

S. no	Description	Resource utilization (adjoint method)	Resource utilization (Gauss-elimination method)
1	Clock: Tmax (target)	10.00 ns	10.00 ns
2	Clock: Tmax (achieved)	8.741 ns	9.100 ns
3	Latency (max)	**2134 cycles**	*2491 cycles*
4	BRAM	82/624 ~13%	78/624 ~12%
5	DSP	*131/1728 ~7%*	**28/1728 ~1%**
6	FF	*15648/460800 ~3%*	**8405/460800 ~1%**
7	LUTs	*21261/230400 ~9%*	**14636/230400 ~6%**

Table 2 Resource utilization for integrated EKF hardware accelerator

S. no	Description	Resource utilization (integrated hardware) (adjoint method)	Resource utilization (integrated hardware) (Gauss-elimination method)
1	Clock: Tmax (target)	10.00 ns	10.00 ns
2	WNS (setup slack)	4.229 ns	4.712 ns
3	WHS (Hold Slack)	0.010 ns	0.010 ns
4	BRAM	38/312 ~12.18%	36/312 ~11.54%
5	DSP	101/1728 ~5.84%	22/1728 ~1.26%
6	FF	22264/460800 ~4.83%	15480/460800 ~3.36%
7	LUTs	24475/230400 ~10.62%	14852/230400 ~6.5%

the EKF Hardware Accelerator and AXI DMA, contribute 13% to the overall power consumption.

3.2 *Python-Based DSL Implementation*

The Python DSL converts the code written in DSL to Verilog code [2]. The EKF implemented using this tool primarily comprises *SimpleBRAMs* and *SizedFIFOs* to handle data, enabling effective use of parallelism and pipelining through DSL constructs. Since the current version of the tool lacks support for floating-point data types, a custom floating-point unit was designed using Python DSL, utilizing the IEEE 754 single-precision floating-point format. Figure 2 depicts the architecture for EKF implementation in Python DSL. This architecture includes various *SizedFI-*

Table 3 Power utilization for integrated EKF hardware accelerator

S. no	Description	Resource utilization (integrated hardware) (adjoint method)	Resource utilization (integrated hardware) (Gauss-elimination method)
1	Total power consumption	3.794 W	3.741 W
2	Dynamic power	3.099 W ~82%	3.047 W ~81%
3	Static power	0.695 W ~18%	0.694 W ~19%
Dynamic power breakup			
4	BRAM	0.053 W ~2%	0.071 W ~2%
5	DSP	0.050 W ~2%	0.016 W ~1%
6	ARM processor	2.660 W ~85%	2.660 W ~87%
7	Logic	0.120 W ~4%	0.121 W ~4%
8	Clock	0.064 W ~2%	0.049 W ~2%

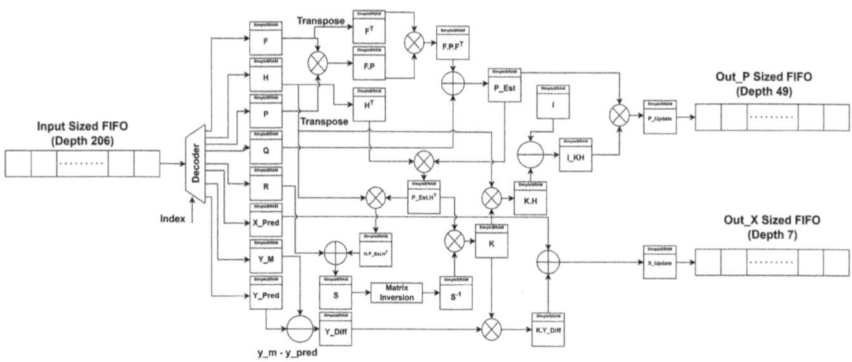

Fig. 2 Architecture of EKF hardware in Python DSL

FOs, *SimpleBRAMs*, floating-point matrix multipliers, matrix adders, and a matrix inversion module.

3.2.1 Matrix Multiplication

Here we will discuss the algorithm used for implementing matrix multiplication. The multiplication involves two matrices, $A_{N \times Q}$ and $B_{Q \times M}$, resulting in a product matrix $C_{N \times M}$. The algorithm utilizes *SimpleBRAMs* and one Floating Point FMA module.

- Initialization of *SimpleBRAMs*: Matrices $A_{N \times Q}$ and $B_{Q \times M}$ are fed to *SimpleBRAMs*, while *SimpleBRAM C* is simultaneously initialized with zeros as shown in Fig. 3.

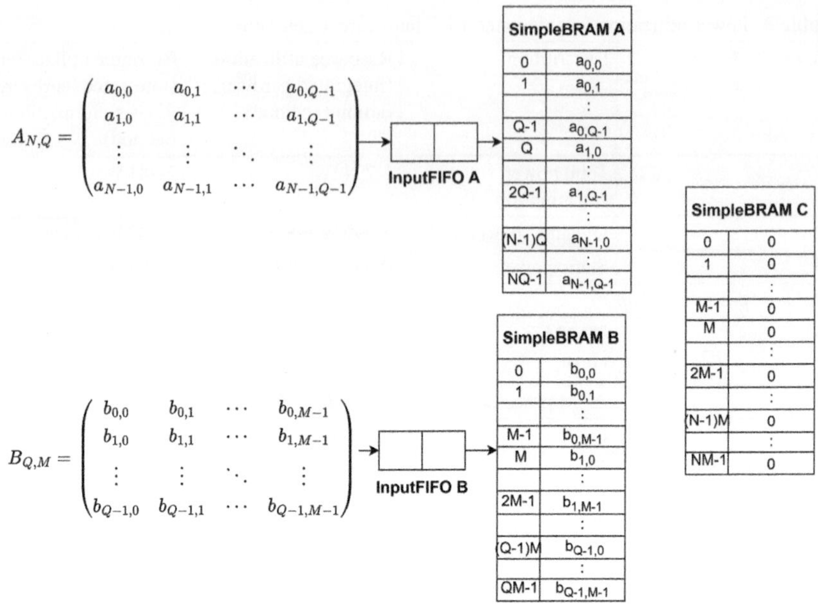

Fig. 3 Initialization of SimpleBRAMs

- Computation Step: After initializing the *SimpleBRAMs*, data from the *SimpleBRAMs* will be accessed at specific addresses and enqueued into the Floating Point FMA. The result from the FMA will then be stored in *SimpleBRAM C*. The algorithm given below will be followed:

Algorithm 3 Matrix Multiplication

Require: Initialized A, B, and C *SimpleBRAMs*

```
for k = 0 to Q − 1 do
    for i = 0 to N − 1 do
        for j = 0 to M − 1 do
            C[i * M + j] ← A[i * Q + k] · B[k * M + j] + C[i * M + j]
        end for
    end for
end for
```

3.2.2 Matrix Inversion

To achieve matrix inversion for a 4×4 matrix with a high level of parallelism, we employ the standard method using the adjoint of the matrix. This process requires extensive floating-point computations, including floating-point addition/subtraction (fpa/fps), floating-point multiplication (fpm), and floating-point division (fpd). To optimize performance, we utilize multiple threads or floating-point hardware units. Here, we allocate six threads each for fpa, fps, and fpm, and one thread for fpd. This configuration enables partial parallelism, enhancing computational efficiency. By increasing the number of threads, we can further optimize the process to achieve full parallelism, thereby reducing computation time and improving performance. These floating-point units are not pipelined.

$$A^{-1} = \frac{1}{\det(A)} \operatorname{adj}(A) \tag{17}$$

where $\operatorname{adj}(A)$ represents the adjoint of matrix A, and $\det(A)$ is the determinant of A.

```
with SerialSections ("Determinant_calculation"):
 with LeafSection("SS1"):
  fpm_i0.enqOperands(r_m[10],r_m[15]) #m[10]*m[15]
  fpm_i1.enqOperands(r_m[11],r_m[14]) #m[11]*m[14]
  fpm_i2.enqOperands(r_m[9],r_m[15])  #m[9]*m[15]
  fpm_i3.enqOperands(r_m[11],r_m[13]) #m[11]*m[13]
  fpm_i4.enqOperands(r_m[9],r_m[14])  #m[9]*m[14]
  fpm_i5.enqOperands(r_m[10],r_m[13]) #m[10]*m[13]
 with LeafSection("SS2"):
  fps_i0.enqOperands(fpm_i0.getResult(),fpm_i1.getResult())
  fps_i1.enqOperands(fpm_i2.getResult(),fpm_i3.getResult())
  fps_i2.enqOperands(fpm_i4.getResult(),fpm_i5.getResult())
```

In the given small code snippet of determinant calculation, under *LeafSection("SS1")*, values in registers *r_m* are sent to input FIFOs of six parallel floating-point multiplication (fpm) threads. In *LeafSection("SS2")*, results from these fpm threads are fed into multiple floating-point subtraction (fps) threads as shown in Fig. 4.

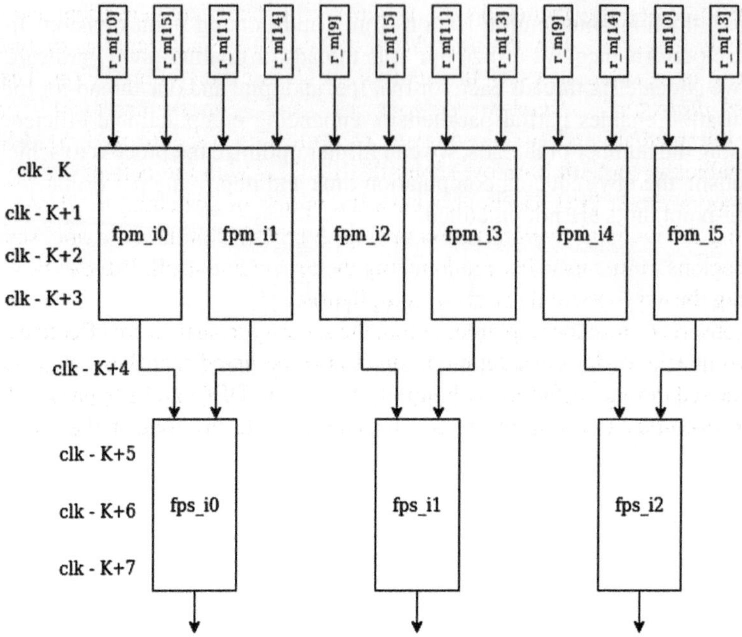

Fig. 4 Running threads in parallel in Python DSL

3.2.3 Matrix Addition

Matrix addition involves accessing corresponding elements from two different *SimpleBRAMs* where matrices are stored. These elements are then fed into a floating-point adder thread, and the resulting sum is stored back into the target *SimpleBRAM*.

```
with ForLoopSection ("mat_add_FPF_T_add_Q", "m", 0, N * N):
 with LeafSection ("mat_add_FPF_T_add_Q_S11"):
  FPF_T_bram.readRequest (m)
  Q_bram.readRequest (m)
 with LeafSection ("mat_add_FPF_T_add_Q_S12"):
  fpa.enqOperands(FPF_T_bram.readResp(),Q_bram.readResp())
 with LeafSection ("mat_add_FPF_T_add_Q_S13"):
  P_est_bram.writeData (m, fpa.getResult())
```

4 Experimental Setup and Results

4.1 Object Tracking Simulation Enviornment

Simulated measurements have been used for this study. Seeker measurements are generated based on the ideal relative positions and velocities with the addition of noise. The EKF uses these measurements to estimate the "true" system states. EKF uses four measurements:

- *Range* (R): This is the distance between two bodies, measured in meters.
- *Range Rate* (RR): It is the rate at which the range reduces, measured in meters/second. It is negative if the bodies are approaching and positive if they are receding.
- *Elevation Angle* (θ_{ele}): Angle of the target with respect to the seeker in the elevation plane, measured in radians.
- *Azimuth Angle* (θ_{azi}): Angle of the target with respect to the seeker in the azimuth plane, measured in radians.

The input for the generation of measurements are the ideal relative positions (r_x, r_y, r_z) and velocities (v_x, v_y, v_z). These are used as inputs for measurement generation. This is done in two steps:

1. *Generation of Ideal Measurements*: Ideal measurements are generated from the ideal values of relative positions. These are the true measurements, generated from input data as given below:

$$R_0 = \sqrt{r_x^2 + r_y^2 + r_z^2} \tag{18}$$

$$\dot{R}_0 = \frac{r_x v_x + r_y v_y + r_z v_z}{R_0} \tag{19}$$

$$\theta_{\text{ele}} = \tan^{-1}\left(\frac{r_x}{\sqrt{r_y^2 + r_z^2}}\right) \tag{20}$$

$$\theta_{\text{azi}} = \tan^{-1}\left(\frac{r_y}{r_z}\right) \tag{21}$$

2. *Addition of Noise to Ideal Measurements*: Noise is added to mimic sensor measurement noise. These measurements are used in the EKF for state estimation. Thus, the measurement is given as

$$z_i = z_{0i} + n_i \tag{22}$$

where

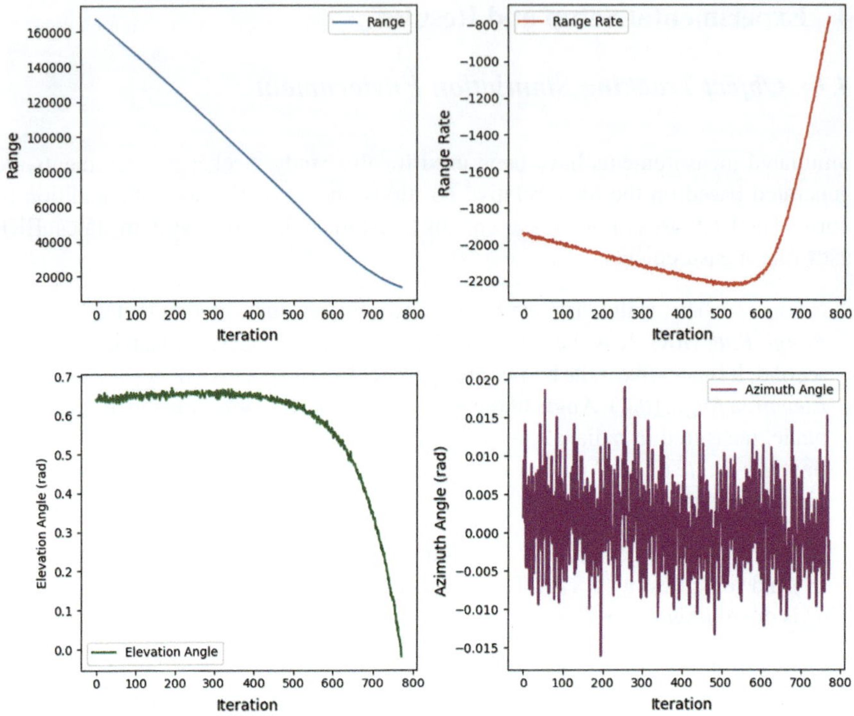

Fig. 5 Noisy measurement plots

z_i: Simulated measurement i, $i = 1, 2, 3, 4$

n_i: Additive Gaussian noise of zero mean and unit variance

Measurements z_1, z_2, z_3, z_4 are used in the EKF. Relative positions and velocities are used as input data to generate the measurements, utilizing six inputs.

Figure 5 shows the measurement plots incorporating Gaussian noise.

4.2 *Hardware Testing Results for Vivado HLS Implementation*

The Vivado HLS-implemented EKF, integrated with the Zynq SoC, was tested on the ZCU104 FPGA board using the PYNQ framework. The EKF ran for 772 time steps, estimating $r_x, r_y, r_z, v_x, v_y, v_z$, and the ballistic coefficient. To measure the accuracy and speedup provided by the designed EKF, a software implementation of the EKF, written in Python, was also executed on the Zynq processor on the ZCU104 board. The speedup was calculated using Eq. 23. $t_{hardware}$ includes the time from when the

Table 4 Comparison of performance

S. no	Description	EKF (integrated hardware) (adjoint method)	EKF (integrated hardware) (Gauss-elimination method)
1	Software execution time	715.600 ms	715.600 ms
2	Hardware execution time	368.218 ms	370.340 ms
3	Speedup	1.9434	1.9417

DMA starts taking inputs from memory until it writes the output of the EKF back to memory, including the EKF computation time.

$$\text{Speed Up} = \frac{t_{\text{software}}}{t_{\text{hardware}}} \tag{23}$$

The designed IP shows speedup in both scenarios, indicating its role as a hardware accelerator. Since the speed is nearly identical in both cases, it can be concluded that the Gaussian elimination method for matrix inversion in the EKF provides the same speed and accuracy as the adjoint method, while using fewer resources (Table 4).

The estimated state plots for both Hardware and Software implementations along with the relative percentage error (calculated using Eq. 24) are presented in Figs. 6, 7, 8, 9, 10, 11, 12, 13, 14, 15, 16, 17, 18, and 19. This is done to evaluate the accuracy of our design.

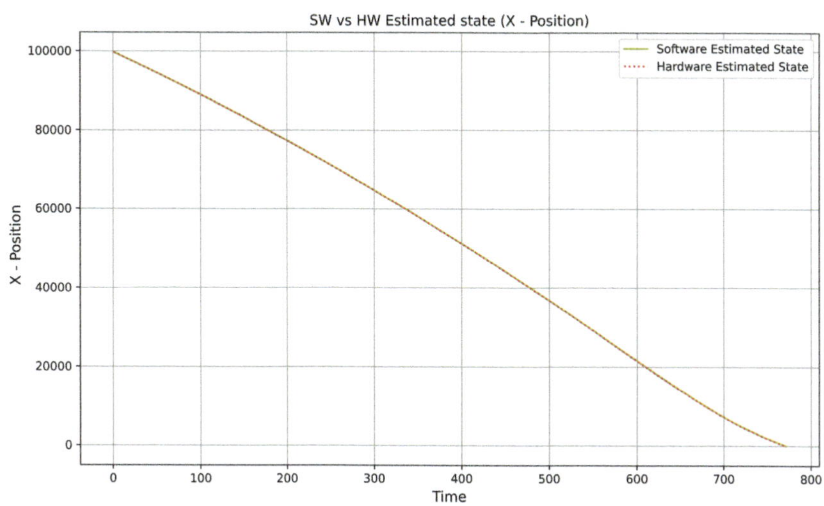

Fig. 6 Relative X-position (r_x)

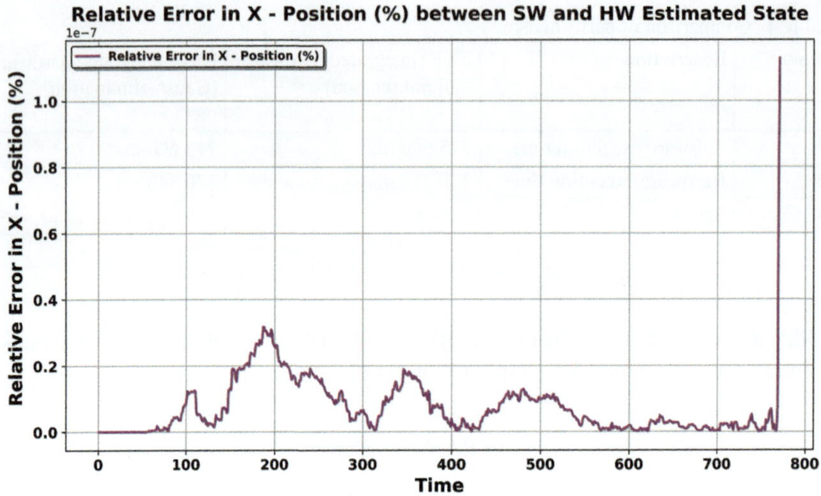

Fig. 7 % error between HW and SW (r_x)

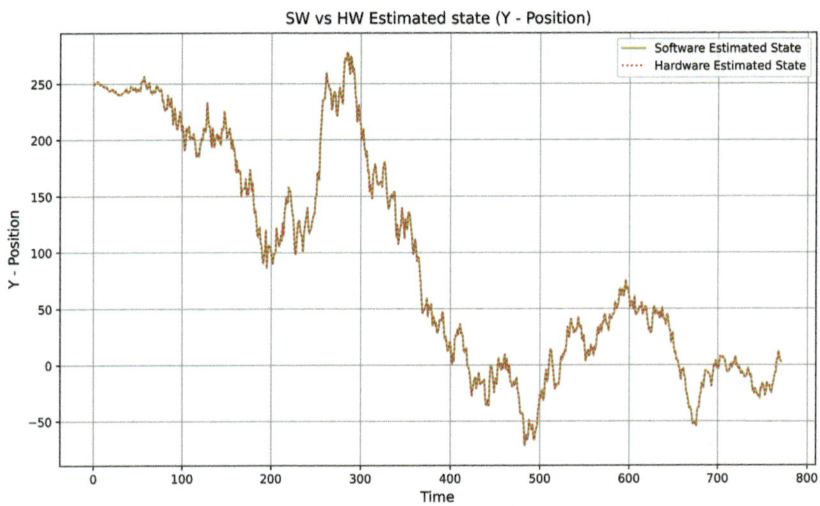

Fig. 8 Relative Y-position (r_y)

$$\text{Relative Error }\% = \left|\frac{X_{\text{software}} - X_{\text{hardware}}}{X_{\text{software}}}\right| \times 100\% \tag{24}$$

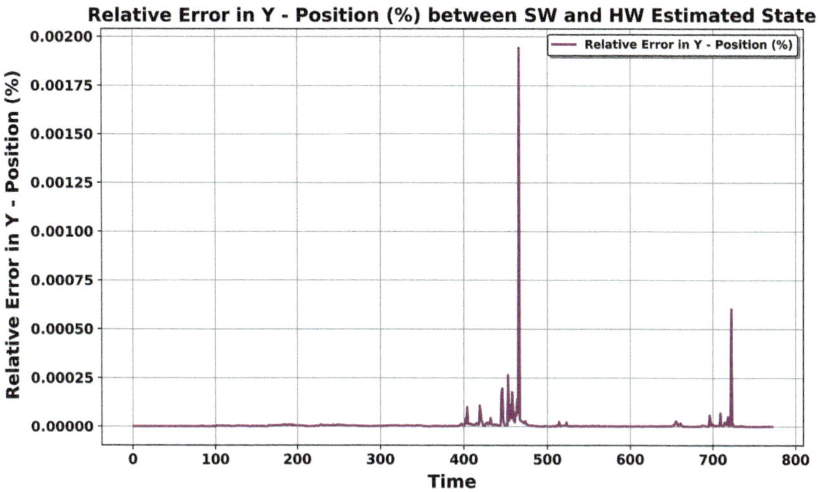

Fig. 9 % error between HW and SW (r_y)

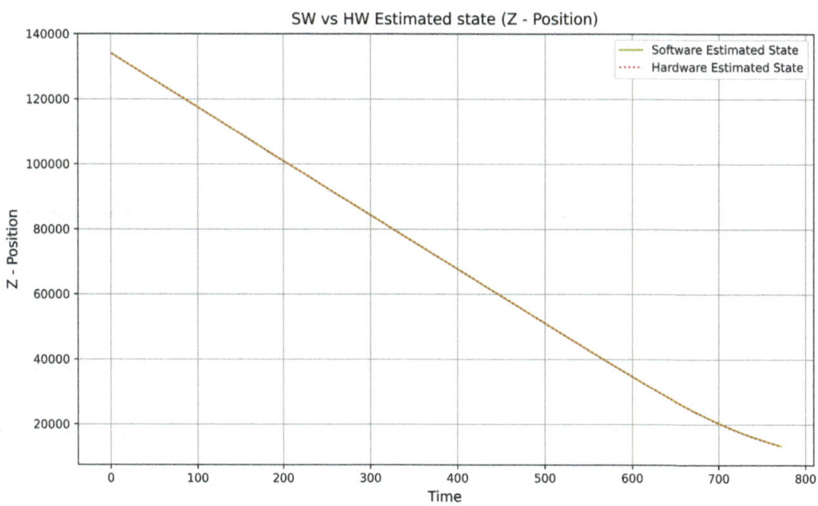

Fig. 10 Relative Z-position (r_z)

4.3 Simulation Results for Python DSL-Based Implementation

Code written in Python DSL for the EKF and the corresponding testbench was converted to RTL Verilog code. The simulation was conducted using Verilator. The EKF was executed for four iterations, and the outputs, $\hat{x}_{k(+)}$ and $P_{k(+)}$, were collected. These outputs from the four iterations were then compared with the EKF imple-

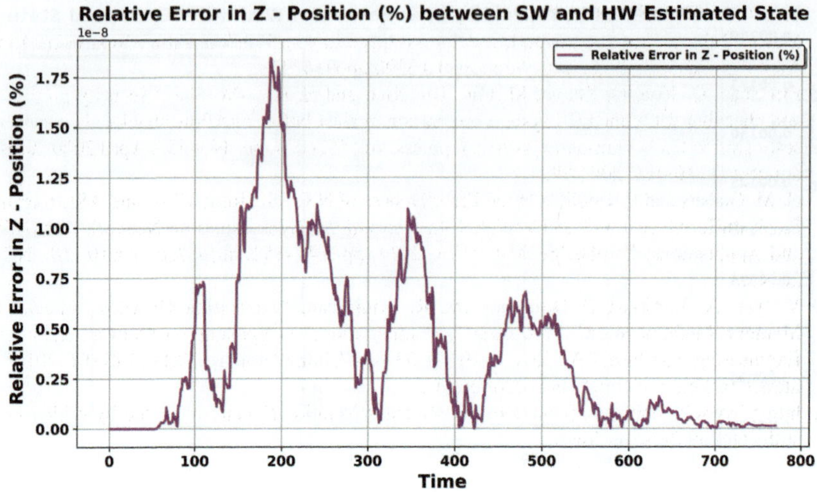

Fig. 11 % error between HW and SW (r_z)

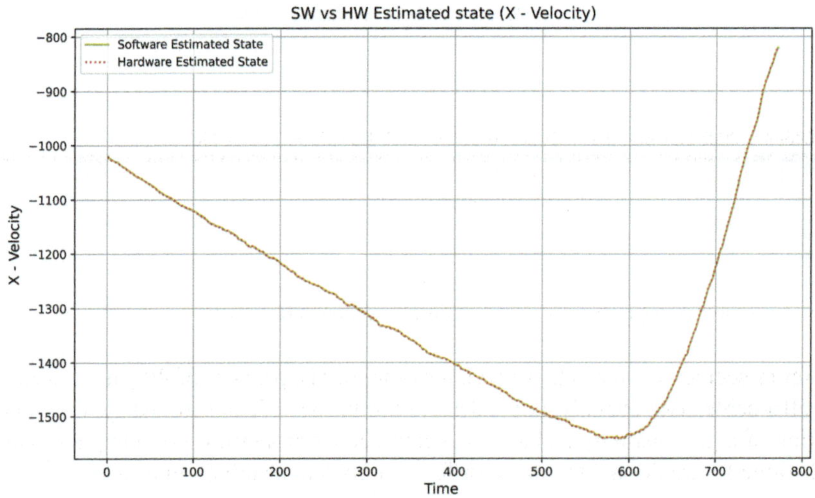

Fig. 12 Velocity in X-direction (v_x)

mentation outputs from Vivado. The relative percentage error for the a posteriori system states and covariance matrix was plotted, as illustrated in Figs. 20 and 21, respectively.

$$\text{Percentage Error} = \left| \frac{X_{\text{PyHLS}} - X_{\text{Vivado}}}{X_{\text{Vivado}}} \right| \times 100\% \tag{25}$$

From Figs. 20 and 21, we can observe that the accuracy of the Python DSL-based implementation of the EKF is comparable to the Vivado-based design. This is

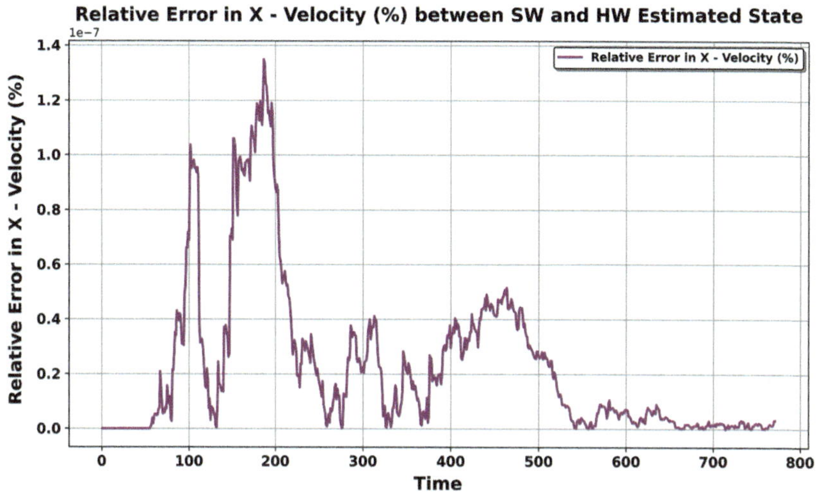

Fig. 13 % error between HW and SW (v_x)

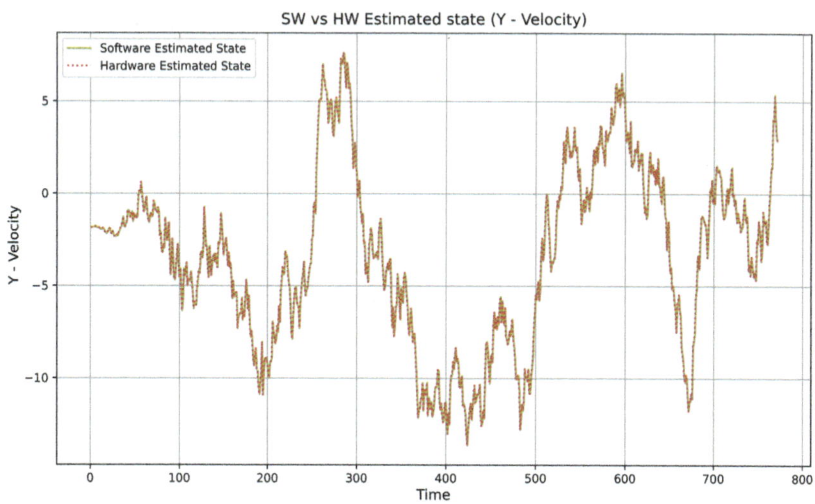

Fig. 14 Velocity in Y-direction (v_y)

noteworthy given that the former implementation uses single-precision floating-point computations, while the latter uses double-precision floating-point computations.

Table 5 displays the clock cycles required for various operations within the Python DSL implementation of the EKF for one iteration. After accounting for input feeding and output extraction, a single iteration of the Python DSL-based EKF consumes a total of 49,932 clock cycles. In contrast, the Vivado HLS implementation completes

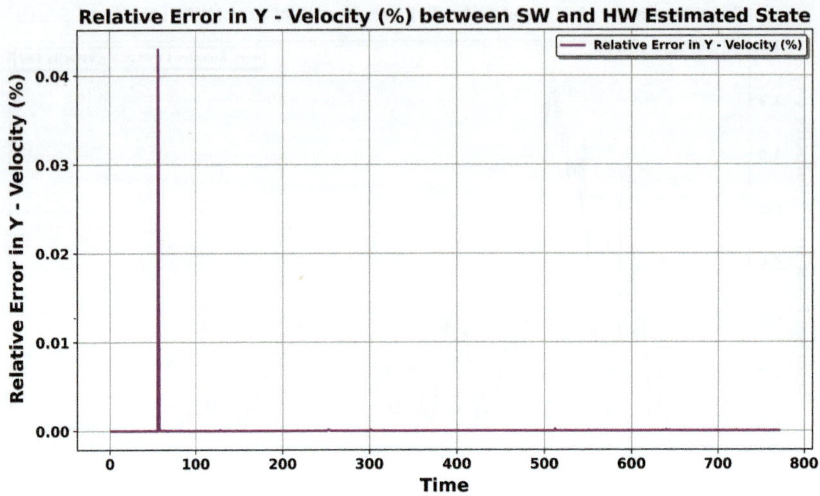

Fig. 15 % error between HW and SW (v_y)

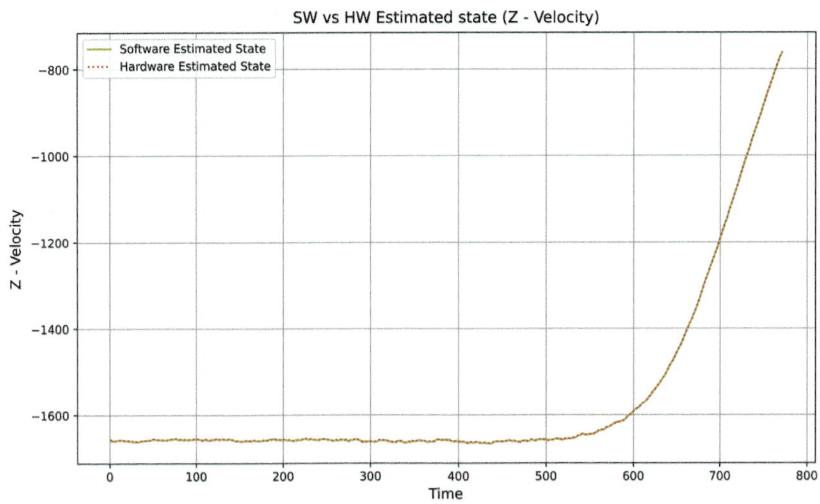

Fig. 16 Velocity in Z-direction (v_z)

the same operation in approximately 2,134 cycles. The higher cycle count in the former is attributed to non-pipelined floating-point units and a memory-dependent architecture.

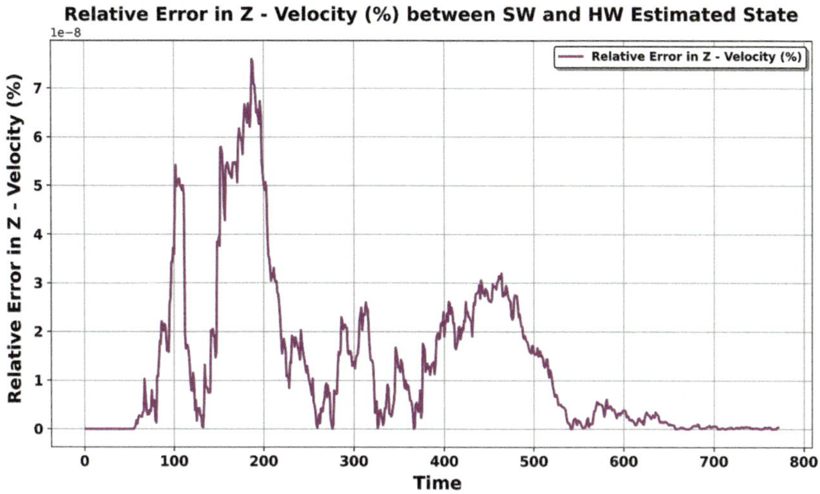

Fig. 17 % error between HW and SW (v_z)

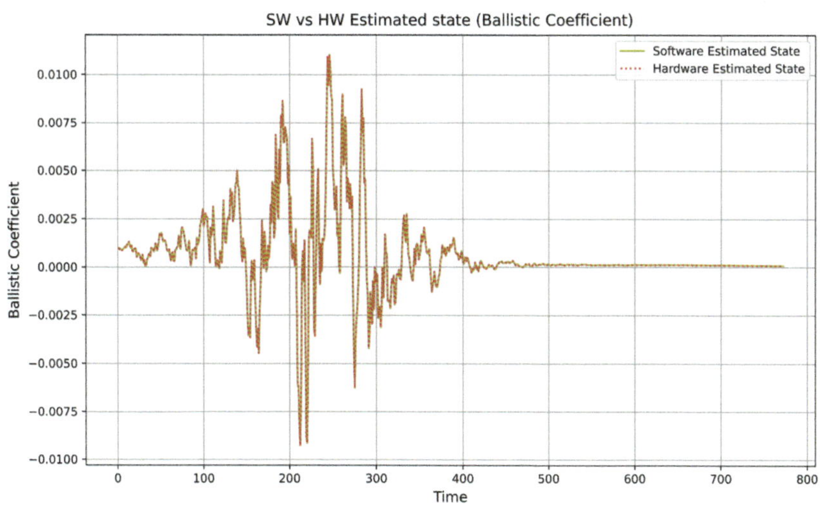

Fig. 18 Ballistic coefficient

5 Future Work

The floating-point units designed in the Python DSL can be pipelined to reduce latency and improve throughput. Greater parallelism in the Python DSL implementation can be achieved by using more threads. Additionally, the application of the EKF can be extended to GPS tracking, which will require modifications to the algorithm.

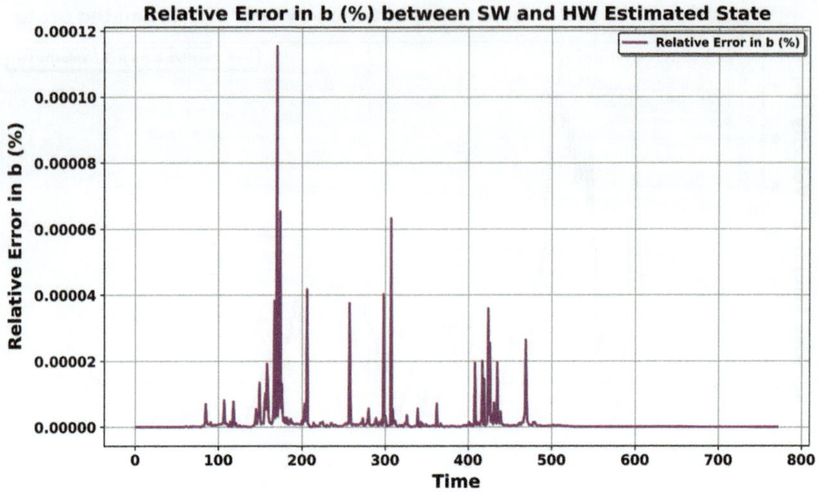

Fig. 19 % error between HW and SW (b)

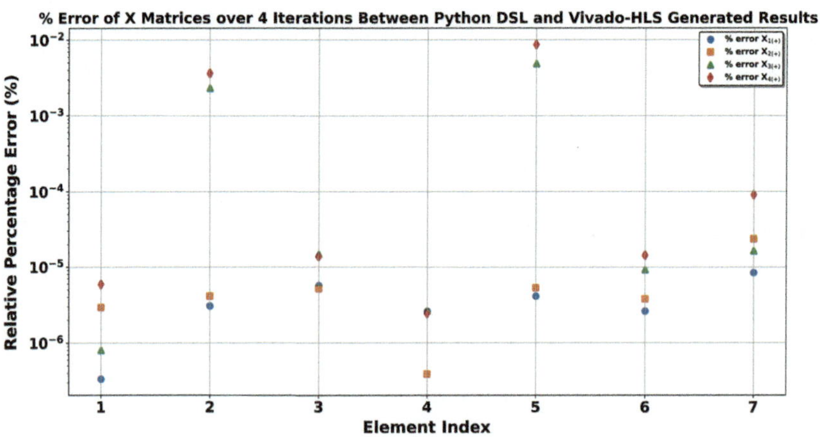

Fig. 20 % Error of X matrices over 4 iterations of EKF

Table 5 Clock cycles taken by computations inside EKF for 1 iteration

S. no	Computation	Clock cycles
1	$F_{7\times7} \cdot P_{7\times7}$ and H^T in parallel	9613
2	F^T	114
3	$(FP)_{7\times7} \cdot (F^T)_{7\times7}$	9613
4	$P\_est_{7\times7} = (FP \cdot F^T)_{7\times7} + Q_{7\times7}$	835
5	$(P\_estH\_T)_{7\times4} = P\_est_{7\times7} \cdot (H^T)_{7\times4}$	5498
6	$(HP\_estH\_T)_{4\times4} = H_{4\times7} \cdot (P\_estH\_T)_{7\times4}$	3285
7	$S_{4\times4} = (HP\_estH\_T)_{4\times4} + R_{4\times4}$	274
8	$Sinv_{4\times4}$	1140
9	$K_{7\times4} = (P\_estH\_H\_T)_{7\times4} \cdot Sinv_{4\times4}$	3151
10	$KH_{7\times7} = K_{7\times4} \cdot H_{4\times7}$ and $ydiff_{4\times1} = ym_{4\times1} - ypred_{4\times1}$ in parallel	5515
11	$(I\_KH)_{7\times7} = I_{7\times7} - KH_{7\times7}$	835
12	$(Out\_P)_{7\times7} = (I\_KH)_{7\times7} \cdot (P\_est)_{7\times7}$ and $KYdiff_{7\times1} = K_{7\times4} \cdot Ydiff_{4\times1}$ in parallel	9624
13	$(Out\_X)_{7\times1} = (X\_pred)_{7\times1} + KYdiff_{7\times1}$	121

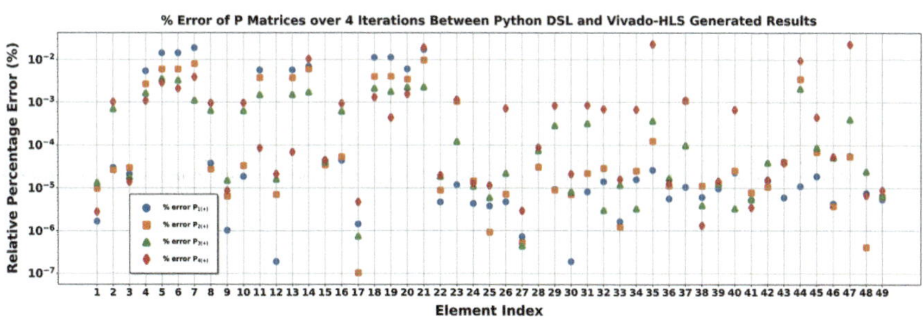

Fig. 21 % Error of P matrices over 4 iterations of EKF

6 Conclusion

In this paper, the EKF is implemented using Vivado HLS. The EKF IP is integrated with the Zynq SoC, and the integrated design is tested on the ZCU104 FPGA Board using the PYNQ overlay. The hardware results are compared with a software implementation in terms of accuracy and performance. Additionally, various single-precision floating-point computational units were designed using a Python-based DSL. These units were then used to design the EKF in the Python DSL, and the RTL simulation results of the Python DSL implementation are compared with the hardware results of the Vivado design in terms of accuracy and performance.

References

1. Kalman R.E.: A New Approach to Linear Filtering and Prediction Problems. Transactions of the ASME–Journal of Basic Engineering **82**(Series D), 35–45 (1960)
2. Datar M., Hegde D.S., Prasad V.D., Prajapati M., Manikanta N., Gupta D., Pavanija J., Pare P., Akash, Gupta S., Patkar S.B.: Python-based DSL for generating Verilog model of Synchronous Digital Circuits (2024). arXiv:2406.09208
3. Kalman Filtering, General Information. In: John Wiley & Sons, Ltd (ed.) Kalman Filtering, 1st edn, vol. 1, pp. 1–29. John Wiley & Sons, Ltd (2008). https://doi.org/10.1002/9780470377819.ch1
4. Kalman Filtering, Linear Dynamic Systems. In: John Wiley & Sons, Ltd (ed.) Kalman Filtering, 1st edn, vol. 2, pp. 31–66. John Wiley & Sons, Ltd (2008). https://doi.org/10.1002/9780470377819.ch2
5. Kalman Filtering, Linear Optimal Filters and Predictors. In: John Wiley & Sons, Ltd (ed.) Kalman Filtering, 1st edn, vol. 4, pp. 131–181. John Wiley & Sons, Ltd (2008). https://doi.org/10.1002/9780470377819.ch4
6. Kalman Filtering, Nonlinear Filtering. In: John Wiley & Sons, Ltd (ed.) Kalman Filtering, 1st edn, vol. 7, pp. 293–353. John Wiley & Sons, Ltd (2008). https://doi.org/10.1002/9780470377819.ch7
7. Naqvi S.R., Gharib S., Khan R., Munir N., Khanam M.: System of Linear Equations, Gaussian Elimination. Global Journal of Computer Science and Technology **15** (2015)
8. Bhale P., Dwivedi P.N., Kumar P., Bhattacharyaa A.: Estimation of Ballistic Coefficient of Reentry Vehicle with Divided Difference Filtering using Noisy RF Seeker Data. In: 9th International Conference on Industrial Technology, pp. 1087–1092. IEEE (2007). https://doi.org/10.1109/ICIT.2006.372311

Enhanced Speed Control for Switched Reluctance Motor a PID and Fuzzy Logic Approach

Gopinath Sengupta, Alok Kanti Deb, Satya Sundar Mishra, and Nikhil Kumar Mishra

Abstract The use of Switched Reluctance Motors (SRMs) has been constrained due to significant speed variations that can lead to instability. To address this challenge, speed control can be effectively managed using PID and fuzzy logic control strategies. This paper exhibits a dual-loop control system for the SRM, consisting of a current loop and a position loop, each controlled independently and concurrently using both PID and fuzzy logic controllers. The performance of these controllers is evaluated and compared to achieve stable speed regulation at the desired level. Simulations are conducted using MATLAB Simulink to demonstrate and analyze the results.

Keywords SRM drive system · Inductance profile · Rotor position sensing · PID controller · FUZZY logic controller

1 Introduction

The Switched Reluctance Motor (SRM) have gained substantial notice in recent years due to their simple mechanical structure, robustness, and fault-tolerant capabilities. Unlike traditional machines, SRMs are characterized by a rotor without windings or

G. Sengupta (✉)
School of Energy Science and Engineering, Indian Institute of Technology Guwahati, Guwahati, India
e-mail: g.sengupta@iitg.ac.in

A. K. Deb
Department of Electrical Engineering, Indian Institute of Technology Kharagpur, Kharagpur, India
e-mail: alokkanti@ee.iitkgp.ernet.in

S. S. Mishra
Department of Electrical Engineering, Trident Academy of Technology, Bhubaneswar, India

N. K. Mishra
Department of Mechanical Engineering, Indian Institute of Technology Guwahati, Guwahati, India
e-mail: nikhil1996@iitg.ac.in

U. S. Dixit et al. (eds.), *Current Progress in Engineering Sciences*,
https://doi.org/10.1007/978-981-96-8753-4_31

permanent magnets, with both the stator, and rotor featuring salient poles, earning them the classification of doubly salient machines [1–3]. This inherent simplicity contributes to their high efficiency and reduced manufacturing costs, making them an attractive option for various applications, including variable speed drives and power generation systems [4–10].

Despite these advantages, SRMs face challenges related to their inherent non-linear characteristics. The relationship linking torque, inductance, and flux linkage varies non-linearly with respect to the stator winding current and position of the rotor [11]. Such non-linearities complicate the design and control of SRM drives, particularly in achieving precise current control, speed regulation, and instantaneous torque control. Torque ripple, a significant issue in SRMs, leads to mechanical vibrations and acoustic noise, adversely affecting speed control performance and overall system reliability [12, 13].

The evolution of power electronics and converter technologies, combined with advanced design methodologies, has revitalized the prospects of SRMs in demanding applications. Notably, SRMs have been explored as alternatives to hydraulic and pneumatic actuators in more electric aircraft (MEA) systems, where robustness and fault tolerance are critical [14, 15]. Additionally, their suitability for electromechanical actuators (EMAs) further highlights their potential in aerospace and other high-reliability domains.

Control strategies play a pivotal role in addressing the drawbacks of SRMs. Researchers have extensively studied current and torque control methods, focusing on minimizing torque ripple and improving speed control performance [16–18]. While conventional proportional–integral (PI) controllers are generally used for speed control, the non-linearity and parameter uncertainty of SRMs necessitate the development of more advanced control techniques. Such efforts aim to enhance the operational efficiency and reliability of SRMs in industrial and commercial applications [19].

In this paper, we present a reliable and straightforward SRM model that has been developed and simulated. The execution of the speed control system, utilizing a dual-loop control approach, is evaluated. The controller design comprises a current loop and a position loop, which are controlled independently and concurrently through PID and fuzzy logic controllers. The responses of these controllers are compared to assess their effectiveness in stabilizing the speed at the desired level.

2 Description of SRM

SRM features a single excitation system with a double salient structure. The rotor is constructed from layers of steel laminations that are shaped and cut to create the desired number of poles. Notably, the rotor does not contain windings, permanent magnets, or squirrel cage bars. The stator, similarly, is made of steel laminations, precisely cut to form the necessary number of poles [20].

The rotor in an SRM moves to a position that minimizes reluctance, which corresponds to the point of maximum inductance, aligning just beneath the energized stator poles. This alignment generates torque, causing the rotor to move in synchronization with the sequence of stator coil excitations until a stable position is reached. When excitation is applied to a pair of diametrically opposite stator poles, the rotor pole nearest to the energized stator poles is attracted, producing torque. Once the rotor poles align with the energized stator poles, the stator poles are de-energized, and the next pair of stator poles is activated, continuing the rotor's movement and producing sustained rotational torque. The torque produced in an SRM is directly influenced by the rate of change of inductance with respect to the rotor's position [1, 2, 19, 20].

3 Mathematical Modeling of SRM

To develop a comprehensive mathematical model for the SRM, a set of differential equations is derived based on machine theory and general electric circuit analysis. The modeling process can be divided into two subsystems: the stator subsystem, described using Kirchhoff's laws to capture the dynamic behavior, and the mechanical subsystem, described using Newton's law to track rotor dynamics and the applied mechanical load. These two subsystems are coupled through energy conversion principles [9]. The inductance profile and the instantaneous torque expression are derived with the assumption of linear magnetic characteristics.

3.1 Electromagnetic Equations

The voltage equations for an SRM are similar to those for other electrical machines, following the principle that the summation of the voltage across the passive components equals the applied voltage. The instantaneous voltage across the terminals of a phase winding in a SRM is governed by Faraday's law, which relates the voltage to the rate of change of flux linkage in the winding and is given by

$$v = iR + V = iR + \frac{d\lambda(\theta, i)}{dt} \tag{1}$$

where V = voltage per phase, λ = flux linkage, i = current per phase, θ = rotor position, R = winding resistance per phase, and t = time (Fig. 1).

Equation (1) can be simplified under certain assumptions: the mutual flux linkage among phase windings is negligible, and the B-H characteristic of the ferromagnetic material is linear. This leads to the expression $\lambda = iL(\theta, i)$. By substituting into Eq. (1) and applying partial derivatives, we obtain

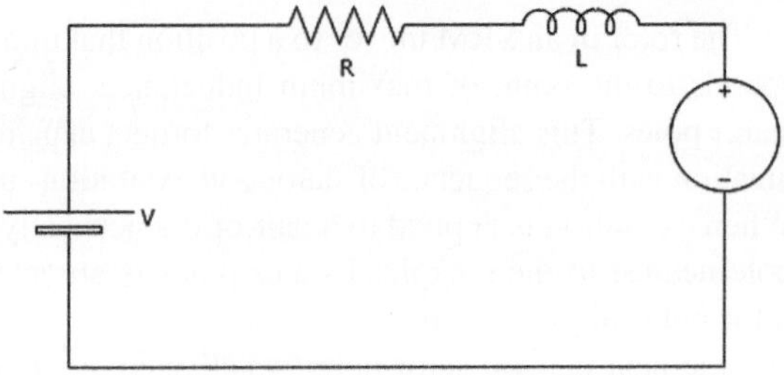

Fig. 1 Equivalent circuit of switched reluctance motor

$$V = iR + V = iR + i\frac{dL(\theta, i)}{dt} + L(\theta, i)\frac{di}{dt} \tag{2}$$

Rearranging gives

$$V = iR + i\frac{dL(\theta, i)}{d\theta}\frac{d\theta}{dt} + L(\theta, i)\frac{di}{dt} \tag{3}$$

Since $d\theta/dt$ represents the angular speed ω, we have

$$V = iR + L(\theta, i)\frac{di}{dt} + i\omega\frac{dL(\theta, i)}{d\theta} \tag{4}$$

In this equation:

- The initial term represents the resistive voltage drop,
- The subsequent term is the inductive voltage drop,
- The third term represents the EMF induced, which can be significant at high speeds.

When the winding is supplied with voltage, the rate of current build-up and its maximum value depend on the excited poles. If the speed is constant, then by considering Eq. (4), the switch should activate the voltage when dL/dt is minimal, leading to a rapid increase in current until the desired value is reached. When $dL/d\theta$ becomes significant and $L(\theta, i)$ increases, di/dt decreases and may even become negative if the EMF due to motion $\omega dL(\theta, i)/d\theta$ overcomes $(V - i\text{R})$. If the voltage is zero, the motional EMF forces the current to zero.

3.2 Torque Equations

The torque and voltage equations for each phase describe the operation of the SRM. For simplicity, let $L(\theta, i)$ be denoted as L. The resistive voltage drop is negligible compared to other voltage drops, allowing us to write the phase voltage as

$$V = L\frac{\mathrm{d}i}{\mathrm{d}t} + i\frac{\mathrm{d}L}{\mathrm{d}t} \tag{5}$$

This can be further modified as

$$V = L\frac{\mathrm{d}i}{\mathrm{d}t} + \frac{i}{2}\frac{\mathrm{d}L}{\mathrm{d}t} + \frac{i}{2}\frac{\mathrm{d}L}{\mathrm{d}t} \tag{6}$$

The expression for the instantaneous power will be

$$Vi = \left[Li\frac{\mathrm{d}i}{\mathrm{d}t} + \frac{i^2}{2}\frac{\mathrm{d}L}{\mathrm{d}t}\right] + \frac{i^2}{2}\frac{\mathrm{d}L}{\mathrm{d}\theta}\frac{\mathrm{d}\theta}{\mathrm{d}t} \tag{7}$$

The first two terms can be rewritten as $\mathrm{d}(Li^2)/2\mathrm{d}t$ and $\mathrm{d}\theta/\mathrm{d}t$ represents the speed of the motor. Thus

$$Vi = \frac{\mathrm{d}}{\mathrm{d}t}\left(\frac{1}{2}Li^2\right) + \omega\frac{i^2}{2}\frac{\mathrm{d}L}{\mathrm{d}\theta} \tag{8}$$

In this equation:

- $\frac{1}{2}Li^2$ represents the increase in stored magnetic energy,
- $\omega\frac{i^2}{2}\frac{\mathrm{d}L}{\mathrm{d}\theta}$ represents the power converted to mechanical work.

The per-phase mechanical power developed is given by

$$w = T\omega \tag{9}$$

where T represents the per-phase torque developed. Comparing the second term in Eqs. (9) and (10), we get

$$T = \frac{1}{2}\frac{\mathrm{d}L}{\mathrm{d}\theta}i^2 \tag{10}$$

This expression shows that the instantaneous torque per phase is a quadratic function of the per-phase current. The torque is positive when $\mathrm{d}L/\mathrm{d}\theta$ is positive, indicating motoring operation, and negative when it is negative, indicating braking. The torque is zero when the phase inductance remains constant.

The total instantaneous torque developed by the SRM is the summation of the torques from each of the phases:

$$T_T = \sum_{k=1}^{q} T = \frac{1}{2}\sum_{k=1}^{q}\frac{\mathrm{d}L(\theta)}{\mathrm{d}\theta}i^2 \tag{11}$$

A motoring torque is generated when the SRM rotates in the positive direction, moving through regions of increasing inductance. During the unaligned zone, the torque is zero, and when passing through regions of decreasing inductance, a braking torque is developed.

3.3 Dynamics Mechanical Subsystem

The differential equation describing the mechanical subsystem can be developed by applying Newton's law, which relates the electromagnetic torque T, load torque T_L, and rotor speed ω:

$$J\frac{d\omega}{dt} = T - T_L - f\omega \tag{12}$$

By taking Laplace Transform, we get

$$\frac{\omega(s)}{T} = \frac{1}{Js + f} \tag{13}$$

where:

J represents the moment of inertia and f represents the viscous friction coefficient.

4 Simulink Modeling of SRM

In industrial applications, a SRM is required to maintain a constant synchronous speed to deliver the desired output. However, the motor often experiences periodic load variations, alternating between high-load and low-load periods. To accommodate these changes, the motor's speed must adjust accordingly. If the speed fluctuates excessively in response to load variations, it can cause disturbances and lead to transient instability in the system. To mitigate this, a feedback control system is employed for precise speed regulation of the SRM. This system incorporates a controller that compares the motor's actual speed with the reference speed set as per the application of the motor. The controller's output is then used to determine the torque reference signal, ensuring stable and optimal operation. The Simulink model of a SRM is built according to the mathematical dynamic equations governing its operation and is designed according to the specification mentioned in Appendix 1. The model is organized into four main blocks: **current block**, **angle block**, **voltage block**, and **torque block**.

4.1 Model Structure

The integrator in the model converts the angular speed into rotor position, which is then applied to the angle block. The angle block transforms the ramp output of the integrator into a periodic function with a period of $\pi/2$ radians, corresponding to the four stator poles of the SRM. This periodic output is fed to the inductance and current blocks, as these are dependent on the position of the rotor.

Current Block: The current block generates chopping currents during the rising inductance region based on rotor position. The voltage in the model is regulated using the hysteresis current control method.

Figure 2 illustrates the current loop control model of SRM. The input current applied to each phase block is controlled by the controller block, which employs PID and FUZZY logic control strategies. The angular speed is divided into two paths:

- A direct feedback path to the angle block via the integrator.
- A path that passes through the controller, where it is multiplied with the reference current to manage the outputs of other blocks.

Voltage Block: As shown in Fig. 4, the difference between the voltage block output and the resistive voltage drop is integrated to compute the product of inductance and current. The inductance block divides this product to obtain the actual current. The actual current is compared with the reference current, and the resulting error is fed back to the voltage block.

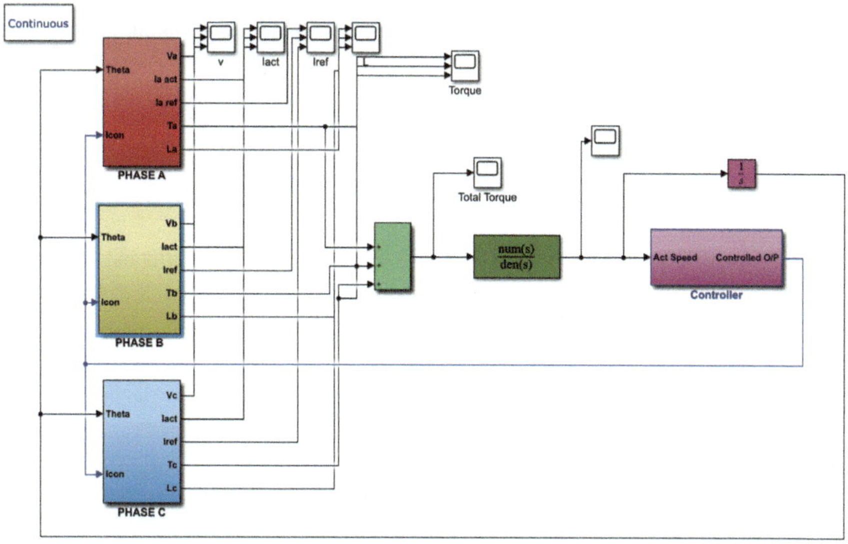

Fig. 2 Current loop control model of SRM using Matlab Simulink

Table 1 Rules used for FUZZY membership function

ER/DER	NB	NM	NS	ZO	PS	PM	PB
NB	NB	NB	NB	NB	NM	NS	ZO
NM	NB	NB	NB	NM	NS	ZO	PS
NS	NB	NB	NM	NS	ZO	PS	PM
ZO	NB	NM	NS	ZO	PS	PM	PB
PS	NM	NS	ZO	PS	PM	PB	PB
PM	NS	ZO	PS	PM	PB	PB	PB
PB	ZO	PS	PM	PB	PB	PB	PB

Table 2 Sequence of reference current with respect to rotor position

Rotor position (Deg.)	Ref. current (I_a)	Ref. current (I_b)	Ref. current (I_c)
0–10	1	0	1
10–30	1	0	0
30–40	1	1	0
40–60	0	1	0
60–70	0	1	1
70–90	0	0	1

Torque Block: The actual current is then supplied to the torque-generating block, which produces the required torque. The combined output from the three-phase torque blocks is passed to a transfer function block derived from the mechanical system equation. The output of this block is the angular speed, which is integrated to determine the rotor position.

4.2 Rotor Position and Phase Current

In SRM operation, current in each phase is applied when its inductance increases with respect to rotor position. The rotor position serves as feedback to the current generator block. For rotor positions ranging from 0° to 90°, the phase currents are observed and tabulated (see Tables 1 and 2).

4.3 Controller Models

In the drive arrangement the motor's speed depends on both the drive current and the angular position, which can be controlled through the current loop and angle loop, respectively. Speed regulation thus has been achieved in three distinct ways:

- By controlling the current loop exclusively.
- By controlling the angle loop exclusively.
- By simultaneously controlling both loops.

In this paper, speed control is first implemented by incorporating the controller within the current loop. The controller processes the output and integrates it into the current loop. Similarly, the same approach is applied to the angle loop. Finally, both loops are controlled simultaneously. The outcomes of each control strategy are then evaluated and compared to determine the most effective approach. Figure 3 presents the both-loop control model of SRM. Here, the input currents to each phase block and the angle block are controlled simultaneously by two controller blocks employing PID and FUZZY logic strategies (Fig. 4).

Physically, the PID and FUZZY controller models share the same structure, with differences only in the control strategy. These are illustrated in Figs. 5 and 6, respectively.

PID Controller: A comparison is made between the reference speed and the actual speed, and the resulting error is scaled using a gain factor to convert it into a per-unit signal. The desired speed is achieved by tuning the PID controller through the Ziegler–Nichols frequency response method.

FUZZY Logic Controller: FUZZY logic controller operates based on weighted inputs and outputs defined by membership functions. Various membership functions—such as trapezoidal, Gaussian, and triangular—can be used. Here, triangular membership functions are chosen for their simplicity and optimal control performance.

The inputs to the Fuzzy Inference System (FIS) are defined as the error and the change in error, while the output is specified as the control signal. Both the input and output variables of the FIS are represented using seven fuzzy sets:

- NB (Negative Big)
- NM (Negative Medium)

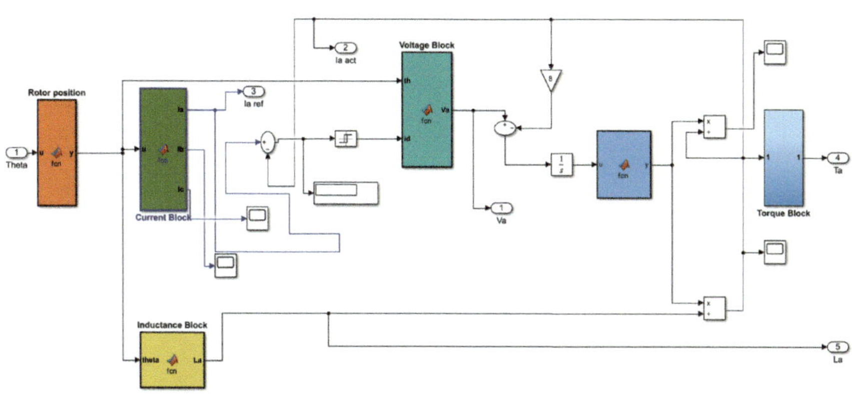

Fig. 3 One Phase of Current loop control model of SRM using Matlab Simulink

- NS (Negative Small)
- Z (Zero)
- PS (Positive Small)
- PM (Positive Medium)
- PB (Positive Big).

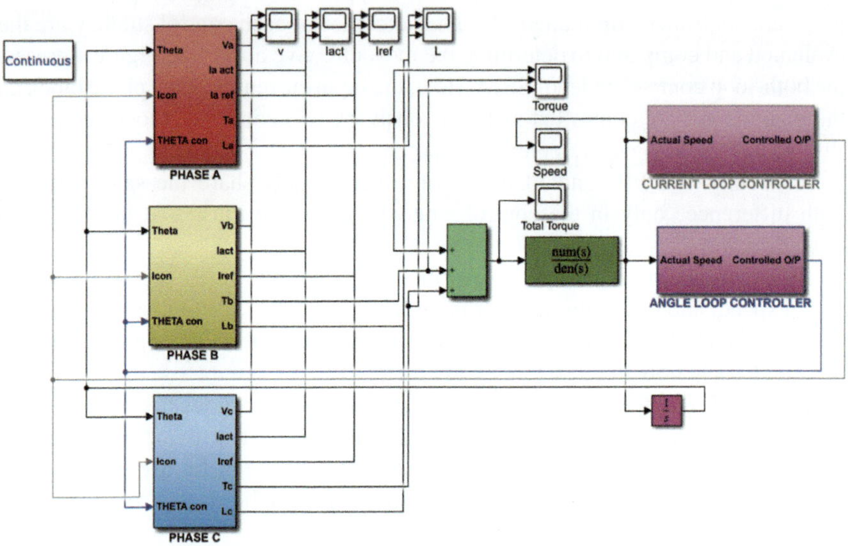

Fig. 4 Both loop(current and angle) control model of SRM using Matlab Simulink

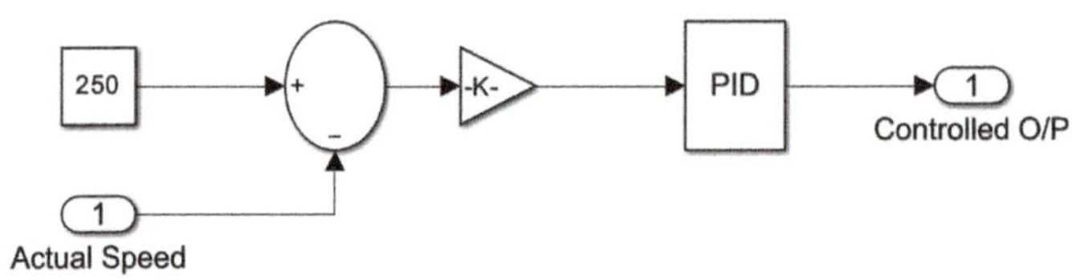

Fig. 5 PID controller

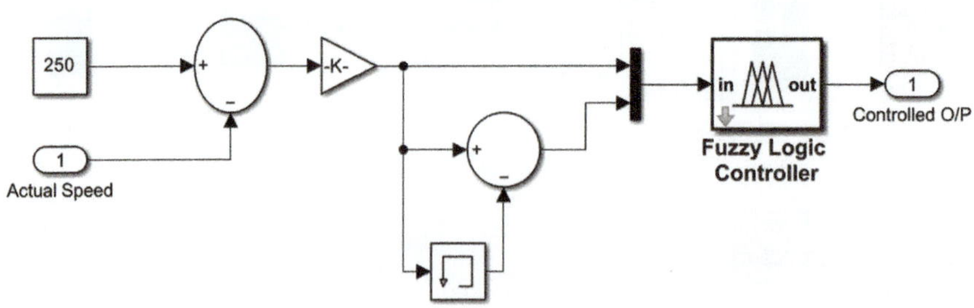

Fig. 6 FUZZY logic controller

5 Results

Open Loop SRM

In the open-loop configuration, the SRM rotor speed exhibits significant variations, as illustrated in Fig. 7. This instability highlights the need for control strategies to maintain the speed at a desired level. The absence of regulation mechanisms leads to fluctuations that make the system unsuitable for practical applications.

Current Loop Control SRM

The implementation of current loop control in the SRM significantly improves system stability. The rotor speed behavior under current loop control is shown in Fig. 8, where both PID and FUZZY logic controllers are employed. A comparison between the two controllers reveals that although the percentage error is similar for both approaches, the FUZZY logic controller achieves a shorter settling time compared to the PID controller.

Time-domain specifications for the current loop control are summarized in Table 3, showing that the delay time, rise time, and peak time for the FUZZY logic controller

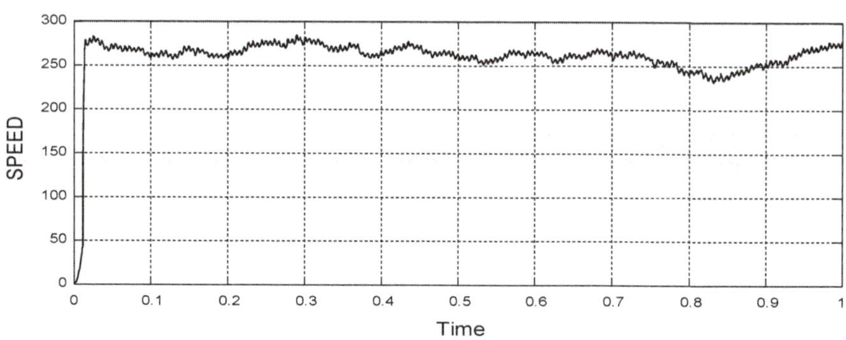

Fig. 7 Rotor speed of open-loop SRM in rad/s

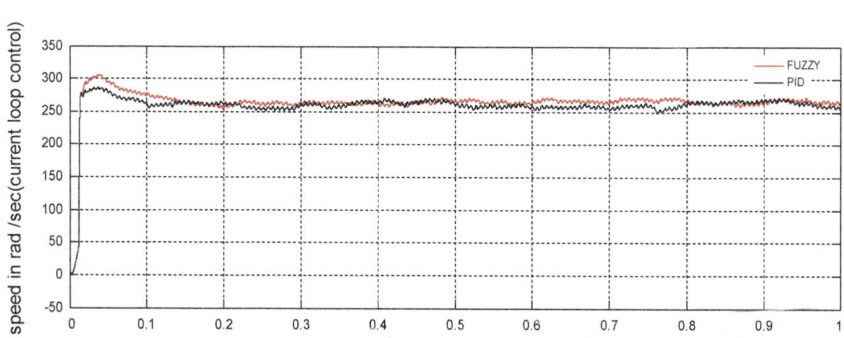

Fig. 8 Rotor speed (in rad/s) for SRM with current loop control

are slightly lower than those of the PID controller. However, the FUZZY logic controller exhibits a larger overshoot of 58 rad/s compared to 38.8 rad/s in the PID controller. Despite this, the FUZZY logic controller provides greater overall stability, reflected in its faster settling time of 0.2617 s compared to 0.3401 s for the PID controller.

Angle Loop Speed Control

When angle loop control is applied to the SRM, the performance of the system is further enhanced. Figure 9 depicts the rotor speed under angle loop control using both PID and FUZZY controllers. Time-domain specifications for this control method, presented in Table 4, indicate notable differences between the two controllers.

The FUZZY logic controller achieves significantly lower overshoot (11.9 rad/s) compared to the PID controller (95.5 rad/s). Furthermore, the FUZZY controller exhibits superior performance in terms of delay time, rise time, and peak time. Its settling time of 0.06408 s with negligible error demonstrates remarkable steadiness and reliability. On the other hand, while the PID controller has a slightly longer settling time of 0.289 s, it maintains an error margin of approximately 5%.

Speed Control of SRM (Both Loop)

The incorporation of both current and angle loop controls further improves the performance of the SRM. Figure 10 illustrates the rotor speed response under this

Table 3 Time domain specifications of current loop control

Specifications	PID	FUZZY
Delay time (in sec)	0.01227	0.01211
Rise time (in sec)	0.013350	0.01341
Peak time (in sec)	0.04236	0.03706
Peak overshoot (in rad/s)	38.8	58
Settling time (in s) (5% error)	0.3401	0.2617

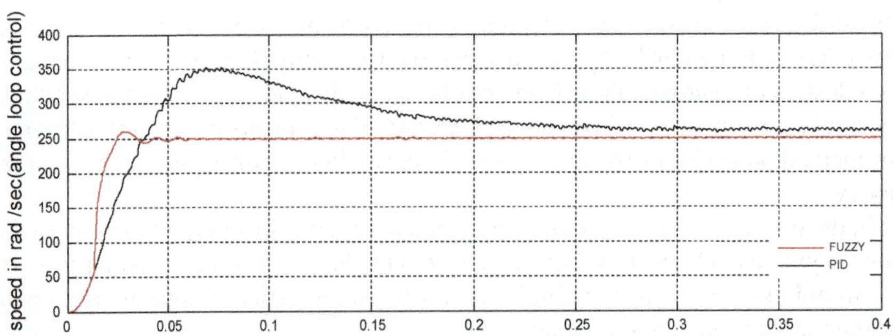

Fig. 9 Rotor speed (in rad/s) for SRM with angle loop control

Table 4 Time domain specifications of angle loop control

Specifications	PID	FUZZY
Delay time (in s)	0.02140	0.01443
Rise time (in s)	0.038	0.02452
Peak time (in s)	0.09113	0.0288
Peak overshoot (in rad/s)	95.5	11.9
Settling time (in s)	0.289 (5% error)	0.06408 (0% error)

configuration, and the corresponding time-domain specifications are provided in Table 5.

In this combined control strategy, the PID controller demonstrates the ability to settle to the final speed without any steady-state error. However, the FUZZY logic controller outperforms the PID controller in terms of response times and overshoot. The FUZZY controller achieves a settling time of 0.0528 s, significantly shorter than the 0.25 s observed for the PID controller. Additionally, the FUZZY controller exhibits minimal overshoot (11 rad/s) compared to the PID controller's 84.2 rad/s.

The combined loop control approach results in noticeable reductions in delay time, rise time, and peak overshoot for both controllers compared to individual current or

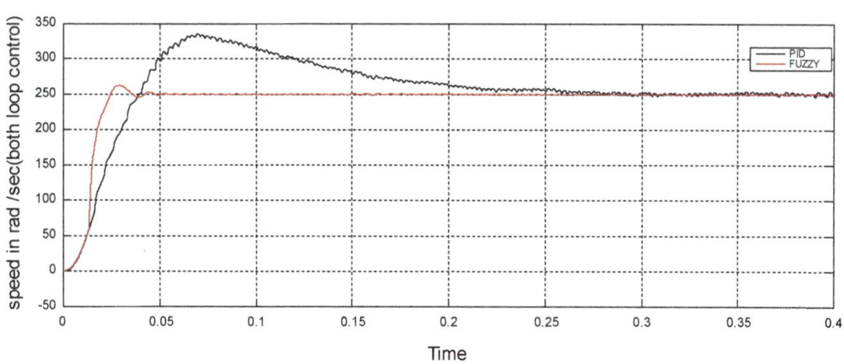

Fig. 10 Rotor speed (in rad/s) for the SRM with simultaneous current and angle loop control

Table 5 Time domain performance specifications of both-loop control SRM

Specifications	PID	FUZZY
Delay time (in s)	0.0198	0.0143
Rise time (in s)	0.03783	0.02446
Peak time (in s)	84.2	11
Peak overshoot (in rad/s)	0.25	0.0528
Settling time (in s)	0.0198	0.0143

angle loop controls. The FUZZY logic controller, in particular, demonstrates superior time-domain performance, making it the more effective option for achieving smooth and stable speed control in SRMs.

6 Conclusion

Switched Reluctance Motors (SRMs) have become increasingly popular in variable-speed drive applications due to their simple structure, robustness, low cost, and the straightforward nature of their associated unipolar power converters. These characteristics make SRMs a competitive alternative to traditional AC and DC drives.

The study demonstrates that the PID controller, when applied in angle loop control, provides more stable speed regulation compared to current loop control. However, the angle loop PID control still exhibits a degree of error. This issue is effectively addressed by incorporating simultaneous control of both the current and angle loops, resulting in improved performance.

A comparison of time-domain specifications, as presented in Tables 3, 4, and 5, highlights the advantages of FUZZY logic controllers over PID controllers. The FUZZY logic controller achieves significantly shorter settling times, reduced overshoot, and near-negligible vibrations, making it a more efficient and reliable choice for speed control in SRMs. These findings underscore the potential of FUZZY logic controllers to enhance the dynamic performance of SRMs, particularly in applications requiring precise and stable speed control.

Appendix 1

See Table 6.

Table 6 Machine specification

Voltage	150 v
Resistance	8Ω
Reference current	5 A
Minimum inductance	8 mH
Maximum inductance	60 mH
Inertia	0.0013
Coefficient of friction	0.0183
Speed	2400 rpm
No. of stator poles	6
No. of rotor poles	4

References

1. Meghana Karunakar M., Dr. G.R.K. Murthy, M. Kiran Kumar: Modeling & Closed Loop Speed Control of 6/4 Pole Switched Reluctance Motor Based on MATLAB/Simulink Environment. International Journal of Electrical, Electronics and Telecommunication Engineering 44(2), 2051–3240 (2021).
2. Vikas S. Wadnerkar, Dr. G. Tulasi Ram Das, Dr. A.D. Rajkumar: Performance Analysis of Switched Reluctance Motor; Design, Modelling and Simulation of 8/6 Switched Reluctance Motor. Journal of Theoretical and Applied Information Technology 2005–2008 JATIT, 1118–1124 (2008).
3. F. Soares, P. J. Costa Branco: Simulation of a 6/4 Switched Reluctance Motor Based on MATLAB/Simulink Environment. IEEE Transactions on Aerospace and Electronic Systems 37(3), 989–1009 (2001).
4. K. Venkataratnam: Special Electric Machine. Universities Press (India) Private Limited, Hyderabad (2008).
5. W. F. Ray, P. J. Lawrenson, R. M. Davis, J. M. Stephenson, N. N. Fulton, R. J. Blake: High-Performance Switched Reluctance Brushless Drives. IEEE Transactions on Industry Applications 22(4), 722–730 (1986).
6. Hamid Ishan Akther, Virendra K. Sharma, A. Chandra: Determination of Optimum Switching Angles for Speed Control of SRM Drive System. Indian Journal of Engineering and Materials Sciences 11(3), 151–168 (2004).
7. A.D. Cheok, Y. Fukuda: A New Torque and Flux Control Method for Switched Reluctance Motor Drives. IEEE Transactions on Power Electronics 17(4), 617–626 (2002).
8. K. M. Rahman, B. Fahimi, G. Suresh, A. V. Rajarathnam, M. Ehsani: Advantages of Switched Reluctance Motor Applications to EV and HEV: Design and Control Issues. IEEE Transactions on Industry Applications 36(1), 111–121 (2000).
9. Krishnan, R.: Switched Reluctance Motor Drives: Modelling, Simulation, Analysis, Design, and Applications. CRC Press, Boca Raton (2001).
10. M. Divandari, B. Rezaie and E. Amiri, "Robust Speed Control of Switched Reluctance Motor Drive Based on Full Order Terminal Sliding Mode Control," 2020 IEEE Applied Power Electronics Conference and Exposition (APEC), New Orleans, LA, USA, 2020, pp. 2423–2428, https://doi.org/10.1109/APEC39645.2020.9124309.
11. A. Dadpour, K. Ansari, M. Divandari, MR. Hosseinkhah "Conversion of shaded-pole induction motor to switched reluctance motor and effects of pole shoe and notch on SRM noise" IEEE conference, Electronics and Nanotechnology, no.5, pp. 344–348, April 2013.
12. B. Amin, Variable Reluctance Machines, Jouve, Paris, 2003.
13. R. Krishnan, Switched Reluctance Motor Drive, CRC Press, New York, 2001.
14. T. Nonaka, Y. Nakazawa, K. Ohyama, H. Fujii, H. Uehara, Y. Hyakutake, Inverter improving motor efficiency of switched reluctance motor for electric vehicle, in: European Power Electronics Conference, EPE, France, 2013, pp. 1–8.
15. H. Hannoun, M. Hilairet, C. Marchand, Design of an SRM speed control strategy for a wide range of operating speeds, IEEE Trans. Ind. Electron. (2010) 2911–2921.
16. Z. Yueying, W. Dafang, Z. Guifan, Y. Dongyu, W. Yu, Research progress of switched reluctance motor drive system, in: International Conference on Mechatronics and Automation, ICMA, Changchun, China, 2009.
17. R. Krishnan, Whither, Motor drives: a case study in switched reluctance motor drives, in: International Conference on Electrical Machines and Systems, ICEMS, Seoul, South Korea, 2007.
18. M.D. Hennen, M. Niessen, C. Heyers, H.J. Brauer, R.W. De Doncker, Development and control of an integrated and distributed inverter for a fault tolerant five-phase switched reluctance traction drive, IEEE Trans. Power Electron. (2010) 547–554.
19. A. Radun, Generating with the switched reluctance motor, in: Proceedings of Applied Power Electronics Conference and Exposition, APEC, vol. 1, 1994, pp. 41–47.

20. P. Asadi, M. Ehsani, B. Fahimi, Design and control characterization of switched reluctance generator for maximum output power, in: Applied Power Electronics Conference and Exposition, APEC, Dallas, USA, 2006.

Portable Health Monitor for Car Drivers Through Contact and Non-contact Sensors

Rahimul I. Mazumdar, Vishakha Chaurasia, Anirban Dasgupta, and Parijat Bhowmick

Abstract This study presents a portable integrated system designed for monitoring the health of automotive drivers using both contact and non-contact sensors to address existing challenges. The contact-based sensors include the MAX30102 for estimating heart rate (HR) and oxygen saturation, a photoplethysmography (PPG) sensor for evaluating heart rate variability (HRV), and a Grove sensor for measuring galvanic skin response (GSR) over time. Additionally, a non-contact thermal camera is employed to estimate core body temperature (CBT). The collected data undergoes comprehensive analysis: the GSR signal is denoised using variational mode decomposition; HRV is computed using peak detection on the PPG data; and parameters such as HR, oxygen saturation, and CBT are estimated via maximum-likelihood estimation. These metrics provide insights into the driver's physical and mental health. For physical health, a high CBT indicates fever, while HR thresholds help diagnose tachycardia or bradycardia. Oxygen saturation levels below normal indicate hypoxia. Mental health is assessed using the GSR signal, where the ratio of tonic to phasic components determines the driver's stress level.

Keywords Heart rate (HR) · Galvanic skin response (GSR) · Core body temperature (CBT) · Maximum-likelihood estimation (MLE) · Variational mode decomposition (VMD)

R. I. Mazumdar (✉) · V. Chaurasia · A. Dasgupta · P. Bhowmick
Department of Electronics and Electrical Engineering, Indian Institute of Technology Guwahati, Guwahati, Assam, India
e-mail: rahimulislam@iitg.ac.in

V. Chaurasia
e-mail: c.vishakha@alumni.iitg.ac.in

A. Dasgupta
e-mail: anirban.dasgupta@iitg.ac.in

P. Bhowmick
e-mail: parijat.bhowmick@iitg.ac.in

U. S. Dixit et al. (eds.), *Current Progress in Engineering Sciences*,
https://doi.org/10.1007/978-981-96-8753-4_32

1 Introduction

Traffic accidents pose a significant problem, leading to fatalities, injuries, and substantial economic losses. Worldwide, approximately 1 million people die annually due to traffic accidents [1], with India having the highest traffic accident rates [2]. The causes of these accidents can be attributed to factors related to the driver, vehicle, and environment [3]. Among all these factors, the driver's health is critical, leading to many accidents [4]. However, this factor is often neglected as a research topic. In this context, the driver's health factors include physical and mental health. Physical health consists of the driver having a fever marked by an increase in metabolism, heart issues such as tachycardia and bradycardia, respiratory issues indicated by hypoxia, etc. A driver with a fever can face accidents due to impaired concentration, slower reaction times, and reduced physical and mental alertness [5]. Tachycardia and bradycardia are dangerous during driving because they can cause symptoms like dizziness, shortness of breath, or chest pain, which can significantly affect the driver's ability to drive safely [6]. Likewise, hypoxia is harmful during driving because it leads to reduced oxygen levels in the blood, which can cause symptoms such as shortness of breath, which can severely impair a driver's ability to operate the vehicle safely [7]. The mental health conditions of the driver include stress. Stress during driving may lead to aggressive driving, increasing the risk of accidents [8]. Identifying these conditions requires understanding these parameters and their sensing techniques.

1.1 Motivation

The motivation of the work is to prevent road accidents by continuously monitoring the physical and mental health of a car driver and alarm him in the event of an anomaly.

1.2 Background

As introduced, some essential conditions that must be monitored during driving include fever, tachycardia, bradycardia, hypoxia, and stress. Fever is a temporary increase in the core body temperature (CBT), often due to an illness. High fever can impair concentration and reaction times. Tachycardia is a condition characterized by an unusually rapid heart rate that can cause dizziness, shortness of breath, and fainting, and can be hazardous when driving. An abnormally slow heart rate known as Bradycardia leads to fatigue, confusion, and fainting. Such type of conditions can also impair driving abilities. Hypoxia occurs when there is a deficiency of oxygen in the blood, resulting in symptoms such as dizziness, confusion, and loss of consciousness. Stress or hypertension can cause elevated emotional tension, leading to reduced concentration, impaired decision-making, and slow reaction times and hence increasing the likelihood of accidents. These conditions are critical for ensuring driver safety. Fever is typically measured using a thermal sensor or non-contact infrared camera that monitors core body temperature. Tachycardia and bradycardia are assessed by measuring the heart rate (HR), which is defined as the number of heartbeats per minute. The HR can be estimated through sensors like the MAX30102 [9]. Hypoxia is detected using pulse oximetry (SPO_2) [10], a method that measures oxygen saturation levels in the blood. Stress can be analyzed through galvanic skin response (GSR) sensors [11] that measure the skin's electrical conductance, which changes due to increased arousal levels due to stress. Hypertension is obtained using heart rate variability (HRV), the variation in time intervals between consecutive heartbeats. HRV is an important measure as it reflects the autonomic nervous system's regulation of the heart and can be estimated by sensors like the MAX86150 [12]. By integrating these sensors into a comprehensive health monitoring system, real-time data can be analyzed to assess the driver's physical and mental state, thereby preventing potential accidents caused by these conditions.

1.3 Related Work

The existing literature describes the use of wearable sensing to obtain signals like electrocardiogram (ECG), photoplethysmograph (PPG), and GSR for monitoring drivers' heart rate and heart rate variability, mainly to analyze the presence of stress or drowsiness during driving. Some prominent works are summarized in Table 1. From the table, we find that in 2010, Shin et al. [13] proposed a drowsiness detection system using ECG and PPG sensors, by observing different HR distributions between normal and drowsy states. Following this, in 2016, Singh et al. [14] employed ECG signals for HR estimation, achieving an impressive accuracy of 2 beats per minute (bpm). They further enhanced the system by sampling ECG data to a central web server, enabling doctors to monitor health remotely. Advancing this field in 2020, Valsalan et al. [15] incorporated sensing of HR and body temperature, to assess a patient's health status based on pulse rate ranges, indicating whether a patient was in a checkup, unwell, had hypothermia, or was healthy. Most recently, in 2023, Yunidar et al. [9] monitored heart rate data using the MAX30102 sensor, comparing the results with pulse oximeter data and achieving an accuracy of 98%. These studies collectively highlight the advancements in health monitoring systems for drivers, showcasing improved methods and increased accuracy over time and summarized in Table 1. However, integrating both wearable and non-wearable sensors for comprehensive and real-time driver health monitoring remains an underexplored area that this work aims to address. This is primarily because most of these sensors communicate via the same protocol, viz., the I2C. However, the microcontroller board such as Arduino or Raspberry Pi has only one or two such available pins.

Table 1 Literature review on health monitoring systems for automotive drivers

Authors, year	Sensors used/parameters	Key findings and contributions
Shin et al. 2010 [13]	ECG and PPG sensor	– Proposed a wireless car driver health monitoring system to address the limitations of existing wired methods – Results showed different HR distributions between normal and drowsy states
Singh et al. 2016 [14]	ECG signal is used for heart rate estimation	– Achieved an accuracy of 2 bpm in heart rate estimation – Sampled ECG data to a central web server for health monitoring by doctors
Valsalan et al. 2020 [15]	Heart rate and CBT, also temperature and humidity inside the room	– Based on pulse rate ranges, the model indicates if a patient is in a checkup, unwell, has hypothermia, or is healthy
Yunidar et al. 2023 [9]	Monitor heart rate data from MAX30102 sensor	– Compared the result of the proposed model concerning pulse oximeter data and got an accuracy of 98%

1.4 Research Gaps

The following research gaps have emerged from the literature reviews listed below:

- The temperature assessments in existing works are not of the core body.
- GSR is often noisy and prone to the nature of contact with the skin.
- No system exists that can simultaneously monitor all the parameters.
- The existing algorithms for parameter estimation are non-precise.
- The existing systems are not tested during real-time driving.

1.5 Contributions

In response to identified research gaps revealed through a thorough literature survey, the contributions of this work are outlined below, showcasing both technical expertise and innovation in addressing critical challenges.

- Devised a strategy to obtain core body temperature rather than the facial surface.
- Proposed denoising of the GSR signal using VMD to enhance data accuracy in varying environmental conditions.

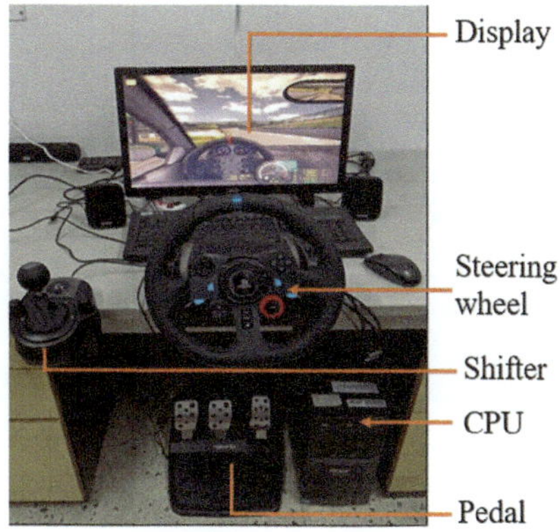

Fig. 1 Driving simulator

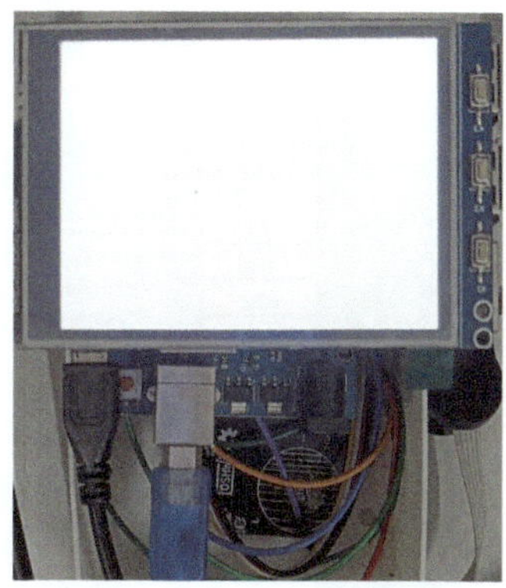

Fig. 2 Portable integrated health monitoring device with driving simulator

- Developed a portable integrated health monitoring system in Fig. 2 utilizing wearable and non-wearable sensors to monitor vital signs such as heart rate, stress levels, and oxygen saturation.
- Improved the precision in estimating the health parameter using MLE.
- Tested the system on a driving simulator in Fig. 1.

2 Interfacing

This work presents a method for interfacing multiple health monitoring sensors with a Raspberry Pi 3, as shown in Fig. 3. The system integrates the MAX30102 for heart rate and SPO_2, the MLX90640 for core body temperature, a thermal camera for skin temperature, the MAX86150 for HRV analysis, and a dedicated GSR sensor. All these sensors utilize the I2C protocol, connecting through the SDA and SCL lines. The SDA represents the serial data, while the SCL denotes the serial clock. The main challenge is that the Raspberry Pi 3 has just two I2C ports, whereas our

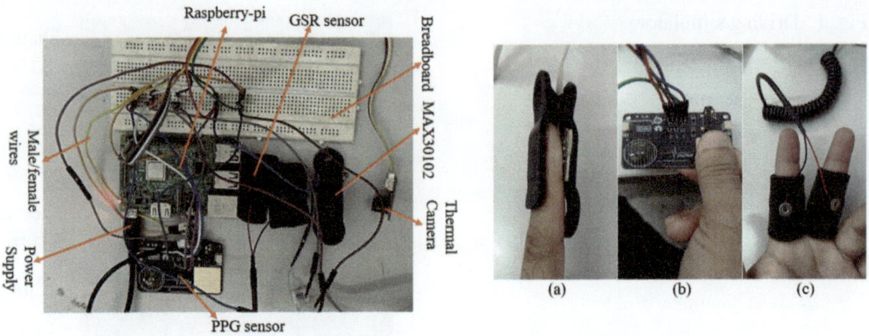

Fig. 3 Assembling of all the sensors, data collection through **a** MAX30102 sensor, **b** PPG sensor, and **c** GSR sensor

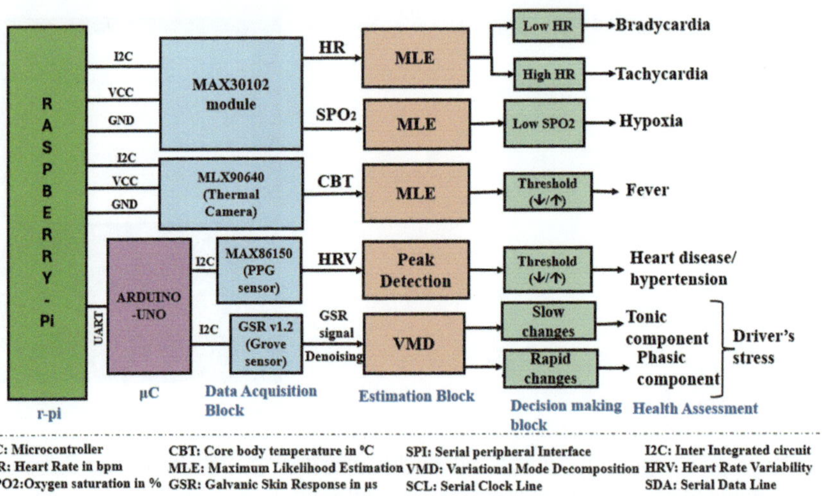

Fig. 4 Framework for health monitoring of a driver with four I2C interfaces using Raspberry Pi and Arduino

application demands four. We utilized an Arduino Uno to address these issues, which offers two additional I2C ports. The Arduino communicates with the Raspberry Pi via UART, which operates at a baud rate of 115200. The Arduino is an arbitrator in this configuration, transferring the data from the sensors to the Raspberry Pi. The Arduino's I2C ports connect the GSRv1.2 and MAX86150 PPG sensors, while the MAX30102 sensor and thermal camera utilize the Raspberry Pi's native I2C ports. The I2C consists of SCL and SDA data lines. Figure 3 illustrates the integration of all sensors with the Raspberry Pi, including data collection from the MAX30102, PPG, and GSR sensors, as outlined in the framework shown in Fig. 4. This configuration was tested on a driving simulator (Fig. 1) and ensures efficient data transmission and synchronization between the sensors, with the Raspberry Pi 3 functioning as the central component of the system.

3 Parameter Estimation

This section explores the estimation methods for HR, SPO_2, CBT, and HRV. These estimated parameters are thresholded to diagnose the appropriate condition. The denoising of the GSR signal and its processing to estimate stress is also discussed. The HR and SPO_2 are obtained from the MAX30102 data. Similarly, the HRV is estimated from the MAX86150 PPG sensor and CBT from MLX90640. The GSR signal from the Grove GSRv1.2 sensor is preprocessed and analyzed for stress assessment. Each of these estimations is explained below.

3.1 Estimating CBT

The proposed method to estimate CBT from thermal images is by focusing on the eye canthus region and applying MLE. Unlike surface temperature measurements, the eye canthus provides a more accurate representation of CBT [16].

The first step is to acquire thermal images using an infrared thermal camera. It is ensured the images capture the face, particularly the eye canthus region. The temperature values from the eye canthus region are marked as a region of interest (ROI) and the average temperature within the ROI is computed to obtain an estimated value of CBT.

Let us assume that the observed temperature T_{obs} in the eye canthus region follows a normal distribution with mean μ (the actual core body temperature) and variance σ^2.

$$T_{\text{obs}} \sim \mathcal{N}(\mu, \sigma^2) \tag{1}$$

The likelihood function for n observations is given by

$$L(\mu, \sigma^2 | T_{\text{obs}}) = \prod_{i=1}^{n} \frac{1}{\sqrt{2\pi\sigma^2}} \exp\left(-\frac{(T_{\text{obs},i} - \mu)^2}{2\sigma^2}\right) \tag{2}$$

The natural logarithm of the likelihood function, known as the log-likelihood, is

$$\ln L(\mu, \sigma^2 | T_{\text{obs}}) = -\frac{n}{2} \ln(2\pi\sigma^2) - \frac{1}{2\sigma^2} \sum_{i=1}^{n} (T_{\text{obs},i} - \mu)^2 \tag{3}$$

The log-likelihood function is maximized to estimate the parameters μ and σ^2. The maximum-likelihood estimate of μ is the sample mean:

$$\hat{\mu} = \frac{1}{n} \sum_{i=1}^{n} T_{\text{obs},i} \tag{4}$$

The maximum-likelihood estimate of σ^2 is the sample variance:

$$\hat{\sigma}^2 = \frac{1}{n}\sum_{i=1}^{n}(T_{\text{obs},i} - \hat{\mu})^2 \tag{5}$$

The estimated CBT, which is derived from temperatures taken from the eye canthus area, is represented by $\hat{\mu}$.

3.2 *Estimating HR and* SPO_2 *from MAX30102 Sensor Data Through MLE*

This section discusses an algorithm for estimating HR and oxygen saturation (SPO_2) from MAX30102 sensor data. MLE is used to enhance the precision. The process starts with data acquisition, which involves the collection of raw infrared (IR) and red-light (RED) data from the MAX30102 sensor. Subsequently, the raw data is subjected to signal preprocessing to remove noise and artifacts.

To preprocess the signal, a low-pass filter with a cutoff frequency of 2.5 Hz is used to eliminate high-frequency noise, while a high-pass filter with a cutoff frequency of 0.5 Hz addresses the baseline drift. For heart rate estimation, peaks in the IR signal corresponding to heartbeats are identified using the Pan–Tompkins peak detection algorithm [17]. The time intervals between successive peaks (RR intervals) are calculated and the heart rate is derived.

$$\text{HR} = \frac{60}{\text{Mean RR Interval (in seconds)}} \tag{6}$$

The SPO_2 is calculated using the ratio of ratios (RoR) method from the RED and IR signals.

$$\text{RoR} = \frac{\text{AC}_{\text{RED}}/\text{DC}_{\text{RED}}}{\text{AC}_{\text{IR}}/\text{DC}_{\text{IR}}} \tag{7}$$

Here AC is the alternating current component (peak-to-peak amplitude), and DC is the direct current component (average value) of the respective signals.

$$\text{SPO}_2 = 110 - 25 \times \text{RoR} \tag{8}$$

MLE is used to refine the estimates of HR and SPO_2 by modeling the observed data. The observed RR intervals RR_{obs} are assumed to follow a normal distribution with mean μ_{RR} (actual mean RR interval) and variance σ^2_{RR}.

$$RR_{\text{obs}} \sim \mathcal{N}(\mu_{RR}, \sigma^2_{RR}) \tag{9}$$

The likelihood function for n observations is given by

$$L(\mu_{RR}, \sigma_{RR}^2 | RR_{\text{obs}}) = \prod_{i=1}^{n} \frac{1}{\sqrt{2\pi\sigma_{RR}^2}} e^{\left(-\frac{(RR_{\text{obs},i} - \mu_{RR})^2}{2\sigma_{RR}^2}\right)} \tag{10}$$

The natural logarithm of the likelihood function, known as the log-likelihood, is

$$\begin{aligned} \ln L(\mu_{RR}, \sigma_{RR}^2 | RR_{\text{obs}}) = & -\frac{n}{2} \ln(2\pi\sigma_{RR}^2) \\ & - \frac{1}{2\sigma_{RR}^2} \sum_{i=1}^{n} (RR_{\text{obs},i} - \mu_{RR})^2 \end{aligned} \tag{11}$$

To estimate μ_{RR} and σ_{RR}^2, the log-likelihood is maximized. The maximum-likelihood estimate of μ_{RR} is the sample mean

$$\hat{\mu}_{RR} = \frac{1}{n} \sum_{i=1}^{n} RR_{\text{obs},i} \tag{12}$$

The maximum-likelihood estimate of σ_{RR}^2 is the sample variance

$$\hat{\sigma}_{RR}^2 = \frac{1}{n} \sum_{i=1}^{n} (RR_{\text{obs},i} - \hat{\mu}_{RR})^2 \tag{13}$$

It is assumed that the observed RoR values follow a normal distribution with mean μ_{RoR} and variance σ_{RoR}^2.

$$\text{RoR}_{\text{obs}} \sim \mathcal{N}(\mu_{\text{RoR}}, \sigma_{\text{RoR}}^2) \tag{14}$$

The likelihood function for n observations is given by

$$L(\mu_{\text{RoR}}, \sigma_{\text{RoR}}^2 | \text{RoR}_{\text{obs}}) = \prod_{i=1}^{n} \frac{1}{\sqrt{2\pi\sigma_{\text{RoR}}^2}} e^{\left(-\frac{(\text{RoR}_{\text{obs},i} - \mu_{\text{RoR}})^2}{2\sigma_{\text{RoR}}^2}\right)} \tag{15}$$

3.3 *Estimating Hypertension*

The estimation of hypertension by analyzing HRV derived from PPG sensor data using peak detection is performed through the following steps. PPG data is acquired using the sensor, the signal is preprocessed to remove noise using a bandpass and moving average filter. In the next step, peaks in the PPG signal are detected using an ensemble of five different methods. To ensure robustness, the peak identified

Table 2 Threshold conditions for health parameter monitoring

Parameters	Threshold conditions	Detected issues
HR	if HR is < 60 bpm	Bradycardia
HR	if HR is > 100 bpm	Tachycardia
SPO_2	if SPO_2 is $< 95\%$	Hypoxia
CBT	if CBT is $> 100\,^\circ$F	Fever
HRV (SDNN)	SDNN ($\geq$150)	Hypertension
GSR signal	Phasic to tonic of the order greater than 10^{-2}	Emotional response (stress)

by the majority of algorithms (at least four out of five) is chosen as the final peak. Subsequently, Heart Rate Variability (HRV) is analyzed by calculating the standard deviation of the time intervals between consecutive R-R peaks (SDNN). The algorithm used for this process is adopted from the work presented in [18].

The SDNN is obtained as

$$\text{SDNN} = \sqrt{\frac{1}{N-1}\sum_{i=1}^{N-1}\left((RR_{i+1} - RR_i) - \overline{RR}_{\text{interval}}\right)^2} \tag{16}$$

Here N is the number of RR peaks and RR_i is the i-th RR peak. $\overline{RR}_{\text{interval}}$ is the mean interval. Typically, high SDNN is associated with a higher risk of hypertension. Threshold values for HRV metrics are determined based on clinical studies and population data.

$$\text{Hypertension Risk} = \begin{cases} \text{Low} & \text{if SDNN} < \text{Threshold}_{\text{SDNN}} \\ \text{High} & \text{if SDNN} \geq \text{Threshold}_{\text{SDNN}} \end{cases} \tag{17}$$

3.4 Estimating Stress from a GSR

To estimate stress from a Galvanic Skin Response (GSR) signal, denoising was incorporated using Variational Mode Decomposition (VMD). The method involves collecting the GSR signal with a dedicated sensor, which often contains noise from motion artifacts, environmental factors, and sensor issues. VMD is subsequently used to decompose the GSR signal into intrinsic mode functions (IMFs), thereby isolating and removing noise components to ensure accurate stress estimation.

The VMD process requires solving the constrained variational problem outlined below:

$$\min_{u_k,\omega_k}\left\{\sum_{k=1}^{K}\left\|\partial_t\left[\left(\delta(t)+\frac{j}{\pi t}\right)*u_k(t)\right]e^{-j\omega_k t}\right\|_2^2\right\} \tag{18}$$

This equation is subjected to the following conditions:

$$\sum_{k=1}^{K} u_k = x. \tag{19}$$

Here, the IMFs are indicated by u_k, the center frequencies of the modes are depicted by ω_k, x is the original GSR signal, and the number of modes is denoted by K. Relevant modes are chosen and combined to recreate a cleaned signal, while noise components are eliminated. The features involving stress are then extracted from the cleaned GSR signal. The slow-changing IMFs of the GSR signal comprise the tonic component, while the rapidly changing IMFs make the phasic component. The ratio of phasic to tonic signal powers is thresholded to determine if the driver has stress or not. To enhance the system's predictive capability, we explored the use of a machine learning model to predict stress levels and health risks based on real-time sensor data. A random forest classifier was trained on GSR, HRV, and CBT data, achieving an accuracy of 92% in distinguishing between low, moderate, and high stress levels in simulated driving conditions. This model provides a more nuanced understanding of driver health than traditional threshold-based approaches, allowing the system to issue alerts before a critical event, such as an impending panic attack or severe hypoxia.

4 Results

We tested the system on a driving simulator. A standard pulse oximeter was used for HR and SPO_2, a Polar H10 heart rate sensor was used for HRV, and a standard digital thermometer was used for CBT as a reference. The results presented in the detailed analysis in Table 3 for 30 participants indicate that the integrated health monitoring system consistently performs close to the accuracy of standard medical devices in various physiological parameters. HR measurements show deviation errors ranging from 1.13% to 3.53%, demonstrating reliable performance in tracking heart conditions such as tachycardia and bradycardia. The oxygen saturation readings exhibit minimal deviation errors, between 1.02% and 1.08%, indicating the system's proficiency in detecting hypoxia with precision comparable to standard pulse oximeters. The HRV data reveal deviation errors from 1.64% to 4.00%, affirming the system's effectiveness in assessing stress and cardiovascular health. The CBT measurements, with deviations ranging from 0. 27% to 0. 80%, underscore the ability of the system to accurately detect fever and other thermal anomalies, closely aligning with the performance of standard medical thermometers. It is also determined whether a driver exhibited tachycardia, bradycardia, hypoxia, fever, or hypertension based on the predefined threshold values shown in Table 2. Table 4 evaluated the three core metrics of the GSR signal: stress level, the ratio of phasic to tonic level, and the difficulty of driving tasks.

Table 3 Comparison of integrated system and standard device measurements for 30 participants

Parameters	HR			SPO_2			HRV			CBT		
Participants	HR (BPM)	Standard HR (BPM)	HR deviation (%)	SPO_2 (%)	Standard SPO_2 (%)	SPO_2 deviation (%)	HRV (ms)	Standard HRV (ms)	HRV deviation (%)	CBT (°F)	Standard CBT (°F)	CBT deviation (%)
1	72	70	2.86	98	97	1.03	55	54	1.85	99.5	99.14	0.54
2	88	85	3.53	95	96	1.04	48	50	4.00	100.4	99.14	0.53
3	76	75	1.39	98	97	1.02	51	51	0.00	99.14	99.14	0.27
4	92	90	2.22	97	97	0.00	60	60	1.67	99.32	98.06	0.54
5	64	65	1.54	96	95	1.05	57	58	1.72	100.22	99.14	0.80
6	83	84	1.25	95	95	1.05	53	54	1.89	98.78	98.6	0.27
7	78	77	1.30	98	97	1.03	54	53	1.89	99.32	99.0	0.54
8	89	87	2.30	95	94	1.06	49	48	2.08	99.68	99.5	0.27
9	82	80	2.50	97	96	1.04	56	55	1.82	98.6	98.0	0.54
10	74	73	1.37	96	95	1.05	53	52	1.92	99.5	98.78	0.54
11	80	78	2.56	99	98	1.02	60	59	1.69	99.86	99.14	0.53
12	86	84	2.38	95	94	1.06	58	57	1.75	99.14	99.14	0.54
13	77	76	1.32	98	97	1.03	51	50	2.00	98.42	98.06	0.54
14	91	89	2.25	96	95	1.05	62	61	1.64	99.5	99.14	0.54
15	79	78	1.28	95	94	1.06	54	53	1.79	100.4	99.14	0.53
16	90	88	2.27	98	97	1.02	50	49	2.04	99.32	99.0	0.54
17	83	82	1.22	94	93	1.08	55	54	1.85	98.6	98.24	0.54
18	86	85	1.16	98	97	1.04	52	51	1.69	99.68	99.14	0.27
19	75	74	1.35	97	96	1.04	52	51	1.96	99.5	98.54	0.54
20	84	83	1.20	96	95	1.05	60	60	1.67	98.78	98.6	0.27
21	88	87	1.15	95	94	1.06	53	52	1.92	99.5	99.14	0.54

(continued)

Table 3 (continued)

Parameters	HR			SPO_2			HRV			CBT		
Participants	HR (BPM)	Standard HR (BPM)	HR deviation (%)	SPO_2 (%)	Standard SPO_2 (%)	SPO_2 deviation (%)	HRV (ms)	Standard HRV (ms)	HRV deviation (%)	CBT (°F)	Standard CBT (°F)	CBT deviation (%)
22	81	80	1.25	99	98	1.02	58	57	1.75	98.24	98.06	0.54
23	73	72	1.39	98	97	1.03	50	49	2.04	100.22	100.06	0.53
24	88	87	1.14	98	97	1.03	51	50	2.00	98.6	98.24	0.54
25	70	69	1.45	95	94	1.05	50	49	2.04	99.32	99.14	0.53
26	87	86	1.16	99	98	1.02	54	53	1.89	99.32	99.0	0.54
27	86	85	1.37	97	96	1.04	50	49	2.04	98.42	98.06	0.54
28	79	78	1.28	95	94	1.06	60	59	1.69	98.78	98.6	0.27
29	86	85	1.18	96	95	1.05	53	52	1.92	99.5	99.14	0.54
30	79	78	1.28	95	94	1.06	60	59	1.69	98.78	98.6	0.27

Table 4 GSR-based stress estimation results

Participants	Stress level	Ratio of phasic/tonic level	Difficulty level of driving
1	High	1.8	High
2	Medium	1.5	Medium
3	High	2.0	High
4	Medium	1.4	Medium
5	Low	1.1	Low
6	Medium	1.6	Medium
7	High	1.9	High
8	Low	1.2	Low
9	Medium	1.5	Medium
10	High	2.1	High
11	Low	1.3	Low
12	Medium	1.4	Medium
13	High	2.0	High
14	Medium	1.5	Medium
15	Low	1.2	Low
16	High	1.7	High
17	Medium	1.4	Medium
18	Low	1.1	Low
19	Medium	1.6	Medium
20	High	1.9	High
21	Low	1.3	Low
22	Medium	1.5	Medium
23	High	2.2	High
24	Medium	1.6	Medium
25	Low	1.2	Low
26	High	2.0	High
27	Medium	1.5	Medium
28	Low	1.3	Low
29	Medium	1.6	Medium
30	High	2.1	High

5 Conclusion and Future Scope

This paper described the integration of a system and subsequent parameter estimation for assessing the physical and mental health of a car driver. The system integrates contact and non-contact sensors to estimate parameters, *viz.*, HR, oxygen saturation, HRV, and CBT. A high CBT indicates fever, reduced oxygen saturation indicates hypoxia, HR gives information on tachycardia and bradycardia, while high HRV from a PPG signal indicates hypertension. The system also analyzes the phasic and tonic components of the GSR signal to determine stress. The novelty of the work lies in solving the shortage of I2C interfaces by using an Arduino as an I2C to UART converter. The varying environmental conditions while GSR capture is also addressed using VMD. The precision of parameters such as CBT, HR, and SPO_2 is improved using MLE. In the future, IoT technology will be incorporated into this system to enable real-time data transmission and analysis on smartphone applications. Such technology will be beneficial to services like Ola and Uber. A better design for wearable sensing can be developed for seamless monitoring in actual driving conditions.

Acknowledgements We want to express our heartfelt gratitude to **Mr. Chinmay Koch**, **Mr. Jyotirmay Ganguly**, intern students from Assam Engineering College, India, and **Mr. Ravindra**, a Research Scholar in the EEE department at IIT Guwahati, for assisting in assembling the hardware components and designing the casing of that hardware using 3D printing. Thank you all for your relentless support and encouragement.

References

1. S. Binder and J. W. Runge, "Road safety and public health: a us perspective and the global challenge," *Injury Prevention*, vol. 10, no. 2, pp. 68–69, 2004.
2. D. Mohan, "Traffic injuries and fatalities in India," *Transportation Research & Injury Prevention Programme, Indian Institute of Technology Delhi*, 2004.
3. Y. Ali, A. Sharma, M. M. Haque, Z. Zheng, and M. Saifuzzaman, "The impact of the connected environment on driving behavior and safety: A driving simulator study," *Accident Analysis & Prevention*, vol. 144, p. 105643, 2020.
4. S. K. Ahmed, M. G. Mohammed, S. O. Abdulqadir, R. G. A. El-Kader, N. A. El-Shall, D. Chandran, M. E. U. Rehman, and K. Dhama, "Road traffic accidental injuries and deaths: A neglected global health issue," *Health science reports*, vol. 6, no. 5, p. e1240, 2023.
5. J. Charest and M. A. Grandner, "Sleep and athletic performance: impacts on physical performance, mental performance, injury risk and recovery, and mental health: an update," *Sleep medicine clinics*, vol. 17, no. 2, pp. 263–282, 2022.
6. P. K. Jha, M. K. Rajendran, P. K. Lenka, A. Acharyya, and A. Dutta, "A fully analog autonomous QRS complex detection and low-complexity asystole, extreme bradycardia, and tachycardia classification system," *IEEE Transactions on Instrumentation and Measurement*, vol. 71, pp. 1–13, 2022.
7. F. Hamza Cherif, L. Hamza Cherif, M. Benabdellah, and G. Nassar, "Monitoring driver health status in real time," *Review of scientific instruments*, vol. 91, no. 3, 2020.
8. V. C. Magaña, W. D. Scherz, R. Seepold, N. M. Madrid, X. G. Pañeda, and R. Garcia, "The effects of the driver's mental state and passenger compartment conditions on driving performance and driving stress," *Sensors*, vol. 20, no. 18, p. 5274, 2020.
9. Y. Yunidar, M. Melinda, K. Khairani, M. Irhamsyah, and N. Basir, "Iot-based heart signal processing system for driver drowsiness detection," *Green Intelligent Systems and Applications*, vol. 3, no. 2, pp. 98–110, 2023.
10. A. Cheung, L. Tu, A. Macnab, B. K. Kwon, and B. Shadgan, "Detection of hypoxia by near-infrared spectroscopy and pulse oximetry: a comparative study," *Journal of biomedical optics*, vol. 27, no. 7, pp. 077001–077001, 2022.

11. R. M. Al Abdi, A. E. Alhitary, E. W. Abdul Hay, and A. K. Al-Bashir, "Objective detection of chronic stress using physiological parameters," *Medical & biological engineering & computing*, vol. 56, pp. 2273–2286, 2018.
12. A. Oktariana, E. Yulianto, Syaifudin, Triwiyanto, H. G. Ariswati, A. M. Maghfiroh, and V. Shankhwar, "Analysis of electrocardiogram and photoplethysmogram signals to detect car driver drowsiness using the threshold method," in *Proceeding of the 3rd International Conference on Electronics, Biomedical Engineering, and Health Informatics: ICEBEHI 2022, 5–6 October, Surabaya, Indonesia*, pp. 681–701, Springer, 2023.
13. H.-S. Shin, S.-J. Jung, J.-J. Kim, and W.-Y. Chung, "Real time car driver's condition monitoring system," in *SENSORS, 2010 IEEE*, pp. 951–954, IEEE, 2010.
14. R. K. Singh, A. Sarkar, and C. Anoop, "A health monitoring system using multiple non-contact ECG sensors for automotive drivers," in *2016 IEEE International Instrumentation and Measurement Technology Conference Proceedings*, pp. 1–6, IEEE, 2016.
15. P. Valsalan, T. A. B. Baomar, and A. H. O. Baabood, "IoT based health monitoring system," *Journal of critical reviews*, vol. 7, no. 4, pp. 739–743, 2020.
16. Q. Wei, H.-J. Park, and J. H. Lee, "Development of a wireless health monitoring system for measuring core body temperature from the back of the body," *Journal of Healthcare Engineering*, vol. 2019, no. 1, p. 8936121, 2019.
17. L. Sathyapriya, L. Murali, and T. Manigandan, "Analysis and detection r-peak detection using modified pan-Tompkins algorithm," in *2014 IEEE International Conference on Advanced Communications, Control and Computing Technologies*, pp. 483–487, IEEE, 2014.
18. S. Heo, S. Kwon, and J. Lee, "Stress detection with single PPG sensor by orchestrating multiple denoising and peak-detecting methods," *IEEE Access*, vol. 9, pp. 47777–47785, 2021.

Classification of Driving Condition Images for Semi-autonomous Vehicles Through Data Augmentation and Transfer Learning

Ravindra Gautam, Anirban Dasgupta, Ayush Datta Jaiswal, and Devansh Sharma

Abstract The accurate classification of driving conditions is crucial for the safe and efficient operation of semi-autonomous vehicles. Even for manually operated vehicles, classification of driving conditions is necessary for improving visibility. Variations in conditions, such as heavy rainfall, dense fog, or darkness, present distinct challenges that directly impact driving dynamics and safety. This initiated research on computer vision techniques to classify driving conditions. However, the complexity of road conditions with different traffic levels, sceneries, etc., poses significant challenges for the classification task. In this paper, we propose the use of ConvNeXT to address this problem. We perform transfer learning by using feature extractions from pre-trained ConvNeXT and train the classification module for categorizing the road images into one of four classes, *viz.* sunny, rainy, foggy, dark. We create training, validation and test sets from the available data from different databases. Due to limited available data, we also employ data augmentations by randomly selected crop regions, random horizontal flips, Gaussian blur with random kernels and standard deviations, and the addition of salt and pepper noise. The performance of ConvNeXT is compared with VGG-16, ResNet-50, DenseNet-121, Inception-v3, Mobilenet-v3, Efficientnet-v2, Xception, and Inception Resnet-v2. ConvNeXT provides the best accuracy of 95.25%, precision of 0.95, and a recall of 0.96, among other competing methods.

R. Gautam (✉) · A. Dasgupta
EEE Department, IIT Guwahati, Guwahati, India
e-mail: ravindra2013@iitg.ac.in

A. Dasgupta
e-mail: anirban.dasgupta@iitg.ac.in

A. D. Jaiswal
Nutanix Technologies, Bangalore, India
e-mail: ayush.jaiswal@alumni.iitg.ac.in

D. Sharma
Oracle India Private Limited, Bangalore, India
e-mail: s.devansh@alumni.iitg.ac.in

U. S. Dixit et al. (eds.), *Current Progress in Engineering Sciences*,
https://doi.org/10.1007/978-981-96-8753-4_33

Keywords Semi-autonomous driving · Driving condition classification · ConvNeXT · Data augmentation · Transfer learning

1 Introduction

Semi-autonomous vehicles are automobiles that have some automated driving features but still require human intervention and oversight [17]. These vehicles typically incorporate advanced driver assistance systems that can handle certain driving tasks under specific conditions, such as highway driving or parking. Still, the driver must remain engaged and ready to take control at necessary times. One such issue in semi-autonomous vehicles is automated driving condition classification. Such classification is required in semi-autonomous vehicles primarily for safety reasons [12]. This classification is also vital for manually operated vehicles, as it determines adverse weather conditions, such as heavy rainfall or fog, to provide automated assistive vision.

1.1 Motivation

The driver's visibility is heavily impacted by adverse driving conditions, as depicted in Fig. 1, which poses challenges for manually operated as well as semi-autonomous vehicles. Therefore, determining driving conditions is crucial for safety, efficiency, and overall driving experience. The motivations can be highlighted as

- *Safety*: Identifying conditions like fog or heavy rain allows for the timely activation of visibility aids, such as fog lights or automated windshield wipers, ensuring better visibility and preventing accidents.
- *Adaptation*: Awareness of current road conditions helps drivers and autonomous systems adjust driving techniques and vehicle settings, ensuring optimal traction and handling on slippery or obscured roads.

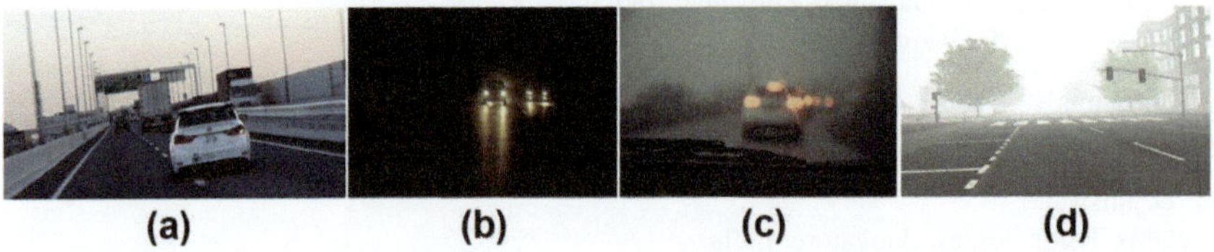

Fig. 1 Road conditions in different weather conditions such as **a** normal bright, **b** dark, **c** rainy, **d** foggy

- *Fuel Efficiency*: Knowledge of weather and road conditions contributes to fuel optimization, as vehicles can adapt their driving patterns to conserve energy when necessary.
- *Real-time Decision Making*: For semi-autonomous and fully autonomous vehicles, real-time classification of driving conditions is vital for safe and efficient navigation. This includes making informed decisions about speed, braking, and maneuvering based on accurate and timely data about the environment.

1.2 Sensing Types

This driving condition classification problem is usually addressed by using sensors and deploying suitable algorithms to detect and classify weather conditions in real-time. The various sensors and technologies to detect and interpret weather conditions are provided in Table 1. Based on the table, the most commonly deployed sensing are Radio Detection and Ranging (RADAR), Light Detection and Ranging (LIDAR), and Thermal and Visual Imaging techniques. LIDAR technology uses laser light to measure distances and create detailed 3D maps of the surroundings, including identifying precipitation particles or fog [23]. RADAR uses radio waves to detect objects and their velocities, helping identify rain or snow [26]. Thermal imagers detect the surface temperature to construct an image. Since night and day will have significantly different temperatures, they can easily be distinguished. Visual sensors capture images and videos, allowing for the recognition of weather patterns like rain, snow, fog as well as road conditions [9, 18]. RADAR and LiDAR systems usually have limited resolution compared to visual imaging. This limitation can impact the ability to detect small-scale weather phenomena accurately, such as thin clouds or fine precipitation particles. Moreover, both RADAR and LiDAR are affected by atmospheric conditions such as humidity, temperature, and particulate matter. Thermal imaging is usually costly, temperature dependent, and has lower resolution, making it inferior to visual image alternatives. Visual range cameras offer detailed visual data facilitating recognition of diverse weather conditions and road surfaces. This contextual information aids in weather classification, including identifying patterns, precipitation intensities, and visibility changes. Given their affordability and accessibility compared to LiDAR or RADAR, we utilize image-based method for the classification task in this work.

1.3 Objectives

In this paper, the main objective is to develop a classification model to indicate an image as sunny, rainy, dark, and foggy, as shown in Fig. 1. We select these four conditions are these are the most predominantly occurring cases during driving.

Table 1 Sensors and methods for weather classification

Author	Year	Sensing	Weather types	Method	Remarks
Lu et al. [11]	2014	Visual imaging	Sunny and cloudy	SVM	Accurately distinguishes sunny and cloudy images
Miething et al. [13]	2019	Thermal imaging	Snow	Sensor fusion	Thermal imaging for effective snow detection
Hong et al. [8]	2020	Radar	Fog	Deep learning	Implemented a novel radar-based method
Yang et al. [27]	2022	LiDAR	Rain, fog, snow	Near-InfraRed	High accuracy in sunny weather
Minhas et al. [14]	2022	Visual imaging	Clear, cloudy, rainy, foggy	CNN	Tested on synthetic data

2 Literature Review

Very few recent works have attempted to address this gap through computer vision techniques for weather classification. These attempts primarily employed neural networks.

One of the prominent recent works is by Guerra et al. [6] in 2018, that focuses on extracting discriminating features specific to rain, fog, and snow conditions. They introduced the Rain Fog Snow (RFS) Dataset comprising 1100 images per category and proposed an algorithm utilizing superpixel delimiting masks for feature extraction. The classification was performed using support vector machines (SVM) and evaluated with ten different Convolutional Neural Network (CNN) architectures. However, as the complexity of the scene increases, the discriminating features reduce.

In 2021, Xiao et al. [25] have used an improved VGG-16 for the classification of weather phenomena images, which are not necessarily regarding driving conditions. Dhananjaya et al. [5] addressed the challenge of achieving robust visual perception in harsh weather and low-light conditions by creating their own data through experiments, encompassing nine weather classifications and achieving 78.15% using a ResNet.

Jokab et al. [21] have proposed a weather classification system using images from a front-view camera processed by a simple CNN in 2022. Their solution demonstrated high accuracy across various weather conditions, with particularly outstanding results in snowy weather detection.

Table 2 Existing road image databases

Database	Description
Weather image recognition data sets [6]	– 6,862 labeled images – 11 classes *viz.* dew, fog/smog, frost, glaze, hail, lightning, rain, rainbow, rime, sandstorm, and snow
Rome weather image classification dataset [22]	– 250 labeled images – Five weathers *viz* sunny, rainy, cloudy, foggy, and snowy scenarios
Daytime and nighttime road images data sets [16]	– 31,567 labeled images – Two classes *viz* 31,567 daylight images and 16,960 nighttime images
Foggy road images dataset [20]	– 102 labeled foggy road images – Three classes *viz* uniform fog, heterogeneous fog, and cloudy fog, captured during both morning and evening times

Recently, the use of DenseNet-121 was proposed by Albelwi et al. [2] in 2022, for classifying weather images on the Rain Fog Snow (RFS) dataset. However, the conditions are generic and do not mimic driving scenarios.

The main issue with these works are that the weather classification datasets are generic and not focussed on driving-related tasks. The driving-condition classification related standard datasets are not available in public. The publicly available image datasets that are related closely to this problem are summarized in Table 2.

2.1 Research Issues

The literature review brings about the following research challenges that need attention while addressing the problem.

- Driving scenarios during the night are diverse in nature, as shown in Fig. 2. Some cases may be too dark, while some contain sufficient street light. The ambient light color also differs significantly, thereby making hue-based features less significant.
- Rainy and foggy conditions present different complexities of the scene depending on the intensity of rain and fog, respectively.
- Often, headlights of other vehicles are on in both dark as well as foggy cases, making these classes difficult to distinguish.
- Variation with different forms of people and vehicles on the road, variations in sceneries, buildings, and posters pose significant challenges [17].
- Discriminating features are difficult to handcraft.
- Real-time implementation on image streams demands high processing throughput in terms of frames per second (fps).

Fig. 2 Variations in night driving conditions

2.2 Contributions

This work has the following two significant contributions.

- Creating a new dataset collecting from a few existing datasets with images of the required labels and using four augmentation methods to increase the sample size.
- Training a ConvNeXT model via transfer learning to classify the four driving scenarios.

3 Dataset Preparation

3.1 Data Collection

We have assembled a dataset featuring images representing four distinct driving conditions *viz.* Foggy, Rainy, Sunny, and Dark. These data were manually sampled pertaining to each of the required classes from Weather Image Recognition data sets, Rome Weather Image Classification Dataset, and Daytime and Nighttime Road Images Data Sets. This dataset was purposefully developed to train deep learning models for classifying these conditions. Three individuals manually performed the labeling, and ambiguous cases were not considered in the dataset. The final dataset encompasses a total of 782 image files, each precisely labeled with one of the four driving conditions. The class distribution is as follows:

- Foggy: 212 images
- Rainy: 202 images
- Sunny: 188 images
- Dark: 180 images.

Some sample images for each class are shown in Fig. 3.

3.2 Data Augmentation

Since the number of images was less, data augmentation techniques were used. Specific augmentation techniques such as rotations or vertical flips were not used, as they will not occur in the context of driving scenarios with respect to the original upright captured road image. The following augmentation techniques were employed.

Fig. 3 Sample images of the training data with the top row having images of sunny class, followed by dark, foggy and rainy

3.2.1 Random Crops

Let W be the width of the image and H be the height of the original image $I(x, y)$. Let (x_1, y_1) denote the coordinates of the top-left corner of the region to be kept, and (x_2, y_2) denote the coordinates of the bottom-right corner. The cropped images $I_k(x, y)$ are obtained as

$$I_k(x, y) = I(x, y) \mid x_1(k) \leq x \leq x_2(k),\ \ y_1(k) \leq y \leq y_2(k)\}\forall k \in [0, 19] \quad (1)$$

Here we use 20 random crops per image. We limit (x_1, y_1) to within 10% of the top-left corner, as given in Eqs. (2) and (3), while (x_2, y_2) is limited to within 10% of the bottom-right corner, as given in Eqs. (4) and (5).

$$x_1 < \lfloor(0.1 \times W) \quad (2)$$

$$y_1 < \lfloor(0.1 \times H) \quad (3)$$

$$\lfloor(0.9 \times W) < x_2 < W \quad (4)$$

$$\lfloor(0.9 \times H) < y_2 < H \quad (5)$$

Here, the $\lfloor$ operation is rounding off to the previous nearest integer. All random variables x_1, x_2, y_1, y_2 have uniform distributions. Some sample images of this technique on an image are shown in Fig. 4.

Fig. 4 Random crops of the original road image for the class 'Dark'

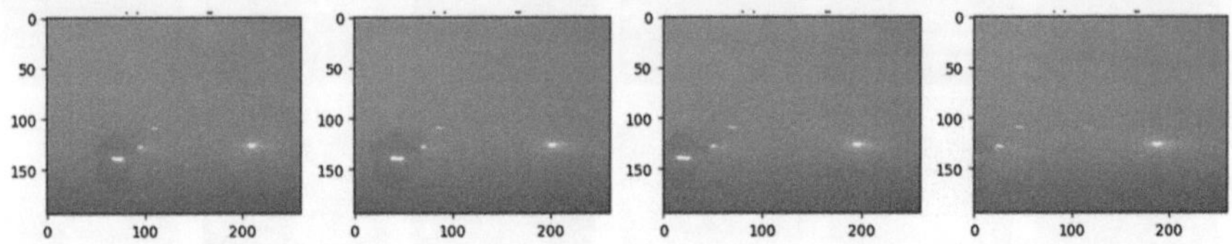

Fig. 5 Random flips of the original road image for the class 'Foggy'

3.2.2 Horizontal Flips

In this augmentation, the image is flipped horizontally on a random vertical axis at p fraction of width W. The new flipped pixel position x' is obtained as

$$x'(k) = \begin{cases} p(k)W - (x - p(k)W) & \text{if } x \leq p(k) \times W \\ x + (W - x) \times (1 - p(k)) & \text{if } x > p(k) \times W \end{cases} \quad (6)$$

Here, $p(k)$ is a random variable sampled from a uniform distribution in the range $[0.4W, 0.6W]$, $\forall k \in [0, 19]$. In this manner, 20 such flips were obtained per image. Figure 5 shows a sample foggy image with some flips about the vertical axis.

3.2.3 Gaussian Blur

This augmentation technique applies Gaussian blur to the image using random kernel sizes and standard deviations. Gaussian blur smoothens the image, helping the model to focus on the essential content of the road image rather than unnecessary details. Additionally, this augmentation mimics real-world scenarios such as motion blur, enhancing the model's robustness.

The blurred images $I_{\text{blurred}}(x, y)$ are obtained by using the following equation.

$$I_{\text{blurred}}(x, y) = \frac{1}{2\pi\sigma^2} \sum_{i=-M}^{M} \sum_{j=-M}^{M} \exp\left(-\frac{i^2 + j^2}{2\sigma^2}\right) \cdot I(x + i, y + j) \quad (7)$$

where σ is the standard deviation controlling the blur intensity, and M determines the kernel size, which defines the extent of the neighboring pixels contributing to the blur effect.

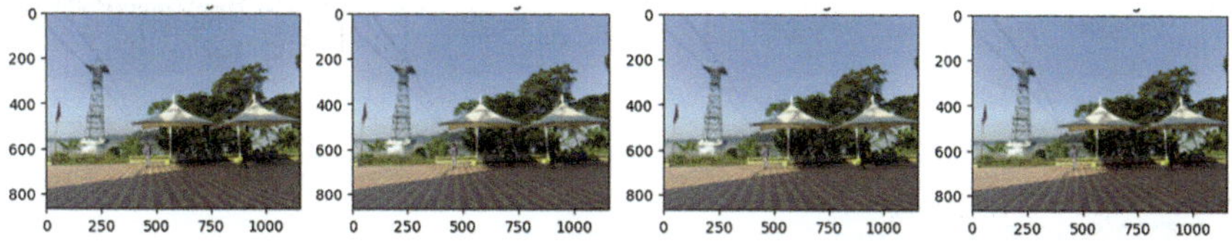

Fig. 6 Gaussian blur cases with random kernel sizes and standard deviations of the original road image for the class 'Sunny'

Fig. 7 Cases of adding salt and pepper noise to an original road image for the class 'Rainy'

In this augmentation, σ is randomly chosen from the range [0.8, 1.2], while M is an integer ($M \in \mathbb{Z}$) selected between 3 and 9 ($3 \leq M \leq 9$). Using this technique, 20 additional augmented images are generated for each input image, providing variability and improving model generalization. A sample case for different blurs for a road image under bright sunny conditions is shown in Fig. 6.

3.2.4 Adding Salt-and-Pepper Noise

The last method is adding salt-and-pepper noise, which helps the model learn to be robust to noise in the input data.

$$I_{\text{noisy}}(x, y) = \begin{cases} 0 & \text{with probability } p_{\text{salt}} \\ 255 & \text{with probability } p_{\text{pepper}} \\ I(x, y) & \text{otherwise} \end{cases} \tag{8}$$

As such, 20 more images were obtained (Fig. 7).

4 Dataset Split

After the augmentation process, each image presented 80 more variants, representing 20 images for four augmentation types. This increased the dataset to 81-fold. The final dataset now becomes as follows:

- Foggy: $212 \times 81 = 17172$ images
- Rainy: $202 \times 81 = 16362$ images
- Sunny: $188 \times 81 = 15228$ images
- Dark: $180 \times 81 = 14580$ images.

These images are further split into training, validation, and test sets with a ratio of 7:2:1.

5 Driving Condition Classification

As explained in the data augmentation process in Sect. 3.2, the size of the data set increased with the increase in data. However, the number of images was still insufficient for the appropriate training of state-of-the-art deep network architectures from scratch. Moreover, the training time would also be significantly large owing to the large number of floating point operations. Hence, we employed transfer learning for training. The transfer learning technique [10] uses the previously learned weights as the starting point for a new model, usually on different data. In this work, we use the feature extraction module from previously learned networks on the Imagenet dataset while training the classification modules on our data.

5.1 Feature Extraction

Feature extraction is performed from the recently introduced ConvNeXT [15] with pretrained weights on the 'ImageNet-21k' dataset with fine-tuning on the 'ImageNet-1k' dataset.

Algorithm 1: ConvNeXT forward pass algorithm

Input: Input image
Output: Feature maps
Pass input image through the first convolution block;
Perform the following operations: Layer normalization;
ConvNeXT block operation;
Layer normalization;
2D convolutions;
Layer normalization;
ConvNeXT block operation;
Layer normalization;
2D convolutions;
Layer normalization;
ConvNeXT block operation;
Layer normalization;
2D convolutions;
Layer normalization;
ConvNeXT block operation;
Perform global average pooling;
Perform ReLU;

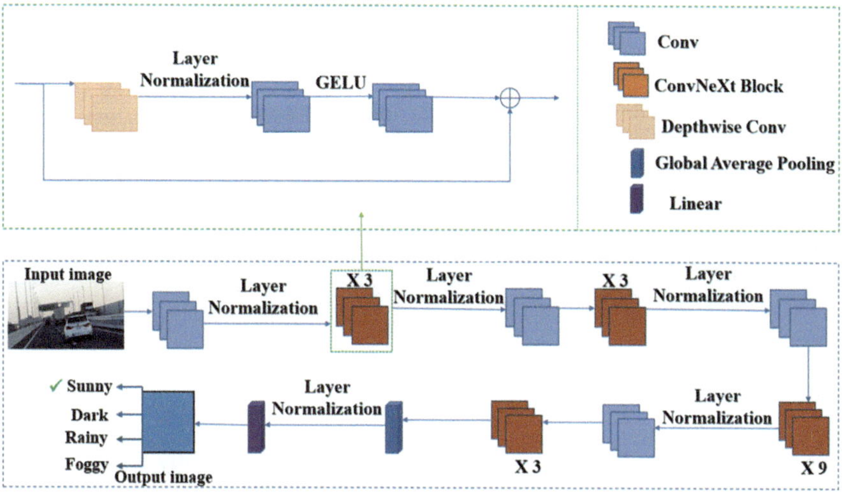

Fig. 8 Block diagram of ConvNeXT architecture

The framework of ConvNext draws inspiration from the vision Transformer model, which pioneered the use of attention mechanisms to enhance the efficiency of model training. Employing an encoder-decoder architecture, convneXT builds upon this foundation. CNNs offer distinct advantages in visual feature extraction, while attention-based models excel at capturing global contextual information with larger receptive fields and robust model representation capabilities. By combining these strengths, the ConvNext architecture was introduced. The feature extraction layers of the ConvNext network are given in Algorithm 1 and depicted in Fig. 8. The first four blocks are convolution, layer normalization, and ConvNext, followed by another layer normalization. This set of four operations is cascaded three more times, followed by the global average pooling and finally with a rectified linear unit (ReLU). These cascades extract features of different scales.

5.2 *ConvNeXT Block*

The ConvNeXT block consists of depthwise convolutions, followed by regular convolutions and a Gaussian Error Linear Unit (GELU) activation function. This activation function GELU(x) is defined as

$$\text{GELU}(x) = x\Phi(x) = 0.5x\left(1 + \text{erf}\left(\frac{x}{\sqrt{2}}\right)\right) \tag{9}$$

In this equation, erf(.) is the Gaussian error function defined as

$$\mathrm{erf}(x) = \frac{2}{\sqrt{\pi}} \int_0^x e^{-t^2} dt \tag{10}$$

Since its computation is complex, it is approximated with the sigmoid function $\sigma(.)$ as

$$\mathrm{erf}(x) = x\sigma(1.702x) \tag{11}$$

This block adopts the structural framework of ResNet. As such, ConvNext embodies a fusion of CNN and attention-based models.

5.3 *Classification*

The features extracted in the previous stage are trained using a fully connected dense network as explained in Algorithm 2. This network is trained with the prepared dataset for 150 epochs, with a batch size of 1.

5.4 *Training*

The categorical cross-entropy loss function $\mathcal{L}$ is used to train this network. This loss $\mathcal{L}$ is defined as

$$\mathcal{L}(y, \hat{y}) = -\sum_i y_i \log(\hat{y}_i) \tag{12}$$

Here, y and $\hat{y}$ represent the true and predicted probability distributions with one-hot encoded vectors. Backpropagation of the loss is performed to compute gradients for each parameter. The model weights are updated using an Adam optimizer based on the gradients.

Algorithm 2: Fully connected layer for 4 classes

Input : Feature vector F from the previous stage
Output: Class predictions

Fully Connected Layer:
F_1 = FullyConnected(2048, 4, Softmax)
Forward Pass:
O = FullyConnectedLayer(F)
Return O

6 Results

This section presents an in-depth analysis of the classification performance of our model across various weather classes. Table 3 provides a comprehensive overview of the classification metrics calculated for each feature extraction method, including accuracy, precision, recall, and F1-Score. We evaluate the performance of ConvNeXT using transfer learning from various pre-trained models *viz.* VGG-16, DenseNet-121, ResNet50, Inception v3, MobileNet v3, EfficientNet v2, Xception, and Inception ResNet v2. We utilize the pre-trained weights on the ImageNet dataset for feature extraction and subsequently train the fully connected classification module. VGG-16 [3], characterized by its deep architecture primarily composed of 3×3 convolutional layers with max-pooling, extracts hierarchical features. However, its high parameter count demands significant computational resources. DenseNet [28] employs dense connectivity patterns, efficiently aggregating features from all preceding layers to enhance parameter efficiency and gradient flow throughout the network. ResNet [7] introduces residual connections, mitigating the vanishing gradient problem and facilitating training of very deep networks. It excels at capturing fine-grained details. GoogleNet [1], featuring inception modules, utilizes multiple filter sizes within the same layer to effectively capture multi-scale features while maintaining computational efficiency. MobileNet [19] employs depthwise separable convolutions and linear bottlenecks to achieve good accuracy with reduced model size and processing power requirements compared to traditional networks. EfficientNet [4] utilizes a compound scaling method that uniformly scales the depth, width, and resolution of the network, resulting in high-performance models that are smaller and faster than competing approaches. Xception [24] relies on depthwise separable convolutions, breaking down standard convolutions into more efficient steps. Inception ResNet builds upon the Inception architecture by incorporating residual connections. This

Table 3 Performance comparison of different architectures

Architecture	Metrics			
	Accuracy	Precision	Recall	F1-score
CNN [21]	81.78	0.79	0.78	0.78
VGG16 [25]	92.04	0.97	0.86	0.91
ResNet50 [5]	91.72	0.84	0.92	0.88
DenseNet121 [2]	90.49	0.91	0.88	0.89
Inceptionv3	90.82	0.88	0.92	0.899
Mobilenetv3	90.07	0.93	0.84	0.88
Efficientnet v2	90.48	0.92	0.87	0.89
Xception	93.56	0.92	0.96	0.939
Inception Resnetv2	94.23	0.92	0.96	0.939
ConvNeXT	95.25	0.95	0.96	0.955

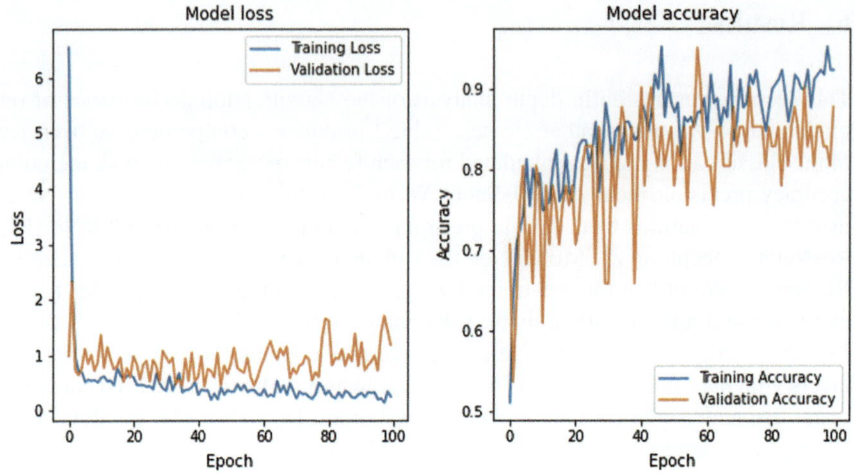

Fig. 9 The plots show the accuracy and loss of the model for 100 epochs

Table 4 Confusion matrix for four classes

		Predicted			
		Sunny	Dark	Rainy	Foggy
4 * Actual	Sunny	0.94	0.03	0.01	0.02
	Dark	0.04	0.93	0.03	0.01
	Rainy	0.01	0.01	0.96	0.02
	Foggy	0.01	0.003	0.007	0.98

technique allows the network to learn from the original input alongside transformed features, enhancing gradient flow and overall performance compared to the base Inception model.

The training and validation plots for the accuracies and losses are shown in Fig. 9. The table reveals that the model exhibits robust precision, recall, and F1 scores. The confusion matrix for the model is given in Table 4.

7 Conclusion

In this research, the problem being addressed is the classification of driving conditions into sunny, rainy, dark, and foggy. We have used visual images over LiDAR, RADAR, and thermal data, owing to specific advantages as discussed. We propose to utilize a network based on ConvNeXt as the backbone. We create our training and test data by selecting appropriate images from the Weather Image Recognition dataset, Rome Weather Image Classification Dataset, and Daytime and Nighttime Road Images Dataset. The data augmentation techniques employed to enhance model robustness include random crops, horizontal flips, Gaussian blur, and salt-and-pepper noise injection. The research compared the performance of popular architectures. Looking ahead, future research routes include the reversal of adverse driving conditions using techniques like pix2pix and cycle GANs, aimed at enhancing visibility and further improving driving safety. This research underscores the potential of deep learning in revolutionizing driving assistance systems, paving the way for safer and more efficient transportation solutions. When developing a weather classification model for deployment on resource-constrained devices like Raspberry Pi, obtaining a trade-off between accuracy and speed will be important. In the final model, the images will typically be captured from a camera mounted on a car, and the road scenery will be viewed outward.

References

1. Al-Qizwini, M., Barjasteh, I., Al-Qassab, H., Radha, H.: Deep learning algorithm for autonomous driving using Googlenet. In: 2017 IEEE intelligent vehicles symposium (IV). pp. 89–96. IEEE (2017)
2. Albelwi, S.A.: Deep architecture based on DenseNet-121 model for weather image recognition. International Journal of Advanced Computer Science and Applications **13**(10) (2022)
3. Alippi, C., Disabato, S., Roveri, M.: Moving convolutional neural networks to embedded systems: the AlexNet and VGG-16 case. In: 2018 17th ACM/IEEE International Conference on Information Processing in Sensor Networks (IPSN). pp. 212–223. IEEE (2018)
4. Atila, Ü., Uçar, M., Akyol, K., Uçar, E.: Plant leaf disease classification using EfficientNet deep learning model. Ecological Informatics **61**, 101182 (2021)
5. Dhananjaya, M.M., Kumar, V.R., Yogamani, S.: Weather and light level classification for autonomous driving: Dataset, baseline and active learning. In: 2021 IEEE International Intelligent Transportation Systems Conference (ITSC). pp. 2816–2821. IEEE (2021)
6. Guerra, J.C.V., Khanam, Z., Ehsan, S., Stolkin, R., McDonald-Maier, K.: Weather classification: A new multi-class dataset, data augmentation approach and comprehensive evaluations of convolutional neural networks. In: 2018 NASA/ESA Conference on Adaptive Hardware and Systems (AHS). pp. 305–310. IEEE (2018)
7. He, F., Liu, T., Tao, D.: Why ResNet works? Residuals generalize. IEEE transactions on neural networks and learning systems **31**(12), 5349–5362 (2020)
8. Hong, Z., Petillot, Y., Wang, S.: Radarslam: Radar based large-scale slam in all weathers. In: 2020 IEEE/RSJ International Conference on Intelligent Robots and Systems (IROS). pp. 5164–5170. IEEE (2020)
9. Kang, L.W., Chou, K.L., Fu, R.H.: Deep learning-based weather image recognition. In: 2018 International Symposium on Computer, Consumer and Control (IS3C). pp. 384–387. IEEE (2018)
10. Liu, T., Alibhai, S., Wang, J., Liu, Q., He, X., Wu, C.: Exploring transfer learning to reduce training overhead of HPC data in machine learning. In: 2019 IEEE International Conference on Networking, Architecture and Storage (NAS). pp. 1–7. IEEE (2019)
11. Lu, C., Lin, D., Jia, J., Tang, C.K.: Two-class weather classification. In: Proceedings of the IEEE Conference on Computer Vision and Pattern Recognition. pp. 3718–3725 (2014)
12. Marti, E., De Miguel, M.A., Garcia, F., Perez, J.: A review of sensor technologies for perception in automated driving. IEEE Intelligent Transportation Systems Magazine **11**(4), 94–108 (2019)
13. Miethig, B., Liu, A., Habibi, S., Mohrenschildt, M.v.: Leveraging thermal imaging for autonomous driving. In: 2019 IEEE Transportation Electrification Conference and Expo (ITEC). pp. 1–5. IEEE (2019)

14. Minhas, S., Khanam, Z., Ehsan, S., McDonald-Maier, K., Hernández-Sabaté, A.: Weather classification by utilizing synthetic data. Sensors **22**(9), 3193 (2022)
15. Qi, J., Nguyen, M., Yan, W.Q.: Waste classification from digital images using ConvNeXt. In: Pacific-Rim Symposium on Image and Video Technology. pp. 1–13. Springer (2022)
16. RAMAN: Daytime and nighttime road images. https://www.kaggle.com/datasets/raman77768/day-time-and-night-time-road-images (2020)
17. Sharma, T., Chehri, A., Fofana, I., Jadhav, S., Khare, S., Debaque, B., Duclos, N., Arya, D.: Deep learning-based object detection and classification for autonomous vehicles in different weather scenarios of Quebec, Canada. IEEE Access (2024)
18. Shehata, M.S., Cai, J., Badawy, W.M., Burr, T.W., Pervez, M.S., Johannesson, R.J., Radmanesh, A.: Video-based automatic incident detection for smart roads: The outdoor environmental challenges regarding false alarms. IEEE Transactions on Intelligent Transportation Systems **9**(2), 349–360 (2008)
19. Sinha, D., El-Sharkawy, M.: Thin MobileNet: An enhanced MobileNet architecture. In: 2019 IEEE 10th annual ubiquitous computing, electronics & mobile communication conference (UEMCON). pp. 0280–0285. IEEE (2019)
20. Tarel, J.P., Hautiere, N., Cord, A., Gruyer, D., Halmaoui, H.: Improved visibility of road scene images under heterogeneous fog. In: 2010 IEEE intelligent vehicles symposium. pp. 478–485. IEEE (2010)
21. Triva, J., Grbić, R., Vranješ, M., Teslić, N.: Weather condition classification in vehicle environment based on front-view camera images. In: 2022 21st International Symposium INFOTEH-JAHORINA (INFOTEH). pp. 1–4. IEEE (2022)
22. Vaz, R.: Rome weather image classification. https://www.kaggle.com/datasets/rogeriovaz/rome-weather-classification (2022)
23. Wallace, A.M., Halimi, A., Buller, G.S.: Full waveform lidar for adverse weather conditions. IEEE transactions on vehicular technology **69**(7), 7064–7077 (2020)
24. Wu, X., Liu, R., Yang, H., Chen, Z.: An Xception based convolutional neural network for scene image classification with transfer learning. In: 2020 2nd international conference on information technology and computer application (ITCA). pp. 262–267. IEEE (2020)
25. Xiao, H., Zhang, F., Shen, Z., Wu, K., Zhang, J.: Classification of weather phenomenon from images by using deep convolutional neural network. Earth and Space Science **8**(5), e2020EA001604 (2021)
26. Xu, Z.W., Wu, J., Wu, Z.S.: A survey of ionospheric effects on space-based radar. Waves in Random Media **14**(2), S189 (2004)
27. Yang, T., Li, Y., Ruichek, Y., Yan, Z.: Performance modeling a near-infrared ToF lidar under fog: A data-driven approach. IEEE Transactions on Intelligent Transportation Systems **23**(8), 11227–11236 (2021)
28. Zhang, K., Guo, Y., Wang, X., Yuan, J., Ding, Q.: Multiple feature reweight DenseNet for image classification. IEEE Access **7**, 9872–9880 (2019)

How Multilingual LLMs Are! A Case Study of LLMs Using Multilingual Sentiment Analysis in Indian Language

Saurabh Kumar, Shifali Agrahari, Sanasam Ranbir Singh, Arpit Saikia, Rhitwik Choudhury, and Anubhav Bhattacharya

Abstract Large Language Models (LLMs) are computational models known for their ability to achieve general-purpose language understanding and generate text by training the model's billion parameters on massive amounts of text data. Multilingual LLMs are language models that support multiple languages and can understand and generate text in different languages. In the current scenario, there is a lack of Multilingual LLMs that have dedicated support for Indian languages. This paper has tried to assess the current performance of LLMs in Indian languages by hand-picking a few multilingual LLMs (like XLM-R, mBERT, MuRIL, Navarasa, IndicBERT v2, mT5) that support these Indian languages and fine-tuning them using the technique of prompt tuning and few-shot learning for multilingual Sentiment Analysis. Sentiment Analysis involves processing text to determine whether the emotional tone of the message is positive, negative, or neutral. It requires a model to understand and interpret the context, tone, and implicit meaning of text, which provides a robust test of the model's natural language understanding capabilities.

Keywords LLMs · Multilingual LLMs · Sentiment analysis · Prompt tuning · Accuracy

---

S. Kumar · S. Agrahari (✉) · S. R. Singh
Department of CSE, Indian Institute of Technology Guwahati, Guwahati, Assam, India
e-mail: a.shifali@iitg.ac.in

S. Kumar
e-mail: saurabh1003@iitg.ac.in

S. R. Singh
e-mail: ranbir@iitg.ac.in

A. Saikia
Department of CSE, National Institute of Technology, Silchar, Assam, India

R. Choudhury
Department of EE, National Institute of Technology, Silchar, Assam, India

A. Bhattacharya
Department of CSE, Rajiv Gandhi Institute of Petroleum Technology, Jais, Uttar Pradesh, India

U. S. Dixit et al. (eds.), *Current Progress in Engineering Sciences*,
https://doi.org/10.1007/978-981-96-8753-4_34

1 Introduction

The rise of multilingual model architectures has transformed natural language processing (NLP). This transformation has been spearheaded by the introduction of multilingual pre-trained language models such as mBERT [7], mT5 [19]. These models harness extensive unsupervised textual data from various languages, enabling zero-shot and few-shot cross-lingual transfer for a wide range of downstream NLP tasks. However, their performance can vary significantly depending on the specific task and language pair [10].

Along with these advancements, large language models (LLMs) such as GPT-3 [5] have gained significant attention. These models are renowned for their ability to achieve general-purpose language understanding and generate text by training billions of parameters on vast amounts of textual data. Research has demonstrated the impressive performance of LLMs in few-shot in-context learning, particularly in public English sentiment analysis tasks [20]. Despite being predominantly pre-trained on English corpora, some research has revealed fascinating multilingual capabilities in both public and proprietary LLMs [14].

However, the effectiveness of these multilingual LLMs in handling Indian languages, which vary significantly in syntax, script, and cultural context, remains underexplored. Our study aims to assess the performance of LLMs in Indian languages, using sentiment analysis as a case study. By evaluating these models within the context of India's linguistic diversity, we seek to gain insights into their strengths and limitations, identify areas for improvement, and propose directions for future research. Ultimately, this will contribute to the development of more inclusive and accurate NLP technologies. This study evaluates the performance of several multilingual LLMs, including XLM-R, mBERT, MuRIL, Navarasa, and IndicBERT v2, specifically for sentiment analysis in Indian languages. Our methodology involves fine-tuning these models using techniques like prompt tuning and few-shot learning. By showcasing the effectiveness of these models and pinpointing areas for improvement, we aim to drive the development of multilingual LLMs that better accommodate Indian languages, thereby enhancing their applicability in diverse linguistic contexts.

To evaluate the LLMs effectively, we have employed a combination of few-shot learning, fine-tuning, and prompt-based tuning. Specifically, we have tested these models with varying proportions of training data: 1%, 2%, and 10%. Few-shot learning enables the models to perform well with limited examples, while fine-tuning and prompt-based tuning allow us to adapt the models more closely to the specific nuances of Indian languages. This approach will help us assess how these models handle sentiment analysis with minimal training data and determine the most effective strategies for enhancing their performance.

2 Literature Review

Any LLMs analysis like sentiment analysis faces unique challenges when applied to Indian languages, primarily due to two major issues: the scarcity of low-resource datasets and the lack of support for Indian languages in most Large Language Models (LLMs). Because most of the LLMs trained on English language due to high resource language, whereas Indian low-resource languages face a deficit of annotated data.

2.1 Multilingual Pre-trained Language Models

In recent years, the rise of multilingual model architectures has been notable, particularly following the introduction of multilingual pre-trained language models like mBERT [7], XLM-R [4], mT5 [19], and BLOOM [11]. These models harness the power of extensive unsupervised textual data spanning numerous languages, which facilitates zero-shot and few-shot cross-lingual transfer from one language to another across a variety of downstream NLP tasks, although the performance can vary [10].

More recently, large language models (LLMs) such as GPT-3 [5], Llama-2 [15], and Llama-3 [21] have gained significant attention due to their outstanding performance in text generation. Research, such as [20], has highlighted the strong capabilities of LLMs in few-shot in-context learning for public English sentiment analysis tasks. While most LLMs are pre-trained on corpora primarily in English, some studies have revealed intriguing multilingual capabilities in both public and proprietary LLMs [14]. However, to the best of our knowledge, the cross-lingual transfer abilities of these LLMs have not been thoroughly explored for sentiment analysis tasks. It remains unclear how these models compare to existing multilingual pre-trained models in terms of cross-lingual transfer.

2.2 Multilingual Case Study

Vilares et al. [18] introduced the first code-switching corpus with sentiment labels, showing the robustness of a multilingual approach. Akhtar et al. [1], Uppal et al. [16] have contributed to the sentiment analysis dataset for Hindi.

To confront these challenges, various studies have been conducted focusing on specific sentiment analysis tasks for Indian languages. Regarding sentiment analysis of the "Hinglish" text, Pravalika et al. [13], Cambria et al. [6], Mangat et al. [12] proposed a dictionary-based approach using a movie review dataset in this hybrid language. The method involved creating separate dictionaries for English and Hindi to accommodate word variations and case insensitivity. Feature extraction used tf-idf with unigram, bigram and trigram techniques, while sentiment classification used SVM, Nave Bayes, Neural Network and Logistic Regression algorithms to optimize feature selection and classifier performance for 'Hinglish' text. In [17] an annotated sentiment analysis method was introduced using SVM with a radial basis function for 2,866 Hindi–English code-mixed tweets, achieving an accuracy of 58.2%. In recent research [3], has emerged to address this issue by leveraging machine translation to augment data resources. Khanuja et al. [8] introduced indicSentiment, a dataset spanning 13 Indian languages including Assamese, Bengali, Bodo, Gujarati, Hindi, Kannada, Malayalam, Marathi, Odia, Punjabi, Tamil, Telugu, and Urdu. This data set comprises 1,000 reviews per language, translated and annotated from English reviews collected from various e-Commerce platforms. Recently, in [21], an empirical analysis was conducted to compare the cross-lingual transfer capabilities of public Small Multilingual Language Models (SMLMs) like XLM-R against English-centric LLMs such as Llama-3, in the context of sentiment analysis across English, Spanish, French, and Chinese. Kumar et al. [9] released sentiment analysis datasets for 22 Indian languages.

3 Methodology

The objective of this work is to explore the multilingual transfer capabilities of pre-trained models in the context of sentiment analysis tasks for low-resource languages of India.

3.1 Datasets

The availability of datasets in Indian languages for sentiment analysis is limited. To address this scarcity, we have employed the machine translation method described by Kumar et al. [9] to create a sentiment analysis dataset in various Indian languages. We have used the IIT Patna product review dataset [2] in Hindi (HI) as the source dataset and translated it into several low-resource Indian languages, including Assamese (AS), Bengali (BN), Tamil (TA), Telugu (TE), and Urdu (UR). Additionally, for comparative purposes, we translated the dataset into English (EN). A few samples from the dataset are shown in Table 1. We have designated 80% of the dataset for each language for testing purposes. We have used 1%, 2%, and 10% of the dataset for each language for training the models, following the standard few-shot training approach.

Table 1 Samples from the dataset. Here, review in Hindi is considered as the source language and the rest as the target during translation

Language	Review	Sentiment
Hindi	**लेकिन इसकी बैटरी कमज़ोर है ।**	**Negative**
	इसकी स्क्रीन अभूतपूर्व रंग और अच्छे दृश्य कोण देती है ।	**Positive**
English	But its battery is weak.	Negative
	Its screen gives unprecedented color and good visual angle.	Positive
Assamese	কিন্তু ইয়াৰ বেটাৰী দুৰ্বল।	Negative
	ইয়াৰ পৰ্দাই অসাধাৰণ ৰং আৰু ভাল দৰ্শন কোণ প্ৰদান কৰে।	Positive
Bengali	কিন্তু এর ব্যাটারি দুর্বল।	Negative
	এর স্ক্রিনটি অভূতপূর্ব রঙ এবং ভাল ভিজ্যুয়াল কোণ দেয়।	Positive
Tamil	ஆனால் அதன் பேட்டரி பலவீனமாக உள்ளது.	Negative
	இதன் திரையானது தனித்துவமான வண்ணங்களையும் நல்ல கோணங்களையும் வழங்குகிறது.	Positive
Telugu	కానీ దాని బ్యాటరీ బలహీనంగా ఉంది.	Negative
	దీని స్క్రీన్ అద్భుతమైన రంగులు మరియు మంచి వీక్షణ కోణాలను అందిస్తుంది.	Positive
Urdu	لیکن اس کی بیٹری کمزور ہے۔	Negative
	اس کی سکرین غیر معمولی رنگ اور دیکھنے کے اچھے زاویے فراہم کرتی ہے۔	Positive

3.2 LLM Models Details

We fine-tuned several multilingual LLMs on the IndiSentiment140 dataset:

- **XLM-R (Cross-lingual Language Model—RoBERTa)**: XLM-R is a transformer-based model pretrained on 100 languages, optimized for cross-lingual understanding. It leverages the RoBERTa framework, which is known for its robust performance on monolingual tasks, i.e. the performance remains consistence throughout different domains, and extends this capability to multiple languages. XLM-R is designed to handle a wide variety of languages with different linguistic structures, making it suitable for multilingual tasks such as sentiment analysis.
- **mBERT (Multilingual BERT)**: mBERT is the multilingual version of BERT, pre-trained on the top 104 languages with the largest Wikipedias. It shares the same architecture as BERT but is designed to handle multiple languages, providing a common framework for various language tasks. mBERT's training on diverse languages helps it to understand and generate text across different linguistic contexts, making it a versatile model for sentiment analysis.
- **MuRIL (Multilingual Representations for Indian Languages)**: MuRIL is specifically developed for Indian languages, addressing the unique linguistic properties and challenges of these languages. It is pre-trained on a large corpus of Indian languages, including code-mixed data. MuRIL aims to improve natural language understanding for Indian languages, making it particularly relevant for tasks such as sentiment analysis in the Indian context.
- **Navarasa**: Navarasa is a multilingual language model focused on Indian languages. It is fine-tuned to capture the nuances and variations in sentiment across different Indian languages. The model is named after the nine emotions (Navarasa)

in Indian classical arts, highlighting its emphasis on understanding and classifying emotional content in text.

- **IndicBERT v2**: IndicBERT v2 is an updated version of IndicBERT, optimized for Indian languages. It incorporates improvements based on newer data and training techniques, providing better performance on tasks involving Indian languages. IndicBERT v2 aims to address the resource constraints and linguistic diversity of Indian languages, enhancing its applicability for sentiment analysis.

Each model was fine-tuned using prompt tuning and few-shot learning techniques to adapt to the sentiment analysis task in various Indian languages. These techniques involve using a small number of prompt parameters and a limited set of examples to guide the model's adaptation, enabling faster and more efficient training.

4 Results and Analysis

The evaluation of multilingual LLMs for sentiment analysis across various Indian languages, using different percentages of training data (1%, 2%, and 10%), provides several insightful observations. The models considered include XLM-R, mBERT, MuRIL, Navarasa, and Indic BERT v2, and the languages evaluated are English, Hindi, Bengali, Tamil, Telugu, Urdu, and Assamese. The performance in terms of accuracy for different languages considering different LLMs is tabulated in Table 2.

XLM-R demonstrates a consistent improvement in accuracy as the training data increases from 1% to 10%. For example, its accuracy in English improves from 0.553 to 0.653 and in Hindi from 0.591 to 0.722. Other languages such as Bengali, Tamil, Telugu, and Urdu also see steady improvements, with Assamese maintaining a high and stable accuracy of 0.751 across all data percentages. Similarly, mBERT shows significant improvements with increased training data, particularly excelling in English where accuracy jumps from 0.602 to 0.824. While Hindi and Bengali see moderate improvements, Tamil and Telugu show steady accuracy increases. However, Assamese shows a slight decrease in accuracy at 10.

In contrast, MuRIL's performance remains relatively unchanged across different training data percentages, with accuracy staying around 0.491 for most languages and only a slight increase for Tamil at 10%. Like other models, MuRIL maintains a stable high accuracy of 0.751 for Assamese. Navarasa exhibits variability in performance with mixed results across languages. For English, accuracy initially is high at 1% (0.713), drops at 2% (0.314), then recovers at 10% (0.703). Hindi and Bengali see significant improvements at 10%, with accuracy reaching 0.582 and 0.729 respectively. Tamil and Telugu also show strong improvements at 10%, while Assamese performs stably but not as high as some other models.

Indic BERT v2 demonstrates substantial performance improvements with increased training data. English and Hindi see significant jumps in accuracy, reaching 0.756 and 0.758 at 10%. Bengali and Telugu also show noticeable improvements, with accuracy reaching 0.764 and 0.766 respectively. Tamil and Urdu see high accuracy at

Table 2 Accuracy of different LLMs for different Indian languages considering 1%, 2%, and 10% of data samples for training the model using prompt tuning

LLM model	Train data (%)	Languages						
		EN	HI	BN	TA	TE	UR	AS
XLM-R	1	0.553	0.591	0.538	0.546	0.599	0.594	0.751
	2	0.620	0.665	0.563	0.636	0.632	0.621	0.751
	10	0.653	0.722	0.567	0.689	0.678	0.719	0.751
mBERT	1	0.602	0.511	0.586	0.553	0.491	0.509	0.751
	2	0.690	0.613	0.625	0.604	0.615	0.528	0.751
	10	0.824	0.676	0.720	0.664	0.698	0.703	0.742
MuRIL	1	0.491	0.510	0.562	0.489	0.491	0.509	0.751
	2	0.491	0.490	0.491	0.489	0.491	0.491	0.751
	10	0.491	0.490	0.491	0.511	0.491	0.491	0.751
Navarasa	1	0.713	0.518	0.364	0.575	0.539	0.704	0.579
	2	0.314	0.519	0.581	0.466	0.554	0.623	0.455
	10	0.703	0.582	0.729	0.714	0.702	0.705	0.731
Indic BERT v2	1	0.747	0.665	0.650	0.283	**0.766**	0.251	0.554
	2	**0.756**	0.242	0.236	**0.763**	0.659	0.662	0.262
	10	**0.756**	**0.758**	**0.764**	0.762	**0.766**	**0.763**	**0.766**

10%, indicating good adaptability, while Assamese shows consistent improvements, reaching 0.766 at 10%.

Overall, increasing the training data from 1% to 10% generally leads to better performance across all models, highlighting the importance of training data volume in fine-tuning LLMs for sentiment analysis. XLM-R and mBERT stand out as the most effective models, showing substantial improvements across most languages. Indic BERT v2 also performs well, especially with larger training data percentages. MuRIL's performance remains stable but unimpressive compared to other models, indicating potential limitations in adapting to sentiment analysis with limited training data variation. Some models like Navarasa show variability across different languages, suggesting that certain LLMs may be more suited to specific linguistic contexts. Almost all models maintain high accuracy for Assamese, potentially due to the dataset characteristics or the nature of the language data used.

5 Conclusion

This study underscores the effectiveness of multilingual LLMs in sentiment analysis across Indian languages. XLM-R and mBERT emerged as robust performers i.e. the performance remains consistence throughout different datasets, particularly in high-resource languages like Hindi and Bengali. However, challenges remain in optimizing performance for low-resource languages. Future research should focus on enhancing model adaptability and performance in diverse linguistic contexts. IndicBERT v2 demonstrated superior performance compared to other language models, primarily due to its specialized training on Indian languages. This targeted training allowed Indic BERT v2 to effectively capture and leverage the nuances and complexities inherent to Indian linguistic structures and contexts.

References

1. Akhtar, M.S., Ekbal, A., Bhattacharyya, P.: Aspect based sentiment analysis in Hindi: resource creation and evaluation. In: Proceedings of the tenth international conference on language resources and evaluation (LREC'16). pp. 2703–2709 (2016)
2. Akhtar, M.S., Kumar, A., Ekbal, A., Bhattacharyya, P.: A hybrid deep learning architecture for sentiment analysis. In: Matsumoto, Y., Prasad, R. (eds.) Proceedings of COLING 2016, the 26th International Conference on Computational Linguistics: Technical Papers. pp. 482–493. The COLING 2016 Organizing Committee, Osaka, Japan (Dec 2016), https://aclanthology.org/C16-1047
3. Araújo, M., Pereira, A., Benevenuto, F.: A comparative study of machine translation for multilingual sentence-level sentiment analysis. Information Sciences **512**, 1078–1102 (2020)
4. Barbieri, F., Espinosa Anke, L., Camacho-Collados, J.: XLM-T: Multilingual language models in Twitter for sentiment analysis and beyond. In: Proceedings of the Thirteenth Language Resources and Evaluation Conference. European Language Resources Association, Marseille, France (Jun 2022), https://aclanthology.org/2022.lrec-1.27
5. Brown, T., Mann, B., Ryder, N., Subbiah, M., Kaplan, J.D., Dhariwal, P., Neelakantan, A., Shyam, P., Sastry, G., Askell, A., et al.: Language models are few-shot learners. Advances in neural information processing systems **33**, 1877–1901 (2020)
6. Cambria, E., Das, D., Bandyopadhyay, S., Feraco, A.: Affective computing and sentiment analysis. A practical guide to sentiment analysis pp. 1–10 (2017)
7. Devlin, J., Chang, M.W., Lee, K., Toutanova, K.: Bert: Pre-training of deep bidirectional transformers for language understanding. arXiv:1810.04805 (2018)
8. Khanuja, S., Bansal, D., Mehtani, S., Khosla, S., Dey, A., Gopalan, B., Margam, D.K., Aggarwal, P., Nagipogu, R.T., Dave, S., et al.: Muril: Multilingual representations for Indian languages. arXiv:2103.10730 (2021)
9. Kumar, S., Sanasam, R., Nandi, S.: IndiSentiment140: Sentiment analysis dataset for Indian languages with emphasis on low-resource languages using machine translation. In: Duh, K., Gomez, H., Bethard, S. (eds.) Proceedings of the 2024 Conference of the North American Chapter of the Association for Computational Linguistics: Human Language Technologies (Volume 1: Long Papers). pp. 7682–7691. Association for Computational Linguistics, Mexico City, Mexico (Jun 2024), https://aclanthology.org/2024.naacl-long.425
10. Lauscher, A., Ravishankar, V., Vulić, I., Glavaš, G.: From zero to hero: On the limitations of zero-shot cross-lingual transfer with multilingual transformers. arXiv:2005.00633 (2020)

11. Le Scao, T., Fan, A., Akiki, C., Pavlick, E., Ilić, S., Hesslow, D., Castagné, R., Luccioni, A.S., Yvon, F., Gallé, M., et al.: Bloom: A 176b-parameter open-access multilingual language model (2023)
12. Mangat, V., et al.: Dictionary based sentiment analysis of Hinglish text. International Journal of Advanced Research in Computer Science **8**(5) (2017)
13. Pravalika, A., Oza, V., Meghana, N., Kamath, S.S.: Domain-specific sentiment analysis approaches for code-mixed social network data. In: 2017 8th international conference on computing, communication and networking technologies (ICCCNT). pp. 1–6. IEEE (2017)
14. Qin, L., Chen, Q., Zhou, Y., Chen, Z., Li, Y., Liao, L., Li, M., Che, W., Yu, P.S.: Multilingual large language model: A survey of resources, taxonomy and frontiers. arXiv:2404.04925 (2024)
15. Touvron, H., Martin, L., Stone, K., Albert, P., Almahairi, A., Babaei, Y., Bashlykov, N., Batra, S., Bhargava, P., Bhosale, S., et al.: Llama 2: Open foundation and fine-tuned chat models. arXiv:2307.09288 (2023)
16. Uppal, S., Gupta, V., Swaminathan, A., Zhang, H., Mahata, D., Gosangi, R., Shah, R., Stent, A.: Two-step classification using recasted data for low resource settings. In: Proceedings of the 1st Conference of the Asia-Pacific Chapter of the Association for Computational Linguistics and the 10th International Joint Conference on Natural Language Processing. pp. 706–719 (2020)
17. Vijay, D., Bohra, A., Singh, V., Akhtar, S.S., Shrivastava, M.: Corpus creation and emotion prediction for Hindi–English code-mixed social media text. In: Proceedings of the 2018 conference of the North American chapter of the Association for Computational Linguistics: student research workshop. pp. 128–135 (2018)
18. Vilares, D., Alonso, M.A., Gómez-Rodríguez, C.: Sentiment analysis on monolingual, multilingual and code-switching twitter corpora. In: WASSA 2015, the 6th Workshop on Computational Approaches to Subjectivity, Sentiment and Social Media Analysis. pp. 2–8. Association for Computational Linguistics (2015)
19. Xue, L., Constant, N., Roberts, A., Kale, M., Al-Rfou, R., Siddhant, A., Barua, A., Raffel, C.: mt5: A massively multilingual pre-trained text-to-text transformer. arXiv:2010.11934 (2020)
20. Zhu, W., Liu, H., Dong, Q., Xu, J., Huang, S., Kong, L., Chen, J., Li, L.: Multilingual machine translation with large language models: Empirical results and analysis. arXiv:2304.04675 (2023)
21. Zhu, X., Gardiner, S., Roldán, T., Rossouw, D.: The model arena for cross-lingual sentiment analysis: A comparative study in the era of large language models. arXiv:2406.19358 (2024)

A Review of Treatment Technologies and Their Effects on Browning of Garlic

Ajay Kumar Shakya and Chandan Das

Abstract Scientific research on garlic bioactive chemicals usually mentions their health benefits. Due to its many health benefits, garlic has become popular and a focus of scientific investigation. Garlic has been processed using sustainable methods to improve its biological, physicochemical, and bioactive qualities. Most popular processing methods are examined in this study to determine their effects on browning of garlic. Food safety and shelf life are often achieved through thermal processing. However, nutritional value and taste may diminish. Non-thermal technologies, including high hydrostatic pressure, pulsed electric field, and ultrasounds, promise to reduce costs and enable more ecologically friendly black garlic production while keeping its nutritional and sensory characteristics. These technologies increase phytochemical content by improving bioaccessibility and simplifying processing-related biosynthesis.

Keywords Garlic · Black garlic · Bioactive compounds · Treatment technologies

1 Introduction

Since ancient times, garlic (*Allium sativum*) has been recognized for its prophylactic and therapeutic properties. Its numerous beneficial effects include antioxidant and antimicrobial activities and reducing cancer risk and cardiovascular disease risk factors [5]. Among plant foods rich in organosulfur compounds, garlic is particularly notable. Garlic stands out due to its widespread consumption and substantial research focused on its organosulfur compounds, even while cruciferous vegetables like broccoli, cabbage, and cauliflower, as well as *Allium* vegetables like garlic, onion, and leek, are essential sources [37]. Garlic has been shown to have antimicrobial, antioxidant, antidiabetic, and cancer-preventive qualities [39]. A broad category of characteristics components has been mentioned for years back, including lipids

A. K. Shakya · C. Das (✉)
Department of Chemical Engineering, Indian Institute of Technology Guwahati, Guwahati, Assam, India
e-mail: cdas@iitg.ac.in

U. S. Dixit et al. (eds.), *Current Progress in Engineering Sciences*,
https://doi.org/10.1007/978-981-96-8753-4_35

[45], carbohydrates [11], proteins [12], organic sulfur compounds [29], polyphenols [7], and garlic entophytes [36]. Garlic bulbs are generally stored for several months after harvest to guarantee year-round availability. The quality of the bulbs must be preserved, and proper storage conditions are essential for extending their shelf life. Garlic bulbs are commercially stored at 0–3 °C to avoid spoiling problems such as surface discoloration, withering, sprouting, and mold growth. Temperatures of about 2 °C have been shown to minimize molds growth and sprouting, while 3 °C has been found to preserve firmness and taste for up to 9 months [18].

Interestingly, black garlic, a processed form of garlic, was discovered in the cargo of a medieval shipwreck from the fifteenth century near the port of Gdańsk (southern Baltic Sea, Poland) [1]. This discovery revealed that black garlic, well-preserved for about 500 years, has an extended shelf life as well as has the following benefits as shown in Fig. 1. Black garlic is produced through thermal processing that reduces water activity and increases Maillard reaction (MR) products, contributing to its extended shelf life and enhanced organoleptic properties [43]. Moreover, non-thermal techniques has been primarily used to prepare black garlic in order to boost food safety and prolong the shelf life of garlic [3].

Nevertheless, this article offers a thorough overview of many procedures designed to improve the quality of garlic by augmenting its bioactive elements. The article

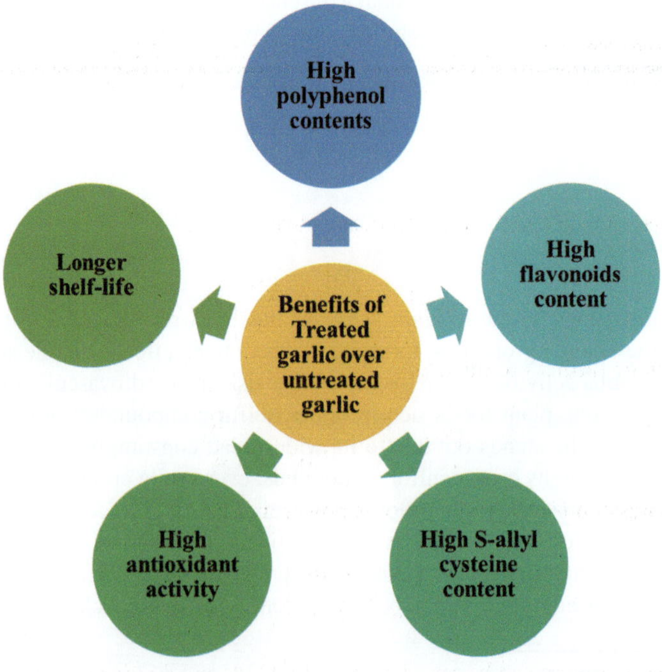

Fig. 1 Benefits of treated garlic over untreated garlic

discusses several processing methods, these approaches aim to enhance the concentration and effectiveness of critical bioactive chemicals. The study provides valuable insights into studying different treatments to improve garlic's health benefits and flavor profile by examining their effects on garlic's browning. This data will aid researchers in choosing the most efficient techniques for handling garlic, enhancing its overall quality.

2 Effect of Treatment Technologies on the Browning and Its Related Compounds

2.1 Effect of Thermal Treatment

Garlic that has undergone varying temperatures was shown to have a more significant allicin concentration than raw garlic. Additionally, the garlic samples had higher total flavonoid content (TFC) and total phenolic content (TPC) levels. However, the primary identified phenolic acids varied at different stages of processing as reported by the authors were mainly hydroxycinnamic acid derivatives (Chlorogenic acid, Caffeic acid, p-Coumaric acid, Ferulic acid, m-Coumaric acid, and o-Coumaric acid). Out of all the major flavonoid subgroups identified in garlic, flavonols (Catechin, Epicatechin, and Epigallocatechin gallate) were reported to be present in the highest levels, flavanones (Such as Naringenin), and flavones (Quercitrin and apigenin) being the next highest, although not in fresh garlic [21]. However, according to [46], the heating process enhances phenolic content by breaking down bound forms (such as esterified and glycosylated phenolics), thereby increasing the free forms. On the other hand, the decrease or suppression of the enzymatic oxidation of antioxidant components in raw garlic is responsible for the rise in phenolic content in heated samples [14].

In addition, [13] have explained that converting unstable garlic components into stable soluble forms results in the generation of solid antioxidant molecules. In contrast, Gorinstein et al. [16] have explained that the breakdown of high-molecular-weight polyphenols releases new low molecular weight phenolics, increasing the TPC and TFC [16]. However, Corzo-Martínez et al. [13] reported that producing several bioactive chemicals during heating, including TPC, TFC, and intermediate MR products, is the outcome of these intricate interactions. These compounds enhance the bioactivity of black garlic, particularly its antioxidant capacity [13].

Furthermore, researchers have studied not only the effect of heat on the garlic but also, they have varied the humidity of the environment around the garlic with heat. For instance, elevated temperature and humidity levels resulted in black garlic with enhanced antioxidant properties and higher TPC. The greatest increases in TPC in black garlic kept in KCl containers—which have a slightly greater relative humidity than NaCl—illustrated the impact of relative humidity. This variation in TPC was more than double the initial amount. Therefore, these findings indicate that higher relative humidity at the same temperature promotes the formation of phenolic compounds [35]. In addition, Medina et al. [28] reported the same observations in their study that the TPC and antioxidant capacity of garlic that was heat-treated at different temperatures and kept at 90% relative humidity increased significantly [28]. According to [13], heat treatment changes the unstable molecules in raw garlic into stable, soluble compounds with vigorous antioxidant activity.

Furthermore, temperature influences allicin degradation and accelerates increases in total acids, TPC, and hydroxymethylfurfural (HMF) levels. At higher temperatures, the MRs occur more rapidly, leading to a more significant accumulation of HMF in black garlic [41, 49]. In addition, black garlic was found to contain 98 various chemical compounds, including lipids and derivatives, organosulfur compounds, saccharides and their derivatives, and amino acids and derivatives among them most concentrated compounds have been shown in Fig. 2. However, analysis of the volatile profile during black garlic fermentation showed a reduction in sulfur volatiles and a notable increase in sweet and roasted volatiles, particularly furfural and its derivatives, due to the enhanced MR's [30]. By analyzing variations in the MRs, the heat process clarifies the mechanism underlying the production of black garlic and helps enhance quality control. Garlic turns black because of the MRs between fructose or glucose and amino acids [48]. The process of heating lipids and carbohydrates with amino acids found in garlic and other foods results in the production of a variety of intermediates and products by the MRs. Food gets its distinctive dark brown color from melanoidin and other brown polymer compounds created during the MRs [27].

Additionally, [47] have demonstrated that black garlic extracts had more potent antioxidant and ACE inhibitory qualities. The study demonstrates that black garlic extract contained particularly high concentrations of Amadori chemicals, including N-(1-deoxy-D-fructose-1-yl)-L-arginine (Fru-Arg) [47]. But heat processing and

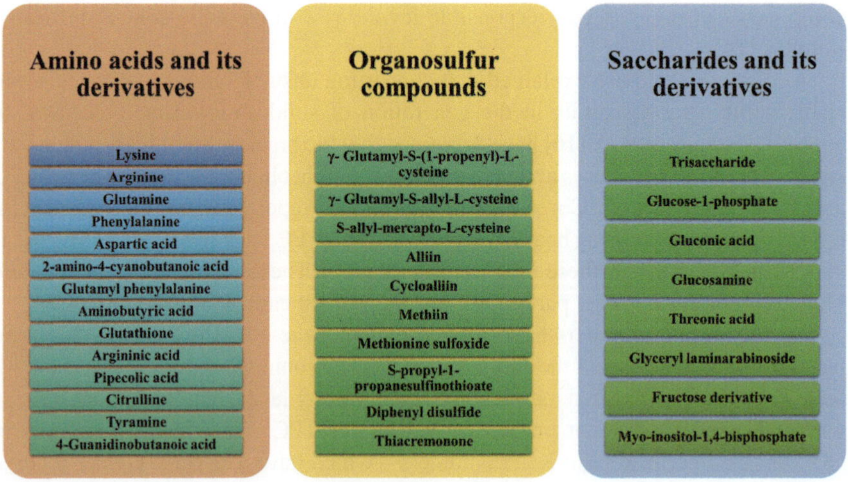

Fig. 2 Chemical compounds found in black garlic

aging, which liberate free sugars and amino acids and promote MRs and the synthesis of complex compounds, are probably the cause of this increase [40]. Moreover, the elevated levels of amino acids, as shown in Fig. 3, in the 5-day black garlic are probably due to the breakdown of proteins or peptides. This breakdown may be attributed to enzymatic or non-enzymatic processes, such as pyrolysis, as reported by [26].

Moreover, temperature significantly influences water evaporation, the activity of internal garlic enzymes, and the pace of chemical reactions involving the various compounds present in garlic [38]. Generally, the formation of black color in garlic is caused by non-enzymatic browning, which is highly dependent on the heating temperature [38]. In addition, [6] also reported the same observation; according to them, heated garlic showed increasing browning with temperature, with higher temperatures causing the browning to occur more quickly. The degree of blackness in black garlic is determined by the level of MRs, phenol oxidation, and polysaccharide degradation under regulated temperature incubation [2].

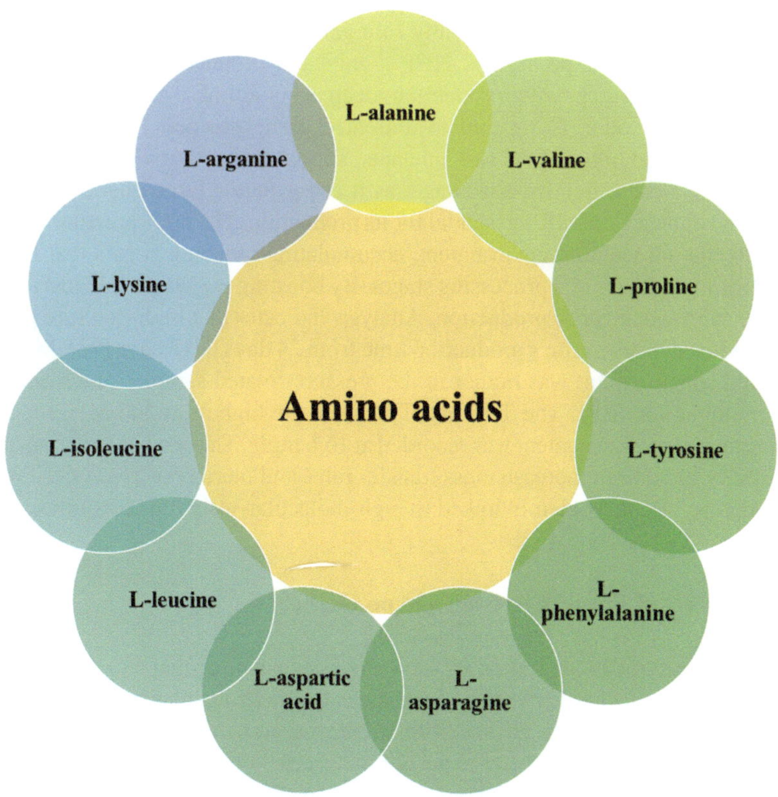

Fig. 3 Presence of amino acids in 5-day black garlic

[20] reported that the black garlic manufacturing process can be improved by applying steaming pretreatment at different incubation intervals. Their research indicates that breaking down the cell wall and destroying the cell structure can accelerate the formation of a dark brown color and improve the synthesis of bioactive components and antioxidant activity [32]. This promotes the Maillard browning reaction by facilitating the easy release of constituents like amino-N content, reduced sugar, and phenolic compounds [25]. Furthermore, the researchers found that the color shift that occurs in black garlic with aging is largely influenced by both enzymatic and non-enzymatic browning processes [19].

2.2 *Effect of Non-thermal Treatments on Garlic*

2.2.1 Effect of High Hydrostatic Pressure on Garlic

HHP can produce a microbiologically safe garlic powder with a milder flavor and improved antioxidant properties. HHP significantly reduced the diallyl disulfide content during garlic powder production, decreasing alliinase activity and reducing garlic's pungent odor. Compared to untreated garlic powder, HHP-treated garlic powder exhibited higher TPC and enhanced antioxidant activity [33]. Additionally, HHP offers a cost-effective advantage, as it can generate large quantities of black garlic and enhance the MR's essential for its production. The high-pressure treatment disrupts the intracellular environment, accumulating reducing sugars that facilitate the MRs during the initial processing stages. By boosting the MRs, HHP increases the efficiency of black garlic production. Analysis indicates that high-pressure pretreatment reduced black garlic's production time from 24 days to 15 days [23]. Moreover, the total sugar content was higher in the pressure-treated samples and reached the maximum at 300 MPa. The TPC and TFC were the highest at 300 MPa for 5 min, while the melanoidin content was recorded at 16.6 mg/g. This study has demonstrated the ability of HHP to increase mass transfer rates and thereby increase the TPC and TFC that were noncovalently linked to high-molecular-weight melanoidins [50].

2.2.2 Effect of Pulsed Electric Field on Garlic

The PEF technique is a non-thermal food processing method that was recently developed [31]. This technology uses varying electric field intensities and quick pulse waves when processing food. According to the researchers, there was an increase in both TPC and TFC [8]. A brief exposure to a pulsed electric field can improve the characteristics of garlic by boosting the bioactive chemicals present in garlic.

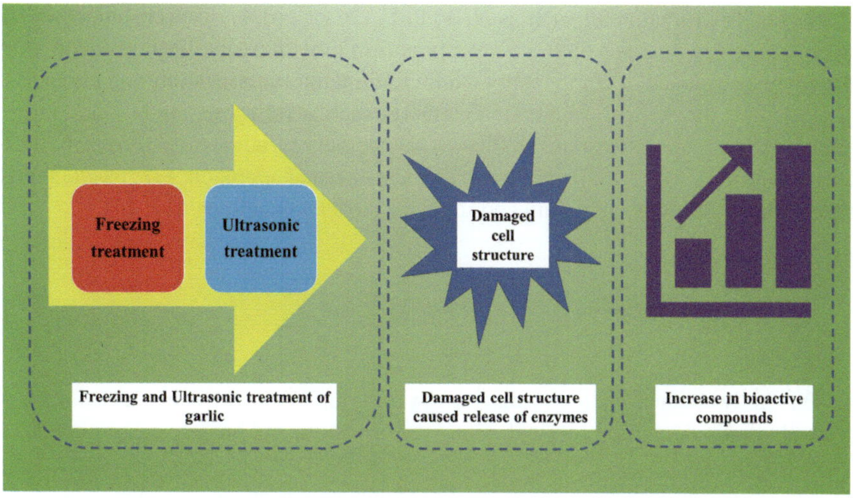

Fig. 4 Effect of freezing and ultrasonic treatment on garlic

2.2.3 Effect of Ultrasound on Garlic

Ultrasonic pretreatment shortens processing times but often requires multiple applications in the food industry to boost bioactive compounds in garlic, all while maintaining the garlic's overall appearance and color. In addition, combining ultrasonic and freezing pretreatments has enhanced garlic's bioactive compound levels [9].

According to [24], freezing pretreatment disrupts garlic cell structures and promotes enzymatic reactions, thereby enhancing the production of functional compounds. [4] reported that ultrasonic pretreatment causes more severe cell damage. This increased cell damage allows enzymes to interact more frequently with substrates, leading to a higher rate of bioactive compounds (Fig. 4) [34].

3 Role of Treatment Technologies in a Sustainable Food System

The bioactive components in garlic are greatly affected by thermal and non-thermal treatments, which impact its nutritional and therapeutic characteristics. Thermal treatments can deactivate enzymes such as alliinase, enhance the stability of garlic's flavor, and facilitate the production of advantageous chemicals such as TPC, TFC, and sulfur derivatives (SAC, DAS, DADS, DATS, and ajoene) [2, 17, 44]. Nevertheless, elevated temperatures can deteriorate delicate components such as allicin and diminish volatile chemicals that contribute to garlic's aroma and medicinal benefits [42].

Non-thermal treatment, such as HHP, PEF, ultrasound, and fermentation, effectively preserves heat-sensitive chemicals and improves the quality of garlic with the enhanced bioactive components while minimizing the browning color of garlic [22]. On the other hand, fermentation methods, such as the formation of black garlic, greatly enhance the antioxidant qualities and generate new advantageous metabolites. Despite the need for specialized equipment, non-thermal approaches are fundamental due to their capacity to maintain and improve bioactive chemicals [9, 15].

4 Summary

- This study updates the knowledge on how different treatment technologies, specifically thermal and non-thermal, affect the browning and bioactive compounds of garlic.
- Treated garlic, especially black garlic, shows enhanced physicochemical properties, including higher levels of reducing sugar, TPC, TFC, organosulfur compounds, and total protein compared to fresh garlic.
- While the Maillard reaction produces distinct chemical species, it also alters antioxidant characteristics, which are favorably associated with browning development.

Acknowledgements Any thoughts, writing, or summaries presented in this article are solely those of the authors and do not significantly affect IIT GUWAHATI perspectives.

Ajay Kumar Shakya: Conceptualization, Investigation, Writing—original draft. Dr. Chandan Das: Conceptualization, Supervision, Writing—review and editing.

References

1. Badura, M., Możejko, B., Ossowski, W., 2013. Bulbs of onion (Allium cepa L.) and garlic (*Allium sativum L.*) from the 15th-century Copper Wreck in Gdańsk (Baltic Sea): a part of victualling? J. Archaeol. Sci. 40, 4066–4072. https://doi.org/10.1016/j.jas.2013.05.026
2. Bae, S.E., Cho, S.Y., Won, Y.D., Lee, S.H., Park, H.J., 2014. Changes in S-allyl cysteine contents and physicochemical properties of black garlic during heat treatment. LWT - Food Sci. Technol. 55, 397–402. https://doi.org/10.1016/j.lwt.2013.05.006
3. Barba, F.J., Criado, M.N., Belda-Galbis, C.M., Esteve, M.J., Rodrigo, D., 2014. Stevia rebaudiana Bertoni as a natural antioxidant/antimicrobial for high pressure processed fruit extract: Processing parameter optimization. Food Chem. 148, 261–267. https://doi.org/10.1016/j.foodchem.2013.10.048
4. Barekat, S., Soltanizadeh, N., 2017. Improvement of meat tenderness by simultaneous application of high-intensity ultrasonic radiation and papain treatment. Innov. Food Sci. Emerg. Technol. 39, 223–229. https://doi.org/10.1016/j.ifset.2016.12.009
5. Bayan, L., Koulivand, P.H., Gorji, A., 2014. Garlic: a review of potential therapeutic effects. Avicenna J. phytomedicine 4, 6.

6. Benzing-Purdie, L.M., Ripmeester, J.A., Ratcliffe, C.I., 1985. Effects of temperature on Maillard reaction products. J. Agric. Food Chem. 33, 31–33. https://doi.org/10.1021/jf00061a009
7. Bunea, A., Andjelkovic, M., Socaciu, C., Bobis, O., Neacsu, M., Verhé, R., Camp, J. Van, 2008. Total and individual carotenoids and phenolic acids content in fresh, refrigerated and processed spinach (Spinacia oleracea L.). Food Chem. 108, 649–656. https://doi.org/10.1016/j.foodchem.2007.11.056
8. Chang, C.-K., Tsai, S.-Y., Tsai, M.-S., Tu, A.-T., Hou, C.-Y., Cheng, K.-C., Zhu, W.-L., Mulyani, R., Hsieh, C.-W., 2024. Promoting the Aging Process and Enhancing the Production of Antioxidant Components of Garlic through Pulsed Electric Field Treatments. Antioxidants 13, 374. https://doi.org/10.3390/antiox13030374
9. Chen, Y.T., Chen, Y.A., Lee, C.H., Wu, J.T., Cheng, K.C., Hsieh, C.W., 2020. A strategy for promoting γ-glutamyltransferase activity and enzymatic synthesis of S-allyl-(L)-cysteine in aged garlic via high hydrostatic pressure pretreatments. Food Chem. 316, 126347. https://doi.org/10.1016/j.foodchem.2020.126347
10. Chen, Y., Lee, C., Chen, Yi-an, Wu, J., Tsai, M., Cheng, K., Hsieh, C., 2020. LWT - Food Science and Technology Preparation of S -allyl cysteine-enriched garlic by two-step processing 124.
11. Cheong, K.L., Yan, F., Huang, X., 2012. Enymologic characterization of garlic fructan exohydrolase. J. Food Biochem. 36, 248–253. https://doi.org/10.1111/j.1745-4514.2010.00532.x
12. Choi, I., Cha, H., Lee, Y., 2014. Physicochemical and Antioxidant Properties of Black Garlic. Molecules 19, 16811–16823. https://doi.org/10.3390/molecules191016811
13. Corzo-Martínez, M., Corzo, N., Villamiel, M., 2007. Biological properties of onions and garlic. Trends Food Sci. Technol. 18, 609–625. https://doi.org/10.1016/j.tifs.2007.07.011
14. Dewanto, V., Wu, X., Adom, K.K., Liu, R.H., 2002. Thermal Processing Enhances the Nutritional Value of Tomatoes by Increasing Total Antioxidant Activity. J. Agric. Food Chem. 50, 3010–3014. https://doi.org/10.1021/jf0115589
15. Fine, F., Gervais, P., 2004. Efficiency of pulsed UV light for microbial decontamination of food powders. J. Food Prot. 67, 787–792. https://doi.org/10.4315/0362-028X-67.4.787
16. Gorinstein, S., Drzewiecki, J., Leontowicz, H., Leontowicz, M., Najman, K., Jastrzebski, Z., Zachwieja, Z., Barton, H., Shtabsky, B., Katrich, E., Trakhtenberg, S., 2005. Comparison of the bioactive compounds and antioxidant potentials of fresh and cooked Polish, Ukrainian, and Israeli garlic. J. Agric. Food Chem. 53, 2726–2732. https://doi.org/10.1021/jf0404593
17. Harmita, H., Suryadi, H., Syarif, M., Liksasa, L.D., 2020. Gas Chromatography Analysis of Diallyl Disulphide and Diallyl Trisulphide and Antioxidant Activity in Black Garlic. Int. J. Pharm. Investig. 10, 17–23. https://doi.org/10.5530/ijpi.2020.1.4
18. Ichikawa, M., Ide, N., Ono, K., 2006. Changes in Organosulfur Compounds in Garlic Cloves during Storage. J. Agric. Food Chem. 54, 4849–4854. https://doi.org/10.1021/jf060083o
19. Kang, O.-J., 2016. Evaluation of Melanoidins Formed from Black Garlic after Different Thermal Processing Steps. Prev. Nutr. Food Sci. 21, 398–405. https://doi.org/10.3746/pnf.2016.21.4.398
20. Karnjanapratum, S., Supapvanich, S., Kaewthong, P., Takeungwongtrakul, S., 2021. Impact of steaming pretreatment process on characteristics and antioxidant activities of black garlic (Allium sativum L.). J. Food Sci. Technol. 58, 1869–1876. https://doi.org/10.1007/s13197-020-04698-7
21. Kim, J.S., Kang, O.J., Gweon, O.C., 2013. Comparison of phenolic acids and flavonoids in black garlic at different thermal processing steps. J. Funct. Foods 5, 80–86. https://doi.org/10.1016/j.jff.2012.08.006
22. Kim, K.W., Kim, Y.T., Kim, M., Noh, B.S., Choi, W.S., 2014. Effect of high hydrostatic pressure (HHP) treatment on flavor, physicochemical properties and biological functionalities of garlic. LWT - Food Sci. Technol. 55, 347–354. https://doi.org/10.1016/j.lwt.2013.08.027
23. Li, F., Cao, J., Liu, Q., Hu, X., Liao, X., Zhang, Y., 2020. Acceleration of the Maillard reaction and achievement of product quality by high pressure pretreatment during black garlic processing. Food Chem. 318, 126517. https://doi.org/10.1016/j.foodchem.2020.126517

24. Li, N., Lu, X., Pei, H., Qiao, X., 2015. Effect of Freezing Pretreatment on the Processing Time and Quality of Black Garlic. J. Food Process Eng. 38, 329–335. https://doi.org/10.1111/jfpe.12156
25. Li, N., Lu, X., Pei, H., Qiao, X., Al, N.L.I.E.T., 2014. Effect of freezing pretreatment on the processing time and quality of black garlic. https://doi.org/10.1111/jfpe.12156
26. Liang, T., Wei, F., Lu, Y., Kodani, Y., Nakada, M., Miyakawa, T., Tanokura, M., 2015. Comprehensive NMR analysis of compositional changes of black garlic during thermal processing. J. Agric. Food Chem. 63, 683–691. https://doi.org/10.1021/jf504836d
27. Manzocco, L., Calligaris, S., Mastrocola, D., Nicoli, M.C., Lerici, C.R., 2000. Review of non-enzymatic browning and antioxidant capacity in processed foods. Trends Food Sci. Technol. 11, 340–346. https://doi.org/10.1016/S0924-2244(01)00014-0
28. Medina, T., Angeles, M., Jesús, P., Rafael, M., Tania, M., 2016. Evolution of some physicochemical and antioxidant properties of black garlic whole bulbs and peeled cloves. FOOD Chem. 199, 135–139. https://doi.org/10.1016/j.foodchem.2015.11.128
29. Milner, J.A., 2001. Recent Advances on the Nutritional Effects Associated with the Use of Garlic as a Supplement A Historical Perspective on Garlic and Cancer 1 1027–1031.
30. Molina-Calle, M., de Medina, V.S., Priego-Capote, F., de Castro, M.D.L., 2017. Establishing compositional differences between fresh and black garlic by a metabolomics approach based on LC–QTOF MS/MS analysis. J. Food Compos. Anal. 62, 155–163. https://doi.org/10.1016/j.jfca.2017.05.004
31. Naliyadhara, N., Kumar, A., Girisa, S., Daimary, U.D., Hegde, M., Kunnumakkara, A.B., 2022. Pulsed electric field (PEF): Avant-garde extraction escalation technology in food industry. Trends Food Sci. Technol. 122, 238–255. https://doi.org/10.1016/j.tifs.2022.02.019
32. Nugrahedi, P.Y., Dekker, M., Widianarko, B., Verkerk, R., 2016. Quality of cabbage during long term steaming; phytochemical, texture and colour evaluation. LWT 65, 421–427. https://doi.org/10.1016/j.lwt.2015.08.034
33. Park, I., Kim, J.U., Shahbaz, H.M., Jung, D., Jo, M., Lee, K.S., Lee, H., Park, J., 2019. Original article High hydrostatic pressure treatment for manufacturing of garlic powder with improved microbial safety and antioxidant activity 325–334. https://doi.org/10.1111/ijfs.13937
34. Park, S.H., Lee, H., Kim, H.S., Kim, Y.R., Noh, S.H., 2014. Optimum conditions for S-allyl-(L)-cysteine accumulation in aged garlic by RSM. Food Sci. Biotechnol. 23, 717–722. https://doi.org/10.1007/s10068-014-0097-1
35. Pires, L. de S., Todisco, K.M., Janzantti, N.S., Mauro, M.A., 2019. Black garlic: Effects of the processing on the kinetics of browning and moisture transfer and on antioxidant properties. J. Food Process. Preserv. 43, 1–15. https://doi.org/10.1111/jfpp.14133
36. Qiu, Z., Lu, X., Li, N., Zhang, M., Qiao, X., 2018. Characterization of garlic endophytes isolated from the black garlic processing. Microbiologyopen 7, 1–11. https://doi.org/10.1002/mbo3.547
37. Ramirez, D.A., Locatelli, D.A., González, R.E., Cavagnaro, P.F., Camargo, A.B., 2017. Analytical methods for bioactive sulfur compounds in Allium: An integrated review and future directions. J. Food Compos. Anal. 61, 4–19. https://doi.org/10.1016/j.jfca.2016.09.012
38. Rapusas, R.S., Driscoll, R.H., 1995. Kinetics of non-enzymatic browning in onion slices during isothermal heating. J. Food Eng. 24, 417–429. https://doi.org/10.1016/0260-8774(95)90054-F
39. Reuter, H.D., 1995. Allium sativum and Allium ursinum: Part 2 pharmacology and medicinal application. Phytomedicine 2, 73–91. https://doi.org/10.1016/S0944-7113(11)80052-8
40. Ryu, J.H., Kang, D., 2017. Physicochemical Properties, Biological Activity, Health Benefits, and General Limitations of Aged Black Garlic: A Review. Molecules 22, 919. https://doi.org/10.3390/molecules22060919
41. Shakya, A.K., Das, C., 2023. A Review of Bioactive Compounds and Maillard Reaction-Based Products Generated During the Thermal Treatment of Garlic, in: Agro and Food Processing Technologies. Springer Nature Singapore, Singapore, pp. 35–55. https://doi.org/10.1007/978-981-19-9704-4_3
42. Singh, A., 2014. Effect of Drying Characteristics of Garlic-A Review. J. Food Process. Technol. 05. https://doi.org/10.4172/2157-7110.1000318

43. Toledano Medina, M.Á., Merinas-Amo, T., Fernández-Bedmar, Z., Font, R., del Río-Celestino, M., Pérez-Aparicio, J., Moreno-Ortega, A., Alonso-Moraga, Á., Moreno-Rojas, R., 2019. Physicochemical Characterization and Biological Activities of Black and White Garlic: In Vivo and In Vitro Assays. Foods 8, 220. https://doi.org/10.3390/foods8060220
44. Torres-Palazzolo, C., Ramirez, D., Locatelli, D., Manucha, W., Castro, C., Camargo, A., 2018. Bioaccessibility and permeability of bioactive compounds in raw and cooked garlic. J. Food Compos. Anal. 70, 49–53. https://doi.org/10.1016/j.jfca.2018.03.008
45. Tsiaganis, M.C., Laskari, K., Melissari, E., 2006. Fatty acid composition of Allium species lipids. J. Food Compos. Anal. 19, 620–627. https://doi.org/10.1016/j.jfca.2005.06.003
46. Xu, G., Ye, X., Chen, J., Liu, D., 2007. Effect of Heat Treatment on the Phenolic Compounds and Antioxidant Capacity of Citrus Peel Extract. J. Agric. Food Chem. 55, 330–335. https://doi.org/10.1021/jf062517l
47. Yu, J., Shan, Y., Li, S., Zhang, L., 2020. Potential contribution of Amadori compounds to antioxidant and angiotensin I converting enzyme inhibitory activities of raw and black garlic. Lwt 129. https://doi.org/10.1016/j.lwt.2020.109553
48. Yuan, H., Sun, L., Chen, M., Wang, J., 2018. An analysis of the changes on intermediate products during the thermal processing of black garlic. Food Chem. 239, 56–61. https://doi.org/10.1016/j.foodchem.2017.06.079
49. Zhang, X., Li, N., Lu, X., Liu, P., Qiao, X., 2016. Effects of temperature on the quality of black garlic. J. Sci. Food Agric. 96, 2366–2372. https://doi.org/10.1002/jsfa.7351
50. Zhao, Y., Jiang, Y., Ding, Y., Wang, D., Deng, Y., 2019. High hydrostatic pressure-assisted extraction of high-molecular-weight melanoidins from black garlic: Composition, structure, and bioactive properties. J. Food Qual. 2019. https://doi.org/10.1155/2019/1682749

[illegible]

Recent Advancement in Monitoring of Environmental Pollutants Using Structured Nano-catalyst Integrated Electrochemical Sensing Technique: A Review

Chandra Bhan, Ravi Ravi, and Animes Kumar Golder

Abstract The presence of pesticides, pharmaceuticals, heavy metals, and other contaminants in environmental matrices is a growing global concern. One of the primary challenges in detecting environmental pollutants at trace levels is achieving high sensitivity and selectivity in complex matrices such as contaminated soil and water. Therefore, in-situ sensing and quantifying these compounds is crucial for assessing contamination severity at its source. Vertically aligned nanostructures (VANs) on the functioning electrode's surface offer a promising solution for monitoring pollutants due to their tailored surface properties, including high surface area (increase more attachment sites for the electroactive species), increased surface roughness (expand more nucleation sites), fast charge transfer kinetics (electron transfer inside and/or outside from nanostructure surface to electrolyte), and high stability. However, conventional methods for fabricating VANs involve many environmentally aggressive chemicals. Consequently, there is a growing concern about developing nature-inspired VANs using phytochemicals derived from plants and their organs for efficient sensing of environmental pollutants. This review comprehensively discusses the synthesis of powder and VANs using conventional and nature-inspired methods and explores the pivotal role of nanostructures in enhancing the sensing performance of sensors. This report also documents the comparison of immobilized (using binder) and VANs onto working electrode-based sensors for detecting environmental pollutants. The main focus of this review is on the nature-inspired

C. Bhan · A. K. Golder (✉)
Department of Chemical Engineering, Indian Institute of Technology Guwahati, Guwahati, Assam, India
e-mail: animes@iitg.ac.in

C. Bhan
e-mail: bchandra@iitg.ac.in

R. Ravi · A. K. Golder
Centre for the Environment, Indian Institute of Technology Guwahati, Guwahati, Assam, India
e-mail: r.vashist@iitg.ac.in

U. S. Dixit et al. (eds.), *Current Progress in Engineering Sciences*,
https://doi.org/10.1007/978-981-96-8753-4_36

development of VANs based miniaturized three-electrode system device for on-site and real-time monitoring of environmental pollutants with high stability and reusability.

Keyword Nanomaterials · Environmental pollutants · Sensing techniques · Electrochemical sensing device

1 Introduction

In recent years, there has been increasing global concern about environmental pollutants and the health implications caused by harmful contaminants originating from intensive industries, agricultural operations, and fast urbanization. Furthermore, polluted drinking water is an acute global concern for half of the world's population [1]. Exposure to these pollutants can cause a variety of health problems, including mental and physical dysfunctions, neurological abnormalities, and a weaker immune system, which can lead to death in many situations [2, 3]. The aforementioned challenges have highlighted the importance of developing reliable, highly stable, user-friendly, cost-effective, and eco-friendly equipment for in-situ sensing and quantification of environmental pollutants in complex matrices such as contaminated soil and water at low concentration levels to protect the environment and human health. Therefore, in-situ detection and quantification of these compounds is crucial for assessing contamination severity at its source.

The existence of environmental pollutants can be determined by gas chromatography–mass spectrometry, high-performance liquid chromatography, atomic absorption spectroscopy, mass spectrometry, etc. [4–6]. However, these analytical measurements have been time-consuming, costly, need extensive skills to operate, challenging to transport owing to the large equipment size, and high installation cost triggering high measurement costs. A cost-effective, highly stable, sensitive, flexible, and eco-friendly device is tremendously imperious for in-situ sensing and quantification of environmental pollutants.

Electrochemical sensors stand out with high potential for in-situ sensing and quantification of environmental pollutants, including pesticides, nitroaromatics, and heavy metals. [7]. They are advantageous due to their ease, low cost, excellent sensing performance (detection limit and sensitivity), enhanced stability, and high selectivity during pollutant sensing. Furthermore, modifying functioning electrode surface by diversely shaped nanomaterials using binder (Nafion, polyvinylidene fluoride, polytetrafluoroethylene, polytetrafluoroethylene, carboxymethyl cellulose, and polypyrrole), for instance, ZnO, V_2O_5, $BiVO_4$, Bi_2S_3, CuS, PtNPs, and AgNPs, could enhance the sensor's performance in a great extent owing to their tailored surface properties, increased active adsorption sites for targeted analytes, improved electron transfer kinetics, and high chemical stability [7–10]. These features enable efficient detection of environmental residues even at low concentrations, ensuring high stability and making nanomaterials promising candidates for sensitive voltammetric sensing platforms. Simultaneously, nanostructures have garnered significant research interest for a variety of applications, such as wastewater treatment, water splitting, CO_2 reduction, and corrosion inhibition [11–17]. Therefore, a major drawback is the fast flaking off nanostructures from the functioning electrode surface, which can limit the sensor's lifespan and reproducibility [18]. Additionally, the binder acts as a matrix on the working electrode's surface, which could potentially interfere during sensing. It also increases the preparation step (immobilization of nanostructures onto the support electrode's surface), and the cost of the sensor. However, growing nanostructure directly in vertical configuration onto the functioning electrode surface could remove the problem associated with fast flaking off nanostructures, complete surface modification step, need of the binder, and decreased cost of the fabricated sensors [19].

In current findings, it has been documented that the directly growing nanostructure of metal, metal oxide, and metal sulfide in vertical configuration onto the functioning electrode surface could boost the sensor's performance by providing high electroactive area (increase more attachment sites for the electroactive species), increased surface roughness (expand more nucleation sites), fast charge transfer kinetics (electron transfer inside and/or outside from nanostructure surface to electrolyte solution), and high stability during the course of environmental pollutant sensing [10, 19]. However, conventional methods for growing nanomaterials in vertical configuration predominantly involve environmentally aggressive chemicals that are not eco-friendly [20]. Consequently, there is a growing concern to develop nature-inspired formation of nanostructures in vertical configuration using phytochemicals derived from plants and their organs, and their utilization in the miniaturized three-electrode device on patterned FTO for efficient sensing of environmental pollutants at their point of origin is need of the hour.

Miniaturized three-electrode system devices offer several advantages over large traditional devices. Their compact size and lightweight nature enhance portability and ease of use, making them ideal for applications requiring mobility, such as field testing or integration into wearable technology [21]. Additionally, miniaturized systems often provide superior sensitivity and faster response times due to their smaller electrode areas, leading to better signal-to-noise ratios, rapid measurements, and more precision [22]. This integration facilitates advanced analytical capabilities that are difficult to achieve with larger, bulkier devices. Overall, the miniaturized three-electrode systems represent a significant advancement in analytical technology, offering increased efficiency, convenience, and cost-effectiveness compared to traditional devices [23]. The miniaturization of electrochemically active nature-inspired vertically aligned nanostructures (VANs) on portable devices has opened new possibilities for on-site and real-time monitoring of environmental pollutants. This integration not only enhances the functionality of the device but also allows for more complex and precise analytical processes in a reduced dimension compared to conventional, bulkier devices. Fabrication of electrochemical sensing devices employs diverse techniques such as physical methods (electroplating, evaporation, sputtering), chemical processes (sol–gel method, chemical vapor deposition), and printing techniques (screen printing, inkjet printing) [24]. Advances in materials science, electrochemical sensor design, and nanostructure integration on portable platforms continue to improve the sensitivity and reliability of environmental pollutant detection systems, significantly contributing to environmental health protection.

The prime object of this review is to summarize a complete overview of the nature-inspired development of VANs based miniaturized three-electrode system configuration onto the surface of patterned FTO glass substrate for in-situ sensing and quantifying environmental pollutants. This review also provides the synthesis methodology of powder and VANs by conventional and bio-based routes, and their functionality to enhance the electrochemical sensor's performance by immobilizing or directly growing nanostructures on the surface of the working electrode. Furthermore, this review documents the comparison of immobilized and direct growth of nanostructures onto working electrode-based sensors for detecting environmental pollutants.

2 Synthesis Route Impacts the Functionality of Nanomaterials

In the case of electrochemical sensors, nanostructures are synthesized directly onto the functioning electrode surface or powder form (immobilized onto the surface of the base electrode with the help of binder) of various morphology using conventional and nature-inspired routes for insitu detection and quantification of environmental pollutants (Fig. 1). Additionally, nanostructures of diverse morphology are utilized in the various applications, for instance, CO_2 conversion, wastewater purification, H_2 production, and corrosion inhibition.

2.1 Hydrothermal Method

The hydrothermal method is a versatile technique for synthesizing nanostructures, involving chemical reactions in aqueous solutions at elevated temperatures and pressures within a sealed autoclave. This method is a versatile and effective technique for producing high-quality nanostructures with controlled size and shape, making it valuable for a wide range of scientific and industrial applications. Gopi et al. synthesized $BiVO_4$ nanostructures in a hydrothermal route at 180 °C for 4 h. Subsequently, prepared nanostructures were embedded onto the surface of screen-printed carbon electrode (SPCE) for electrochemical sensing of paraoxon. The fabricated sensor exhibited a limit of detection of (LOD) of 0.03 μM and a sensitivity of 0.345 $\mu A\ \mu M^{-1}\ cm^{-2}$ within the concentration range of 0.05–860 μM [25]. Nair et al. prepared $FeTiO_3$ nanoplates using the hydrothermal method at 180 °C for 24 h for electrocatalytic monitoring of nitrofurazone [26]. Chellappa et al. synthesized MnO_2

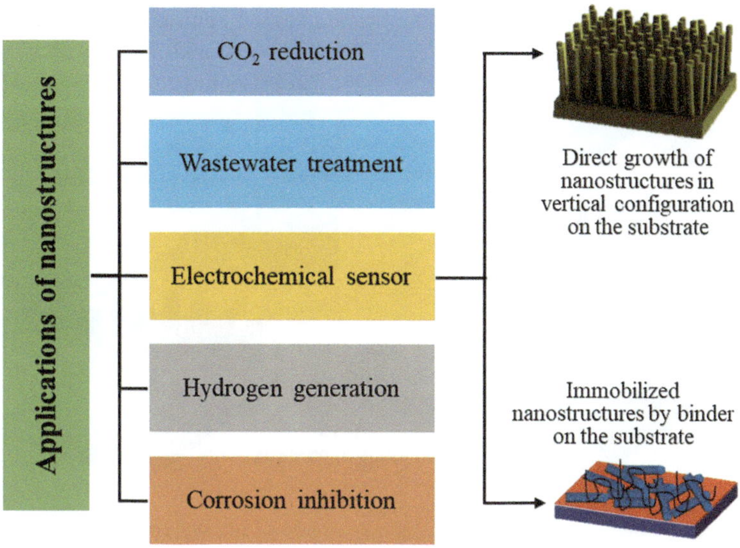

Fig. 1 Application of nanostructures in different fields

nanorods in a hydrothermal route at 180 °C for 3 h for electrochemical sensing of environmental anthropogenic pollutants and nitrobenzene [27].

Ahmad et al. grew palm trunk-like ZnO nanoparticles in the vertical configuration on the surface of the Ag electrode using a hydrothermal route at 95 °C for 20 h for detection and quantification of glucose [19]. Wang et al. developed the vertically aligned $Ni_5P_4/Ni_2P/C$ nanosheet onto the surface of ITO glass substrate using hydrothermal method at 120 °C for 8 h for sensing of Cu^{2+} and paracetamol [28]. Ahmad et al. proposed a glucose sensor based on ZnO nanorods aligned in vertical configuration modified with CuO by hydrothermal method, as shown in Fig. 2 [29]. Step-by-step synthesis of nanostructures aligned in the vertical configuration onto the surface of the bare substrate by hydrothermal route, as illustrated in Fig. 3.

2.2 *Microwave-Assisted Method*

The microwave-assisted technique for nanostructures synthesis is a versatile and efficient approach that utilizes microwave radiation to rapidly heat the reaction mixture, leading to the formation of nanostructures. This technique merited faster reaction time, uniform heating, improved yield, and enhanced control over particle size and distribution, resulting in homogeneous nucleation and growth conditions that produce high-quality nanostructures with reduced aggregation. Its adaptability allows for the synthesis of a wide variety of nanostructures, including metals, oxides, sulfides,

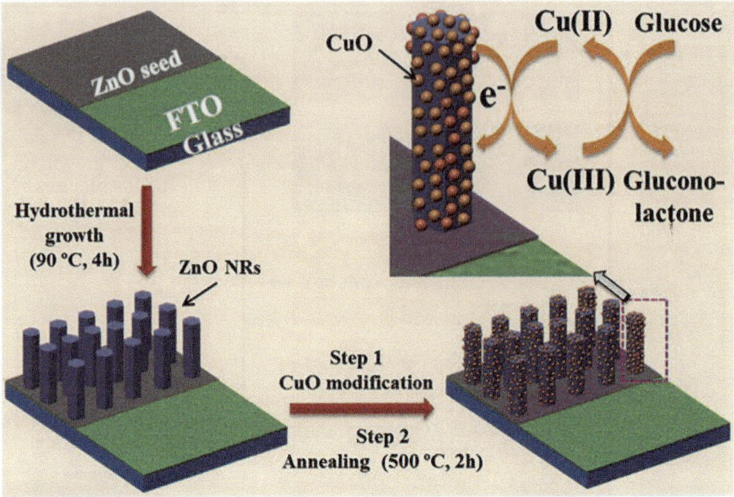

Fig. 2 Step-by-step fabrication of ZnO nanorods modified CuO aligned in vertical configuration onto the surface of FTO glass substrate and their utilization in electrochemical sensing of glucose. Reprinted with permission from ref. [29]. Copyright 2017 Nature

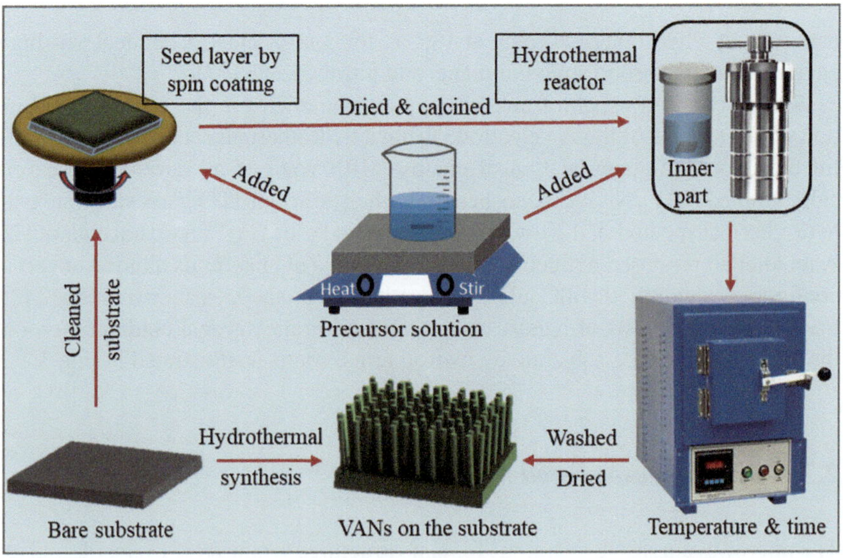

Fig. 3 Schematic illustration of the synthesis of nanostructures aligned in vertical configuration onto the surface of the bare substrate by hydrothermal route

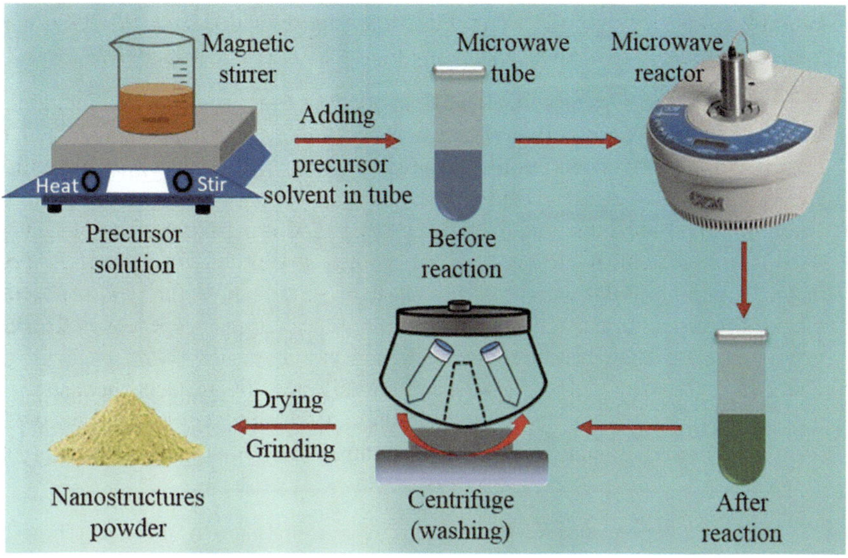

Fig. 4 Graphical representation of the synthesis of nanostructures by the microwave-assisted method

and composite materials, making it a valuable technique for synthesizing nanostructures. Gawal and Golder synthesize CQDs/CdS nanocomposite at room temperature (12 h) for application in the photocatalytic conversion of CO_2 to methanol with 1060.52 μmol/g·h production. In CQDs/CdS nanocomposite, CdS and CQDs were prepared by microwave-assisted (100 °C for 5 min) and hydrothermal methods (180 °C for 12 h), respectively [30]. In another study, they also prepared In_2O_3/CdS nanocomposite by microwave-assisted method at 100 °C for 5 min in the application of photocatalytic CO_2 conversion to formic acid and carbon monoxide [31]. Zhan et al. synthesized ZnO nanorods in vertical configuration onto the surface of a stainless-steel substrate using the microwave-assisted method at 10 W (5 min) for sensing glucose and H_2O_2 [32]. Figure 4 demonstrates the steps involved in the preparation of nanostructures using the microwave-assisted method.

2.3 *Electrochemical Deposition Method*

The electrochemical deposition route for the preparation of nanostructures involves immersing a substrate (working electrode) in an electrolyte solution containing metal ions, leading to the reduction of metal ions from an electrolyte solution onto a conductive substrate by applying a controlled potential between the working electrode and a counter electrode. This method allows precise control over the deposition parameters, such as current density, potential, electrolyte composition, and deposition time,

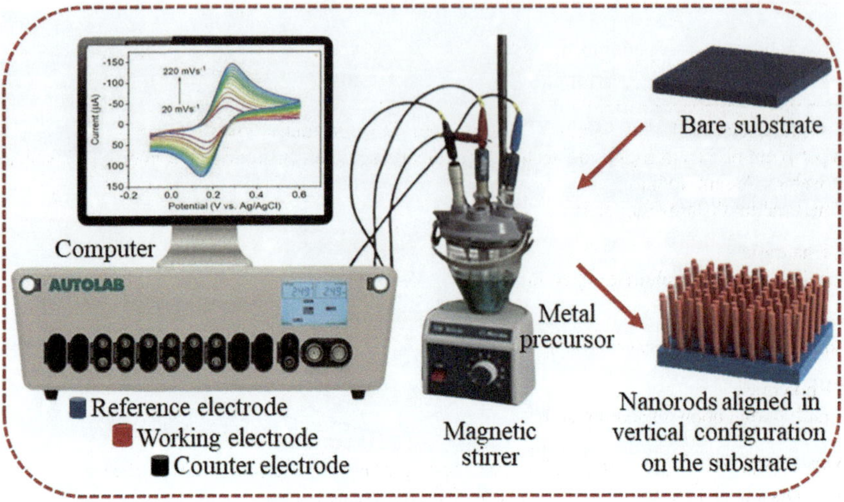

Fig. 5 Schematic representation of the synthesis of nanostructures by the electrochemical deposition method

enabling the synthesis of nanostructures with tailored size, shape, and distribution. This method is widely used in applications such as catalysis, sensors, and electronic devices due to its simplicity, scalability, and ability to produce high-purity nanoparticles. Figure 5 shows the synthesis of nanostructures by electrochemical deposition techniques. Ma et al. utilized a gradual electrodeposition process for modification of glassy carbon electrode (GCE) via deposition of PtNPs followed by ZIF-8 film under constant applied potential and current density. This deposition results in improved electrochemically active surface area and sensing performance of electrochemical sensor for the detection of ascorbic acid [33]. In another study, carbon nano-pipettes (CNPs) were decorated with metals such as Au, Ag, and Pt by an electro-deposition technique. Further, Au@CNPs, Ag@CNPs, and Pt@CNPs were utilized for glucose sensing, Cl^- sensing, and H^+ and H_2O_2 sensing, respectively [34]. Kathiresan et al. deposited GO/MWCNTs/Py over the GCE using an electro-deposition process regulated by cyclic voltammetry and used to detect dopamine [35]. In another study, Zn NPs were deposited over TiO_2 nanotubes through an electrodeposition process performed at a constant potential of -1.6 V [36].

2.4 *Chemical-Based Synthesis Method*

Based on the origin of reducing and capping agents, the nanomaterial synthesis process can be classified into chemical-based and bio-based methods. In chemical-based methods, intensive chemicals are utilized to reduce and capping agents to

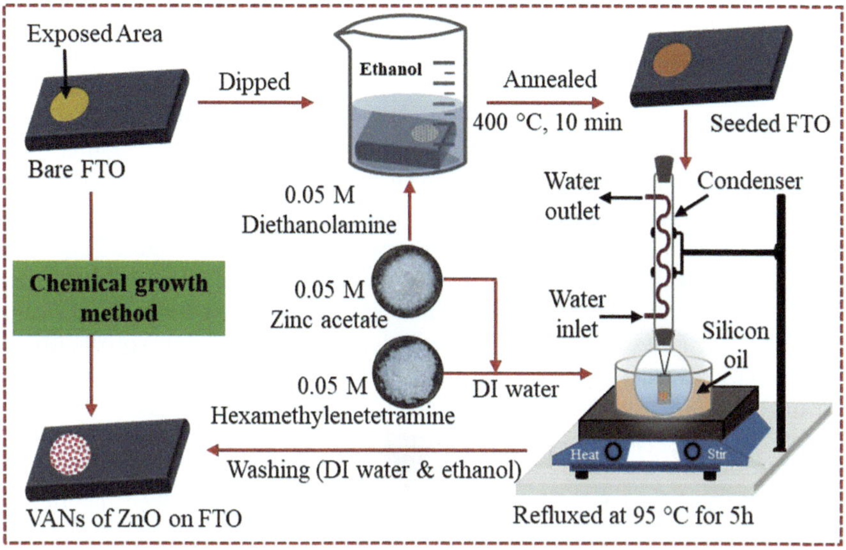

Fig. 6 Synthesis of ZnO nanorods aligned in vertical configuration on the surface of bare FTO glass substrate by chemical growth method [37]

synthesize nanomaterials. The chemical-based synthesis method produces nanomaterial with more precise and controlled physiochemical characteristics such as high surface area, uniform size and shape, high crystallinity, high conductivity, and stability. Mali et al. developed vertically aligned CdS-coated ZnO using a chemical method and observed uniformly distributed nanorods with an average diameter of 60 nm [37]. In another study, chemical-based synthesis of Ni-SnO_2 for electrochemical sensing of ascorbic acid in aqueous solution [38]. Zhai et al. chemically synthesized Ag-doped ZnO with an average diameter of 50–60 nm for electrochemical detection of phosphothioate pesticides [39]. A systematic diagram of the synthesis of vertically aligned nanorods of ZnO on FTO glass is shown in Fig. 6.

2.5 *Bio-based Synthesis Method*

Recently, the bio-based route for nanomaterials synthesis has gained massive attention due to its eco-friendly, high efficiency, and abundance in nature. Bio-based synthesis methods employ bio-analysts present in organisms and their parts, such as plants and microorganisms. The plant-based bio-extract works as a one-pot system, providing reducing and capping agents for nanomaterial synthesis. Additionally, numerous studies have reported bio-based synthesized nanomaterials for various applications such as electrochemical sensing, CO_2 reduction, organic pollutant degradation, and water splitting. Chowdhury et al. synthesized bio-based SnO_2 nanoparticles using *S. edule* extract, which is a rich source of ascorbic acid and other bioanalytes. The synthesized catalysts were spherical in shape with an average size of 4.5 nm with the existence of oxygen vacancies and surface functionalization. These properties could enhance its electrochemical activity [15]. In another study, *S. edule* extract was also utilized to synthesize bio-based 1D Bi_2S_3 nanorods with a diameter of 12.4–81.8 nm for electrocatalytic CO_2 conversion to formate. Figure 7 shows the FETEM and HRTEM images of Bi_2S_3 nanorods synthesized using *S. edule* extract for electrochemical CO_2 reduction to formate [17]. Ali et al. synthesized CuO nanostructures using *F. elastic* extract with flower-like particles with a size of 20–107 nm. The synthesized CuO and PANI-modified carbon paste electrode showed efficient electrochemical sensing of Pb^{2+}, Hg^{2+}, and Cd^{2+} [40]. In another study, *A. nervosa* extract-based ZnO nanoclusters were utilized for electrochemical sensing of vitamin C [41].

Conclusively, chemical-based and bio-based methods have been reported for the synthesis of electrochemically active nanostructures. The utilization of vegetal extract minimizes the use of intensive chemicals for nanomaterials synthesis and

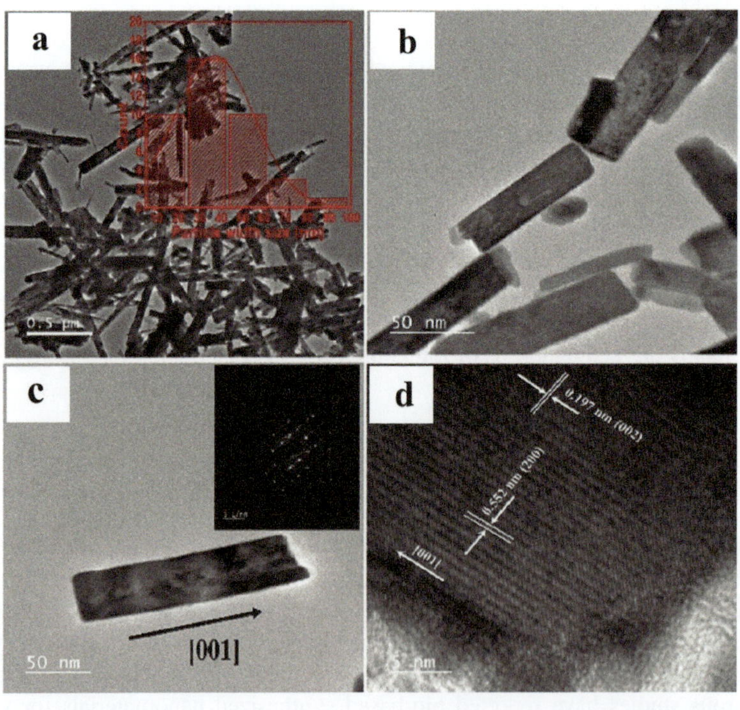

Fig. 7 **a–c** FETEM and **d** HRTEM image of Bi_2S_3 nanorods synthesized by the nature-inspired method using vegetal extract of *S. edule* fruit (Insert micrograph in Fig. 7c is the SAED pattern of Bi_2S_3 nanorod). Reprinted with permission from ref. [17]. Copyright 2023 Elsevier

also diminishes the generation of harmful by-products. Bio-based nanomaterials have several phytochemical advantages over chemical-based nanomaterials, such as high surface area, high conductivity, prominent oxygen vacancies, metallic defects, surface functionalization, and high hydrophilicity, resulting in higher electrocatalytic efficiency for sensing of environmental pollutants.

3 Role of Nanomaterials in Enhancing Sensing Performance

Nanomaterials have demonstrated significant potential in improving the sensing performance of electrochemical sensors during the course of environmental pollutant sensing due to their distinctive physicochemical characteristics. The higher surface area of nanoparticles offers a larger number of adsorption sites for targeted electroactive species, which could improve the sensor's performance. Additionally, nanoparticles' electrical, magnetic, and optical characteristics can be tailored by altering their size, morphology, and precursor concentration, providing an effective sensing platform for in-situ sensing and quantifying environmental pollutants. Based on the functioning of the nanostructures on the working electrode surface (immobilized using binder or direct growth of VANs), nanoparticles can enhance the sensing performance (LOD and sensitivity), selectivity, and life of the electrochemical sensor. The role of diverse nanomaterials in improving electrochemical sensors' performance is discussed in the subsequent sections.

3.1 Immobilized Nanomaterials onto the Base Electrode

In this type of electrochemical sensor, nanoparticles of different materials, such as ZnO, SnO_2, $BiVO_4$, Bi_2S_3, CdS, AgNPs, and AuNPs functionalized onto the surface of working electrode (FTO, ITO, screen-printed carbon, glassy carbon, boron-doped diamond, Au, and Pt) using binder (Nafion, polyvinylidene fluoride, polytetrafluoroethylene, polytetrafluoroethylene, carboxymethyl cellulose, and polypyrrole) for sensing of targeted analytes. Additionally, binder form a matrix on the functioning electrode surface to affix nanostructures and enhance their electrocatalytic interactions with target electroactive species. This type of sensor's widely utilized in sensing pesticides, pharmaceuticals, heavy metals, etc. Das et al. synthesized 1D Ag nanorods by an ascorbic acid-assisted route. Subsequently, Ag nanorods functionalized onto the surface of GCE using Nafion as a binder for electrochemical sensing of aniline, as illustrated in Fig. 8. The fabricated sensor showed the LOD of 0.032 μM and sensitivity of 1.4841 $\mu A \mu M^{-1}$ cm^{-2} [18]. Ameen et al. proposed an electrochemical sensor based on ZnO nanoglobules modified GCE for voltammetry sensing of piperidine. The sensor exhibited the LOD of 3.3 μM with a sensitivity of

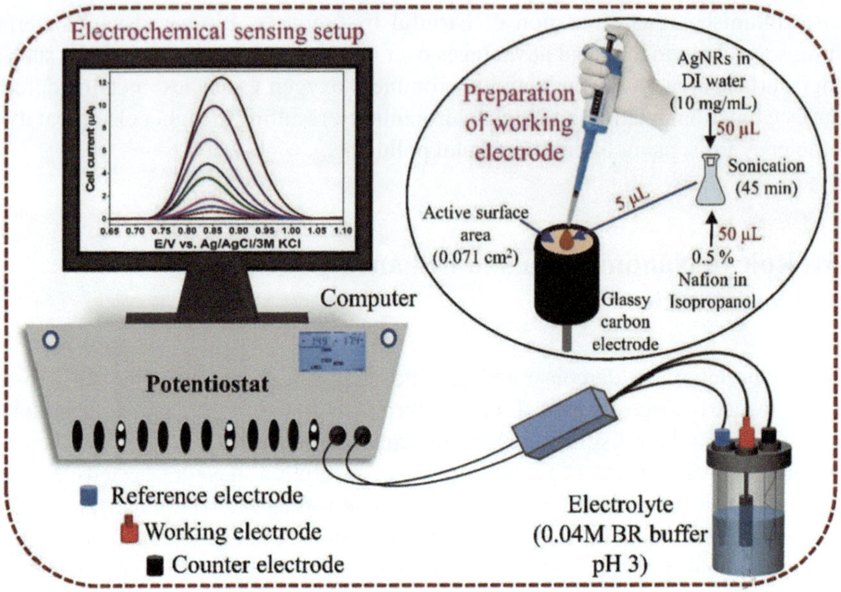

Fig. 8 Graphical representation of AgNRs functionalized on the surface of GCE using Nafion as a binder by the drop-casting technique for electrochemical determination of aniline in tap and river water samples. Reprinted with permission from ref. [18]. Copyright 2024 Elsevier

0.059 $\mu A \mu M^{-1}$ cm^{-2} [42]. Dash et al. prepared Ag@Pt core–shell nanostructures using phytochemicals present in *P. guajava* leaves for electrocatalytic sensing of Pb(II). The proposed sensor demonstrates the LOD of 0.0008 μM with a sensitivity of 307.26 $\mu A \mu M^{-1}$ cm^{-2} [43]. Teshome et al. fabricated an electrochemical sensor based on Fe-doped ZnO nanostructures modified GCE for voltammetry sensing of endrin. The LOD of the sensor was 0.019 μM within the concentration range of 0.1–70 μM [44].

3.2 *Direct Growth of VANs onto the Base Electrode*

In this type of sensor, VANs directly grow on the surface of the working electrode (FTO, ITO, silicon wafers, and glass substrate) for electrochemical sensing of environmental pollutants. VANs provided enhanced surface area (increased adsorption sites for sensing analytes), more roughness (maximum nucleation site), improved electron transport kinetics, uniformity, and density control (ensuring reproducibility and consistency), resulting in enhanced sensing performance of electrochemical sensor during the course of environmental pollutant sensing. Ahmad et al. fabricated a sensor based on ZnO nanorods aligned in a vertical configuration onto the

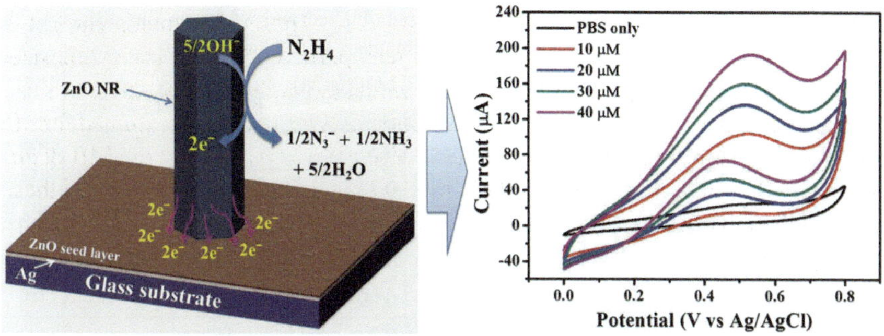

Fig. 9 Schematic diagram of ZnO nanorods aligned in a vertical configuration onto the surface of Ag electrode for electrocatalytic sensing of hydrazine. Reprinted with permission from ref. [10]. Copyright 2017 Elsevier

surface of the Ag electrode for electrocatalytic sensing of hydrazine, as demonstrated in Fig. 9. The LOD of the sensor was 0.05 μM with the sensitivity of 105.5 $\mu A \mu M^{-1}$ cm^{-2} [10]. Zhang et al. modified a carbon fiber paper with binder-free vertically aligned AgNPs-embedded ZnO nanorods and utilized for electrochemical detection of baicalein, daidzein, and luteolin. The developed electrode showed high stability and reusability along with low LODs of 0.0034, 0.0066, and 0.0034 μM, respectively [45]. In another study, binder-free VANs of ZnO were developed over the FTO glass substrate and exhibited 10.18 $\mu A \mu M^{-1}$ cm^{-2} sensitivity and 0.5 μM LOD for p-nitroaniline sensing [46]. Weng et al. synthesized the VANs of rGO-modified ZnO for electrochemical sensing of paeoniflorin [47]. The comparison of sensing performance (LOD and sensitivity) of immobilized and VANs on the surface of working electrode for sensing of environmental pollutants, as shown in Table 1.

In conclusion, VANs demonstrate good sensing performance (LOD and sensitivity) in comparison to the immobilized nanostructures onto the base electrode' surface using binder for electrocatalytic sensing of environmental pollutants owing to favorable morphology of nanostructures, which typically feature a maximum surface area (improves the interaction between the sensor and target analytes). Additionally, binder could act as an interference or potentially block adsorption sites during the course of sensing. It could increase the sensor's cost, nanomaterials delamination onto the host electrode surface, and limit the life of the sensor. Alternatively, the high specific surface area of VANs allows for more active sites for reactions or binding events, which is crucial for effective sensing. Nevertheless, published synthesis routes of VANs on the substrate's surface utilize aggressive chemicals that are not environmentally friendly. Therefore, exploring nature-inspired synthesis of VANs on the working electrode surfaces offers a promising avenue for future research, particularly in the electrochemical sensing of environmental pollutants at their source. This approach could offer environmentally benign alternatives while maintaining the performance benefits of VANs during the application in environmental remediation.

Table 1 Comparison of sensing performance of immobilized and VANs on the surface of the working electrode for sensing of environmental pollutants

Electrode	Sensor type	Analytes	LOD (μM)	Sensitivity (μAμM^{-1} cm^{-2})	Ref.
ZnO/FTO	Binder-free	p-Nitroaniline	0.5	10.18	[46]
Ce/TiO$_2$/rGO	Binder-based		0.35	49.12	[48]
C@ZnO/Alloy	Binder-free	Hydrazine	0.1	9.4	[49]
ZnO/Ag			0.0039	59.18	[50]
ZnO/Ag			0.005	105.5	[10]
ZnO/Nafion/Au	Binder-based		0.2	8.56	[51]
Au/SWCNHs/GCE			1.1	–	[52]
SnO$_2$/CeO$_2$/GCE			0.179	81	[53]
ZnO/FTO	Binder-free	Piperidine	0.06	0.53	[20]
ZnO/GCE	Binder-based		3.3	0.059	[42]
ZnO/GCE			9.4	0.238	[54]
CuO	Binder-free	H$_2$O$_2$	-	0.03	[55]
3.5-NH$_3$-PM	Binder-based		0.25	31.3	[56]
Ag-doped ZnO/GO/GCE	Binder-free	Carbamate	0.34×10^{-3}	–	[57]
CMO/B- rGO/GCE	Binder-based		0.002	4.06	[58]

4 Fabrication of Miniature Three-Electrode Device of VANs on FTO

Nature-inspired development of VANs onto the FTO glass substrate's surface can be used to construct miniature three-electrode devices for in-situ sensing and quantifying environmental pollutants at their point of origin. For manufacturing a miniature three-electrode system device, the initial step contains pattering of FTO (discontinue the F-doped SnO$_2$ layer) in three sections, one for the counter electrode, second for the working electrode, and third for the reference electrode, followed by ultrasonically cleaned in acetone, ethanol, and DI water. A bare FTO glass substrate is used as a counter electrode, and for the reference electrode, a layer of Ag/AgCl ink is drop-casted by masking scotch tape on the other sections to stop the nanostructures from spreading to undesired regions. In the middle section, the seed layer is deposited on the surface of the bare FTO, followed by nature-inspired formation of VANs on the surface of the seeded FTO glass substrate by masking other sections of the substrate, which serves as the working electrode, as shown in Fig. 10.

Dash et al. [24] fabricated a miniaturized three-electrode system device by pattering on the FTO glass substrate followed by coating of Ag/AgCl ink (left section

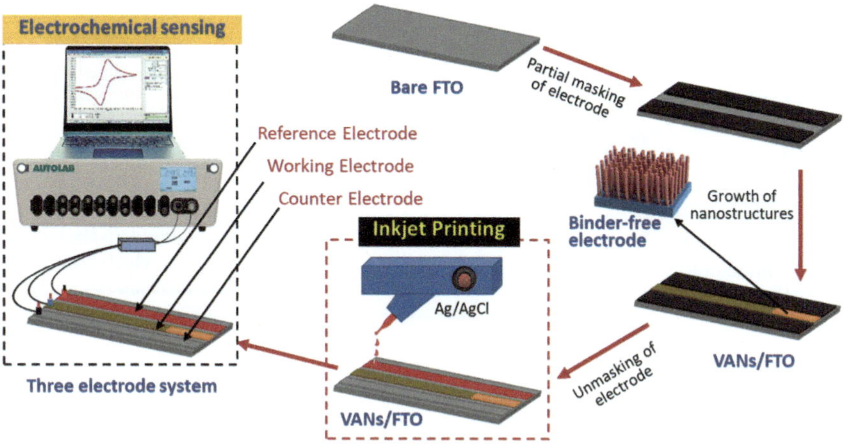

Fig. 10 Fabrication of miniature three-electrode device of VANs on FTO as working electrode, bare FTO as a counter electrode, and inkjet printed Ag/AgCl as a reference electrode for electrochemical sensing of environmental pollutants

acts as reference electrode and right section, bare FTO used as counter electrode), and immobilization of chemically synthesized Cu@AgNRs (mid-section, utilized as working electrode) using Nafion as binder for simultaneous electrochemical sensing of heavy metals in tap and river water samples, as illustrates in Figs. 11a and b [24]. However, the synthesis of Cu@AgNRs nanostructures predominantly involves environmentally aggressive chemicals that are not eco-friendly.

To the extent of our knowledge, Bhan and Golder appear to be the first published nature-inspired VANs on the surface of FTO glass substrate using vegetal extract of *S. edule* for electrochemical sensing of p-Chloroaniline in tap and river water samples [59]. Figure 12 illustrates the graphical representation of the nature-inspired synthesis of VANs of ZnO onto the surface of the FTO glass substrate. The fabricated sensor exhibited the LOD of 0.021 μM and a sensitivity of 0.194 $\mu A \mu M^{-1}$ cm^{-2} within the concentration range of 0–200 μM [59]. Moreover, the advancement of nature-inspired VANs of various nanomaterials (SnO_2, CuO, $BiVO_4$, Bi_2S_3, CdS, AgNPs, AuNPs, etc.) on the support electrode surface, offer significant insight into the future of electrochemical sensor, which enhances sensor's performance during the course of environmental pollutant sensing. Therefore, nature-inspired formation of VANs on FTO substrate-based miniaturized three-electrode system device demonstrates potential utilization in sensing and quantification of environmental pollutants at their source of origin owing to binder-free, simple, and environmentally friendly method with excellent sensing performance.

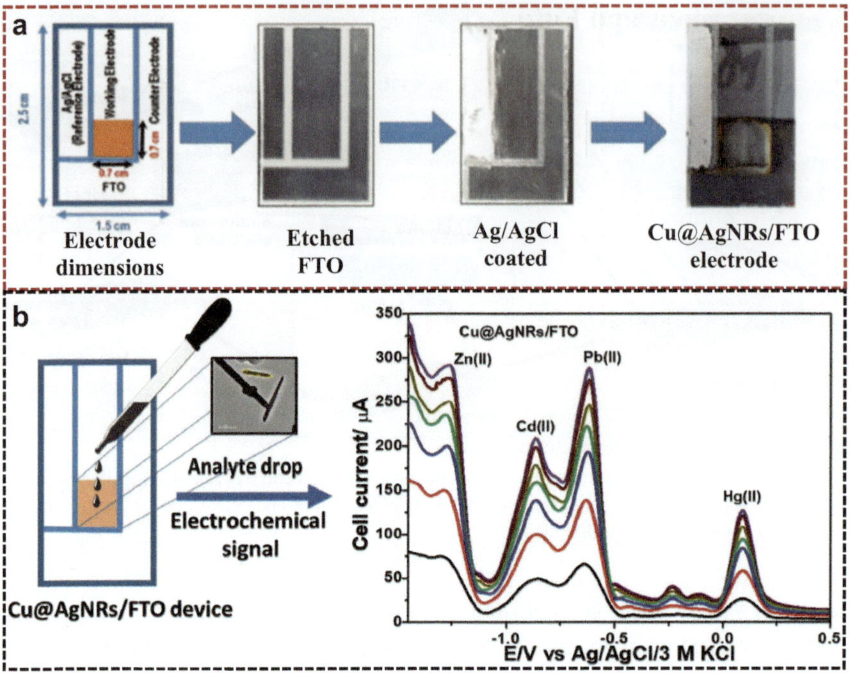

Fig. 11 **a** and **b** demonstrate the steps in fabricating three-electrode device configurations for simultaneous electrochemical sensing of heavy metals in tap and river water samples. Reprinted with permission from ref. [24]. Copyright 2024 Elsevier

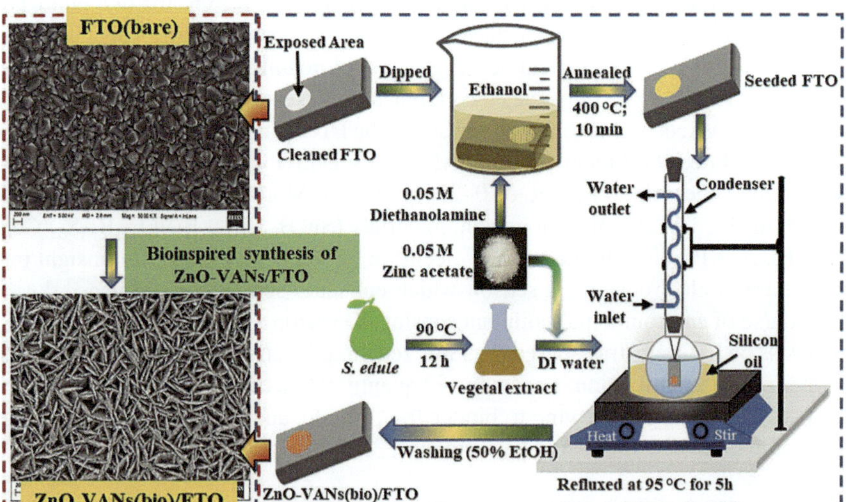

Fig. 12 Graphical representation of the nature-inspired synthesis of VANs of ZnO onto the surface of FTO glass substrate using vegetal extract of *S. edule* fruit. Reprinted with permission from ref. [59]. Copyright 2023 Elsevier

5 Conclusions and Future Perspective

This report provided a brief discussion on the synthesis of powder and VANs by conventional (using environmentally aggressive chemicals) and nature-inspired (using phytochemical presents in the plants and their organs) methods and explored the crucial role of nanostructures in improving the sensing performance of the electrochemical sensor by immobilizing and directly growing nanostructure in vertical configuration on the host electrode surface for environmental pollutants sensing. VANs exhibited excellent sensing performance (LOD, sensitivity, stability, and reproducibility) in comparison to the immobilized nanostructures on the base electrode's surface owing to the high electroactive surface area, high surface roughness, fast charge transfer kinetics, and efficient substrate infiltration shape of nanostructures. In addition, nanostructures aligned in the vertical configuration could eliminate the fast flaking off nanostructures from the base electrode's surface, the binder's need, and the step of nanostructures functionalization on the base electrode surface using a binder. Additionally, in functionalized nanostructures-based electrochemical sensors, the binder acts as a matrix that could block the adsorption sites for targeted analytes and potentially interfere during sensing. This review also provided the direction for the fabrication of the nature-inspired VANs based miniaturized three-electrode system on the FTO-based device for sensing and quantification of environmental pollutants at their source of origin with high selectivity, sensitivity, and stability to protect human and environmental health.

A critical future objective in electrochemical sensors is to bridge the gap between real-world analytical applications and emerging concepts in electrochemical sensing. Achieving this will enable various analytical measurement scenarios where electrochemical sensors can perform at their highest potential. The primary challenges in electrochemical sensing involve developing sustainable methods for the direct growth of nature-inspired nanostructures aligned in vertical configuration on the working electrode surfaces to enhance their efficiency and fabricating a miniaturized three-electrode system-based device with high stability and sensitivity for electrochemical sensing of environmental pollutants. VANs exhibited high surface area, resulting in improved electrochemical sensor performance. However, high surface area could increase surface fouling and nonspecific adsorption of interference species, decreasing electrochemical sensor performance. Progress in this area involves preventing surface fouling, reducing nonspecific adsorption of interference, developing novel strategies to enhance signal-to-noise ratios, and developing new nanomaterials. These advancements aim to address the persistent issue of improving charge transfer characteristics. In addition, further modifying nature-inspired VANs by other nanomaterials also provides valuable insight into the advancement of electrochemical sensors. This approach has the potential to significantly enhance the performance of electrochemical sensors in-situ sensing and quantifying environmental pollutants.

References

1. Wang, Y.: Applications of Carbon Nanotubes and Graphene for Electrochemical Sensing of Environmental Pollutants, J. Nanosci. Nanotechnol. 16 7852–7872 (2016).
2. Ravi, R.: Pharmaceutically Active Compounds' (PhACs) Threat: An Environmental Prospective, in: Sustain. Environ., Springer Nature Singapore, Singapore. 51–66 (2023).
3. Ravi, R.: A tuneable bioinspired process of Pt-doping in TiO_2 for improved photoelectrochemical and photocatalytic functionalities, Colloids Surfaces A Physicochem. Eng. Asp. 663 131034 (2023).
4. Su, G.: A chemometric-assisted method for automatic, rapid and non-targeted detection of multi-pesticides in plant-derived foods by gas chromatography–mass spectrometry, Food Chem. 443 138573 (2024).
5. Li, T.: Dihydroxyl modified UiO-66 as dispersive solid-phase extraction sorbent coupled with ultra-high performance liquid chromatography tandem mass spectrometry for detection of neonicotinoid insecticides, J. Chromatogr. A 1727 464970 (2024).
6. Golder, A.K.: Synthesis of low-cost bentonite/Duranta erecta's fruit powder imbedded alginate beads and its application in surfactant removal, Environ. Sci. Pollut. Res. 28 58945–58957 (2021).
7. Sangamithirai, D.: Electrochemical sensing platform for the detection of nitroaromatics using g-C_3N_4/V_2O_5 nanocomposites modified glassy carbon electrode, Electrochim. Acta 434 141308 (2022).
8. Ravi: Enhanced adsorption capacity of designed bentonite and alginate beads for the effective removal of methylene blue, Appl. Clay Sci. 169 102–111 (2019).
9. Farhat, Z.: Fabrication of used-tea embedded alginate beads for cationic dye remediation: Synergistic effect of surface adsorption and intraparticle diffusion, Surfaces and Interfaces 51 104601 (2024).
10. Ahmad, R.: Highly stable hydrazine chemical sensor based on vertically-aligned ZnO nanorods grown on electrode, J. Colloid Interface Sci. 494 153–158 (2017).
11. Ravi, R.: Bio-based Au Doping with Dominant Oxygen Vacancies and Ti^{3+} Defects on Photocatalytic Functionalities of TiO_2, Ind. Eng. Chem. Res. 62 20702–20715 (2023).
12. Kumar, A.: A novel eco-friendly inhibitor of chayote fruit extract for mild steel corrosion in 1 M HCl: Electrochemical, weight loss studies, and the effect of temperature, Sustain. Chem. Pharm. 36 101261 (2023).
13. Ravi, R.: Efficient recovery of TiO_2 and Pt-doped TiO_2 photocatalysts in wastewater treatment using a pilot-scale cross-flow ultrafiltration membrane system, Sep. Purif. Technol. 339 126710 (2024).
14. Kumar, A.: Corrosion inhibition of mild steel by Praecitrullus fistulosus (tinda fruit and peel) extracts, Sci. Total Environ. 929 172569 (2024).
15. Chowdhury, A.: A tunable bioinspired process of SnO_2 NPs synthesis for electrochemical CO_2-into-formate conversion, J. CO_2 Util. 66 102263 (2022).
16. Bhan, C.: Utilizing bioinspired AgNPs as an antibacterial agent to enhance ceramic membrane performance, J. Environ. Chem. Eng. 11 110283 (2023).
17. Chowdhury, A.: A simple template-free bioinspired route of 1D Bi_2S_3 nanorods synthesis for electrochemical CO_2 reduction to formate, J. Ind. Eng. Chem. 127 138–148 (2023).
18. Das, D.: Improved selectivity of electrochemical aniline sensing using one-dimensional silver nanorods with high aspect ratio synthesized by ascorbic acid assisted method, Anal. Chim. Acta 1310 342697 (2024).
19. Ahmad, R.: Facile fabrication of palm trunk–like ZnO hierarchical nanostructure–based biosensor for wide-range glucose detection, Chem. Eng. J. 492 152432 (2024).
20. Khan, U.: Fabrication of ultra-sensitive piperidine chemical sensor with a direct grown well-aligned ZnO nanorods on FTO substrate as a working electrode, Solid State Sci. 97 105986 (2019).
21. Kim, J.H.: A fully microfabricated carbon nanotube three-electrode system on glass substrate for miniaturized electrochemical biosensors, Biomed. Microdevices 14 613–624 (2012).

22. Wang, Q.: Miniaturized Electrochemical Sensing Platforms for Quantitative Monitoring of Glutamate Dynamics in the Central Nervous System, Angew. Chemie Int. Ed. 202406867 (2024).
23. de Moraes, N.C.: Design of novel, simple, and inexpensive 3D printing-based miniaturized electrochemical platform containing embedded disposable detector for analytical applications, Electrophoresis 41 278–286 (2020).
24. Dash, S.R.: Ultrasensitive electrochemical sensors based on Cu and Cu@Ag nanorods for simultaneous heavy metal detection, Mater. Chem. Phys. 318 129255 (2024).
25. Gopi, P.K.: High-performance electrochemical sensing of hazardous pesticide Paraoxon using $BiVO_4$ nano dendrites equipped catalytic strips, Chemosphere 288 132511 (2022).
26. Nair, V.S.: Hydrothermal synthesis of iron titanate hexagonal nanoplates for electrochemical detection of nitrofurazone, Microchim. Acta 191 245 (2024).
27. Chellappa, V.: Hydrothermal synthesis of MnO_2 nanorods for efficient electrochemical detection of environmental anthropogenic pollutants and nitrobenzene, Inorg. Chem. Commun. 160 112015 (2024).
28. Wang, X.: Growth of $Ni_5P_4/Ni_2P/C$ nanosheet arrays on ITO as dual-function electrochemical sensor for efficient Cu^{2+} and paracetamol detection, Ceram. Int. 50 18584–18593 (2024).
29. Ahmad, R.: Highly Efficient Non-Enzymatic Glucose Sensor Based on CuO Modified Vertically-Grown ZnO Nanorods on Electrode, Sci. Rep. 7 5715 (2017).
30. Gawal, P.M.: Vegetal route for synthesis of CQDs/CdS nanocomposites for photocatalytic reduction of CO_2 to methanol under visible light, Colloids Surfaces A Physicochem. Eng. Asp. 683 133068 (2024).
31. Gawal, P.M.: Plant-Based Phytochemicals for Synthesis of Z-Scheme In_2O_3/CdS Heterostructures: DFT Analysis and Photocatalytic CO_2 Reduction to HCOOH and CO, Langmuir 40 13538–13549 (2024).
32. Liu, W.: Rapid synthesis of vertically-aligned zinc oxide nanorods on stainless steel for non-enzymatic glucose and H_2O_2 photoelectrochemical sensor, Appl. Surf. Sci. 480 341–348 (2019).
33. Ma, Y.: An electrochemical sensor based on the modification of platinum nanoparticles and ZIF-8 membrane for the detection of ascorbic acid, Talanta 226 122105 (2021).
34. Wang, Y.: Electrodeposition of Metal Nanoparticles inside Carbon Nanopipettes for Sensing Applications, Anal. Chem. 94 16987–16991 (2022).
35. Kathiresan, V.: A simple one-step electrochemical deposition of bioinspired nanocomposite for the non-enzymatic detection of dopamine, J. Anal. Sci. Technol. 12 5 (2021).
36. Wang, K.: Zinc Nanoparticles Electrodeposited on TiO_2 Nanotube Arrays Using Deep Eutectic Solvents for Implantable Electrochemical Sensors, ACS Appl. Nano Mater. 6 8238–8249 (2023).
37. Mali, S.S.: Chemically grown vertically aligned 1D ZnO nanorods with CdS coating for efficient quantum dot sensitized solar cells (QDSSC): a controlled synthesis route, Dalt. Trans. 42 16961 (2013).
38. Singh, A.: Electrochemical sensing of ascorbic acid (AA) from human sweat using Ni–SnO_2 modified wearable electrode, Inorg. Chem. Commun. 152 110718 (2023).
39. Zhai, X.: A highly selective and recyclable sensor for the electroanalysis of phosphothioate pesticides using silver-doped ZnO nanorods arrays, Anal. Chim. Acta 1152 338285 (2021).
40. Slimane, D.: Green synthesis of copper oxide nanoparticles using Ficus elastica extract for the electrochemical simultaneous detection of Cd^{2+} , Pb^{2+} , and Hg^{2+}, RSC Adv. 13 18734–18747 (2023).
41. Singh, P.: Bioinspired triangular ZnO nanoclusters synthesized by Argyreia nervosa nascent leaf extract for the efficient electrochemical determination of vitamin C, RSC Adv. 11 25752–25763 (2021).
42. Ameen, S.: An electrochemical sensing platform based on hollow mesoporous ZnO nanoglobules modified glassy carbon electrode: Selective detection of piperidine chemical, Chem. Eng. J. 270 564–571 (2015).

43. Dash, S.R.: Synthesis of highly structured spherical Ag@Pt core-shell NPs using bio-analytes for electrocatalytic Pb(II) sensing, Sensors Actuators B Chem. 314 128062 (2020).
44. Teshome, T.: Electrochemical determination of endrin from fruit juice samples based on Fe-doped ZnO nanomaterial modified glassy carbon electrode, J. Appl. Electrochem. 54 865–877 (2024).
45. Zhang, Z.: In situ fabrication of Ag nanoparticles modified vertically-grown ZnO nanorods on carbon fiber paper electrode for simultaneous detection of luteolin, daidzein and baicalein, Colloids Surfaces A Physicochem. Eng. Asp. 694 134118 (2024).
46. Ahmad, R.: Development of highly-stable binder-free chemical sensor electrodes for p-nitroaniline detection, J. Colloid Interface Sci. 494 300–306 (2017).
47. Weng, N.: Vertically well-aligned ZnO nanorods based biosensor for determination of paeoniflorin as an antidepressant medicine, Alexandria Eng. J. 86 471–480 (2024).
48. Kumar, N.: Hydrothermally synthesized polycrystalline Ce-TiO_2@rGO compound for sensing of para-nitroaniline, Bull. Mater. Sci. 45 87 (2022).
49. Liu, J.: C@ZnO nanorod array-based hydrazine electrochemical sensor with improved sensitivity and stability, Dalt. Trans. 39 8693 (2010).
50. Ahmad, R.: Highly sensitive hydrazine chemical sensor based on ZnO nanorods field-effect transistor, Chem. Commun. 50 1890 (2014).
51. Umar, A.: Zinc oxide nanonail based chemical sensor for hydrazine detection, Chem. Commun. 7345 166–168 (2008).
52. Zhao, S.: A high-performance hydrazine electrochemical sensor based on gold nanoparticles/single-walled carbon nanohorns composite film, Appl. Surf. Sci. 369 36–42 (2016).
53. Mohammad, A.: A sensitive electrochemical detection of hydrazine based on SnO_2/CeO_2 nanostructured oxide, Microchem. J. 171 106784 (2021).
54. Ameen, S.: Spindles shaped ZnO modified glassy carbon electrode for the selective monitoring of piperidine, Mater. Lett. 148 188–191 (2015).
55. Jia, W.: Vertically Aligned CuO Nanowires Based Electrode for Amperometric Detection of Hydrogen Peroxide, Electroanalysis 20 2153–2157 (2008).
56. Candia-Onfray, C.: Nanostructured Fe-N-C pyrolyzed catalyst for the H_2O_2 electrochemical sensing, Electrochim. Acta 387 138468 (2021).
57. Wang, C.: Development of Electrochemical Sensor Based on Graphene Oxide Electrode Modified by Silver-Doped ZnO Nanorods for Detection of Carbamate Pesticide in Food, Int. J. Electrochem. Sci. 15 5623–5631 (2020).
58. Alagarsamy, S.: An ultrasensitive carbamate pesticide detection sensor based on metal molybdate encapsulated with boron doped reduced graphene oxide nanocomposite, Colloids Surfaces A Physicochem. Eng. Asp. 673 131830 (2023).
59. Bhan, C.: Bio-based hierarchical vertically aligned 2D ZnO nanostructures for ultra selective electrochemical sensing of p-Chloroaniline, Chem. Eng. J. 475 146122 (2023).

Natural Analytes for Formation of Silver Nanostructures for Electrocatalytic Sensing of Carbendazim

Chandra Bhan, Smruti Ranjan Dash, and Animes Kumar Golder

Abstract Recently, the spreading of agricultural pesticides in the food chain, waterbodies, and soil has become a major concern, creating enormous challenges for regulatory bodies. Agricultural pesticide utilization has increased for improved crop production to fulfill the demands of a huge population, resulting in pesticide pollution throughout the world. Developing a highly efficient in-situ system for the sensing and quantifying of pesticides is the need of the hour to control pesticide pollution in India and globally. Utilization of intensive chemicals in conventional nanomaterial synthesis could produce toxic by-products. Bio-based methods could be highly efficient and eco-friendly alternatives for nanoparticle synthesis. Herein, we report the formation of silver nanoparticles (AgNPs) using a vegetal extract of *Citrus Limon* by a microwave-assisted method for electrochemical sensing of carbendazim. The average particle size of highly crystalline and mesoporous bio-based AgNPs was 10.33 ± 2.3 nm with a BET surface area of 392.5 m^2/g. AgNPs/GCE decreases the charge transfer resistance from 2830 to 474 KΩ and increases the electroactive surface area by 2.83 times that of bare GCE. The detection limit, quantification limit, sensitivity, and linear concentration range of the AgNPs/GCE were 0.279 μM, 0.931 μM, 0.668 $\mu A.\ \mu M^{-1}.cm^{-2}$, and 0–50 μM, respectively. The proposed sensor showed a 3.53% decline in the peak current up to the 5th cycle of carbendazim sensing. Moreover, AgNPs/GCE exhibited excellent sensing performance with high stability for electrocatalytic sensing of carbendazim.

Keywords Bio-based synthesis · Microwave-assisted method · Silver nanoparticles · Electrochemical sensing · Carbendazim fungicide

C. Bhan · A. K. Golder (✉)
Department of Chemical Engineering, Indian Institute of Technology Guwahati, Guwahati, Assam, India
e-mail: animes@iitg.ac.in

C. Bhan
e-mail: bchandra@iitg.ac.in

S. R. Dash · A. K. Golder
Centre for the Environment, Indian Institute of Technology Guwahati, Guwahati, Assam, India
e-mail: d.smruti@iitg.ac.in

U. S. Dixit et al. (eds.), *Current Progress in Engineering Sciences*,
https://doi.org/10.1007/978-981-96-8753-4_37

1 Introduction

In recent decades, the rising demand for agricultural products to fulfill the needs of a huge population has led to the increased production of high-quality crops. Consequently, numerous pesticides have been utilized to prevent plant diseases and pest attacks, thereby boosting agricultural productivity. However, the overuse of pesticides has resulted in significant pollutants in both the food chain and aquatic water bodies [1]. Carbendazim, a fungicide commonly used to protect crops like rice, vegetables, fruits, lettuce, legumes, cucurbits, and sunflowers from fungal diseases such as scorch, blight, powdery mildew, spots, and rot, is of particular concern [2]. The presence of carbendazim in food, water, and soil poses serious challenges due to its harmful impacts on the ecosystem, especially on humans, invertebrates, and aquatic animals. Carbendazim exposure can cause mutagenic effects, alter cell proliferation, inhibit embryonic growth, lead to liver disease, infertility, hematological abnormalities, inhibit antioxidant enzymes, and disrupt thyroid gland function [3]. Therefore, the on-site detection and monitoring of these pesticides are crucial for initiating effective eradication measures.

Various methods, such as chromatographic techniques, chemical methods, electrochemical sensing, and spectroscopy, have been employed to detect pesticides. Chromatographic techniques, while precise, are time-consuming and pose challenges for on-site detection [4]. Spectroscopy techniques are economical and fast but suffer from low sensitivity and high detection limits [5]. Chemical methods lack precision, require intensive chemicals, and are typically qualitative. Among these methods, electrochemical sensing stands out as a quick, precise, economical, and highly sensitive technique for pesticide detection. However, developing highly efficient and eco-friendly electrocatalysts remains challenging. Conductive electrodes such as platinum (Pt), graphite, glassy carbon (GC), and gold (Au) electrodes have been used for electrochemical sensing. However, they often exhibit low sensitivity and selectivity, high electrical resistance, and a low electrochemically active surface area. Modifying these electrodes with metal oxide and sulfide nanostructures, such as ZnO, AuNPs, AgNPs, SnO_2, CuO, Co_3O_4, and Bi_2S_3, can enhance their electrochemical activity [1, 6–8]. These modifications provide a larger active surface area, higher conductivity, improved charge transfer, and increased surface interaction with targeted analytes [7, 9]. AgNPs, in particular, possess antimicrobial, electrical, and high catalytic properties, making them a potential candidate for electrode modification to ultimately enhance electrochemical activity [10].

Conventional nanostructure synthesis methods often utilize intensive chemicals, which can generate toxic by-products, further exacerbating environmental issues. In contrast, bio-based methods offer an efficient and eco-friendly alternative for electrocatalyst synthesis, utilizing plant parts such as leaves, fruits, and seeds as sources of reducing and capping agents [11–13]. Plant extracts primarily contain ascorbic acid, gallic acid, citric acid, flavonoids, and polyphenols. Moreover, the surface functionalization of electrocatalysts by these organic compounds enhances electrochemical activity by facilitating better interaction with pollutants [14]. For instance, in our previous work, ZnO nanostructures were synthesized onto the surface of FTO glass substrates using *Sechium edule* extract for the electrochemical detection of p-chloroaniline. The growth of bio-based ZnO on the FTO glass substrate increased the effective surface area by 1.2-fold and decreased the electron transfer resistance by 2.58 times compared to bare FTO [1]. In another study, bio-based carbon dots (CDs) were synthesized using a vegetal extract of *Psidium guajava leaves* and immobilized on a GCE for the electrochemical sensing of chlorpyrifos. The nanostructures immobilized electrode exhibited better sensitivity and repeatability during chlorpyrifos detection, with results comparable to HPLC analysis [15].

Lemon juice, containing citric acid, amino acids, phenolic compounds, and vitamins, has also been reported as a suitable medium for nanostructure synthesis [16]. Utilizing abundantly available and eco-friendly plant resources for synthesizing highly sensitive and efficient nanostructures for the electrochemical detection of pesticides represents a promising new research domain for environmental sustainability. Herein, we synthesized silver nanoparticles (AgNPs) using *Citrus limon* extract in a microwave reactor. The prepared Ag nanostructures were characterized for their elemental, morphological, and physical properties. Subsequently, the Ag nanostructures were immobilized on a GCE for electrochemical sensing of the carbendazim. The electrochemical analysis was conducted in a three-electrode system. The detection limit, quantification limit, and sensitivity were calculated, and the stability was analyzed over 10 consecutive sensing cycles.

2 Materials and Methods

2.1 Reagents

$AgNO_3$ (CAS No: 7761-88-8), NaCl (CAS No: 7647-14-5), $(C_6H_9NO)_n$ (PVP K-30, CAS No: 9003-39-8), $C_9H_9N_3O_2$ (CAS No: 10605-21-7), $Ru(NH_3)_6Cl_2$ (CAS No: 15305-72-3), KCl (CAS No: 7447-40-7), Na_2HPO_4 (CAS No: 7558-79-4) and NaH_2PO_4 (CAS No: 13472-35-0) were purchased from Merck, Mumbai, India. Nafion D-521® (CAS No: 31175-20-9) was procured from Alfa Aesar, Haverhill, USA.

2.2 Experimental Methodology

Bio-based microwave-assisted synthesis of AgNPs

Fresh lemon was bought from the vegetable market of IIT Guwahati, India. The lemon was cut into halves and squeezed to obtain juice, and filtered using a 0.45 μm nylon filter. The synthesis of AgNPs was carried out using the microwave-assisted method, as illustrated in Fig. 1. For this, a reaction solution containing 2 mL of 0.1 M $AgNO_3$

solution, 200 mg polyvinylpyrrolidone (PVP) K-30, and 10 mg NaCl was stirred for proper mixing and transferred to a 5 mL cylindrical glass tube with a circular pressure cap. Further, 2 mL of freshly prepared *Citrus limon* extract was added, and the mixture was incubated in a microwave reactor at 90 W power and 100 °C for a duration of 5 min, following cool-down to room temperature and centrifugation at 13,500 rpm for 10 min with consecutive washing with 50% ethanol for 5 cycles to eradicate the contaminations. Further, the obtained pellet was dispersed into 2 mL of DI water for characterization and electrochemical sensing activities.

Preparation of AgNPs modified GCE

The GCE with a diameter of 0.3 cm was polished using alumina powder (0.3 μm) to get a smooth surface, followed by ultrasonic cleaning for 5 min in DI water. Further, it was washed in DI water by dipping it and drying it at room temperature. Simultaneously, 100 μL of prepared AgNPs (10 mg/mL) was dispersed in 900 μL DI water, 100 μL Nafion, and 900 μL isopropanol solution and ultrasonicated for 1 h. Further, 5 μL of the mixture solution was drop-casted over a cleaned GEC surface, followed by drying at room temperature before use for electrochemical characterization and sensing of carbendazim.

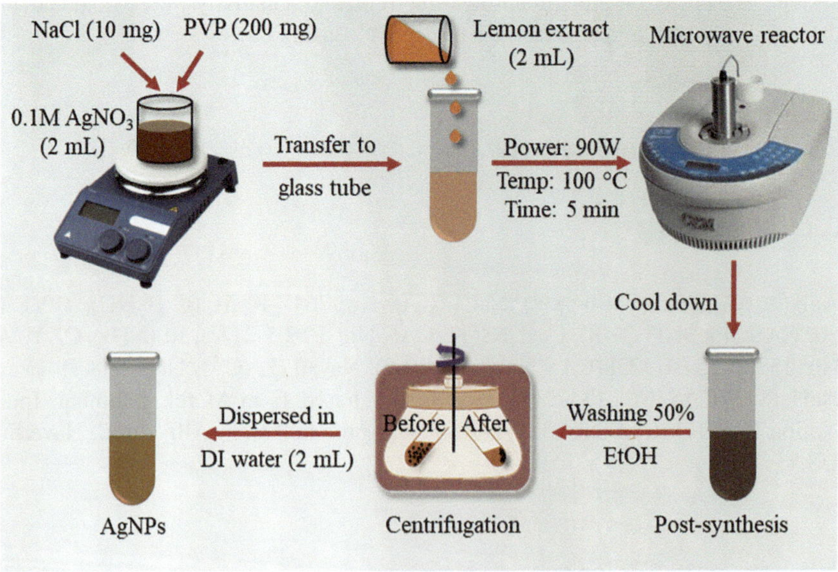

Fig. 1 Graphical illustration of bio-based microwave-assisted synthesis of AgNPs

2.3 Characterization

Field emission scanning electron microscope (FESEM, Carl Zeiss, Jena, Germany) and Field emission transmission electron microscopy (FETEM, JEM-2100 F, JEOL, USA) were utilized for morphological analysis. The electron diffraction X-ray (EDX, Sigma, Carl Zeiss, Jena, Germany) was performed for elemental analysis. X-ray diffraction (XRD) pattern of AgNPs was performed using the XRD analyzer (D2 Advance, Bruker, Karlsruhe, Germany). The thermal stability of the AgNPs was analyzed by a thermogravimetric analyzer (TGA 209 F1, Netzsch, Germany) from a temperature of 30–900 °C with a 10 °C min^{-1} heating rate in the N_2 atmosphere. The BET surface area analysis was performed in a pore size analyzer (Autosorb-IQ MP, Quantachrome, Florida, USA) with degassing at 180 °C for 2 h.

2.4 Electrochemical Studies and Carbendazim Sensing

The electrochemical characterization and sensing of carbendazim were performed in a Potentiostat/Galvanostat (PGSTAT 302 N, Metrohm, Switzerland). This electrochemical setup contains three-electrode cell arrangements, namely, reference electrode (Ag/AgCl/3 M KCl), counter electrode (platinum wire, 1 mm diameter), and functioning electrode (bare GCE and AgNPs/GCE). The electrochemical impedance spectrometry (EIS) determined the charge transfer resistance of bare and AgNPs immobilized GCE. The electroactive surface area (A_{eff}) of the working electrode (bare and AgNPs immobilized GCE) was also determined by cyclic voltammetry (CV) at different scan rates (5–500 mV/s). The calibration curve of varying concentrations of carbendazim (0–200 μM) was plotted through square wave voltammetry (SWV) analysis using AgNPs/GCE. SWV was also performed to determine the stability of AgNPs/GCE at a carbendazim concentration of 50 μM.

3 Results and Discussion

3.1 Morphological and Elemental Analysis

Figure 2a and b illustrate the FESEM and FETEM images of synthesized bio-based AgNPs using bio-extract of *Citrus limon* by microwave-assisted route. The obtained results confirmed that the AgNPs are irregularly spherical shaped with an average diameter of 10.33 ± 2.3 nm with a normal distribution (inset of Fig. 2b). The size of non-agglomerated particles was measured in ImageJ software for the size distribution profile. Figure 2c shows the HRTEM and SAED pattern (inset of Fig. 2c) images of AgNPs. The HRTEM image exhibited a 0.233 nm d-spacing value, aligning with XRD results. Additionally, the SAED pattern confirms the polycrystalline nature

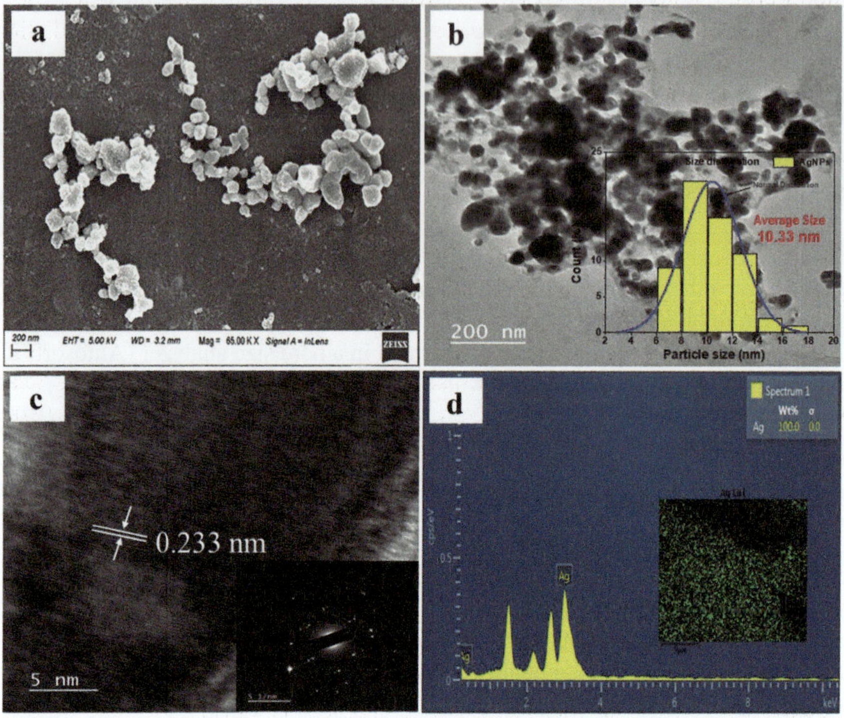

Fig. 2 **a** FESEM image, **b** FETEM image and particle size distribution (inset), **c** HRTEM and SAED pattern (inset) images, and **d** Elemental spectra and image mapping (inset) of AgNPs

of AgNPs. Fig. 2d illustrates the elemental analysis and image mapping (inset of Fig. 2d), confirming the uniform distribution of AgNPs. However, additional peaks were also observed, corresponding to Al and Au from the aluminum foil used for sample preparation and the Au coating before sample analysis, respectively.

3.2 XRD Analysis

Figure 3 illustrates the XRD spectrum of prepared bio-based AgNPs and exhibits sharp peaks at $2\theta = 38.5$ (111), 45.1 (200), 65.4 (220), 78.5 (311), and 81.8° (222), corresponding to AgNPs, according to JCPDS No. 07-090-2671. The sharp peaks indicate the high crystallinity of AgNPs, supporting their high electrocatalytic activity for sensing organic analytes. The d-spacing value of the most prominent peak at $2\theta = 38.5°$ was found to be 0.233 nm, which is in line with the HRTEM result.

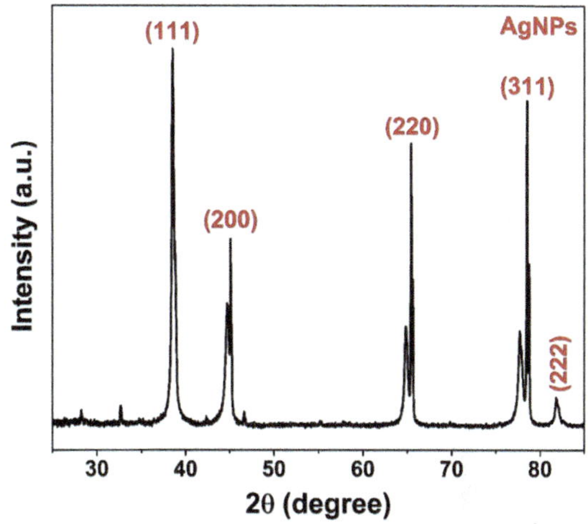

Fig. 3 XRD pattern of bio-based AgNPs

3.3 *TGA and BET Analyses*

The TGA profile of AgNPs is illustrated in Fig. 4a. The obtained results confirmed about 2.5% mass loss till 200 °C, corresponding to the vaporization of surface water. Further, the mass loss increased to ~27.5% at 700 °C, which could be due to evaporation of internal water and degradation of loosely bound functional groups. Additionally, about ~53% mass was lost up to 900 °C, indicating further degradation of tightly bound polyphenolic groups. The reported study also observed a similar TGA profile [17].

Figure 4b illustrates the N_2 adsorption–desorption curve and BET parameters of AgNPs. The BET surface area of AgNPs was 392.5 m^2/g, while the average pore volume and size were 0.6 cc/g and 6.26 nm, respectively. The N_2 adsorption–desorption hysteresis loop confirmed that AgNPs follow the Type-IV isotherm, indicating the mesoporous nature of the AgNPs [13]. This mesoporous nature could

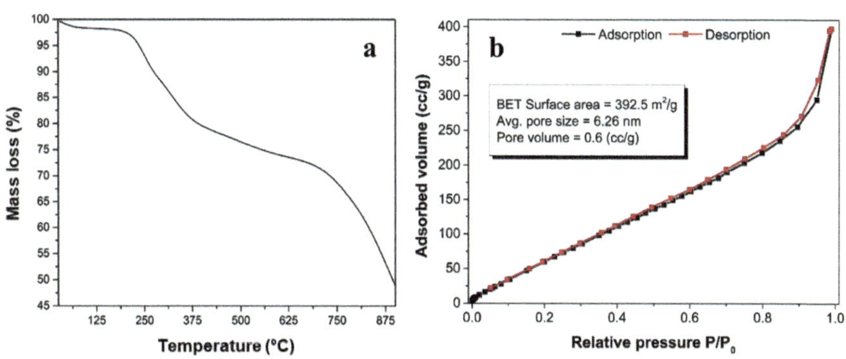

Fig. 4 **a** TGA profile and **b** BET parameters and N_2 adsorption–desorption curve of AgNPs

be regulated by surface-adsorbed functional groups. AgNPs with higher surface area could provide a large number of active sites to interact with targeted analytes, which ultimately increases the electrochemical sensing efficiency [18].

3.4 *Electrochemical Characterization and Carbendazim Sensing*

EIS analysis

To determine the charge, transfer resistance (R_p) and electrolyte solution resistance (R_s) of bare GCE and AgNPs/GCE, EIS analysis was performed in 100 mM KCl electrolyte using 1 mM $Ru(NH_3)_6Cl_2$ as an analyte between the frequency range of 0.01–10^5 Hz (Fig. 5). R_p and R_s of bare GCE and AgNPs/GCE were determined from the inserted Randles circuit in Fig. 5 by fitting the Nyquist plots data [19]. The R_p of bare GCE and AgNPs/GCE were 2830 KΩ (R_s = 129 Ω) and 474 kΩ (R_s = 130 Ω), respectively. A notable reduction (83.25%) in the R_p was observed after the immobilization of AgNPs onto the surface of bare GCE, and no significant changes were noted in the value of R_s [6].

Electroactive surface area of functioning electrode

The electroactive surface area of functioning electrodes (bare and AgNPs immobilized GCE) was determined via CV conducted at different scan rates (5–500 mV/s) using 1 mM $[Ru(NH_3)_6]Cl_2$ as an analyte in 0.1 M KCl electrolyte. The obtained cyclic voltammograms are illustrated in Fig. 6. The obtained data was fitted to the Randles–Sevcik equation (Eq. 1), and the slope of the linear plot of I_{pa} versus $v^{1/2}$ was used to calculate effective surface area (A_{eff}, cm^2) [1].

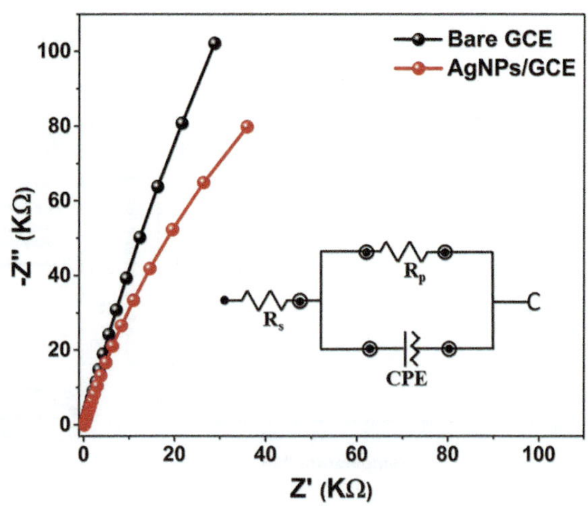

Fig. 5 EIS analyses of bare GCE and AgNPs/GCE with Randles circuit

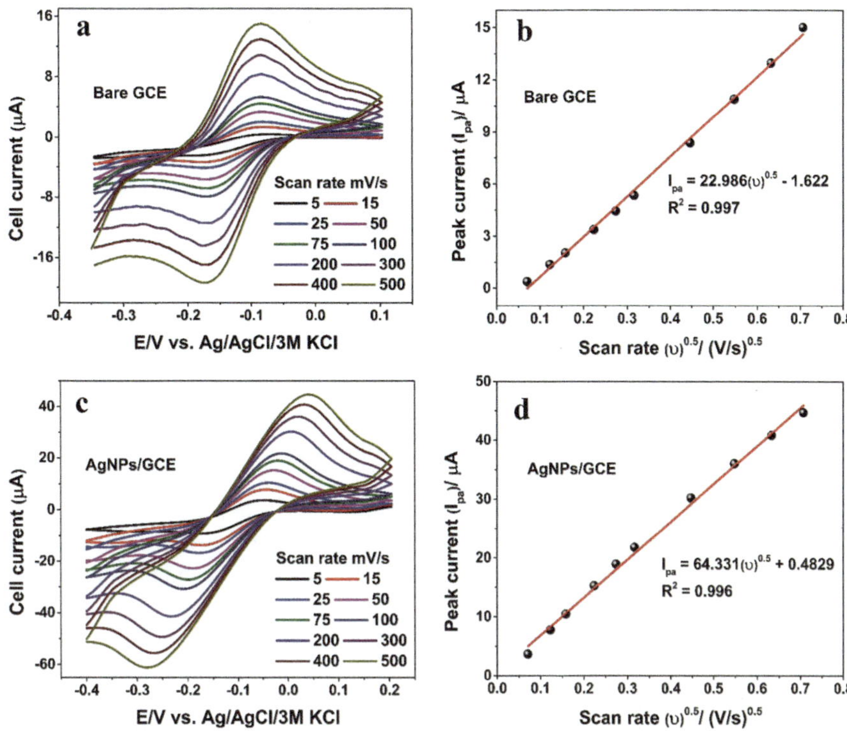

Fig. 6 **a** CV and **b** I_{pa} versus $v^{1/2}$ of bare GCE. **c** CV and **d** I_{pa} versus $v^{1/2}$ of AgNPs/GCE

$$I_{Pa} = \left(2.69 \times 10^5\right) n^{3/2} A_e D^{1/2} v^{1/2} C_0 \tag{1}$$

where I_{pa}, C_0, D, v, and n represent anodic peak current (A), the concentration of analyte (mol/cm^3), diffusion coefficient (8.43×10^{-6} cm^2/s), scan rate (V/s), and number of electrons involved ($n = 1$). The electrocatalytic surface area of bare GCE was determined to be 0.029 cm^2, which increased to 0.082 cm^2 after the immobilization of AgNPs onto the surface of GCE. The increase in the effective surface area enhanced electrochemical sensing performance due to the presence of more binding sites for targeted electrochemical species.

Electrochemical detection of carbendazim using AgNPs/GCE

The electrocatalytic sensing performance of carbendazim based on AgNPs/GCE was investigated by SWV and corresponding calibration curve recorded in different carbendazim concentrations (0–200 μM) using 0.1 M PBS electrolyte at pH 7 (Fig. 7a). The graph of I_{pa} and carbendazim concentration exhibited two linear relationships between the concentration range of 0–50 and 50–200 μM owing to the profusion of deposited carbendazim ions onto the surface of AgNPs/GCE altered the sensing reaction from diffusion to surface control regime (Fig. 7b). The detection and quantification limit for carbendazim sensing using AgNPs/GCE were investigated by Eqs. 2 and 3 [1].

$$\text{Detection Limit} = \frac{3S_d}{m} \tag{2}$$

$$\text{Quantification Limit} = \frac{10S_d}{m} \tag{3}$$

where m and S_d are the slope of the I_{pa} versus carbendazim concentration graph and the blank run standard deviation (0.0051), respectively. The I_{pa} versus carbendazim concentration curve slope was found to be 0.0548 (R^2 = 0.982) and 0.0177 (R^2 = 0.994) μA. μM^{-1} within the concentration range of 0–50 and 50–200 μM, respectively (Fig. 7b). The detection limit was 0.279 (quantification limit 0.931 μM) and 0.864 μM (quantification limit 2.881 μM) within the concentration range of 0–50 and 50–200, respectively. The sensitivity of the AgNPs/GCE was determined by the slope of the I_{pa} versus carbendazim concentration curve and electroactive surface area (0.082 cm^2) obtained from Fig. 6d. The sensitivity of carbendazim detection over AgNPs/GCE was 0.668 and 0.216 μA. $\mu M^{-1}.cm^{-2}$ between the concentration range of 0–50 and 50–100 μM, respectively.

Stability of AgNPs/GCE during carbendazim detection

The stability of AgNPs/GCE was investigated by SWV performed in 0.1 M PBS electrolyte (pH 7) at a fixed concentration of carbendazim (50 μM). The change in peak current was recorded after each cycle during the course of carbendazim sensing, as shown in Fig. 8. The decrease in the peak response of 3.53 and 15.67% was recorded up to the 5th and 10th cycles of carbendazim sensing, respectively. The decrease in the peak current could be due to the profusion of carbendazim ions onto the surface of AgNPs/GCE [1, 6]. Table 1 illustrates the comparison of

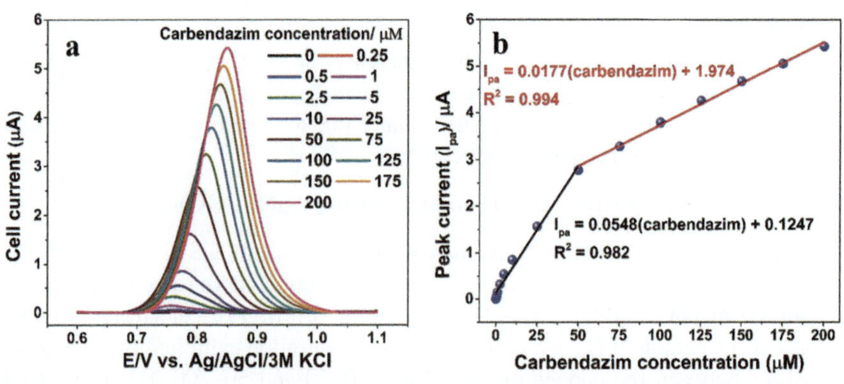

Fig. 7 **a** SWV of carbendazim sensing over AgNPs/GCE in the working range of 0–200 μM using 0.1 M PBS electrolyte at pH 7 and **b** I_{pa} versus carbendazim concentration curve

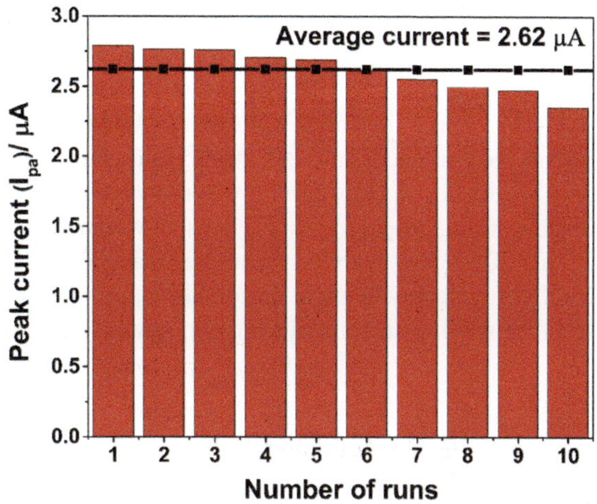

Fig. 8 Stability of AgNPs/GCE during the course of carbendazim sensing

Table 1 Performance of AgNPs/GCE in electrochemical sensing of carbendazim and its comparison with the recently reported sensors

Working electrodes	Technique	LOD (μM)	Sensitivity (μAμM^{-1} cm^{-2})	Linear range (μM)	Sources
AgNPs/GCE	**SWV**	**0.279**	**0.668**	**0–50**	**Present work**
		0.864	**0.216**	**50–200**	
NPG/GCE	DPV	0.24	0.485	3–120	[20]
LIG/Pt	SWV	0.67	–	1–40	[21]
GCE/RGO-Pt	DPV	2.96	–	25–115	[22]
TCP/CPE	ASV	0.3	–	0.5–10	[23]
MOB/CPE	DPV	0.12	–	0.29–6	[24]
Rhg/GCE	DPV	0.013	22.54	0.1–100	[25]

the sensing performance of carbendazim with the recently reported sensors. The fabricated sensing platform clearly outperformed several previously reported studies.

4 Conclusion

We have successfully synthesized bio-based AgNPs using natural analytes present in *Citrus Limon* extract by a microwave-assisted method for electrochemical sensing of carbendazim. The morphological analysis confirmed nearly spherical-shaped nanoparticles with an average size of 10.33 ± 2.3 nm with a normal distribution of particles. The BET surface area of mesoporous AgNPs was 392.5 m^2/g with the following type IV N_2 adsorption–desorption isotherm. However, TGA analysis reported a mass loss of ~53% up to 900 °C. The electron transfer resistance was decreased by 5.97-fold, and the effective surface area was increased by 64.63% with AgNPs immobilization. The proposed AgNPs/GCE exhibited a sensitivity of 0.668 $\mu A. \mu M^{-1}.cm^{-2}$, detection limit of 0.279 μM, and quantification limit of 0.931 μM within the carbendazim concentration ranging from 0 to 50 μM. Only a 3.53 and 15.67% decline in peak current was recorded up to the 5th and 10th cycle of carbendazim sensing, respectively. The present work clearly outperformed several previous studies. Therefore, AgNPs synthesized using a nature-inspired method are a potential catalyst for electrochemical sensor fabrication with high sensitivity and low detection limit for sensing and quantifying agricultural pesticides.

References

1. Bhan, C.: Bio-based hierarchical vertically aligned 2D ZnO nanostructures for ultra selective electrochemical sensing of p-Chloroaniline, Chem. Eng. J. 475 146122 (2023).
2. Liang, Y.: Fluorescence resonance energy transfer-based analytical methods for determination of food contaminants: A review, Food Control 152 109892 (2023).
3. Sharma, M.: Carbendazim toxicity in different cell lines and mammalian tissues, J. Biochem. Mol. Toxicol. 36 (2022).
4. Ravi, R.: Efficient recovery of TiO_2 and Pt-doped TiO_2 photocatalysts in wastewater treatment using a pilot-scale cross-flow ultrafiltration membrane system, Sep. Purif. Technol. 339 126710 (2024).
5. Ravi, R.: Enhanced adsorption capacity of designed bentonite and alginate beads for the effective removal of methylene blue, Appl. Clay Sci. 169 102–111 (2019).
6. Das, D.: Improved selectivity of electrochemical aniline sensing using one-dimensional silver nanorods with high aspect ratio synthesized by ascorbic acid assisted method, Anal. Chim. Acta 1310 342697 (2024).
7. Chowdhury, A.: A tunable bioinspired process of SnO_2 NPs synthesis for electrochemical CO_2-into-formate conversion, J. CO_2 Util. 66 102263 (2022).
8. Chowdhury, A.: A simple template-free bioinspired route of 1D Bi_2S_3 nanorods synthesis for electrochemical CO_2 reduction to formate, J. Ind. Eng. Chem. 127 138–148 (2023).
9. Dash, S.R.: Synthesis of highly structured spherical Ag@Pt core-shell NPs using bio-analytes for electrocatalytic Pb(II) sensing, Sensors Actuators B Chem. 314 128062 (2020).
10. Bhan, C.: Utilizing bioinspired AgNPs as an antibacterial agent to enhance ceramic membrane performance, J. Environ. Chem. Eng. 11 110283 (2023).
11. Kumar, A.: A novel eco-friendly inhibitor of chayote fruit extract for mild steel corrosion in 1 M HCl: Electrochemical, weight loss studies, and the effect of temperature, Sustain. Chem. Pharm. 36 101261 (2023).
12. Gawal, P.M.: Plant-Based Phytochemicals for Synthesis of Z-Scheme In_2O_3/CdS Heterostructures: DFT Analysis and Photocatalytic CO_2 Reduction to HCOOH and CO, Langmuir 40 13538–13549 (2024).
13. Ravi, R.: Pharmaceutically Active Compounds' (PhACs) Threat: An Environmental Prospective, in: Sustain. Environ., Springer Nature Singapore, Singapore. 51–66 (2023).
14. Ravi, R.: Bio-based Au Doping with Dominant Oxygen Vacancies and Ti^{3+} Defects on Photocatalytic Functionalities of TiO_2, Ind. Eng. Chem. Res. 62 20702–20715 (2023).
15. Dash, S.R.: Carbon Dots Derived from Waste Psidium Guajava Leaves for Electrocatalytic Sensing of Chlorpyrifos, Electroanalysis 34 1141–1149 (2022).
16. Mahiuddin, M.: Green synthesis of crystalline bismuth nanoparticles using lemon juice, RSC Adv. 11 26683–26686 (2021).

17. Sreelekha, E.: A Comparative Study on the Synthesis, Characterization, and Antioxidant Activity of Green and Chemically Synthesized Silver Nanoparticles, Bionanoscience 11 489–496 (2021).
18. Golder, A.K.: Synthesis of low-cost bentonite/Duranta erecta's fruit powder imbedded alginate beads and its application in surfactant removal, Environ. Sci. Pollut. Res. 28 58945–58957 (2021).
19. Kumar, A.: Corrosion inhibition of mild steel by Praecitrullus fistulosus (tinda fruit and peel) extracts, Sci. Total Environ. 929 172569 (2024).
20. Gao, X.: Electroactive nanoporous gold driven electrochemical sensor for the simultaneous detection of carbendazim and methyl parathion, Electrochim. Acta 310 78–85 (2019).
21. Wang, L.: Electrochemical Sensor Based on Laser-Induced Graphene for Carbendazim Detection in Water, Foods 12 2277 (2023).
22. Pham, T.S.H.: Graphene Nanocomposites Based Electrochemical Sensing Platform for Simultaneous Detection of Multi-drugs, Electroanalysis 34 435–444 (2022).
23. Ashrafi, A.M.: Trace Determination of Carbendazim Fungicide Using Adsorptive Stripping Voltammetry with a Carbon Paste Electrode Containing Tricresyl Phosphate, Int. J. Electrochem. Sci. 7 9717–9731 (2012).
24. Silva, J. de O.S.: Simultaneous determination of carbendazim and carbaryl pesticides in water bodies samples using a new voltammetric sensor based on Moringa oleifera biochar, Chemosphere 347 140707 (2024).
25. Yang, Y.: Electrochemical Sensors Based on Reduced Holey Graphene for Detection of Carbendazim, Phys. Status Solidi 219 (2022).

Vegetable Oils to Polyurethane Sheets: Green Chemistry Synthesis Pathways

Mangal Mangal, Ramesh N. Goswami, and Tamal Banerjee

Abstract Due to the growing concern over the depletion of the global crude oil stock and environmental issues, the replacement of petroleum-based polymers with environmentally benign alternatives derived from renewable resources has attracted great attention. In this study, a variety of bio-based polyurethanes were effectively made from polyols produced by the thiol-ene reaction of vegetable oils. Crude oils were hydroxylated with performic acid, created on-site by the reaction of hydrogen peroxide and formic acid to create the polyol. The resulting polyols were then used in the synthesis of polyurethanes. The characteristics of these bio-based polyurethanes can be determined by thermogravimetric analysis (TGA), differential scanning calorimetry (DSC), dynamic mechanical analysis (DMA), and tensile testing to understand their mechanical properties. Additionally, the mechanical and thermal properties of these plastics were enhanced by incorporating reinforcing materials such as cellulose nanofibers. These environmentally friendly polyurethanes, synthesized from various vegetable oils, exhibit potential for diverse applications, including packaging, construction, automotive, and biomedical fields, demonstrating their versatility and sustainability.

Keywords Bio-polyol · Renewable resources · Sustainable synthesis pathways · Bio-derived polyurethane

M. Mangal · T. Banerjee (✉)
Department of Chemical Engineering, Indian Institute of Technology Guwahati, Guwahati, Assam, India
e-mail: tamalb@iitg.ac.in

M. Mangal
e-mail: m.mangal@iitg.ac.in

R. N. Goswami
Center for Environment, Indian Institute of Technology Guwahati, Guwahati, Assam, India

U. S. Dixit et al. (eds.), *Current Progress in Engineering Sciences*,
https://doi.org/10.1007/978-981-96-8753-4_38

1 Introduction

In recent years, the quest for sustainable materials has gained considerable momentum due to the pressing need to address environmental concerns and reduce the world's reliance on non-renewable resources. Within this context, polyurethane, a versatile and widely used polymer, has garnered significant attention as a potential eco-friendly material. Traditional polyurethane production heavily relies on petroleum-based feedstocks, contributing to the depletion of fossil fuel reserves and generating harmful emissions during its synthesis [1, 2]. Researchers have explored alternative routes to overcome these challenges and foster a more sustainable future, leading to the development of bio-based polyurethanes. Among various renewable resources, vegetable oils have emerged as a promising candidate for the synthesis of bio-based polyurethane sheets [3]. Vegetable oils, derived from plants, exhibit remarkable properties that make them an attractive bioresource for polymeric materials. It is rich in unsaturated fatty acids, such as linoleic and linolenic acids, which possess reactive functional groups capable of polymerization and crosslinking [4, 5]. These features make vegetable oils an excellent precursor for producing polyurethane sheets through an eco-friendly, cost-effective process [6].

The synthesis of vegetable oil-based polyurethane sheets, apart from castor oil-based sheets, involves a multi-step process, starting with the extraction and purification of linseed oil. Subsequently, the linolenic acid present in the oil is chemically modified to introduce reactive functional groups [7]. These functionalized vegetable oil derivatives are then combined with different diisocyanates, such as toluene diisocyanate (TDI) or methylene diphenyl diisocyanate (MDI), to form a polyurethane prepolymer. Finally, the prepolymer is cured and crosslinked to produce robust and flexible polyurethane sheets. The chemical reaction between these components leads to the formation of a polymeric network with tunable mechanical and thermal properties, making it suitable for a wide range of applications, from coatings and adhesives to packaging materials and biomedical devices [8]. The general synthesis process of polyurethane from polyol and polyisocyanate is depicted in Fig. 1. In this section, we present a comprehensive investigation into the synthesis of vegetable oil-based polyurethane sheets, highlighting its environmentally friendly aspects, unique material properties, and potential applications.

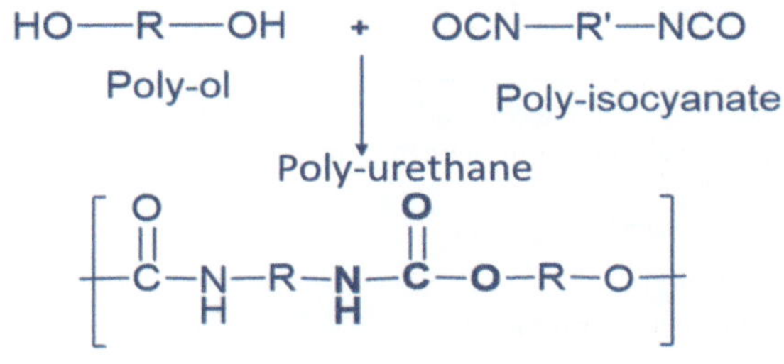

Fig. 1 General reaction mechanism for polyurethane formation

2 Selection Criteria of Polyols and Poly-isocyanates

Polyurethanes are versatile polymers synthesized from the reaction between polyols and diisocyanates [9]. In the pursuit of more sustainable materials, bio-based polyols, often derived from vegetable oils, have become a key focus. These polyols provide the "soft segments" in polyurethane, imparting flexibility, elasticity, and, in some cases, biodegradability [3, 10]. The corresponding "hard segments" are formed by diisocyanates, which contribute to the material's rigidity, mechanical strength, and durability. Polyols, the primary hydroxyl-containing components in polyurethane synthesis, are crucial for defining the polymer properties [11]. Vegetable oils, rich in unsaturated fatty acids, offer a renewable source of polyols that can be modified to suit different applications. The selection of polyols depends on the vegetable oil's chemical composition, which affects polyurethane properties. Soybean oil, rich in linoleic acid, is ideal for soft, flexible applications like foams and elastomers [12, 13]. Castor oil, with natural hydroxyl groups, simplifies synthesis and is used in rigid or semi-rigid products [14]. Linseed oil, high in linolenic acid, enhances flexibility and biodegradability [15]. The different polyols that can be employed in the subsequent synthesis are listed in Fig. 2. Each oil imparts specific characteristics, making the resulting polyurethanes suitable for applications like flexible foams, coatings, adhesives, and biodegradable materials.

Diisocyanates are the second essential component in polyurethane synthesis, determining the hard segments in the polymer and influencing the overall structure and properties [16, 17]. They are categorized into two main types: aromatic and aliphatic. Aromatic diisocyanates, such as toluene diisocyanate (TDI) and methylene diphenyl diisocyanate (MDI), are widely used in polyurethane production. They produce rigid and durable materials with excellent mechanical strength and thermal stability. TDI is available in two isomeric forms, 2, 4 and 2, 6, while MDI is offered as either pure MDI or polymeric MDI. Both types are preferred for applications requiring durability, such as foams, elastomers, and coatings [18]. Aliphatic diisocyanates, including hexamethylene diisocyanate (HDI) and isophorone diisocyanate

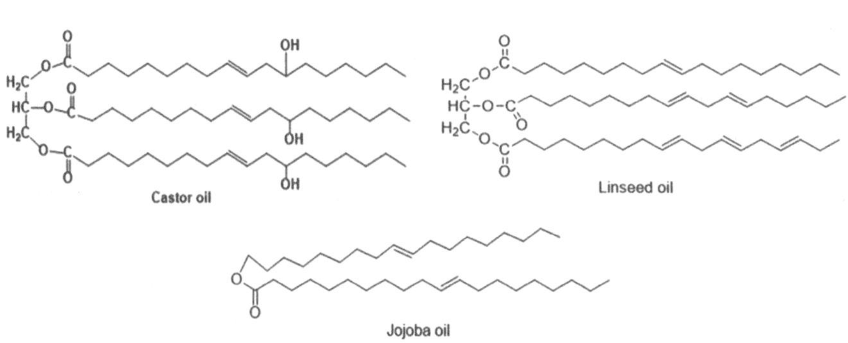

Fig. 2 Comprehensive overview of different types of bio-polyols utilized in the synthesis of sustainable polyurethane materials

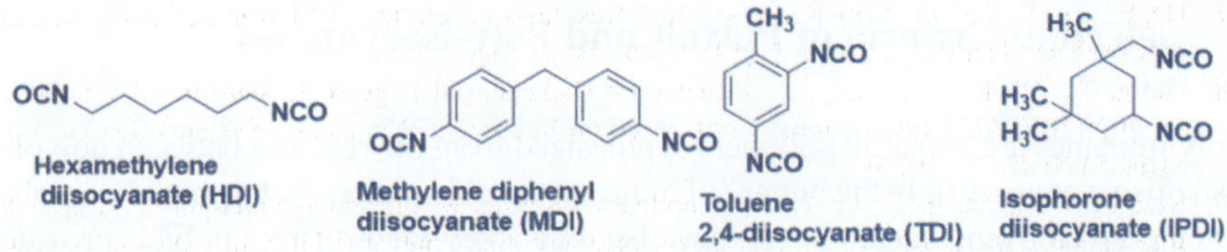

Fig. 3 Various kinds of aliphatic and aromatic diisocyanates

(IPDI), are used in specialized applications that demand UV resistance and light stability. These diisocyanates are crucial for coatings that need to maintain clarity and color under light exposure and for elastomers requiring specific flexibility and mechanical properties [19, 20]. A list of different aliphatic and aromatic diisocyanates is provided in Fig. 3. This combination allows for tailoring polyurethane properties to suit various applications, from flexible foams to durable coatings and elastomers.

The careful selection of polyols and diisocyanates allows for precise tuning of polyurethane properties [21]. Polyols with a higher hydroxyl functionality typically lead to polyurethanes with greater rigidity and higher thermal stability, whereas polyols with lower functionality result in more flexible and elastic materials [22]. Furthermore, the type of diisocyanate plays a pivotal role in determining the material's mechanical and chemical stability. Aromatic diisocyanates typically impart greater strength and durability, while aliphatic diisocyanates offer enhanced flexibility and UV stability [23]. This ability to tailor the properties of polyurethanes based on the chosen polyol and diisocyanate combination makes them highly versatile for applications ranging from flexible foams to durable coatings, elastomers, and biodegradable packaging materials [24, 25].

3 Different Routes of Polyurethane Synthesis

3.1 Direct Polyol Utilization

Direct polyol utilization involves the direct reaction of polyols, such as polyester or polyether polyols, with isocyanates to produce polyurethanes [26]. Polyols, which contain multiple hydroxyl (-OH) groups, react with the reactive -NCO groups of isocyanates to form urethane linkages (-NHCOO-). This process typically takes place in a solvent-free environment, making it more environmentally friendly [27]. The reaction begins by mixing polyols and isocyanates, often in the presence of catalysts such as tin-based compounds (e.g., dibutyltin dilaurate) or amine-based catalysts, which accelerate the reaction and ensure proper polymer formation shown in Fig. 4. Depending on the desired properties of the final polyurethane, various additives can be introduced during the reaction. For example, blowing agents create a cellular structure for foam production, while fillers like silica or calcium carbonate enhance

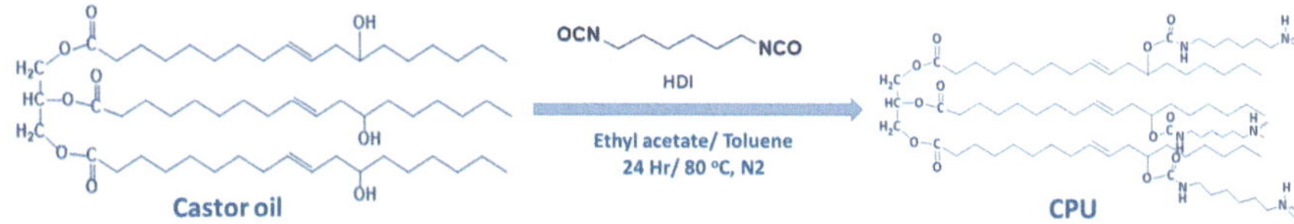

Fig. 4 Detailed mechanistic pathway for synthesis of polyurethane from castor oil-based polyols via polyaddition

the mechanical properties of the polymer [9]. A key advantage of this solvent-free process is its sustainability, as it reduces the use of volatile organic compounds (VOCs), lowering the environmental impact [28]. The polyols and isocyanates are typically mixed and reacted in processes such as mold casting, foaming, or coating, depending on the specific end-use product. The process allows for the customization of polyurethane properties by adjusting the type of polyol and isocyanate, the concentration of catalysts, and the inclusion of specific additives [29]. This flexibility results in versatile applications, ranging from flexible and rigid foams to durable elastomers and protective coatings, making direct polyol utilization a valuable method for industrial polyurethane production. In the synthesis of polyurethanes from castor oil, the process involves reacting the hydroxyl groups of castor oil with diisocyanates. This reaction occurs in an inert atmosphere at temperatures between 50 and 100 °C for 24 h, with toluene or ethyl acetate used as the solvent [30]. Notably, no catalyst is required for this reaction. Once the reaction is complete, the mixture is poured into a Teflon mold and left in an open atmosphere for 3–4 days to allow the solvent to evaporate. This results in the formation of solid polyurethane sheets [31]. This method highlights the potential of using renewable, bio-based materials like castor oil as a sustainable alternative to traditional petrochemical-based polyols in polyurethane production.

3.2 Epoxidation and Hydroxylation

To synthesize bio-polyurethanes from vegetable oils, a crucial initial step is the epoxidation of these oils. This process is essential for oils that do not naturally contain hydroxyl groups [32]. Unlike castor oil, which has inherent hydroxyl groups, other oils such as jojoba, soybean, linseed, palm, and sunflower oils require conversion of their carbon–carbon double bonds (C=C) into epoxy groups. This conversion is achieved by reacting the oils with an oxidizing agent, typically performic acid [33]. The epoxidation process involves introducing epoxy groups into the double bonds of the oil's fatty acid chains. Following this, a ring-opening reaction occurs where the epoxy groups are converted into hydroxyl groups. These newly formed hydroxyl groups are then available to react with diisocyanates, leading to the production of polyurethanes. This multi-step process transforms the initial oil into a functional material suitable for polyurethane synthesis. The same epoxidation and subsequent ring-opening reactions can be applied to various oils, including soybean and sunflower oils, not just linseed and jojoba oils. The epoxidation of the double bonds in these oils is typically carried out using peracids or hydrogen peroxide in the presence of an acid catalyst [34]. The resulting epoxidized oils are then polymerized to form polyurethanes, which exhibit desirable mechanical and thermal properties. This approach not only enhances the performance of the final products but also ensures their environmental friendliness and sustainability [35].

Jojoba oil has a mix of saturated and unsaturated fatty acids; the epoxidation involves reacting its C=C bonds with a peracid, often catalyzed by formic acid [33]. The epoxidized groups in jojoba oil then undergo ring-opening reactions in the presence of existing hydroxyl groups [36]. Figure 5 illustrates the process flows for the synthesis of jojoba oil-based polyurethane via epoxidation and ring-opening. The formic acid catalyst facilitates this process, resulting in the formation of tertiary hydrogen atoms next to the ether linkages [37]. As the number of epoxy groups decreases, polyols with varying distributions of hydroxyl groups are formed. These polyols significantly influence the structure of the resulting polyurethanes, affecting their crosslink density and mechanical properties.

Linseed oil, which is rich in polyunsaturated fatty acids, undergoes epoxidation where its C=C bonds are converted into epoxy groups [38]. This conversion is achieved by reacting the double bonds with a peracid, typically formed in situ from hydrogen peroxide and a carboxylic acid like formic acid [39]. The process flow for the synthesis of linseed oil-based polyurethane via epoxidation and ring-opening is shown in Fig. 6. Formic acid acts as a catalyst, promoting the formation of epoxy groups. After epoxidation, the epoxy groups in linseed oil can undergo ring-opening reactions, particularly in the presence of hydroxyl groups originally present in the oil [40]. The formic acid catalyst helps in facilitating the ring-opening reaction by forming tertiary hydrogen atoms adjacent to the newly created ether

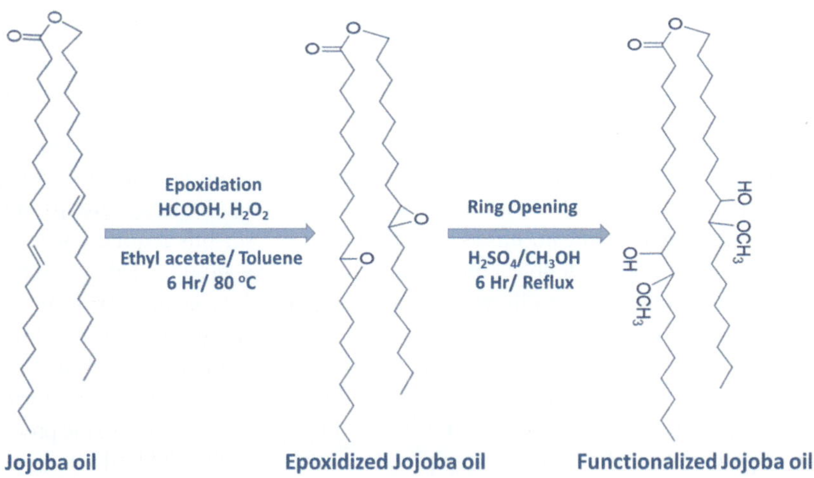

Fig. 5 Detailed mechanistic multi-step pathway for jojoba oil-based polyurethane via epoxidation and ring opening

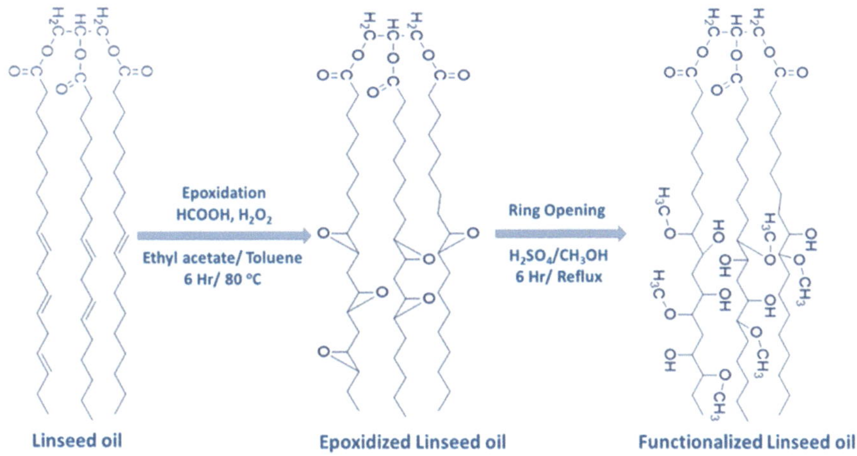

Fig. 6 Process flow for linseed oil-based polyurethane via epoxidation and ring opening

linkages. This reaction reduces the number of epoxy groups and results in the formation of polyols with evenly distributed hydroxyl groups. These polyols influence the crosslink density and mechanical properties of the final polyurethane [41, 42].

3.3 *Transesterification Processes*

Transesterification processes are pivotal in modifying ester structures to produce polyols that are suitable for polyurethane synthesis shown in Fig. 6. This method involves the exchange of ester groups between different esters or between an ester and an alcohol [43]. By altering the ester structure through transesterification, it is possible to create polyols with desired properties for use in polyurethane formulations. The transesterification process typically involves reacting an ester with an alcohol, or in some cases, another ester, in the presence of a catalyst. This reaction leads to the formation of new ester and alcohol products as shown in Fig. 7 [44, 45]. The choice of catalyst, often a base or acid, influences the rate and efficiency of the reaction. During the reaction, the original ester group is exchanged with the new alcohol group, resulting in the formation of a different ester product and a new alcohol [46]. These newly formed esters and alcohols can then be used as polyols in the synthesis of polyurethanes. The ability to tailor the properties of these polyols through transesterification makes this method highly valuable in producing polyurethanes with specific characteristics, such as improved mechanical strength or enhanced flexibility.

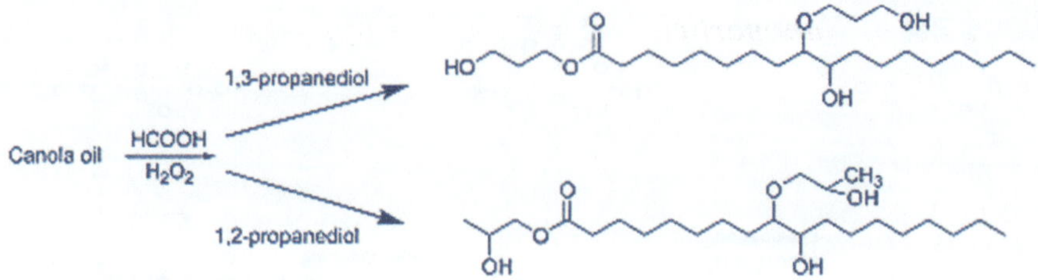

Fig. 7 Transesterification processes [47]

4 The Influence of Fillers on the Biodegradability and Mechanical Properties of Polyurethane Materials

4.1 Types of Fillers Used in Bio-Based Polyurethanes

In bio-based polyurethanes, the inclusion of various fillers is essential to enhance their mechanical properties and biodegradability. These fillers play a pivotal role in modifying the material's performance and environmental impact. Among the commonly used fillers are cellulose, cellulose nanofibers, lignin, chitosan, and starch [48]. Cellulose and cellulose nanofibers are particularly valued for their ability to significantly improve the strength and rigidity of the polyurethane matrix. Cellulose, being a fundamental structural component of plant cell walls, provides reinforcement by integrating into the polymer network. Cellulose nanofibers, with their high aspect ratio and nanometer-scale dimensions, offer enhanced mechanical reinforcement and increase the tensile strength of the final product [49, 50]. Lignin, a complex aromatic polymer found in plant cell walls, is another filler that contributes to the enhanced mechanical performance and thermal stability of bio-based polyurethanes. Its presence improves the rigidity of the material and helps in resisting degradation under thermal stress [51]. Chitosan, derived from the chitin found in crustacean exoskeletons, is noted for its biodegradability and antimicrobial properties. Its incorporation not only improves the biodegradability of the polyurethane but also provides an additional layer of protection against microbial attack, which can be beneficial for various applications. Starch, a polysaccharide obtained from a wide range of plant sources, is used to improve the mechanical properties and reduce the cost of bio-based polyurethanes. Starch enhances the tensile strength of the material and can be utilized to adjust the cost-effectiveness of the production process [52].

4.2 *Role of Nanoparticles and Fibers in Enhancing Properties*

Nanoparticles and fibers are integral to optimizing the properties of bio-based polyurethanes. Nanoparticles, such as nanoclays, metal oxides, or carbon nanotubes, are known for their ability to significantly enhance mechanical strength, thermal stability, and barrier properties [53, 54]. The high surface area-to-volume ratio of nanoparticles facilitates strong interactions with the polymer matrix, which leads to improvements in overall performance characteristics. For instance, nanoparticles can enhance the thermal resistance of the material, making it more suitable for high-temperature applications [55, 56]. Fibers, especially cellulose nanofibers, provide essential reinforcement to the polyurethane matrix. These fibers, characterized by their high aspect ratio, integrate well within the polymer structure, offering substantial improvements in tensile strength and impact resistance. The strong bonding between the fibers and the polymer matrix helps distribute stress more evenly across the material, thus enhancing its overall mechanical performance. The use of fibers also contributes to the structural integrity of the polyurethane, making it more durable and resistant to deformation under stress [57, 58].

4.3 *Biodegradability Improvement Through Filler Incorporation*

The incorporation of bio-based fillers is a crucial strategy for enhancing the biodegradability of polyurethanes. Fillers such as cellulose, lignin, chitosan, and starch are inherently biodegradable and contribute to reducing the environmental impact of the material [59]. Cellulose and its derivatives enhance the biodegradability of polyurethane due to their organic and plant-based origins, as seen clearly from its molecular structure shown in Fig. 8. The hydroxyl groups in cellulose's structure allow it to bond with isocyanates in polyurethane, strengthening the material while still facilitating its eventual breakdown. Additionally, the flexibility of the cellulose chains and their crosslinking ability help stabilize the polyurethane matrix, improving its mechanical properties. The presence of cellulose in the polyurethane matrix facilitates the breakdown of the material by microbial action, leading to a more sustainable end product [60]. Lignin and chitosan also play significant roles in improving the biodegradability of bio-based polyurethanes. Lignin's complex structure helps to slow down the degradation process, but when used in combination with other biodegradable fillers, it supports a gradual breakdown. Chitosan's natural biodegradability and its ability to support microbial growth further enhance the overall decomposition of the polyurethane material [61]. Starch, when incorporated into the formulation, accelerates the degradation process by providing a readily decomposable component. Its polysaccharide nature is conducive to microbial action, which helps in breaking down the polyurethane more efficiently [62].

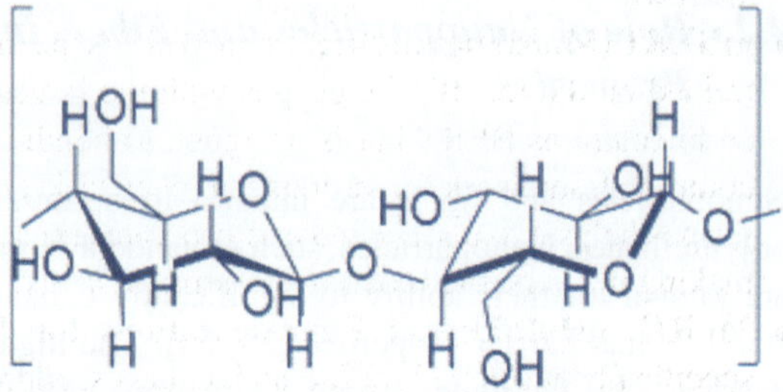

Fig. 8 Cellulose network used in polymer composites

5 Impact of -NCO/-OH Molar Ratio

The NCO/OH molar ratio is a crucial determinant of the physicochemical properties of polyurethane materials, as it reflects the balance between isocyanate (NCO) and hydroxyl (OH) groups during polymerization [63]. An important factor influencing the ultimate strength of polyurethane materials formed is the NCO/OH molar ratio. This ratio, which shows how well isocyanate (NCO) and hydroxyl (OH) groups are balanced, establishes how much crosslinking occurs inside the polymer network [64]. In general, a greater NCO/OH ratio leads to a higher crosslink density, which improves the polyurethane sheets' mechanical strength, hardness, and thermal stability. On the other hand, a lower ratio could result in a material that is weaker overall but more flexible and less brittle. In order to customize the material qualities to particular applications and make sure that the polyurethane sheets fulfill the required performance standards for resilience, durability, and load-bearing capacity, it is imperative to optimize this ratio [65]. This ratio is calculated using the weights of the cross-linker (chain extender), polyol, and filler materials, denoted as WL, WPolyol, and WF, respectively. The calculations for this ratio are detailed in Eqs. 1–4. The molar content of isocyanate groups in the cross-linker is represented by [NCO]L, while [OH] polyol and [OH] filler denote the molar content of hydroxyl groups in the polyol and filler materials [66]. By incorporating these values into the equation, the NCO/OH molar ratio is determined, guiding the formulation to achieve desired material properties. A well-balanced ratio ensures optimal crosslinking, affecting the material's mechanical strength, flexibility, and stability, while also influencing its thermal and chemical resistance [67, 68].

$$\frac{\mathrm{NCO}}{\mathrm{OH}} = \frac{\mathrm{W_L[NCO]_L}}{\mathrm{W_{polyol}[OH]_{polyol} + W_F[OH]_F}} \tag{1}$$

$$[\mathrm{NCO}]_\mathrm{L} = \frac{\text{number of NCO groups in linker}}{\text{Molecular weight of linker}}\ \mathrm{mmol/g} \tag{2}$$

$$[\mathrm{OH}]_\mathrm{CO} = \frac{\text{number of OH groups in polyol}}{\text{Molecular weight of polyol}}\ \mathrm{mmol/gm} \tag{3}$$

$$[OH]_F = \frac{\text{number of OH groups in filler}}{\text{Molecular weight of filler}}\ \text{mmol/gm} \tag{4}$$

6 Applications and Challenges

Bio-polyurethanes are gaining traction across various industries due to their eco-friendly attributes. In packaging, they offer a biodegradable alternative to conventional plastics, mitigating environmental impact. The automotive sector uses bio-polyurethanes in interior components and cushioning, balancing durability with a lower carbon footprint. In construction, they enhance thermal insulation in applications like sealants and coatings, promoting sustainability. Furniture makers replace traditional foams with bio-polyurethanes for similar comfort with less environmental harm. They also find use in sustainable footwear and textiles, providing eco-friendly options for soles, coatings, and finishes. Household products such as cushions and mattresses also benefit from their sustainability. Challenges include higher production costs compared to traditional polyurethanes, variability in performance due to raw material differences, and supply chain issues related to bio-based feedstocks. Meeting regulatory standards and convincing consumers to switch from traditional materials can be complex. Future developments will focus on improving bio-polyurethane performance, enhancing cost-efficiency through advanced feedstocks, and integrating recycling and circular economy principles. As production technologies advance and costs decrease, bio-polyurethanes are expected to penetrate new markets, driving broader adoption and sustainability.

7 Conclusion

Polyurethanes (PUs) are highly versatile materials used in a wide range of products, including paints, coatings, elastomers, insulation, fibers, and flexible foams. The diversity of polyols, diisocyanates, and other agents used in PU synthesis allows for customized materials tailored to specific applications. Traditionally, petroleum-based polyols were the primary source, but rising costs, energy demands, and environmental concerns have increased the need for sustainable alternatives. Bio-based materials, such as cellulose fibers, are now gaining attention for their potential to improve the environmental impact and performance of PUs. The selection of polyols and poly-isocyanates plays a key role in determining PU properties. Polyols sourced from petroleum or renewable bio-based sources are chosen based on factors like hydroxyl number, molecular weight, and functionality. The type of poly-isocyanates, such as MDI or TDI, affects the reactivity and final material characteristics. Different routes for PU synthesis, such as direct polyol utilization, epoxidation and hydroxylation, and transesterification processes, offer sustainable methods for creating bio-based polyols from natural oils and fats. These alternative synthesis routes help reduce reliance on fossil fuels and contribute to more eco-friendly PU formulations.

Apart from synthesis materials, fillers also contain a significant impact on both the biodegradability and mechanical properties of PUs. Natural fillers, such as cellulose and starch, enhance biodegradability, while inorganic fillers like clay and silica improve strength and stability. The NCO/OH molar ratio is essential in determining the final properties of PU. Achieving the right balance ensures optimal flexibility and durability without compromising biodegradability. Bio-based PUs, incorporating sustainable polyols, fillers, and innovative synthesis routes, are more environmentally friendly and economical than conventional polymers. However, challenges like cost, processing, and maintaining performance remain, though the potential for sustainable PU systems is significant.

Acknowledgements The author (Mangal) would like to acknowledge a PMRF (Prime Minister's Research Fellowship) fellowship from the Government of India to conduct the research. The authors would also like to acknowledge the financial support from the DST-TDT scheme via Project DST/TDT/WMT /Ag Waste/2021/16(G).

References

1. Gama, N.V., A. Ferreira, and A. Barros-Timmons, Polyurethane foams: Past, present, and future. Materials, 2018. 11(10): p. 1841.
2. Mori, R., Replacing all petroleum-based chemical products with natural biomass-based chemical products: a tutorial review. RSC Sustainability, 2023. 1(2): p. 179–212.
3. Campana, F., et al., Green polyurethanes from bio-based building blocks: recent advances and applications. Green Synthesis and Catalysis, 2024.
4. Tremblay-Parrado, K.-K., C. García-Astrain, and L. Avérous, Click chemistry for the synthesis of biobased polymers and networks derived from vegetable oils. Green Chemistry, 2021. 23(12): p. 4296–4327.
5. Zubair, M., et al., Recent advances in lipid derived bio-based materials for food packaging applications. Macromolecular Materials and Engineering, 2021. 306(7): p. 2000799.
6. Zafar, F., et al., A review on cleaner production of polymeric and nanocomposite coatings based on waterborne polyurethane dispersions from seed oils. Progress in Organic Coatings, 2019. 131: p. 259–275.
7. Maurya, A.K., F.M. de Souza, and R.K. Gupta, P olyurethane and Its Composites: Synthesis to Application, in Polyurethanes: Preparation, Properties, and Applications Volume 1: Fundamentals. 2023, ACS Publications. p. 1–20.
8. Wendels, S. and L. Avérous, Biobased polyurethanes for biomedical applications. Bioactive Materials, 2021. 6(4): p. 1083–1106.
9. Akindoyo, J.O., et al., Polyurethane types, synthesis and applications–a review. Rsc Advances, 2016. 6(115): p. 114453–114482.
10. Kaikade, D.S. and A.S. Sabnis, Polyurethane foams from vegetable oil-based polyols: a review. Polymer Bulletin, 2023. 80(3): p. 2239–2261.
11. Tran, M.H. and E.Y. Lee, Production of polyols and polyurethane from biomass: a review. Environmental Chemistry Letters, 2023. 21(4): p. 2199–2223.
12. Sawpan, M.A., Polyurethanes from vegetable oils and applications: a review. Journal of Polymer Research, 2018. 25(8): p. 184.

13. Phung Hai, T.A., et al., Renewable polyurethanes from sustainable biological precursors. Biomacromolecules, 2021. 22(5): p. 1770–1794.
14. Prasain, K. and D.H. Hua, Design, Synthesis, Property, and Application of Plant Oil Polymers. Renewable Polymers: Synthesis, Processing, and Technology, 2011: p. 23–67.
15. Chowdhury, A. and P. Anthony, A review on biodegradable polymeric materials derived from vegetable oils for diverse applications. Int J Sci Res, 2016. 5: p. 1786–1791.
16. Yilgör, I., E. Yilgör, and G.L. Wilkes, Critical parameters in designing segmented polyurethanes and their effect on morphology and properties: A comprehensive review. Polymer, 2015. 58: p. A1-A36.
17. Asensio, M., et al., The role of diisocyanate structure to modify properties of segmented polyurethanes. Materials, 2023. 16(4): p. 1633.
18. Schupp, T. and P.M. Plehiers, Absorption, distribution, metabolism, and excretion of methylene diphenyl diisocyanate and toluene diisocyanate: Many similarities and few differences. Toxicology and Industrial Health, 2022. 38(9): p. 500–528.
19. De Pariza, X.L., et al., Polythiourethanes: Synthesis, applications, and opportunities. Progress in Polymer Science, 2023. 145: p. 101735.
20. Dodge, J., Polyurethanes and polyureas. Synthetic Methods in Step-Growth Polymers, 2003: p. 197–263.
21. Wienen, D., et al., An overview of polyurethane biomaterials and their use in drug delivery. Journal of Controlled Release, 2023. 363: p. 376–388.
22. Furtwengler, P. and L. Avérous, Renewable polyols for advanced polyurethane foams from diverse biomass resources. Polymer Chemistry, 2018. 9(32): p. 4258–4287.
23. Xie, F., et al., Degradation and stabilization of polyurethane elastomers. Progress in Polymer Science, 2019. 90: p. 211–268.
24. Mouren, A. and L. Avérous, Sustainable cycloaliphatic polyurethanes: from synthesis to applications. Chemical Society Reviews, 2023. 52(1): p. 277–317.
25. Ghosh, A., et al., Polyurethane chemistry for the agricultural applications–Recent advancement and future prospects, in Polyurethanes: Preparation, Properties, and Applications Volume 3: Emerging Applications. 2023, ACS Publications. p. 1–36.
26. Ten Dam, J. and U. Hanefeld, Renewable chemicals: dehydroxylation of glycerol and polyols. ChemSusChem, 2011. 4(8): p. 1017–1034.
27. Yuen, A., et al., Room temperature synthesis of non-isocyanate polyurethanes (NIPUs) using highly reactive N-substituted 8-membered cyclic carbonates. Polymer Chemistry, 2016. 7(11): p. 2105–2111.
28. Wegener, G., et al., Trends in industrial catalysis in the polyurethane industry. Applied Catalysis A: General, 2001. 221(1–2): p. 303–335.
29. Menon, A., et al., Polyurethane (PU): Structure, properties, and applications, in Handbook of Thermosetting Foams, Aerogels, and Hydrogels. 2024, Elsevier. p. 67–92.
30. Gurunathan, T., S. Mohanty, and S.K. Nayak, Isocyanate terminated castor oil-based polyurethane prepolymer: Synthesis and characterization. Progress in Organic Coatings, 2015. 80: p. 39–48.
31. Stirna, U., et al., Structure and properties of the polyurethane and polyurethane foam synthesized from castor oil polyols. Journal of cellular plastics, 2012. 48(6): p. 476–488.
32. Olcay, H., E.D. Kocak, and Z. Yıldız, Sustainability in polyurethane synthesis and bio-based polyurethanes. Sustainability in the Textile and Apparel Industries: Sourcing Synthetic and Novel Alternative Raw Materials, 2020: p. 139–156.
33. Moser, B.R., et al., A review of fatty epoxide ring opening reactions: Chemistry, recent advances, and applications. Journal of the American Oil Chemists' Society, 2022. 99(10): p. 801–842.
34. Danov, S., et al., Recent advances in the field of selective epoxidation of vegetable oils and their derivatives: a review and perspective. Catalysis Science & Technology, 2017. 7(17): p. 3659–3675.
35. Ghasemlou, M., et al., Polyurethanes from seed oil-based polyols: A review of synthesis, mechanical and thermal properties. Industrial Crops and Products, 2019. 142: p. 111841.

36. Abdel-Hameed, H.S., et al., Chemical transformation of Jojoba oil and Soybean oil and study of their uses as bio-lubricants. Industrial Crops and Products, 2022. 187: p. 115256.
37. Nie, R., et al., Recent advances in catalytic transfer hydrogenation with formic acid over heterogeneous transition metal catalysts. ACS Catalysis, 2021. 11(3): p. 1071–1095.
38. Mahendran, A.R., et al. Synthesis and characterization of a bio-based resin from linseed oil. in Macromolecular Symposia. 2012. Wiley Online Library.
39. Campanella, A., C. Fontanini, and M.A. Baltanas, High yield epoxidation of fatty acid methyl esters with performic acid generated in situ. Chemical Engineering Journal, 2008. 144(3): p. 466–475.
40. Omonov, T.S. and J.M. Curtis, Plant oil-based epoxy intermediates for polymers. Bio-based plant oil polymers and composites, 2016: p. 99–125.
41. Desai, S., et al., Effect of polyols and diisocyanates on thermo-mechanical and morphological properties of polyurethanes. European Polymer Journal, 2000. 36(4): p. 711–725.
42. Maji, P.K. and A.K. Bhowmick, Influence of number of functional groups of hyperbranched polyol on cure kinetics and physical properties of polyurethanes. Journal of Polymer Science Part A: Polymer Chemistry, 2009. 47(3): p. 731–745.
43. Desroches, M., et al., From vegetable oils to polyurethanes: synthetic routes to polyols and main industrial products. Polymer reviews, 2012. 52(1): p. 38–79.
44. Khan, Z., et al., Current developments in esterification reaction: A review on process and parameters. Journal of Industrial and Engineering Chemistry, 2021. 103: p. 80–101.
45. Leung, D.Y., X. Wu, and M.K.H. Leung, A review on biodiesel production using catalyzed transesterification. Applied energy, 2010. 87(4): p. 1083–1095.
46. Bondioli, P., The preparation of fatty acid esters by means of catalytic reactions. Topics in Catalysis, 2004. 27(1): p. 77–82.
47. Kong, X., G. Liu, and J.M. Curtis, Novel polyurethane produced from canola oil based poly (ether ester) polyols: Synthesis, characterization and properties. European Polymer Journal, 2012. 48(12): p. 2097–2106.
48. Członka, S., et al., Bio-based polyurethane composite foams with improved mechanical, thermal, and antibacterial properties. Materials, 2020. 13(5): p. 1108.
49. Saba, N., et al., Mechanical, morphological and structural properties of cellulose nanofibers reinforced epoxy composites. International journal of biological macromolecules, 2017. 97: p. 190–200.
50. Shaghaleh, H., X. Xu, and S. Wang, Current progress in production of biopolymeric materials based on cellulose, cellulose nanofibers, and cellulose derivatives. RSC advances, 2018. 8(2): p. 825–842.
51. Quinsaat, J.E.Q., et al., Preparation of mechanically robust bio-based polyurethane foams using depolymerized native lignin. ACS Applied Polymer Materials, 2021. 3(11): p. 5845–5856.
52. Zia, F., et al., Starch based polyurethanes: A critical review updating recent literature. Carbohydrate polymers, 2015. 134: p. 784–798.
53. Arik, N., N. Horzum, and Y.B. Truong, Development and characterizations of engineered electrospun bio-based polyurethane containing essential oils. Membranes, 2022. 12(2): p. 209.
54. Hassan, T., et al., Functional nanocomposites and their potential applications: A review. Journal of Polymer Research, 2021. 28(2): p. 36.
55. Zou, H., S. Wu, and J. Shen, Polymer/silica nanocomposites: preparation, characterization, properties, and applications. Chemical reviews, 2008. 108(9): p. 3893–3957.
56. Zhao, X., et al., Polymer-supported nanocomposites for environmental application: a review. Chemical engineering journal, 2011. 170(2–3): p. 381–394.
57. Nurazzi, N., et al., A review on mechanical performance of hybrid natural fiber polymer composites for structural applications. Polymers, 2021. 13(13): p. 2170.
58. Chattopadhyay, D.K. and K. Raju, Structural engineering of polyurethane coatings for high performance applications. Progress in polymer science, 2007. 32(3): p. 352–418.
59. Gutiérrez, T.J. and V.A. Alvarez, Cellulosic materials as natural fillers in starch-containing matrix-based films: A review. Polymer Bulletin, 2017. 74: p. 2401–2430.

60. Pathak, V.M., Review on the current status of polymer degradation: a microbial approach. Bioresources and Bioprocessing, 2017. 4(1): p. 1–31.
61. Lizundia, E., et al., Multifunctional lignin-based nanocomposites and nanohybrids. Green Chemistry, 2021. 23(18): p. 6698–6760.
62. Jiang, T., et al., Starch-based biodegradable materials: Challenges and opportunities. Advanced Industrial and Engineering Polymer Research, 2020. 3(1): p. 8–18.
63. Prisacariu, C., Polyurethane elastomers: from morphology to mechanical aspects. 2011: Springer Science & Business Media.
64. Caraculacu, A. and S. Coseri, Isocyanates in polyaddition processes. Structure and reaction mechanisms. Progress in Polymer Science, 2001. 26(5): p. 799–851.
65. Somarathna, H., et al., The use of polyurethane for structural and infrastructural engineering applications: A state-of-the-art review. Construction and Building Materials, 2018. 190: p. 995–1014.
66. Tilly, J.C., et al., Spectroscopic and rheological cross-analysis of polyester polyol cure behavior: Role of polyester secondary hydroxyl content. ACS omega, 2019. 4(1): p. 932–939.
67. Shao, J., et al., Robust polyurethane materials with synergistic ionic and metal-ligand crosslinking double networks for specific Cu-ion removal. Chemical Engineering Journal, 2023. 474: p. 145719.
68. Zhang, K., et al., Structurally dynamic hydrogels for biomedical applications: pursuing a fine balance between macroscopic stability and microscopic dynamics. Chemical Reviews, 2021. 121(18): p. 11149–11193.

Carbon Quantum Dots for Photocatalytic Application: A Mini Review

Pramod Madhukar Gawal and Animes Kumar Golder

Abstract Carbon quantum dots (CQDs) are emerging as a promising class of nanomaterials with unique optical, electronic, and chemical properties, making them highly suitable for photocatalytic applications. This mini-review provides a thorough overview of CQD synthesis methods, functionalization techniques, and their photocatalytic uses in environmental remediation and energy conversion. CQDs are particularly effective due to their high surface area, tunable photoluminescence, and efficient charge separation. They show potential in degrading organic pollutants, enhancing hydrogen production via photocatalytic water splitting, and CO_2 reduction. However, challenges such as stability, scalability, and cost-effective synthesis remain. The review discusses these issues and suggests future research directions to optimize CQD performance and scalability, aiming to address global environmental and energy challenges with sustainable solutions.

Keyword Carbon quantum dots · Photocatalysis · Environmental application · Energy conversion

1 Introduction

Photocatalysis, a process that harnesses light energy to drive chemical reactions, holds significant promise for addressing pressing environmental issues and energy shortage[1]. Traditional photocatalysts like TiO_2 and ZnO have been widely studied and utilized. However, these materials face limitations such as poor visible light absorption and high recombination rates of photogenerated electron (e^-)/hole(h^+)

P. M. Gawal · A. K. Golder (✉)
Department of Chemical Engineering, Indian Institute of Technology Guwahati, Guwahati, Assam, India
e-mail: animes@iitg.ac.in

P. M. Gawal
e-mail: g.pramod@iitg.ac.in

U. S. Dixit et al. (eds.), *Current Progress in Engineering Sciences*,
https://doi.org/10.1007/978-981-96-8753-4_39

pairs, which reduce their overall efficiency [2, 3]. Consequently, there is a continuous search for new materials that can overcome these drawbacks and enhance photocatalytic performance.

CQDs have emerged as a new class of nanomaterials with unique and advantageous properties, making them particularly suitable for photocatalytic applications. CQDs are typically less than 10 nm in size and consist of a graphitic core with various functional groups on the surface [4]. This structure endows them with remarkable photostability, tunable photoluminescence, low toxicity, and ease of synthesis. These characteristics position CQDs as a promising alternative to traditional photocatalysts [5].

The synthesis of CQDs can be generally classified into bottom-up and top-down approaches. Bottom-up methods involve assembling small molecular precursors into CQDs, while Top-down methods involve fragmenting larger carbon structures to produce CQDs. Each synthesis method has its advantages and limitations, with the selection typically based on the desired properties and intended applications of the CQDs [6, 7].

In addition to their favorable structural, optical, and electronic properties, CQDs exhibit excellent photocatalytic activity, making them suitable for various environmental and energy-related applications [5]. They have been explored for the degradation of organic pollutants, photocatalytic water splitting, CO_2 reduction, and improving the efficiency of photocatalytic fuel cells (Fig. 1). The ability of CQDs to act as both electron donors and acceptors facilitates efficient charge separation and transport, which is crucial for enhancing photocatalytic performance [8–11].

Despite their promising properties, CQD-based photocatalysts face several challenges that need to be addressed for practical applications. These challenges include ensuring long-term stability, developing cost-effective and scalable synthesis methods, and enhancing photocatalytic efficiency through a better understanding and optimization of CQD properties [12, 13].

This mini-review offers a detailed overview of the synthesis methods, functionalization techniques, and photocatalytic applications of CQDs. It also delves into the mechanisms underlying their photocatalytic activity, current challenges, and future research directions. By addressing these aspects, this review aims to highlight the potential of CQDs as versatile materials for environmental remediation and energy conversion and to spur further research and development in this exciting field.

2 Properties of Carbon Quantum Dots

CQDs have attracted considerable interest as a novel class of nanomaterials due to their unique and advantageous properties. These properties make CQDs especially suitable for photocatalytic applications, offering promising solutions in environmental remediation and energy conversion. This section explores the structural, optical, and electronic properties of CQDs that enhance their effectiveness in these applications.

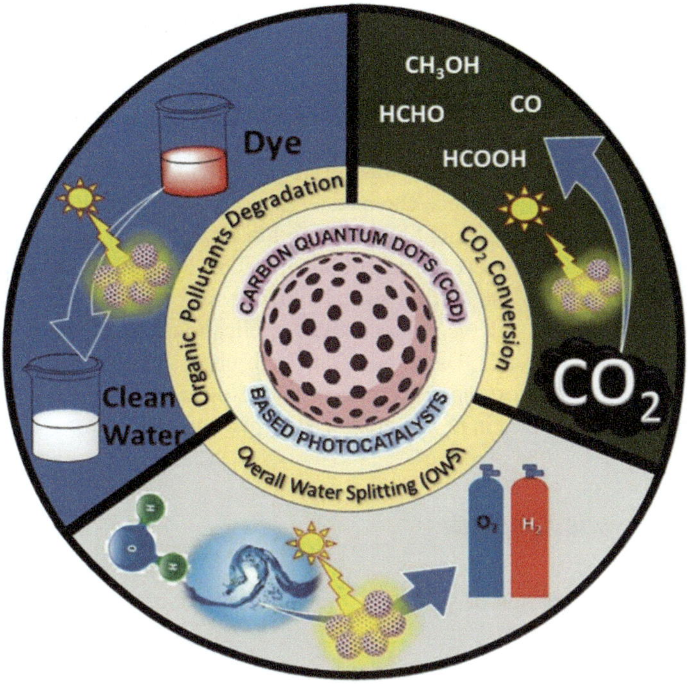

Fig. 1 Photocatalytic application of CQD-based photocatalysts in organic pollutant degradation, CO_2 conversion, and water splitting

2.1 *Structural Properties*

Structurally, CQDs are nanoscale particles, typically smaller than 10 nm in diameter, featuring a graphitic core with various functional groups on their surface. This small size and graphitic structure confer several benefits [4, 5]:

- **Surface Area**: The high surface area of CQDs facilitates enhanced interaction with reactants during photocatalytic processes, which is crucial for improving catalytic efficiency.
- **Surface Functionalization**: The diverse functional groups on the surface of CQDs enable facile functionalization. This capability enables the tuning of their chemical properties and interactions with different reactants, which can be exploited to enhance specific catalytic processes.
- **Shape and Size Control**: CQDs can be synthesized in various shapes and sizes, which can be controlled through different synthesis methods. This control over physical properties is essential for optimizing their photocatalytic performance for specific applications.

2.2 Optical Properties

A key characteristic of CQDs is their optical properties, notably strong and tunable photoluminescence [5, 6]:

- **Photoluminescence**: CQDs exhibit strong photoluminescence, which is dependent on their size and surface chemistry. This property is particularly useful for applications requiring efficient light absorption and emission.
- **Tunable Emission Spectra**: The emission spectra of CQDs can be tailored by modifying their size and surface states, allowing for customization according to the requirements of specific photocatalytic applications.
- **High Photostability**: CQDs demonstrate high photostability, maintaining their luminescent properties under prolonged light exposure. This stability is critical for their long-term use in photocatalytic systems.

2.3 Electronic Properties

The electronic properties of CQDs are another key factor in their suitability for photocatalytic applications[5]:

- **Conductivity**: CQDs possess excellent electrical conductivity, which facilitates efficient charge separation and transfer during photocatalytic processes. This property is essential for reducing recombination rates of e^-/h^+ pairs, thereby enhancing photocatalytic efficiency.
- **Tunable Bandgap**: The bandgap of CQDs can be tuned by varying their size and surface states. This tunability allows CQDs to absorb a broad spectrum of light, including both UV and visible regions, which is advantageous for utilizing solar energy effectively.
- **Charge Carrier Dynamics**: CQDs can function as both electron donors and acceptors, facilitating efficient charge separation and transport. This dual capability enhances their catalytic efficiency, as it facilitates the involvement of more reactive sites in photocatalytic reactions.

2.4 Surface Chemistry

The surface chemistry of CQDs is crucial in determining their photocatalytic performance [4, 5]:

- **Functional Groups**: The surface functional groups of CQDs can engage in redox reactions, enhancing their reactivity. These groups can also be modified to improve the interaction with specific reactants, thus tailoring the photocatalytic activity for targeted applications.

- **Surface Passivation**: Surface passivation, through the introduction of passivating agents or doping with other elements, can significantly enhance the quantum yield of CQDs. This improvement makes them more efficient in capturing and converting light energy.

2.5 Comparison with Traditional Photocatalysts

When compared to traditional photocatalysts like TiO_2 and ZnO, CQDs offer several advantages [4–6]:

- **Visible Light Absorption**: Unlike TiO_2 and ZnO, which primarily absorb UV light, CQDs can absorb a broad spectrum of light, including visible light. This property makes them more effective in utilizing sunlight, which contains a significant portion of visible light.
- **Lower Recombination Rates**: The efficient charge separation and transfer capabilities of CQDs result in lower recombination rates of e^-/h^+ pairs, enhancing photocatalytic efficiency.
- **Environmental and Economic Benefits**: CQDs can be synthesized from inexpensive and abundant carbon sources, making them cost-effective and environmentally friendly alternatives to traditional photocatalysts.

3 Synthesis of Carbon Quantum Dots

CQDs synthesis can be broadly classified into bottom-up and top-down approaches (Fig. 2). Each method presents specific advantages, limitations, and potential for scalability, depending on the desired properties and applications of the CQDs [6, 7, 12].

3.1 Top-Down Methods

Top-down methods entail fragmenting larger carbon structures to produce CQDs. Common techniques include [6, 7, 12].

Laser Ablation:

- **Process**: Laser irradiation of a carbon source, such as graphite, to generate CQDs.
- **Advantages**: Provides precise control over the size and properties of CQDs.
- **Limitations**: Requires sophisticated equipment and can be costly.

Electrochemical Oxidation:

- **Process**: Oxidizing graphite rods in an electrochemical cell to produce CQDs.

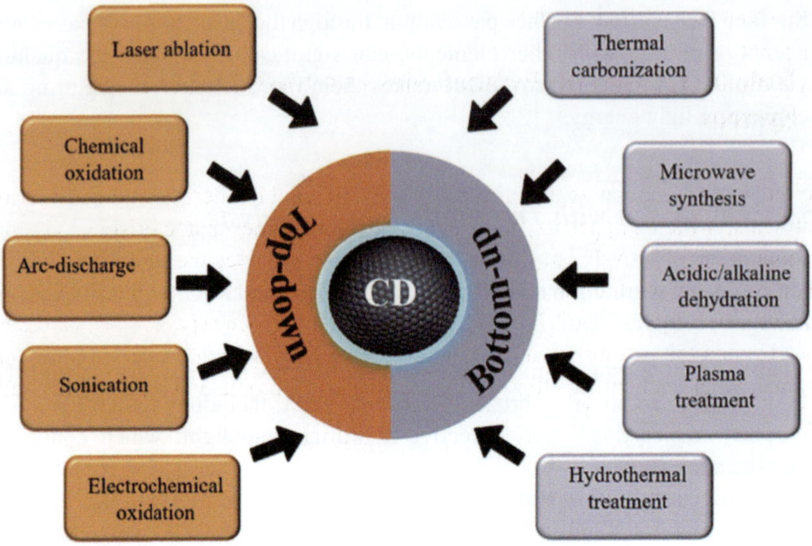

Fig. 2 Various synthesis methods for carbon quantum dots (CQDs)

- **Advantages**: Allows tuning of CQD properties by adjusting reaction parameters.
- **Limitations**: It may involve complex setup and control of reaction conditions.

Arc Discharge:

- **Process**: Generating CQDs through arc discharge between graphite electrodes in an inert atmosphere.
- **Advantages**: Suitable for producing large quantities of CQDs.
- **Limitations**: Requires high-energy input and specialized equipment.

3.2 Bottom-Up Methods

Bottom-up methods involve assembling small molecular precursors into CQDs. Techniques include [6, 7, 12].

Hydrothermal/Solvothermal Synthesis:

- **Process**: Subjecting organic precursors to high temperature and pressure to form CQDs.
- **Advantages**: Allows control over the size and functionalization of CQDs.
- **Limitations**: Requires high-pressure reactors and precise temperature control.

Microwave-Assisted Synthesis:

- **Process**: Using microwave irradiation to rapidly heat carbon precursors, producing CQDs.

- **Advantages**: Efficient and rapid synthesis method.
- **Limitations**: Requires microwave reactors and careful control of reaction parameters.

Combustion/Synthesis:

- **Process**: Controlled combustion of organic materials to yield CQDs.
- **Advantages**: Simple and scalable method.
- **Limitations**: It may produce impurities that require additional purification steps.

3.3 Comparison of Synthesis Techniques

Each synthesis method offers distinct benefits and challenges. Top-down methods provide precise control over the size and morphology of CQDs but often require sophisticated equipment and conditions. In contrast, bottom-up methods can achieve higher yields and better functionalization, making them more suitable for large-scale production. The synthesis of CQDs presents a versatile and promising approach for enhancing photocatalytic applications. Both top-down and bottom-up methods offer unique advantages and potential for optimization (Table 1). By addressing the challenges and exploring innovative synthesis techniques, the full potential of CQDs in environmental remediation and energy conversion can be realized. This comprehensive overview of CQD synthesis methods highlights the ongoing advancements and future directions in this exciting field, encouraging further research and development. These methods can produce high yields and allow for easy functionalization, making them more suitable for large-scale production.

Table 1 Comparison of the synthesis method of CQDs

Synthesis method	Advantages	Limitations
Laser ablation	Precise control over size and properties	Costly and requires sophisticated equipment
Electrochemical oxidation	Tunable properties by adjusting parameters	Complex setup and reaction control
Arc discharge	Suitable for large quantities	High-energy input and specialized equipment
Hydrothermal/Solvothermal	Control over size and functionalization	Requires high-pressure reactors
Microwave-assisted	Efficient and rapid synthesis	Requires microwave reactors
Combustion	Simple and scalable	Potential impurities needing purification

4 Photocatalytic Applications of CQDs

CQDs have been extensively studied for their potential in various photocatalytic applications, with a particular focus on environmental remediation and energy conversion. This section delves into the specific roles CQDs play in these domains, highlighting their performance, mechanisms, and advantages.

4.1 Environmental Remediation

Environmental pollution, particularly from organic contaminants and water pollutants, poses a significant threat to ecosystems and human health. Photocatalytic degradation with CQDs presents a promising solution, leveraging their high surface area, efficient charge separation, and strong light absorption [7].

4.1.1 Degradation of Organic Pollutants

CQDs have shown remarkable efficiency in degrading various organic pollutants, including dyes, pharmaceuticals, and pesticides, under visible light irradiation. The large surface area of CQDs offers abundant active sites for pollutant adsorption and degradation. Additionally, the effective separation of e^-/h^+ pairs in CQDs reduces recombination losses, thus enhancing photocatalytic activity [9, 13–15]

- **Dyes**: CQDs can effectively degrade industrial dyes such as methylene blue, rhodamine B, and methyl orange, which are commonly used in textile industries and are notorious for their persistence in the environment. Under visible light, CQDs produce reactive oxygen species such as superoxide anions ($O_2\bullet^-$) and hydroxyl radicals (•OH), which attack the dye molecules, breaking them down into less harmful substances.
- **Pharmaceuticals**: The presence of pharmaceutical residues in water bodies is a growing concern. CQDs can photocatalytically degrade various pharmaceuticals, including antibiotics and analgesics, reducing their environmental impact. The process involves the adsorption of the pharmaceutical molecules onto the CQD surface, followed by their degradation through ROS generated during light irradiation.
- **Pesticides**: Pesticides, due to their extensive use in agriculture, often contaminate water sources. CQDs can degrade pesticides like atrazine, dichlorvos, and parathion through a similar mechanism involving ROS, ensuring that these harmful chemicals are broken down into less toxic by-products.

4.1.2 Water Splitting

Photocatalytic water splitting is a promising approach for producing hydrogen, a clean and renewable energy source. CQDs enhance the efficiency of this process due to their excellent light absorption and charge carrier dynamics [11].

- **Hydrogen Production**: CQD-based photocatalysts exhibit significant improvements in hydrogen production rates compared to traditional photocatalysts. The strong light-harvesting capability of CQDs allows them to absorb a broad spectrum of sunlight, including visible light, which is essential for optimizing solar energy utilization. Upon light absorption, CQDs generate e^-/h^+ pairs; the e^- reduces water molecules to produce hydrogen, while h^+ oxidizes water to form oxygen.
- **Charge Carrier Dynamics**: The nanoscale size and large surface area of CQDs enable efficient separation and transport of photogenerated e^-/h^+ pairs. This minimizes the recombination of e^-/h^+ pairs, thereby enhancing the overall photocatalytic efficiency. Moreover, surface functionalization and doping of CQDs can further improve their charge transfer properties and catalytic activity.

4.2 Energy Conversion

In addition to environmental remediation, CQDs are highly effective in energy conversion applications, particularly in reducing CO_2 and improving the efficiency of photocatalytic fuel cells [10].

4.2.1 CO_2 Reduction

The conversion of carbon dioxide (CO_2) into value-added chemicals is a crucial step toward mitigating climate change and developing sustainable energy sources. CQD-based catalysts are particularly effective in this application due to their tunable electronic properties and strong light-harvesting capabilities [8, 16]

- **Value-Added Chemicals**: CQDs can photocatalytically reduce CO_2 into valuable chemicals such as methanol, ethanol, formic acid, carbon monoxide, and methane. The process involves the activation of CO_2 molecules adsorbed on the CQD surface, followed by their reduction through photogenerated electrons. The tunable bandgap of CQDs allows for the optimization of light absorption and charge transfer properties, making them highly efficient for CO_2 reduction.
- **Light-Harvesting Capabilities**: CQDs possess excellent light absorption properties, enabling them to utilize a wide range of the solar spectrum. This characteristic is crucial for the efficient conversion of CO_2 under natural sunlight. Furthermore,

the ability to engineer the surface chemistry of CQDs through doping and functionalization can enhance their catalytic performance and selectivity for specific reduction products.

4.2.2 Photocatalytic Fuel Cells

Photocatalytic fuel cells (PFCs) are an innovative approach to converting light energy into electrical energy. CQDs can significantly improve the efficiency and performance of PFCs by enhancing light absorption and charge transfer processes [17].

- **Efficiency Enhancement**: The incorporation of CQDs in PFCs leads to higher power output and better overall performance. CQDs enhance the light absorption capacity of the photocatalysts, ensuring more efficient utilization of solar energy. Additionally, the excellent charge separation and transport properties of CQDs reduce electron–hole recombination, resulting in more effective photogenerated charge carrier utilization.
- **Charge Transfer**: CQDs act as both electron donors and acceptors, facilitating efficient charge transfer within the PFC system. This dual functionality is critical for maintaining high photocatalytic activity and achieving superior power conversion efficiency. Surface modifications of CQDs can further improve their interaction with other components of the PFC, optimizing the overall device performance.

Table 2 summarizes the recent studies on photocatalytic applications using CQDs or CQDs-based catalysts.

5 Mechanism of Photocatalytic Activity

The photocatalytic performance of CQDs is primarily driven by their ability to absorb light and generate e^-/h^+ pairs. Under light irradiation, CQDs absorb photons, exciting electrons from the valence band to the conduction band. This process creates e^-/h^+ pairs that participate in redox reactions, facilitating various photocatalytic applications. The efficiency of these processes is influenced by several key factors, including light absorption, charge carrier dynamics, and surface reactions.

5.1 Light Absorption

Light absorption plays a crucial role in governing the photocatalytic efficiency of CQDs. One of the most advantageous properties of CQDs is their tunable absorption

Table 2 Recent studies on photocatalytic application by CQDs or CQDs-based catalysts

Catalyst	Synthesis method	Role of CQDs	Application	Source
CQDs	Hydrothermal treatment	–	Methylene blue degradation	[18]
Nitrogen-doped carbon dots	Hydrothermal treatment	–	Methylene blue dye degradation	[19]
CQDs/CdS	Simple chemical deposition	Electron reservoir	Hydrogen production	[20]
CQDs/TiO_2	Hydrothermal	Electron reservoir, photosensitizer	Hydrogen production	[21]
CQDs/CdS	Simple deposition method	Electron reservoir	CO_2 reduction	[1]
CQDs/Cu_2O	Simple wet impregnation method	Electron reservoir	CO_2 reduction	[22]
TiO_2/CQDs	–	Electron reservoir	Photocatalytic fuel cell	[23]
$BiVO_4$/CQDs	–	Electron reservoir	Photocatalytic fuel cell	[24]

capability, which allows them to absorb a broad spectrum of light, including ultraviolet (UV) and visible regions. This broad absorption spectrum is crucial for maximizing the utilization of solar energy, making CQDs highly effective in photocatalytic applications [4–6].

- **Tunable Absorption Properties**: The ability to tune the absorption properties of CQDs is achieved through the control of their size, shape, and surface chemistry. Smaller CQDs tend to have higher energy band gaps, enabling them to absorb higher energy UV light. In contrast, larger CQDs can absorb lower energy visible light. This tunability ensures that CQDs can be optimized for specific photocatalytic applications depending on the light source available.
- **Broad Light Spectrum Absorption**: The capability of CQDs to absorb both UV and visible light significantly enhances their photocatalytic performance. In natural sunlight, which contains a mixture of UV, visible, and infrared light, CQDs can effectively harness a larger portion of the solar spectrum compared to traditional photocatalysts, which often have limited absorption ranges.

5.2 *Charge Carrier Dynamics*

Efficient separation and transport of photogenerated e^-/h^+ pairs are vital for achieving high photocatalytic activity. CQDs exhibit several characteristics that facilitate these processes, thereby enhancing their overall photocatalytic performance [10, 11, 17].

- **Efficient Charge Separation**: Upon light absorption, CQDs generate $^-/h^+$ pairs. The nanoscale size and large surface area of CQDs promote efficient separation of charge carriers. The large surface area provides numerous active sites for the reactant adsorption and the formation of electron–hole pairs, while the small size reduces the distance that the carriers need to travel, minimizing recombination losses.
- **Excellent Electron Mobility**: CQDs possess excellent electron mobility, which is crucial for the efficient transport of photogenerated electrons. This property ensures that electrons can quickly move to the reaction sites where they participate in redox reactions. The high mobility of electrons in CQDs is attributed to their graphitic core structure, which provides a conductive pathway for electron transport.
- **Charge Carrier Lifetime**: The lifetime of photogenerated charge carriers is another important factor. Longer carrier lifetimes allow for more efficient participation in photocatalytic reactions. Surface passivation and doping of CQDs can further enhance the charge carrier lifetime by reducing recombination rates and introducing trap states that facilitate charge separation.

5.3 *Surface Reactions*

Surface reactions are crucial in determining the photocatalytic performance of CQDs. The active sites on the surfaces of CQDs are crucial for the adsorption of reactants and the initiation of redox reactions. Functionalization and doping of CQDs can introduce additional active sites, enhancing their reactivity and selectivity.

- **Active Sites**: The surface of CQDs is abundant in functional groups, including carboxyl, hydroxyl, and carbonyl groups. These functional groups serve as active sites for the reactant adsorption and the generation of ROS, which are essential for photocatalytic degradation processes.
- **Functionalization**: Functionalization involves modifying the surface of CQDs with various chemical groups to enhance their photocatalytic properties. For example, doping CQDs with heteroatoms such as nitrogen, sulfur, or phosphorus can introduce additional active sites and alter the electronic properties of CQDs, improving their light absorption and charge carrier dynamics.
- **Enhanced Reactivity and Selectivity**: By tailoring the surface chemistry of CQDs, their reactivity and selectivity in photocatalytic reactions can be significantly enhanced. Functionalized CQDs can exhibit improved adsorption of

specific reactants, leading to higher photocatalytic efficiency and selectivity for desired products.

6 Challenges and Future Perspectives

Despite their promising properties, CQD-based photocatalysts encounter several challenges that must be overcome to fully unlock their potential for practical applications. These challenges primarily revolve around stability, scalability, and efficiency. Addressing these issues through targeted research and development efforts is crucial for the commercial viability and widespread adoption of CQDs in photocatalytic applications.

Stability

A key challenge in the practical application of CQD-based photocatalysts is maintaining their long-term stability. Stability is crucial for maintaining consistent photocatalytic performance over extended periods and under various operational conditions.

- **Photostability**: CQDs must resist photobleaching, a process where prolonged exposure to light degrades the material, reducing its effectiveness. Photostability can be improved through surface passivation, doping, and the development of more robust CQD structures that can withstand continuous light exposure without significant degradation.
- **Environmental Stability**: CQDs must also maintain their structural integrity and functional properties in diverse environmental conditions, including varying temperatures, pH levels, and the presence of potentially reactive species. Research should focus on understanding how these factors affect CQD stability and on developing strategies to enhance their resilience.

Scalability

For CQDs to be used on an industrial scale, it is essential to develop cost-effective and scalable synthesis methods. Current laboratory-scale production techniques may not be suitable for large-scale applications due to high costs, complexity, and low yield.

- **Cost-Effective Synthesis Methods**: New synthesis methods that are both cost-effective and environmentally friendly are needed. Techniques such as microwave-assisted synthesis, hydrothermal methods, and electrochemical approaches offer potential pathways for scalable production. These methods need to be optimized for higher yield and lower costs while maintaining the desired properties of CQDs.
- **Environmental Impact**: The environmental impact of the synthesis process must be minimized. This includes reducing the use of hazardous chemicals, lowering energy consumption, and ensuring that the production process generates minimal waste.

Efficiency

Enhancing the photocatalytic efficiency of CQDs is essential for their effective use in practical applications. This involves a comprehensive understanding of the fundamental mechanisms that govern their photocatalytic activity and optimizing their properties accordingly.

- **Fundamental Understanding**: Gaining a deeper understanding of the mechanisms involved in light absorption, charge separation, and surface reactions in CQDs will help in designing more efficient photocatalysts. Advanced characterization techniques and theoretical modeling offer valuable insights into these processes at the atomic and molecular scales.
- **Material Optimization**: Exploring new materials for doping, composite formation, and surface modification can significantly enhance the photocatalytic efficiency of CQDs. Doping with heteroatoms (e.g., nitrogen, sulfur, phosphorus) can introduce additional active sites and improve electronic properties, while forming composites with other nanomaterials can create synergistic effects that boost performance.
- **Surface Functionalization**: Surface functionalization techniques can be used to introduce specific chemical groups that enhance the reactivity and selectivity of CQDs in photocatalytic reactions. Functionalized CQDs can exhibit improved adsorption of reactants and facilitate more efficient charge transfer processes.

6.1 Stability Improvement

Research aimed at improving the stability of CQDs should focus on several key areas:

- **Surface Passivation**: Encapsulating CQDs with stable, inert materials can shield them from photobleaching and environmental degradation.
- **Robust Structural Design**: Developing CQDs with robust structural features that can withstand harsh operational conditions will enhance their long-term stability.
- **Understanding Degradation Mechanisms**: Investigating the mechanisms underlying CQD degradation under various conditions can help in devising strategies to mitigate these effects.

6.2 Cost-Effective Synthesis

Developing synthesis methods that are both cost-effective and environmentally friendly involves:

- **Optimization of Existing Methods**: Refining current synthesis techniques to improve yield, reduce costs, and minimize environmental impact.

- **Innovation in Synthesis Techniques**: Exploring new and innovative synthesis methods that offer scalability and cost-effectiveness, such as green chemistry approaches that use sustainable materials and processes.

6.3 Enhancing Photocatalytic Efficiency

Enhancing the photocatalytic efficiency of CQDs requires the following:

- **Doping and Composites**: Investigating the effects of different dopants and composite materials on the photocatalytic properties of CQDs to identify combinations that offer the best performance.
- **Surface Modification**: Utilizing surface modification techniques to introduce functional groups that enhance reactivity and selectivity in photocatalytic reactions.
- **Mechanistic Studies**: Conducting detailed mechanistic studies to understand the interactions between light, CQDs, and reactants, which will inform the design of more efficient photocatalysts.

7 Conclusions

In conclusion, CQDs offer significant advancements in photocatalysis due to their exceptional properties, including high surface area, tunable photoluminescence, and efficient charge separation. These attributes make CQDs highly effective for environmental remediation, such as degrading organic pollutants and enhancing hydrogen production through photocatalytic water splitting, as well as for energy conversion, including CO_2 reduction and improved photocatalytic fuel cell performance. Despite their potential, practical challenges such as stability, scalability, and cost-effective synthesis need to be addressed. Future research should focus on optimizing these aspects to fully realize CQDs' capabilities, leading to sustainable solutions for global environmental and energy challenges.

Acknowledgements The authors would like to acknowledge the Department of Chemical Engineering and the IIT Guwahati for providing all the facilities for this work.

References

1. P. M. Gawal and A. K. Golder, Colloids Surfaces A Physicochem. Eng. Asp. **683**, 133068 (2024).
2. N. Shehzad, M. Tahir, K. Johari, T. Murugesan, and M. Hussain, J. CO2 Util. **26**, 98 (2018).
3. C. B. Ong, L. Y. Ng, and A. W. Mohammad, Renew. Sustain. Energy Rev. **81**, 536 (2018).

4. J. Wang, J. Jiang, F. Li, J. Zou, K. Xiang, H. Wang, Y. Li, and X. Li, Green Chem. **25**, 32 (2022).
5. S. J. Phang and L. L. Tan, Catal. Sci. Technol. **9**, 5882 (2019).
6. Y. Yao, H. Zhang, K. Hu, G. Nie, Y. Yang, Y. Wang, X. Duan, and S. Wang, J. Environ. Chem. Eng. **10**, 107336 (2022).
7. Q. S. Si, W. Q. Guo, H. Z. Wang, B. H. Liu, and N. Q. Ren, Chinese Chem. Lett. **31**, 2556 (2020).
8. Y. H. Park, G. Murali, J. K. R. Modigunta, I. In, and S. Il In, Front. Chem. **9**, 1 (2021).
9. S. Sharma, V. Dutta, P. Singh, P. Raizada, A. Rahmani-Sani, A. Hosseini-Bandegharaei, and V. K. Thakur, J. Clean. Prod. **228**, 755 (2019).
10. K. A. S. Fernando, S. Sahu, Y. Liu, W. K. Lewis, E. A. Guliants, A. Jafariyan, P. Wang, C. E. Bunker, and Y. P. Sun, ACS Appl. Mater. Interfaces **7**, 8363 (2015).
11. M. Sharma, S. R. Kafle, A. Singh, A. Chakraborty, and B. S. Kim, ChemCatChem **202400056** (2024).
12. H. Jung, V. S. Sapner, A. Adhikari, B. R. Sathe, and R. Patel, Front. Chem. **10**, 1 (2022).
13. A. Mei, Z. Xu, X. Wang, Y. Liu, J. Chen, J. Fan, and Q. Shi, Environ. Res. **214**, 114160 (2022).
14. S. M. Chaudhari, P. M. Gawal, P. K. Sane, S. M. Sontakke, and P. R. Nemade, Res. Chem. Intermed. **44**, 3115 (2018).
15. S. Gengan, H. C. Ananda Murthy, M. Sillanpää, and T. Nhat, Results Chem. **4** (2022).
16. P. M. Gawal and A. K. Golder, Langmuir **40**, 13538 (2024).
17. Z. Yan, W. Su, W. Xu, Q. Mao, and L. Xue, Chinese Chem. Lett. 110217 (2024).
18. G. S. Das, J. P. Shim, A. Bhatnagar, K. M. Tripathi, and T. Y. Kim, Sci. Rep. **9**, 1 (2019).
19. V. Ramar, S. Moothattu, and K. Balasubramanian, Sol. Energy **169**, 120 (2018).
20. D. Gogoi, R. Koyani, A. K. Golder, and N. R. Peela, Sol. Energy **208**, 966 (2020).
21. H. Yu, Y. Zhao, C. Zhou, L. Shang, Y. Peng, Y. Cao, L. Z. Wu, C. H. Tung, and T. Zhang, J. Mater. Chem. A **2**, 3344 (2014).
22. T. Kulandaivalu, S. Abdul Rashid, N. Sabli, and T. L. Tan, Diam. Relat. Mater. **91**, 64 (2019).
23. S. Zhang, J. Tang, W. Chen, T. Qian, X. Li, Z. Feng, J. He, R. Zhang, Z. Yang, H. Li, P. Pan, K. Zhang, L. Zheng, and D. Feng, Appl. Energy **357**, 122526 (2024).
24. X. Ma, H. Du, M. Tan, J. Qian, M. Deng, D. Hao, Q. Wang, and H. Zhu, Sep. Purif. Technol. **339**, 126644 (2024).

Nature-Based Synthesis and Characterization of CdS from *Aegle Marmelos* Leave Extract for Efficient Visible-Light-Driven Methylene Blue Degradation

Pramod Madhukar Gawal and Animes Kumar Golder

Abstract This study demonstrates the implication of natural resources for the synthesis of CdS nanoparticles (NPs) using *Aegle marmelos* leaf extract, promoting an eco-friendly and sustainable approach to NPs production. The synthesis method utilizes the reducing and stabilizing properties of natural analytes existing in this leaf extract, and the resultant CdS nanoparticles were characterized using FTIR, SEM, XRD, TEM, UV-DRS, and EDX techniques. The absorption spectrum revealed a threshold at 530 nm, with a bandgap energy of 2.4 eV. XRD analysis confirmed a cubic structure with a crystal size of 7 nm, while SEM and TEM analyses revealed an average particle size of 28 nm. EDX analysis confirmed the composition of CdS NPs with sulfur and cadmium contents of 51.80 and 48.20%, respectively. Elemental mapping verified the uniform distribution of these elements. The catalytic activity of the synthesized catalyst was assessed for methylene blue dye degradation under visible light illumination. The CdS(bio) catalyst achieved an MB degradation of 98.76% within 120 min. This green synthesis method minimizes environmental impact while also demonstrating potential applications in photocatalysis for dye degradation.

Keywords CdS nanoparticles · Methylene blue · Nature-inspired synthesis · *Aegle marmelos* leaf extract · Dye degradation

---

P. M. Gawal · A. K. Golder (✉)
Department of Chemical Engineering, Indian Institute of Technology Guwahati, Guwahati, Assam, India
e-mail: animes@iitg.ac.in

P. M. Gawal
e-mail: g.pramod@iitg.ac.in

U. S. Dixit et al. (eds.), *Current Progress in Engineering Sciences*,
https://doi.org/10.1007/978-981-96-8753-4_40

1 Introduction

Synthetic dye pollution is a major environmental concern due to its persistence and resistance to biodegradation [1]. Methylene blue (MB), a widely used cationic dye in the textile, paper, and pharmaceutical industries, is a common wastewater pollutant [2]. Traditional treatment methods, such as adsorption and chemical oxidation, often fail to achieve complete removal or lead to secondary pollution [3]. Photocatalysis has emerged as an effective and sustainable solution for degrading organic contaminants like MB under light irradiation [2].

The synthesis of semiconductor nanoparticles (NPs), such as cadmium sulfide (CdS), has attracted significant interest due to their unique structural, optical, and electronic properties. These characteristics make them ideal for various applications, including photovoltaics, photocatalysis, and optoelectronic devices [4–6]. Conventional chemical methods for synthesizing CdS NPs often involve hazardous reagents and severe conditions, which present environmental and health hazards. Consequently, there is a growing demand for greener and more sustainable synthesis approaches [7–9].

In recent years, bio-based synthesis methods have emerged as a promising alternative, utilizing aqueous plant-extracts as the sources of reducing and stabilizing agents. These methods offer several advantages, including eco-friendliness, cost-effectiveness, simplicity, and availability of plentiful bio-source [10, 11]. The North-eastern parts of India are well known for their resource reservation and the indigenous bio-analytes that exist in various plants that could efficiently serve as reducing and capping agents for the eco-friendly synthesis of semiconductor NPs [2]. The rate of formation of NPs in a bio-based process can further be accelerated by several folds with a short microwave irradiation.

Herein, we have synthesized bio-based CdS NPs with phytochemicals found in *Aegle marmelos* leaf extract via microwave irradiation. *Aegle marmelos*, commonly known as Bael, is a medicinal plant widely known for its rich phytochemical profile, including flavonoids, phenolic acids, and alkaloids. These compounds can act as reducing agents, facilitating the formation of nanoparticles [4, 12]. The synthesized nanoparticles were analyzed with a range of characterization techniques to evaluate their structural and optical properties. The photocatalytic activity was studied for MB degradation under visible light illumination. The CdS(bio) achieved the MB degradation of 98.76% within 120 min. The implications of this nature-based synthesis approach for environmental sustainability and potential applications in photocatalytic dye degradation.

2 Material and Method

2.1 *Materials*

Thiourea (99.9%) and ethanol (EtOH) (99.99%) were obtained from Merck, India. Cadmium acetate dihydrate (99.50%) was sourced from Himedia, India. Methylene blue was purchased from SRL, India. All experimental work was carried out using deionized (DI) water from the water purifier (Millipore, USA).

2.2 *Phytochemical Extraction from* Aegle Marmelos

Fresh Aegle marmelos leaves (10 g) were first rinsed thoroughly with tap water for 5 min, then with DI water. The cleaned leaves were chopped into small pieces and placed in a beaker containing 100 mL of 50% EtOH. The mixture was stirred at 350 rpm using a magnetic stirrer at 70 °C until the solvent turned dark green. The resulting extract was cooled, filtered through a nylon cloth, and then stored for future use [3].

2.3 *Bioinspired Synthesis of CdS*

A volume of 5 mL of *Aegle marmelos* extract was added to 10 mL cadmium acetate (0.1 M) solution. Then, 10 mL of thiourea (0.1 M) was added dropwise under magnetic stirring [12]. The mixture was transferred to a microwave reactor and irradiated at 100 °C for 5 min. The obtained precipitate was centrifuged at 5000 rpm for 5 min, washed multiple times with deionized (DI) water followed by 50% ethanol (EtOH), and then dried in an oven at 65 °C for 15 h to yield CdS NPs.

2.4 *Characterizations Techniques*

The crystal structure and size of samples were analyzed by X-ray diffractometer (XRD, Smart lab, Rigaku, Japan) using Cu Kα radiation over the 2θ range of 10°–80°. UV-diffuse reflectance spectra (DRS) and absorption spectra were obtained using a spectrophotometer (UV-2600, Shimadzu, Japan) within the 200–700 nm range with $BaSO_4$ as reference. The morphological features of catalysts, including shape and size, were examined using a scanning electron microscope (SEM, model: Sigma 300, make: Zeiss, Germany) and transmission electron microscope (TEM, JEM2100F, JEOL, Japan). Surface elemental analysis of the samples was performed via energy-dispersive X-ray spectroscopy (EDS, Sigma 300, Zeiss, Germany). Fourier transform infrared spectroscopy (FTIR, I.R. Affinity 1, Shimadzu, Japan) was conducted to identify the relevant functional groups.

2.5 *Photocatalytic Degradation Studies*

The photocatalytic degradation of MB was studied by a desktop photoreactor under visible light irradiation. A 250 W metal halide (MH) lamp (\(\lambda > 420\) nm, Halonix Technologies, India) was used, providing a light intensity of 3.51 W/m^2. For the experiment, 25 mg of catalyst was mixed with 10 ppm of MB of 50 mL and stirred in the dark for 30 min to achieve adsorption equilibrium. The mixture was then exposed to visible light irradiation. Samples were collected every 10 min for analysis. The solution concentration was measured based on its absorbance at λmax = 665 nm. The MB degradation % was determined using Eq. 1.

$$\%\ \text{MB degradation} = \frac{C_0 - C}{C_0} \times 100 \tag{1}$$

where C_0 represents the initial concentration of MB, and C is the concentration after the irradiation period.

3 Results and Discussion

3.1 *UV–visible Spectrum*

Figure 1a presents the absorption spectra of CdS(bio) NPs. The absorption threshold was observed at 530 nm, aligning with previous findings [13], indicating that the NPs operate within the visible range. The bandgap energy was calculated using the Tauc equation (Eq. 2).

$$\alpha = \frac{(hv - E_g)^n}{hv} \tag{2}$$

where hv represents the incident photon energy, α is the absorption coefficient, and Eg denotes the bandgap energy, with n being a constant (1/2 for direct allowed transitions). Figure 1b presents the graph of $(\alpha hv)^{1/2}$ versus energy. The bandgap of the CdS(bio) NPs was determined to be 2.4 eV, consistent with previous reports [14].

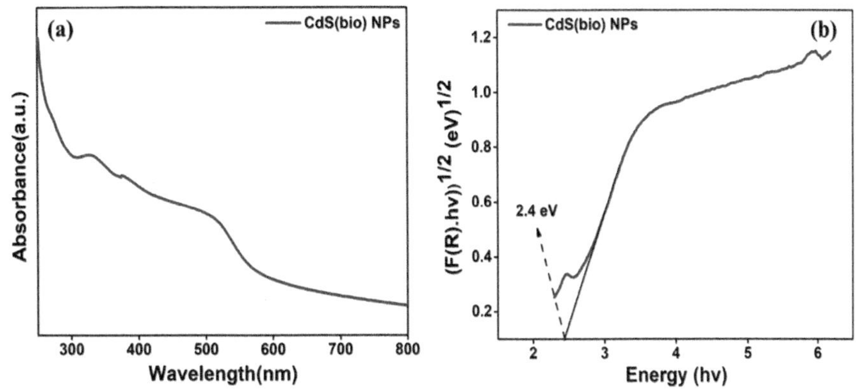

Fig. 1 **a** UV–vis absorption spectra and **b** bandgap calculation using Tauc's plot of CdS(bio) NPs

3.2 XRD Analysis

Figure 2a displays the XRD pattern of bio-synthesized CdS NPs. The peak at 2θ values of 26.25(111), 43.6(220), and 51.6(311) confirm the cubic structure of CdS (JCPDF database: 00-002-0454). These peak positions match well with the reported literature [12]. The crystal size was calculated using Debye Scherer's equation (Eq. 3) [2].

$$D = \frac{K\lambda}{\beta \mathrm{Cos}\theta} \tag{3}$$

where λ represents X-ray radiation wavelength (0.1540 nm), β represents FWHM, D is the average crystal size, θ is the diffraction angle, and K is a dimensionless factor (0.89). The crystal size of CdS(bio) NPs was determined to be 7 nm (Table 1).

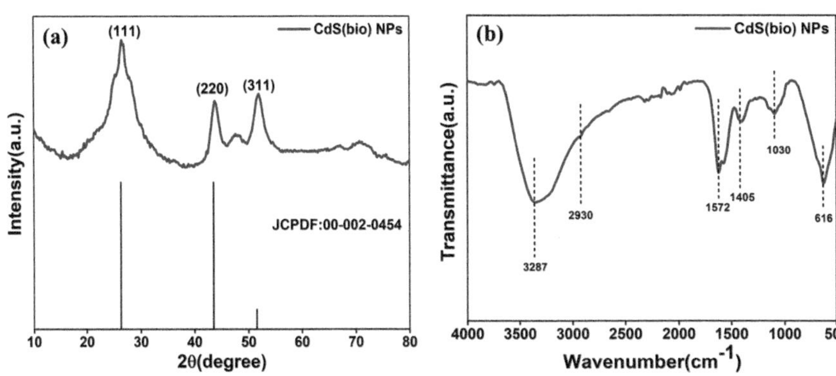

Fig. 2 **a** X-ray diffraction pattern and **b** FTIR spectrum of CdS(bio) NPs

Table 1 Physical characteristics of CdS(bio) NPs

Catalyst	Bandgap (eV)	Crystallite size (nm)	Average size (nm)
CdS(bio) NPs	2.4	7	28

3.3 FTIR Analysis

Figure 2b presents the FTIR spectrum of CdS(bio) NPs. The wide peak at 3290 cm^{-1} indicates the O–H stretching vibrations of H_2O. Peaks at 1572 and 1405 cm^{-1} correspond to the C=C stretching of cyclic alkenes and C–C bending in aromatic compounds, respectively. The weak band at 1030 cm^{-1} corresponds to C–O stretching, while the band at 616 cm^{-1} indicates CdS stretching. The weak band at 2930 cm^{-1} is attributed to Cd–O stretching. These peaks align with the reported literature [1].

3.4 SEM, TEM, and EDX Analyses

Figure 3a presents SEM images of CdS(bio) NPs, which exhibit well-defined shapes successfully formed through the bio-based method. The average size of CdS(bio) was 28 nm, consistent with values reported in the literature[4]. Figure 3b displays the TEM images of CdS(bio), consistent with SEM analyses. Figure 3c shows the EDX analysis of CdS(bio) composites revealed Cd and S contents of 51.80 and 48.20%, respectively, confirming the successful synthesis of CdS(bio)NPs. Figure 3d displays the elemental mapping of CdS(bio) NPs.

4 Photocatalytic Reduction Methylene Blue

The synthesized catalyst was studied for MB dye degradation under visible light illumination. The calibration of methylene blue is presented in Fig. 4a. The methylene dye degradation using CdS(bio) NPs is illustrated in Fig. 4b. The 98.76% degradation of methylene blue was achieved in 120 min.

5 Conclusions

The sustainable synthesis route of CdS NPs using *Aegle marmelos* leaf extract has been successfully demonstrated involving benchtop studies. The characterizations of synthesized materials reveal an optical absorption threshold at 530 nm, indicating the NPs function well within the visible range, with a bandgap energy of 2.4 eV.

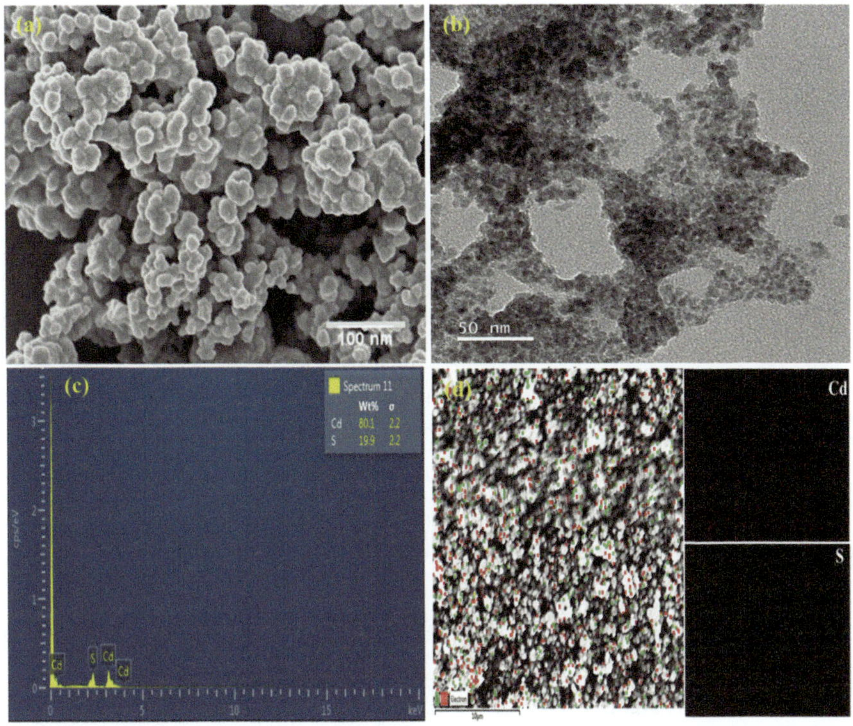

Fig. 3 **a** SEM image, **b** TEM image, **c** EDX spectrum, and **d** elemental mapping of CdS(bio) NPs

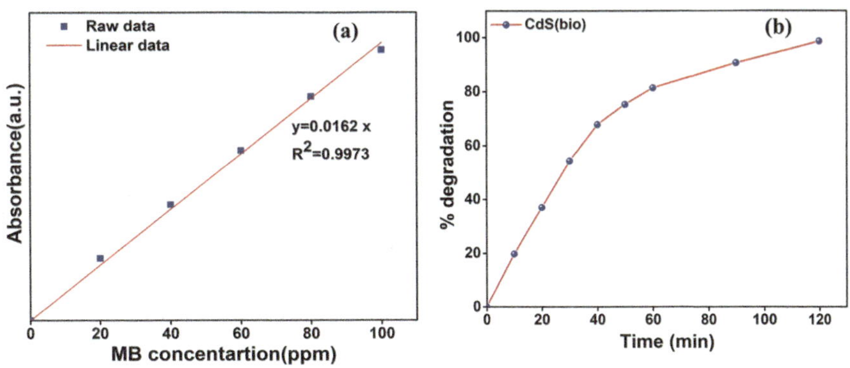

Fig. 4 **a** Calibration graph of methylene blue standard, and **b** photocatalytic degradation of MB with reaction time using CdS(bio) NPs

XRD analysis confirms the cubic structure of the CdS NPs, with a crystal size of 7 nm. FTIR spectra showed characteristic functional groups, such as CdS, C=C, O–H, C–C, C–O, and Cd–O stretching vibrations. SEM and TEM analyses revealed an average particle size of 28 nm, while EDX analysis of CdS NPs displayed sulfur and cadmium contents (approx.) 51.80 and 48.20%, respectively. Elemental mapping further confirms the uniform distribution of elements across the nanoparticle surface. The catalyst demonstrated 98.76% degradation of MB in 120 min under visible light illumination. In a nutshell, *Aegle marmelos* leaf is effective in functioning as the source of stabilizing and reducing agents for the nature-based synthesis of CdS NPs, providing a sustainable alternative for nanoparticle production with potential applications of dye degradation.

Acknowledgements The authors express their gratitude to the Central Instruments Facility and the Analytical Laboratory of the Department of Chemical Engineering at IIT Guwahati.

References

1. B. Lellis, C. Z. Fávaro-Polonio, J. A. Pamphile, and J. C. Polonio, Biotechnol. Res. Innov. **3**, 275 (2019).
2. S. M. Chaudhari, P. M. Gawal, P. K. Sane, S. M. Sontakke, and P. R. Nemade, Res. Chem. Intermed. **44**, 3115 (2018).
3. P. Kumari and A. Kumar, Results in Surfaces and Interfaces **11**, 100122 (2023).
4. P. M. Gawal and A. K. Golder, Langmuir **40**, 13538 (2024).
5. J. Wang and M. Isshiki, in edited by S. Kasap and P. Capper (Springer US, Boston, MA, 2007), pp. 829–842.
6. K. Veerathangam, M. S. Pandian, and P. Ramasamy, Mater. Sci. Semicond. Process. **108**, 104869 (2020).
7. B. Srinivasa Goud, Y. Suresh, S. Annapurna, A. K. Singh, and G. Bhikshamaiah, Mater. Today Proc. **3**, 4003 (2016).
8. S. Naranthatta, P. Janardhanan, R. Pilankatta, and S. S. Nair, ACS Omega **6**, 8646 (2021).
9. H. Dabhane, S. Ghotekar, P. Tambade, S. Pansambal, H. C. A. Murthy, R. Oza, and V. Medhane, Environ. Chem. Ecotoxicol. **3**, 209 (2021).
10. R. Ravi and A. K. Golder, Ind. Eng. Chem. Res. **62**, 20702 (2023).
11. C. Bhan and A. Kumar Golder, Chem. Eng. J. **475**, 146122 (2023).
12. P. M. Gawal and A. K. Golder, Colloids Surfaces A Physicochem. Eng. Asp. **683**, 133068 (2024).
13. Pramod M Gawal and Animes K Golder, *Photocatalytic Reduction of CO_2 to Value-Added Chemicals Catalyzed by CdS/CQDs Nanocomposites* (Guwahati, 2022).
14. P. M. Gawal and A. K. Golder, *Plant-Based Phytochemicals for Synthesis of Z-Scheme In_2O_3 /CdS Heterostructure for Photocatalytic CO_2 Reduction to Fuels* (Guwahati, 2023).

Recent Advances in Bio-based Synthesis of Nanomaterials for Photocatalytic Applications

Ravi Ravi, Chandra Bhan, and Animes Kumar Golder

Abstract Recently, nanotechnology has emerged as a rapidly growing field focused on the fabrication, characterization, and application of nanomaterials. In the realm of wastewater treatment, nanomaterials play a crucial role in various processes, including adsorption, antimicrobial activity, photocatalysis, stabilization, water splitting, etc. Among these, photocatalysis is considered the most efficient and eco-friendly technique. Physical and chemical methods are widely employed for nanoparticle synthesis. However, they have drawbacks such as high labor requirements, high cost, and the use of extensive chemicals. In contrast, bio-based methods offer an eco-friendly and economical approach to producing photocatalytically active nanomaterials with improved properties. Bio-based methods utilize biological units, such as plant extracts, as sources of reducing and capping agents in a one-pot synthesis system. This study reviewed reported studies on the bio-based synthesis of photocatalyst nanomaterials for photocatalytic applications. Bio-based synthesis has been employed to create native and non-metal/metal-doped photocatalysts, heterojunctions, and nanocomposites using plant extracts. The primary bio-analysts present in these extracts include ascorbic acid, gallic acid, citric acid, phenolic compounds, flavonoids, and other chelating agents. These diverse bio-analysts produce superior photocatalysts compared to those synthesized using chemical methods. This review summarizes the stands of bio-based routes over chemical-based synthesis methods, highlighting improvements in bandgap, surface hydrophilicity, oxygen vacancies, and photocatalytic degradation efficiency. Additionally, it extensively covers nanomaterial synthesis methods, bio-based synthesized photocatalyst nanomaterials, and their photocatalytic applications. Conclusively, adopting bio-based synthesis routes

R. Ravi · A. K. Golder (✉)
Centre for the Environment, IIT Guwahati, Guwahati, Assam, India
e-mail: animes@iitg.ac.in

R. Ravi
e-mail: r.vashist@iitg.ac.in

C. Bhan · A. K. Golder
Department of Chemical Engineering, IIT Guwahati, Guwahati, Assam, India
e-mail: bchandra@iitg.ac.in

U. S. Dixit et al. (eds.), *Current Progress in Engineering Sciences*,
https://doi.org/10.1007/978-981-96-8753-4_41

not only improves the photocatalytic properties of photocatalysts but also reduces synthesis costs and minimizes the use of harmful chemicals.

Keywords Photocatalyst nanomaterials · Vegetal extract · Bio-based synthesis · Photocatalytic applications

1 Introduction

Recently, nanotechnology has emerged as a significant field focused on the fabrication, characterization, and application of nanomaterials. Nanomaterials are materials with at least one dimension in the nanoscale range, typically between 1 and 100 nm [1]. Their exceptional properties, such as high active surface area, remarkable optical characteristics, inertness, high catalytic activity, and strong mechanical and electrical properties, have garnered significant attention in a short period [2]. Nanoparticles have shown several applications in wastewater treatment, including adsorption, photocatalysis, water splitting, antimicrobial activity, stabilization, reduction of harmful gases, and waste management. Beyond waste treatment, nanomaterials are extensively used in drug delivery devices, sensing, paints, sports equipment, analytical instruments, energy storage, and electronic devices [3–7]. Figure 1 provides a graphical representation of different nanomaterials and their applications. Among these, photocatalytic nanomaterials are reported to be the most efficient and eco-friendly for the complete degradation of organic pollutants through mineralization. Photocatalytic nanomaterials utilize light energy to degrade organic pollutants via catalytic reactions that generate free radicals [8]. For complete photodegradation, the organic pollutants are mineralized into CO_2 and H_2O. However, synthesizing these nanomaterials with precise characteristics is challenging. Various methods have been employed to synthesize nanomaterials, with their properties controlled by regulating the synthesis method and reaction parameters such as pH, temperature, mixing rate, and concentration. Conventional physical and chemical synthesis methods have several disadvantages. Nonetheless, bio-based methods could be a good alternative for synthesizing nanomaterials [5].

This review extensively covers the synthesis methods of nanomaterials, focusing on bio-based synthesized photocatalyst nanomaterials and their photocatalytic applications. Additionally, bio-based photocatalytic nanomaterials are compared with their chemically synthesized counterparts. The review examines studies published in the last decade on the bio-based synthesis of photocatalyst nanomaterials for wastewater treatment, water splitting, and CO_2 reduction applications. Relevant keywords such as "green synthesis of nanomaterials," "photocatalytic degradation," "photocatalytic reduction of CO_2," and "photocatalytic water splitting" were used to search databases, including Scopus, Web of Science, Google Scholar, PubMed, and ResearchGate.

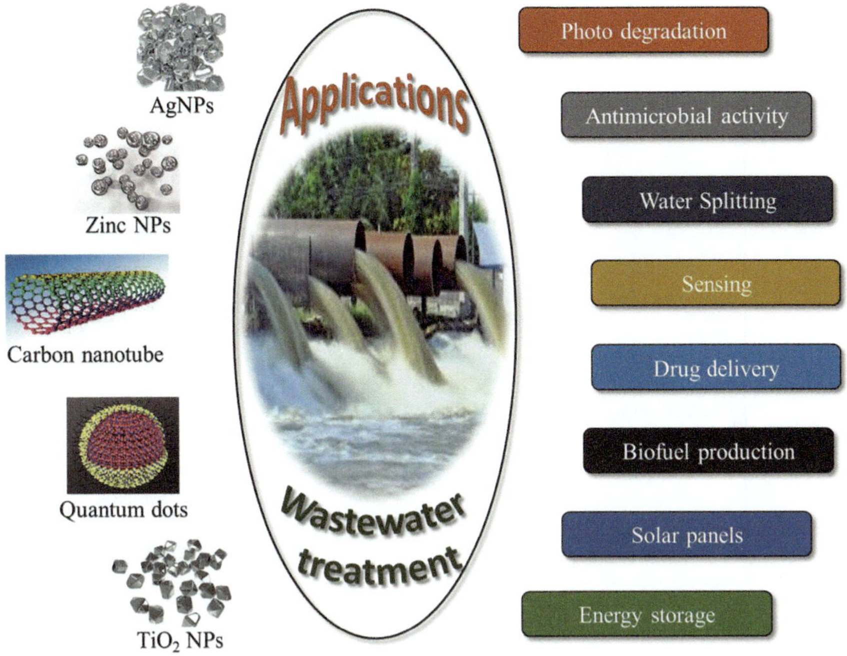

Fig. 1 Graphical illustration of various nanomaterials and their applications in numerous fields

2 Tactics for Synthesis of Nanomaterials

Fabricating nanomaterials with specific properties is achievable by precisely controlling optimized parameters during synthesis. The physicochemical attributes of nanomaterials are significantly influenced by the chosen synthesis method. Variations in synthesis parameters, such as pH, temperature, concentration, volume, reaction time, and the surrounding environment, can significantly alter the properties of nanomaterials. Based on the precursor and direction of nanomaterial growth, two widely utilized tactics for nanomaterial synthesis are the bottom-up and top-down approaches [9]. The bottom-up approach includes physical and chemical methods, while the top-down approach primarily involves physical methods. A visual representation of these approaches is depicted in Fig. 2. The synthesis method plays a crucial role in determining photocatalytic nanomaterials' optical, physical, and chemical properties.

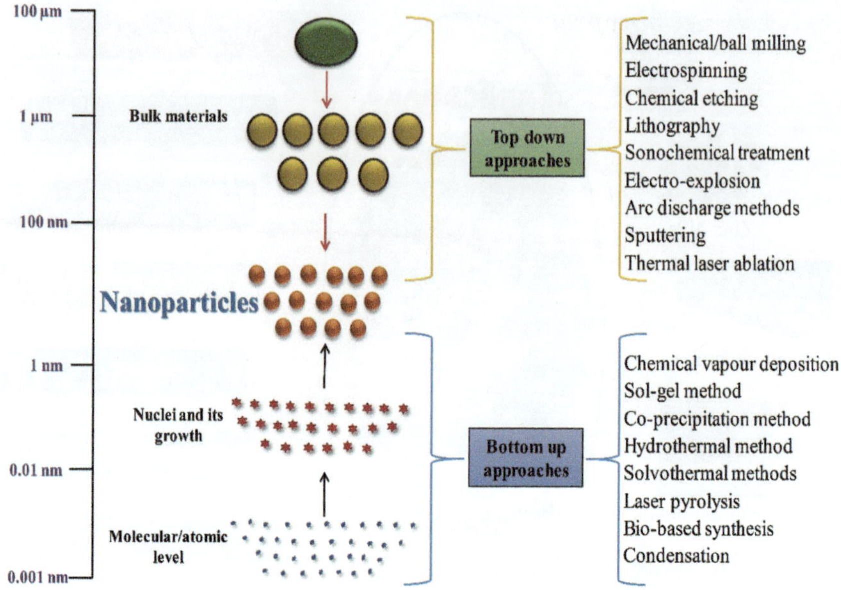

Fig. 2 Pictorial representation of top-down and bottom-up approaches for nanomaterials synthesis

2.1 Top-Down Approach

In a top-down approach, using various techniques, bulky materials are broken down into nanoscale particles. This approach extensively synthesizes nanostructured materials from naturally available raw materials, such as plant parts (leaves, stems, bark, seeds, etc.), soil particles, alloys, metal sheets, and industrial and agricultural waste. Techniques for synthesizing nanomaterials with a top-down approach include mechanical/ball milling, electrospinning, chemical etching, lithography, sonochemical treatment, electro-explosion, arc discharge methods, acid/base treatment, sputtering, and thermal laser ablation [9]. Nanomaterials produced through this approach are comparatively easier to shape with desired properties due to minimal impact on physicochemical parameters. Additionally, nanomaterials formed using the top-down approach are extensively applied in the adsorption of pollutants from contaminated water. However, synthesizing photocatalytic nanomaterials is less commonly undertaken using the top-down approach. Compared to the bottom-up approach, nanomaterials produced via top-down methods are less functional for many applications [10].

2.2 Bottom-Up Approach

Bottom-up approaches, also called self-assembly, are extensively employed to synthesize nanomaterials from their ionic state. This process involves assembling the smallest units at the ionic level through physical or chemical forces to form nanostructured materials [9]. Compared to the top-down approach, the bottom-up approach allows for synthesizing a diverse range of nanomaterials under controlled conditions. However, even slight variations in synthesis parameters can lead to significant changes in the structure and properties of the synthesized nanomaterials.

Nanomaterials produced through this approach have broad applications, including adsorption, photocatalysis, scaffold formation, drug delivery devices, water splitting, CO_2 reduction, biofuel production, and antimicrobial activity [11]. For enhanced photocatalytic activity under visible light irradiation, it is desirable to target photocatalytic nanomaterials with high surface area, small particle size, low bandgap, delayed recombination, and efficient charge transfer [12]. Numerous methods have been employed to synthesize photocatalytic nanomaterials from their precursor solutions (ionic form), primarily following the bottom-up approach. Common methods used for the bottom-up synthesis of photocatalytic nanomaterials include chemical vapor deposition (CVD), sol-gel, co-precipitation, hydrothermal, and solvothermal methods.

Chemical vapor deposition (CVD)

CVD techniques were initially employed for depositing solid material onto a surface through the chemical interaction of vaporized precursors with a heated solid substrate. The resulting deposited material can be thin films, crystals, or powder. Figure 3 provides a graphical illustration of the CVD deposition mechanism. CVD techniques can be categorized based on reaction conditions, including metal-organic CVD, laser CVD, low and high-pressure CVD, plasma CVD, and chemical and photochemical CVD. Initially, CVD was extensively used to synthesize high-quality thin-film coatings with specific properties. However, in the modern era, CVD has been adapted for forming advanced materials, such as photocatalyst nanocomposites, photocatalyst/nanomaterial-coated membranes, nanostructures, heat and corrosion-resistant coatings, dielectrics, and 2D and 3D polymeric materials [13]. The physical, topographic, and chemical properties in CVD methods can be easily controlled by regulating temperature, pressure, gas flow rate, and substrate type, and varying the composition of gases and deposited material.

Chemical and thermal CVD methods have been employed to form native photocatalytic materials and their nanocomposites under specific pressure and temperature conditions. Subramaniam et al. utilized the CVD technique to fabricate polyaniline-coated titania nanotubes in a rotary bed plasma chamber with plasma power ranging from 25 to 75 W. The thickness of the resulting uniform film decreased from 8.5 to 1.2 nm as the plasma power increased. The prepared nanotubes exhibited a bandgap of 2.54 eV and demonstrated a 33% increase in the photocatalytic degradation of RB 5 dye under visible light [13]. In a study by Lee et al., a gradually structured TiO_2

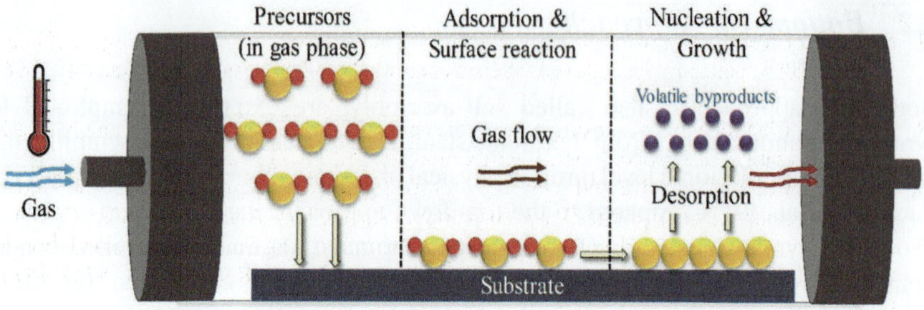

Fig. 3 The schematic representation of thin-film coating of the desired compound on a substrate surface by the chemical vapor deposition method

thin-film with a uniform thickness of 0.5–0.7 μm was developed, showing excellent photocatalytic degradation of bromothymol blue dye. The thickness of the TiO_2 film increased with deposition time [14]. Cherevan et al. (2018) utilized high-temperature CVD to develop a hard CNT template for synthesizing Ta_2O_5 [16]. Additionally, Kim et al. employed CVD techniques to create a photocatalytic graphene column for the photodegradation of phenolic compounds [15].

Sol–gel method

The sol-gel method is widely used for synthesizing photocatalytic materials because it produces materials with well-controlled physicochemical properties, high surface area, and desired morphology. This method has been utilized for synthesizing various non-silica photocatalysts, including ZnO, SnO_2, TiO_2, iron oxide, heterojunctions, and other metal oxides [16]. The sol-gel method typically involves five main steps, as illustrated in Fig. 4. The hydrolysis and condensation steps are crucial in determining the fate of nanomaterials by forming interconnected porous structures, which are further stabilized and shaped through aging, drying, and calcination at high temperatures [17].

In a study by El-Dafrawy et al., a series of ZnO and ZnO/PVA photocatalysts were synthesized using the sol-gel method. Characterization techniques revealed the formation of spherical-shaped ZnO with the wurtzite phase. The synthesized photocatalysts demonstrated effective photodegradation against MB and MO dyes under UV light and reduced blood glucose levels in rats [16]. In another study, Dara et al. developed a nanostructured Tb_2FeMnO_6 photocatalyst using the sol-gel method, with maleic acid and pomegranate paste employed as fuel for the combustion process. The bandgaps were found to be 3.3 and 2.96 eV, respectively. The photodegradation results confirmed up to 80 and 50% discoloration of erythrosine and malachite green dyes, respectively [18]. Furthermore, Salavati-Niasari et al. synthesized $ZnTiO_3$ nanostructures using benzene-1, 3, 5-tricarboxylic acid and achieved approximately 70% degradation of methyl orange dye within 1 h [17].

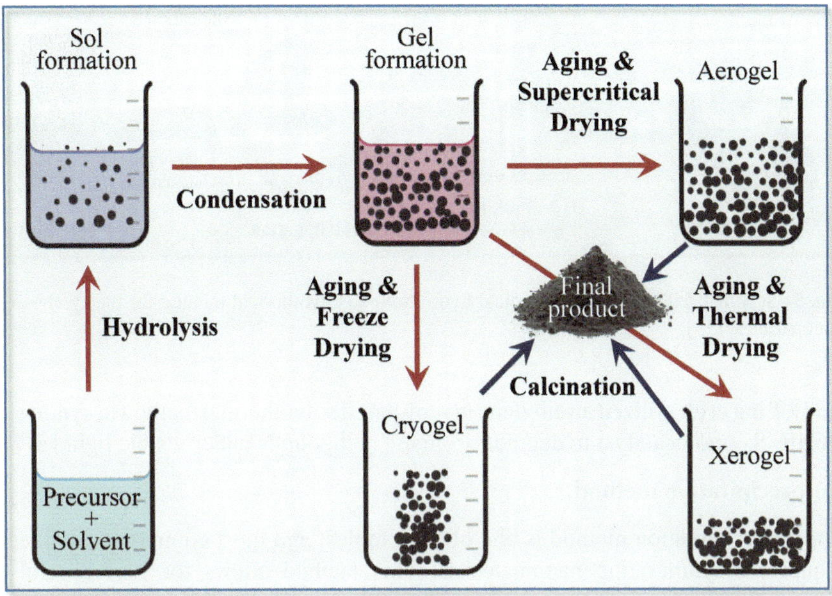

Fig. 4 The steps involved in synthesizing nanomaterials using the sol–gel method [18]

Solvothermal and hydrothermal methods

Solvothermal and hydrothermal methods are well-known methods used for the synthesis of nanoparticles. These processes involve chemical reactions in an aqueous phase under elevated pressure and temperature conditions. While hydrothermal and solvothermal methods are similar, the distinction lies in using water in hydrothermal and other solvents (rather than water) in solvothermal synthesis. Solvothermal methods are highly used to synthesize nanorods, nanowires, metal oxides, vertically aligned nanoparticles, etc. Thermal autoclave and microwave-assisted solvothermal/hydrothermal methods have garnered significant attention due to their ability to create high pressure and temperature in a closed chamber [19]. A schematic description of the solvothermal method utilized for synthesizing TiO_2-GO nanocomposite is given in Fig. 5.

In a study by Yadav and Kim, a TiO_2 and GO nanocomposite was synthesized using solvothermal methods at 130 °C for 4 h, with ethanol as a reaction solvent. Characterization techniques confirmed the presence of uniformly distributed anatase TiO_2 and graphene oxide in the prepared nanocomposite. Furthermore, photocatalytic degradation of benzene gases was performed under UV light irradiation, and almost complete benzene degradation was observed within 50 minutes [19]. Kamble and Ling developed a visible-light-driven hexagonal bi-pyramidal-shaped $BiVO_4$ using an ethylene glycol-based solvothermal method at 250 °C for 24 h. The photocatalytic degradation of methylene blue (MB) dye was performed, and approximately 91% degradation efficiency was achieved under visible light irradiation [20]. In another

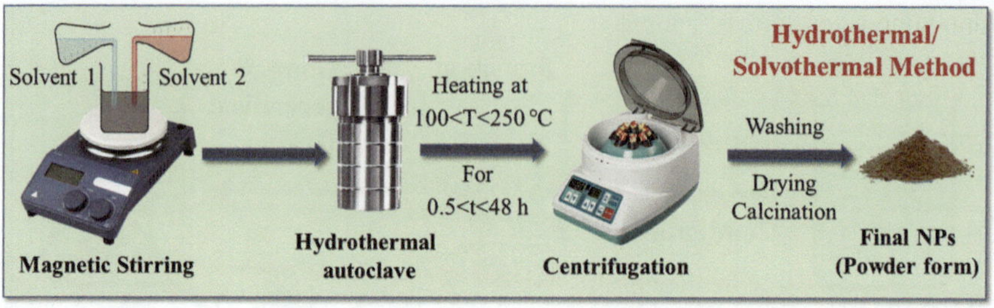

Fig. 5 Schematic illustration of a typical hydrothermal/solvothermal method for the synthesis of nanoparticles [19]

study, Ling et al. utilized an ethylene glycol-based solvothermal method to synthesize a $CdIn_2S_4$ photocatalyst to degrade aromatic compounds under visible light [21].

Co-precipitation method

The co-precipitation method is one of the simplest and most commonly used techniques for synthesizing nanomaterials. This method allows for easy control of physicochemical parameters, enabling the tailoring of desired material characteristics. The co-precipitation method involves simultaneous nucleation, growth, and precipitation, where ions are reduced and aggregate to form crystals. The size of the formed particles is influenced by various physicochemical parameters, including the reducing and capping agent, pH, temperature, mixing speed, reaction time, and volume [22]. Figure 6 provides a schematic representation of the synthesis of WO_3/$CoWO_4$ nanocomposites through the co-precipitation method.

In a study by Thilagavathi et al., the co-precipitation method was employed to synthesize 20 nm-sized WO_3/$CoWO_4$ nanocomposites at 80 °C for 2 h. The prepared

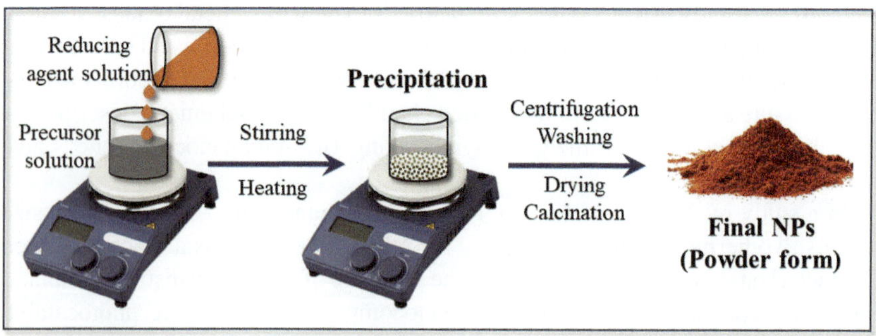

Fig. 6 Schematic representation of the typical synthesis of nanomaterials by co-precipitation method

nanocomposites, with a bandgap of 2.51 eV, exhibited excellent photocatalytic degradation of methylene blue (MB) dye, achieving an efficiency of 86.5% [22]. Pramothkumar et al. synthesized $CuMn_2O_4$ using the co-precipitation method at room temperature for 12 h. The resulting flake-like photocatalyst with a 50–100 nm size and a bandgap of 2.54 eV showed ~86 and 92% photodegradation of MB and malachite green dye, respectively [23]. In another study, Gupta et al. explored the effect of various metals (Co, Ni, Cu, and Zn) on the optical and physical properties of Fe_2O_4. The bandgap was found to be in the range of 1.11–1.81 eV, and each variant exhibited high degradation efficiency against several toxic dyes [24].

3 Physical, Chemical, and Bio-based Routes of Photocatalytic Nanomaterials Synthesis

3.1 Physical and Chemical-Based Synthesis Route

The physicochemical properties of photocatalytic nanomaterials are significantly influenced by the synthesis method. Nanomaterials can be synthesized through physical, chemical, and biological routes. A graphical illustration of these synthesis methods is presented in Fig. 7. Each route has its own specializations and drawbacks for synthesizing specific nanoparticles. Physical methods extensively used for synthesizing advanced nanomaterials include sputtering, ball milling, electrochemical methods, electrospinning, sonochemical methods, and condensation, and are primarily based on the top-down approach [9]. Despite their highly controlled synthesis mechanisms, physical methods are often limited by high costs, high power consumption, the generation of significant solid waste from the top-down approach, the need for sophisticated machinery, and the requirement for highly skilled operators.

Among these methods, chemical routes are highly favored for nanoparticle synthesis due to the availability of various chemicals as reducing agents, rapid synthesis, minimal solid waste generation, ease of handling, and the lack of need for expensive machinery, which allows for the development of highly precise nanomaterials. However, using non-biodegradable and toxic chemicals in these processes has raised environmental concerns, diminishing the popularity of chemical-based synthesis methods. Additionally, generating harmful by-products during synthesis further reduces the appeal of chemical methods [5]. Both chemical and bio-based routes follow a bottom-up approach to nanomaterial synthesis [8]. In bio-based synthesis, chemicals used in chemical routes are replaced by bio-based agents in plant extracts and microbes.

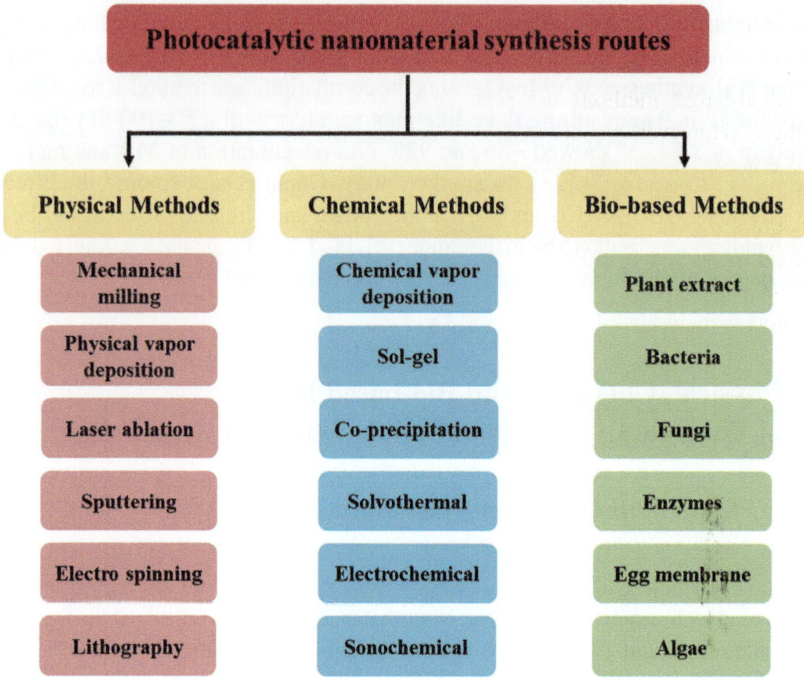

Fig. 7 Graphical illustration of nanomaterial synthesis methods following physical, chemical, and bio-based routes

3.2 *Bio-based (Green) Synthesis Route*

The bio-based (green) synthesis technique presents a promising alternative to physical and chemical methods, addressing many drawbacks. Bio-based synthesis methods are cost-effective, eco-friendly, abundant, and biocompatible, and offer precise control over the size and shape of nanomaterials [25]. Various biological entities, including plant extracts, enzymes, bacteria, fungi, and algae, have been utilized for bio-based nanomaterial synthesis. Among these, plant extracts are particularly favored due to their advantages, such as the availability of both reducing and capping agents, ease of extraction, minimal waste generation, and abundance in nature [26]. Common analytes in plant extracts include reducing agents (e.g., gallic acid, ascorbic acid, citric acid), flavonoids, chelating agents, and polyphenols. These bio-analytes facilitate a one-pot system for the controlled synthesis of nanomaterials. Additionally, the presence of numerous functional groups in plant extracts enables surface functionalization, enhancing photocatalytic activity through improved surface interactions and increased generation of free radicals. In recent decades, bio-based synthesized nanomaterials have gained significant attention for various photocatalytic applications due to their advantages over chemical methods.

3.3 Relative Stands of Bio-Based and Chemical Routes

Chemical-based methods are widely employed for photocatalytic nanomaterial synthesis. However, bio-based routes for synthesizing photocatalytic nanomaterials have emerged as more efficient and eco-friendly alternatives. Table 1 compares chemical-based and bio-based photocatalytic nanomaterials regarding their bandgap and photocatalytic activity for degrading organic pollutants such as dyes and pharmaceutical waste. Bio-based methods offer an eco-friendly alternative to chemical-based methods and provide additional advantages related to specific characteristics. These advantages include:

- **Narrower bandgap**: Bio-based photocatalysts often have a narrower bandgap, enhancing their activity under visible light.
- **Improved surface hydrophilicity**: Natural chelating agents and bio-analytes in plant extracts improve the surface hydrophilicity of bio-based photocatalysts.
- **Generation of oxygen vacancies and lattice defects**: Bio-based methods can create oxygen vacancies and metallic defects, increasing photocatalytic efficiency by generating free electrons and electron traps.
- **Enhanced visible-light-driven activity**: These characteristics' combined effects synergistically enhance the materials' photocatalytic activity under visible light.

For instance, bio-based Pt-doped TiO_2 and Au-doped TiO_2 have demonstrated a lower bandgap, improved charge transfer, delayed recombination, and better surface hydrophilicity compared to chemical-based counterparts. Additionally, bio-based photocatalysts exhibited approximately 1.25 and 1.18 times higher photocatalytic degradation of ciprofloxacin (CIP) and chloroquine (CLQ), respectively [8, 12]. Alsalmah reported that bio-based Cu-doped Bi2O3 exhibited superior photocatalytic activity to chemical-based counterparts due to delayed recombination and enhanced surface functionalization [27]. Two independent studies observed. Similar results for

Table 1 Comparison of bio-based and chemical-based photocatalysts in terms of their bandgap and photocatalytic degradation efficiency

Catalyst	Bandgap (eV)		Photocatalytic degradation (%)			Ref.
	Chemical	Bio-based	Pollutant	Chemical	Bio-based	
ZnO	3.25	3.17	RhB dye	70.9	80.2	[28]
Pt-doped TiO_2	3.14	3.02	CIP	68.3	85.5	[12]
Au-doped TiO_2	2.73	2.45	CLQ	80.5	94.6	[8]
$Bi_2 - xCu_xO_3$	2.38	-	TB dye	73.1	85.5	[27]
Mo-doped ZnO	3.47	3.43	MO dye	81.0	85.0	[29]

 # Rhodamine B dye; CIP: Ciprofloxacin; CLQ: Chloroquine; RhB; MO: Methyl orange; TB: Trypan Blue dye

ZnO and Mo-doped ZnO [28, 29]. In conclusion, while both bio-based and chemical-based photocatalysts have advantages, bio-based methods provide significant environmental and cost benefits and superior photocatalytic performance under certain conditions. This review summarizes various photocatalysts synthesized using bio-based methods, highlighting their optical and physicochemical characteristics and their photocatalytic efficiency.

4 Photocatalytic Applications of Bio-based Nanomaterials

Bio-based nanomaterials have been employed for various photocatalytic applications, including the degradation of organic pollutants (wastewater treatment), CO_2 reduction, and water splitting (green H_2 generation) under light energy. The photocatalytic process involves the generation of free radicals from electron-hole pairs produced after the excitation of electrons from the valence band to the conduction band under light irradiation [8]. A typical photocatalytic mechanism is illustrated in Fig. 8.

Table 2 lists various bio-based photocatalysts, including information on their shape, size, bandgap, applications, and the source of vegetal extract used in the synthesis. In a study, Likasari et al. utilized a bio-based method to synthesize NiO (size: 13.9–20.2 nm) using *T. erecta* L leaf vegetal extract. The resulting NiO photocatalyst demonstrated a reasonable photodegradation efficiency of 53% in 2 h against MB dye and exhibited good glucose sensing and antimicrobial activity [30]. In another study, Arumugam et al. synthesized spherical-shaped ZnO (size: 35 nm) through a bio-mediated method using *F. carica* leaf extract as a reducing and capping agent. The prepared material exhibited excellent antimicrobial activity and degraded ~95.3% of MB within 70 min under sunlight irradiation [31]. Munyai et al. developed CdS using *S. frutescens* leaf extract. The resulting spherical-shaped photocatalysts (size: 12–32 nm) exhibited a bandgap of 3.55 eV and demonstrated ~91 and 68%

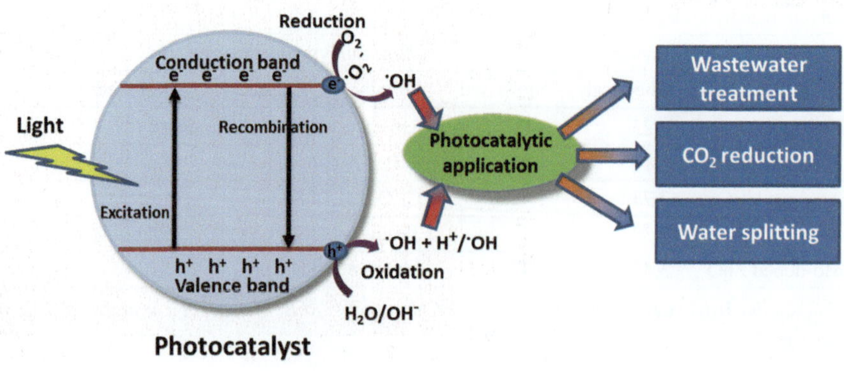

Fig. 8 Mechanism of the photocatalytic process under light energy

photodegradation of dyes and sulfisoxazole, respectively, within 2 h [32]. Additionally, Haseena et al. synthesized a magnetic Fe_2O_3 photocatalyst using *L. aspera* and J. *podagrica* leaf extract, with bandgaps of 2.42 and 2.50 eV, respectively, for the removal of textile waste [33]. Yu et al. prepared an AgNPs-biochar photocatalyst using *E. japonica* leaf extract and demonstrated about 99.9% RhB dye degradation within 10 min [34]. *A. marmelos* leaf extract-based TiO_2 exhibited a bandgap of 3.38 eV and degraded 66.2% ornidazole (ODZ) [35]. Fatimah et al. utilized *P. pinnata* leaf extract to synthesize SnO_2 (size: 8–20 nm) and exhibited good antimicrobial activity [36].

Gawal and Golder synthesized bio-based CQDs/CdS nanocomposites using *A. marmelos* vegetal extract for CO_2 reduction and observed the generation of 1.06 mmol/g·h methanol under visible light [26]. Bio-based In2O3/CdS also converted CO_2 to formic acid with a yield of 1.06 mmol/g·h [25]. In another study, bio-based ZnO was synthesized using *S. erianthum* extract for CO_2 reduction and produced formic acid (0.19 mmol/g·h), methanol (0.015 mmol/g·h), and ethanol (0.012 mmol/g·h) [37]. A bio-based Cu_2O NPs synthesized using *P. guajava* were utilized for photocatalytic reduction of CO_2 and water splitting applications under light energy. The obtained results confirmed the generation of 0.10 mmol/g·h methanol and 7.8 mmol H_2/g·h (hydrogen) [38]. In another study, *A. linearis*-based RuO_2/Cu_2O deposited over the ITO surface exhibited 0.029 mmol H_2 generation (per h) per gram of catalysts [39].

Various plant extracts have been reported to efficiently synthesize different photocatalysts, serving as both reducing and capping agents. Most synthesized nanomaterials were less than 50 nm in size, although some were around 100 nm. The development of photocatalysts with diverse morphologies, such as spherical, nanorods, and flower shapes, demonstrates the versatility of bio-based synthesis methods in tailoring nanomaterial structures. The prevalence of spherical-shaped photocatalysts may be due to specific interactions and conditions provided by the plant extracts during the synthesis process. Although none of the studies explicitly mentioned post-synthesis difficulties, researchers often encounter challenges related to washing, drying, and grinding during the synthesis and characterization of nanomaterials. Proper washing steps, such as using ethanol and water mixtures, are essential for removing impurities and unreacted precursors from the synthesized nanoparticles. Additionally, techniques like sonication and calcination may be employed to disperse particles and address issues related to agglomeration.

5 Conclusions

In conclusion, this review has highlighted the various routes for synthesizing photocatalytic nanomaterials, summarizing them based on the direction of nanoparticle formation (top-down vs. bottom-up) and the nature of the routes (physical, chemical, and bio-based). Bio-based methods present an environmentally friendly and

Table 2 Bio-based photocatalysts with specific characteristics such as size, shape, and bandgap, source of vegetal extract, and their photocatalytic applications and efficiencies

Catalysts	Extract source	Size (nm); shape; bandgap (eV)	Photocatalytic application (efficiency)	Source
NiO	*Tagetes erecta* L	13.9 and 20.2 nm; spherical; 3.6 eV	MB photodegradation (53%); glucose sensing; and antimicrobial activity	[30]
ZnO	*F. carica*	35 nm; spherical; 3.8 eV	Antimicrobial activity; photodegradation of MB (95.3%)	[31]
CdS	*S. frutescens*	12–32 nm; spherical; 3.55 eV	Photodegradation of dyes and sulfisoxazole	[32]
Fe_2O_3	*L. aspera* & *J. podagrica*	59.1 and 44.7 nm; spherical; 2.42 and 2.5 eV	Photodegradation of dyes	[33]
AgNPs-biochar	*E. japonica*	<35 nm; uniformly spherical; 1.2 eV	Photodegradation of RhB dye	[34]
TiO_2	*A. marmelos*	<150 nm; spherical; 3.38 eV	UV light-driven degradation of ornidazole	[35]
SnO_2	*P. pinnata*	8–20 nm; flower-like shape; 3.5 eV	Antimicrobial activity; bromophenol blue photodegradation	[36]
CQDs/CdS composite	*A. marmelos*	73 nm; rod-shaped; 2.32 eV	CO_2 reduction to Methanol (1.06 mmol/g·h)	[26]
In_2O_3/CdS	*A. marmelos*	25.1 nm; spherical-shaped; 2.18 eV	CO_2 reduction to formic acid (0.51 mmol/g·h)	[25]
ZnO	Egg membrane	~500 nm; macron shape; 2.9 eV	CO_2 reduction to formic acid (0.19 mmol/g·h)	[37]
	S. erianthum	<50 nm; HOLLOW spherical; 3.0 eV	CO_2 reduction to methanol (0.015 mmol/g·h) and ethanol (0.012 mmol/g·h)	
Cu_2O NPs	*P. guajava*	~1500 nm; octahedral; 1.8 eV	CO_2 reduction to methanol (0.10 mmol/g·h)	[38]
			Water splitting (7.8 mmol H_2/g·h)	
CuO NPs	*Zea mays*	Agglomerated; spherical; 3.55 eV	Water splitting (0.14 mmol H_2/g·h)	[40]
RuO_2/Cu_2O/ITO	*A. linearis*	~2.25 nm; not defined; 2.1 eV	Water splitting (0.029 mmol H_2/g·h)	[39]
ZnO NPs	*M. oleifera*	20–150 nm; spherical; 3.12 eV	Water splitting (0.34 mmol H_2/g·h)	[41]

efficient alternative to chemical and physical routes for synthesizing photocatalysts. These studies underscore the ongoing efforts to optimize bio-based synthesis methods to produce well-defined and well-dispersed photocatalytic nanomaterials. Bio-based photocatalysts demonstrate superior optical and physicochemical properties and enhanced photocatalytic activity compared to their chemical counterparts. Additionally, several bio-based photocatalysts have been successfully employed in photocatalytic applications such as wastewater treatment, CO_2 reduction to value-added chemicals, and water splitting for green H_2 fuel generation. There is considerable potential for further exploration of bio-based routes to synthesize modified photocatalysts, including metal and non-metal-doped photocatalysts, heterojunctions, and nanocomposites. Given the vast diversity of Earth's ecosystem, many plant-derived bio-analytes remain unexplored for photocatalyst synthesis. Ongoing research in bio-based synthesis methods highlights the adaptability and versatility of these approaches in tailoring photocatalytic nanomaterials' properties. Expanding the range of plant extracts and fostering interdisciplinary collaborations could further advance the photocatalytic field and uncover new opportunities for sustainable and efficient photocatalyst synthesis.

Acknowledgements The authors would like to acknowledge the Department of Science and Technology, Government of India, for providing financial support to carry out this research (under grant no. IMP/2018/000241). The authors are also thankful to the Centre for the Environment, Department of Chemical Engineering, and Central Instruments Facility, Indian Institute of Technology Guwahati, for providing all the facilities for this work.

References

1. Bhan C.: Bio-based hierarchical vertically aligned 2D ZnO nanostructures for ultra-selective electrochemical sensing of p-Chloroaniline, Chemical Engineering Journal, 146122 (2023).
2. Ravi R., Golder A.K.: Pharmaceutically Active Compounds'(PhACs) Threat: An Environmental Prospective, in: North-East Research Conclave, Springer, pp. 51–66 (2022).
3. Arora S.: Higher-order turbulence statistics and multiscale characterization of morphodynamics in a riverbank section with an upstream mining pit, Physics of Fluids, 35 (2023).
4. Golder A.K.: Synthesis of low-cost bentonite/Duranta erecta's fruit powder imbedded alginate beads and its application in surfactant removal, Environmental Science and Pollution Research, 28, 58945–58957 (2021).
5. Bhan C.: Utilizing Bioinspired AgNPs as an Antibacterial Agent to Enhance Ceramic Membrane Performance, Journal of Environmental Chemical Engineering, 110283 (2023).
6. Ravi, Enhanced adsorption capacity of designed bentonite and alginate beads for the effective removal of methylene blue, Applied Clay Science, 169, 102–111 (2019).
7. Kumar A.: A novel eco-friendly inhibitor of chayote fruit extract for mild steel corrosion in 1 M HCl: Electrochemical, weight loss studies, and the effect of temperature, Sustainable Chemistry and Pharmacy, 36 101261 (2023).
8. Ravi R.: A tuneable bioinspired process of Pt-doping in TiO_2 for improved photoelectrochemical and photocatalytic functionalities, Colloids and Surfaces A: Physicochemical and Engineering Aspects, 663 131034 (2023).
9. Baig N.: Nanomaterials: a review of synthesis methods, properties, recent progress, and challenges, Materials Advances, 2, 1821–1871 (2021).

10. Calvo-de la Rosa J.: Optimization of the synthesis of copper ferrite nanoparticles by a polymer-assisted sol–gel method, ACS omega, 4, 18289–18298 (2019).
11. Ravi R.: Efficient recovery of TiO_2 and Pt-doped TiO_2 photocatalysts in wastewater treatment using a pilot-scale cross-flow ultrafiltration membrane system, Separation and Purification Technology, 126710 (2024).
12. Ravi R.: Bio-based au doping with dominant oxygen vacancies and Ti^{3+} defects on photocatalytic functionalities of TiO_2, Industrial & Engineering Chemistry Research, 62, 20702–20715 (2023).
13. Subramaniam M.: Synthesis of titania nanotubes/polyaniline via rotating bed-plasma enhanced chemical vapor deposition for enhanced visible light photodegradation, Applied Surface Science, 484, 740–750 (2019).
14. H. Lee, S.H. Park, Y.-K. Park, S.-J. Kim, B.-H. Kim, J. Lee, S.-C. Jung, Photocatalytic activity of hierarchically structured TiO_2 films synthesized by chemical vapor deposition, International Journal of Photoenergy, 2014 (2014).
15. Kim J.: Photocatalytic degradation of phenol using chemical vapor desposition graphene column, Catalysts, 10, 1251 (2020).
16. El-Dafrawy S.M.: Synthesis, photocatalytic and antidiabetic properties of ZnO/PVA nanoparticles, Scientific Reports, 11, 1–11 (2021).
17. Salavati-Niasari M.: Synthesis, characterization, and morphological control of $ZnTiO_3$ nanoparticles through sol-gel processes and its photocatalyst application, Advanced Powder Technology, 27, 2066–2075 (2016).
18. Dara M.: Sol–gel auto combustion synthesis, characterization, and application of Tb_2FeMnO_6 nanostructures as an effective photocatalyst for the discoloration of organic dye contaminants in wastewater, RSC Advances, 11, 26844–26854 (2021).
19. Yadav H.M.: Solvothermal synthesis of anatase TiO_2-graphene oxide nanocomposites and their photocatalytic performance, Journal of Alloys and Compounds, 688, 123–129 (2016).
20. Kamble G.S.: Solvothermal synthesis of facet-dependent $BiVO_4$ photocatalyst with enhanced visible-light-driven photocatalytic degradation of organic pollutant: assessment of toxicity by zebrafish embryo, Scientific reports, 10, 1–11 (2020).
21. Ling C.: Solvothermal synthesis of CdIn 2 S 4 photocatalyst for selective photosynthesis of organic aromatic compounds under visible light, Scientific reports, 7, 1–16 (2017).
22. Thilagavathi T.: WO_3/$CoWO_4$ nanocomposite synthesis using a facile co-precipitation method for enhanced photocatalytic applications, Journal of Physics and Chemistry of Solids, 154, 110066 (2021).
23. Pramothkumar A.: Flake-like $CuMn_2O_4$ nanoparticles synthesized via co-precipitation method for photocatalytic activity, Physica B: Condensed Matter, 572, 117–124 (2019).
24. Gupta N.K.: Photocatalytic degradation of organic pollutants over MFe_2O_4 (M= Co, Ni, Cu, Zn) nanoparticles at neutral pH, Scientific reports, 10, 1–11 (2020).
25. Gawal P.M.: Plant-Based Phytochemicals for Synthesis of Z-Scheme In_2O_3/CdS Heterostructures: DFT Analysis and Photocatalytic CO_2 Reduction to HCOOH and CO, Langmuir, (2024).
26. Gawal P.M.: Vegetal route for synthesis of CQDs/CdS nanocomposites for photocatalytic reduction of CO_2 to methanol under visible light, Colloids and Surfaces A: Physicochemical and Engineering Aspects, 683, 133068 (2024).
27. Alsalmah H.A.: Green synthesis of copper doped bismuth oxide: A novel inorganic material for photocatalytic mineralization of Trypan blue dye, Inorganic Chemistry Communications, 163, 112270 (2024).
28. Khalid N.: Green synthesis and characterizations of bi-functional Mo-doped ZnO nanostructures for antimicrobial and photocatalytic applications, Materials Chemistry and Physics, 296, 127306 (2023).
29. Sudha A.: Assessment of de-toxicity of treated dye solution via seed germination test: Photocatalytic treatment by Morinda citrifolia leaf extract mediated ZnO: Mo/g-C_3N_4 nanocomposite, Materials Science in Semiconductor Processing, 178, 108437 (2024).

30. Likasari I.D.: NiO nanoparticles synthesized by using *Tagetes erecta* L leaf extract and their activities for photocatalysis, electrochemical sensing, and antibacterial features, Chemical Physics Letters, 780, 138914 (2021).
31. Arumugam J.: Green synthesis of zinc oxide nanoparticles using Ficus carica leaf extract and their bactericidal and photocatalytic performance evaluation, Chemical Physics Letters, 783, 139040 (2021).
32. Munyai S.: Green synthesis of Cadmium Sulphide nanoparticles for the photodegradation of Malachite green dye, Sulfisoxazole and removal of bacteria, Optik, 247, 167851 (2021).
33. Haseena S.: Bio-Synthesize of Photocatalytic Fe_2O_3 Nanoparticles Using Leucas aspera and Jatropha podagrica Leaf Extract for an Effective Removal of Textile Dye Pollutants, Optik, 168275 (2021).
34. Yu C.: Green synthesized nanosilver-biochar photocatalyst for persulfate activation under visible-light illumination, Chemosphere, 131237 (2021).
35. Ahmad W.: Green synthesis of photocatalytic TiO_2 nanoparticles for potential application in photochemical degradation of ornidazole, Journal of Inorganic and Organometallic Polymers and Materials, 31 (2021) 614–623.
36. Fatimah I.: Flower-like SnO_2 Nanoparticle Biofabrication Using Pometia pinnata Leaf Extract and Study on Its Photocatalytic and Antibacterial Activities, Nanomaterials, 11, 3012 (2021).
37. Garay-Rodríguez L.F.: Renewable formic acid production from CO_2 reduction using green ZnO nanoarchitectures, Materials Science in Semiconductor Processing, 161, 107458 (2023).
38. Torres-Arellano S.: Optimized CO_2 photoreduction using cuprous oxide (Cu_2O) nanoparticles synthesized using *Psidium guajava* extract, Energy, Ecology and Environment, 1–13 (2024).
39. Ismail E.: RuO_2 nanoparticles by a novel green process via Aspalathus linearis natural extract & their water splitting response, Journal of Alloys and Compounds, 662, 283–289 (2016).
40. Torres-Arellano S.: Biologically mediated synthesis of CuO nanoparticles using corn COB (Zea mays) ash for photocatalytic hydrogen production, Materials Chemistry and Physics, 301, 127640 (2023).
41. Archana B.: Enhanced photocatalytic hydrogen generation and photostability of ZnO nanoparticles obtained via green synthesis, International Journal of Hydrogen Energy, 42, 5125–5131 (2017).

A Colorimetric and Photoresistive Sensor for Quantification of Fluoride in Water

Shrikant Kashyap, Purushotam Shivshakti Mahan, and Tapas K. Mandal

Abstract Fluoride contamination of potable water has become a global concern. Fluoride poisoning can escalate musculoskeletal and dental disorders; therefore, monitoring of fluoride in drinking water is of utmost importance for healthcare applications. This article reports the application of ferric chloride ($FeCl_3$) and ammonium thiocyanate (NH_4SCN) to form a colorimetric complex with fluoride (F^-), which is used for colorimetric sensing and quantification of F^- present in water. The formation of this complex was corroborated by a UV–vis spectroscopy with a prominent peak at a wavelength (λ_{max}) of 455 nm. Akin to UV–vis absorbance, the electrical resistance of the colorimetric complex was measured using a developed prototype consisting of a light-dependent resistor (LDR) and a light-emitting diode (LED). The prototype was integrated with a digital multimeter. The resistances of the spiked samples of the fluoride complex were measured to prepare a calibration plot. The calibration plot exhibited good linearity with an R^2 value of 0.978. The prototype detected F^- concentration below the recommended threshold value of 1.5 mgL^{-1}. The response of the prototype was validated against the UV–vis spectrophotometer, and an absolute average percentage error of 6.23% was reported. This reasonable performance prototype proves its worthiness to be translated into a Point of Care Testing (POCT) device for on-site sensing and quantification of F^- to ensure safe drinking water quality.

Keywords Healthcare · Prototype · POCT · Fluoride · Sensor · LED-LDR

S. Kashyap · P. S. Mahan · T. K. Mandal (✉)
Department of Chemical Engineering, Indian Institute of Technology Guwahati, Guwahati, Assam, India
e-mail: tapasche@iitg.ac.in

S. Kashyap
e-mail: shrikant2016@iitg.ac.in

P. S. Mahan
e-mail: p.shivshakti@iitg.ac.in

U. S. Dixit et al. (eds.), *Current Progress in Engineering Sciences*,
https://doi.org/10.1007/978-981-96-8753-4_42

1 Introduction

The essence of life is water, and access to clean and safe drinking water is the fundamental right for all. Yet, it remains a challenge for humankind as the water available is contaminated by various anthropogenic and natural sources. Among these contaminants, fluoride poisoning poses a significant risk to public health worldwide. At optimum concentration, fluoride benefits health [1]; however, ingested in excess amounts leads to dental and skeletal fluorosis [2, 3]. To mitigate these health issues, the World Health Organisation (WHO) has advised a threshold limit of fluoride in drinking water at 1.5 ppm [4].

Therefore, there is a need to monitor the level of fluoride poisoning in natural and municipal water supply sources for healthcare applications. The detection and quantification of fluoride in water have been extensively studied, with numerous methodologies developed over the past few decades. Fluoride monitoring can be done using traditional methods, such as ion-selective electrodes (ISEs), fluorescence spectroscopy, UV–vis spectroscopy (UV–vis), atomic absorption spectroscopy (AAS), inductively coupled plasma mass spectroscopy (ICP-MS), and ion chromatography (IC) [5, 6]. Due to their high accuracy and robustness, these techniques are considered the gold standard.

ISE techniques offer high accuracy; however, they require rigorous calibration, expensive equipment, and skilled personnel, and they are unsuitable for on-site analysis. Also, the selectivity is compromised using this technique [7]. Similarly, ion chromatography and other spectroscopy techniques, due to their robustness, are more accurate as well as selective, but the requirement for high-end instrumentation, skilled personnel, and rigorous sample preparation methods restricts their use to well-equipped laboratories and limits their accessibility for routine and on-site water quality monitoring only [5, 7, 8].

Over the years, sensor-based applications have garnered significant interest in healthcare due to their potential for low-cost, portability, and user-friendly detection [9]. Colorimetric sensors, which rely on visible color changes in fluoride, have been promising due to their naivety and naked-eye detection. For example, sensors utilizing metal complexes and dyes, such as zirconium-alizarin complex and Eriochrome Cyanine R, have demonstrated effective colorimetric responses to fluoride [10]. However, these systems often suffer from limitations in sensitivity and selectivity, necessitating further optimization [11]. Photoresistive sensors offer another promising approach. These sensors change their electrical resistance in response to light. These can be coherently integrated with colorimetric sensors to provide a quantifiable change in the electrical signal corresponding to the color intensity [12]. The integration of photoresistive sensors with colorimetric assays has been explored for various analytes, though their application for fluoride detection remains relatively underexplored [13].

This paper presents the development of a novel colorimetric and photoresistive sensor for quantifying fluoride in water. For colorimetric sensing and quantification, fluoride (F^-) in water, ferric chloride ($FeCl_3$), and ammonium thiocyanate

(NH_4SCN) have been employed as reagents to generate a colorimetric complex with fluoride (F^-). The proposed sensor prototype leverages the unique properties of colorimetric reagents and a photoresistor to produce a detectable signal corresponding to changes in fluoride levels. Integrating these two sensing mechanisms aims to enhance sensitivity, selectivity, and usability, providing a cost-effective POCT kit for on-site monitoring of fluoride in remote and resource-limited areas.

2 Materials and Methods

2.1 Materials

Ammonium thiocyanate (NH_4SCN), ferric chloride ($FeCl_3$), and sodium fluoride (NaF) were procured from Himedia, India. All the chemicals were of analytical grade and used without any further purification. Milli-Q grade water was used throughout the experiment unless stated otherwise. Digital multimeter, light-emitting diode (LED), light-dependent resistors (LDR), connecting wires, resistor, and 9 V battery were purchased from the online store Robu, India. Quartz cuvettes for spectrophotometric analysis were purchased from Axiva, India.

2.2 Reaction Scheme

An equimolar mixture of $FeCl_3$ and NH_4SCN was mixed in equal amounts to give a red color solution called $Fe(SCN)^{+3}$, which decolorizes upon the addition of F^- solution. The decolorization was spectrophotometrically observed to be proportional to F^- concentration and obeyed Beer-Lambert's law. The detailed reaction scheme is described in Fig. 1. It is well-established that F^- on interaction with iron (iii) thiocyanate gives colorless Fe (III) fluoride, as shown in Fig. 1. In this regard, the specific reaction can be written as [14, 15],

$$Fe(SCN) + 6NaF \rightarrow [FeF_6]_3^- + 6Na^+ + 3(SCN)^-$$

2.3 Design and Development of the Portable LED-LDR Sensor

The development of the portable POCT device necessitates using an LED light, LDR, cuvette, 20 × 20 × 20 mm lightproof box, 9 V battery, and a digital multimeter. To create the LED-LDR sensor, first, the LED and LDR were attached precisely opposite

Fig. 1 Schematic of the reaction

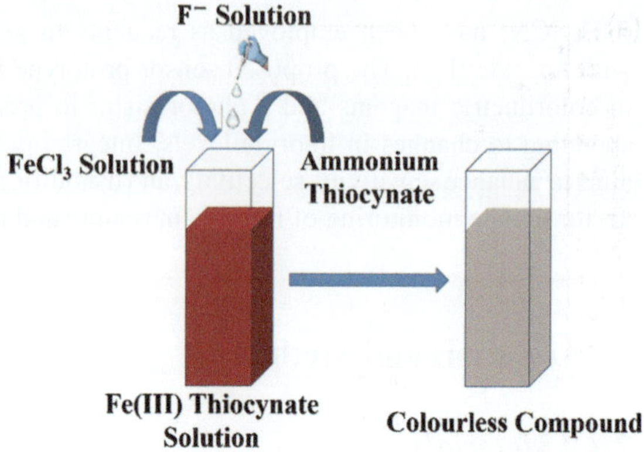

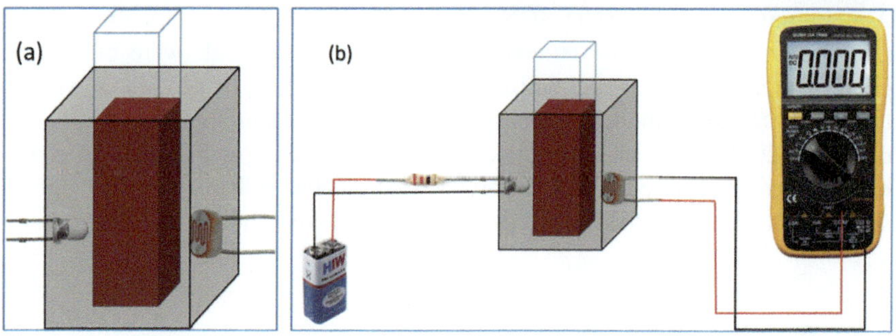

Fig. 2 Design of LED-LDR sensor

and facing each other within the lightproof box (Fig. 2a), leaving a 10 mm × 10 mm space for the cuvette to be inserted, as illustrated in Fig. 2 [16, 17]. A 220 Ω resistor connects the LED light to a 9 V battery to provide a steady power supply. The transmitted light intensity from the cuvette was measured by connecting the LDR to a multimeter.

3 Results and Discussion

3.1 Spectroscopic Study of the Colorimetric Complex

The color fading of Fe(SCN) corroborated the development of a colorimetric complex. The colorimetric complex decolorizes as we increase the concentration of fluoride ions, which is proportional to absorbance and obeys Beer-Lambert's law with a negative slope. The complex so formed was characterized spectroscopically using a UV–vis spectrophotometer. The study revealed a decrease in absorbance

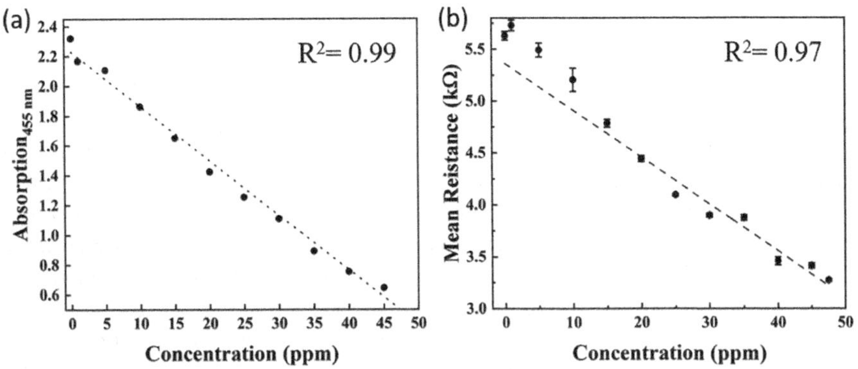

Fig. 3 Calibration plot of the POCT sensor; **a** calibration using a UV–vis spectrophotometer, **b** calibration using the POCT kit

with an increase in concentration. The maximum absorbance was reported at 455 nm [18], and calibration was prepared, as shown in Fig. 3a, for validation purposes of the POCT kit.

3.2 Sensor Calibration

Different fluoride concentrations from 0 to 50 ppm with equal increments of 5 ppm were prepared for the sensor calibration. The cuvette containing the sample was placed inside the lightproof box. The LED was illuminated, and the light intensity was sensed by the LDR sensor, which transferred the signal to the digital multimeter, and the resistance corresponding to it was measured. All the readings were taken in triplicate, and the average value was reported. Plot was drawn w.r.t concentration versus resistance (Fig. 3b). The sensor exhibited good linearity in the 1–45 ppm concentration range with an R^2 value of 0.97. The upper limit of detection was reported to be 45 ppm, the highest value documented in the literature, thus covering the maximum detection range. The minimal magnitude of the error bars reported in Fig. 3b justifies the present prototype's better sensitivity, accuracy, and repeatability. In support of our claim, a detailed critical comparison of the present prototype with the available fluoride sensor is presented in Table 1.

3.3 Interference Study

Experiments were performed to establish the sensor's selectivity for real-world applications. Its interference with other potential co-existing ions such as Cl^-, CO_3^{2-}, SO_4^{2-}, NO_3^-, Fe^{2+}, Fe^{3+}, Ca^{2+}, and Mg^{2+} heavy metal ions was investigated. The

Table 1 Comparison of the present work with available fluoride sensors

Make	Parameters		
	Detection range (ppm)	Sensitivity and response time	Ref.
Aquasol (AE210)	0.1–2 and 1–20	Sensitivity is not mentioned	[19]
Hanna- Hi739	20–20	Sensitivity ≥ 0.1 mgL^{-1} and detection time ≥ 1 min	[20]
Prerana laboratories	N.A	Sensitivity is not mentioned, and 10 min ≤ detection time ≤ 30 min	[21]
Merck (colorimetric)	0–0.8	Sensitivity is not mentioned	[22]
Merck (photometric)	0.1–2.5	Sensitivity is not mentioned, and detection time ≥ 5 min	[23]
Himedia	0–2.5	Sensitivity is not mentioned, and the detection time is not mentioned	[24]
Proposed sensor prototype	0–47.5	Good, can identify samples with a sensitivity of ≥ 0.1mg/L, detection time ~ 20 s	

study shows that these ions do not significantly interfere with the sensor's response, as evidenced by a minimal change in the absorbance of interfering ions compared to the reagent blank (Fig. 4). Figure 4 validates the selectivity of F^- over other interfering ions and the applicability and worthiness of the sensor in the real-world scenario, ensuring reliable F^- detection in complex sample matrices.

Fig. 4 Interference study

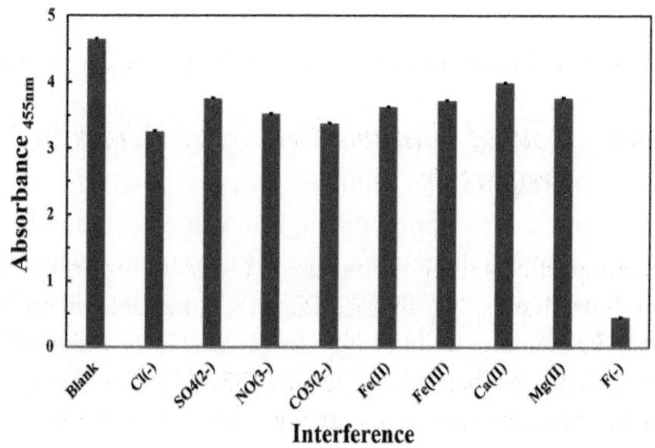

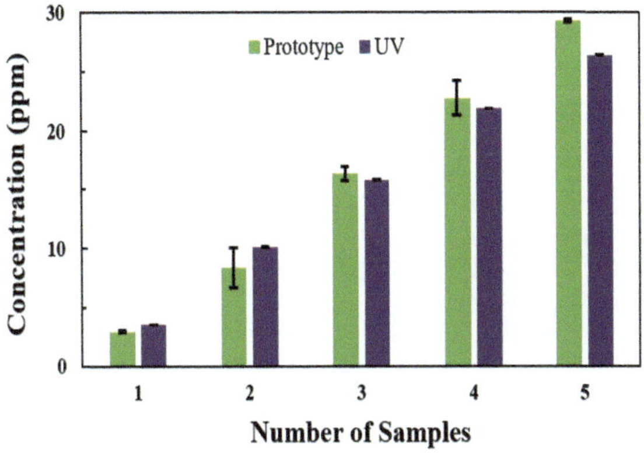

Fig. 5 Validation of the prototype performance w.r.t. UV–vis spectrophotometer

3.4 Assessment of Performance of POCT Kit Using Spiked Samples and Validation with UV–vis Results

The concentration values for the POCT kit were obtained from the calibration plot and compared with those obtained from the absorbance of the UV–vis spectrophotometer (Fig. 5). It was observed that the in-house developed POCT sensor exhibits good accuracy with an absolute average percentage error of 6.23% against the UV–vis spectrophotometer. Therefore, it is clear that the developed prototype is easy to use and very effective for determining the content of unknown F^-.

4 Conclusion

In summary, we present the development of a highly sensitive and cost-effective prototype for rapid detection and quantification of F^- in water samples. The core of the sensor is a colorimetric reaction that results in the decolorization of a red-colored FeSCN complex. An LED-LDR sensor was created to measure changes in color intensity, and this sensor was integrated into a digital multimeter. A calibration plot was prepared, and regression analysis exhibited good linearity with an R^2 value of 0.97. Testing with spiked water samples demonstrated good similarity with UV–vis absorbance with an absolute percentage error of 6.23%. The results indicate the developed sensor is user-friendly and has the potential to be translated into a handheld point-of-care testing (POCT) device for on-site detection and quantification of F^-.

References

1. Zhang, L., Gao, X., Chen, X., Zhao, M., Wu, H., and Liu, Y., *A smartphone integrated ratiometric fluorescent sensor for point-of-care testing of fluoride ions*. Anal Bioanal Chem, 2022. **414**(13): p. 3999–4009, https://doi.org/10.1007/s00216-022-04046-y
2. Borysewicz-Lewicka, M. and Opydo-Szymaczek, J., *Fluoride in Polish Drinking Water and the Possible Risk of Dental Fluorosis*. Polish Journal of Environmental Studies, 2016. **25**(1): p. 9–15, https://doi.org/10.15244/Pjoes/60352
3. Ullah, R., Zafar, M.S., and Shahani, N., *Potential fluoride toxicity from oral medicaments: A review*. Iran J Basic Med Sci, 2017. **20**(8): p. 841–848, https://doi.org/10.22038/IJBMS.2017.9104
4. Organization, W.H., *Guidelines for drinking-water quality*. 2002: World Health Organization. 9241545356
5. Dhillon, A., Nair, M., and Kumar, D., *Analytical methods for determination and sensing of fluoride in biotic and abiotic sources: a review*. Analytical Methods, 2016. **8**(27): p. 5338–5352, https://doi.org/10.1039/c6ay01534d
6. Vanarse, V.B., Thakur, S., and Bandyopadhyay, D., *Recirculatory Solvotaxis of a Nematic Droplet on Water Surface Enabling Miniaturization*. Langmuir, 2025, https://doi.org/10.1021/acs.langmuir.4c04647
7. Medvedev, D.A., *Current drawbacks of proton-conducting ceramic materials: How to overcome them for real electrochemical purposes*. Current Opinion in Green and Sustainable Chemistry, 2021. **32**: p. 100549, https://doi.org/10.1016/J.Cogsc.2021.100549
8. Vanarse, V., Ravi, B., De, S., Dubey, S., and Bandyopadhyay, D., *Controlled Micro–Nano-Scale Droplet Generation via Spin Dewetting*. Processes, 2024. **12**(8): p. 1562, https://doi.org/10.3390/pr12081562
9. Florkowski, C., Don-Wauchope, A., Gimenez, N., Rodriguez-Capote, K., Wils, J., and Zemlin, A., *Point-of-care testing (POCT) and evidence-based laboratory medicine (EBLM)–does it leverage any advantage in clinical decision making?* Critical reviews in clinical laboratory sciences, 2017. **54**(7–8): p. 471–494, https://doi.org/10.1080/10408363.2017.1399336
10. Yamamura, S., Wade, M., and Sikes, J., *Direct Spectrophotometric Fluoride Determination*. Analytical Chemistry, 1962. **34**(10): p. 1308–1312, 0003–2700
11. Kim, Y. and Lee, B., *Propectives of Environmental Colorimetric-Sensors*. Korean Chemical Engineering Research, 2011. **49**(4): p. 393–399, https://doi.org/10.9713/kcer.2011.49.4.393
12. Ravula, R. and Mandal, T.K., *A photoresistor-based portable digital sensor for rapid colorimetric detection of Arsenic*. Microchemical Journal, 2024. **196**: p. 109574, https://doi.org/10.1016/j.microc.2023.109574
13. Doulos, L., Tsangrassoulis, A., Bouroussis, C., and Topalis, F., *Reviewing drawbacks of conventional photosensors: are CCD/CMOS sensors the next generation*. Lux Europa, 2013: p. 1–6
14. Mandal, N., Mitra, S., and Bandyopadhyay, D., *Paper-Sensors for Point-of-Care Monitoring of Drinking Water Quality*. Ieee Sensors Journal, 2019. **19**(18): p. 7936–7941, https://doi.org/10.1109/Jsen.2019.2919269
15. Foster, M.D., *Colorimetric determination of fluoride in water using ferric chloride*. Industrial & Engineering Chemistry Analytical Edition, 1933. **5**(4): p. 234–236, 0096–4484
16. Kashyap, S. and Mandal, T.K., *Prototype of a POCT Device for Colorimetric Sensing of Chromium(VI) for Healthcare Applications*. Journal of Electronic Materials, 2025: p. 1–8, https://doi.org/10.1007/s11664-025-11768-z
17. Kashyap, S., Joshi, S., Mahan, P.S., Kayal, S., and Mandal, T.K., *A highly efficient indigenous portable optical sensor for measurement of total chromium concentration in environmental and biological samples: Design, development, and validation with spectroscopic results*. Microchemical Journal, 2025: p. 113098, https://doi.org/10.1016/j.microc.2025.113098
18. Boivin, G., Dupuis, J., and Meunier, P.J., *Fluoride and osteoporosis*. World Rev Nutr Diet, 1993. **73**: p. 80–103, https://doi.org/10.1159/000422461

19. Aquasol, https://www.rakiro.net/product-details/aquasol-fluoride-test-kit-ae210, (26 December 2024)
20. HI-739, . https://hannainst.in/product/fluoride-hr-checker-hc-hi739/?srsltid=AfmBOopgCK560mpYOhsGm_9Iu0LyDkHoe6z2GyM12nvJHrr0K6knQ0Zgwmg&gQT=1, (26 December 2024)
21. Fluoide kits, https://www.watertestingkits.co.in/water-testing-kits.html, (26 December 2024)
22. Colorimetric, https://www.sigmaaldrich.com/IN/en/product/mm/118771?srsltid=AfmBOopwDb2Y68oTGR7VQ9v1914SmyTH0qJE0508hIFdbUpUvjrZY3Nz, (26 December 2024)
23. photometric, https://www.sigmaaldrich.com/IN/en/product/mm/117243#product-documentation, (26 December 2026)
24. Fluoride testing kit, https://www.himedialabs.com/in/fluoride-testing-kit.html, (26 December 2024)

Carbon Dioxide (CO_2) Sequestration Using Stable Microbubble Dispersions

Stuti Dubey and Subrata Kumar Majumder

Abstract Previous studies have primarily focused on microbubble dispersions using synthetic surfactants, leaving a gap in understanding the behavior of natural non-ionic surfactants, particularly in optimizing foam stability for effective CO_2 sequestration. This study investigates microbubble dispersions stabilized with Acacia concinna (Shikakai) surfactant under varying conditions of surfactant concentration, pH, and brine concentration. The Sauter mean bubble diameter, ranging from 61 to 114 μm, and the high gas volume fraction of 81–87% demonstrate the surfactant's effectiveness in creating highly aerated dispersions, suitable for enhanced oil recovery (EOR) applications. Moderate salt concentrations (1000 ppm NaCl) enhance foam stability, increasing the half-life duration to 160 min due to reduced electrostatic repulsion and improved micelle formation. Conversely, excessive salt concentrations (3000 ppm NaCl) decrease foam stability, with the half-life duration dropping to 104 min, attributed to the salting-out effect and reduced surfactant solubility. Higher surfactant concentrations improve stability across pH values from 5 to 10, with non-ionic surfactants showing less sensitivity to pH changes. This work addresses the complex interactions at the gas–liquid interface, particularly how pH and brine concentration influence foam stability in the presence of CO_2. The study provides valuable insights for optimizing surfactant formulations in EOR, enhancing CO_2 sequestration efficiency, improving oil recovery rates, and contributing to global efforts to mitigate climate change through sustainable resource management and environmental protection.

Keywords Acacia concinna · CO_2 sequestration · Enhanced oil recovery · Liquid drainage · Microbubble dispersion

---

S. Dubey (✉) · S. K. Majumder
Department of Chemical Engineering, Indian Institute of Technology Guwahati, Guwahati, Assam, India
e-mail: stuti_dubey@iitg.ac.in

S. K. Majumder
e-mail: skmaju@iitg.ac.in

U. S. Dixit et al. (eds.), *Current Progress in Engineering Sciences*,
https://doi.org/10.1007/978-981-96-8753-4_43

Nomenclature

d_{bs}	Sauter mean bubble diameter
φ_g	Gas volume fraction
μ_s	Viscosity of solution
V_t	Volume of drained liquid at time (t)
V_{max}	Volume of final drained liquid at time (t)
$t_{1/2}$	Half-life for liquid drainage
n	Sigmoidal character of the drainage curve
R_d	Rate of liquid drainage

1 Introduction

Under sustainable resource development scenarios, it was predicted that global enhanced oil recovery (EOR) output will increase to over 4 million barrels per day by 2040. EOR using carbon dioxide (CO_2) injection combined with carbon capture and storage can help produce more oil while storing CO_2 more affordably. In recent times, interest in CO_2-EOR has increased significantly due to increased support for carbon capture, utilization, and storage (CCUS) work. In geologically favorable circumstances, CO_2-EOR offers a compelling way to lower the emission intensity of oil delivered to the market and, hence, lessen its environmental effects [1]. This technology offers dual benefits of cost reduction and increased oil production [2]. It also emphasizes the importance of CO_2-EOR strategies in reducing carbon emissions and maximizing oil production, especially through methods like water-alternating gas with CO_2 (WAG-CO_2) flooding [3]. EOR utilizes CO_2 injection to extract more oil from reservoirs, aiding in both environmental and economic benefits [4].

The issue of high gas mobility in porous medium can be resolved by CO_2 foam. When compared to gas injection, injected foam seems to be a more viscous fluid, which helps to increase the displacement efficiency [5]. The potential of foam EOR is reduced by the great heterogeneity and high viscosity of crude oil. Furthermore, one of the primary difficulties in the foam EOR process is the negative impact of crude oil on foam stability. Despite this technology being relatively developed, the primary issues preventing its widespread use are related to geological storage stability [6]. The remarkable rise in research on the special characteristics of micro and nanobubbles (MNB) has brought attention to the uses of micro and nanosized materials in several cutting-edge scientific and technological domains, such as colloidal gas aphrons (CGS), CO_2-EOR, medical, and food domains [6]. Injecting CO_2 as a microbubble could maximize capillary trapping, which is one of the main CO_2 trapping mechanisms, and is another benefit of using microbubbles to help sequester carbon [7]. In deep reservoirs with high temperatures and pressures, chemical agents like surfactants, polymers, and nanoparticles are tested for such conditions and are used to create microbubbles [8].

Microbubble dispersions drain in two stages: first, liquid drains under gravity; second, foam breaks down as a result of thinner films between bubbles. This process has been explained in terms of microbubbles rising [9]. Prior research on microbubbles mostly focused on analyzing the stability of microbubbles under various circumstances, including temperature, surfactant concentration, pressure, and polymer concentrations. Images of the CGAs were recorded throughout time to study the size of the CGA bubble and the pace of drainage [10]. The rate of drainage was measured as a function of many factors, including the impact of viscosity, to determine the half-life of CGAs and examine the variations in the apparent viscosity and microbubble size of the CGA-based drilling fluids. They demonstrated that when microbubbles are exposed to a depressurization process after pressure, they become unstable (that is, the bubbles enlarge beyond their initial size [11]. Research has shown that the interaction between nanoparticles and surfactants is affected by salinity and pH, with higher nanoparticle concentrations enhancing colloidal stability in high-salinity brine at low pH [12]. Additionally, the use of nanoparticles in combination with surfactants can significantly increase foam stability, reduce foam drainage, and improve oil recovery in reservoir conditions [13]. Furthermore, the stability of CO_2 microbubbles is primarily dependent on xanthan gum polymer concentration, with higher surfactant concentrations and stirring rates increasing stability but salinity decreasing it, showcasing significant interactions between these factors [14]. These findings highlight the complex interplay of surfactants, nanoparticles, and environmental conditions in determining the stability of CO_2 microbubbles.

Previous studies have focused on using microbubble dispersions using synthetic surfactants. Studies with microbubble dispersion using a natural non-ionic surfactant to optimize foam stability for effective CO_2 sequestration have been limited. The research gap it addresses involves the limited understanding of how pH and brine concentration specifically influence foam stability in the presence of CO_2, particularly the complex interactions at the gas–liquid interface. The overall objective of this study lies in its potential to improve oil recovery rates while simultaneously reducing atmospheric CO_2 levels, aligning with global efforts to mitigate climate change. This work not only enhances the effectiveness of CO_2-EOR processes but also contributes to sustainable resource management and environmental protection.

2 Materials and Methodology

2.1 Materials Required

A non-ionic, naturally derived surfactant, Acacia concinna (sourced from Northeast Nursery), was used in conjunction with sodium chloride (NaCl) brine (obtained from Sigma Aldrich). Hydrochloric acid (HCl) and Sodium hydroxide (NaOH) (from Sigma Aldrich) were diluted for pH adjustment, generated with deionized water (DI) of 100 ml solutions.

2.2 Preparation of Solutions for Stability

Surfactant solutions were prepared using vigorous agitation with a magnetic stirrer. A 100 mL volume of surfactant at concentrations of 500, 1000, and 4000 parts per million (ppm) was mixed with deionized water (DI) at a rotational speed of 2000 revolutions per minute (rpm) for 30 min under ambient conditions (25 °C, atm press). Sodium chloride (NaCl) was added to generate brine solutions of concentrations 1000 and 3000 ppm, respectively. The rheology of the resulting solutions was determined using a Rheometer (Anton-Paar) at a constant temperature of 25 °C.

2.3 Measurement of Microbubble Size

The size of the microbubbles within the column was analyzed using an LED light source. A Nikon COOLPIX P520 camera (1080P Full HD, 3.2-inch ultrahigh-resolution) was used to capture images and videos for the stability experiments. Bubble diameter measurements were then obtained using image analysis software. Typically, 150–300 bubbles were marked and analyzed to obtain a statistically relevant representation of the size distribution. The Sauter mean diameter, representing the average bubble size based on volume and surface area, was the primary metric used to characterize bubble dimensions. The mathematical expression for the Sauter mean diameter is given by:

$$d_{bs} = \frac{\sum_{i=1}^{n} n_i d_{bi}^3}{\sum_{i=1}^{n} n_i d_{bi}^2} \tag{1}$$

2.4 Stability Analysis

To assess liquid separation from the microbubble dispersion, samples were transferred to 100-ml graduated cylinders. The volumes of the foam phase (V_f) and drained liquid (V_l) were recorded over time. Upon complete foam collapse, the maximum volume of drained liquid (V_{max}) was measured. All experiments were replicated thrice for data reliability. The gas volume fraction, which is the ratio of gas volume (100 ml-V_{max}) to the total dispersion volume, provides a measure of entrapped gas quantity. The half-life time ($t_{1/2}$) is the time required for half of the overall liquid volume to drain from the dispersion.

3 Results and Discussions

3.1 Physical Properties of Microbubble Dispersion

The physical properties of the microbubble dispersion are measured through the analysis of bubble diameter (d_{sb}), gas fraction (φ_g), and viscosity of solutions (μ_s) at different conditions. Within the 61–114 µm range, the Sauter mean bubble diameter of the microbubble dispersion is comparable to the values published in the literature [15]. Additionally, Natawijaya et al. found that microbubbles with a dominant size of 10–50 µm led to higher oil recovery compared to larger microbubbles, showcasing the influence of microbubble size on EOR efficiency under different brine conditions [16]. The gas volume fraction was measured in the range of 80.89–87.77. A gas volume fraction of 81–87% means that the dispersion is highly aerated, which is typical in applications where a high degree of gas hold-up is desired, such as in enhanced oil recovery (EOR) that benefits from increased gas–liquid contact. These figures align with the 60–70% range that has been documented in the literature [17, 18]. The interfacial properties of surfactant are measured through surface tension, which shows a reduction in the values from 72 to 30.11 mN/m. Rheology plays a vital role in the efficiency of surfactant mixtures at different pH and brine concentrations. The viscosity of the solution was reduced from 4.83 to 1.15 cP. Reducing viscosity in surfactant mixtures at different pH and brine enhances CO_2-in-brine foam stabilization, benefiting enhanced carbon capture and storage efficiency.

3.2 Kinetics of Foam Drainage

The foam drainage process in the microbubble column follows a sigmoidal trend due to the sequential phases of liquid movement. Initially, gravity-driven drainage (Phase I) causes liquid accumulation at the base, leading to a rapid increase in drainage rate. Simultaneously, capillary forces drive liquid from the thin films between bubbles to the plateau borders. As drainage progresses, gas diffusion from smaller to larger bubbles alters the foam structure, slowing down the process. Eventually, in Phase II, the dispersion loses its colloidal nature, transforming into a wet foam where liquid drainage is controlled by capillary suction and molecular interactions. The relationship between the liquid drainage from the foam dispersion and time is expressed empirically. The liquid drainage curve of the microbubble dispersion has a profile that resembles an "S" shape with an early rise in drainage rate, as shown in Fig. 1.

$$V_t = V_{\max} \frac{t^n}{t_{1/2}^n + t^n} \tag{2}$$

The sigmoidal model accurately represents this dynamic transition, capturing the interplay of gravitational, capillary, and molecular forces. The parameters V_{max},

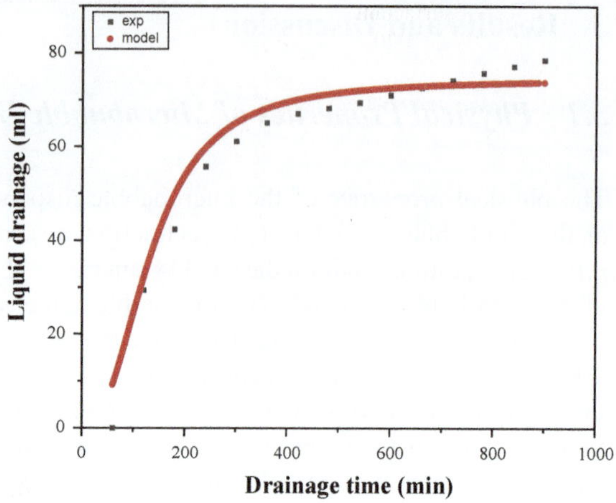

Fig. 1 Graphical representations of experimental and model values of maximum drained liquid volume for the microbubble dispersion (AC concentration-500 ppm, pH 6)

K, and n were determined using nonlinear curve fitting, providing a quantitative framework for understanding the kinetics of foam stability.

This sigmoidal model is chosen because it accurately captures the three-stage drainage process: (i) an initial slow drainage phase, (ii) a rapid drainage phase, and (iii) a plateau phase where drainage stabilizes. These stages correspond to gravitational drainage, capillary effects, and bubble coalescence. The experimental data showed a strong correlation in Eq. (2), with R^2 values ranging from 0.97 to 0.99.

3.3 *Effect of Surfactant Concentration on Microbubble Stability*

The stability curves for acacia concinna solution dispersions of concentration ranging from 500, 1000, and 4000 ppm with pH values of 5, 6, 8, and 10 have been prepared. Surfactant concentration affects foam stability at different pH levels. The stability of microbubbles dispersion increased as the concentration shifted toward 500–4000 ppm at a pH of 8, as shown in Fig. 2. The experiments were repeated three times for consistency, and the calculated percentage error due to repeatability ranged between 0.5 and 2% ensuring the reliability of the obtained data. Higher surfactant concentration enhances colloidal stability, crucial for reducing gas mobility and enhancing oil recovery in reservoir conditions [12].

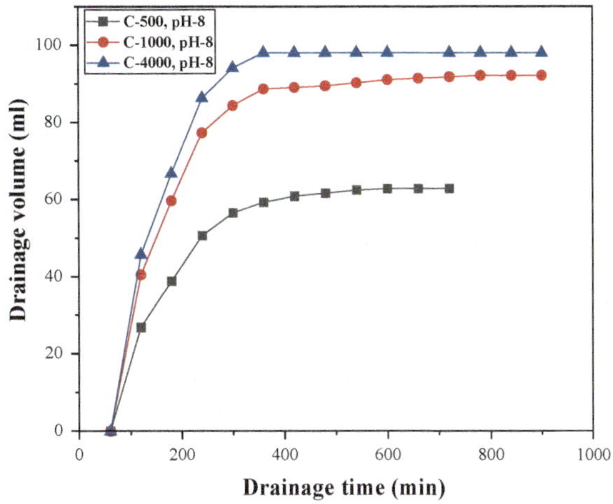

Fig. 2 Effect of surfactant concentration (500, 1000, and 4000 ppm) on microbubble dispersion at pH 8

3.4 Effect of pH on Microbubble Stability

The pH level plays a crucial role in determining the optimal surfactant concentration for CO_2 foam stability. Non-ionic surfactants do not carry any charge and are less affected by pH changes. It shows that solution stability is higher at pH values beyond 6 than at lower pH values (around 5), suggesting that pH changes have a good impact on dispersion stability with an error percentage of >1%. With a half-life of around 146 min, the stability of the dispersion from the non-ionic surfactant remains the same. Lowering the pH of surfactant solution can have a detrimental effect on foam stability due to repulsive interactions between amine headgroups, especially in the presence of high salinity [19].

3.5 Effect of NaCl on Stability

It was observed that the stability of the dispersion decreases as the concentration of salt (NaCl) increases, with the half-life duration first rising from 113 to 160 min at pH 6 with 1000 ppm NaCl concentration, then decreasing to 104 min at 3000 ppm NaCl concentration, indicating a complex interaction between the surfactant and the salt concentration. The calculated percentage error due to repeatability was in the range of 0.5–2%. The initial increase in half-life duration suggests that a moderate amount of salt improves the foam stability. The presence of salt can reduce the critical micelle concentration (CMC), leading to more efficient micelle formation and better surfactant performance at the interface. The decrease in half-life duration at higher salt concentrations suggests that excessive salt destabilizes the foam. High concentrations of salt can lead to a "salting-out" effect, where the surfactant becomes less soluble in water. This reduces the concentration of surfactant molecules available to stabilize the foam. Despite earlier findings of reduced dispersion stability due to higher electrolyte concentrations [20]. Research by Haowei et al. demonstrated that the conformance control performance of non-chemical CO_2 microbubbles declined with an increase in brine salinity, highlighting the sensitivity of microbubbles to changes in water salinity [21]. The overall effect of brine on surfactant-stabilized foam is concentration-dependent. Moderate concentrations of brine (1000 ppm NaCl) enhance foam stability, while high concentrations (3000 ppm NaCl) lead to decreased stability.

4 Conclusions

The study of microbubble dispersions stabilized with Acacia concinna (Shikakai) surfactant under varying conditions of surfactant concentration, pH, and brine concentration reveals critical insights into the factors affecting foam stability. The Sauter mean bubble diameter within the 61–114 μm range and the high gas volume fraction of 81–87% indicate the effectiveness of the surfactant in creating highly aerated dispersions, suitable for enhanced oil recovery (EOR) applications. The results demonstrate that moderate salt concentrations (1000 ppm NaCl) improve foam stability, as evidenced by an increase in half-life duration to 160 min, likely due to reduced electrostatic repulsion and enhanced micelle formation. However, excessive salt concentrations (3000 ppm NaCl) lead to a decrease in foam stability, with a half-life duration dropping to 104 min, attributed to the salting-out effect and reduced surfactant solubility. The impact of pH on foam stability further underscores the importance of optimizing surfactant concentration, as higher concentrations enhance stability across a range of pH values from 5 to 10. These findings highlight the complex interplay between surfactant concentration, brine salinity, and pH in determining the stability of microbubble dispersions, providing valuable guidelines for the formulation of effective surfactant systems in EOR and other petroleum applications.

Acknowledgements The authors thank the Department of Chemical Engineering at the Indian Institute of Technology Guwahati for providing various research facilities.

References

1. K. Novak Mavar, N. Gaurina-Međimurec, L. Hrnčević, Significance of enhanced oil recovery in carbon dioxide emission reduction, Sustain. 13 (2021) 1–27. https://doi.org/10.3390/su1304 1800.
2. A. Bera, S. Satapathy, J. Daneti, Perspectives of CO_2 Injection Strategies for Enhanced Oil Recovery and Storage in Indian Oilfields, Energy and Fuels 38 (2024) 10613–10633. https:// doi.org/10.1021/acs.energyfuels.4c01374.

3. M. Al-Ghnemi, E. Ozkan, K. Amini, H. Kazemi, Numerical Modeling Assessment of CO_2 - EOR and Sequestration Potential in a Light-Oil Carbonate Reservoir, Proc. - SPE Symp. Improv. Oil Recover. 2024-April (2024). https://doi.org/10.2118/218163-MS.
4. R. Umrani, R. Jones, C. Ware, D. Janise, R. Ozah, N. Joshi, T. Energy, OTC-32212-MS Evaluating an Area of Review for CO_2 Sequestration and Storage : A Subsurface Modelling Workflow Tied to Regulatory Guidelines, (2023). https://doi.org/10.4043/32212-MS.
5. R.M. Sunmonu, M. Onyekonwu, Enhanced oil recovery using foam injection; a mechanistic approach, Soc. Pet. Eng. - 37th Niger. Annu. Int. Conf. Exhib. NAICE 2013 - To Grow Africa's Oil Gas Prod. Required Policy, Funding, Technol., Tech. Capab. 2 (2013) 1061–1073. https://doi.org/10.2118/167589-ms.
6. X. Li, B. Peng, Q. Liu, J. Liu, L. Shang, Micro and nanobubbles technologies as a new horizon for CO_2-EOR and CO_2 geological storage techniques: A review, Fuel 341 (2023). https://doi.org/10.1016/j.fuel.2023.127661.
7. D. Zhang, J. Song, Mechanisms for geological carbon sequestration, Procedia IUTAM 10 (2014) 319–327. https://doi.org/10.1016/j.piutam.2014.01.027.
8. A. Telmadarreie, A. Doda, J.J. Trivedi, E. Kuru, P. Choi, CO_2 microbubbles - A potential fluid for enhanced oil recovery: Bulk and porous media studies, J. Pet. Sci. Eng. 138 (2016) 160–173. https://doi.org/10.1016/j.petrol.2015.10.035.
9. Y.L. Yan, C.T. Qu, N.S. Zhang, Z.G. Yang, L. Liu, A study on the kinetics of liquid drainage from colloidal gas aphrons (CGAs), Colloids Surfaces A Physicochem. Eng. Asp. 259 (2005) 167–172. https://doi.org/10.1016/j.colsurfa.2005.02.028.
10. Y. Dai, T. Deng, Stabilization and characterization of colloidal gas aphron dispersions, J. Colloid Interface Sci. 261 (2003) 360–365. https://doi.org/10.1016/S0021-9797(03)00056-0.
11. N. Bjorndalen, E. Kuru, Stability of microbubble based drilling fluids under downhole conditions, Can. Int. Pet. Conf. 2006, CIPC 2006 (2006) 1–11. https://doi.org/10.2118/2006-079.
12. X. Zhang, G. Zhang, J. Ge, Y. Wang, CO_2 foam stabilized with switchable surfactants and modified nanoparticles effected by PH and salinity, Soc. Pet. Eng. - SPE Eur. Featur. 82nd EAGE Conf. Exhib. 1 (2020) 1–10. https://doi.org/10.2118/200584-MS.
13. B.R.S. de Azevedo, B.G. Alvarenga, A.M. Percebom, A. Pérez-Gramatges, Interplay of Interfacial and Rheological Properties on Drainage Reduction in CO_2 Foam Stabilised by Surfactant/Nanoparticle Mixtures in Brine, Colloids and Interfaces 7 (2023). https://doi.org/10.3390/colloids7010002.
14. N.N.H. Le, Y. Sugai, K. Sasaki, Investigation of stability of CO_2 microbubbles—colloidal gas aphrons for enhanced oil recovery using definitive screening design, Colloids and Interfaces 4 (2020). https://doi.org/10.3390/colloids4020026.
15. R.W. Alves, P. Jauregi, A.A.U. De Souza, S.M.A.G.U. De Souza, Recovery of Norbixin Using Colloidal Gas Aphrons, (2001) 1–9.
16. M.A. Natawijaya, Y. Sugai, F. Anggara, CO_2 microbubble colloidal gas aphrons for EOR application: the generation using porous filter, diameter size analysis and gas blocking impact on sweep efficiency, J. Pet. Explor. Prod. Technol. 10 (2020) 103–113. https://doi.org/10.1007/s13202-019-0680-3.
17. P. Jauregi, J. Varley, Lysozyme separation by colloidal gas aphrons, Prog. Colloid Polym. Sci. 100 (1996) 362–367. https://doi.org/10.1007/bfb0115809.
18. K. Matsushita, A.H. Mollah, D.C. Stuckey, C. del Cerro, A.I. Bailey, Predispersed solvent extraction of dilute products using colloidal gas aphrons and colloidal liquid aphrons: Aphron preparation, stability and size, Colloids and Surfaces 69 (1992) 65–72. https://doi.org/10.1016/0166-6622(92)80239-X.
19. L. Le, R. Ramanathan, H. Nasr-El-Din, Evaluation of an ethoxylated amine surfactant for CO_2-foam stability at high salinity conditions, Soc. Pet. Eng. - Abu Dhabi Int. Pet. Exhib. Conf. 2019, ADIP 2019 (2019). https://doi.org/10.2118/197515-ms.
20. P. Jauregi, S. Gilmour, J. Varley, Characterisation of colloidal gas aphrons for subsequent use for protein recovery, Chem. Eng. J. 65 (1997) 1–11. https://doi.org/10.1016/S1385-8947(96)03154-3.

21. H. Jia, H. Yu, H. Tang, J. Shi, E. Tang, Y. Wang, Experimental Investigation of Non-Chemical CO_2 Microbubbles EOR Performance in Low-Permeability Resevoirs, Int. Pet. Technol. Conf. IPTC 2024 (2024). https://doi.org/10.2523/IPTC-24500-MS.

GPSR Compliance

The European Union's (EU) General Product Safety Regulation (GPSR) is a set of rules that requires consumer products to be safe and our obligations to ensure this.

If you have any concerns about our products, you can contact us on ProductSafety@springernature.com

In case Publisher is established outside the EU, the EU authorized representative is:

Springer Nature Customer Service Center GmbH
Europaplatz 3
69115 Heidelberg, Germany

Batch number: 10584684

Printed by Printforce, the Netherlands